Intelligent Sensor Networks

The Integration of Sensor Networks, Signal Processing and Machine Learning

OTHER TELECOMMUNICATIONS BOOKS FROM AUERBACH

Ad Hoc Mobile Wireless Networks: Principles, Protocols, and Applications
Subir Kumar Sarkar, T.G. Basavaraju, and C. Puttamadappa
ISBN 978-1-4200-6221-2

Communication and Networking in Smart Grids
Yang Xiao (Editor)
ISBN 978-1-4398-7873-6

Delay Tolerant Networks: Protocols and Applications
Athanasios V. Vasilakos, Yan Zhang, and Thrasyvoulos Spyropoulos
ISBN 978-1-4398-1108-5

Emerging Wireless Networks: Concepts, Techniques and Applications
Christian Makaya and Samuel Pierre (Editors)
ISBN 978-1-4398-2135-0

Game Theory in Communication Networks: Cooperative Resolution of Interactive Networking Scenarios
Josephina Antoniou and Andreas Pitsillides
ISBN 978-1-4398-4808-1

Green Communications: Theoretical Fundamentals, Algorithms and Applications
Jinsong Wu, Sundeep Rangan, and Honggang Zhang
ISBN 978-1-4665-0107-2

Green Communications and Networking
F. Richard Yu, Xi Zhang, and Victor C.M. Leung (Editors)
ISBN 978-1-4398-9913-7

Green Mobile Devices and Networks: Energy Optimization and Scavenging Techniques
Hrishikesh Venkataraman and Gabriel-Miro Muntean (Editors)
ISBN 978-1-4398-5989-6

Handbook on Mobile Ad Hoc and Pervasive Communications
Laurence T. Yang, Xingang Liu, and Mieso K. Denko (Editors)
ISBN 978-1-4398-4616-2

IP Telephony Interconnection Reference: Challenges, Models, and Engineering
Mohamed Boucadair, Isabel Borges, Pedro Miguel Neves, and Olafur Pall Einarsson
ISBN 978-1-4398-5178-4

LTE-Advanced Air Interface Technology
Xincheng Zhang and Xiaojin Zhou
ISBN 978-1-4665-0152-2

Media Networks: Architectures, Applications, and Standards
Hassnaa Moustafa and Sherali Zeadally (Editors)
ISBN 978-1-4398-7728-9

Multihomed Communication with SCTP (Stream Control Transmission Protocol)
Victor C.M. Leung, Eduardo Parente Ribeiro, Alan Wagner, and Janardhan Iyengar
ISBN 978-1-4665-6698-9

Multimedia Communications and Networking
Mario Marques da Silva
ISBN 978-1-4398-7484-4

Near Field Communications Handbook
Syed A. Ahson and Mohammad Ilyas (Editors)
ISBN 978-1-4200-8814-4

Next-Generation Batteries and Fuel Cells for Commercial, Military, and Space Applications
A. R. Jha, ISBN 978-1-4398-5066-4

Physical Principles of Wireless Communications, Second Edition
Victor L. Granatstein, ISBN 978-1-4398-7897-2

Security of Mobile Communications
Noureddine Boudriga, ISBN 978-0-8493-7941-3

Smart Grid Security: An End-to-End View of Security in the New Electrical Grid
Gilbert N. Sorebo and Michael C. Echols
ISBN 978-1-4398-5587-4

Transmission Techniques for 4G Systems
Mário Marques da Silva
ISBN 978-1-4665-1233-7

Transmission Techniques for Emergent Multicast and Broadcast Systems
Mário Marques da Silva, Americo Correia, Rui Dinis, Nuno Souto, and Joao Carlos Silva
ISBN 978-1-4398-1593-9

TV Content Analysis: Techniques and Applications
Yiannis Kompatsiaris, Bernard Merialdo, and Shiguo Lian (Editors)
ISBN 978-1-4398-5560-7

TV White Space Spectrum Technologies: Regulations, Standards, and Applications
Rashid Abdelhaleem Saeed and Stephen J. Shellhammer
ISBN 978-1-4398-4879-1

Wireless Sensor Networks: Current Status and Future Trends
Shafiullah Khan, Al-Sakib Khan Pathan, and Nabil Ali Alrajeh
ISBN 978-1-4665-0606-0

Wireless Sensor Networks: Principles and Practice
Fei Hu and Xiaojun Cao
ISBN 978-1-4200-9215-8

Intelligent Sensor Networks

The Integration of Sensor Networks, Signal Processing and Machine Learning

Edited by

FEI HU

QI HAO

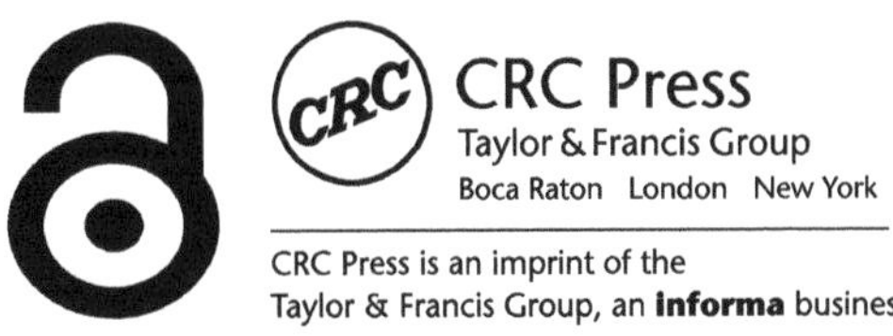

CRC Press
Taylor & Francis Group
6000 Broken Sound Parkway NW, Suite 300
Boca Raton, FL 33487-2742

Printed in the United States of America on acid-free paper
Version Date: 20121023

International Standard Book Number: 978-1-1381-9974-3 (Paperback)
International Standard Book Number: 978-1-4398-9281-7 (Hardback)

Library of Congress Cataloging-in-Publication Data

Intelligent sensor networks : the integration of sensor networks, signal processing, and machine learning / editors, Fei Hu, Qi Hao.
p. cm.
Summary: "In the last decade, wireless or wired sensor networks have attracted much attention. However, most designs target general sensor network issues including protocol stack (routing, MAC, etc.) and security issues. This book focuses on the close integration of sensing, networking, and smart signal processing via machine learning. Based on their world-class research, the authors present the fundamentals of intelligent sensor networks. They cover sensing and sampling, distributed signal processing, and intelligent signal learning. In addition, they present cutting-edge research results from leading experts"-- Provided by publisher.
Includes bibliographical references and index.
ISBN 978-1-4398-9281-7 (hardback)
1. Wireless sensor networks. I. Hu, Fei, 1972- II. Hao, Qi, 1973-

TK7872.D48I4685 2012
681'.2--dc23 2012032571

Visit the Taylor & Francis Web site at
http://www.taylorandfrancis.com

and the CRC Press Web site at
http://www.crcpress.com

To Yang Fang and Gloria (Ge Ge)

Contents

PART II Intelligent Sensor Networks: Signal Processing

PART III Intelligent Sensor Networks: Sensors and Sensor Networks

Preface

Nowadays, various sensors are used to collect the data of our environment, including RFID, video cameras, pressure, acoustic, etc. A typical sensor converts the physical energy of the object under examination into electrical signals to deduce the information of interest. Intelligent sensing technology utilizes proper prior knowledge and machine learning techniques to enhance the process of information acquisition. Therefore, the whole sensing procedure can be performed at three levels: data, information, and knowledge. At the data level, each sample represents a measure of target energy within a certain temporal-spatial volume. For example, the pixel value of a video camera represents the number of photons emitted within a certain area during a specific time window. The information acquired by a sensor is represented by a probabilistic belief over random variables. For example, the information of target position can be represented by a Gaussian distribution. The knowledge acquired by a sensor is represented by a statistical model describing relations among random variables. For example, the behavior of a target can be represented by a hierarchical hidden Markov model; a situation can be represented by a hierarchical Bayesian network.

The unprecedented advances in wireless networking technology enable the deployment of a large number of sensors over a wide area without limits of wires. However, the main challenge for developing wireless sensor networks is limited resources: power supply, computing complexity, and communication bandwidth. There are several possible solutions to overcome these obstacles: (1) development of "smart" sensing nodes that can reduce the data volume for information representation as well as energy consumption; (2) development of distributed computing "intelligence" that allows data fusion, state estimation, and machine learning to be performed in a distributed way; and (3) development of ad hoc networking "intelligence" that can guarantee the connectivity of sensor systems under various conditions. Besides, energy harvest technologies and powerless sensor networks have been developed to relax the limits of power supply.

The "intelligence" of sensor networks can be achieved in four ways: (1) spatial awareness, (2) data awareness, (3) group awareness, and (4) context awareness. Spatial awareness refers to an intelligent sensor network's capability of knowing the relative geometric information of its members and targets under examination. This awareness is implemented through sensor self-calibration and target state estimation. Data awareness refers to an intelligent sensor network's capability of reducing the data volume for information representation. This awareness is implemented through data prediction, data fusion, and statistical model building. Group awareness refers to an intelligent sensor network's capability of all members knowing each other's states and adjusting each member's behavior according to other members' actions. For example, distributed estimation and data learning are performed through the collaboration of a group of sensor nodes; ad hoc networking techniques maintain the connectivity of the whole network. Context awareness refers to an intelligent sensor network's capability of changing its operation modes based on the knowledge of situations and resources to achieve maximum efficiency of sensing performance. This awareness is implemented through proper context representation and distributed inference.

Book Highlights

So far there are no published books on intelligent sensor network from a machine learning and signal processing perspective. Unlike current sensor network books, this book has contributions from world-famous sensing experts that deal with the following issues:

1. *Emphasize "intelligent" designs in sensor networks*: The intelligence of sensor networks can be developed through distributed machine learning or smart sensor design. Machine learning of sensor networks can be performed in supervised, unsupervised, or semisupervised ways. In Part I, Chapter 1—Machine Learning Basics—we have provided a comprehensive picture of this area. Machine learning technology is a multidisciplinary field that includes probability and statistics, psychology, information theory, and artificial intelligence. Note that sensor networks often operate in very challenging conditions and need to accommodate environmental changes, hardware degradation, and inaccurate sensor readings. Thus they should learn and adapt to the changes in their operation environment. Machine learning can be used to achieve intelligent learning and adaptation. In Part I we have also included some chapters that emphasize these "intelligence" aspects of sensor networks. For example, we have explained the components of the intelligent sensor (transducer) interfacing problem. We have also discussed how the network can intelligently choose the "best" assignment from the available sensors to the missions to maximize the utility of the network.
2. *Detail signal processing principles in intelligent sensor networks*: Recently, a few advanced signal processing principles have been applied in sensor networks. For example, compressive sensing is an efficient and effective signal acquisition and sampling framework for sensor networks. It can save transmittal and computational power significantly at the sensor node. Its signal acquisition and compression scheme is *very simple*, so it is suitable for inexpensive sensors. As another example, a Kalman filter can be used to identify the sensor data pollution attacks in sensor networks.
3. *Elaborate important platforms on intelligent sensor networks*: The platforms of intelligent sensor networks include smart sensors, RFID-assisted nodes, and distributed self-organization architecture. This book covers these platforms. For example, in Part III we have included two chapters on RFID-based sensor function enhancement. The sensor/RFID integration can make the sensor better identify and trace surrounding objects.
4. *Explain interesting applications on intelligent sensor networks*: Intelligent sensor networks can be used for target tracking, object identification, structural health monitoring, and other important applications. In most chapters, we have clearly explained how those "intelligent" designs can be used for realistic applications. For example, in structural health monitoring applications, we can embed the sensors in a concrete bridge. Thus, a bridge fracture can be detected in time. Those embedded sensors for field applications can be powered through solar-cell batteries. In healthcare applications, we can use medical sensors and intelligent body area sensor networks to achieve low-cost, 24/7 patient monitoring from a remote office.

Targeted Audience

This book is suitable for the following types of readers:

1. *College students*: This book can serve as a textbook or reference book for college courses on intelligent sensor networks. The courses could be offered in computer science, electrical and computer engineering, information technology and science, or other departments.
2. *Researchers*: Because each chapter is written by leading experts, the contents will be very useful for researchers (such as graduate students and professors) who are interested in the application of artificial intelligence and signal processing in sensor networks.
3. *Computer scientists*: We have provided many computing algorithms on machine learning and data processing in this book. Thus, computer scientists could refer to those principles in their own design.
4. *Engineers*: We have also provided useful intelligent sensor node/sensor network design examples. Thus, company engineers could use those principles in their product design.

Book Architecture

This book includes three parts as follows:

Part I—Machine Learning: This part describes the application of machine learning and other artificial intelligence principles in sensor network intelligence. It covers the basics of machine learning, including smart sensor/transducer architecture and data representation for intelligent sensors, modal parameter–based structural health monitoring based on wireless smart sensors, sensor-mission assignment problems in which the objective is to maximize the overall utility of the network under different constraints, reducing the amount of communication in sensor networks by means of learning techniques, neurodisorder patient monitoring via gait sensor networks, and cognitive radio-based sensor networks.

Part II—Signal Processing: This part describes the optimization of sensor network performance based on digital signal processing (DSP) techniques. It includes the following important topics: cross-layer integration of routing and application-specific signal processing, on-board image processing in wireless multimedia sensor networks for intelligent transportation systems, and essential signal processing and data analysis methods to effectively handle and process the data acquired with the sensor networks for civil infrastructure systems. It also includes a paradigm for validating the extent of spatiotemporal associations among data sources to enhance data cleaning in sensor networks, a sensor stream reduction application, a basic methodology that is composed of four phases (characterization, reduction tools, robustness, and conception), discussions on how the compressive sensing (CS) can be used as a useful framework for the sensor networks to compress and acquire signals and save transmittal and computational power in sensors, and the use of Kalman filters for attack detection in a water system sensor network that consists of water level sensors and velocity sensors.

Part III—Networking: This part focuses on detailed network protocol design in order to achieve an intelligent sensor networking scenario. It covers the following topics: energy-efficient opportunistic routing protocol for sensor networking; multi-agent-driven wireless sensor cooperation for limited resource allocation; an illustration of how distributed event detection can achieve both high accuracy and energy efficiency; blanket/sweep/barrier coverage issues in sensor networks; linear state-estimator, locally and additionally, to perform management procedures that support the network of state-estimators to establish self-organization; low-power solution for wireless passive sensor network; the fusion of pre/post RFID correction techniques to reduce anomalies; RFID systems and sensor integration for tele-medicine; and the new generation of intrusion detection sensor networks.

Disclaimer: We have tried our best to provide credits to all cited publications in this book. We sincerely thank all authors who have published materials on intelligent sensor network and who have directly/indirectly contributed to this book through our citations. If you have questions on the contents of this book, please contact the editors Fei Hu (fei@eng.ua.edu) or Qi Hao (qh@eng.ua.edu). We will correct any errors and thus improve this book in future editions.

MATLAB® is a registered trademark of The Mathworks, Inc. For product information, please contact:

The MathWorks, Inc.
3 Apple Hill Drive
Natick, MA, 01760-2098 USA
Tel: 508-647-7000
Fax: 508-647-7001
E-mail: info@mathworks.com
Web: www.mathworks.com

Editors

Dr. Fei Hu is currently an associate professor in the Department of Electrical and Computer Engineering at The University of Alabama, Tuscaloosa, Alabama. He received his PhDs from Tongji University (Shanghai, China) in the field of signal processing (in 1999) and from Clarkson University (New York) in the field of electrical and computer engineering (in 2002). He has published over 150 journal/conference papers and book chapters. Dr. Hu's research has been supported by U.S. National Science Foundation, Cisco, Sprint, and other sources. His research expertise can be summarized as *3S—security, signals*, and *sensors*: (1) security, which includes cyberphysical system security and medical security issues; (2) signals, which refers to intelligent signal processing, that is, using machine learning algorithms to process sensing signals; and (3) sensors, which includes wireless sensor network design issues.

Dr. Qi Hao is currently an assistant professor in the Department of Electrical and Computer Engineering at The University of Alabama, Tuscaloosa, Alabama. He received his PhD from Duke University, Durham, North Carolina, in 2006, and his BE and ME from Shanghai Jiao Tong University, China, in 1994 and 1997, respectively, all in electrical engineering. His postdoctoral training in the Center for Visualization and Virtual Environment at The University of Kentucky was focused on 3D computer vision for human tracking and identification. His current research interests include smart sensors, intelligent wireless sensor networks, and distributed information processing. His research has been supported by U.S. National Science Foundation and other sources.

Contributors

Sameh Abdel-Naby
Center for Sensor Web Technologies
University College Dublin
Dublin, Ireland

Stephan Adler
Department of Mathematics and Computer Science
Computer Systems and Telematics
Freie Universität Berlin
Berlin, Germany

Andre L.L. Aquino
Computer Institute
Federal University of Alagoas
Alagoas, Brazil

Ville Autio
RF Media Laboratory
Centria University of Applied Sciences
Ylivieska, Finland

Qutub Ali Bakhtiar
Network Laboratory
Technological University of America
Coconut Creek, Florida

Gianluca Bontempi
Machine Learning Group
Department of Computer Science
Université Libre de Bruxelles
Brussels, Belgium

Xiaojun Cao
Department of Computer Science
Georgia State University
Atlanta, Georgia

David Cartes
Center for Advanced Power Systems
Florida State University
Tallahassee, Florida

F. Necati Catbas
Department of Civil, Environmental and Construction Engineering
University of Central Florida
Orlando, Florida

Jae-Gun Choi
Department of Information and Communications
Gwangju Institute of Science and Technology
Gwangju, South Korea

Peter Darcy
Information and Communication Technology
School of Information and Communication Technology
Institute for Integrated and Intelligent Systems
Griffith University
Brisbane, Queensland, Australia

Hristo Djidjev
Information Sciences
Los Alamos National Laboratory
Los Alamos, New Mexico

Norman Dziengel
Department of Mathematics and Computer Science
Computer Systems and Telematics
Freie Universität Berlin
Berlin, Germany

Touria El-Mezyani
Center for Advanced Power Systems
Florida State University
Tallahassee, Florida

Paulo R.S. Silva Filho
Computer Institute
Federal University of Alagoas
Alagoas, Brazil

Rama Murthy Garimella
International Institute of Information Technology
Hyderabad, India

Marco Ghibaudi
Real-Time Systems Laboratory
Scuola Superiore Sant'Anna
and
Scuola Superiore Sant'Anna Research Unit
National Inter-University Consortium for Telecommunications
Pisa, Italy

Mustafa Gul
Department of Civil and Environmental Engineering
University of Alberta
Edmonton, Alberta, Canada

Qi Hao
Department of Electrical and Computer Engineering
The University of Alabama
Tuscaloosa, Alabama

Fei Hu
Department of Electrical and Computer Engineering
The University of Alabama
Tuscaloosa, Alabama

S. Sitharama Iyengar
School of Computer and Information Sciences
Florida International University
Miami, Florida

Vasanth Iyer
School of Computer and Information Sciences
Florida International University
Miami, Florida

Joni Jamsa
RF Media Laboratory
Centria University of Applied Sciences
Ylivieska, Finland

Krasimira Kapitanova
Department of Computer Science
University of Virginia
Charlottesville, Virginia

Zakaria Kasmi
Department of Mathematics and Computer Science
Computer Systems and Telematics
Freie Universität Berlin
Berlin, Germany

David Keller
Center for Advanced Power Systems
Florida State University
Tallahassee, Florida

Abdelmajid Khelil
Department of Computer Science
Technical University of Darmstadt
Darmstadt, Germany

Jerry Krill
Applied Physics Laboratory
Johns Hopkins University
Laurel, Maryland

Sumit Kumar
International Institute of Information Technology
Hyderabad, India

Yann-Aël Le Borgne
Machine Learning Group
Department of Computer Science
Université Libre de Bruxelles
Brussels, Belgium

Heung-No Lee
Department of Information and Communications
Gwangju Institute of Science and Technology
Gwangju, South Korea

Xuefeng Liu
Department of Computing
Hong Kong Polytechnic University
Kowloon, Hong Kong, People's Republic of China

Mika Luimula
Faculty of Telecommunication and e-Business
Turku University of Applied Sciences
Turku, Finland

Mohammadreza Mahmudimanesh
Department of Computer Science
Technical University of Darmstadt
Darmstadt, Germany

Kia Makki
Technological University of America
Coconut Creek, Florida

Kebina Manandhar
Department of Computer Science
Georgia State University
Atlanta, Georgia

Shruti Mantravadi
Army College of Dental Sciences
Dr. NTR University of Health Sciences
Secunderabad, India

Marlin H. Mickle
Department of Electrical and Computer Engineering
University of Pittsburgh
Pittsburgh, Pennsylvania

Konstantin Mikhaylov
Oulu Southern Institute
University of Oulu
Ylivieska, Finland

Conor Muldoon
Center for Sensor Web Technologies
University College Dublin
Dublin, Ireland

Michael O'Driscoll (retired)
Applied Physics Laboratory
Department of Research and Exploratory Development
Johns Hopkins University
Laurel, Maryland

Ajay Ogirala
Department of Electrical and Computer Engineering
University of Pittsburgh
Pittsburgh, Pennsylvania

Gregory O'Hare
Center for Sensor Web Technologies
University College Dublin
Dublin, Ireland

Paolo Pagano
National Laboratory of Photonic Networks
National Interuniversity Consortium for Telecommunications
and
Real-Time Systems Laboratory
Scuola Superiore Sant'Anna
Pisa, Italy

Zoltan Papp
Instituut voor Toegepast Natuurkundig Onderzoek
Oude Waalsdorperweg
Den Haag, the Netherlands

Sang-Jun Park
Department of Information and Communications
Gwangju Institute of Science and Technology
Gwangju, South Korea

Matteo Petracca
National Laboratory of Photonic Networks
National Interuniversity Consortium for Telecommunications
and
Real-Time Systems Laboratory
Scuola Superiore Sant'Anna
Pisa, Italy

Niki Pissinou
School of Computer and Information Sciences
Florida International University
Miami, Florida

Miodrag Potkonjak
Department of Computer Science
University of California
Los Angeles, California

Prapassara Pupunwiwat
Information and Communication Technology
School of Information and Communication Technology
Institute for Integrated and Intelligent Systems
Griffith University
Brisbane, Queensland, Australia

Wanzhi Qiu
Victoria Research Laboratory
National ICT Australia
and
Department of Electrical and Electronic Engineering
The University of Melbourne
Parkville, Victoria, Australia

Ricardo A. Rabelo
Department of Computer
Federal University of Ouro Preto
Minas Gerais, Brazil

Hosam Rowaihy
Department of Computer Engineering
King Fahd University of Petroleum and Minerals
Dhahran, Saudi Arabia

Vyasa Sai
Department of Electrical and Computer Engineering
University of Pittsburgh
Pittsburgh, Pennsylvania

Claudio Salvadori
Real-Time Systems Laboratory
Scuola Superiore Sant'Anna
and
Scuola Superiore Sant'Anna Research Unit
National Inter-University Consortium for Telecommunications
Pisa, Italy

Jochen Schiller
Department of Mathematics and Computer Science
Computer Systems and Telematics
Freie Universität Berlin
Berlin, Germany

Joris Sijs
Instituut voor Toegepast Natuurkundig Onderzoek
Oude Waalsdorperweg
Den Haag, the Netherlands

Deepti Singhal
International Institute of Information Technology
Hyderabad, India

Efstratios Skafidas
Victoria Research Laboratory
National ICT Australia
and
Department of Electrical and Electronic Engineering
The University of Melbourne
Parkville, Victoria, Australia

Sang H. Son
Department of Information and Communication Engineering
Daegu Gyeongbuk Institute of Science and Technology
Daegu, Korea

Sanjeev Srivastava
Center for Advanced Power Systems
Florida State University
Tallahassee, Florida

Bela Stantic
Information and Communication Technology
School of Information and Communication Technology
Institute for Integrated and Intelligent Systems
Griffith University
Brisbane, Queensland, Australia

Qingquan Sun
Department of Electrical and Computer Engineering
The University of Alabama
Tuscaloosa, Alabama

Neeraj Suri
Department of Computer Science
Technical University of Darmstadt
Darmstadt, Germany

Shaojie Tang
Department of Computer Science
Illinois Institute of Technology
Chicago, Illinois

Jouni Tervonen
Oulu Southern Institute
University of Oulu
Ylivieska, Finland

Elizabeth F. Wanner
Departamento de Computacao
Centro Federal Tecnologica de Minas Gerais
Minas Gerais, Brazil

Georg Wittenburg
Laboratoire d'Informatique de L'École Polytechnique
Institut National de Recherche en Informatique et en Automatique
Palaiseau, France

Xiaohua Xu
Department of Computer Science
Illinois Institute of Technology
Chicago, Illinois

Ping Yi
School of Information Security Engineering
Shanghai Jiao Tong University
Xuhui, Shanghai, People's Republic of China

Ting Zhu
Department of Computer Science
Binghamton University
Binghamton, New York

Marco Ziegert
Computer Systems and Telematics
Freie Universität Berlin
Berlin, Germany

Olga Zlydareva
Center for Sensor Web Technologies
University College Dublin
Dublin, Ireland

I

INTELLIGENT SENSOR NETWORKS: MACHINE LEARNING APPROACH

Chapter 1

Machine Learning Basics

Krasimira Kapitanova and Sang H. Son

Contents

The goal of machine learning is to design and develop algorithms that allow systems to use empirical data, experience, and training to evolve and adapt to changes that occur in their environment. A major focus of machine learning research is to automatically induce models, such as rules and patterns, from the training data it analyzes. As shown in Figure 1.1, machine learning combines

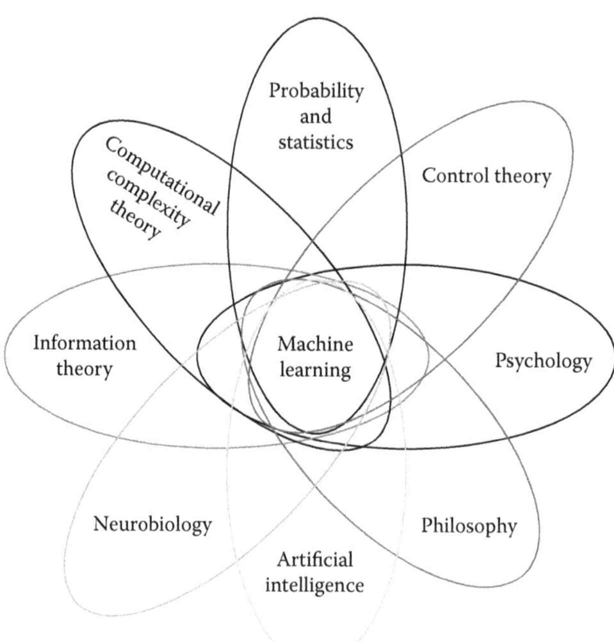

Figure 1.1 Machine learning is a broad discipline, combining approaches from many different areas.

techniques and approaches from various areas, including probability and statistics, psychology, information theory, and artificial intelligence.

Wireless sensor network (WSN) applications operate in very challenging conditions, where they constantly have to accommodate environmental changes, hardware degradation, and inaccurate sensor readings. Therefore, in order to maintain sufficient operational correctness, a WSN application often needs to learn and adapt to the changes in its running environment. Machine learning has been used to help address these issues. A number of machine learning algorithms have been employed in a wide range of sensor network applications, including activity recognition, health care, education, and for improving the efficiency of heating, ventilating, and air conditioning (HVAC) system.

The abundance of machine learning algorithms can be divided into two main classes: *supervised* and *unsupervised* learning, based on whether the training data instances are labeled. In supervised learning, the learner is supplied with labeled training instances, where both the input and the correct output are given. In unsupervised learning, the correct output is not provided with the input. Instead, the learning program must rely on other sources of feedback to determine whether or not it is learning correctly. A third class of machine learning techniques, called *semi-supervised learning*, uses a combination of both labeled and unlabeled data for training. Figure 1.2 shows the relationship between these three machine learning classes.

In this chapter, we have surveyed machine learning algorithms in sensor networks from the perspective of what types of applications they have been used for. We give examples from all three machine learning classes and discuss how they have been applied in a number of sensor network applications. We present the most frequently used machine learning algorithms, including clustering, Bayes probabilistic models, Markov models, and decision trees. We also analyze the

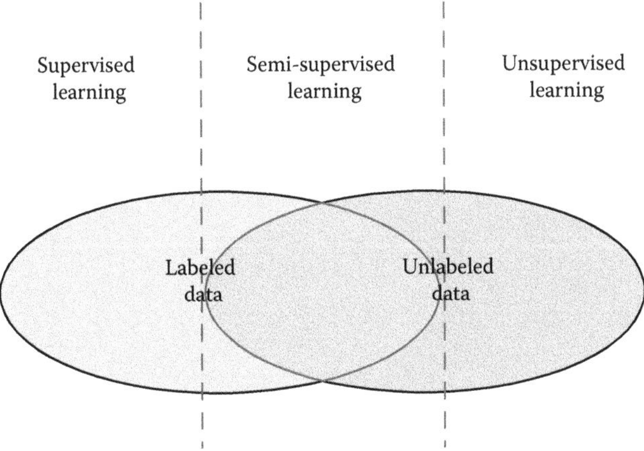

Figure 1.2 Machine learning algorithms are divided into supervised learning, which used labeled training data, and unsupervised learning, where labeled training data is not available. A third class of machine learning technique, semi-supervised learning, makes use of both labeled and unlabeled training data.

challenges, advantages, and drawbacks of using different machine learning algorithms. Figure 1.3 shows the machine learning algorithms introduced in this chapter.

1.1 Supervised Learning

In supervised learning, the learner is provided with labeled input data. This data contains a sequence of input/output pairs of the form $\langle x_i, y_i \rangle$, where x_i is a possible input and y_i is the correctly labeled output associated with it. The aim of the learner in supervised learning is to learn the mapping from inputs to outputs. The learning program is expected to learn a function f that accounts for the input/output pairs seen so far, $f(x_i) = y_i$, for all i. This function f is called a *classifier* if the output is discrete and a *regression function* if the output is continuous. The job of the classifier/regression function is to correctly predict the outputs of inputs it has not seen before. For example, the inputs can be a set of sensor firings and the outputs can be the activities that have caused those sensor nodes to fire.

The execution of a supervised learning algorithm can be divided into five main steps (Figure 1.4).

Step 1 is to determine what training data is needed and to collect that data. Here we need to answer two questions: "What data is necessary?" and "How much data do we need?" The designers have to decide what training data can best represent real-world scenarios for the specific application. They also need to determine how much training data should be collected. Although the more training data we have, the better we can train the learning algorithm, collecting training data and providing correct labels can often be expensive and laborious. Therefore, an application designer always strives to maintain the size of the training data not only large enough to provide sufficient training but also small enough to avoid any unnecessary costs associated with data collection and labeling.

Step 2 is to identify the feature set, also called *feature vector*, to be used to represent the input. Each feature in the feature set represents a characteristic of the objects/events that are being classified. There is a trade-off between the size of the feature vector and the classification accuracy of the machine learning algorithm. A large feature vector significantly increases the complexity of the

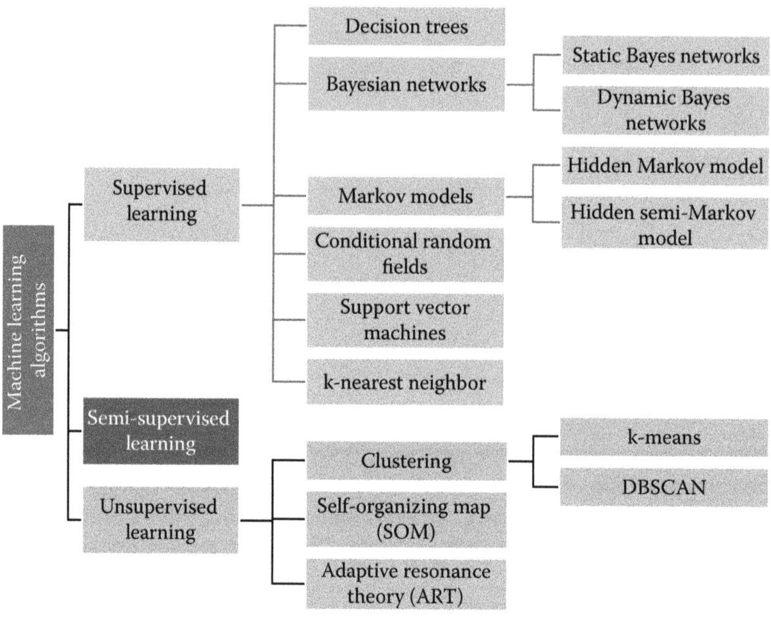

Figure 1.3 Classification of the machine learning algorithms most widely used in WSN applications.

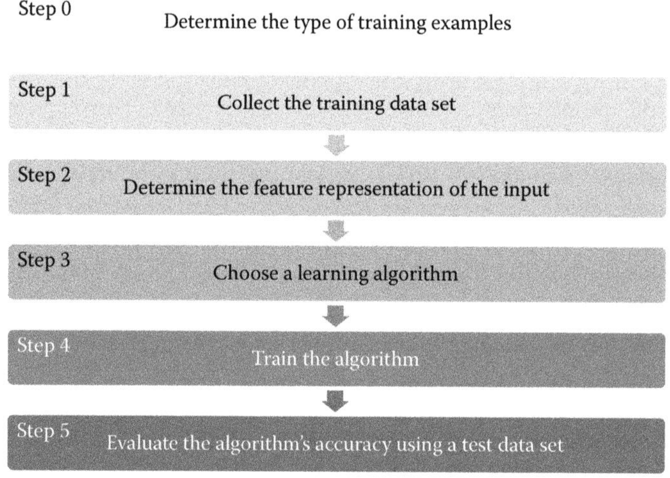

Figure 1.4 The stages of supervised machine learning.

classification. However, using a small feature vector, which does not contain sufficient description of the objects/events, could lead to poor classification accuracy. Therefore, the feature vector should be sufficiently large to represent the important features of the object/event and small enough to avoid excessive complexity.

Step 3 is to select a suitable learning algorithm. A number of factors have to be considered when choosing a learning algorithm for a particular task, including the content and size of the training dataset, noise in the system, accuracy of the labeling, and the heterogeneity and redundancy of the input data. We also have to evaluate the requirements and characteristics of the sensor network application itself. For example, for an activity recognition application, the duration of sensor use plays a significant role in determining the activity being executed. Therefore, to achieve high activity recognition accuracy, we would prefer to use machine learning algorithms that can explicitly model state duration.

The most frequently used supervised machine learning algorithms include support vector machines, naive Bayes classifiers, decision trees, hidden Markov models, conditional random field, and k-nearest neighbor algorithms. There are also a number of approaches that have been applied to improve the performance of the chosen classifiers, such as bagging, boosting, and using classifier ensembles. Each of the algorithms has its advantages and disadvantages, which make it suitable for some types of applications but inappropriate for others.

Step 4 is to train the chosen learning algorithm using the collected training data. In this step, the algorithm learns the function that best matches the input/output training instances.

Step 5 is evaluation of the algorithm's accuracy. We assess the accuracy of the learned function with the help of testing dataset, where the testing dataset is different from the training dataset. In this step, we evaluate how accurately the machine learning algorithm classifies entries from the testing set based on the function it has learned through the training dataset.

Different supervised learning algorithms have been used and evaluated experimentally in a variety of sensor network applications. In the rest of this section, we describe some of the algorithms that are most frequently used in WSN applications.

1.1.1 Decision Trees

Decision trees are characterized by fast execution time, ease in the interpretation of the rules, and scalability for large multidimensional datasets (Cabena et al. 1998, Han 2005). The goal of decision tree learning is to create a model that predicts the value of the output variable based on the input variables in the feature vector. Each node corresponds to one of the feature vector variables. From every node, there are edges to children, where there is an edge per each of the possible values (or range of values) of the input variable associated with the node. Each leaf represents a possible value for the output variable. The output variable is determined by following a path that starts at the root and is guided by the values of the input variables.

Figure 1.5 shows an example decision tree for a sensor network activity detection application. In this scenario, we assume that there are only two events of interest in the kitchen: *cooking* and *getting a drink*. The decision tree uses sensor node firings to distinguish between these two activities. For example, if there is movement in the kitchen and the stove is being used, the algorithm determines that the residents must be cooking. However, if there is movement in the kitchen, the stove is not being used, and somebody opens the cups cupboard, the algorithm decides that the activity being performed at the moment is *getting a drink*. This is a simple example illustrating how decision trees can be applied to sensor network applications. In reality, the decision trees that are learned by real applications are much more complex.

The C4.5 algorithm is one of the well-known, top-down, greedy search algorithms for building decision trees (Quinlan 1993). The algorithm uses entropy and information gain metrics to induce a decision tree. The C4.5 algorithm has been used for *activity recognition* in the PlaceLab project at MIT (Logan et al. 2007). The authors of the project monitored a home deployed with over 900 sensors, including wired reed switches, current and water flow inputs, object and person motion detectors, and radio frequency identification (RFID) tags. They collected data for 43 typical house activities, and C4.5 was one of the classifiers used by their activity recognition approach.

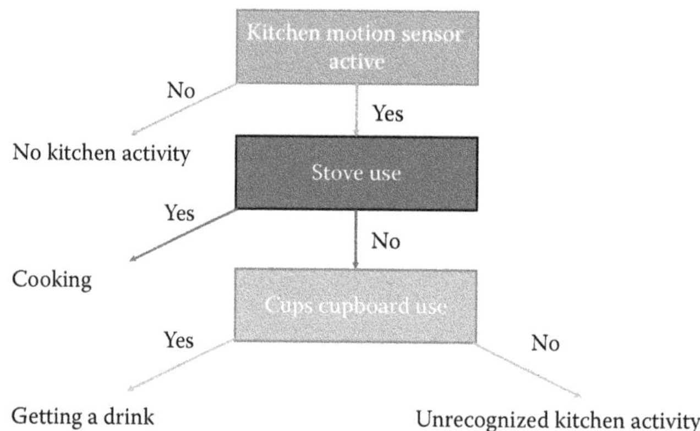

Figure 1.5 Example decision tree for an activity detection application. In this scenario, we are interested only in two of the kitchen activities: cooking and getting a drink. The decision tree is used to determine which one of these activities is currently occurring based on the sensor nodes that are firing in the kitchen.

C4.5 was used for *target recognition* in an underwater wireless sensor surveillance system (Cayirci et al. 2006). Each node in the network was equipped with multiple microsensors of various types, including acoustic, magnetic, radiation, and mechanical sensors. The readings from these sensors were used by the decision tree recognition algorithms to classify submarines, small delivery vehicles, mines, and divers.

C4.5 was also used as part of an algorithm to *automatically recognize physical activities and their intensities* (Tapia et al. 2007). The algorithm monitors the readings of triaxial wireless accelerometers and wireless heart rate monitors. The approach was evaluated using datasets consisting of 10 physical gymnasium activities collected from a total of 21 people.

1.1.2 Bayesian Network Classifiers

Bayesian probability interprets the concept of probability as *degree of belief.* A Bayesian classifier analyzes the feature vector describing a particular input instance and assigns the instance to the most likely class. A Bayesian classifier is based on applying Bayes' theorem to evaluate the likelihood of particular events. Bayes' theorem gives the relationship between the *prior* and *posterior* beliefs for two events. In Bayes' theorem, P(A) is the prior initial belief in A. P(A|B) is the posterior belief in A, after B has been encountered, i.e., the conditional probability of A given B. Similarly for B, P(B) is the prior initial belief in A, and P(B|A) is the posterior belief in B given A. Assuming that $P(B) \neq 0$, Bayes' theorem states that $P(A|B) = \dfrac{P(B|A)P(A)}{P(B)}$.

The Bayesian network is a probabilistic model that represents a set of random variables and their conditional dependencies via a direct acyclic graph. For example, a Bayesian network could represent the probabilistic relationships between activities and sensor readings. Given a set of sensor readings, the Bayesian network can be used to evaluate the probabilities that various activities are being performed.

Bayesian networks have a number of advantages. Since a Bayes network relates only nodes that are probabilistically related by a causal dependency, an enormous saving of computation can result. Therefore, there is no need to store all possible configurations of states. Instead, all that needs to be stored is the combinations of states between sets of related parent–child nodes. Also Bayes networks are extremely adaptable. They can be started off small, with limited knowledge about the domain, and grow as they acquire new knowledge.

Bayes networks have been applied to a variety of *sensor fusion problems*, where data from various sources must be integrated in order to build a complete picture of the current situation. They have also been used in *monitoring and alerting applications* where the application should recognize whether specific events have occurred and decide if an alert or a notification should be sent. Further, they have been applied to a number of *activity recognition* applications and evaluated using numerous single- and multiple-resident home deployments.

Bayesian networks can be divided into two groups, *static* and *dynamic*, based on whether they are able to model temporal aspects of the events/activities of interest. We introduce an example classifier for each of these two classes: static naive Bayes classifier and dynamic naive Bayes classifier.

1.1.2.1 Static Bayesian Network Classifiers

A very commonly used representative of the static Bayesian networks is the static *naive Bayes classifier*. Learning Bayesian classifiers can be significantly simplified by making the naive assumption that the features describing a class are independent. The classifier makes the assumption that the presence or absence of a feature of a class is unrelated to the presence or absence of any of the other features in the feature vector. The naive Bayes classifier is one of the most practical learning methods, and it has been widely used in many sensor network applications, including *activity recognition* in residence for elders (van Kasteren and Kröse 2007), activity recognition in the PlaceLab project at MIT (Logan et al. 2007), *outlier detection* (Janakiram et al. 2006), and *body sensor networks* (Maurer et al. 2006).

Figure 1.6 shows a naive Bayesian model for the recognition of an activity. In this scenario, the activity at time t, $activity_t$, is independent of any previous activities. It is also assumed that the sensor data R_t is dependent only on the $activity_t$.

Naive Bayes classifiers have the following advantages:

1. They can be trained very efficiently.
2. They are very well suited for categorical features.
3. In spite of their naive design and the independence assumptions, naive Bayes classifiers have performed very well in many complex real-world situations. They can work with more than 1000 features.
4. They are good for combining multiple models and can be used in an iterative way.

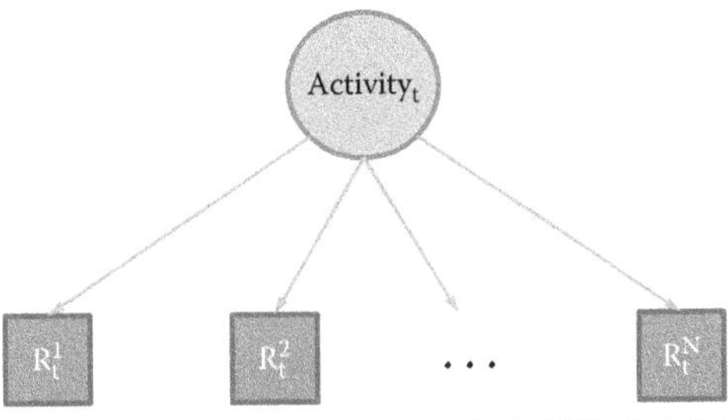

Figure 1.6 Static Bayesian network: $activity_t$ denotes the activity being detected at time t, and R_t^i represents the data from sensor i at time t.

A disadvantage of naive Bayes classifiers is that, if conditional independence is not true, i.e., there is dependence between the features of the analyzed classes, they may not be a good model. Also naive Bayes classifiers assume that all attributes that influence a classification decision are observable and represented. Despite these drawbacks, experiments have demonstrated that naive Bayes classifiers are very accurate in a number of problem domains. Simple naive Bayes networks have even been proved comparable with more complex algorithms, such as decision trees (Tapia 2004).

1.1.2.2 Dynamic Bayesian Network Classifiers

Another disadvantage of static Bayesian networks is that they cannot model the temporal aspect of sensor network events. Dynamic Bayesian networks, however, are capable of representing a sequence of variables, where the sequence can be consecutive readings from a sensor node. Therefore, dynamic Bayesian networks, although more complex, might be better suited for modeling events and activities in sensor network applications.

Figure 1.7 shows a naive dynamic Bayesian model, where the activity_{t+1} variable is directly influenced only by the previous variable, activity_t. The assumption with these models is that an event can cause another event in the future, but not vice versa. Therefore, directed arcs between events/activities should flow forward in time and cycles are not allowed.

Dynamic models have been used in *activity recognition* applications. A naive dynamic Bayes classifier is compared to a naive static Bayes classifier using two publicly available datasets (van Kasteren and Kröse 2007). The dynamic Bayes classifier is shown to achieve higher activity recognition accuracy than the static model. A dynamic Bayesian filter was successfully applied to the *simultaneous tracking and activity recognition* problem, which exploits the synergy between location and activity to provide the information necessary for automatic health monitoring (Wilson and Atkenson 2005).

1.1.3 Markov Models

A process is considered to be Markov if it exhibits the Markov property, which is the lack of memory, i.e., the conditional probability distribution of future states of the process depends only on the present state, and not on the events that preceded it. We discuss two types of Markov models: hidden Markov model (HMM) and hidden semi-Markov model (HSMM).

1.1.3.1 Hidden Markov Model

An HMM can be viewed as a simple dynamic Bayesian network. When using an HMM, the system is assumed to be a Markov process with unobserved (hidden) states. Even though the sequence of

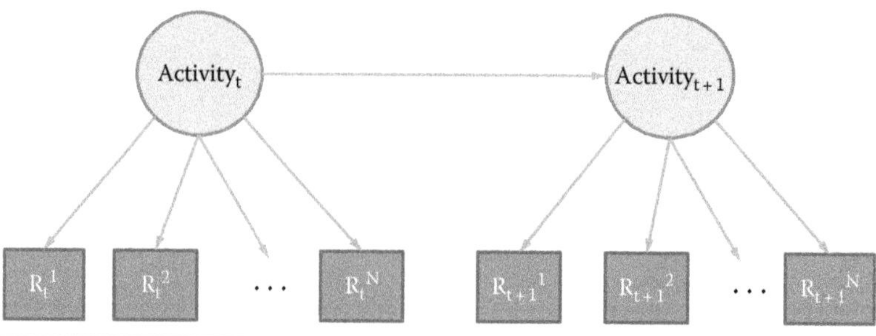

Figure 1.7 An example of a naive dynamic Bayesian network.

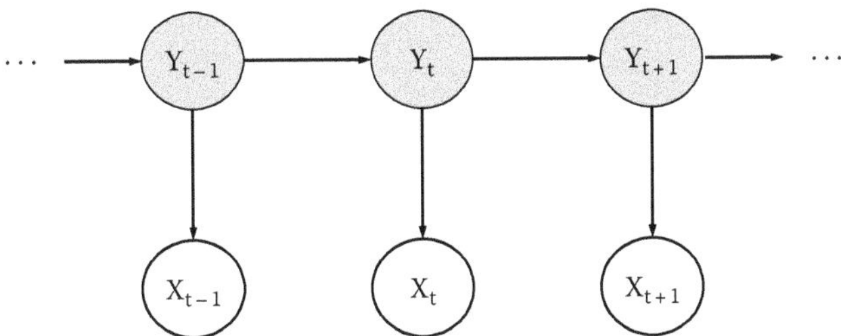

Figure 1.8 Hidden Markov model example. The states of the system Y_i are hidden, but their corresponding outputs X_i are visible.

states is hidden, the output, which is dependent on the state, is visible. Therefore, at each time step, there is a hidden variable and an observable output variable. In sensor network applications, the hidden variable could be the event or activity performed and the observable output variable is the vector of sensor readings.

Figure 1.8 shows an example HMM, where the states of the system Y are hidden, but the output variables X are visible. There are two dependency assumptions that define this model, represented by the directed arrows in the figure.

1. Markov assumption: the hidden variable at time t, namely Y_t, depends only on the previous hidden variable Y_{t-1} (Rabiner 1989).
2. The observable output variable at time t, namely X_t, depends only on the hidden variable Y_t.

With these assumptions, we can specify an HMM using the following three probability distributions:

1. *Initial-state distribution*: the distribution over initial states $p(Y_1)$
2. *Transition distribution*: the distribution $p(Y_t|Y_{t+1})$, which represents the probability of going from one state to the next
3. *Observation distribution*: the distribution $p(X_t|Y_t)$, which indicates the probability that the hidden state Y_t would generate observation X_t

Learning the parameters of these distributions corresponds to maximizing the joint probability distribution p(X, Y) of the paired observation and label sequences in the training data. Modeling the joint probability distribution p(X, Y) makes HMMs a *generative model*.

HMMs have been extensively used in many sensor network applications. Most of the earlier work on *activity recognition* used HMMs to recognize the activities from sensor data (Patterson et al. 2005, Wilson and Atkenson 2005, van Kasteren et al. 2008). An HMM is also used in the smart thermostat project (Lu et al. 2010). The smart thermostat technology automatically *senses the occupancy and sleep patterns in a home* and uses these patterns to automatically operate the heating, ventilation, and air cooling (HVAC) system in the home. The authors employ an HMM to estimate the probability of the home being in each of three states: unoccupied, occupied and the residents are active, and occupied with the residents sleeping. HMMs were also applied in a *biometric identification* application for multi-resident homes (Srinivasan et al. 2010). In this project, height sensors were mounted above the doorways in a home and an HMM was used to identify the location of each of the residents.

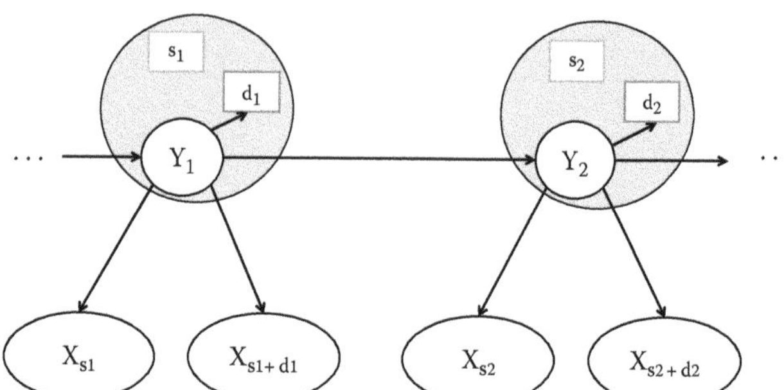

Figure 1.9 Hidden semi-Markov model. Each hidden state y_i is characterized by start position s_i and a duration d_i. This means that the system is in state y_i from time s_i to time $s_i + d_i$.

A weakness of conventional HMMs is their lack of flexibility in modeling state durations. With HMMs, there is a constant probability of changing state, given that the system is in its current state of the model. This, however, limits the modeling capability. For example, the activity preparing dinner typically spans at least several minutes. To prepare dinner in less than a couple of minutes is not very usual. The geometric distribution used by HMMs to represent time duration cannot be used to represent event distributions where shorter durations are less possible.

1.1.3.2 Hidden Semi-Markov Models

An HSMM differs from an HMM in that HSMMs explicitly model the duration of hidden states. This means that the probability of there being a change in the hidden state depends on the amount of time that has elapsed since entry into the current state (Figure 1.9).

A number of projects have used HSMMs to *learn and recognize human activities* of daily living (Zhang et al. 2008, Duong et al. 2009, van Kasteren et al. 2010). HSMMs were also applied to *behavior understanding* from video streams in a nursing center (Chung and Liu 2008). The proposed approach infers human behaviors through three contexts: spatial, activities, and temporal. HSMM were also used in a *mobility tracking* application for cellular networks (Mark and Zaidi 2002).

The activity recognition accuracy achieved by HSMM is compared to that of HMM (van Kasteren et al. 2010). The authors evaluate the recognition performance of these models using two fully annotated real-world datasets consisting of several weeks of data. The first dataset was collected in a three-room single-resident apartment and the second dataset was from a six-room single-resident house. The results show that HSMM consistently outperforms HMM. This indicates that accurate duration modeling is important in real-world activity recognition applications as it can lead to significantly better performance. The use of duration in the classification process helps especially in scenarios where the sensor data does not provide sufficient information to distinguish between activities.

1.1.4 Conditional Random Fields

Conditional random fields (CRFs) are often considered an alternative to HMMs. The CRF is a statistical modeling method, which is a type of an undirected probabilistic graphical model that defines a single log-linear distribution over label sequences given a particular observation sequence. It is used to encode known relationships between observations and construct consistent interpretations.

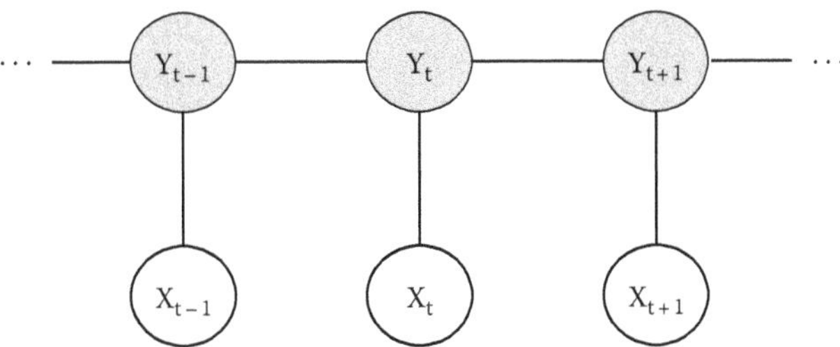

Figure 1.10 A linear-chain CRF model. Similar to an HMM, the states of the system Y_i are hidden, but their corresponding outputs X_i are visible. Unlike the HMM model, however, the graph represented by the CRF model is undirected.

The CRF model that most closely resembles an HMM is the linear-chain CRF. As Figure 1.10 shows, the model of a linear-chain CRF is very similar to that of an HMM (Figure 1.8). The model still contains hidden variables and corresponding observable variables at each time step. However, unlike the HMM, the CRF model is undirected. This means that two connected nodes no longer represent a conditional distribution. Instead we can talk about potential between two connected nodes. In comparison with HMM, the two conditional probabilities, *observation* probability $p(X_t|Y_t)$ and *transition* probability $p(Y_t|Y_{t+1})$, have been replaced by the corresponding potentials. The essential difference lies in the way we learn the model parameters. In the case of HMMs, the parameters are learned by maximizing the *joint* probability distribution p(X, Y). CRFs are *discriminative models*. The parameters of a CRF are learned by maximizing the *conditional* probability distribution p(Y|X), which belongs to the family of exponential distributions (Sutton and McCailum 2006).

CRF models have been applied to *activity recognition* in the home from video streams, in which primitive actions, such as go-from-A-to-B, are recognized in a laboratory-like dining room and kitchen setup (Truyen et al. 2005). The results from these experiments show that CRFs perform significantly better than the equivalent generative HMMs even when a large portion of the data labels are missing. CRFs were also used for *modeling concurrent and interleaving activities* (Hu et al. 2008). The authors perform experiments using one of the MIT PlaceLab datasets (Logan et al. 2007), PLA1, which consists of 4 hours of sensor data.

T. van Kasteren et al. use four different datasets, two bathroom datasets and two kitchen datasets, to compare the performance of HMM to that of CRF (van Kasteren et al. 2010). The experiments show that, when applied to activity recognition tasks, CRF models achieve higher accuracy than HMM models. The authors contribute the results to the flexibility of discriminative models, such as CRF, in dealing with violations of the modeling assumptions. However, the higher accuracy achieved by CRF models comes at a price.

1. Training discriminative models takes much longer than training their generative counterparts.
2. Discriminative models are more prone to *overfitting*. Overfitting occurs when a model describes random noise instead of the underlying relationship. This happens when the model is trained to maximize its performance on the training data. However, a model's efficiency is determined not by how well it performs on the training data but by its generality and how it performs on unseen data.

Whether the improved recognition performance of CRFs is worth the extra computational cost depends on the application. The data can be modeled more accurately using an HSMM, which

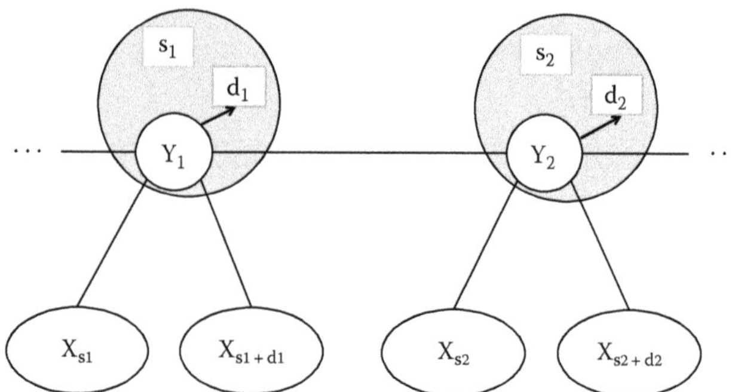

Figure 1.11 An example semi-Markov CRF. Similar to an HSMM model, each of the hidden states y_i is characterized by start position s_i and a duration d_i. However, unlike an HSMM, the HMCRF graph is undirected.

allows both speedy learning and good performance, and is less prone to overfitting. However, it does result in slower inference and depends on correct modeling assumptions for the durations.

1.1.4.1 Semi-Markov Conditional Random Fields

Similar to HMMs, which have their semi-Markov variants, CRFs also have a semi-Markov variant: semi-Markov conditional random fields (SMCRFs). An example SMCRF model is shown in Figure 1.11. The SMCRF inherits features from both semi-Markov models and CRFs as follows:

1. It models the duration of states explicitly (like HSMM).
2. Each of the hidden states is characterized by a start position and duration (like HSMM).
3. The graph of the model is undirected (like CRF).

Hierarchical SMCRFs were used in an *activity recognition* application on a small laboratory dataset from the domain of video surveillance (Truyen et al. 2008). The task was to recognize indoor trajectories and activities of a person from the noisy positions extracted from the video. The data had 90 sequences, each of which corresponded to one of three possible activities: preparing a short meal, preparing a normal meal, and having a snack. The hierarchical SMCRF outperformed both a conventional CRF and a dynamic CRF.

SMCRFs were also used for activity recognition by van Kasteren et al. (2010). The results show that unlike the big improvement achieved by using HSMMs over HMMs, SMCRFs only slightly outperform CRFs. The authors attribute this result to the fact that CRFs are more robust in dealing with violations to the modeling assumptions. Therefore, allowing to explicitly model duration distributions might not have the same significant benefits as seen with HSMM.

1.1.5 Support Vector Machines

A support vector machine (SVM) is a non-probabilistic binary linear classifier. The output prediction of an SVM is one of the two possible classes. Given a set of training instances, each marked as belonging to one of the two classes, an SVM algorithm builds an N-dimensional hyperplane model that assigns future instances into one of the two possible output classes.

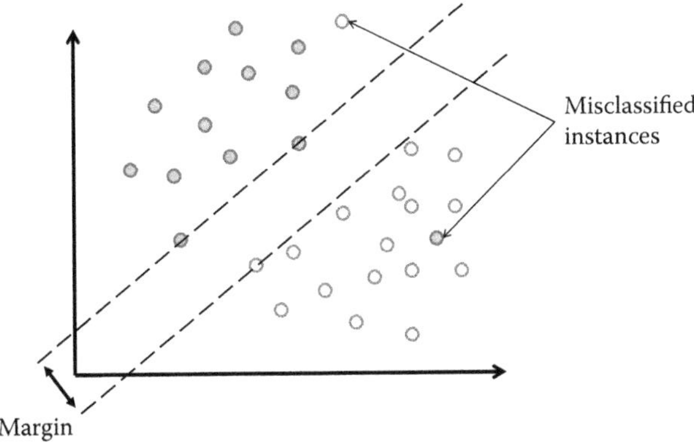

Figure 1.12 A two-dimensional SVM model. The instances of the two possible classes are divided by a clear gap.

As shown in Figure 1.12, an SVM model is a representation of the input instances as points in space, mapped so that the instances of the separate classes are divided by a clear gap. New examples are then mapped into that same space and predicted to belong to a class based on which side of the gap they fall on. In other words, the goal of the SVM analysis is to find a line that separates the instances based on their class. There are an infinite number of possible lines, and one of the challenges with SVM models is finding the optimal line.

SVMs have been applied to a large number of sensor network applications. Sathik et al. use SVMs in *early forest fire detection* applications (Mohamed Sathik et al. 2010). SVMs were also applied to *target classification* applications for distributed sensor networks (Li et al. 2001). The experiments were performed on real seismic and acoustic data. SVMs are compared to a k-nearest neighbor algorithm and a maximum likelihood algorithm and are shown to achieve the highest target classification accuracy. Tran and Nguyen use SVMs to achieve accurate *geographic location estimations* for nodes in a WSN, where the majority of nodes do not have effective self-positioning functionality (Tran and Nguyen 2008). SVMs were also applied to investigating the possibility of *recognizing visual memory recall* (Bulling and Roggen 2011). The project aims to find if people react differently to images they have already seen as opposed to images they are seeing for the first time.

1.1.6 k-Nearest Neighbor Algorithms

The k-nearest neighbor (k-NN) algorithm is among the simplest of machine learning algorithms, yet it has proven to be very accurate in a number of scenarios. The training examples are vectors in a multidimensional feature space, each with a class label. The training phase of the algorithm consists only of storing the feature vectors and class labels of the training samples. A new instance is classified by a majority vote of its neighbors, with the instance being assigned the class that is most common among its k nearest neighbors (Figure 1.13).

The best choice of k depends upon the data. k must be a positive integer and it is typically small. If $k = 1$, the new instance is simply assigned to the class of its nearest neighbor. Larger values of k reduce the effect of noise on the classification but make boundaries between classes less distinct. A good k can be selected by various heuristic techniques, for example, cross-validation.

Although the k-NN algorithm is quite accurate, the time required to classify an instance could be high since the algorithm has to compute the distances (or similarity) of that instance to all

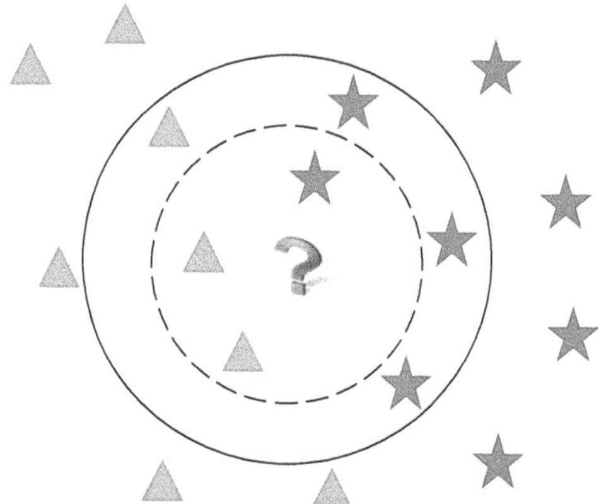

Figure 1.13 Example of k-nearest algorithm classification. The question mark is the test sample, and it should be classified as either a star or a triangle. If k = 3, the test sample is assigned to the class of triangles because there are two triangles and one star inside the inner circle. If k = 7, the test sample is assigned to the class of stars since there are four stars and three triangles in the outer circle.

the instances in the training set. Therefore, the classification time of k-NN is proportional to the number of features and the number of training instances.

k-NN algorithms have been applied to a wide variety of sensor network applications. Ganesan et al. propose the use of k-NN for *spatial data interpolation* in sensor networks (Ganesan et al. 2004). Due to its simplicity, k-NN allows the sampling to be done in a distributed and inexpensive manner. A disadvantage with this approach, however, is that k-NN interpolation techniques might perform poorly in highly irregular settings. Winter et al. also analyze the application of k-NN queries for *spatial data queries* in sensor networks (Winter et al. 2005). They design two algorithms based on k-NN, which are used to intelligently prune off irrelevant nodes during query propagation, thus reducing the energy consumption while maintaining high query accuracy. Duarte and Hu evaluate the accuracy of k-NN in the context of *vehicle classification* (Duarte and Hu 2004). The authors collect a real-world dataset and analyze both the acoustic and the seismic modality. The results show that in this application scenario, k-NN algorithms achieve comparable accuracy to that of SVMs.

1.2 Unsupervised Learning

Collecting labeled data is resource and time consuming, and accurate labeling is often hard to achieve. For example, obtaining sufficient training data for activity recognition in a home might require 3 or 4 weeks of collecting and labeling data. Further, labeling is difficult not only for remote areas that are not easily accessible, but also for home and commercial building deployments. For any of those deployments, someone has to perform the data labeling. In a home deployment, the labeling can be done by the residents themselves, in which case they have to keep a log of what they are doing and at what time. Previous experience has shown that these logs are often incomplete and inaccurate. An alternative solution is to install cameras throughout the house and monitor the activities of the residents. However, this approach is considered to be privacy-invasive and therefore not suitable.

In unsupervised learning, the learner is provided with input data, which has not been labeled. The aim of the learner is to find the inherent patterns in the data that can be used to determine the correct output value for new data instances. The assumption here is that there is a structure to the input space, such that certain patterns occur more often than others, and we want to see what generally happens and what does not. In statistics, this is called *density estimation.*

Unsupervised learning algorithms are very useful for sensor network applications for the following reasons:

- Collecting labeled data is resource and time-consuming.
- Accurate labeling is hard to achieve.
- Sensor networks applications are often deployed in unpredictable and constantly changing environments. Therefore, the applications need to evolve and learn without any guidance, by using unlabeled patterns.

A variety of unsupervised learning algorithms have been used in sensor network applications, including different clustering algorithms, such as k-means and mixture models; self-organizing maps (SOMs); and adaptive resonance theory (ART). In the rest of this section, we describe some of the most commonly used unsupervised learning algorithms.

1.2.1 Clustering

Clustering, also called *cluster analysis,* is one form of unsupervised learning. It is often employed in pattern recognition tasks and activity detection applications. A clustering algorithm partitions the input instances into a fixed number of subsets, called *clusters,* so that the instances in the same cluster are similar to one another with respect to some set of metrics (Figure 1.14).

Cluster analysis itself is not one specific algorithm, but the general task to be solved. The clustering can be achieved by a number of algorithms, which differ significantly in their notion of what constitutes a cluster and how to efficiently find them. The choice of appropriate clustering algorithms and parameter settings, including values, such as the distance function to use, a density threshold, or the number of expected clusters, depends on the individual dataset and intended use of the results.

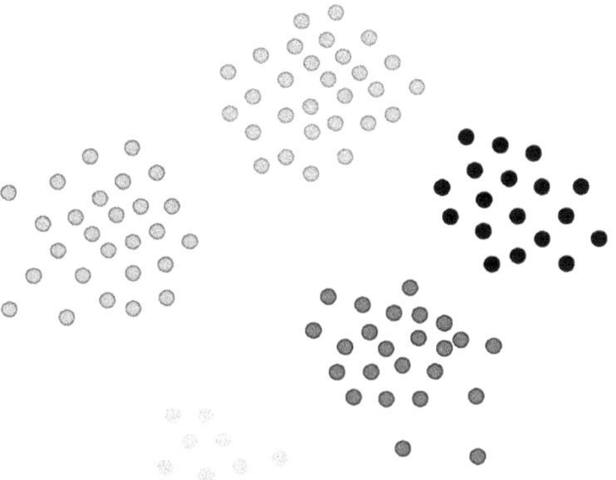

Figure 1.14 A clustering algorithm divides the set of input data instances into groups, called clusters. The instances in the same group are more similar to each other than to those in other clusters.

The notion of a *cluster* varies between algorithms and the clusters found by different algorithms vary significantly in their properties. Typical cluster models include the following:

- *Connectivity models*: An example of a connectivity model algorithm is hierarchical clustering, which builds models based on distance connectivity.
- *Centroid models*: A representative of this set of algorithms is the k-means algorithm. With this algorithm, each cluster is represented by a single mean vector.
- *Distribution models*: Clusters are modeled using statistics distributions.
- *Density models*: An example of density model clustering is density-based spatial clustering for applications with noise (DBSCAN). In this type, clusters are identified as areas with higher density than non-clusters.
- *Group models*: These clustering algorithms are not able to provide a refined model for the results. Instead, they can only generate the group information.

We discuss in more detail two of the most common clustering algorithms used in sensor network applications: k-means clustering and DBSCAN clustering.

1.2.1.1 k-Means Clustering

The goal of k-means clustering is to partition the input instances into k clusters, where each instance belongs to the cluster with the nearest mean. Since the problem is NP-hard, the common approach is to search only for approximate solutions. There are a number of efficient heuristic algorithms that can quickly converge to a local optimum, such as the Lloyd's algorithm (Lioyd 1982). Since the algorithms find only local optimums, they are usually run multiple times with different random initializations.

An advantage of the k-means algorithm is that it is simple and converges quickly when the number of dimensions of the data is small. However, k-means clustering also has a number of drawbacks. First, k must be specified in advance. Also the algorithms prefer clusters of approximately similar sizes. This often leads to incorrectly cut borders in between clusters, which is not surprising since, being a centroid model algorithm, k-means optimizes for cluster center rather than cluster borders.

Figure 1.15 shows a clustering example, where $k = 2$ and k-means is not able to accurately define the borders between the two clusters. There are two density clusters in that figure. One of them is much larger and contains circles. The other one is smaller and consists of triangles. Since k-means optimizes for cluster center and tends to produce clusters with similar sizes, it incorrectly splits the data instances into a dark and a light cluster. These two clusters, however, do not overlap with the original density clusters of the input data.

k-Means clustering has been used in a number of WSN applications. A k-means algorithm is used in the *fingerprint and timing-based snooping (FATS) security attack* to cluster together sensors that are temporally correlated (Srinivasan et al. 2008). This allows the attack to identify sensors that fire together and hence identify sensors that are located in the same room. k-Means clustering has also been used to address the *multiple sink location problem* in large-scale WSNs (Oyman and Ersoy 2004). In large-scale networks with a large number of sensor nodes, multiple sink nodes should be deployed not only to increase the manageability of the network but also to prolong the lifetime of the network by reducing the energy dissipation of each node. Al-Karaki et al. apply k-means clustering to *data aggregation* and more specifically to finding the minimum number of aggregation points in order to maximize the network lifetime (Al-Karaki et al. 2004). The results

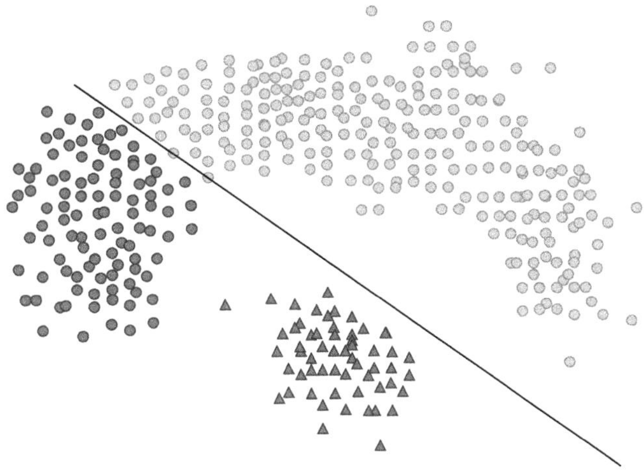

Figure 1.15 k-Means clustering might incorrectly cut the borders between density-based clusters.

from their experiments show that, compared to a number of other algorithms, such as a genetic algorithm and a simple greedy algorithm, k-means clustering achieves the highest network lifetime extension.

1.2.1.2 DBSCAN Clustering

The DBSCAN is the most popular density-based clustering algorithm. In density-based clustering, clusters are defined as areas of higher density than the remainder of the dataset. DBSCAN requires two parameters: distance threshold (Eps-neighborhood of a point) and minimum number of points required to form a cluster (MinPts) (Ester et al. 1996). DBSCAN is based on connecting points within a certain distance of each other, i.e., points that are in the same Eps-neighborhood. However, in order to make a cluster, DBSCAN requires that for each point in the cluster, there are at least MinPts in the Eps-neighborhood. Figure 1.16 shows an example of DBSCAN clustering. The dataset is the same as that in Figure 1.15, but since a density-based clustering algorithm has been used, the data is clustered correctly.

An advantage of DBSCAN is that, unlike many other clustering algorithms, it can form clusters of any arbitrary shape. Another useful property of the algorithm is that its complexity is fairly low and it will discover essentially the same clusters in each run. Therefore, in contrast to k-means clustering, DBSCAN can be run only once rather than multiple times. The main drawback of DBSCAN is that it expects sufficiently significant density drop in order to detect cluster borders. If the cluster densities decrease continuously, DBSCAN might often produce clusters whose borders look arbitrary.

In sensor network applications, DBSCAN has been used as part of the *FATS security attack* to identify the function of each room, such as bathroom, kitchen, or bedroom (Srinivasan et al. 2008). DBSCAN generates temporal activity clusters, each of which forms a continuous temporal block with a relatively high density of sensor firings. Experiments show that DBSCAN performs very well because it automatically leaves out outliers and computes high-density clusters. However, when DBSCAN is applied to the step of identifying which sensors are in the same

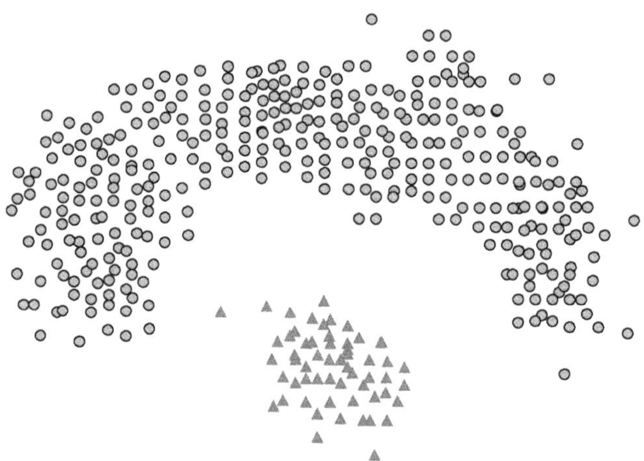

Figure 1.16 An example density-based clustering with DBSCAN.

room, k-means clustering performs much better. This is especially true for scenarios where all devices are highly correlated temporally and there is no significant density drop on the boundary of clusters.

Apiletti et al. also apply DBSCAN to *detecting sensor correlation* (Apiletti et al. 2011). The authors perform experiments using data collected from a sensor network deployed in university laboratories. The results show that DBSCAN is able to identify different numbers of clusters based on which day of the week it is analyzing. This allows it to construct more accurate models for the sensor use patterns in the laboratories. DBSCAN also successfully detects noisy sensors.

1.2.2 Self-Organizing Map

SOMs provide a way of representing multidimensional data in much lower dimensional spaces—typically one or two dimensions. The process of reducing the dimensionality of the feature vectors is a data compression technique known as *vector quantization*. SOMs, as indicated by their name, produce a representation of the compressed feature space, called a map. An extremely valuable property of these maps is that the information is stored in such a way that any topological relationships within the training set are maintained.

A SOM contains components called nodes. Each node is associated with (1) a position in the map space and (2) a vector of weights, where the dimension of this vector is the same as that of the input data instances. The nodes are regularly spaced in the map, which is typically a rectangular or a hexagonal grid. A typical example of SOMs is a color map (Figure 1.17). Each color is represented by a three-dimensional vector containing values for red, green, and blue. However, the color SOM represents the colors in a two-dimensional space.

The procedure of placing an input data instance onto the map is the following:

1. Initialize the weights of the nodes on the map.
2. Choose an input training instance.
3. Find the node with the closest vector to that of the input instance. This node is called the best matching unit (BMU).

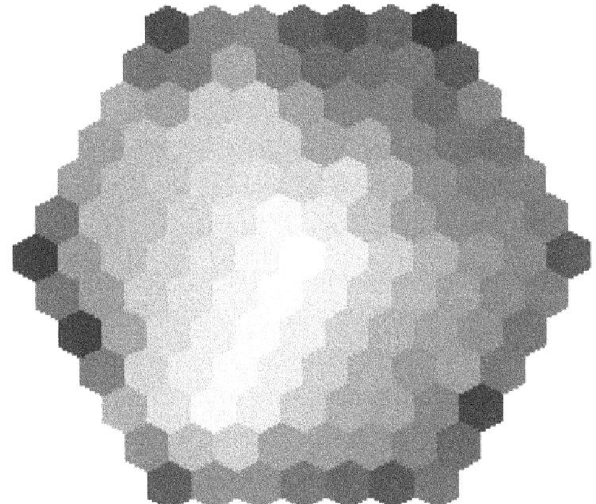

Figure 1.17 An example SOM representation for colors.

4. Calculate the radius of the BMU's neighborhood. This value is often set to the radius of the whole map, but it decreases at each time step. Any node found within this radius is considered to be inside the BMU's neighborhood.
5. Once the BMU is located, it is assigned the values from the vector of the input instance. In addition, the weights of the nodes close to the BMU are also adjusted towards the input vector. The closer a neighbor node is to the BMU, the more its weight is altered.

In sensor networks, SOMs have been applied to *anomaly detection* caused by faulty sensors and unusual phenomenon, such as harsh environmental conditions (Siripanadorn et al. 2010). Paladina et al. have also used SOMs for *node localization* (Paladina et al. 2007). Their localization technique is based on a simple SOM implemented on each of the sensor nodes. The main advantages of this approach are the limited storage and computing cost. However, the processing time required by the SOMs increases with the size of the input data. Giorgetti et al. have also applied SOMs to addressing node localization (Giorgetti et al. 2007). Their SOM-based algorithm computes *virtual* coordinates that are used in location-aided routing. If the location information for a few anchor nodes is available, the algorithm is also able to compute the *absolute* positions of the nodes. The results from the experiments further show that the SOM-based algorithm performs especially well for networks with low connectivity, which tend to be harder to localize, and in the presence of irregular radio patterns or anisotropic deployment. A variation of a SOM, called growing self-organized map, is employed to achieve accurate *detection of human activities* of daily living within smart home environments (Zheng et al. 2008).

1.2.3 Adaptive Resonance Theory

Most existing learning algorithms are either *stable* (they preserve previously learned information) or *plastic* (they retain the potential to adapt to new input instances indefinitely). Typically, algorithms that are stable cannot easily learn new information and algorithms that are plastic tend to forget the old information they have learned. This conflict between stability and plasticity is called the *stability–plasticity dilemma* (Carpenter and Grossberg 1987).

The ART architectures attempt to provide a solution to the stability–plasticity dilemma. ART is a family of different neural architectures that address the issue of how a learning system can preserve its previously learned knowledge while keeping its ability to learn new patterns. An ART model is capable of distinguishing between familiar and unfamiliar events, as well as between expected and unexpected events.

An ART system contains two functionally complementary subsystems that allow it to process familiar and unfamiliar events: *attentional subsystem* and *orienting subsystem*. Familiar events are processed within the attentional subsystem. This goal of this subsystem is to constantly establish even more precise internal representations of and responses to familiar events. By itself, however, the attentional subsystem is unable to simultaneously maintain stable representations of familiar categories and to create new categories for unfamiliar events. This is where the orienting subsystem helps. It is used to reset the attentional subsystem when an unfamiliar event occurs. The orienting subsystem is essential for expressing whether a novel pattern is familiar and well represented by an existing recognition code, or unfamiliar and in need of a new recognition code.

Figure 1.18 shows the architecture of an ART system. The attentional system has two successive stages, F_1 and F_2, which encode patterns of activation in short-term memory. The input pattern is received at F_1, and the classification is performed at F_2. Bottom-up and top-down pathways between the two stages contain adaptive long-term memory traces. The orienting subsystem measures the similarity between the input instance vector and the pattern produced by the fields in the attentional subsystem. If the two are similar, i.e., if the attentional subsystem has been able to recognize the input instance, the orienting subsystem does not interfere. However, if the two patterns are significantly different, the orienting subsystem resets the output of the recognition layer. The effect of the reset is to force the output of the attentional system back to zero, which allows the system to search for a better match.

A drawback of some of the ART architectures is that the results of the models depend significantly on the order in which the training instances are processed. The effect can be reduced to some extent by using a slower learning rate, where differential equations are used and the degree of training on

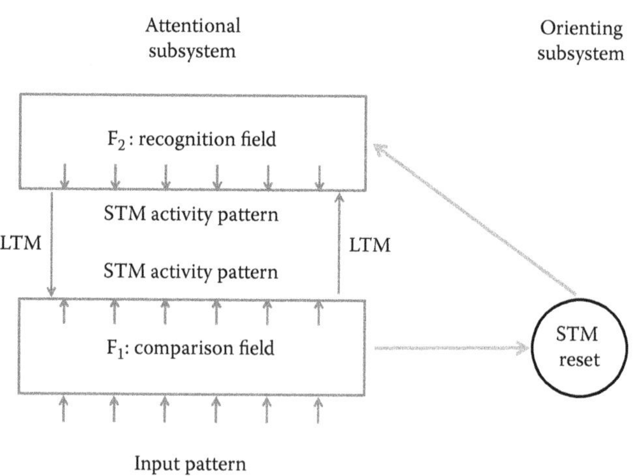

Figure 1.18 The architecture of an ART system has two subsystems: attentional, responsible for processing familiar events, and orienting, which helps reset the attentional subsystem when an unfamiliar event occurs. The attentional subsystem contains a comparison field, where the input is received, and a recognition field, which assigns the input to a category. Both short-term memory (STM) and long-term memory (LTM) are employed.

an input depends on the time the input is available. However, even with slow training, the order of training still affects the system regardless of the size of the input dataset.

ART classifiers have been applied to WSN applications to address *anomaly detection problems in unknown environments* (Li et al. 2010). A fuzzy ART classifier is used to label multidimensional sensor data into discrete classes and detect sensor-level anomalies. An ART classification is also employed by an *intruder detection* system that uses a WSN and mobile robots (Li and Parker 2008). The sensor network uses an unsupervised fuzzy ART classifier to learn and detect intruders in a previously unknown environment. Upon the detection of an intruder, a mobile robot travels to investigate the position where the intruder is supposed to be. Kulakov and Davcev incorporate ART into a technique used for *detection of unusual sensor events and sensor failures* (Kulakov and Davcev 2005). Through simulation, where one of the input sensor nodes is failed on purpose, the authors show the improvement in data robustness achieved by their approach.

1.2.4 Other Unsupervised Machine Learning Algorithms

There is a wide variety of unsupervised learning algorithms, in addition to k-means clustering, DBSCAN, SOM, and ART, which have been often applied to WSN application. The SmartHouse project uses a system of sensors to monitor a person's activities at home (Barger et al. 2005). The goal of the project is to recognize and detect different behavioral patterns. The authors use *mixture models* to develop a probabilistic model of the behavioral patterns. The mixture model approach serves to cluster the observations with each cluster considered to be a different event type.

A number of activity recognition projects have developed unsupervised learning algorithms that *extract models from text corpora or the web*. The Guide project uses unsupervised learning methods to detect activities using RFID tags placed on objects (Philipose et al. 2003). This method relies on data mining techniques to extract activity models from the web in an unsupervised fashion. For this project, the authors have mined the temporal structure of about 15,000 home activities.

Gu et al. develop another unsupervised approach based on RFID-tagged object-use fingerprints to recognize activities without human labeling (Gu et al. 2010). The activity models they use are built based on object-use fingerprints, which are sets of contrast patterns describing significant differences in object use between any two activity classes. This is done by first mining a set of object terms for each activity class from the web and then mining contrast patterns among object terms based on emerging patterns to distinguish between any two activity patterns.

Wyatt et al. also employ generic mined models from the web (Wyatt et al. 2005). Given an unlabeled trace of object names from a user performing their activities of daily living, they use the generic mined models to segment the trace into labeled instances of activities. After that, they use the labeled instances to learn custom models of the activity from the data. For example, they learn details such as order of object use, duration of use, and whether additional objects are used.

Tapia et al. develop a similar approach where they extract relevant information on the functional similarity of objects automatically from WordNet, which is an online lexical reference system for the English language (Tapia et al. 2006). The information about the functional similarity among objects is represented in a hierarchical form known as *ontology*. This ontology is used to help mitigate the problem of model incompleteness, which often affects the techniques used to construct activity recognition models.

An unsupervised approach based on detecting and analyzing the sequence of objects that are being used by the residents is described in Wu et al. (2007). The activity recognition method is based on RFID object use correlated with video streams and information collected from how-to websites such as about.com. Since video streams are used, the approach provides high-grained activity recognition. For example, it can differentiate between making tea and making coffee. However, as previously mentioned, collecting video data of home activities is difficult due to privacy concerns.

Dimitrov et al. develop a system that relies on unsupervised recognition to identify activities of daily living in a smart home environment (Dimitrov et al. 2010). The system utilizes *background domain knowledge* about the user activities, which is stored in a self-updating probabilistic knowledge base. The system aims to build the best possible explanation for the observed stream of sensor events.

1.3 Semi-Supervised Learning

Semi-supervised learning algorithms use both labeled and unlabeled data for training. The labeled data is typically a small percentage of the training dataset. The goal of semi-supervised learning is to (1) understand how combining labeled and unlabeled data may change the learning behavior and (2) design algorithms that take advantage of such a combination. Semi-supervised learning is a very promising approach since it can use readily available unlabeled data to improve supervised learning tasks when the labeled data is scarce or expensive.

There are many different semi-supervised learning algorithms. Some of the most commonly used ones include the following:

- *Expectation–maximization (EM) with generative mixture models*: EM is an iterative method for finding maximum likelihood estimates of parameters in statistical models, where the models depend on unobserved latent variables (Dempster et al. 1977). Each iteration of the algorithm consists of an expectation step (e-step) followed by a maximization step (m-step). EM with generative mixture models is suitable for applications where the classes specified by the application produce well-clustered data.
- *Self-training*: Self-training can refer to a variety of schemes for using unlabeled data. Ng and Cardie implement self-training by *bagging* and majority voting (Ng and Cardie 2003). An ensemble of classifiers is trained on the labeled data instances and then the classifiers are used to classify the unlabeled examples independently. Only those examples, for which all classifiers assign the same label, are added to the labeled training set, and the classifier ensemble is retrained. The process continues until a stop condition is met.

  A single classifier can also be self-trained. Similar to the ensemble of classifiers, the single classifier is first trained on all labeled data. Then the classifier is applied to the unlabeled instances. Only those instances that meet a selection criterion are added to the labeled set and used for retraining.
- *Co-training*: Co-training requires two or more *views* of the data, i.e., disjoint feature sets that provide different complementary information about the instances (Blum and Mitchell 1998). Ideally, the two feature sets for each instance are conditionally independent. Also each feature set should be sufficient to accurately assign each instance to its respective class. The first step in co-training is to use all labeled data and train a separate classifier for each view. Then, the most confident predictions of each classifier are used on the unlabeled data to construct additional labeled training instances. Co-training is a suitable algorithm to use if the features of the dataset naturally split into two sets.
- *Transductive SVMs*: Transductive SVMs extend general SVMs in that they could also use partially labeled data for semi-supervised learning by following the principles of *transduction* (Gammerman et al. 1998). In inductive learning, the algorithm is trained on specific training

instances, but the goal is to learn general rules, which are then applied to the test cases. By contrast, transductive learning is reasoning from specific training cases to specific testing cases.

- *Graph-based methods*: These are algorithms that utilize the graph structure obtained by capturing pairwise similarities between the labeled and unlabeled instances (Zhu 2007). These algorithms define a graph structure where the nodes are labeled and unlabeled instances, and the edges, which may be weighted, represent the similarity of the nodes they connect.

In sensor networks, semi-supervised learning has been applied to *localization* of mobile objects. Pan et al. develop a probabilistic semi-supervised learning approach to reduce the calibration effort and increase the tracking accuracy of their system (Pan et al. 2007). Their method is based on semi-supervised CRFs, which effectively enhance the learned model from a small set of training data with abundant unlabeled data. To make the method more efficient, the authors employ a Generalized EM algorithm coupled with domain constraints. Yang et al. use a semi-supervised manifold learning algorithm to estimate the locations of mobile nodes in a WSN (Yang et al. 2010). The algorithm is used to compute a subspace mapping function between the signal space and the physical space by using a small amount of labeled data and a large amount of unlabeled data.

Wang et al. develop a semi-supervised learning algorithm based on SVM (Wang et al. 2007). The algorithm has been applied to *target classification*, and the experimental results show that it can accurately classify targets in sensor networks.

Semi-supervised learning has also been applied to *object detection and recognition* of commonly displaced items. Xie et al. propose a dual-camera sensor network that can be used as memory assistant tool (Xie et al. 2008). Their approach extracts the color features of every new object and then uses a semi-supervised clustering algorithm to classify the object. The user is provided with the option to review the results of the classification algorithm and label images that have been mislabeled, thus providing real-time feedback to the system to refine the data model of the semi-supervised clustering.

1.4 Summary

Machine learning has been steadily entering the area of sensor network applications. Since its application to routing problems in wireless networks as early as 1994 (Cowan et al. 1994), it has been used to address problems, such as activity recognition, localization, sensor fusion, monitoring and alerting, outlier detection, energy efficiency in the home, to name a few. Future work will further extend both the application domains and the set of machine learning techniques that are used.

References

Al-Karaki, J. N., R. Ul-Mustafa, and A. E. Kamal. Data aggregation in wireless sensor networks - Exact and approximate algorithms. *Workshop on High Performance Switching and Routing*, Phoenix, AZ, 2004, pp. 241–245.

Apiletti, D., E. Baralis, and T. Carquitelli. Energy-saving models for wireless sensor networks. *Knowledge and Information Systems* (Springer London) 28(3), 2011: 615–644.

Barger, T., D. Brown, and M. Alwan. Health status monitoring through analysis of behavioral patterns. *IEEE Transactions on Systems, Man, and Cybernetics, Part A: Systems and Humans* 35(1), 2005: 22–27.

Blum, A. and T. Mitchell. Combining labeled and unlabeled data with co-training. *11th Annual Conference on Computational Learning Theory*, Madison, WI, 1998, pp. 92–100.

Bulling, A. and D. Roggen. Recognition of visual memory recall processes using eye movement analysis. *Ubiquitous Computing (UbiComp '11)*, New York, 2011, pp. 455–469.

Cabena, P., P. Hadjinian, R. Stadler, J. Verhees, and A. Zanasi. *Discovering Data Mining: From Concept to Implementation*. Upper Saddle River, NJ: Prentice-Hall, Inc., 1998.

Carpenter, G. and S. Grossberg. A massively parallel architecture for a self-organizing neural pattern recognition machine. *Journal on Computer Vision, Graphics, and Image Processing* 37(1), January 1987: 54–115.

Cayirci, E., H. Tezcan, Y. Dogan, and V. Coskun. Wireless sensor networks for underwater surveillance systems. *Ad Hoc Networks* 4(4), July 2006: 431–446.

Chung, P.-C. and C.-D. Liu. A daily behavior enabled hidden Markov model for human behavior understanding. *Pattern Recognition* 41(5), May 2008: 1589–1597.

Cowan, J., G. Tesauro, and J. Alspector. Packet routing in dynamically changing networks: A reinforcement learning approach. *Advances in Neural Information Processing Systems* 6, 1994: 671–678.

Dempster, A. P., N. M. Laird, and D. B. Rubin. Maximum likelihood from incomplete data via the EM algorithm. *Journal of the Royal Statistical Society. Series B (Methodological)* 39(1), 1977: 1–38.

Dimitrov, T., J. Pauli, and E. Naroska. Unsupervised recognition of ADLs. In *Proceedings of the 6th Hellenic conference on Artificial Intelligence: Theories, Models and Applications (SETN'10)*, Stasinos Konstantopoulos, Stavros Perantonis, Vangelis Karkaletsis, Constantine D. Spyropoulos, and George Vouros (Eds.). Springer-Verlag, Berlin, Heidelberg, 2010, pp. 71–80. DOI=10.1007/978-3-642-12842-4_11 http://dx.doi.org/10.1007/978-3-642-12842-4_11

Duarte, M. and Y. H. Hu. Vehicle classification in distributed sensor networks. *Journal of Parallel and Distributed Computing* 64(7), July 2004: 826–838.

Duong, T., D. Phung, H. Bui, and S. Venkatesh. Efficient duration and hierarchical modeling for human activity recognition. *Artificial Intelligence* 173(7–8), 2009: 830–856.

Ester, M., H.-P. Kriegel, J. Sander, and X. Xu. A density-based algorithm for discovering clusters in large spatial databases with noise. *International Conference on Knowledge Discovery in Databases and Data Mining (KDD '96)*, Portland, OR, 1996, pp. 226–231.

Gammerman, A., V. Vovk, and V. Vapnik. Learning by transduction. In *Proceedings of the Fourteenth conference on Uncertainty in Artificial Intelligence (UAI'98)*, Gregory F. Cooper and Serafín Moral (Eds.). Morgan Kaufmann Publishers Inc., San Francisco, CA, USA, 1998: 148–155.

Ganesan, D., S. Ratnasamy, H. Wang, and D. Estrin. Coping with irregular spatio-temporal sampling in sensor networks. *ACM SIGCOMM Computer Communication Review* 34, 1 (January 2004), 2004: 125–130. DOI=10.1145/972374.972396 http://doi.acm.org/10.1145/972374.972396

Giorgetti, G., S. K. S. Gupta, and G. Manes. Wireless localization using self-organizing maps. *6th International Conference on Information Processing in Sensor Networks (IPSN '07)*, Cambridge, MA, 2007, pp. 293–302.

Gu, T., S. Chen, X. Tao, and J. Lu. An unsupervised approach to activity recognition and segmentation based on object-use fingerprints. *Data and Knowledge Engineering* 69, 2010: 533–544.

Han, J. *Data Mining: Concepts and Techniques*. San Francisco, CA: Morgan Kaufmann Publisher Inc., 2005.

Hu, D. H., S. J. Pan, V. W. Zheng, N. N. Liu, and Q. Yang. Real world activity recognition with multiple goals. *Proceedings of the 10th International Conference on Ubiquitous Computing (UbiComp '08)*, Seoul, South Korea, 2008, pp. 30–39.

Janakiram, D., V. AdiMallikarjuna Reddy, and A. V. U. Phani Kumar. Outlier detection in wireless sensor networks using Bayesian belief networks. *First International Conference on Communication System Software and Middleware*, New Delhi, India, 2006, pp. 1–6.

van Kasteren, T., G. Englebienne, and B. Kröse. Activity recognition using semi-Markov models on real world smart home datasets. *Journal of Ambient Intelligence and Smart Environments* 2(3), 2010: 311–325.

van Kasteren, T. and B. Kröse. Bayesian activity recognition in residence for elders. *3rd International Conference on Intelligent Environments*, Ulm, Germany, 2007, pp. 209–212.

van Kasteren, T., A. Noulas, G. Englebienne, and B. Kröse. Accurate activity recognition in a home setting. *10th International Conference on Ubiquitous Computing (UbiComp '08)*, New York, 2008, pp. 1–9.

Kulakov, A. and D. Davcev. Tracking of unusual events in wireless sensor networks based on artificial neural-networks algorithms. *International Conference on Information Technology: Coding and Computing (ITCC '05)*, Las Vegas, NV, 2005, pp. 534–539.

Li, Y. Y. and L. E. Parker. Intruder detection using a wireless sensor network with an intelligent mobile robot response. *Proceedings of the IEEE Southeast Conference*, 2008, Huntsville, AL, pp. 37–42.

Li, Y. Y., M. Thomason, and L. E. Parker. Detecting time-related changes in wireless sensor networks using symbol compression and probabilistic suffix trees. *IEEE International Conference on Intelligent Robots and Systems (IROS '10)*, Taipei, 2010, pp. 2946–2951.

Li, D., K. D. Wong, Y. H. Hu, and A. M. Sayeed. Detection, classification, and tracking of targets. *IEEE Signal Processing Magazine* 19, March 2001: 17–29.

Lioyd, S. Least squares quantization in PCM. *IEEE Transactions on Information Theory* 28(2), March 1982: 129–137.

Logan, B., J. Healey, M. Philipose, and E. M. Tapia. A long-term evaluation of sensing modalities for activity recognition. *UbiComp.*, Innsbruck, Austria, 2007, pp. 483–500. New York: Springer-Verlag.

Lu, J. et al. The smart thermostat: using occupancy sensors to save energy in homes. *8th ACM Conference on Embedded Networked Sensor Systems (SenSys '10)*, New York, 2010, pp. 211–224.

Mark, B. L. and Z. R. Zaidi. Robust mobility tracking for cellular networks. *IEEE International Conference on Communications*, New York, 2002, pp. 445–449.

Maurer, U., A. Smailagic, D. P. Siewiorek, and M. Deisher. Activity recognition and monitoring using multiple sensors on different body positions. *International Workshop on Wearable and Implantable Body Sensor Networks*, Washington, DC , 2006, pp. 113–116. Los Alamitos, CA: IEEE Computer Society.

Mohamed Sathik, M., M. Syed Mohamed, and A. Balasubramanian. Fire detection using support vector machine in wireless sensor network and rescue using pervasive devices. *International Journal on Advanced Networking and Applications* 2(2), 2010: 636–639.

Ng, V. and C. Cardie. Bootstrapping coreference classifiers with multiple machine learning algorithms. *Conference on Empirical Methods in Natural Language Processing (EMNLP '03)*, Stroudsburg, PA, USA, 2003, pp. 113–120.

Oyman, E. I. and C. Ersoy. Multiple sink network design problem in large scale wireless sensor networks. *IEEE International Conference on Communications*, Paris, France, 2004, pp. 3663–3667.

Paladina, L., M. Paone, G. Jellamo, and A. Puliafito. Self-organizing maps for distributed localization in wireless sensor networks. *IEEE Symposium on Computers and Communications*, Aveiro, Portugal, 2007, pp. 1113–1118.

Pan, R. et al. Domain-constrained semi-supervised mining of tracking models in sensor networks. *13th ACM SIGKDD International Conference on Knowledge Discovery and Data Mining (KDD '07)*, San Jose, CA, 2007, pp. 1023–1027.

Patterson, D. J., D. Fox, H. Kautz, and M. Philipose. Fine-grained activity recognition by aggregating abstract object usage. *9th IEEE International Symposium on Wearable Computers (ISWC '05)*, Osaka, Japan, 2005, pp. 44–51.

Philipose, M., K. Fiskin, D. Fox, H. Kautz, D. Patterson, and M. Perkowitz. Guide: Towards understanding daily life via auto-identification and statistical analysis, In Proc. of the Int. Workshop on Ubiquitous Computing for Pervasive Healthcare Applications. *UbiHealth Workshop at UbiComp*, 2003.

Quinlan, J. R. Induction of Decision Trees. Mach. Learn. 1, 1 (March 1986), 81–106. DOI=10.1023/A:1022643204877 http://dx.doi.org/10.1023/A:1022643204877.

Quinlan, R. *C4.5: Programs for Machine Learning*. San Mateo, CA: Morgan Kaufmann Publishers Inc., 1993.

Rabiner, L. R. A tutorial on hidden Markov models and selected applications in speech recognition. *Proceedings of the IEEE* 77, San Francisco, CA, USA, February 1989, pp. 257–286.

Siripanadorn, S., N. Hattagam, and N. Teaumroong. Anomaly detection in wireless sensor networks using self-organizing map and wavelets. *International Journal of Communications* 4(3), 2010: 74–83.

Srinivasan, V., J. Stankovic, and K. Whitehouse. Protecting your daily in-home activity information from a wireless snooping attack. *10th International Conference on Ubiquitous Computing (UbiComp '08)*, Seoul, Korea, 2008, pp. 202–211.

Srinivasan, V., J. Stankovic, and K. Whitehouse. Using height sensors for biometric identification in multi-resident homes. In *Proceedings of the 8th international conference on Pervasive Computing (Pervasive'10)*, Patrik Floréen, Antonio Krüger, and Mirjana Spasojevic (Eds.). Springer-Verlag, Berlin, Heidelberg, 337–354. DOI=10.1007/978-3-642-12654-3_20 http://dx.doi.org/10.1007/978-3-642-12654-3_20

Sutton, C. and A. McCailum. An introduction to conditional random fields for relational learning. In *Introduction to Statistical Relational Learning*, by B. Taskar and L. Getoor (eds.). Cambridge, MA: MIT Press, 2006, pp. 93–129.

Tapia, E. Activity recognition in the home using simple and ubiquitous sensors. In *Pervasive Computing*, A. Ferscha and F. Mattern. Berlin/Heidelberg, Germany: Springer, 2004, pp. 158–175.

Tapia, E. M., T. Choudhury, and M. Philipose. Building reliable activity models using hierarchical shrinkage and mined ontology. *Pervasive*. New York: Springer-Verlag, 2006.

Tapia, E. M. et al. Real-time recognition of physical activities and their intensities using wireless accelerometers and a heart rate monitor. *11th IEEE International Symposium on Wearable Computers (ISWC '07)*, Washington, DC, 2007, pp. 1–4.

Tran, D. A. and T. Nguyen. Localization in wireless sensor networks based on support vector machines. *IEEE Transactions on Parallel Distributed Systems* 19(7), 2008: 981–994.

Truyen, T. T., H. H. Bui, and S. Venkatesh. Human activity learning and segmentation using partially hidden discriminative models. *Workshop on Human Activity Recognition and Modelling (HAREM '05)*, Oxford, U.K., 2005, pp. 87–95.

Truyen, T. T., D. Q. Phung, H. H. Bui, and S. Venkatesh. Hierarchical semi-Markov conditional random fields for recursive sequential data. *Neural Information Processing Systems (NIPS '08)*, Vancouver, B.C., Canada, 2008.

Wang, X., S. Wang, D. Bi, and L. Ding. Hierarchical wireless multimedia sensor networks for collaborative hybrid semi-supervised classifier learning. *Sensors* 7(11), 2007: 2693–2722.

Wilson, D. H. and C. Atkenson. Simultaneous tracking and activity recognition (STAR) using many anonymous, binary sensors. *Pervasive*, Munich, Germany, 2005, 62–79.

Winter, J., Y. Xu, and W.-C. Lee. Energy Efficient Processing of K Nearest Neighbor Queries in Location-aware Sensor Networks. In *Proceedings of the The Second Annual International Conference on Mobile and Ubiquitous Systems: Networking and Services(MOBIQUITOUS '05)*. IEEE Computer Society, Washington, DC, USA, 281–292. DOI=10.1109/MOBIQUITOUS.2005.28 http://dx.doi.org/10.1109/MOBIQUITOUS.2005.28

Wu, J., A. Osuntogun, T. Choudhury, M. Philipose, and J. M. Rehg. A scalable approach to activity recognition based on object use. *IEEE 11th International Conference on Computer Vision (ICCV '07)*, Rio de Janeiro, Brazil, 2007, 1–8.

Wyatt, D., M. Philipose, and T. Choudhury. Unsupervised activity recognition using automatically mined common sense. *20th National Conference on Artificial Intelligence - Volume 1 (AAAI '05)*, Pittsburgh, PA, 2005, pp. 21–27.

Xie, D., T. Yan, D. Ganesan, and A. Hanson. Design and implementation of a dual-camera wireless sensor network for object retrieval. *7th International Conference on Information Processing in Sensor Networks (IPSN '07)*, St-Louis, MO, 2008, pp. 469–480.

Yang, B., J. Xu, J. Yang, and M. Li. Localization algorithm in wireless sensor networks based on semi-supervised manifold learning and its application. *Cluster Computing* 13(4), December 2010: 435–446.

Zhang, W., F. Chen, W. Xu, and Y. Du. Learning Human Activity Containing Sparse Irrelevant Events in Long Sequence. In *Proceedings of the 2008 Congress on Image and Signal Processing*, Vol. 4 - Volume 04 (CISP '08), Vol. 4. IEEE Computer Society, Washington, DC, USA, 211–215. DOI=10.1109/CISP.2008.283 http://dx.doi.org/10.1109/CISP.2008.283

Zheng, H., H. Wang, and N. Black. Human activity detection in smart home environment with self-adaptive neural networks. *IEEE International Conference on Networking, Sensing, and Control (ICNSC '08)*, Sanya, China, 2008, pp. 1505–1510.

Zhu, X. Semi-supervised learning literature survey. Madison, WI: University of Wisconsin, Computer Science TR 1530, 2007.

Chapter 2

Modeling Unreliable Data and Sensors

Using Event Log Performance and F-Measure Attribute Selection

Vasanth Iyer, S. Sitharama Iyengar, and Niki Pissinou

Contents

2.1 Introduction

During preprocessing event log, training samples are sorted by their area of fire damage according to high, medium, small, and accidental small fire classes. The preprocessing step allows studying the probability distribution and in our case the samples are highly skewed, giving an estimate that accidental small fires are more likely compared to large fires. Statistically we are interested in the many factors that influence accidental small fires; from the training samples, it follows a normal distribution. Graphically describing normal distribution, it takes the form of a bell-shape curve, which is also known as the Gaussian function. This is a first approximation for a real-valued random variable, which tends to cluster around a single mean value. The sensor model needs to learn the expected ranges for the baseline attributes being measured, giving better density estimation with increasing samples count. The baseline discrete parameters capture only the sensor ranges, making event prediction function hard to train with a Gaussian density function, without specific temporal understanding of the datasets. The dynamic features present in a sequence of patterns are localized and used to predict events, which otherwise may not be an attributing feature to the static data mining algorithm.

We have studied the spatial features and baseline discrete sensor measurements and all the attributes available that have a high classification error [1], which can sometimes amount to 50% of the error in the case of accidental small fire category. To further investigate factors that can cause such fires, we include data pertaining to human specific temporal attributes such as number of visitors and traffic patterns coming into the forest area, thus filtering events with local significance. Temporal attributes is a better estimator, given the type of training samples, which are difficult to calibrate and any approximation may induce false alarms. Relevance-based ranking function is highly suitable to order higher bound sample chosen by domain experts as ideal estimates and still maintaining the desired low false alarm rates. The method of ranking uses the function of sensor precision and event relevance weights, which are then linearly added to represent data from fire activity logs. The rest of the chapter is organized as follows. Sections 2.2 and 2.3 provide related work and state of the art. Section 2.4 defines sensor measurements and fire activity to model the data and algorithm computational complexity. Section 2.5 defines the additive ranking weight to classify accidental small fires using precision and relevance of the event collection. Section 2.6 discusses the performance of the large fire ranking function in terms of Fire Weather Index (FWI) [2] attributes and uses precision and discusses error rate in terms of false alarms. Sections 2.7 and 2.8 discuss preliminary information retrieval (IR) [3] choices with extensive simulation using Machine Learning (ML) probabilistic algorithms. Section 2.9 uses WEKA [4,5] tool to automatically do attribute selection ranking, this allows to a good statistical knowledge for weather data, and in Section 2.10, the constraints related to machine learning model with respect to weather data and how to use Bayes net are discussed. Future work in Section 2.11 discusses massive datasets and how to computationally use a new framework and port ML algorithms. The chapter concludes in Section 2.12 with summary of results performance.

2.2 Background

The machine learning repository [6] provides collection of supervised databases that are used for the empirical analysis of event prediction algorithms with unsupervised datasets from distributed wireless sensor networks. Sensor network generate huge amounts of data that need to be validated for its relevance; keeping only necessary data helps avoid high computational overloads due to data redundancies. Calibration of sensors may not be always possible and the data aggregating algorithms need to have domain rules to detect any outliers from the datastream, given other parameters are kept constant. Given a dataset of forest fire events for a region, the training algorithm will be able to transform correlated attributes from sensor networks datastream to validate and classify the events and reject the outliers reliably. The preliminary work models the empirical data with a ranking function without spatial information to predict the likelihood of different events. The concept of IR, such as precision and relevance, is used in context to sensor networks, which not only allow to understand the domain topics but also add high reliability to the large dataset. In this case study, we use forest fire and environmental conservation as the theme and study which environmental factors attribute to such events, for example, temporal attributes such as humans, peak weather conditions, surface fuel buildups, and wind spread factors. The Burnt Area (BA), which is the ground truth is broadly studied with respect to small and large fires. The framework needs to extend queries in topics that are spatially aware, making sensing an essential source of discovering information.

2.3 State of the Art

The sensitivity of a sensor network not only depends on the accuracy of individual sensors but also on the spatial distribute collaborative sensing of neighboring sensors. Now we can define the data mining criteria, which is a standard measure to evaluate queries in a sensor networks.

Precision How close a features measurement match at every independent value, when measured in time.

Accuracy How redundant a basic feature sampling measurement assumption are, due to the fact the independent and identically distributed (i.i.d) statistical for measuring correlated spatially arranged sensors do not hold good for medium to large weather data sets (e.g. Confusion matrix shown Table 2.1 and the formula for ranking precision and accuracy are given below). The more standard form of precision and recall is given in Ranking section in 2.1.1.

Table 2.1 Confusion Matrix of Classifier

	Positive	*Negative*
Positive	tp	fn
Negative	fp	tn

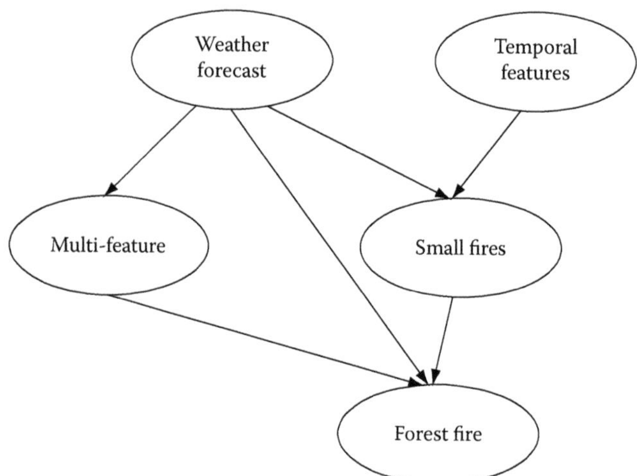

Figure 2.1 Bayes Net representation.

$$\text{accuracy} = \frac{t_p + t_n}{n}$$

$$\text{precision} = \frac{t_p}{t_p + f_p}$$

$$F_\beta = \frac{2 \times \text{Precision} \times \text{Recall}}{\text{Precision} + \text{Recall}}$$

The first criteria, which is precision, cannot be generalized in the case of weather data, as weather data are not independent and identically distributed (i.i.d.), but happens to be highly correlated feature. As most of the generative models like Naive Bayes assume i.i.d. strictly, one needs to find alternative models, which better approximate the observed samples. A better model using conditional probabilities compared to the basic Nave Bayes is its network dependent version called Bayes Net. This model allows to specify conditional relationships that are dependent and independent of the observed phenomena respectively. A typical Bayes Net is illustrated in Figure 2.1, which identifies entities such as small, large forest fires and its dependencies. The multi-feature allows to analyze the general pattern in weather data, and local events are further studied by including temporal dependencies in the form of human traffic patters. The enhanced model not only improves the statistical assumptions [7], it also performs much better in error handling, which is one of the major disadvantage of large low-resource energy constrained sensor networks. In this work, one of the goals is to better estimate the performance of the underlying algorithms used. To illustrate the need for such a metric we use three algorithms and measure its precision and accuracy as shown in Table 2.2. The average is calculated by combining precision and accuracy. Algorithm III has the highest average even though it has very low precision; to overcome this limitation we adapt a weighted average, which is also called the F-score. The calculated value of F-score is shown in the final column for all the three algorithms, and now using F-score the weights do properly rank the algorithm III as the lowest due to its low precision value.

Adding sensor data to the manual forest fire event logs allows the study of automated correlated real-time information, which properly trained allows estimation and classification of future sensor outputs from the same geographical region. Event logs contain spatial information such as GPS

Table 2.2 Mean Comparison of Precision and Recall

	Precision	*Recall*	*Average*	*F-score*
Algorithm I	0.5	0.4	0.45	0.444
Algorithm II	0.7	0.1	0.4	0.175
Algorithm III	0.02	1.0	0.51	0.0392

and the area of the fire-damaged region but lack correlated information which could lead to better estimation of the fire event under study. Moreover, the spatial information is collected manually and takes considerable amount of time to classify it, while the sensor measurements are measured in real-time and can be approximated by machine learning algorithms to classify events and notify if a fire alarm condition has reached.

Ranking functions allows filtering unrelated data and present only relevant information to the user's query [8–10]. In the IR domain, there are many efficient ways to rank the relevance of a document in a collection given a user query. Similarly, we like to rank the order of an event occurring given the precision and relevance of the prior fire probabilities and a hypothesis. The terms defined as precision and relevance are inversely proportional. In problems where the recorded evidence is small and rare, one can use a precision scale instead to rank the evidence and making other correlated events as relevant balancing out the summed weights used in events ranking. We rank the precision weights higher whenever fire events occur along with higher alarm conditions. The high alarm condition is always true for precision ranking and holds across accidental small, medium, and large fires equally, unlike relevance, which accounts for majority of fires including only the accidental small fire ones. The precision ranking uses ground truth such as actual fire evidence, further eliminating any possibility of errors due to outliers and weak evidences. These ground truths are well-established naturally occurring phenomenon, which occur rarely and leave significant evidence of the BAs in meter square of the forest. A relevance ranking weight does an exhaustive search of all the prior fire events, which makes it unique to the natural habitat of the particular geographical area, where the concept is machine learnt.

2.4 Sensor Measurement and Fire Activity

The samples of the fire events are sorted in terms of BA in hectares(ha), the frequency distribution versus BA is shown in Figure 2.2. We can model the BA in terms of a function as shown in Equation 2.1, which allows us to study the behavior of fire activities over time and predict newer events reliably.

$$BA = f(x) \tag{2.1}$$

2.4.1 Sorting by BA (ha)

The histogram shows that the BA is skewed with large number of small fires and very few large fires, making likelihood of small fires more predictable. We further classify fires into four categories that allows to select the performance of the system in terms of precision and relevance. Figure 2.2b shows the new classes without any other weather attribute that is likely to cause the events.

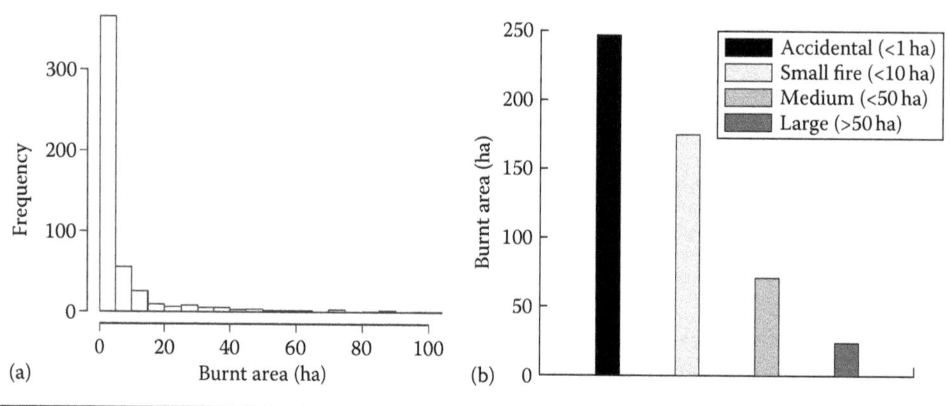

Figure 2.2 Histograms of empirical samples. (a) Empirical log collection and (b) four-class classification.

2.4.2 Sorting by BA with Temporal Attributes

$$\mathrm{V_{train}}(\mathrm{FireEvent_{Day\ of\ week}}) \leftarrow \tag{2.2}$$

$$\hat{V}(\mathrm{FireEvent_{Day\ of\ week}})_{\mathrm{Temporal\ Variable}} +$$

$$\hat{V}(\mathrm{FireEvent_{Day\ of\ week}})_{\mathrm{Correlated\ measurements}}$$

$$\text{Temporal Variables} = \text{Month of the year} + \text{Day of the week} \tag{2.3}$$

$$\text{Correlated measurement} = \text{temperature} + \text{humidity} + \text{wind} + \text{rain} \tag{2.4}$$

$$\text{Classifiers} = \{\text{accidental; small, medium, large}\} \tag{2.5}$$

2.4.3 Estimating Training Values with Sample Data

Sample datasets are based on 517 Fire Location rules from UCI forest fire repository to classify fire activity for a geographical area. The equation representing the target function of BA from the empirical data is given in Equation 2.1. In our case, the hypothesis to be maximized in terms of the temporal attributes are given in Equation 2.2 for a four-class classification, as given in Equation 2.5 and according to Equation 2.1. The assumption here is that the training set D is an unbiased representation to learn the concept c and can estimate the inputs x_i. The previously defined dependent variable **Fire Location**, which is used to estimate given the independent correlated measurements and its relation to the temporal attributed are given in Equations 2.4 through 2.6. The target concepts are present in the training samples and we like to see the influence of adding unlabelled sensor measurements to further accurately learn the concepts of the human-induced accidental small fires versus the more natural accruing types of the medium and large fires. For the sake of clarity of machine learning domain, we convert the correlated sensor data to ordinal [11,12] types, as illustrated in the following:

$$\text{temperature} = \{\text{cool; mild; hot}\} \tag{2.6}$$

$$\text{humidity} = \{\text{normal; high}\} \tag{2.7}$$

$$\text{wind} = \{\text{true; false}\} \tag{2.8}$$

The model estimation of the the target function with weights w_1, w_2 as shown allows to minimize the training error, where x_1, x_2 are temporal and correlated measurements.

$$\hat{V} = w_1 x_1 + w_2 x_2 \tag{2.9}$$

The learning algorithm needs to define the best fit for the given hypothesis and adjust the weights to minimize the error-and-misclassifications.

$$E \equiv \sum (V_{\text{train}}(\text{FireEvent}) - \hat{V}(\text{FireEvent}))^2 \tag{2.10}$$

2.4.4 Algorithm Complexity

Search space consists of all the possible patterns of the features, given our data model, $4 \times 3 \times 3 = 36$ possibilities for each rule when using attributes temperature, humidity, and wind. As there are 517 rules from the collected dataset instances, each rule can have 36 possibilities and the complete search space will have 36^{517} different possibilities. To minimize the complexity of search space, we can further cut down on the sample instances by using spatial clustering and removing any redundancies in similar features. Given the $\langle X, Y \rangle$ positions, we can cluster into groups the possible fire types into accidental small fires and others that have medium and larger BA as large fires. As measuring ambient phenomena are correlated we expect clustering would be best suited. Let us take five clusters to contain all the samples, then the search space reduces to $36^5 = 60 \times 10^6$ possible rule sets. These methods are used with preprocessing to reduce redundancies in the model and are very practical optimizations of machine learning algorithms. To judge the effectiveness of the model and the classification effectiveness, we initially rely on real-valued numeric model such as [13] to estimate the errors. In contrast to the previous approach, we use ordinal values as defined in Equations 2.8 through 2.11 to build a tree classifier and further reduce errors.

2.5 Alarm Ranking Function for Accidental Small Fires

In our example event data log such as forest fires, which may be incomplete and how do we infer knowledge from the missing datasets. Relevance factor of fire event learning concept can be defined as all fire events that are tagged in the log, as we are interested in reporting fire occurrences. The inverse concept precision is seen from the histogram plot in Figure 2.2a where there are very few large fires. When reporting on major fire events, the highest ranked samples are retrieved that has a higher corresponding rank leading to a precision learning concept close to 1.

2.5.1 Ranking Function

The design of a good ranking function needs to balance the relevance and precision of the events in a way to express a summable numerical quantity, which signifies the importance of the new sample and how reliable the prior probabilities were, as shown in Table 2.1 with false positives. As the ranking functions are evaluated for a given query, we define the query criteria for retrieving accidental small fires and large wild fires for our collection.

$$\text{Precision} = \frac{\text{number of relevant forest fire events retrieved}}{\text{number of forest fires retrieved in query}}$$

$$\text{Relevance} = \frac{\text{number of relevant forest fire events retrieved}}{\text{number of relevant forest fires classified}}$$

2.5.2 F-Measure

A measure which combines precision and recall for a small dataset is F-measure and is the weighted harmonic mean of precision and relevance. The traditional F-measure or balanced F-score$=\dfrac{\text{Precision} \times \text{Recall}}{2 \times (\text{Precision} + \text{Recall})}$, is because recall and precision are evenly weighted. The general formula for nonnegative real β is that it is based on van Rijsbergen's effectiveness E given by

$$\text{Alarm}_{\text{rank}} = 1 - \frac{1}{\alpha\frac{1}{p} + (1-\alpha)\frac{1}{r}}$$

We further look into accidental small fires, as they are very probabilistic and any conceptual link to the attributes may lead to rank the idea of precision and relevance of the collection. α− the alarm weight are calculated based on the reliability of the ground truth, higher precision weightage is given to large and medium fires compared to accidental small fires. The precision weight factor for small fires can then be evaluated using $\dfrac{\alpha}{1-\alpha}$.

2.5.3 Performance of Fire Topics Classification Using Temporal Ranking

$$BA_{\text{Days\%}} < \mathbf{50\ ha} = \alpha + \beta \ln(NF) \tag{2.11}$$

where $\alpha = 2.895$ and $\beta = 1.265$, which is the variance of the BA data versus fire activity showing logarithmic $O(\lg(BA)$ complexity as shown in Figure 2.3a.

The broad classification of topics in the training samples collection [14] are reflected in two categories, accidental small fires and large wild fires. To evaluate the performance of the two we use weighted precision versus relevance to estimate the ranking information F@(0.5) for the previous equation . The F-scores are calculated and weighted for high reliability by using F@(0.5), which is twice the precision compared to its equivalent relevance scale. Reliability and precision are proportionally weighted while relevance is inversely proportional. The performance scores show

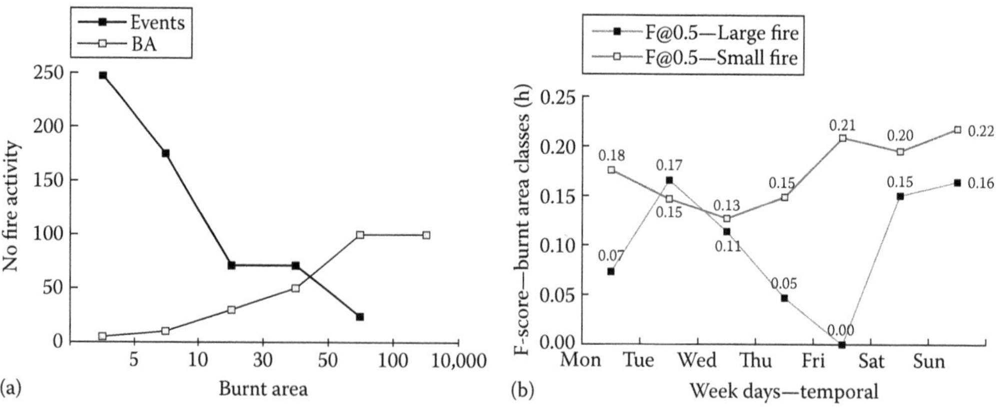

Figure 2.3 Fire activity plot and its F-score transform. (a) Fire activity vs. burnt area yields a logarithmic relation and (b) twice the precision vs. relevance.

Table 2.3 Performance of Ranking with False Alarm Rates

Other Test	*Yellow Region*	*Blue Region*	*Temporal*	*FWI*
Alarm	H = Hit	H = Hit	H = 311	H = 69
No alarm	M = Miss	M = Miss	M = 111	M = 3
False alarm	F = False alarm	Z = Null hypothesis	F = 8	F = 307

that query evaluation for accidental small fires is 1.7 times higher when compared to queries for large wild fires or from the same collection. The plot in Figure 2.3b shows queries for accidental and small fires for the temporal attribute days of the week, where the accidental small fire F-scores have much higher values.

2.5.4 *Ranking Accidental Small Fires*

Accidental small wild fires are possible all through the year, making them a viable application for automated sensor measurements. The measurements such as temperature, humidity, and wind gust are automated, while temporal attributes such as human traffic and day of the week are used to study the small fire events (Table 2.3). The peaks of the plot in Figure 2.4 suggest high alarm during weekends followed by only one high alarm day during the normal week. The weightage of the ranking suggest that small fires events are caused due to temporal attributes such as human traffic and vehicular routes more than any observed correlated sensor measurements.

2.6 Alarm Ranking Function for Large Fires

In the previous section we used BA and data from sensor networks to classify forest fires into four classes. In this section we will instead use domain knowledge to precisely predict fires by using

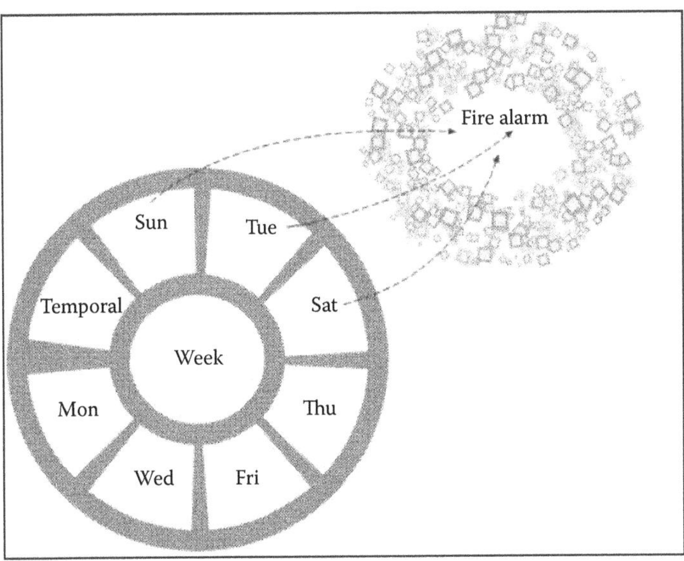

Figure 2.4 Fire alarm days ranked using *F*@(0.5).

Table 2.4 FWI Classes

FWI Types	*Measure (Normalized)*
Low (AF)	0–8
Medium (SF)	8–13
High (MF)	13–32
Very high (LF)	$32 > FWI < 80$

FWIs {low; moderate; high; very high} as shown in Table 2.4. FWI is calculated using ISI and BUI, where ISI represents the Initial Spread Index and BUI represents the Build Up Index, which indicate fire behavior and respectively represent rate of fire spread, fuel consumption, and fire intensity. All FWI indexes are significantly correlated with the number of fires and the burned area, especially when $BA > 100$ is the area burnt by large fires. The average FWI index variation during the year is shown in Figure 2.5a and b. It increases during the month of May and peaks in August and September and starts reducing in the month of October. The following equation shows a numerical representation of BA when using FWI classes. Which is mean daily burned area per month and the mean daily number of fire events per month (NF).

$$BA_{FWI} > \mathbf{50\ ha} = (BUI) + (ISI)^x \tag{2.12}$$

Where ISI and BUI are calculated from the environment for a given fuel type, x is the estimated geometrical fire spread factor. The FWI index is highly correlated with the number of fires and the BA. The plot in Figure 2.5 shows the correlated region when $FWI > 33$ (very high) in the case of large fires.

2.6.1 Performance of Fire Topics Classification Using FWI Ranking

Large fire occurrences damage more than 50 ha in total amounts for majority of the BA (ha). It is a high priority to avoid large fire incidents and help forest conservation. As they are hard to detect and have a varying threshold, it is also a cause of false alarms [15] in an automated system. Plotting all the correlated FWI components that relate to fire activity, plot from Figure 2.5a shows that peak months have a gradual increase of FWI index and are also correlated with large fire incidents. The area of high correlation is shown in yellow with lowest false alarm for a given FWI threshold. The lower bound conditions for fire activity are shown in blue, which have higher false alarm due to valid area above the yellow region.

The temporal correlation for large fires using the FWI F-score is plotted in Figure 2.5b. It shows that large fires are invariant to temporal changes and perform better than small fires for a given precision and recall measure.

2.6.2 Misclassification and Cost of False Alarms

$$\begin{pmatrix} & \multicolumn{2}{c}{Decisionmade} \\ & LargeFire & SmallFire \\ LargeFire & 0 & 1000 \\ SmallFire & 1 & 0 \end{pmatrix}$$

Cost function matrix for misclassification

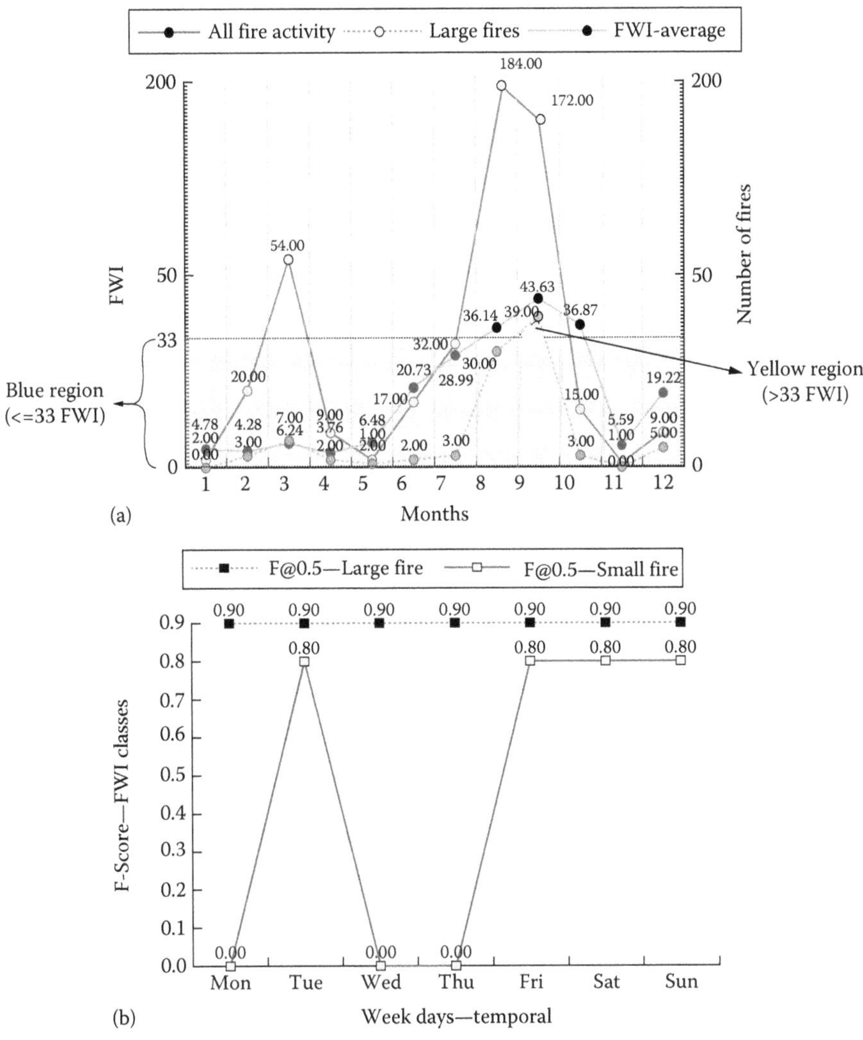

Figure 2.5 Lower bound range of FWI<= 33 (blue) and higher bound range FWI>33 (yellow) show all large fires (dotted line) fall into the yellow region. (a) FWI-based classification and (b) FWI for large fires shows invariance

While testing the performance of the algorithm error rates are simulated, which allows to find the sensitiveness of the system to false positives. False positives have more significance with higher bound values such as large fires, which are very rare and hard to classify. We define a hit in terms of precision to avoid false alarms [15] and Table 2.1. From the given cost matrix we show that using FWI the alarms are very precise, while using temporal the alarms are more accurate, which includes large amounts of false positives.

2.7 Machine Learning Algorithms

Probabilistic algorithms when used with density estimation and class classification yields lowest error. This allows to provide a baseline analysis of the system attributes being used.

2.7.1 Naive Bayes

One can use Naive Bayes [16] that by design presumes the class densities a priori, which have been determined and accurate. The model calculates the class conditional probabilities of the input feature vectors. To understand the underlying skewed structure of the dataset, we further create thresholds for accidental small fires compared to medium and large fires as shown in Table 2.6. So we have the four possible values for the target variable as given in Equation 2.7.

2.7.2 User Query

To validate the model let us predict the fire activity outcome of a peak summer month from the dataset [17]. August has significant number of reported fires compared to other months. Estimating the unknown probabilities (?) using temporal features of fire events given the attribute values for the class.

$$? = \{Month = August; Day = Monday\}$$
$$\{Temperature = Cool; Humidity = High; Wind = True\}$$

The estimated class conditional densities for the independent variables temperature, humidity, and wind conditions are calculated using temporal attributes **month** for the dataset shown in Table 2.5. The datasets further are explored using two temporal variables, **month** and the **day of the week**, as shown in Tables 2.7 and 2.8. The temporal variables introduced into the dataset help gain the

Table 2.5 Posterior Probabilities for Background Weather Data for the Peak Month August

Burnt Area (ha)	*August*	*Monday*	*Temperature*	*Humidity*	*Windy*	*Prior Probability*	*Predictor Variance (%)*
>1 ha	► 0.34	► 0.14	► 0.46	► 0.17	► 0.42	► 0.47	► 57
>1 ha <=10 ha	0.39	0.15	0.35	0.13	0.38	0.33	25.0
>10 ha <=50 ha	0.30	0.14	0.43	0.19	0.50	0.13	17.0
>50 ha	0.33	0.08	0.16	0.08	0.37	0.04	0.02

Table 2.6 Target Variable Occurrences

Fire Types	*Recorded*
Accidental (AF)	247
Small (SF)	175
Medium (MF)	71
Large (LF)	24

Table 2.7 Likelihood of Fires for the Month of August

Fire Type	*Month=August*
Accidental	0.004
Small	0.002
Medium	0.001
Large	0.00004

Table 2.8 Posteriors Probabilities for Temporal Feature Day of the Week

Days	*Accidental*	*Small*	*Medium*	*Large*
Mon	▶ 35	▶ 27	▶ 10	▶ 2
Tue	28	21	11	4
Wed	22	24	5	3
Thu	30	21	9	1
Fri	42	31	12	0
Sat	42	24	11	7
Sun	48	27	13	7
Total	▶ 247	▶ 175	▶ 71	▶ 24

insight of users' dependencies with fire prediction model.

$$g_i(x) = P(\omega_i \| \mathbf{x}) = \frac{p(\mathbf{x} \| \omega_i) P(\omega_i)}{\sum_{i=0}^{i=4} p(x \| \omega_j) P(\omega_j)} \tag{2.13}$$

Substituting the corresponding highlighted values from Tables 2.5 through 2.8 in the Equation 2.13, we get the posterior probability of accidental small fire.

$$\hat{\text{fire}}_{\text{accidental}} = \frac{0.0007547}{0.003565} = 57\% \tag{2.14}$$

$$\hat{\text{fire}}_{\text{Small}} = \frac{0.000333}{0.003565} = 25\% \tag{2.15}$$

$$\hat{\text{fire}}_{\text{medium}} = \frac{0.000223}{0.003565} = 0.17\% \tag{2.16}$$

$$\hat{\text{fire}}_{\text{large}} = \frac{0.000000287}{0.003565} = 0.02\% \tag{2.17}$$

From the posterior probabilities for the month of August for the data collected in Portugal [17], the likelihood of accidental small fires are very high. From cross-validating from the known fact that in summer the likelihood of wild fires is higher, the Bayes rule is able to classify the dataset for accidental and small fires with high accuracy. We use a simulation framework in the following sections to further prove our initial conclusion from the datasets. It is shown that the training time for Naive Bayes scales linearly in both the number of instances and number of attributes.

2.7.3 Tree Classifier

In this section, we focus on the domain rules, which are applicable to the learning system. Tree classifiers lend themselves to use ML rules [11] when searching the hypothesis by further branching on specific attributes. The design of such a classifier needs to sort the weights or entropies [16] of the attributes, which is the basis of its classification effectiveness.

ID3 is a popular tree-classifier algorithm, to implement ID3 as illustrated in Figure 2.6a and Table 2.9 with our attributes. Let (S) be a collection of samples, then using the tree algorithm that uses entropy to split its levels entropy is given by

$$\text{Entropy}(S) = \sum_{i=0}^{i=c} p(i) \log_2 p(i) \tag{2.18}$$

Let us assume a collection (S) has 517 samples [17] with 248, 246, 11, and 12 of *accidental, small, medium, large* fires, respectively. The total entropy calculated from Equation 2.18 is given by

$$\text{Entropy}(S) = \frac{248}{517}\log_2\frac{248}{517} + \frac{246}{517}\log_2\frac{246}{517} + \frac{11}{517}\log_2\frac{11}{517} + \frac{12}{517}\log_2\frac{12}{517} = 1.23$$

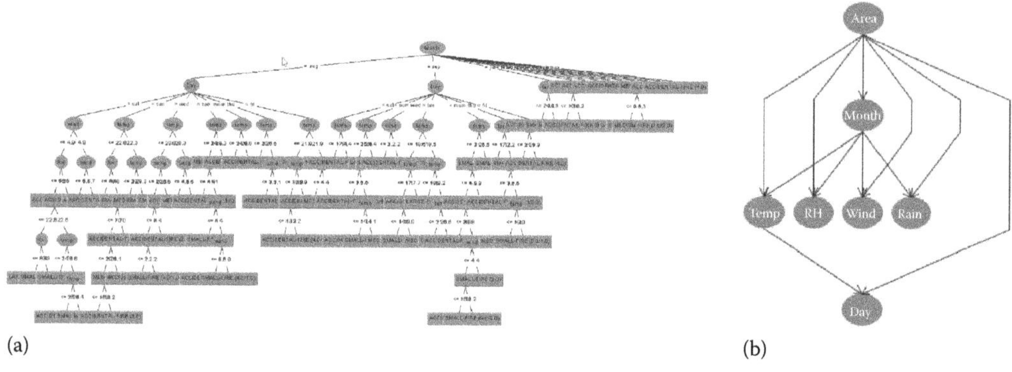

Figure 2.6 Weka algorithm toolkit. (a) Tree classifier and (b) Bayes network.

Table 2.9 Gain Ratio Calculation for Tree Using Entropy

Month	*Temperature*	*Wind*
Not shown	Info: 1.08	Info: 1.20
Not shown	Gain: 1.23 – 1.08 = 0.192	Gain: 1.23 – 1.08 = 0.025

2.7.4 Attribute Selection

ID3 uses a statistical property called information gain to select the best attribute. The gain measures how well the attribute separates training targeted examples, when classifying them into fire events. The measure of purity that we will use is called information and is measured in units called bits. It represents the expected amount of information that would be needed to specify whether a new instance should be classified as accidental, small, medium, or large fires, given that the example reached that node. The gain of an attribute is defined by and illustrated in Table 2.9. Using the calculated attribute for information gain, we show that *temperature* attribute is used before the *wind* attribute to split the tree after the tree root.

$$\text{Gain}(S, A) = \text{Entropy}(S) - \sum_{i=0}^{i=c} \frac{S_v}{|S|}\text{Entropy}(S_v) \tag{2.19}$$

$$\text{Entropy}(S_{\text{Hot}}) = \frac{9}{36}\log_2\frac{9}{36} + \frac{23}{36}\log_2\frac{23}{36} + \frac{3}{36}\log_2\frac{3}{36} + \frac{1}{36}\log_2\frac{1}{36} = 1.282$$

$$\text{Entropy}(S_{\text{Medium}}) = \frac{23}{96}\log_2\frac{23}{96} + \frac{65}{96}\log_2\frac{65}{96} + \frac{3}{96}\log_2\frac{3}{96} + \frac{5}{96}\log_2\frac{5}{96} = 1.175$$

$$\text{Entropy}(S_{\text{Cool}}) = \frac{117}{269}\log_2\frac{117}{269} + \frac{146}{269}\log_2\frac{146}{269} + \frac{2}{269}\log_2\frac{2}{269} + \frac{4}{269}\log_2\frac{4}{269} = 1.05$$

$$\text{Entropy}(\text{temp}) = \frac{43}{517} \times 1.282 + \frac{139}{517} \times 1.175 + \frac{335}{517} \times 1.05 = 1.08$$

$$\text{Gain}(S, \text{temp}) = 1.23 - 1.08 = 0.192$$

$$\text{Entropy}(S_{\text{High}}) = \frac{162}{249}\log_2\frac{162}{249} + \frac{72}{249}\log_2\frac{72}{249} + \frac{8}{249}\log_2\frac{8}{249} + \frac{7}{249}\log_2\frac{7}{249} = 1.1952$$

$$\text{Entropy}(S_{\text{Low}}) = \frac{68}{133}\log_2\frac{68}{133} + \frac{59}{133}\log_2\frac{59}{133} + \frac{2}{133}\log_2\frac{2}{133} + \frac{4}{133}\log_2\frac{4}{133} = 1.24$$

$$\text{Entropy}(\text{wind}) = \frac{361}{517} \times 1.1952 + \frac{156}{517} \times 1.24 = 1.20$$

$$\text{Gain}(S, \text{wind}) = 1.23 - 1.20 = 0.025$$

The internal tree representation for m attributes from n samples will have a complexity of $O(\lg n)$. With increasing inputs given by parameter n the height of the tree will not grow linearly as in the case of Naive Bayes. On the other hand, complexity of building a tree will be $O(mn \lg n)$.

2.8 Simulation

Open-source workbench called WEKA [4] is a useful tool to quantify and validate results, which can be duplicated. WEKA can handle numeric attributes well, so we use the same values for the weather data from the UCI [6] repository datasets. The class variable has to be a nominal one to allow WEKA [4]; we convert all fire types to "0" or "1," where "0" is for accidental small fire and "1" is for large fires making it a two-class classifier; the results are shown as confusion matrix in Tables 2.11 through 2.13. Naive Bayes correctly classifies accidental and small fires (209 out of 247), while the J48 tree classifier does far more, 219 out of 247 and SVM with high precision (235 out of 247).

As WEKA uses kappa [4] statistics internally for evaluating the training sets, a standard score of >60% means training set is correlated. Using J48 simulation we get 53.56% just below and when using SVM we get 0.68 above the correlated index. The comparison on results shows that J48 tree classifier does better than Naive Bayes by 25% and the corresponding SVM does 35% overall showing least bias of the three models. Therefore, using sensor network measurements accidental and small fires can be predicted with high precision using SVM classifier.

2.8.1 Simulation Analysis

WEKA attribute statistics for training set and cross validated testing and its effective correlation kappa score. Tables 2.10 through 2.12 show kappa and other comparison statistics for Naive Bayes, J48 tree, and Support Vector Machine show classifiers for small fires. The experiment is repeated using FWI, which are shown for J48 tree classifier in Table 2.14.

2.8.2 Error Analysis

Equation 2.11 specifies the regression model error and its following confusion matrix from the simulation scores are shown in Tables 2.13 through 2.16, upper bound of small fire (AF+SF) has over 90% precision in SVM, 80% accuracy for J48-Tree, and 61% overall for Naive Bayes. The corresponding baseline performances including all fires categories is 82% for SVM, 72.1% for J48-Tree, and Naive Bayes is 51.64%, which is due to bias toward small fires and only SVM by design is the least biased (Tables 2.17 through 2.19).

When FWI classification [17] is as given in Equation 2.12 for large fires prediction, it is more precise with better precision as shown in the confusion matrix Table 2.20. The percentage of correctly classified is >95%, making it reliable with few false alarms.

Table 2.10 Evaluation on Training Set for Naive Bayes

WEKA Stats	*Results*	*Summary (%)*
Correctly classified instances	267	51.64
Incorrectly classified instances	250	48.35
Kappa statistic	0.1371	
Mean absolute error	0.3022	
Root mean squared error	0.3902	
Relative absolute error	94.86%	
Root relative squared error	97.84%	
Total number of instances	517	

Table 2.11 Evaluation on Training Set for J48 Tree Classifier

WEKA Stats	*Results*	*Summary (%)*
Correctly classified instances	373	72.14
Incorrectly classified instances	144	27.85
Kappa statistic	0.5356	
Mean absolute error	0.1938	
Root mean squared error	0.3113	
Relative absolute error	60.83%	
Root relative squared error	78.04%	
Total number of instances	517	

Table 2.12 Evaluation on Training Set for SVM Linear Classifier

WEKA Stats	*Results*	*Summary (%)*
Correctly classified instances	421	81.43
Incorrectly classified instances	96	18.56
Kappa statistic	0.6893	
Mean absolute error	0.0928	
Root mean squared error	0.3047	
Relative absolute error	29.14%	
Root relative squared error	76.40%	
Total number of instances	517	

Table 2.13 Confusion Matrix for Naive Bayes Using Training Set

	LF	*MF*	*SF*	*AF*
LF	0	1	7	16
MF	0	5	12	54
SF	0	7	53	115
AF	0	0	38	209

Table 2.14 Confusion Matrix on Training Set for J48 Tree Classifier

	LF	*MF*	*SF*	*AF*
LF	7	0	7	10
MF	0	29	15	27
SF	1	7	118	49
AF	0	5	23	219

Table 2.15 Confusion Matrix on Testing Set for J48 Tree Classifier

	LF	*MF*	*SF*	*AF*
LF	0	2	8	14
MF	2	7	2	40
SF	3	12	71	89
AF	4	15	59	168

Table 2.16 Confusion Matrix on Training Set for SVM Linear Classifier

	LF	*MF*	*SF*	*AF*
LF	7	0	5	12
MF	0	31	15	25
SF	0	1	148	26
AF	0	0	12	235

Table 2.17 Confusion Matrix on Testing Set Using SVM

	LF	*MF*	*SF*	*AF*
LF	0	0	7	17
MF	0	1	17	53
SF	0	3	42	130
AF	0	4	51	192

Table 2.18 Confusion Matrix on Training Set Using Bayes Network

	LF	*MF*	*SF*	*AF*
LF	0	0	5	19
MF	0	4	5	62
SF	0	4	30	141
AF	0	0	23	224

Table 2.19 Confusion Matrix on Testing Set Using Bayes Network

	LF	*MF*	*SF*	*AF*
LF	0	0	5	19
MF	0	2	8	61
SF	0	4	23	148
AF	0	2	28	217

Table 2.20 Confusion Matrix on Training Set for $FWI > 32$

	Very High	*High*
Very high	371	0
High	17	0

2.9 Correlation of Attributes

From statistical point of view, if the attributes have similar values then it creates high bias creating what is called over-fitting error during learning. In our case, **temperature** and **humidly** may have similar values and need to be avoided and substituted with a suitable attribute. To pre-process and analyze, we use all the available sensor measurements in the dataset and WEKA provides the attribute selection as illustrated in Table 2.21.

We use the attribute selection wizard of WEKA to find out the best match. The analysis shows that the Month(100%), Day(10%), and Wind(0%) are highly dependent on the precision. In a two-class classification, the quantitative data are biased toward small fires and SVM does better due to better generalization (Table 2.22). In the qualitative analysis that is based on the frequency of attributes, WEKA picks Month, which is a temporal type. The F-score of small fires is higher, when using temporal attributes as shown in Figure 2.3b, which is also true for WEKA predictions.

Table 2.21 Attribute Selection 10-Fold Cross-Validation (Stratified)

Number of Folds (%)	*No.*	*Attribute*
10(100%)	1	Month
1(10%)	2	Day
0(0%)	3	Temp
0(0%)	4	RH
0(0%)	5	Wind

Table 2.22 Evaluation *FWI* > 32 on Training Set for J48 Tree Classifier

WEKA Stats	*Results*	*Summary (%)*
Correctly classified instances	371	95.61
Incorrectly classified instances	17	4.38
Kappa statistic	463.64	
Mean absolute error	0.0838	
Root mean squared error	0.2047	
Relative absolute error	97.5159%	
Root relative squared error	99.9935%	
Total number of instances	388	

2.10 Better Model for Weather Data

In the error analysis, we show that the true error performance depends on the learning algorithms [1] and how the weights are learnt. As all the data mining algorithms assume samples to be i.i.d., it cannot perform well with highly correlated weather data. In the domain of weather data where features tend to be highly correlated, careful feature selection and a better model are needed to address the shortcomings of earlier models. Bayes network allows to define a better model to define classes and events that are dependent and independent of each other. This model allows to distinguish the overlapping statics using correlated and similar features having the same range and values. The basic Bayes net performance is good while testing with 88% precision as shown in Figure 2.7a and b and has similar accuracy as other models proving its discriminative power for the underlying data.

2.11 Future Work

As machine learning is getting popular, its scalability and study of the family of algorithms need to be studied. Development of a large-scale high-speed system is needed to perform *k*-nearest neighbor

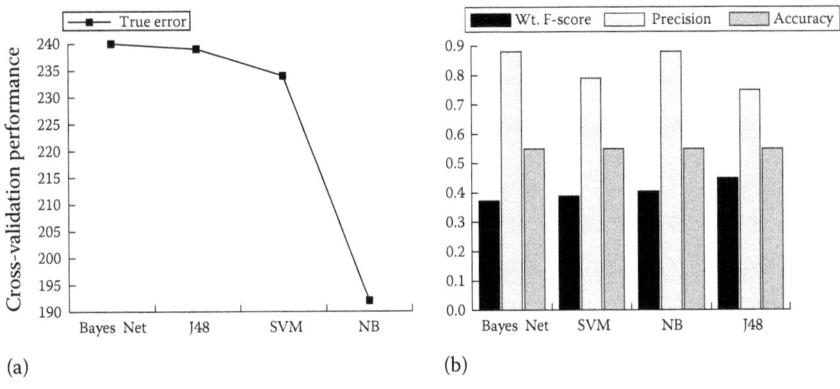

Figure 2.7 Plot of F-score for each model and its performance with weather training data. (a) ML algorithm performance and (b) Bayes network.

searches. The availability of such a system is expected to allow more flexible modeling approaches and much more rapid model turnaround for exploratory analysis. We like to explore Hadoop echo-system to develop machine learning algorithms, data reference set of 20 million profiles to sort the best 1000 records in less than 3 h. Some initial work using the Mahoot machine learning framework is proposed to port the described algorithms in a distributed framework running Hadoop clusters.

2.12 Summary

We use a query log approach of search engines and standard statistical ranking measure to do a baseline analysis and use data from inexpensive sensors to validate against probabilistic ML algorithms. From Table 2.23, we show that precision-⊕ and accuracy-⊙ as denoted, the F-measures matches

Table 2.23 F-Measure Performance for All Tests Compared with WEKA

Experiments	*Model Performance*	*F@0.5 Measure*		*Confusion Matrix*	
	Overall	*Small Fire*	*Large Fire*	*Small Fire*	*Large Fire*
Ranking—Temporal	Bias	0.22⊕	0.17	—	—
Ranking—FWI	Bias	0.1	0.9⊙	—	—
Weka(NB)—Temporal	Generative	0.6⊕	0.1	0.6	0.3
Weka(Bayes Net)—Temporal	Better model	0.88⊕	—	0.88	0.0
Weka(J48)—Temporal	Decision tree	0.7⊕	0.5	0.75	0.3
Weka SVM—Temporal	Generalized	0.87⊕	0.7	0.9⊙	0.3
Weka(J48)—FWI	Invariant	—	0.95⊙	0.1	0.9⊙

the expected WEKA simulation statistics. FWI is able to boost the weak performance of the raw data from sensors, which are typically hard to calibrate. In the qualitative analysis of small fires, the performance of a generalized classifier such as SVM is preferred. Similarly, the qualitative performance for large fires is done by careful attribute selection and we show that an invariant attribute selection such as ($FWI > 50$) yields high classification precision. The use of a better model like Bayes network in the case of weather data helps to increase the performance.

Acknowledgments

The authors thank Shailesh Kumar of Google, Labs, India for suggesting the machine learning framework WEKA and related topics in statistical methods in data mining. The authors would also like to thank Srini Srivathsan for working on the Big Data Analytics and Hadoop framework in context to Machine Learning. Vasanth Iyer appreciates the support from advisors Dr. S. S. Iyengar, Dr. Rama Murthy, and Dr. Rawat in this research effort, which was funded under the FIU, Miami Research Grant.

References

1. Tom M. Mitchell. *Machine Learning*. McGraw-Hill Publications, 1997.
2. Forest fire research and wildland fire safety: *Proceedings of IV International Conference on Forest Fire Research 2002 Wildland Fire Safety Summit*, Luso, Coimbra, Portugal, 18–23 November 2002, pp. 83.
3. Timothy G. Armstrong, A. Moffat, W. Webber, and J. Zobel. EvaluatIR: An online tool for evaluating and comparing IR systems. Computer Science and Software Engineering, University of Melbourne. 2009.
4. Paulo, C. and A. Morais. *A Data Mining Approach to Predict Forest Fires using Meteorological Data.* Department of Information Systems-R&D Algoritmi Centre, University of Minho, Guimaraes, Portugal.
5. Ian H. Witten and E. Frank. *Datamining, Practical Machine Learning*. Elsevier, 2005.
6. WEKA Machine learning software. http://www.cs.waikato.ac.nz/~ml/weka [Accessed May 15, 2011].
7. Mark D. Smuker, J. Allan, and B. Carterette. *A Comparison of Statistical Significance Tests for Information Retrieval Evaluation.* Department of Computer Science, University of Massachusetts, Amherst, MA.
8. Raymond Wan and A. Moffat. Interactive phrase browsing within compressed text. Computer Science and Software Engineering, University of Melbourne. 2001.
9. H. Harb and L. Chen. A Query by example music retrieval algorithm. Maths-Info Department, Ecole Centrale de Lyon. Ecully, France, 2001.
10. Brian Kulis and K. Grauman. Kernelized locality-sensitive hashing for scalable image search, in *Proceedings of the 12th International Conference on Computer Vision* (ICCV), 2009.
11. Vasanth, I., S.S. Iyengar, G. Rama Murthy, N. Parameswaran, D. Singh, and M.B. Srinivas. Effects of channel SNR in mobile cognitive radios and coexisting deployment of cognitive wireless sensor networks. *IEEE 29th International, Performance Computing and Communications Conference (IPCCC)*, pp. 294–300, Albuquerque, NM, 2010.
12. Bhaskar Krishnamachari and S.S. Iyengar. Distributed Bayesian algorithms for fault-tolerant event region detection in wireless sensor networks, *IEEE Trans. Comput.* 53, 3, 2004, pp. 241–250.
13. Vasanth, I., S.S. Iyengar, G. Rama Murthy, and M.B. Srinivas. Modeling unreliable data and sensors: Analyzing false alarm rates using training weighted ranking. ISBN: 978-1-880843-83-3, in *The Proceedings of International Society for Computers and Their Applications*, Honolulu, HI, 2011.

14. David Hawking and S. Robertson. On collection size and retrieval effectiveness. CSIRO Mathematical and Information Sciences, Canberra, ACT, Australia. [Accessed February 22, 2011].
15. Richard R. Brook and S.S. Iyengar. *Robust Distributed Computing and Sensing Algorithm*, ACM, 1996.
16. Frank, A. and A. Asuncion. UCI Machine Learning Repository [http://archive.ics.uci.edu/ml]. Irvine, CA: University of California, School of Information and Computer Science, 2010. [Accessed May 20, 2011].
17. Vasanth, I., S.S. Iyengar, G. Rama Murthy, and M.B. Srinivas. Machine learning and dataming algorithms for predicting accidental forest fires. SENSORCOMM 2011, in *The Fifth International Conference on Sensor Technologies and Applications*, August 21–27, 2011, French Riviera, France, 2011.

Chapter 3

Intelligent Sensor Interfaces and Data Format

Konstantin Mikhaylov, Joni Jamsa, Mika Luimula, Jouni Tervonen, and Ville Autio

Contents

3.1 Introduction

The contemporary world presents novel challenges for distributed measurement and control (DMC) systems. Today's DMC applications are expected to solve complicated tasks in an autonomous mode, which requires a substantial level of intelligence. Thus, effective networking of smart and intelligent sensors is currently an active focus of research and development.

Most of the smart transducers in use today have a structure similar to that shown in Figure 3.1. They generally consist of four major components: the actual transducers (sensors and/or actuators), a signal conditioning and data conversion system, an application processor, and a network communication system (the last two are often considered as a whole—i.e., the so-called network-capable application processor [NCAP]) [1]. To call these types of smart transducers intelligent, these transducers should support one or several intelligent functions such as self-testing, self-identification,

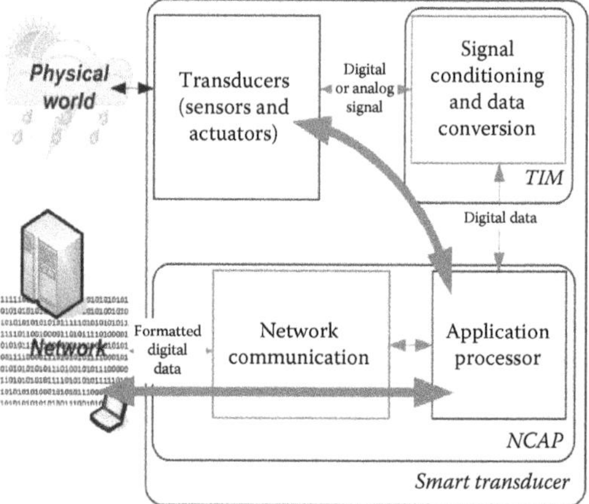

Figure 3.1 Typical structure of a smart transducer.

self-validation, and self-adaptation [2]. Naturally, implementation of these functions introduces special requirements, both for the actual transducers and for the communication interface between the processor (or NCAP) and the transducers.

Besides the actual on-node intelligent function support, networking for an intelligent sensor requires a universal communication mechanism between the nodes and a uniform means of data representation. Indeed, regardless of the size of the sensor network (SN), each node should be able to get in touch with the required sensors, request and receive the required measurement results, understand them, and react accordingly (see Figure 3.2). Additional value the problem of SN's nodes interoperability achieves in conjunction with the Internet Protocol (IP) utilization for SN within Internet of Things concept [3].

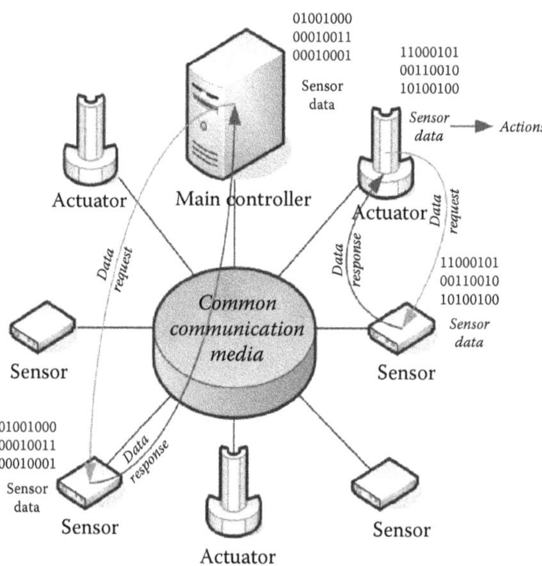

Figure 3.2 Sensor network communication model.

Therefore, the following two major components of the intelligent sensor (transducer) interfacing problem should be considered:

1. The actual physical and electrical interfacing of transducers to processors or NCAPs and implementation of intelligent sensor functions for these
2. NCAP interfacing to other devices within the intelligent SN and data representation

Both of these aspects are discussed in detail in the current chapter.

3.2 Transducer Interfacing to Intelligent Transducer's Processor

As shown in Figure 3.1, a smart transducer is a complex system that is built around an NCAP and the attached transducers. Depending on the application, the interface between the NCAP and the transducers can be implemented as a peer-to-peer (P2P) or a network connection over wired or wireless media [1]. Nonetheless, the intelligent transducer concept introduces some basic requirements for this interface; namely, the support of required intelligent transducer features. The most important capabilities that are required from an intelligent NCAP–transducer interface are [1,4] as follows:

- The standardized physical and electrical connection between the transducer and the NCAP
- The support for detection of transducer connection/disconnection and identification of the transducer by the NCAP
- The support for transducer diagnosis by the NCAP

These features allow connection of the transducers to the NCAP within an intelligent transducer in a simple "plug-and-play" (P&P) mode, where the NCAP automatically detects the attached transducers, calibrates them, and puts them into use. The transducer P&P support also allows a significant simplification of the development and maintenance of the SN application and an increase in system interoperability and adaptability.

We provide a comprehensive picture for the problem of interfacing of a transducer to an NCAP in Sections 3.2.1 through 3.2.3. There, we discuss the plain transducer interfaces in widest use today, the current smart transducer interface standards, and the implementation strategies for the most important intelligent transducer interface features over plain interfaces.

3.2.1 Existing Plain Transducer Interfaces

The presence of a large number of transducer manufacturers has catalyzed the advances in DMC systems in recent years. However, this has resulted in a lack of general agreement on low-level transducer interfaces, which makes the integration of transducers into multivendor networks rather a challenging task [1,4]. The transducers currently available on the market use either analog or a wide range of different digital and quasi-digital interfaces. Therefore, according to Ref. [2] and the International Frequency Sensor Association (IFSA), the proportion of sensors on today's global sensor market is 55%, 30%, and 15% for analog, digital, and quasi-digital output sensors, respectively. Of the digital sensors, the most widespread types for general-purpose applications are presently the ones utilizing I^2C, SPI, and 1-wire interfaces [5]. Illustrations of the most widely used interfaces are presented in Figure 3.3a through f.

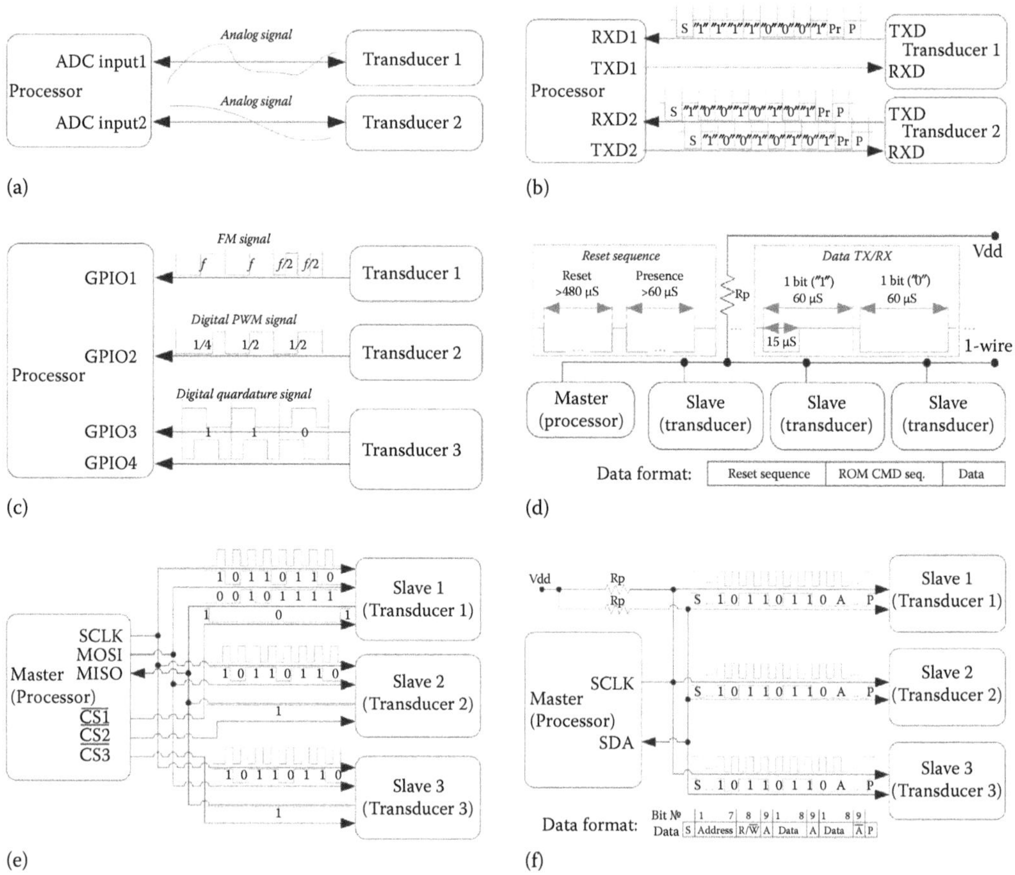

Figure 3.3 The most widely used general-purpose interfaces for transducers. (a) Analog interface. (b) UART interface. (c) Quasi-digital interfaces. (d) 1-wire interface. (e) SPI interface. (f) I^{2}C interface.

The sensors with *analog interfaces* are currently the most widespread [2]. The main advantages of these sensors are their simplicity and low price, although these sensors do require external analog-to-digital converters (ADCs) before the measurements are sent to the digital processing device. If the microcontroller is used as a digital processing device (microcontrollers are currently the most widely used embedded systems [6]), these already have inbuilt ADCs; otherwise, an external ADC should be used. Physical connection for sensors with analog outputs to ADC (see Figure 3.3a) usually utilizes only a single wire (not considering the sensor power supply lines), although this line can require a special shielding to prevent induced noise adding error to the measured value. Typically, several analog sensors would not be connected to ADC over the same physical line at a single time. Obviously, the sensors with analog interfaces cannot provide any mechanisms for their identification.

The *quasi-digital* sensors, as defined in Ref. [7], are "discrete frequency–time domain sensors with frequency, period, duty-cycle, time interval, pulse number, or phase-shift output." Usually, these sensors utilize standardized digital signal voltage levels with special modulation for representing output data. As Ref. [2] reveals, the currently most widespread quasi-digital sensors use frequency modulation (FM) (70% of available on market), pulse-width modulation (PWM) (16%), and duty-cycle (9%) modulation, although some sensors also use pulse number (3%), period (1%), and phase-shift (1%) modulations. Depending on output signal modulation (see, e.g., Figure 3.3c), a quasi-digital interface will require either one (for FM, PWM, duty cycle, pulse number, and period modulation) or two (for phase-shift modulation) physical lines. Similar to analog sensors, the quasi-digital sensors usually have a rather simple structure and low cost, but they are less vulnerable to noise and interference and are capable of providing higher accuracy [2,8]. Another advantage of quasi-digital sensors is that they do not require ADCs, although the measurement reading by processing systems requires some additional effort. Like the analog versions, several quasi-digital sensors usually cannot be connected to the same physical line.

The sensors with *full-featured digital* output convert the measurement results to digital form on-chip. Once converted, digital measurement values can be accessed by the processing device using an appropriate communication interface. Currently, a wide variety of digital interfaces are available for transducers. Among them are the general-purpose ones (e.g., I^2C, SPI, and 1-wire) and a wide range of application-specific ones (e.g., CAN-bus for automotive, Fieldbus for industry, DALI for light control, etc.) [9]. Most of these interfaces utilize serial communication and provide some networking communication capabilities for interconnecting several sensors over the same physical lines. In these cases, the transducers are usually implemented as "slave" devices and can only reply on requests from the "master" device (usually an NCAP). Obviously, digital sensors have more complicated structure compared to analog and quasi-digital sensors, which results in their higher price. Nonetheless, digital sensors have numerous advantages, including connection simplicity and support for some smart features (e.g., sensor self-calibration or inbuilt data processing capabilities) that are of particular value.

We will briefly discuss the communication through I^2C, SPI, 1-wire, and UART interfaces, as these interfaces are the most widespread general-purpose digital communication interfaces and are widely utilized by different sensors [5].

The typical connection of sensors to an NCAP over an *Inter-Integrated Circuit* (I^2C) interface and an I^2C data format are presented in Figure 3.3f. As this figure shows, the I^2C interface uses two common physical lines for the clock (SCLK) and data (SDA), which are pulled up with resistors (Rp) [10]. The I^2C interface can be used to connect multiple slave devices to one master device; therefore, the master device has to initiate the communication by sending the start bit(S) and the address (which usually consists of 7 bits, although addresses of 10 bits are also defined in recent I^2C revisions) of the required slave device. Together with the slave device's address, the master usually transmits the 1-bit Read/Write ($R/\bar{W}$) for defining the communication direction, which would be used until the stop bit (P) closes the current session. The I^2C communication protocol implements per-byte acknowledgments ($A/\bar{A}$). Note that the addresses for I^2C devices are *not unique* and that, depending on the physical connection, one slave device can use multiple (usually 4 or 8) different I^2C addresses. This prevents identification of a single-valued I^2C device based only on its address. The most commonly used data rates for I^2C sensors are between 10 and 400 kbit/s, although recent I^2C revisions also support the rates of 1 and 3.4 Mbit/s.

The *serial peripheral interface (SPI)* is a synchronous serial interface that can operate in full duplex mode [11]. The typical method for connection of several SPI slave devices to the master is presented in Figure 3.3e. As the figure shows, the SPI bus utilizes three common lines for all slave devices: clock (SCLK); master output, slave input (MOSI); master input, slave output (MISO); and a separate chip select ($\overline{CS}$) line for each slave. Therefore, before starting the communication, the SPI master device pulls down the $\overline{CS}$ line of the required slave device to select it. The SPI specification does not define either any maximum data rate (for existing devices it can reach dozens MHz) or any particular addressing scheme or acknowledgment mechanism.

The *1-wire interface* is intended to provide a low data rate communication and power supply over a single physical line [12]. As revealed in Figure 3.3d, the 1-wire network consists of a master device and several slave devices that are connected over the single physical line. The 1-wire line is pulled up with a resistor and can be used for supplying power to the slave devices [12]. The communication over 1-wire bus starts with the reset and synchronization sequence, when the master device first pulls the 1-wire line down for the period of time over 480 μs and then releases it. After that, the connected slave devices have to signal their presence by pulling the 1-wire line down. Thereafter, the master device can start sending or receiving data from the slave devices. For this, the master device first selects the required slave device by sending its unique 64-bit serial number during the read only memory (ROM) command phase and then it starts sending or receiving the actual data. If the master device does not know all of the connected slave devices, it can discover their serial numbers using special procedures. The transmission of each single bit from master to slave, or vice versa, for a 1-wire bus is initialized by the master device by first pulling the line down and then releasing it (for transmitting "1" or receiving data from slave) or keeping it low (for transmitting "0"). The main advantages of a 1-wire bus are its simplicity and the support for device discovery and single-valued identification. The main disadvantage of the 1-wire interface is its low data rate, which usually does not exceed 16 kbit/s, although this low data rate allows implementation of 1-wire networks with cable lengths up to 300 m [13].

The *universal asynchronous receiver/transmitter (UART)* interface is no longer very widely used by sensors, but it is still quite often used by actuators or for inter-processors communication. We are not going to discuss the UART communication in detail, as it is rather well known. Figure 3.3b shows the basics for a UART interface and data formats.

This discussion has covered the interfaces most widely used by a general-purpose plain sensor. As can be seen, the *major portion of the currently existing plain transducer interfaces cannot provide support for the features required for intelligent sensor implementation, such as sensor discovery or single-valued identification.*

3.2.2 Smart Transducer Interfaces and IEEE 1451 Standard

As shown in Section 3.2.1, the majority of the existing general transducer interfaces are unable to provide support for any intelligent sensor functionality. That fact has been the main driving force for development of special smart transducers interfaces since the 1990s. Although several smart transducer interface standards have been proposed in recent years, the most prospective one today is the IEEE 1451 set of standards, which has been also adopted in 2010 as ISO/IEC/IEEE 21451 standard and which is discussed in more detail in the current subsection [14,15].

The IEEE 1451 family of standards includes seven documents that define the set of common communication interfaces for connecting smart transducers to microprocessor-based systems, instruments, and networks in a network-independent environment [1]. As Figure 3.4 reveals, the smart transducer for IEEE 1451 is divided into two components that are interconnected through the transducer-independent interface (TII): the actual NCAP and the transducer interface module (TIM), which contains sensors and actuators, signal conditioning, and data conversion. Although IEEE 1451 does not specify actual physical and media access control (MAC) layers for TII, it provides the interface for different standardized technologies through IEEE 1451.2

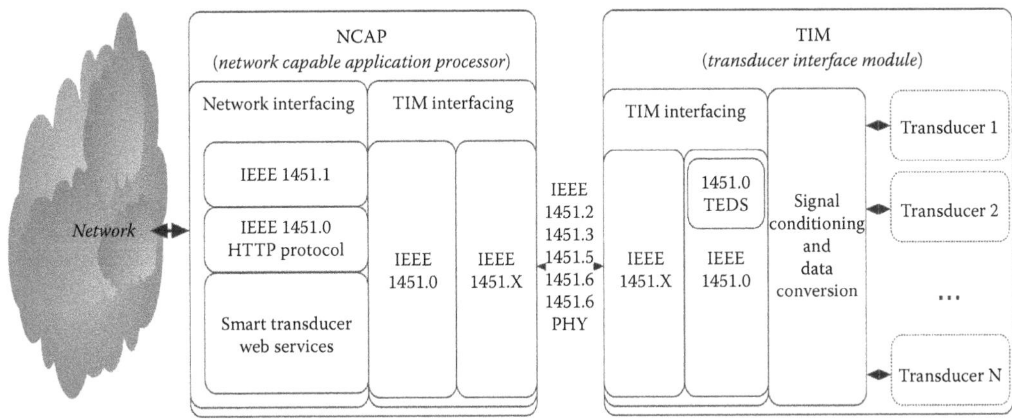

Figure 3.4 The IEEE 1451 family of standards.

(wired point-to-point communication; e.g., SPI, UART, USB), IEEE 1451.3 (wired network; e.g., 1-wire), IEEE 1451.4 (mixed mode—i.e., interfacing transducers with analog output to NCAP), IEEE 1451.5 (wireless communication; e.g., ZigBee, 6LoWPAN, Bluetooth, WiFi), IEEE 1451.6 (CANopen), and IEEE 1451.7 (RFID) [1,14,16]. The discovery, access, and control mechanisms for the transducer supported by the IEEE 1451, both for TIMs connected to NCAPs and for smart transducers within the network, allows achievement of the highest level of network interoperability and implementation of global smart SNs (e.g., see Figure 3.5).

For implementing required smart transducer features, the IEEE 1451 family of standards defines the transducer electronic data sheets (TEDS) for each transducer (or TIM) connectible to the NCAP. The TEDS are memory device attached to the actual transducer (or a memory location accessible by the NCAP—the so-called virtual TEDS) and that stores transducer identification, calibration, correction, and manufacturer-related information (see Figure 3.6a). Depending on the communication interface, the structure of the TEDS can differ slightly. The general TEDS structure, defined by IEEE 1451.0, specifies that the TEDS should include four required and up to six optional components with the specified structure [14,17]. The required parts are (see Figure 3.6a): Meta-TEDS (stores all of the information needed to gain access to any transducer channel and common information for all transducer channels); transducer channel TEDS (provides detailed information about each transducer; e.g., what is measured/controlled and the ranges), user's transducer name TEDS (provides a place to store the name by which this transducer will be known to the system or end users) and PHY TEDS (stores all information about physical communications media between the TIM and the NCAP) [14,18]. The optional TEDS include calibration TEDS, frequency response TEDS, Transfer Function TEDS, text-based TEDS, end user application specific TEDS, and manufacturer defined TEDS. Therefore, the minimum size of the TEDS for, e.g., an IEEE 1451.2 transducer is around 300 bytes, whilst the size of a full-featured TEDS can reach many kilobytes [19]. This is why the simplified—so-called basic—TEDS structure has been provided in IEEE 1451.4 (see Figure 3.6b); this basic structure consists of only 8 bytes that contain the identification information [20,21].

Besides the actual TEDS data, the IEEE 1451 TIM has to contain at least the minimum set of hardware and software to respond to NCAP requests and initiate certain service request messages. In its most basic form, the TIM should [17]

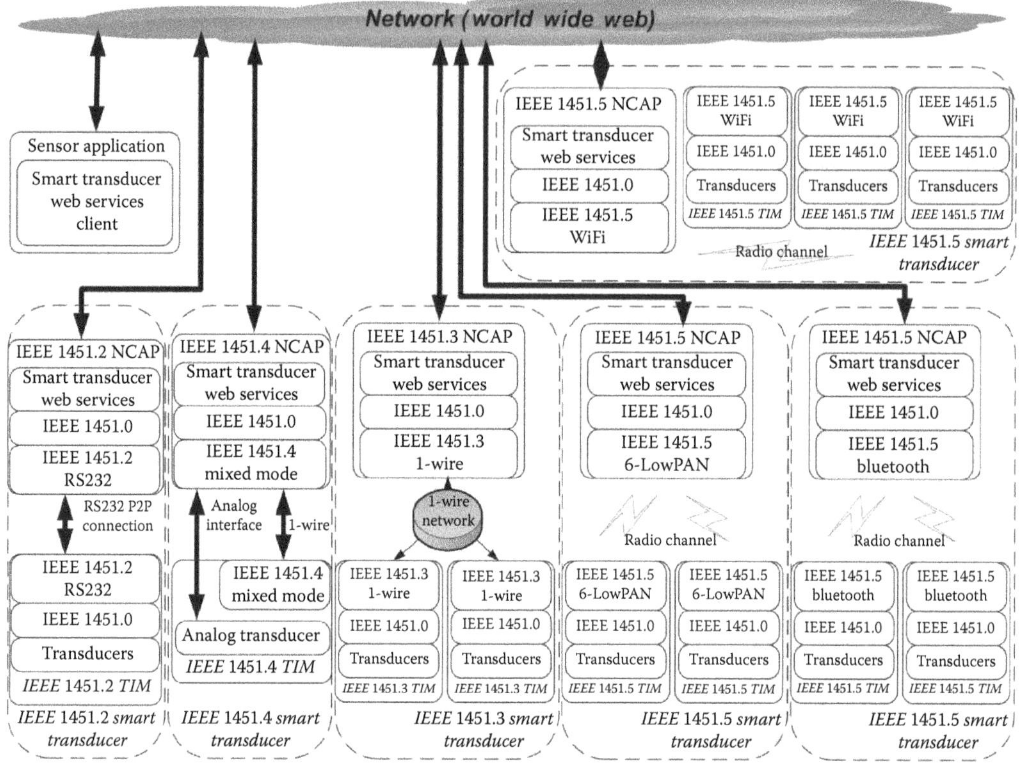

Figure 3.5 A unified web service for IEEE 1451 smart transducers.

- Respond to TIM discovery queries
- Respond to transducer access requests
- Respond to and sometimes initiate transducer management tasks (e.g., sending alerts)
- Respond to and support TEDS management functions

In addition, a TIM may implement various communication, data conversion, signal processing, and calibration algorithms and mechanisms [17].

Although the NCAP network interfaces are not within the scope of the IEEE 1451 standard, some special cases are included in the standard. Therefore, the IEEE 1451.0 describes the communication process between a remote network-based client and an IEEE 1451 NCAP server using hyper text transfer protocol (HTTP). This feature allows implementation of direct communication between NCAPs or allows a single remote web-based client to obtain transducer data from various types of IEEE 1451 transducers connected, e.g., to the World Wide Web [17].

Although IEEE 1451 has numerous advantages, especially its ability to function in a universal way for interfacing any type of transducers (e.g., digital, quasi-digital, and analog ones—see Figure 3.5) to the NCAP (although, most often, this will require addition of an external TIM board with a TEDS block and an IEEE 1451.X communication controller to each transducer) with the support of wide set of intelligent transducer functions, there are some trade-offs for it. The main factors that limit the wide dissemination of the IEEE 1451 standard are its complexity and requirement for additional component usage for its implementation (e.g., memory and a

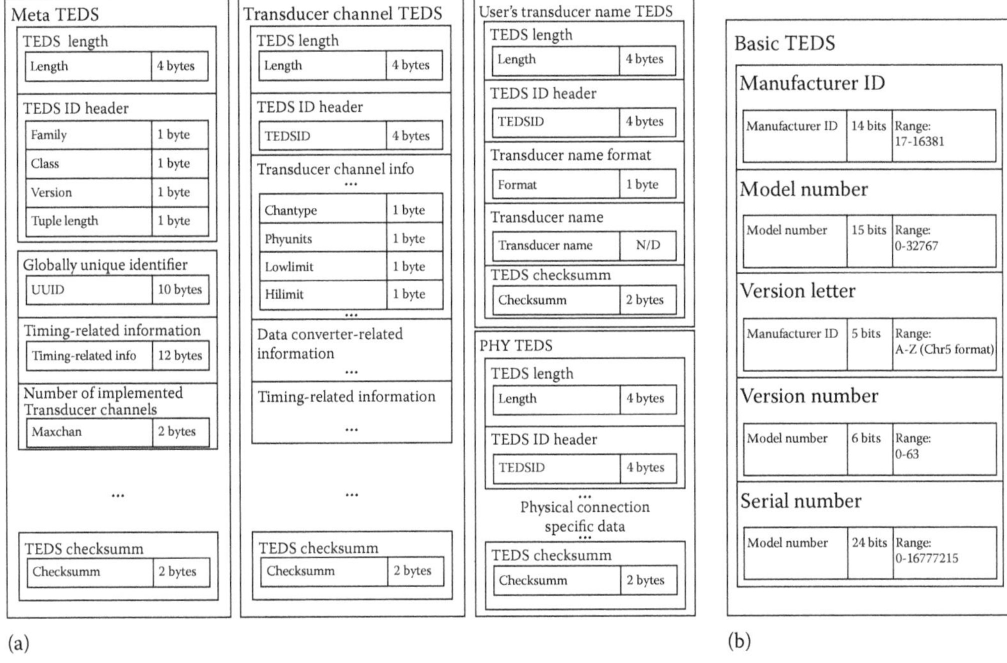

Figure 3.6 IEEE 1451 TEDS formats. (a) IEEE 1451.0 required TEDS. (b) IEEE 1451.4 basic TEDS.

controller for implementing the required TIM), which significantly increases the price of IEEE 1451 compatible solutions [22]. In addition, the absence of IEEE 1451-based SN components, such as transducers, TIMs, and NCAPs on the market is also a negative factor [5].

3.2.3 Implementation of Smart Interface Features over Plain Interfaces in SNs

We have, thus far, discussed the most widespread interfaces for plain sensor and existing smart transducer interfaces. Section 3.2.1 showed that the majority of existing interfaces for plain transducers do not support any intelligent transducer features. Although special smart transducer interfaces exist (e.g., the IEEE 1451 discussed in Section 3.2.2), due to their complexity and high costs, these still have very limited application scope. This makes the problem of intelligent interface features implementation for the sensors with general plain interfaces very real. In the current section, we will show how this can be resolved.

As revealed in Section 3.2.1, the transducers with analog and quasi-digital interfaces have no means either to inform the smart transducer's processor (or NCAP) about their connection or to provide any sort of transducer identification information. For these systems, the only way to detect a sensor connection is to force the processor to perform regular monitoring of the sensor lines. In this case, the sensor connection can be discovered through the analysis of the signal on these lines. Nevertheless, identification of sensors by a standalone processor is usually impossible due to the absence of any identification information in the signal arising from these types of sensors.

The only possible option that can provide some capabilities for identifying the analog or quasi-digital sensors is to use the measurement data from already known sensors on the same or

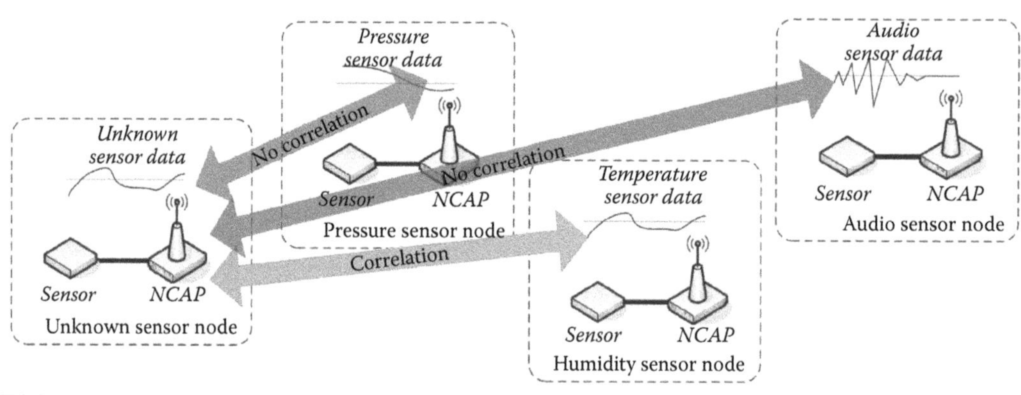

Figure 3.7 Analog sensor identification based on data correlation with neighboring nodes.

neighboring network nodes and to calculate the correlation between their data and the unknown sensor's data (of course, this method assumes the *some correlation exists between the measurements from the different sensors or sensor nodes*)—see Figure 3.7.

Unlike analog and quasi-digital sensors, full-featured digital sensors do not start sending measured data to the processor immediately upon connection but they wait for a *request* from the processor. This makes the connection/disconnection detection for digital sensor rather complicated, especially if several sensors can be connected to processor over the same physical lines [5]. Therefore, the most effective solution for the detection of digital sensor connection would be the use of some external signal generated each time a new sensor is connected. This can be implemented, e.g., through the use of specially designed connectors [5].

In addition to the mechanisms already available for identifying the digital sensors (e.g., address schemes available for I^2C or 1-wire interfaces), the data available from their memory can be used. Sensor identification can be implemented for the widest range of digital interfaces, including the ones that do not have any standardized identification mechanisms, by mechanisms suggested in Ref. [5] based on the following four methods:

- Read from any sensor registers or any other command execution that will return data known in advance (sensor identification is based on the facts of correctness of the physical connection settings, the existence of registers/commands, and the correctness of the retrieved data)
- Write and sequentially read from certain registers with inaccessible bits—i.e., the bits with values that could not be changed (sensor identification is based on the facts of correctness of physical connection settings, the existence of a register, and the position of unchangeable bits)
- Execute a command for which the range of possible return values is known (e.g., a temperature measurement, where sensor identification is based on the facts of the correctness of the physical connection settings, acknowledgment of a command execution, and returned data falling within known limits)
- Write and sequentially read from certain registers or a certain command execution (sensor identification is based only on the facts of correctness of the physical connection settings and acknowledgment of register existence/command execution).

Therefore, this type of sensor identification mechanism can be implemented using the simple tryout algorithm presented in Figure 3.8a. The database, which contains unique request response data for all possible sensors, can be stored either on each node or in a special location within the SN

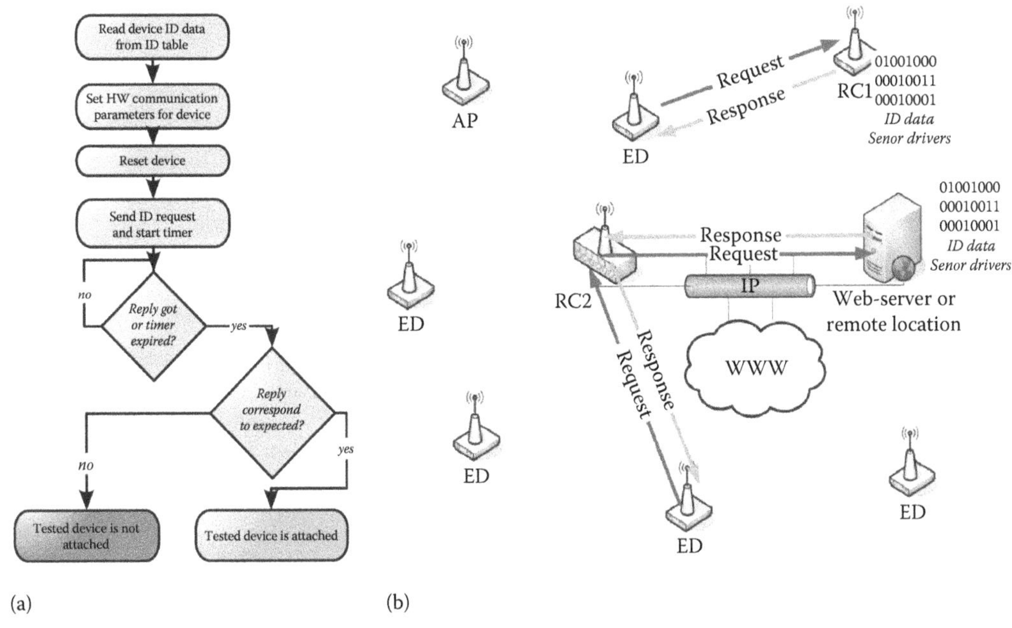

Figure 3.8 Plain digital sensor plug-and-play mechanisms. (a) Plain digital sensor identification algorithm (single device tryout). (b) Network structure with support for plain digital sensor plug-and-play.

(e.g., the resource center [RC] node—see Figure 3.8b). In addition, the RC nodes can be used to provide "sensor drivers"—the pieces of software for sensor node processors that implement the required sensor functionality. This will allow the sensor nodes initially to be free of any sensor-dependent software and will allow downloading of the required sensor drivers from network once the node is attached to it and has identified its available sensors. This mechanism can be used to implement a complete plug-and-play mechanism for sensors with plain digital interfaces. As has been shown in Ref. [5], the suggested identification method can be used for any type of commercially available sensors that utilize plain digital interfaces (i.e., SPI, I^2C, 1-wire, or proprietary ones) *without utilization of any additional components*. Nevertheless, since the suggested mechanism is not based on TEDS or any similar mechanism, proving that the suggested mechanism would allow a single-valued identification for all the existing transducers is not possible.

The networking of the nearby sensor nodes also provides some means for implementing sensor diagnosis or malfunction detection. Similar to the suggested correlation-based analog sensor identification mechanism (see Figure 3.7), the correlation of neighboring node measurements can be used to detect a sensor malfunction. Although this method is not as reliable as the on-node sensor diagnosis of smart sensors, it can partially compensate for the absence of appropriate features for plain sensors.

3.3 Intelligent Networking for Sensor Node

The contemporary sensor networks (SNs) can include numerous heterogeneous nodes manufactured by various vendors [16]. Nonetheless, for proper network operation, each of these nodes should be able to communicate with any other sensor node and with different external devices through some common communication protocol. The success of the transmission control protocol/internet protocol (TCP/IP) over recent years made the TCP/IP protocol stack the de facto standard for large-scale networking [23]. Recent developments have also allowed adoption of the TCP/IP communication for the extreme communication conditions of SN [23,24]. In the current section, we will not focus the hardware implementation of TCP/IP over each specific wired or wireless communication technology that can be utilized by an intelligent SN; instead, we will go one step further and discuss the features of using IP communication within SNs.

One of the main differences in implementing the TCP/IP for SNs, compared to its use in general computer networks, is the very limited resources that are available for SN nodes [16,23]. Indeed, many SN nodes have a very restricted amount of memory that limits the possibility for these nodes to store data on-node. In addition, due to the limited energy and processing capabilities of NCAPs on SN nodes, relaying data processing and analysis on a server device is sometimes more convenient. These considerations, together with the limited amount of energy available on an SN node for transferring data, demonstrate the importance of choosing a format for the message within the SN that will provide sufficiently high compatibility and low energy consumption.

Besides the universal communication and message format, the intelligent SN concept requires the SN nodes (i.e., the NCAPs) to be discoverable and identifiable [25].

The rest of the current subsection is organized as follows: Section 3.3.1 discusses the basic structure of web service stack used for IP-based SN, Section 3.3.2 discusses ways to represent the data in an IP-based SN, and Section 3.3.3 discusses possible ways to implement intelligent functionality in an IP-based SN.

3.3.1 Basic Structure of a Web Service Stack for an IP-Based SN

The structure of a typical web service stack for an SN application can be described as follows. Initially, the web services (i.e., the software system that supports interoperable machine-to-machine interaction over a network) are used to set up the connection between client and server end-points in a platform-independent manner. In SNs, the NCAPs and end-user terminals usually act as client end-points that are connected to a server end-point that runs the database and that contains the measurement data or the data about the sensors (see, e.g., Figure 3.9). Most often in SN, the web services are represented by a TCP/IP protocol stack, on top of which is implemented the HTTP [23,26].

After the initial connection has been established (e.g., consider the case when TCP/IP and HTTP are implemented), the client can send the required request messages to the server. Although this can be done using any HTTP method, in SN GET, POST, PUT, or DELETE are typically used [26]. Finally, the client, using a standardized method, specifies the data that should be placed on the server or requested from server [27] (see Figure 3.10 for an example).

In response, the server informs the client if the request was successful and provides the requested data, if any data have been requested (see Figures 3.9 and 3.10). HTTP is able to carry the data (the "body" of the message) in either direction using any multipurpose internet mail extensions (MIME) type and encoding. The body of the HTTP web service message is typically encoded using extensible markup language (XML) and the format is known both to the client and the server [28]. In order to perform sequences of remote procedure calls (RPCs), the simple object access protocol (SOAP) may be encapsulated into XML body data, although this would inevitably complicate the client and server software [29].

After the sequence of requests is complete, the TCP connection is usually closed (see Figure 3.9).

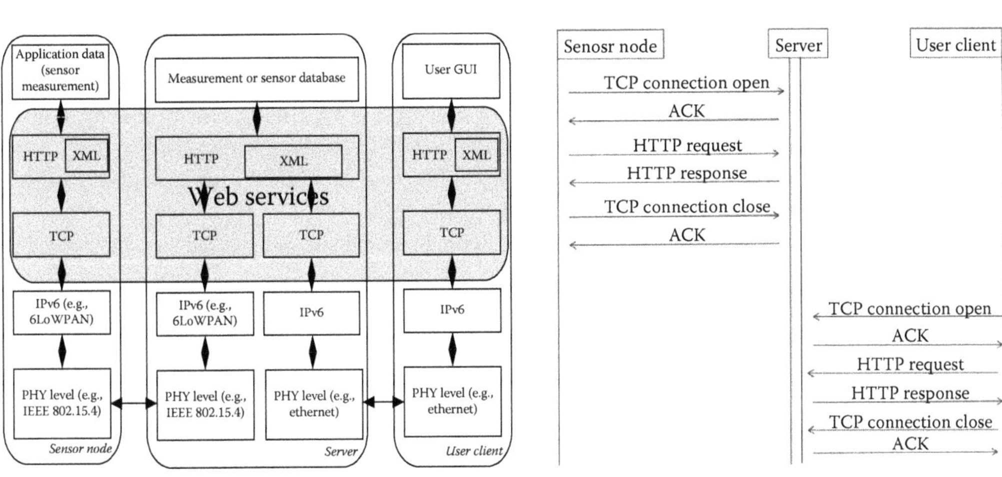

Figure 3.9 Typical client–server interaction and message exchange with SN. (a) Architecture. (b) Messages.

(a)

Client request:

POST/WSNService HTTP/1.1

Host: www.centria.fi

Content–Type: application/json; charset = utf–8

Content–Length: 16

{"temperature–value":8}

Server response:

HTTP/1.1 200 OK

Date: Mon, 5 Dec 2011 21:05:16 GMT

Server: Apache/1.3.3.7 (Unix) (Red–Hat/Linux)

Accept–Ranges: bytes

Connection: close

Content–Type: text/html; charset = UTF–8

(b)

Client request:

GET/index.html HTTP/1.1

Host: www.centria.fi?temperature

Server response:

HTTP/1.1 200 OK

Date: Mon, 5 Dec 2011 22:38:34 GMT

Server: Apache/1.3.3.7 (Unix) (Red–Hat/Linux)

Accept–Ranges: bytes

Connection: close

Content–Type: text/html; charset = UTF–8

{"temperature–value":8}

Figure 3.10 Examples of HTML GET and POST requests and responses within an SN. (a) POST request and response. (b) GET request and response.

One of the advantages of utilizing the HTTP for transferring the data in an SN is that it simplifies the required connection tune-up and especially the set-up of communication through a firewall (see, e.g., Figure 3.11: the TCP/IP-based connection [Figure 3.11a] will most probably require manual firewall rule modification for the server, while the HTTP-based connection [Figure 3.11b] most probably will require no special actions) [30]. The use of HTTP also provides effective authentication, compression, and encryption mechanisms [26].

Today's web services can be implemented either using general web-servers (see e.g., Figure 3.11) or cloud platforms (see Figure 3.12). The implementation of web services using cloud platforms (see Figure 3.12) provides high scalability and cost-efficiency for the application, as it allows dynamic reassignment of the data traffic and required processing between the different resources composing the cloud [31]. This is especially valuable when handling of simultaneous data requests from multiple user clients is required.

Nonetheless, the implementation of HTTP for SN introduces the following challenges:

- High data overhead is caused by the plain text encoding and the verbosity of the HTTP header format.
- TCP Binding (i.e., the association of a node's input or output channels and file descriptors with a transport protocol, a port number, and an IP address) seriously reduces the performance, especially for ad hoc wireless SN with short-lived connections.
- The support for a large number of optional HTTP headers that can have very complex structure and include unnecessary information increases the implementation complexity and SN data traffic overhead.

3.3.2 Methods for Data Representation in IP-Based SNs

Minimization of the total message size and maximization of the packet payload conflicts with the Internet standard but are essential for reducing the energy consumption of SN nodes [32]. Indeed, the standard HTTP headers can be quite voluminous, which usually makes the ratio of payload to whole transferred data rather low [32]. This matters in particular for battery-operated SN nodes that utilize wireless communication, where every used energy Joule and every additional transmitted data byte is meaningful and decreases the node's operation time. As discussed in Section 3.3.1, in SNs, XML is widely used for data encoding [23].

The extensible markup language (XML) [28] is a set of rules for encoding documents in machine-readable form that is widely used by web services for data presentation and exchange. Although the use of XML results in a high level of scalability and interoperability, its use has some drawbacks. The most important of these is that XML opening and closing tags can significantly increase the size of the data (see Figure 3.13). This problem can be partially solved by using the XML compression mechanisms, such as Binary XML.

Binary XML is a set of specifications that defines the compact representation of XML in a binary format. At present, three binary XML compression formats have been introduced, although none of them has yet been widely adopted or accepted as a de facto standard [33].

The first format was suggested by the International Organization for Standardization (ISO) and International Telecommunications Union (ITU) and is called Fast Infoset. This format is proposed as an alternative to the World Wide Web Consortium (W3C) XML syntax that specifies the binary encoding for W3C XML Information Set. The Fast Infoset has been defined as ITU-T (X.891) and ISO (IEC 24824-1) standards [34]. Unlike the other XML representations, the Fast Infoset standard can be beneficial both for low bandwidth systems that require high data compression

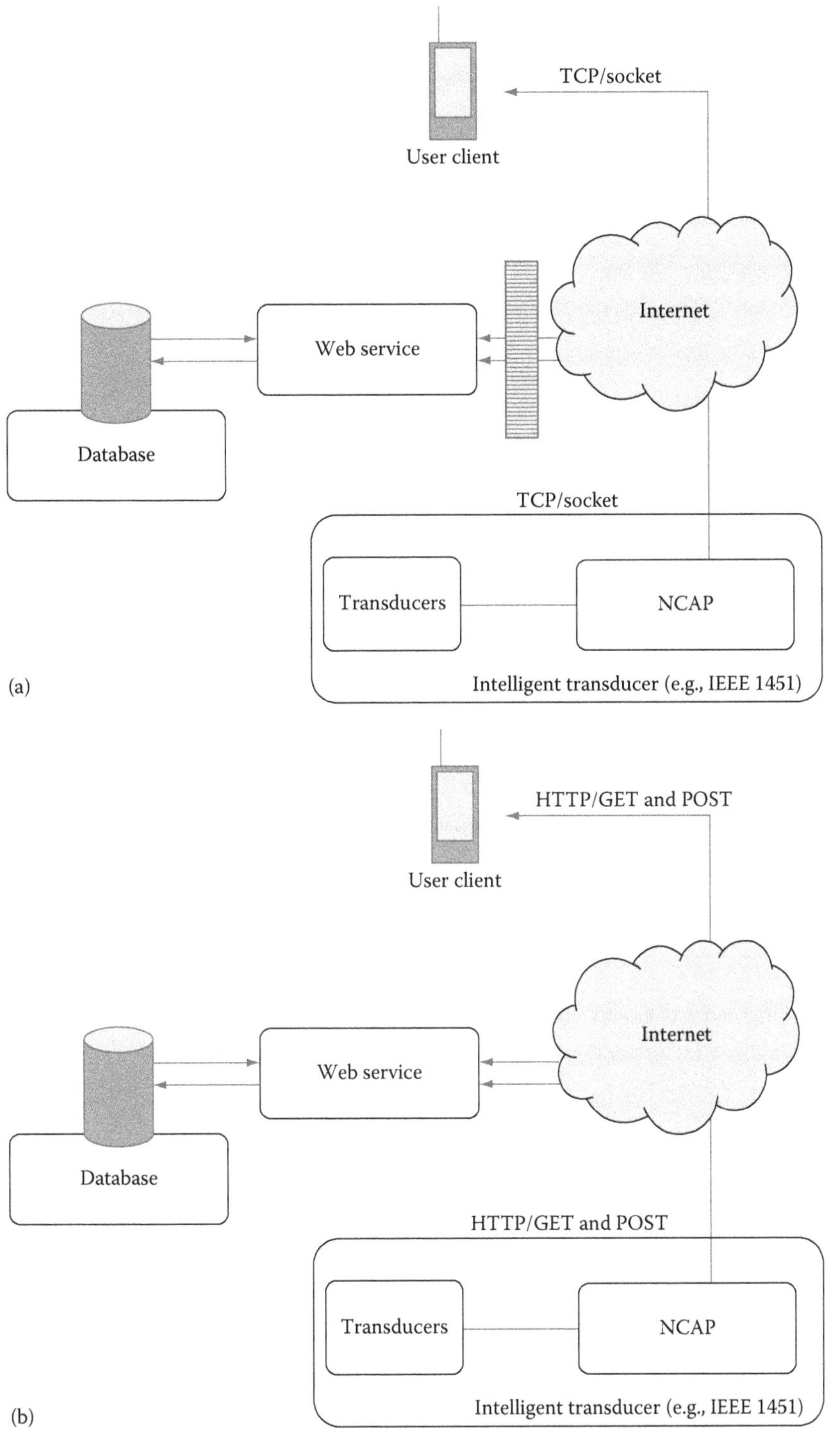

Figure 3.11 Web service layout using a TCP/socket and HTTP in SN. (a) TCP/socket. (b) HTTP.

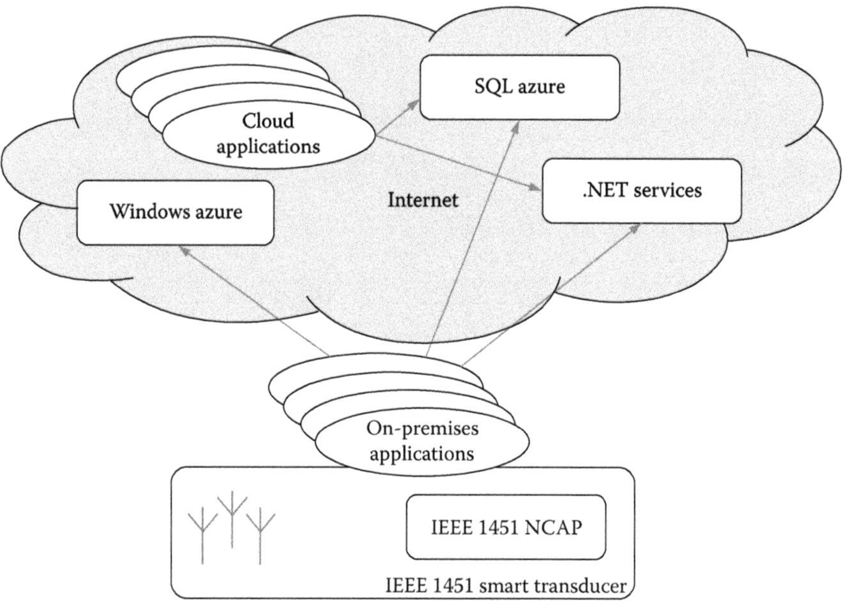

Figure 3.12 Example of web service for an SN using cloud platforms.

```
<?xml version = ″1.0″ encoding = ″utf–8″?>
<root>
  <data–value>8</data–value>
</root>
```

Figure 3.13 Example of an XML file.

(e.g., the connection between an NCAP and a server in SN) and for high performance systems that utilize web services (e.g., end-user client devices in SN) [34].

The second format has been suggested by W3C and is called efficient XML interchange (EXI) format. This is claimed to be a very compact XML representation that can simultaneously optimize performance and utilization of computational resources [35]. The EXI format uses a hybrid approach, drawn from the information and formal language theories, in conjunction with the practical techniques for entropy encoding XML information. Using a relatively simple algorithm, which is amenable to fast and compact implementation, and a small set of data type representations, EXI reliably produces efficient encodings of XML event streams. The grammar production system and format definition of EXI are presented, e.g., in Ref. [35].

The third format is called binary extensible markup language (BXML) and has been suggested by Open Geospatial Consortium (OGC) especially for geosensor networks. This format can be also used for compressing sensor data and has been defined using OGC's geography markup language (GML) geographical features [36,37]. An example of binary XML compression using BXML is presented in Figure 3.14.

The spatial data in this example are very simple and contain only polylines that are stored in an XML file (the data were taken from a Finnish spatial database, Digiroad). This XML file contains a coordinate system, optional data, and collection of coordinates for every object. This file is next converted to binary format based on OGC's BXML (see Figure 3.14). During

XML		*Number of bytes for OGC's BXML:*
<gml:featureMember fid ="1921355">		10
<drd:coordSystem>KKJ3</drd:coordSystem>		9
<drd:NAME>Savelantie</drd:NAME>		17
<drd:ATTRIBUTE>4</drd:ATTRIBUTE>		8
<gml:coordinates		2
decimal = "."		5
cs = " "		5
ts = ","		5
value ="3">		8
3378741 7111164,		16
3378738 7111166,		16
3378712 7111177		13
</gml: coordinates>		1
</gml: featureMember>		1
Overall size:	*269*	*116*

Figure 3.14 Example of binary XML compression using BXML.

this conversion, the traditional XML tags are encoded with bytes; e.g., a GML element with an attribute <gml:featureMember fid = "1921355"> is encoded using a hexadecimal presentation as 0x03 0x01 0x05 0x04 0xF4 0x00 0x1D 0x51 0x4B 0x06, where 0x03 is an element with attributes, 0x01 is a pointer to the element name in the global string table, 0x05 is an attribute start code, 0x04 is a pointer to the attribute's name in the global string table, 0xF4 is a data type (Int32), 0x00 0x1D 0x51 0x4B is the identification number 1921355, and 0x06 marks the end of attributes and the whole element.

3.3.3 Intelligent Functionality for IP-Based SN Nodes

An SN node (i.e., the NCAP of SN node) is made discoverable and identifiable within the SN using either the specifications that have been developed for web services (e.g., SOAP) or the special interface specifications that have been developed especially for sensors (e.g., OpenGIS).

The SOAP [38] relies on XML for its message format and specifies how to exchange structured data for web service implementation. SOAP can be used over multiple connection protocols, including HTTP. The use of SOAP provides the SN node with the possibility for self-identification support, although, as shown in Figure 3.15, the messages that are used by SOAP can be rather voluminous.

The alternative option is the use of the OpenGIS interface standard that has been developed by OGC. According to OGC [37], the OpenGIS interface standard defines OpenLS core services,

```
POST /WSNService HTTP/1.1
Host: www.centria.fi
Content–Type: application/soap + xml; charset = utf–8
Content–Length: 299
SOAPAction: "http://www.w3.org/2003/05/soap–envelope"

<?xml version = "1.0"?>
<soap:Envelope xmlns:soap ="http://www.w3.org/2003/05/soap–envelope">
   <soap:Header>
   </soap:Header>
   <soap:Body>
      <m:InsertSensorValue xmlns:m ="http://www.centria.fi/sensor01">
         <m:value>8</m:value>
      </m:InsertSensorValue>
   </soap:Body>
</soap:Envelope>
```

Figure 3.15 Example of a SOAP POST message.

which forms the services framework for the GeoMobility server (GMS). The core services include directory service; gateway service; location utility service; presentation service, and route service.

GMS, as a location service platform, hosts these services as well as Location Content Databases (e.g., web map server [WMS] and web feature server [WFS]) that are accessed through OGC Interfaces. For geosensor networks, the main part of OGC's work is conducted under sensor web enablement (SWE), which is a suite of specifications related to sensors, sensor data models, and sensor web services. The goal of SWE services is to enable sensors to be accessible and controllable via the Internet [39].

SWE includes the observations and measurements schema (O&M), sensor model language (SensorML), transducer markup language (TML), sensor observations service (SOS), sensor planning service (SPS), sensor alert service (SAS), and web notification service (WNS). XML is a key part of this infrastructure and all of the services and content models are specified using XML schemas [40].

SensorML provides a rich collection of metadata that can be mined and used for discovery of sensor systems and observation process features. These metadata include identifiers, classifiers, constraints (time, legal, and security), capabilities, characteristics, contacts, and references, in addition to inputs, outputs, parameters, and system location. The calibration information, sensor type, and sensor operator could also be included (see Figure 3.16). If required, in addition to SensorML, metadata that provide a functional description of a sensor node can also use TML to provide additional data about the hardware and the information necessary for understanding the data gathering process [41].

The use of SWE and SensorML/TML, together with the backend system that keeps the database containing the data on all known NCAP features (so-called sensor instance registry [SIR]), provides the complete solution for implementing NCAP discovery and identification within SN [42]. In this case, once connected to an SN, the NCAP is required to provide the SensorML/TML data (the metadata transferring mechanism is provided by SOS) to the SIR that reveals the list of sensors connected to this NCAP and all other data about the NCAP (e.g., its location, if known). Once these data are included in the SIR, any user will be able to see this NCAP and its sensors. Using the methods specified in O&M, the user can establish a direct connection and request the measurement data directly from the NCAP. Knowing the NCAP address, the user can also use SAS to subscribe for specific event notifications that will be sent directly to the user by the NCAP.

```
<sml:SensorML xmlns:sml="http://www.opengis.net/sensorML/1.0"
xmlns:swe="http://www.opengis.net/swe/1.0"
xmlns:gml="http://www.opengis.net/gml"
xmlns:xlink="http://www.w3.org/1999/xlink"
xsi:schemaLocation="http://www.opengis.net/sensorML/1.0
http://schemas.opengis.net/sensorML/1.0.0/sensorML.xsd"version="1.0">
<member xlink:role="urn:ogc:def:role:OGC:centriasensor">
<Component gml:id="Centria">
<swe:value>8</swe:value>
</Component>
</member>
</SensorML>
xmlns:swe="http://www.opengis.net/swe/1.0"
xmlns:gml="http://www.opengis.net/gml"
xmlns:xlink="http://www.w3.org/1999/xlink"
xsi:schemaLocation="http://www.opengis.net/sensorML/1.0
http://schemas.opengis.net/sensorML/1.0.0/sensorML.xsd" version="1.0">
<member xlink:role="urn:ogc:def:role:OGC:centriasensor">
<Component gml:id="Centria">
<swe:value>8</swe:value>
</Component>
</member>
</SensorML>
```

Figure 3.16 An example of SensorML data.

In recent years, the OGC's SWE specifications have been effectively adopted by the National Aeronautics and Space Administration (NASA) in their earth observing (EO-1) geosensor networks system [43]. As reported in Refs. [44,45], SWE has also been effectively implemented over the IEEE 1451 NCAPs, which has resulted in the highest level of interoperability and intelligent functionality supported both for sensors on sensor nodes and for sensor nodes inside a network.

3.4 Conclusions

In the current chapter, we have focused on the problem of intelligent sensor network node interfacing from two sides.

In the first place, we have discussed the interface between actual hardware sensors and the processing unit of an intelligent sensor network node. We have also discussed in detail the most widespread plain sensor interfaces (e.g., analog, quasi-digital, and digital ones) and smart sensor interfaces (e.g., IEEE 1451). The plain sensor interfaces most widely used today do not support any of the features that are required for implementing intelligent sensor functionality. The special smart sensor interfaces, although providing the required functionality, have a rather complicated structure that results in increased price and power consumption for this solution. This is especially undesirable for wireless SNs, which often can have very limited resources. Therefore, we have suggested several mechanisms that can be used to implement some intelligent sensor functionalities over plain interfaces using only the processing capabilities of the sensor network nodes and the existing resources of the sensor networks.

In the second place, we have considered the interfacing of an intelligent sensor network node within a network. We have discussed the structure of a web service stack and features for using internet protocol (IP)-based communication within sensor networks. We have also discussed the problems of data representation and compression in these types of networks. Finally, we have shown how the required network-level intelligent functionality can be implemented for IP-based sensor networks.

The chapter shows that much has already been done to enable intelligent sensor networks, but much still remains to do. One of the major factors that presently limit the dissemination of intelligent sensor networks is the high complexity and cost of intelligent sensor network nodes and their components. Although these costs can be reasonable for some life-critical applications (e.g., volcano eruption, earthquake, or tsunami alarm systems), for the majority of everyday life applications, the costs are still too high. This makes the development of simpler solutions for intelligent functionality implementations in sensor networks one of the most important tasks of present-day research.

References

1. E. Song and K. Lee, STWS: A unified web service for IEEE 1451 smart transducers, *IEEE Trans. Instrum. Meas.*, 57, 1749–1756, August 2008.
2. S. Yurish, Smart sensor systems integration: New challenges, IFSA, [Online]. Available: http://www.iaria.org/conferences2011/filesICN11/KeynoteSergeyYurish.pdf, January 2011.
3. The internet of things executive summary, ITU, Geneva, Switzerland, ITU Internet Reports 2005 [Online]. Available: http://www.itu.int/dms_pub/itu-s/opb/pol/S-POL-IR.IT-2005-SUM-PDF-E.pdf
4. K. Lee and R. Schneeman, Distributed measurement and control based on the IEEE 1451 smart transducer interface standards, *IEEE Trans. Instrum. Meas.*, 49, 621–627, June 2000.
5. K. Mikhaylov, T. Pitkaaho, and J. Tervonen, Plug-and-play mechanism for plain transducers with digital interfaces attached to wireless sensor network nodes, submitted for publication.
6. K. Mikhaylov, J. Tervonen, and D. Fadeev, Development of energy efficiency aware applications using commercial low power embedded systems, in embedded systems - Theory and Design Methodology, K. Tanaka, Ed., Rijeka, Croatia: InTech, 2012, pp. 407–430.
7. G. Meijer, *Smart Sensor Systems*. Hoboken, NJ: John Wiley & Sons, Inc., 2008.
8. S. Yurish, Extension of IEEE 1451 Standard to Quasi-Digital Sensors, *Proc. SAS'07*, San Diego, CA, USA, pp. 1–6, February 2007.
9. N. Kirianaki, S. Yurish, N. Shpak, and V. Deynega, *Data Acquisition and Signal Processing for Smart Sensors*. Chichester, U.K.: John Wiley & Sons, 2001.
10. NXP Semiconductors, I2C - bus specification and user manual (UM10204 rev. 4), Eindhoven, Netherlands, 2012. Retrieved from http://www.nxp.com/documents/user_manual/UM10204.pdf
11. Motorola Semiconductor Products Inc., SPI Block Guide (V03.06), Schaumburg, IL, 2003. Retrieved from http://www.ee.nmt.edu/~teare/ee308l/datasheets/S12SPIV3.pdf
12. Maxim Integrated Products, 1-Wire Communication Through Software (AN 126), San Jose, CA, 2002. Retrieved from http://pdfserv.maxim-ic.com/en/an/AN126.pdf
13. Maxim Integrated Products, Book of iButton Standards (AN 937), San Jose, CA, 1997. Retrieved from http://pdfserv.maxim-ic.com/en/an/AN937.pdf
14. IEEE Instrumentation and Measurement Society, IEEE Standard for a Smart Transducer Interface or Sensors and Actuators Common Functions, Communication Protocols, and Transducer Electronic Data Sheet (TEDS) Formats (IEEE Std. 1451.0-2007), New York, NY, 2007. Retrieved from http://ieeexplore.ieee.org/stamp/stamp.jsp?tp=&arnumber=4346346
15. IEEE Instrumentation and Measurement Society, ISO/IEC/IEEE Information technology – Smart transducer interface for sensors and actuators – Common functions, communication protocols, and Transducer Electronic Data Sheet (TEDS) formats (ISO/IEC/IEEE Std. 21450-2010), New York, NY, 2010. Retrieved from http://ieeexplore.ieee.org/stamp/stamp.jsp?tp=&arnumber=5668466

16. M. Kuorilehto, M. Kohvakka, J. Suhonen, P. Hamalainen, M. Hannikainen, and T. Hamalainen, *Ultra-Low Energy Wireless Sensor Networks in Practice: Theory, Realization and Deployment*. Hoboken, NJ: John Wiley & Sons, Inc., 2007.
17. J. Wiczer and K. Lee, A unifying standard for interfacing transducers to networks—IEEE-1451.0, *Proc. Sensors ISA Expo '05*, Chicago, IL, USA, pp. 1–10, 2005.
18. S. Manda and D. Gurkan, IEEE 1451.0 compatible TEDS creation using.NET framework, *Proc. SAS'09*, New Orleans, LA, USA, pp. 281–286, February 2009.
19. J. Wiczer, A summary the IEEE-1451 family of transducer interface standards, *Proc. Sensors Expo'02*, San Jose, CA, USA, pp. 1–9, May 2002.
20. D. Wobschall, IEEE 1451—A universal transducer protocol standard, *Proc. Autotestcon'07*, Baltimore, MD, USA, pp. 359–363, September 2007.
21. IEEE Instrumentation and Measurement Society, ISO/IEC/IEEE Standard for Information technology – Smart transducer interface for sensors and actuators – Part 4: Mixed-mode communication protocols and Transducer Electronic Data Sheet (TEDS) formats (ISO/IEC/IEEE Std. 21451-4-2010), New York, NY, 2010. Retrieved from http://ieeexplore.ieee.org/stamp/stamp.jsp?tp=&arnumber=5668460
22. K. Lee, M. Kim, S. Lee, and H. Lee, IEEE-1451-based smart module for in-vehicle networking systems of intelligent vehicles, *IEEE Trans. Ind. Electron.*, 51, 1150–1158, December 2004.
23. A. Dunkels, Towards TCP/IP for wireless sensor networks, Licentiate thesis, Malardalen University, Eskilstuna, Sweden, 2005.
24. J. Higuera and J. Polo, IEEE 1451 standard in 6LoWPAN sensor networks using a compact physical-layer transducer electronic datasheet, *IEEE Trans. Instrum. Meas.*, 60(8), 2751–2758, August 2011.
25. X. Chu and R. Buyya, Service oriented sensor web, in N.P. Mahalik (Ed.), *Sensor Network and Configuration: Fundamentals, Techniques, Platforms, and Experiments*. Berlin, Germany: Springer-Verlag, 2006.
26. R. Fielding, J. Gettys, J. Mogul, H. Frystyk, L. Masinter, P. Leach, and T. Berners-Lee, Hypertext Transfer Protocol – HTTP/1.1 (RFC 2616), 1999. Retrieved from http://256.com/gray/docs/rfc2616/03.html
27. D. Crockford, RFC4627-The application/json media type for javascript object notation (json), The Internet Society, 2006.
28. T. Bray, J. Paoli, C. Sperberg-McQueen, E. Maler, and F. Yergeau, Extensible markup language (XML) 1.0 (5th edn.), World Wide Web Consortium Std. 2008.
29. M. Olson and U. Ogbuji, The Python web services developer: Messaging technologies compared, IBM developer works, [Online]. Available: http://www-128.ibm.com/developerworks/library/ws-pyth9/, July 2002.
30. M. Strebe and C. Perkins, *Firewalls 24seven*. Alameda, CA: Sybex Inc., 2002.
31. D. Chappell, Introducing the windows azure platform, Microsoft Corporation, [Online]. Available: http://msdn.microsoft.com/en-us/library/ff803364.aspx, October 2010.
32. A. Castellani, M. Ashraf, Z. Shelby, M. Luimula, J. Yli-Hemminki, and N. Bui, Binary WS: Enabling the embedded web, *Proc. Future Network and Mobile Summit'10*, Florence, Italy, pp. 1–8, June 2010.
33. J. Chang, i-Technology viewpoint: The performance woe of binary XML, [Online]. Available: http://soa.sys-con.com/node/250512, August 2008.
34. Telecommunication Standardization Sector of International Telecommunication Union, Series X: Data Networks, Open System Communications And Security: Information technology – Generic applications of ASN.1: Fast infoset (ITU-T Recommendation X.891), Geneva, Switzerland, 2005. Retrieved from http://www.itu.int/rec/dologin_pub.asp?lang=e&id=T-REC-X.891-200505-I!!PDF-E&type=items

35. J. Schneider and T. Kamiya (Eds.), Efficient XML interchange (EXI) format 1.0 (W3C Recommendation, 10 March 2011), World Wide Web Consortium, 2011. Retrieved from http://www.w3.org/TR/2011/REC-exi-20110310/
36. Binary extensible markup language (BXML) encoding specification (OGC 03-002r9), Open Geospatial Consortium Inc., C. Bruce (Ed.), 2006. Retrieved from http://portal.opengeospatial.org/files/?artifact_id=13636
37. M. Botts (Ed.), Open GIS sensor model language (SensorML) Implementation Specification (OGC 05-086r2), Open Geospatial Consortium Inc., 2006. Retrieved from http://portal.opengeospatial.org/files/?artifact_id=13879
38. D. Box, D. Ehnebuske, G. Kakiwaya, A. Layman, N. Mendelson, H.F. Nielsen, S. Tathe, and D. Winer, Simple object access protocol (SOAP) 1.1, World Wide Web Consortium, 2000. Retrieved from http://www.w3.org/TR/2000/NOTE-SOAP-20000508/
39. A. Sheth, C. Henson, and S.S. Sahoo, Semantic sensor web, *IEEE Internet Comput.*, 12(4), 78–83, July 2008.
40. M. Botts, G. Percivall, C. Reed, and J. Davidson, OGC Sensor web enablement: Overview and high level architecture, in S. Nittel, A. Labrinidis, and A. Stefanidis (Eds.), *GeoSensor Networks*, 4540, 175–190, December 2007.
41. A. Walkowski, Sensor web enablement—An overview, in M. Grothe and J. Kooijman (Eds.), *Sensor Web Enablement*, 45, 69–72, 2008.
42. S. Jirka, A. Bröring, and C. Stach, Discovery mechanism on sensor web, *Sensors*, 6, 2661–2681, 2009.
43. S. Chien et al., Lights out autonomous operation of an earth observing sensorweb, *Proc. RCSGSO'07*, Moscow, Russia, pp. 1–8, June 2007.
44. E. Song and K. Lee, Integration of IEEE 1451 smart transducers and OGC-SWE using STWS, *Proc. SAS'09*, New Orleans, LA, USA, pp. 298–303, February 2009.
45. S. Fairgrieve, J. Makuch, and S. Falke, PULSENet^{TM}: An implementation of sensor web standards, *Proc. CTS'09*, Baltimore, MD, USA, pp. 64–75, May 2009.

Chapter 4

Smart Wireless Sensor Nodes for Structural Health Monitoring

Xuefeng Liu, Shaojie Tang, and Xiaohua Xu

Contents

Because of the low cost, high scalability, and ease of deployment, wireless sensor networks (WSNs) are emerging as a promising sensing paradigm that the structural engineering field has begun to consider as a substitute for traditional tethered structural health monitoring (SHM) systems. For a WSN-based SHM system, particularly the one used for long-term purpose, to provide real-time information about the structure's healthy status and in the meantime, to avoid high cost of streaming the raw data wirelessly, embedding SHM algorithms within the network is usually necessary. However, unlike other monitoring applications of WSNs, such as environmental and habitat monitoring where embedded algorithms are as simple as average, max, min, etc., many SHM algorithms are centralized and computationally intensive. Implementing SHM algorithms within WSNs hence becomes the main roadblock of many WSN-based SHM systems. This chapter mainly focuses on designing and implementing effective and energy-efficient SHM algorithms in resource-limited WSNs. We first give a summary review of the recent advances of embedding SHM algorithms within WSNs. Then the modal analysis, a classic SHM algorithm widely adopted in SHM, is chosen as an example. How this technique can be embedded within a WSN is described in a step-by-step manner. At last, we propose a WSN-Cloud system architecture, which we believe is a promising paradigm for the future SHM.

4.1 Introduction

Civil structures such as dams, long-span bridges, skyscrapers, etc., are critical components of the economic and industrial infrastructure. Therefore, it is important to monitor their integrity and detect/pinpoint any possible damage before it reaches to a critical state. This is the objective of SHM [1].

Traditional SHM systems are wire-based and centralized. In a typical SHM system, different types of sensors, such as accelerometers, and strain gauges, are deployed on the structure under monitoring. These sensor nodes collect the vibration and strain of the structure under different locations and transmit the data through cables to a central station. Based on the data, SHM algorithms are implemented to extract damage-associated information to make corresponding decisions about structural condition [1].

According to the duration of deployment, SHM systems can be largely divided into two categories: short- and long-term monitoring. Short-term SHM systems are generally used in routine annual inspection or urgent safety evaluation after some unexpected events such as earthquake, overload, or collisions. These short-term systems are usually deployed on structures for a few hours to collect enough amounts of data for off-line diagnosis afterward. Examples of short-term SHM systems can be found in the Humber Bridge of United Kingdom [2], and the National Aquatic Centre in Beijing, China [3]. The second category of SHM systems is those used for long-term monitoring. Sensor nodes in these systems are deployed on structures for months, years, or even decades to monitor the structures' healthy condition. Different from short-term monitoring systems where data are processed off-line by human operators, most of the long-term SHM systems require the healthy condition of the structure be reported in a real-time or near real-time manner. Examples of long-term monitoring SHM systems can be found in the Tsing Ma Bridge and Stonecutters Bridge in Hong Kong [4].

The main drawback of traditional wire-based SHM systems is the high cost. The high cost mainly comes from the centralized data acquisition system (DAC), long cables, sensors, and in-field servers. Particularly for DAC, its price increases dramatically with the number of channels it can accept. As a result, the cost of a typical wire-based SHM system is generally high. For example, the costs of the systems deployed on the Bill Emerson Memorial Bridge and Tsing Ma Bridge reach \$1.3 and \$8 million, respectively [4].

In addition, deploying a wire-based SHM system generally takes a long period of time. This drawback is particularly apparent in SHM systems used for short-term purpose. Considering the length of cables used in an SHM system deployed on a large civil infrastructure can reach thousands or even tens of thousand meters, deployment can take hours or even days to obtain measurement data just for a few minutes. Moreover, constrained by the number of sensor nodes and the capability of DAC, it is quite common that an SHM system is repeatedly deployed in different areas of a structure to implement measurement. This dramatically increases the deployment cost. We have collaborated with civil researches to deploy a wire-based SHM system on the Hedong Bridge in Guangzhou, China (see Figure 4.1). The DAC system we used can only support inputs from

(a) (b)

Figure 4.1 A wired-based SHM system deployed on the Hedong Bridge, China: (a) Hedong Bridge and (b) deploying a wired system.

seven accelerometers simultaneously. To measure the vibration at different locations across the whole bridge, the systems were hence moved to 15 different areas of the bridge to implement measurement, respectively. For each deployment, it took about 2 h for sensor installation, cable deployment, and initial debugging.

Recent years have witnessed a booming advancement of WSNs and an increasing interest of using WSNs for SHM. Compared with the traditional wire-based SHM systems, wireless communication eradicates the need for wires and therefore represents a significant cost reduction and convenience in deployment. A WSN-based SHM system can achieve finer grain of monitoring, which potentially increases the accuracy and reliability of the system.

However, SHM is different in many aspects from most of the existing applications of WSNs. Table 4.1 summarizes the main differences between SHM and a typical application of WSNs, environmental monitoring, in terms of sensing and processing algorithms. Briefly speaking, sensor nodes in an SHM system implement synchronized sensing with relatively high sampling frequency. Moreover, SHM algorithms to detect damage are based on a bunch of data (in a level of thousands and tens of thousands) and are usually centralized and complicated.

Table 4.1 Difference between SHM and Environmental Monitoring

	SHM	*Environmental Monitoring*
Sensor type	Accelerometers, strain gauges	Temperature, light, humidity
Sampling pattern	Synchronous sampling round by round	Not necessarily synchronized
Sampling frequency	*X*00 – *X*000/s	*X*/s,min
Processing algorithms	On a bunch of data (>*X*0000) centralized, computationally intensive	Simple, easy to be distributed

Moreover, the difficulty of designing a WSN-based SHM system is different for short- and long-term applications. Designing a WSN for short-term SHM is relatively easy. Generally speaking, short-term SHM systems only need to address synchronized data sampling and reliable data collection. The former task can be realized using various time synchronization protocols [5] and resampling techniques [6]. In addition, considering the high cost of wireless transmissions in WSNs, wireless sensor nodes in a short-term SHM system can be equipped with local storage device, such as a μSD card or USB to save the measured data in a real-time manner. The locally stored data in wireless sensor nodes can be retrieved afterward by human operators.

On the contrary, designing a WSN for long-term SHM is much more challenging. A long-term SHM system not only needs to have a longer system lifetime and higher system reliability but embedding SHM algorithms within the network becomes a necessity. This task is difficult mainly due to the following two factors:

First, although there exist some SHM algorithms that are intrinsically distributed, most of the traditional SHM algorithms are centralized, which means that their implementation requires the availability of the raw data from all the deployed sensor nodes. However, considering the high cost of transmitting raw data in a wireless environment, it is desirable that deployed wireless sensor nodes use their local information only or at most, exchange the information only with their nearby neighbors. To distribute these centralized SHM algorithms is a challenging task.

Moreover, different from many applications of WSNs where simple aggregation functions such as average, max, min, etc. are widely used, most of the classic SHM algorithms involve complex matrix computational techniques such as singular value decomposition (SVD), eigen value decomposition, as well as other time domain or frequency domain signal processing methods. Some of these algorithms can be computationally very intensive and require a large auxiliary memory space for computation. For example, it was reported in [6] that implementing the SVD on a small 48-by-50 data matrix that includes data only from a few sensor nodes would take 150 s in Imote2 running at 100 MHz. Further, considering the time complexity of SVD on a data matrix $\mathbf{H} \in \mathbb{R}^{n \times n}$ is $\Theta(n^3)$ [7], the SVD on an $\mathbf{H}$ including data from a large number of sensor nodes is essentially infeasible for most of the available off-the-shelf wireless sensor nodes. How to modify these resource-consuming SHM algorithms and make them lightweight is also a challenging task.

In this chapter, we target the WSN-based SHM systems used for long-term monitoring purposes and mainly focus on how to design and implement SHM algorithms in resource-limited wireless sensor nodes. We first give a summary review of the recent efforts of embedding SHM algorithms within the WSNs. We then select an SHM algorithm that is widely used in civil engineering field, modal analysis, and describe how to implement it within WSNs.

4.2 Related Works

What distinguishes WSNs from traditional tethered structural monitoring systems is that the wireless sensor nodes are "smart" and able to process the response measurements they collected or received from others. Autonomous execution of damage detection algorithms by the wireless sensor represents an important step toward automated SHM.

There exist a number of SHM algorithms developed by civil engineers and they have shown advantages in different structures and environmental conditions. However, based on the difficulties to be implemented in a typical WSN, they can be largely divided as (1) inherently distributed and lightweight, (2) inherently distributed but computationally intensive, and (3) centralized.

Some SHM algorithms can be implemented on a WSN directly without any modification. These algorithms share two properties: (1) They are inherently distributed, which means that each sensor node, based on its own measured data, can make a decision on the condition of the structure; and (2) the complexity of the algorithms is low. For example, to detect damage, some SHM algorithms rely on examining the change of a vibration characteristic called natural frequencies. Natural frequency is a global parameter of structures, and under some assumptions on the input excitation and environmental noise they can be estimated based on time history data from each sensor node [8]. One rough but simple approach to extract natural frequencies is peak-picking [9]. In the peak-picking method, the power spectral density (PSD) of the measured time history from a sensor node is calculated using the fast Fourier transform, and then the some "peaks" on the PSD are selected whose locations are selected as the identified natural frequencies. In a WSN-based SHM system implementing this strategy, any wireless sensor node is able to identify a set of natural frequencies using peak-picking without sharing data with each other. The peak-picking method itself is lightweight. However, a drawback of using this peak-picking method is that it can only give approximate estimation of natural frequencies. An example of such a WSN system can be found in [10].

Some SHM algorithms, although inherently distributed, cannot be directly implemented in a WSN due to the high computational complexity and large memory space required. Examples of these algorithms include the auto-regressive and auto-regressive exogenous inputs (AR-ARX) method [11], the damage localization assurance criterion (DLAC) method [12,13], and the wavelet method [14,15]. The AR-ARX method is based on the premise that if there was damage in a structure, the prediction model previously identified using the undamaged time history would not be able to reproduce the newly obtained time series. In the AR-ARX method, a sensor node (1) first identifies an AR model based on its collect data and (2) then searches through a database that stores the AR models of the structure under a variety of environmental and operational conditions to find a best match and then based on which, (3) identifies an ARX model to obtain the decision on the healthy status. Except for the task in the first stage, the last two tasks are computationally intensive and require large memory space. To address this problem, Lynch et al. [16] modified the AR-ARX method and made it applicable for WSNs. The basic idea is very simple: After a sensor node identifies its AR model, it will send the corresponding parameters to a central server and let the server finish the two remaining cumbersome tasks.

The DLAC method is also a distributed SHM algorithm. In the DLAC, each sensor node collects its own data, calculates its PSD, identifies natural frequencies, and obtains damage information by comparing identified natural frequencies with the reference ones. In the DLAC, the natural frequencies are identified using the rational fraction polynomial (RFP) method [17] instead of the aforementioned peak-picking method, since the RFP can provide more accurate estimation. However, implementing the DLAC is much more time consuming than the peak-picking and hence not applicable for most of the off-the-shelf wireless sensor nodes. To address this problem, the DLAC is tailored for WSNs and within which, the most time-consuming task of the DLAC, the RFP is offloaded to a central server. After the server has finished the RFP, the natural frequencies are transmitted back to the sensor nodes for the remaining tasks.

The wavelet transform (WT) or the wavelet packet transform (WPT) of the time histories collected from individual sensor nodes have also been used for damage detection [14,15]. Wavelet-based approaches are based on the assumption that the signal energy at some certain frequency spectrum bands extracted from the WT/WPT will change after damage. However, traditional WT and WPT are computational intensive, requiring large auxiliary memory space and thus are not suitable for WSNs. To address this problem, the lifting scheme wavelet transform is proposed in [18], which has the advantages of fast implementation, fully in-place calculation without auxiliary memory, and integer-to-integer mapping. This modification on WT and WPT has proven to be very effective to improve the efficiency of WSNs using WT/WPT to detect damage.

Different from the distributed SHM algorithms mentioned earlier by which decision can be made based on data from individual sensor nodes, a large percentage of SHM algorithms are centralized. They require the raw data from all the deployed sensor nodes. Embedding centralized SHM algorithms within a WSN is not an easy task. In this chapter, we give an example of how centralized SHM algorithms can be made distributed in a WSN. Since there exist a large variety of algorithms for SHM that have been proposed by civil engineers, we select one technique called modal analysis. Modal analysis is one of the most fundamental techniques in SHM. Using modal analysis, structural vibration characteristics, called as the modal parameters, are identified that will in turn give damage-associated information.

4.3 Background: Modal Analysis

In this section, we first give some basic background associated with modal analysis and then describe, in a step-by-step manner, how this can be embedded within wireless sensor nodes.

4.3.1 Modal Parameters

Every structure has tendency to oscillate with much larger amplitude at some frequencies than others. These frequencies are called *natural frequencies*. (This concept was mentioned in Section 4.2.) When a structure is vibrating under one of its natural frequencies, the corresponding vibrational pattern it exhibits is called a *mode shape* for this natural frequency.

For example, for a structure with n-degrees of freedom (DOFs), its natural frequency set and mode shapes are denoted, respectively, as:

$$\mathbf{f} = [f^1, f^2, \ldots f^n]' \tag{4.1}$$

$$\Phi = [\Psi^{\mathbf{1}}, \Psi^{\mathbf{2}}, \ldots, \Psi^{\mathbf{n}}] = \begin{bmatrix} \phi_1^1 & \phi_1^2 & \cdots & \phi_1^n \\ \phi_2^1 & \phi_2^2 & \cdots & \phi_2^n \\ \vdots & \vdots & \ddots & \vdots \\ \phi_n^1 & \phi_n^2 & \cdots & \phi_n^n \end{bmatrix} \tag{4.2}$$

where

f^k $(k = 1, \ldots, n)$ is the kth natural frequency

$\Psi^{\mathbf{k}}(k = 1, \ldots, n)$ is the mode shape corresponding to f^k

$\phi_i^k(i = 1, 2, \ldots, n)$ is the value of $\Psi^{\mathbf{k}}$ at the ith DOF

For convenience, f^k and $\Psi^{\mathbf{k}}$ are also called *modal parameters* corresponding to the kth mode of a structure. As an example, Figure 4.2 illustrates the first three mode shapes of a typical cantilevered beam, extracted from the measurements of the deployed 12 sensor nodes. Each mode shape corresponds to a certain natural frequency of this cantilever beam.

Modal parameters are determined only by the physical property of structure (i.e., mass, stiffness, damping, etc.). When damage occurs on a structure, its internal property will be changed, and consequently, modal parameters will be deviated from those corresponding to this structure in the healthy condition. Therefore, by examining the changes in these modal parameters, damage on the

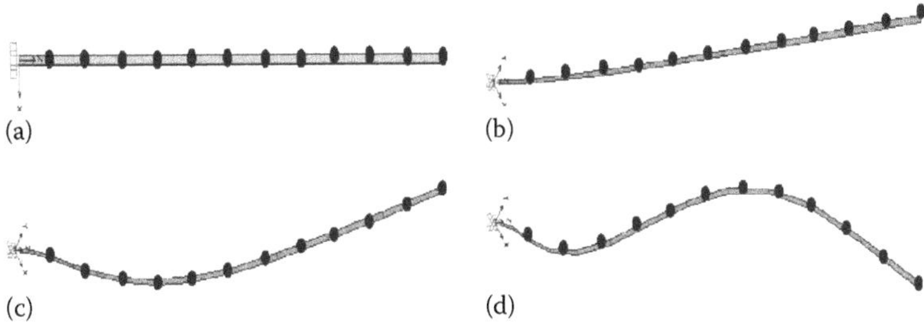

Figure 4.2 Mode shapes of a typical cantilevered beam: (a) original beam, (b) mode shape 1, (c) mode shape 2, and (d) mode shape 3.

structure can be roughly detected and located. Modal parameters can also be used as the inputs for finite element model (FEM) updating [19], which is able to precisely locate and quantify structural damage.

It should also be noted that different from natural frequency vector $\mathbf{f}$, mode shape vector $\Psi^{\mathbf{k}}$ has an element corresponding to each sensor node. Moreover, elements in $\Psi^{\mathbf{k}}$ only represent the relative vibration amplitudes of structure at corresponding sensor nodes. In other words, two mode shape vectors $\Psi^{\mathbf{k}}$ and $\Psi^{\mathbf{j}}$ are the same if there exists a nonzero scalar ζ, which satisfies $\Psi^{\mathbf{k}} = \zeta\Psi^{\mathbf{j}}$. This property leads to one of the important constraints when designing distributed modal analysis. Details about this constraint will be given in the next section.

To identify modal parameters, civil engineers have developed a larger number of classic modal analysis algorithms including stochastic subspace identification (SSI) [20], the eigensystem realization algorithm (ERA) [21], the frequency domain decomposition [22], and the enhanced frequency domain decomposition [23]. In this chapter, we choose the ERA for modal parameter identification and briefly introduce how the modal parameters are identified using the ERA.

4.3.2 The ERA

In this section, we briefly introduce the ERA. The ERA is able to give accurate modal parameter estimate using output data-only and has been widely used by civil engineers for many years.

Assume a total of m sensor nodes are deployed on a structure and the collected data are denoted as $\mathbf{y}(k) = [y^1(k), y^2(k), \ldots, y^m(k)]'$ $(k = 1, \ldots, N_{ori})$, where $y^i(k)$ is the data sampled by the ith sensor at kth time step and N_{ori} is the total number of data points collected in each node. To obtain modal parameters, the ERA first identifies, from measured responses $\mathbf{y}(k)$, a series of parameters $\mathbf{Y}(k)$ called *Markov parameters*. The Markov parameters $\mathbf{Y}(k)$ are calculated as the cross-correlation function (CCF) of the measurement $\mathbf{y}$ and a reference signal y^{ref}:

$$\mathbf{Y}(k) = CCF_{\mathbf{y}y^{ref}}(k) = \begin{bmatrix} CCF_{y^1y^{ref}}(k) \\ CCF_{y^2y^{ref}}(k) \\ \vdots \\ CCF_{y^my^{ref}}(k) \end{bmatrix} \tag{4.3}$$

where $CCF_{y^iy^{ref}}$ is the CCF between the ith measurement y^i and the reference y^{ref}. Generally speaking, measured signal from any deployed sensor node can be selected as y^{ref}. To accurately estimate $CCF_{y^iy^{ref}}$, we first use the Welch's averaged periodogram method [24] to calculate the cross-spectral density (CSD) between y^i and y^{ref}, then inverse fast Fourier transform (ifft) is implemented on the CSD to obtain the CCF.

In the Welch's method, to calculate the CSD of two signals x and y, x and y are first divided into n_d number of overlapping segments. The CSD of x and y, denoted as G_{xy}, is then calculated as

$$G_{xy}(\omega) = \frac{1}{n_d N} \sum_{i=1}^{n_d} X_i^*(\omega)\, Y_i(\omega) \tag{4.4}$$

where

$X_i(\omega)$ and $Y_i(\omega)$ are the Fourier transforms of the ith segment of x and y
"*" denotes the complex conjugate

N is data points in each segment of x (or y) as well as the obtained $G_{xy}(\omega)$. N is generally taken as a power of two values 1024 or 2048 to give reasonable results. To decrease the noise, n_d practically ranges from 10 to 20.

After obtaining the CSD of y^{ref} with each response in $\mathbf{y}$, the Markov parameters $\mathbf{Y}(k)$ are then calculated as the ifft of the obtained CSD:

$$\mathbf{Y}(k) = \begin{bmatrix} ifft(G_{y^1 y^{ref}}) \\ ifft(G_{y^2 y^{ref}}) \\ \vdots \\ ifft(G_{y^m y^{ref}}) \end{bmatrix} (k) \tag{4.5}$$

Having obtained the Markov parameters $\mathbf{Y}(1), \mathbf{Y}(2), \ldots$, the ERA begins by forming the Hankel matrix composed of these Markov parameters and implement the SVD to obtain modal parameters. The detailed procedure of the ERA is summarized in Figure 4.3. It can be seen that the ERA can

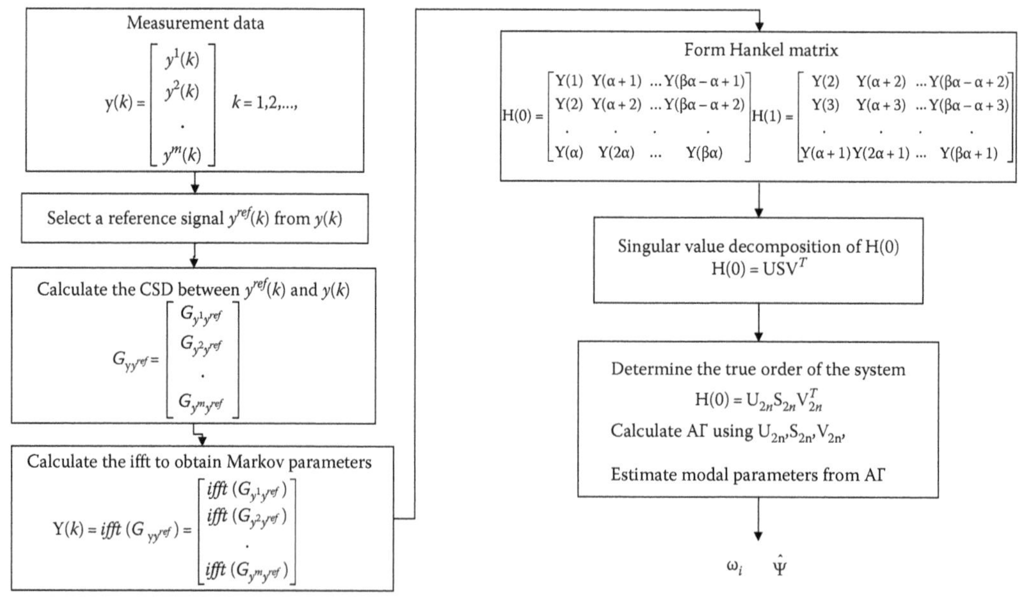

Figure 4.3 Procedures of the ERA algorithm.

be largely divided into two stages. In the first stage, the Markov parameters are identified. These Markov parameters are then used to identify the modal parameters in the second stage.

In the following three sections, we will introduce, in a step-by-step manner, how these centralized modal analysis algorithms are tailored for WSNs.

4.4 Distributed Modal Analysis

4.4.1 *Stage 1: Try to Distribute the Initial Stage of Modal Analysis Algorithms*

To design distributed version of centralized modal analysis algorithms, the detailed procedures in the ERA should be analyzed. It can be seen from Figure 4.3 that the CSD estimation between the time history of a reference sensor and that of each sensor is first calculated. Therefore, if the CSDs can be calculated in a way suitable for WSNs, the efficiency of these algorithms in a WSN can be significantly improved. Nagayama and Spencer [6] proposed a decentralized approach illustrated in Figure 4.4a to calculate the CSDs without necessitating the collection of all the measured data. In this strategy, the reference node broadcasts its measured time history record to all the remaining nodes. After receiving the reference signal, each node calculates a CSD estimation and then transmits it back to a sink node where the remaining portions of the algorithms are implemented. Considering the amount of data in the CSDs is much smaller than the original time history record, the amount of transmitted data in this approach is much smaller than the traditional one where all the raw data are transmitted to the sink. Moreover, part of the computation load that was concentrated at the sink node (i.e., the one responsible for calculating the CSD) is partially off-loaded to the other nodes, which is favorable for a homogeneous WSN in which no "super nodes" exist in the network.

This decentralized approach is further improved in [25] where the decentralized random decrement technique (RDT) is adopted to calculate the CSDs. With the help of the RDT, the reference node does not even need to send all the measured time history record, only some trigger points in the time history found by the RDT need to be broadcast. Once the trigger points are received, each node calculates the CSDs that are subsequently collected at the sink node to continue the

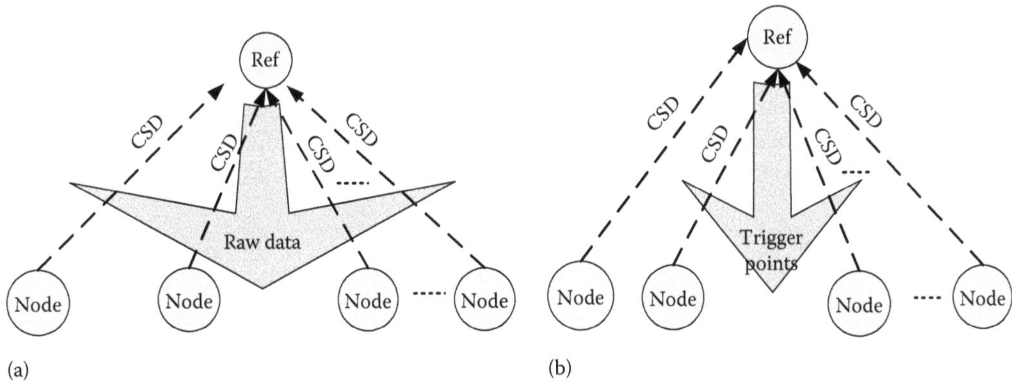

Figure 4.4 Two approaches of calculating the CSDs in a distributed way: (a) the approach proposed in [6] and (b) the approach proposed in [25].

remaining damage identification procedures. Considering the trigger information is in general much shorter than the time history record broadcast by the reference node, this RDT-based decentralized approach can considerably reduce wireless data transmissions. This approach is illustrated in Figure 4.4b.

4.4.2 Stage 2: Divide and Conquer

If only the CSD estimations in the modal analysis algorithms are made to be distributed, there remain some problems since the CSDs of all the nodes still need to be wirelessly transmitted to a sink where the remaining steps of the ERA are finished. First, transmitting the CSDs of all the sensor nodes to the sink is a challenging task considering the CSD of each node contains thousands of points that are usually in a double-precision floating-point format. In addition, in a large civil structure, the CSDs usually need to be transmitted in a multi-hop manner, which considerably downgrades the performance of the system. The second problem is associated with computation. When the sink node receives the CSDs for the deployed sensor nodes, the computational resources required to identify modal parameters usually exceed the capacity of the most existing off-the-shelf wireless sensor nodes, especially when the number of sensor nodes is large. Therefore, a PC is generally used as the sink node that can increase the system cost and difficulties in deployment.

To address the aforementioned problems, instead of using data from all the sensor nodes in a batch manner, we can divide the deployed sensor nodes into clusters and implement the ERA in each cluster. We then obtain a set of natural frequencies and "local" mode shapes, and these cluster-based modal parameters will be "merged" together afterward. This is very similar to the "divide and conquer" strategy widely adopted by computer scientists to solve various mathematical problems. A minor difference might be that in the original "divide and conquer" algorithms, the original problem is solved by dividing the original problem in a *recursive* way, while in this cluster-based ERA, the division of a WSN needs to be carried out only once.

This cluster-based ERA is illustrated in Figure 4.5. In this approach, the whole network is partitioned into a number of clusters. A cluster head (CH) is designated in each cluster to perform intra-cluster modal analysis using traditional centralized modal analysis algorithms. The identified modal parameters in all clusters are then assembled together to obtain the modal parameters for the whole structure. Compared with the centralized approach, the cluster-based approach has at least two advantages. The first advantage of this cluster-based approach is associated with the wireless communication. By dividing sensor nodes into single-hop clusters in which sensor nodes in each cluster are within single-hop communication with their CH, we can avoid multi-hop relay and thus reduce the corresponding wireless communications.

Second, compared with the centralized approach, the computational resources required in each cluster to compute the modal parameters is significantly decreased. By reducing the computational complexity, it is possible to use common wireless sensor nodes instead of PC to implement modal analysis algorithms.

The third advantage of this approach is that by dividing sensor nodes into clusters, the computation of the ERA can be made parallel. All the CHs can work at the same time, thus the overall computation time is decreased.

However, clustering must satisfy some constraints. First, clusters must overlap with each other. This constraint is a prerequisite for the local mode shapes to be stitched together. As we have introduced in Section 4.3, mode shape vectors identified using the ERA only represent the relative vibration amplitudes at sensor nodes involved. Therefore, mode shapes identified in different clusters cannot be directly assembled. This can be demonstrated in Figure 4.6a, where the deployed

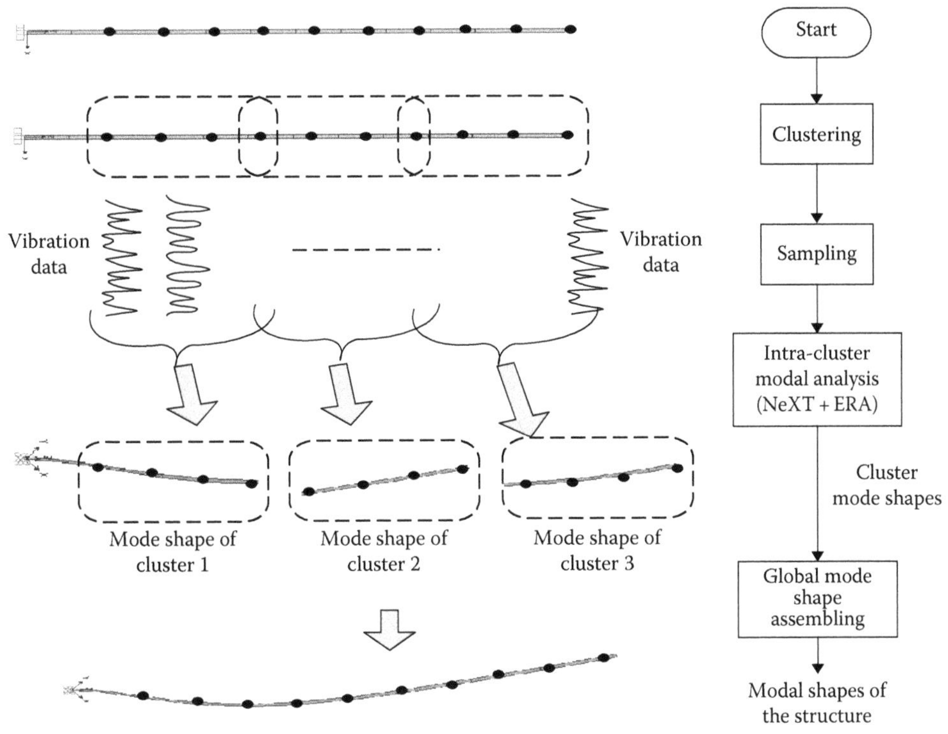

Figure 4.5 Overview of cluster-based modal analysis process.

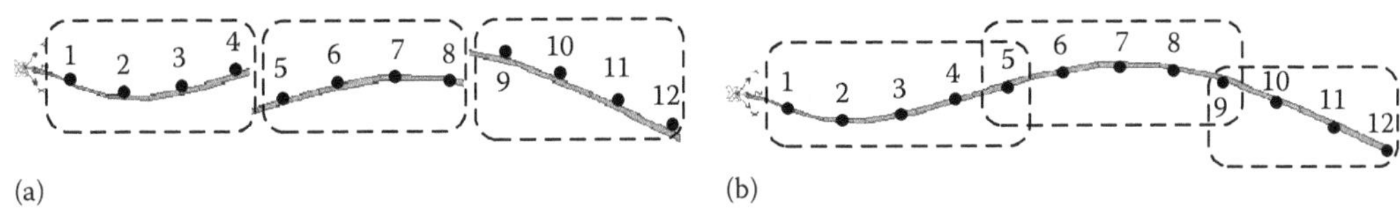

Figure 4.6 Mode shape assembling when (a) clusters do not overlap and (b) clusters overlap.

12 sensor nodes in Figure 4.2 are partitioned into three clusters to identify the third mode shape. Although the mode shape of each cluster is correctly identified, we still cannot obtain the mode shapes for the whole structure. The key to solve this problem is overlapping. We must ensure that each cluster has at least one node which also belongs to another cluster and all the clusters are connected through the overlapping nodes. For example, in Figure 4.6b, mode shapes identified in each of the three clusters can be assembled together with the help of the overlapping nodes 5 and 9. This requirement of overlapping must be satisfied when dividing sensor nodes into clusters.

Another constraint is the number of sensor nodes in a cluster. To avoid the under-determined problem, the ERA also requires that the number of sensor nodes in each cluster should be larger than the number of modal parameters to be identified.

Given a WSN, different clustering strategies will generate clusters with different sizes and network topologies and therefore can result in different energy consumption, wireless bandwidth consumed, delay, etc. Correspondingly, clustering can be optimized according to different objective functions that may vary for different hardware, wireless communication protocols, and other specific scenarios of WSN-based SHM systems. For example, for a WSN in which wireless sensor nodes are battery powered, energy efficiency is an important issue. Therefore, how to divide the deployed sensor nodes such that the energy consumption is minimized is important. This optimal clustering problem is studied in [26]. Besides energy consumption, other possible objective functions for clustering can be wireless transmissions, load-balance, delay, etc.

4.5 WSN-Cloud SHM: New Possibility toward SHM of Large Civil Infrastructures

In the SHM algorithms described earlier, we introduced how to use wireless sensor nodes to estimate modal parameters. However, to obtain damage location and further quantify damage severity, we generally still have one step to go. The estimated modal parameters will be sent to a server where the FEM of the structure under monitoring is updated. This procedure is called model updating [19]. In FEM updating, parameters of the structure's FEM, which are directly associated with the physical property of the structure, are adjusted to reduce a penalty function based on residuals between the modal parameters estimated from measured data and the corresponding FEM predictions. The updated FEM directly provides information of damage location and severity. However, for a large civil infrastructure where an accurate FEM can contain tens of thousands or even hundreds of thousands of small "structural elements," model updating is extremely resource demanding and can take hours or even days even for a powerful PC.

To alleviate the computational burden of the server as well as to decrease the associated delay, civil engineers have proposed a scheme called a multi-scale SHM [27,28]. In this strategy, two different FEMs, one coarse and one refined, are established for a given structure. The former FEM consists of smaller number of large-sized structural elements and the latter contains small-sized but large number of elements. Correspondingly, the updating of the coarse FEM takes much less computation time than the latter. Initially, estimated modal parameters are used to update the coarse FEM. Only when damage is detected on this coarse FEM, the refined FEM is updated for the detailed damage localization and quantification. This multi-scale strategy can significantly decrease the computational load for the server. Moreover, in this strategy, the updating of the coarse FEM only requires the "coarse" modal parameters, whose identification does not need all the deployed sensor nodes. Therefore, it is possible that only part of the deployed sensor nodes need to work. This can increase the lifetime of the WSN.

However, the server of the SHM systems using this multi-scale strategy still needs to be powerful enough to handle the task of updating the refined FEM when the damage is suspected to occur. Considering most of the time, the server is running coarse-FEM updating where the computational load is low; it is a waste to purchase a powerful server that is under-loaded most of the time.

Cloud computing, being able to provide dynamically scalable resources, can be a perfect substitute for the server used in the aforementioned SHM system. Instead of purchasing a powerful server, we can buy the computational resources from cloud provider and only pay for the resources we have used. This "pay as you go" business pattern can dramatically reduce the total cost of SHM systems. The property of dynamic scaling of multi-scaled SHM makes cloud computing a perfect platform in this application.

A future SHM system is envisioned as shown in Figure 4.7. A large number of wireless sensor nodes are deployed on different locations of the structure under monitoring, and a gateway node, serving as in-field commander, is able to communicate with both the WSN and the Internet.

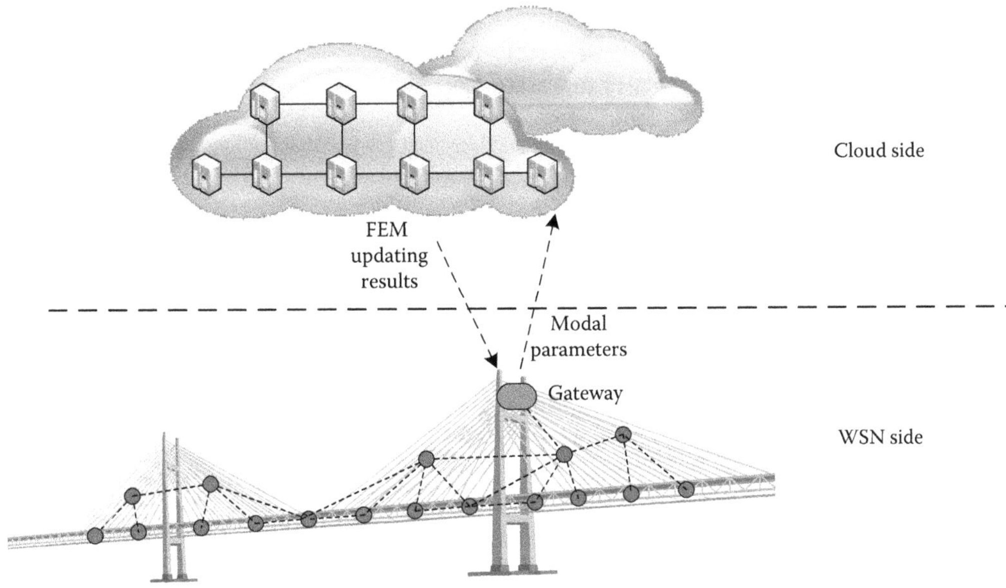

Figure 4.7 Architecture of WSN-Cloud SHM.

Initially, under the command of the gateway node, part of the wireless sensor nodes are activated to sample and compute the modal parameters. The modal parameters are sent to the gateway and are then forwarded to the cloud servers where the model updating is implemented. The updating will be transmitted back to the gateway. If damage is not detected, the aforementioned procedures are repeated for every predetermined period of time. Once damage is found on the coarse FEM, more wireless sensor nodes will be activated by the gateway nodes and a refined modal parameters will then send from the gateway to the cloud side to implement FEM updating on the refined FEM. We call this hybrid architecture as "WSN-Cloud SHM."

This leaves much space for us to explore and to realize a practical WSN-Cloud SHM system. For example, most of the existing applications of Cloud computing, particularly web-based applications, can use "MapReduce" programming model [29]. However, different from web-associated applications such as text tokenization, indexing, and search, implementing FEM updating in the form of MapReduce is not straightforward and needs in-depth investigation. Moreover, besides the cloud user point of view, how to provide different levels of cloud-based SHM services for infrastructure owners is an interesting question. The answers to these questions will lead to great economic benefit in the future.

References

1. C. Farrar and K. Worden, An introduction to SHM, *Philosophical Transactions of the Royal Society*, 365(1851), 303, 2007.
2. J. Brownjohn, M. Bocciolone, A. Curami, M. Falco, and A. Zasso, Humber bridge full-scale measurement campaigns 1990-1991, *Journal of Wind Engineering and Industrial Aerodynamics*, 52, 185–218, 1994.
3. J. Ou and H. Li, Structural health monitoring in mainland china: Review and future trends, *Structural Health Monitoring*, 9(3), 219, 2010.

4. K. Wong, Instrumentation and health monitoring of cable-supported bridges, *Structural Control and Health Monitoring*, 11(2), 91–124, 2004.
5. M. Maróti, B. Kusy, G. Simon, and Á. Lédeczi, The flooding time synchronization protocol, in *Proceedings of the Second International Conference on Embedded Networked Sensor Systems*. ACM, 2004, pp. 39–49.
6. T. Nagayama and B. Spencer Jr, Structural health monitoring using smart sensors, *N.S.E.L. Series 001*, 2008.
7. G. Golub and C. Van Loan, Matrix computations (Johns Hopkins studies in mathematical sciences), 1996.
8. C. Farrar and S. Doebling, An overview of modal-based damage identification methods, in *Proceedings of DAMAS Conference*, 1997.
9. D. Ewins, Modal testing: theory, practice and application. Research Studies Press, Ltd., 2000.
10. E. Monroig and Y. Fujino, Multivariate autoregressive models for local damage detection using small clusters of wireless sensors, in *Proceedings of the Third European Workshop on Structural Health Monitoring*, Granada, Spain, 2006.
11. H. Sohn and C. Farrar, Damage diagnosis using time series analysis of vibration signals, *Smart Materials and Structures*, 10, 446, 2001.
12. A. Messina, I. Jones, and E. Williams, Damage detection and localization using natural frequency changes, in *Proceedings of Conference on Identification in Engineering Systems*, 1996, pp. 67–76.
13. A. Messina, E. Williams, and T. Contursi, Structural damage detection by a sensitivity and statistical-based method, *Journal of Sound and Vibration*, 216(5), 791–808, 1998.
14. Z. Hou, M. Noori, and R. St Amand, Wavelet-based approach for structural damage detection, *Journal of Engineering Mechanics*, 126(7), 677–683, 2000.
15. C. Chang and Z. Sun, Structural damage assessment based on wavelet packet transform, *Journal of Structural Engineering*, 128, 1354, 2002.
16. J. Lynch, A. Sundararajan, K. Law, A. Kiremidjian, and E. Carryer, Embedding damage detection algorithms in a wireless sensing unit for operational power efficiency, *Smart Materials and Structures*, 13, 800, 2004.
17. M. Richardson and D. Formenti, Parameter estimation from frequency response measurements using rational fraction polynomials, in *Proceedings of the 1st International Modal Analysis Conference*, Vol. 1, 1982, pp. 167–186.
18. Y. Zhang and J. Li, Wavelet-based vibration sensor data compression technique for civil infrastructure condition monitoring, *Journal of Computing in Civil Engineering*, 20, 390, 2006.
19. M. Friswell and J. Mottershead, *Finite Element Model Updating in Structural Dynamics*. Springer, Berlin, Germany, Vol. 38, 1995.
20. B. Peeters and G. De Roeck, Reference-based stochastic subspace identification for output-only modal analysis, *Mechanical Systems and Signal Processing*, 13(6), 855–878, 1999.
21. J. Juang and R. Pappa, An eigensystem realization algorithm for modal parameter identification and model reduction, *Journal of Guidance*, 8(5), 620–627, 1985.
22. R. Brincker, L. Zhang, and P. Andersen, Modal identification from ambient responses using frequency domain decomposition, in *Proceedings of the 18th International Modal Analysis Conference*, 2000, pp. 625–630.
23. N. Jacobsen, P. Andersen, and R. Brincker, Using enhanced frequency domain decomposition as a robust technique to harmonic excitation in operational modal analysis, in *Proceedings of ISMA2006: International Conference on Noise & Vibration Engineering*, Leuven, Belgium, 2006, pp. 18–20.
24. P. Welch, The use of fast fourier transform for the estimation of power spectra: A method based on time averaging over short, modified periodograms, *IEEE Transactions on Audio and Electroacoustics*, 15(2), 70–73, 1967.

25. S. Sim, J. Carbonell-Márquez, B. Spencer Jr, and H. Jo, Decentralized random decrement technique for efficient data aggregation and system identification in wireless smart sensor networks, *Probabilistic Engineering Mechanics*, 26(1), 81–91, 2011.
26. X. Liu, J. Cao, S. Lai, C. Yang, H. Wu, and Y. Xu, Energy efficient clustering for WSN-based structural health monitoring, in *IEEE INFOCOM*, Vol. 2. Citeseer, 2011, pp. 1028–1037.
27. Z. Li, T. Chan, Y. Yu, and Z. Sun, Concurrent multi-scale modeling of civil infrastructures for analyses on structural deterioration. Part i: Modeling methodology and strategy, *Finite Elements in Analysis and Design*, 45(11), 782–794, 2009.
28. T. Chan, Z. Li, Y. Yu, and Z. Sun, Concurrent multi-scale modeling of civil infrastructures for analyses on structural deteriorating. Part ii: Model updating and verification, *Finite Elements in Analysis and Design*, 45(11), 795–805, 2009.
29. J. Dean and S. Ghemawat, Mapreduce: Simplified data processing on large clusters, *Communications of the ACM*, 51(1), 107–113, 2008.

Chapter 5

Knowledge Representation and Reasoning for the Design of Resilient Sensor Networks

David Kelle, Touria El-Mezyani, Sanjeev Srivastava, and David Cartes

Contents

In this chapter, we propose a new approach for the design of resilient sensor networks using a knowledge representation (KR) formalism that will facilitate reasoning and collaboration among sensors. This approach utilizes abstract simplicial complexes (ASCs), which serve as the building blocks of this KR, permitting the encoding of more complex relationships than models based on graph theory due to their additional structure. These mathematical constructions exist in the framework of combinatorial algebraic topology (CAT), which uses discrete methods to formalize how algebraic equations relate to the structure of a space. Its algebraic nature allows formulation of algorithms and automation of decision making. Electrical power systems are a natural choice for an application of this KR because of their large size, distributed information, and complex dynamics. The developed approach can be used to design resilient sensor networks that can assist the power management, control, and monitoring systems in inferring and predicting the system state, as well as determining the health of all components, including sensors. This approach is illustrated on an IEEE 14 bus power system.

5.1 Introduction

Sensing and measurement of large-scale system variables such as weather conditions (wind speed, temperature, etc.), power flow, and market prices are essential for improving the monitoring and control of engineered systems. This information can be gathered by means of a distributed heterogeneous network of sensors. The heterogeneous nature of knowledge (e.g., electrical, weather, etc.) results in significant challenges involving control and monitoring systems. Usually, engineers design the sensing and data processing for control and monitoring by considering one particular type of knowledge for a given physical domain. Advanced sensing techniques will collect valuable information about the system conditions throughout the network for multiple physical domains. By considering these multiple forms of knowledge in concert, one can improve problem solving.

Using new communications such as the Internet and reliable media such as wireless, broadband power lines, or fiber optics, a network of distributed sensors will have the ability to communicate quickly in order to infer valuable information. For example, in large-scale systems, sensors will have the ability to interact and coordinate with control and monitoring systems for optimal system control and decision making.

While the study of sensor network design for the state estimation problem is fairly widespread [1,13], their use in distributed, large-scale systems for managing complexity and uncertainty, as well as real-time decentralized control and monitoring is seen as an increasingly important avenue of research. As such, our objective in this research is to develop tools that assist sensors to infer the system state, detect and diagnose failures, as well as to learn and adapt to their changing environment. In order to achieve these objectives, we introduce an approach based on a sensor management system (SMS). This SMS will be able to quickly collect important data and process it as well as monitor the health of sensors and restore erroneous or missing data. This system will inform (1) the controller about the state of the system, and (2) the monitoring system about the health of sensors and assist decisions. By optimizing the communication pathway between sensors and the control or monitoring system, as well as minimizing the number of sensors required for performing a task, we can improve system resilience and avoid problems, such as high maintenance cost, as well as latency and congestion.

KR is the key to the design of an intelligent SMS because it provides the basic cognitive structure of reasoning and machine learning. This can enable an SMS to predict possible outcomes and required actions from its perception of a situation. KR and reasoning is the study of thought as a computational process. It is concerned with how a system uses its knowledge to infer solutions and make decisions. While logical reasoning produces answers to problems based on fixed rules, analogical reasoning produces conclusions by extending a known comparison, strategy, or concept in one domain called the source, to another domain, called the target. Analogies can help gain understanding of new concepts by comparing them with past experiences. There are many references such as [3,7–10] that discuss the theory of analogical reasoning in the context of the human mind. Hadamard in [10] suggests that human creative insight comes from the application of an analogy from a different domain of knowledge, and Collins in [3] explores its role in individuals with incomplete knowledge of the source domain. However, most work done to represent analogical reasoning focuses on graph-theoretic representations which are limited in the dimension of allowed relations.

The goal of this chapter is to develop tools for analogical reasoning and problem solving using CAT. We present a method to form solution strategies as well as make analogies, applying successful strategies to new situations. These methods use ASCs as a formalism for KR. An ASC is an extension of a graph, where the idea of an edge is generalized to arbitrary dimensions. This allows much richer structures such as the mutual adjacency of any number of vertices. Models based on simplicial complexes can encode more complex relationships than the traditional graph-theoretic models due to their additional structure, making ASCs a good choice for the building blocks of an analogy-capable KR. In [5], the authors developed a computer model for analogy solving using simplicial complexes. The source and the target analogs are represented as simplices and the analogy is modeled as topological deformations of these simplices along a chain sequence according to a set of rules. However, this proposed approach was adapted to the specific problem of solving IQ test questions, and was not applied to general intelligent systems. In order to apply this approach to engineering systems we need to develop a basic mathematical language of reasoning. This will be used to (1) model the knowledge collected about the system (e.g., mathematical constraints, etc.), (2) use the KR technique, and (3) develop reasoning process using analogies. The reasoning process will then analyze system conditions, build strategies for communication and interaction between sensors, and initiate necessary actions in the case of missing data or loss of sensors. The benefit of the proposed sensing systems offers the ability of fast data collection, processing, and reliability and resilience in the presence of failures. This new intelligent sensing system promise to increase the system resilience and improve performance of system monitoring and control.

5.2 Background

In this section, we present an overview and background of the existing mathematical tools applied in this chapter. Simplicial complexes form the backbone of the proposed knowledge structure. We first define some important concepts that involve ASCs and relations [4], and then discuss how ASCs can been used to represent knowledge [5].

5.2.1 Simplicial Complexes

Simplicial complexes are abstract structures developed in algebraic topology [14]. Geometrically, simplices are an extension of triangles and tetrahedra to arbitrary dimensions.

Definition 5.1 (Simplex, face, facet) An m-simplex $S = \{x_1, \dots, x_{m+1}\}$ is the smallest convex set in $\mathbb{R}^m$ containing the given set of $m + 1$ points (Figure 5.1). A simplex formed from a subset of the points of S is called a face of S, and a maximal simplex under inclusion is called a facet.

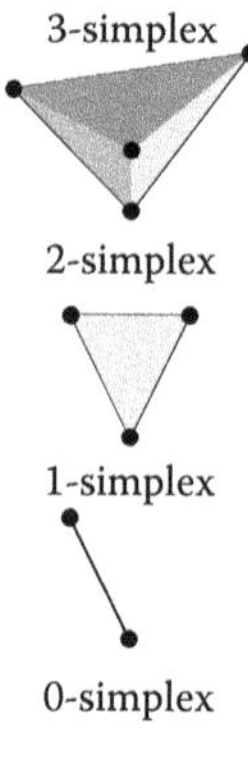

Figure 5.1 ***m*-simplexes for $0 \leq m \leq 3$.**

Simplices may be glued together along their faces. Such a geometric structure is called a *simplicial complex* Δ if every face of every simplex is also considered to be a simplex in Δ. A simplicial complex that is contained in another is called a *sub-complex*. However, we are only interested in the combinatorial properties of a simplicial complex. With this in mind, we consider an *ASC*.

Definition 5.2 (Abstract simplicial complexes) An *ASC* is a collection Δ of subsets of a set of vertices $V = \{x_1, \ldots, x_n\}$, such that if S is an element of Δ then so is every nonempty subset of S [11,14]. The dimension of an ASC is the dimension of its largest facet. An ASC is called *pure* if all of its facets have the same dimension [6].

5.2.2 Predicate Relations on Simplicial Complexes

Dowker provides a way to represent binary relations as a simplicial complex [4]. Let $R \subset A \times B$ be a binary relation between sets A and B. As a subset of $A \times B$, $(a, b) \in R$ means that a is related to b by the relation R, that is, aRb. For each fixed $b \in B$, $\Delta_A(B)$ describes what elements of A are related to b and vice versa. In this way, two dual simplicial complexes,* $\Delta_A(B)$ and $\Delta_B(A)$, are associated to R. This can be taken further by replacing the arbitrary set B with a set of predicates $P = \{p_1, \ldots, p_q\}$. This allows us to define R so that $(a, p_i) \in R$ if and only if a satisfies the predicate p_i, that is, $p_i(a)$ is true. $\Delta_A(P)$ tells what elements of A satisfy each predicate $p_i \in P$, whereas $\Delta_P(A)$ tells all the predicates satisfied by each $a \in A$. The use of simplicial complexes to represent data is part of a larger theory called Q-analysis [2].

Example 5.1: 1-ary predicate

The belief *"Per, Hans, and Leif are male"* is represented as follows:

$$(Per, male),\ (Hans, male),\ (Leif, male) \in R$$

where *male* is the predicate *"is male."* The simplex $\{Per,\ Hans,\ Leif\}$ is associated to the predicate *male*,† as shown in Figure 5.2.

* In the rest of the chapter, we drop the reference to the predicate and object sets and just refer to the complex as Δ.
† From here the simplex will just be referred to as the predicate itself.

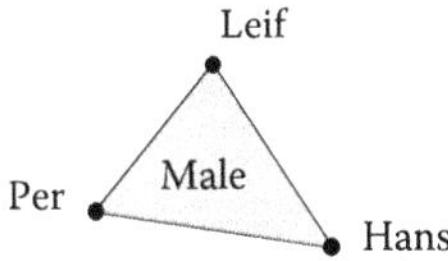

Figure 5.2 1-ary predicate.

5.2.3 Path Algebra

An algebra is a collection of mathematical objects along with specific rules for their interactions. Algebras can provide languages in which problems of different types can be discussed. A path algebra in graph theory is a set of paths equipped with two binary operations that satisfy certain requirements. Path algebras are used to solve problems like finding optimal paths between two vertices [12], finding all paths emanating from a source or converging to a target [17], determining shortest and longest paths, etc. In [5], the authors applied simplicial complexes to the problem of solving IQ test analogy questions by selecting deformations that had fewer steps and preserved the most properties, but did not develop a rigorous algebra of these deformations or use them for reasoning. We define this algebra by extending path algebras from graph theory to ASC so that it can be used to solve problems with analogies in a simplicial KR.

First, we present some background notions from [12] about path algebras. A path algebra is defined to be a set P, with two binary operations $\vee$ and $\circ$ that satisfy the following properties:

- $\vee$ is idempotent, commutative, and associative. That is, for all $a, b, c \in P$,

$$a \vee a = a,$$
$$a \vee b = b \vee a,$$
$$(a \vee b) \vee c = a \vee (b \vee c)\,.$$

- $\circ$ is associative as well as left and right distributive. For all $a, b, c \in P$,

$$(a \circ b) \circ c = a \circ (b \circ c)\,,$$
$$a \circ (b \vee c) = (a \circ b) \vee (a \circ c)\,,$$
$$(a \vee b) \circ c = (a \circ c) \vee (b \circ c)\,.$$

- There exist elements $\epsilon, \lambda \in P$ such that for any $a \in P$,

$$\epsilon \circ a = a = a \circ \epsilon,$$
$$\lambda \vee a = a = a \vee \lambda,$$
$$\lambda \circ a = \lambda.$$

The operation $\vee$ is called the join and $\circ$ is called the product. The elements ϵ and λ are the units of the product and join operations, respectively. By defining the join operator to select the shortest path and the product as path concatenation, Manger shows that the closure (Definition 5.4) of the adjacency matrix of a graph yields the shortest path from node i to node j in its i,jth entry [12]. The join and product of two matrices is defined in terms of the join and product of paths.

Definition 5.3 (Join and product) Let A, B be $n \times n$ matrices over P. Define the operations $\vee$ and $\circ$ on the matrices A, B:

$$A \vee B = \left[a_{ij} \vee b_{ij}\right]$$

$$A \circ B = \left[\bigvee_{k=1}^{n} a_{ik} \circ b_{kj}\right]$$

The powers of the matrix A are defined inductively with the product operation.

Definition 5.4 (Stability index and closure) A square matrix A is said to be *stable* if for some integer q,

$$\bigvee_{k=1}^{q} A^k = \bigvee_{k=1}^{q+1} A^k$$

The smallest such q is called the stability index of A, and the join of the matrices A^k as k ranges from 1 to q is called the (weak) closure of A, written $\hat{A}$:

$$\hat{A} = \bigvee_{k=1}^{q} A^k$$

5.3 Tools for Representation

Our goal is to extend the binary relation used in [4,5] in order to describe relationships that are more complex than the sharing of common properties.

Example 5.2: Necessity of extension

The belief *"Per, Hans, and Leif are brothers"* cannot be encoded as in *Example 5.1* by just replacing the predicate *m*; the statements *"Per is a brother, Leif is a brother, and Hans is a brother"* do not have the intended meaning.

Instead, we extend the set A to a product of sets $A_1 \times A_2 \times \cdots \times A_m$, and define the predicate as follows.

Definition 5.5 (m-Ary predicate) Let m be the minimal number of variables required to express a relationship. We define the m-ary predicate p on m variables to be the map from the product of sets $A_1 \times A_2 \times \cdots \times A_m$ to the set $\{True, False\}$:

$$p : A_1 \times A_2 \times \cdots \times A_m \rightarrow \{True,\ False\}.$$

Predicates with $m = 1$ are called properties.

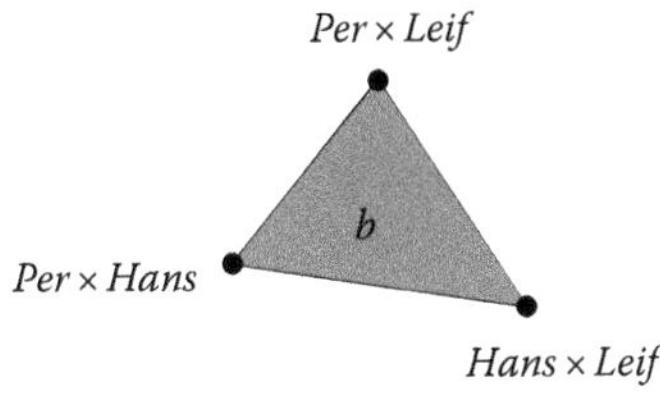

Figure 5.3 Binary predicate.

Example 5.3: Binary predicates

Per, Hans, and *Leif* are brothers. The following beliefs imply the desired belief:

1. *Per* and *Hans* are brothers.
2. *Per* and *Leif* are brothers.
3. *Leif* and *Hans* are brothers.

Figure 5.3 shows the topological representation of the brotherhood of *Per, Hans,* and *Leif*.*

5.4 Tools for Reasoning

The use of analogy in problem solving allows an intelligent system to connect different knowledge domains via method, strategy, or context to an unfamiliar problem.

In this section, we develop an algebra for reasoning by analogy by defining a concept similar to a path called a facet chain. We discuss how this algebra can be used to find a "best" facet chain, and then introduce an algorithm for reasoning.

5.4.1 Facet Chain Algebra

We begin by creating some basic terminology about facet chains and define the operations necessary for creating an algebra.

Definition 5.6 (Links and size) Let $\mathcal{F}$ be the collection of facets of a finite ASC. We define a link to be a set containing two facets $L_{i,j} = \{F_i, F_j\}$. A connected link is a link whose facets are adjacent. The size $|L_{i,j}|$ of the link $L_{i,j}$ is the number of vertices in the intersection $F_i \cap F_j$. The link $\{F_i, F_i\} = \{F_i\} = \epsilon_i$ is called a trivial link. For convenience, we sometimes write $L_{i,j}^{|L_{i,j}|}$ to denote the link along with its size.

Consider an alphabet $\mathcal{L}$ as the set of links of a finite ASC. Then a word over $\mathcal{L}$ is a sequence of links; the set of words over $\mathcal{L}$ is denoted $\mathcal{L}^*$.

* Together, the statements exhibit logical dependencies; e.g., (1), (2) $\Rightarrow$ (3).

Definition 5.7 (Facet chains and length) We refer to an element $\Gamma \in \mathcal{L}^*$ as a facet chain.* A connected facet chain is a facet chain whose links are all connected. The length of a facet chain is written as $\|\Gamma\|$ and is equal to the number of its nontrivial links. We write a facet chain from F_i to F_j as $\Gamma(i,j)$ and the set of facet chains from F_i to F_j as $C_{i,j}$.

Definition 5.8 (Critical size) Let $\Gamma \in \mathcal{L}^*$. We refer to the link in Γ with the smallest size as the critical link of Γ and its size as the critical size $|\Gamma|_c$:

$$|\Gamma|_c = min\{|L_{i,i+1}| : i = 1, \ldots, ||\Gamma|| - 1\}$$

Definition 5.9 (Aspect ratio) Let $\Gamma \in \mathcal{L}^*$. Define the aspect ratio AR of Γ to be the ratio of its length to its critical size:

$$AR(\Gamma) = \frac{\|\Gamma\|}{|\Gamma|_c}$$

If the denominator is zero, define the aspect ratio to be ∞.

Depending on the ASC, there can be many facet chains connecting two facets. We develop the chain operations join and product along with a norm on chains to select the desired facet chain.

Definition 5.10 (Chain norm) Let α be a positive rational number, $\alpha \in \mathbb{Q}^+$. Define the function ρ_α to assign to every chain Γ in $\mathcal{F}^*$ a value in $\mathbb{Q}^+ \cup \{\infty\}$:

$$\rho_\alpha(\Gamma) = \alpha \cdot AR(\Gamma)$$

The chain norm allows the join to select chains based on two requirements: short length and large critical size. For low aspect ratios, it prioritizes the length requirement while prioritizing the size for high aspect ratios. The constant α determines the aspect ratio at which this distinction occurs.

Remark 5.1 (Connected facet chains) The norm $\rho_\alpha(\Gamma)$ is finite if and only if Γ is connected and has finite length.

Remark 5.2 ($\mathcal{F}$ is a metric space) The set of facets $\mathcal{F}$ is a metric space with distance between two facets defined as $d(F_i, F_j) = \min\{\rho(\Gamma(i,j))|\Gamma(i,j) \in C_{i,j}\}$.

Now we can define the operations join and product for facet chains.

---

* Our definition of a facet chain differs from a chain in algebraic topology, which is defined as a linear combination of simplices.

Definition 5.11 (Join and product of facet chains) Let $\Gamma(i,j), \Gamma(k,l)$ be two facet chains in $\mathcal{F}^*$. Define the join of $\Gamma(i,j)$ and $\Gamma(k,l)$ to be the one with the smallest norm:

$$\Gamma_{i,j} \vee \Gamma_{k,l} = \begin{cases} \Gamma(i,j): & \rho_\alpha(\Gamma(i,j)) < \rho_\alpha(\Gamma(k,l)) \\ \Gamma(k,l): & \rho_\alpha(\Gamma(k,l)) < \rho_\alpha(\Gamma(i,j)) \\ \Gamma(i,j): & \rho_\alpha(\Gamma(k,l)) = \rho_\alpha(\Gamma(i,j)), \\ & i < k \end{cases}$$

Now, we define the product of $\Gamma(i,j)$ and $\Gamma(k,l)$ to be concatenation of the two facet chains when the last link of $\Gamma(i,j)$ equals the first link of $\Gamma(k,l)$:

$$\Gamma(i,j) \circ \Gamma(k,l) = \begin{cases} \Gamma(i,j) * \Gamma(k,l): & j = k \\ \lambda: & \text{otherwise} \end{cases}$$

The set of chains $\mathcal{F}^*$ along with the operations $\circ$ and $\vee$ form the facet chain algebra.

5.4.2 Finding the Best Facet Chain

Recall that Definition 5.3 extends the operations of join and product to square matrices. Applied to the aforementioned definitions, the join operation $\bigvee_{k=1}^{n} \Gamma(i,k) \circ \Gamma(k,j)$ operates on the set of chains in $C_{i,j}$. From Definition 5.4, the closure of the adjacency matrix can yield useful information about paths from one vertex to another. We now define two matrices that allow us to quickly compute the best chain between two facets; the first plays a similar role as the adjacency matrix in graph theory, and the second encodes the instructions for following the chain.

Definition 5.12 (Link matrix) Let $\mathcal{I}$ be the square matrix of connected links:

$$\mathcal{I} = \left[L_{i,j}^{|L_{i,j}|} \right]$$

We can easily think of a chain $\Gamma(i,j)$ as transforming the facet F_i into F_j. However, this transformation requires knowledge of all the facets involved. In a situation where a system must learn from an analogy between two domains, we might not have this complete knowledge in the target domain. By considering a sequence of vertex permutations that correspond to the links in the chain $\Gamma(i,j)$, we are able to obtain an explicit sequence of instructions that is independent of any facet knowledge. If the size of a link is one less than the dimension of the facets, then there is only one possible permutation as each facet has only one vertex that is not in their intersection. If it is smaller than this, then there are several permutations that change the facets as desired. However, since we are only dealing with combinatorial information, each facet is viewed as a set and sets are invariant under ordering. We consider two permutations to be equivalent if they both correspond to the same link. Each equivalence class is represented by one of its elements.

Definition 5.13 (Permutation matrix) To a link $L_{i,j}$ between facets F_i and F_j, associate the permutation $\sigma_{i,j} = (i_1 j_1)(i_2 j_2) \ldots (i_n j_n)$, where $n = |F_j \setminus (F_i \cap F_j)|$ and the vertices $i_1, \ldots, i_n \in F_i$ and $j_1, \ldots, j_n \in F_j$.

Define the matrix $\mathcal{P} = [\sigma_{i,j}]$ to be the matrix of these permutations.

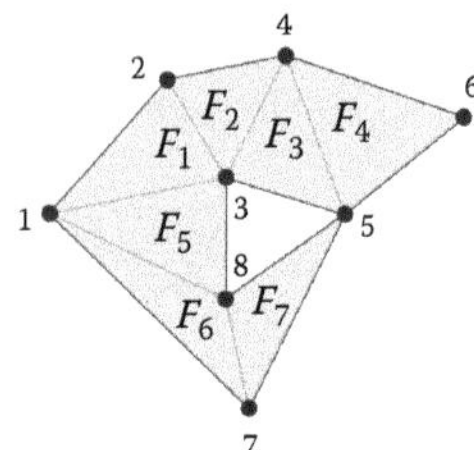

Figure 5.4 Example ASC with seven facets.

Proposition 5.1 The i, jth entry of the matrix $\hat{\mathcal{I}}$ is the chain from F_i to F_j with the smallest chain norm, and the i, jth entry of $\hat{\mathcal{P}}$ is a sequence of permutations on vertices that when applied to the facet F_i will sequentially transform it into F_j. The ith column of $\hat{\mathcal{I}}$ represents all possible best facet chains originating at F_i and the jth row of $\hat{\mathcal{I}}$ represents all best facet chains terminating at F_j.

Therefore, the matrix $\hat{\mathcal{I}}$ tells us, for every two facets in the ASC, which facet chain takes the shortest number of steps and preserves the most properties during each step; the matrix $\hat{\mathcal{P}}$ tells us how to actually perform the corresponding transformation.

Example 5.4:

Let Δ be the simplicial complex shown in Figure 5.4. The matrix $\mathcal{I}$ is given as follows:

$$\begin{bmatrix} \epsilon_1 & L^2_{1,2} & L^1_{1,3} & 0 & L^2_{1,5} & L^1_{1,6} & 0 \\ L^2_{2,1} & \epsilon_2 & L^2_{2,3} & L^1_{2,4} & L^1_{2,5} & 0 & 0 \\ L^1_{3,1} & L^2_{3,2} & \epsilon_3 & L^2_{3,4} & L^1_{3,5} & 0 & L^1_{3,7} \\ 0 & L^1_{4,2} & L^2_{4,3} & \epsilon_4 & 0 & 0 & L^1_{4,7} \\ L^2_{5,1} & L^1_{5,2} & L^1_{5,3} & 0 & \epsilon_5 & L^2_{5,6} & L^1_{5,7} \\ L^1_{6,1} & 0 & 0 & 0 & L^2_{6,5} & \epsilon_6 & L^2_{6,7} \\ 0 & 0 & L^1_{7,3} & L^1_{7,4} & L^1_{7,5} & L^2_{7,6} & \epsilon_7 \end{bmatrix} = \mathcal{I}$$

Similarly, we have the matrix $\mathcal{P}$ shown in Equation 5.1:

$$\mathcal{P} = \begin{bmatrix} e & (1\,4) & (1\,5)(2\,4) & \cdots & (2\,8) & (2\,7)(3\,8) & \cdots \\ (4\,1) & e & (2\,5) & (2\,6)(3\,5) & (2\,8)(4\,1) & \cdots & \cdots \\ (1\,5)(2\,4) & (2\,5) & e & (3\,6) & (4\,8)(5\,1) & \cdots & (4\,7)(3\,8) \\ \cdots & (2\,6)(3\,5) & (6\,3) & e & \cdots & \cdots & (4\,7)(6\,8) \\ (2\,8) & (4\,8)(5\,7) & (4\,8)(5\,7) & \cdots & e & (3\,7) & (3\,5)(1\,7) \\ (2\,7)(3\,8) & \cdots & \cdots & \cdots & (3\,7) & e & (1\,5) \\ \cdots & (4\,7)(3\,8) & (4\,7)(3\,8) & (4\,7)(6\,8) & (3\,5)(1\,7) & (1\,5) & e \end{bmatrix} \tag{5.1}$$

The elements of $\mathcal{P}$ written as three dots are permutations corresponding to nonconnected links called *teleportations*.* These are the links of zero size, which are suppressed for clarity.

* Rodriguez refers to disconnected paths in graph theory as teleportations, and we extend the terminology here [17].

Teleportation chains have potential utility for reasoning with hypothetical knowledge of uncertain or future events.

Remark 5.3 (Norm of teleportations) The norm of a teleportation is always ∞. This can be easily seen as the critical size of a teleportation is zero. Since teleportations have infinite norms, they are never selected by the join when a connected link is available.

There can be many facet chains connecting a pair of facets. For example, $\Gamma_a(1,4), \Gamma_b(1,4)$ are two facet chains* connecting F_1 to F_4:

$$\Gamma_a = L^2_{1,2}, L^2_{2,3}, L^2_{3,4}$$

$$\Gamma_b = L^2_{1,2}, L^1_{2,4}$$

Since the aspect ratios of the two facet chains are greater than one, the chain norm for $\alpha = 1$ gives priority to the size of the critical link. The corresponding chain norms are shown as follows:

$$\rho_1(\Gamma_a) = \frac{3}{2}$$

$$\rho_1(\Gamma_b) = \frac{2}{1}$$

The chain Γ_a has the smallest chain norm of the three, with length three and norm two, whereas Γ_b has length two and norm one. This yields $\rho_1(\Gamma_a) < \rho_1(\Gamma_b)$, so $\Gamma_a \vee \Gamma_b = \Gamma_a$.

5.4.3 Reasoning Algorithm

The reasoning process starts with the ASC representing the target domain called the *target complex*. The target complex is where the problem statement can be represented. The system may not have complete knowledge of the target domain and may only know the initial target facet and a few vertices of the final target facet. The next step is to find a suitable source complex that is rich enough to contain strategies relevant to the problem. The answers to questions posed in the target complex should then be inferred from reasoning in the source complex (see Figure 5.5). The problem of analogy-finding can be formulated as follows: the source complex, initial target facet, and the correspondence ϕ between the two complexes are known, as well as a partial vertex set of the final target facet; we want to determine the best facet chain that connects the initial target facet with a final target facet containing the partial vertex set. In this work, we assume ϕ is a bijection from the target complex to a subset of the source complex.

In practice, α can be chosen based on the specifics of the ASC. For example, choosing α to be smaller than the aspect ratios of all facet chains in the system would result in reasoning that prefers keeping as many connections as possible at the expense of using more steps. This reasoning is more cautious in a sense. Selecting α to be large would result in more hasty reasoning that prefers quick solutions with few steps, at the expense of maintaining fewer connections.

* From here, we drop the reference to the facets F_1 and F_4.

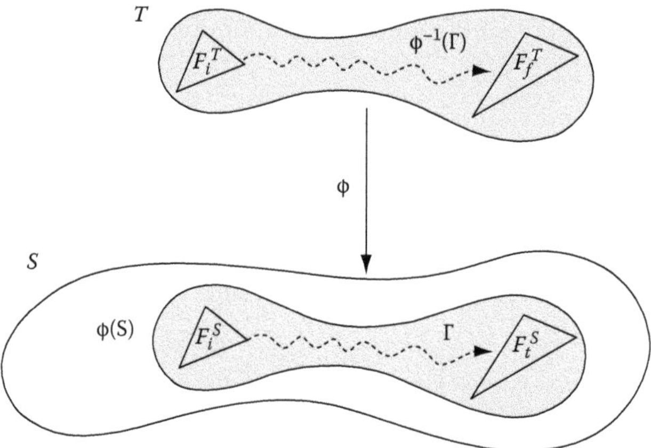

Figure 5.5 The inverse map ϕ^{-1} sends the solution Γ from the source to the target.

The analogy-finding algorithm consists of four steps*:

1. Map initial target facet and the partial final target vertex set, respectively F_i^T and $V_f^{\prime T}$, to the initial source facet and partial final source vertex set, respectively:
$$\phi\left(F_i^T\right) = F_i^S$$
$$\phi\left(V_f^{\prime T}\right) = V_f^{\prime S}$$
2. Find the set of facet chains $C_{if}^{\prime S}$ that start at F_i^S and end at a facet containing $V_f^{\prime S}$. This is done by taking entries of the ith column of $\hat{\mathcal{I}}$ whose last facets contain $V_f^{\prime S}$. The join of the facet chains in $\hat{\mathcal{I}}$ is the desired facet chain $\Gamma(F_i^S, F_f^S) \in S$.
3. Map the sequence of permutations corresponding to the chain from Step 2 to the target complex:
$$\phi^{-1}\left(\sigma_{if}^S\right) = \sigma_{if}^T$$
4. Apply the permutations in order to the initial target facet until the resulting facet contains $V_f^{\prime T}$. This sequence of facets is the desired facet chain in T.

5.5 Sensing System Design for Power Systems

The integration of distributed energy resources in power grids and the development of smart grids have raised several challenges such as power grid monitoring and control. Sensing systems play a vital role in solving these challenges and in improving the power grid reliability. When sensors experience hardware or software failures, the resulting contingencies and missing data can cause problems with the data processing algorithms utilized in power systems [16]. We refer to the sensors with missing or erroneous data as lost sensors.

* If the final target facet is known, use F_f^T in place of $V_f^{\prime T}$.

Sensors can be the primary sources of failure. Missing or erroneous data sent to the relays or controls can lead to false tripping of circuit breakers or other switching devices, which may lead to blackouts. Sequences of fault events like lost sensors or line sags can be recorded before the blackout happens [15]. If these events are identified quickly and sensor failures are corrected efficiently, major blackouts could be avoided.

To handle this rapid identification of events, we propose to integrate an intelligent SMS with the control and monitoring systems, allowing tasks or services to be accomplished as well as the selection of optimal sensor teams needed. We consider a service to be the monitoring and detection of fault events, and the correction and reconstruction of erroneous or missing data. The SMS should feature communication among sensors with the goal of identifying fault events and correcting errors in data that lead to contingencies.

The integration of the SMS with the sensing system will increase the system resilience and reliability. The SMS can be developed as a software layer in the sensing system, which selects the teams of sensors needed for tasks, checks the health of sensors, corrects and reconstructs data, and infers decisions to controllers.

In the following section, we illustrate the reasoning process with the IEEE 14 bus system (Figure 5.6). We have chosen this case as a simple example for illustration purposes. Although there are simpler techniques that can accomplish the computations shown, our contribution is that this structure of reasoning can be extended to much more difficult problems where those methods fail and flexible reasoning is needed.

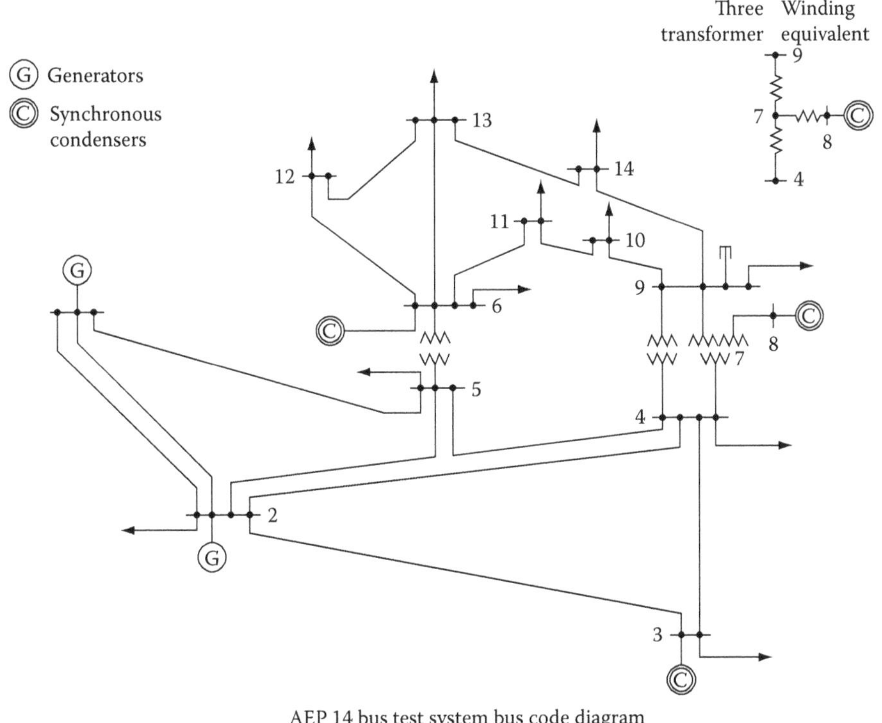

Figure 5.6 IEEE 14 bus system.

5.5.1 Development of Knowledge

In this application, we develop sensing services performed by a given team T_i of sensors that are able to detect fault events and reconstruct data from lost sensors. The proposed KR contains information about the services needed for the identification of sensor failures. These services use teams of sensors in a given power line and its adjacent lines. Each service records data to be processed in order to detect and locate the failure.

5.5.1.1 Redundancy and Agreement among Sensors

In order to monitor the sensor network, sensors are checked using their redundancy. The verification of sensor values is done using Kirchhoff's Current Law (KCL) or Voltage Law (KVL). For the examples in this chapter, we use KCL. Consider the sensors at buses $1, \ldots, 5$ in the IEEE-14 bus system as shown in Figure 5.7. At each transmission line $L_{i,j}$, two sensors are placed in the head and the tail respectively. A sensor $s_{i,j}$ is indexed such that i corresponds to the nearest bus and j corresponds to the remaining adjacent bus. The current measured by sensor $s_{i,j}$ is written $I_{i,j}$ and the current measured by sensor $s_{j,i}$ is $I_{j,i}$. The reference direction of current is defined to be toward the higher numbered bus, toward the load, or away from the generator. Since sensors $s_{1,2}$ and $s_{2,1}$ are on the same line, a redundancy check can be done without any computation and simply amounts to checking if $I_{i,j} = I_{j,i}$.

Definition 5.14 (Directly and indirectly redundant teams) Assuming non-fault conditions, we refer to a team of sensors as a directly redundant team if the sensors are all duplicates or backups of each other. For current measurements, this is true only for sensors that are all on the same line. Applying KCL or KVL to a network yields a system of linear equations. If a team of sensors satisfies exactly one of these equations, it is called an indirectly redundant team.

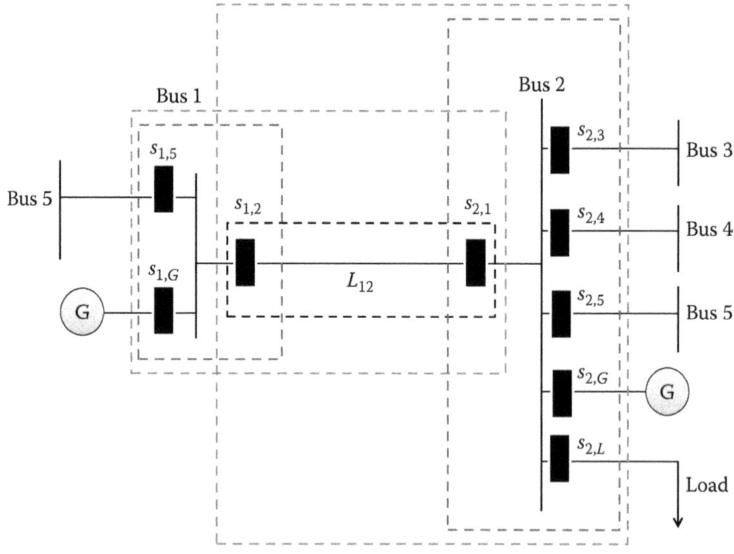

Figure 5.7 Example of team coverage and redundancy.

Example 5.5: Directly and indirectly redundant teams

In Figure 5.7, the team of current sensors $T_1 = \{s_{1,2}, s_{2,1}\}$ is directly redundant, while the team $T_2 = \{s_{2,1}, s_{2,3}, s_{2,4}, s_{2,5}, s_{2,L}, s_{2,G}\}$ is indirectly redundant. The second case can be seen from applying KCL to bus 2 in Figure 5.7. We have

$$I_{2,1} + I_{2,G} = I_{2,3} + I_{2,4} + I_{2,5} + I_{2,L}. \tag{5.2}$$

The variable $I_{2,1}$ is the current measured by sensor $s_{2,1}$ and is defined to flow from bus 1 to bus 2.

Definition 5.15 (Composite redundancy) We say a team satisfies a composite redundancy if it is a union of directly and indirectly redundant teams.

Example 5.6:

In Figure 5.7, the team $T_3 = \{s_{1,2}, s_{2,1}, s_{2,3}, s_{2,4}, s_{2,5}, s_{2,L}, s_{2,G}\}$ is composite as it is a union of T_1 and T_2. However, $T_4 = \{s_{1,5}, s_{1,2}, s_{2,1}, s_{2,5}, s_{2,4}\}$ is not composite. Every sensor in a composite team can be validated by and contribute to a redundancy check. In order to make T_4 composite, we would either have to add $s_{2,3}$ or remove $s_{2,4}$ and $s_{2,5}$.

5.5.1.2 Representation of Knowledge

We begin by specifying the predicates of the KR and creating facets that satisfy these predicates. Since we are interested in the satisfaction of redundancies, we choose the two predicates*: "The team of sensors is directly redundant" and "The team of sensors is indirectly redundant." We refer to their respective simplexes as direct and indirect.

Example 5.7:

Take the two sensor teams $\{s_{1,2}, s_{2,1}\}$ and $\{s_{1,G}, s_{1,5}, s_{1,2}\}$ from the network in Figure 5.7. The two simplexes associated to these teams are shown in Figure 5.8.

Let $s_{i,j}$ and $s_{j,i}$ denote the sensors placed on $L_{i,j}$ respectively at the head and tail of the line. In order to design the ASC of sensors in power networks, we first assume that one or more sensors are placed in every transmission line. This allows the power network to have full sensor coverage. The ASC can then be developed considering three steps: (1) At the first step, we consider only the

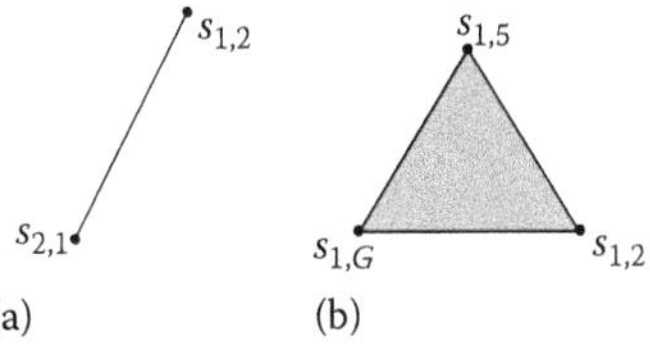

Figure 5.8 (a) Simplex for a directly redundant team and (b) simplex for an indirectly redundant team.

* The resulting equations such as Equation 5.2 can be seen as a modification of the facet ideal in [6].

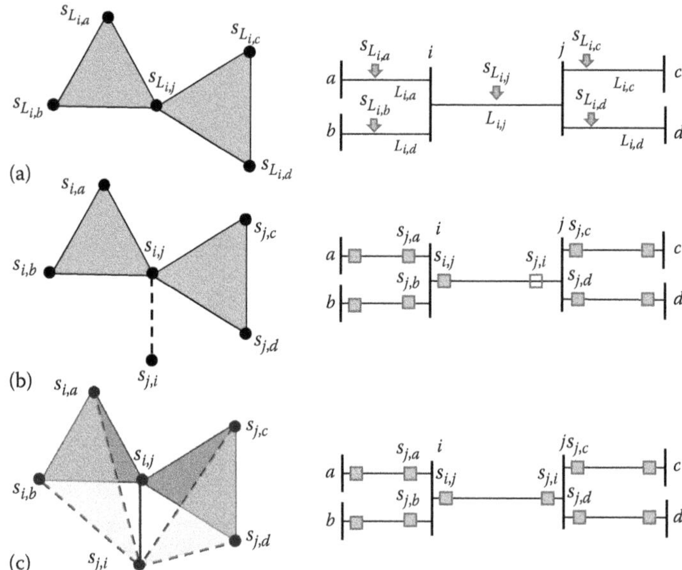

Figure 5.9 Constructing the ASC for full coverage. (a) Construction of indirect simplices and line identification. (b) Direct simplices are added. (c) ASC is completed by adding remaining indirect simplices.

indirectly redundant teams and at each line $L_{i,j}$ we consider a possible measurement* (Figure 5.9a); (2) at the second step, we add sensors in directly redundant teams as shown in Figure 5.9b; and finally (3) we add simplices by connecting the new vertex to the rest of the vertices (Figure 5.9c) so that each of the added simplices is itself an indirectly redundant team.

The result of step (1) is shown in Figure 5.10. Step (3) produces the ASC in Figure 5.11 whose simplices represent algebraic substitution of the direct redundancy into the indirect redundancy.

5.5.1.3 Strategy Chains for Redundancy Checks

Here we use facet chains to represent the strategy of checking sensors for contingency from the knowledge of their redundancy. Each of the predicates in the KR represented in Figure 5.11 can be used for redundancy checking, providing strategies for the monitoring service to detect sensor failures. Consider s_i as the service of the agreement check that is executed by the team $T_2 = s_{2,1}$, $s_{2,5}$, $s_{2,4}$, $s_{2,3}$, $s_{2,L}$, $s_{2,G}$ of indirectly redundant sensors.

Given a lost sensor to recover data from, form a chain with the following steps:

1. Delete vertices from the SC that do not correspond to existing sensors in the network.
2. If a simplex has more than one vertex removed, delete the simplex.
3. Compute the transitive closure of the link matrix of the ASC.
4. Pick the row of the transitive closure matrix corresponding to a simplex that contains the lost sensor vertex. From this row, select the entry with the smallest chain norm. Call this entry the optimal chain.

The resulting chain identifies the sensors whose data are needed for the reconstruction.

* In the case of KCL, each simplex will correspond to a bus of the network.

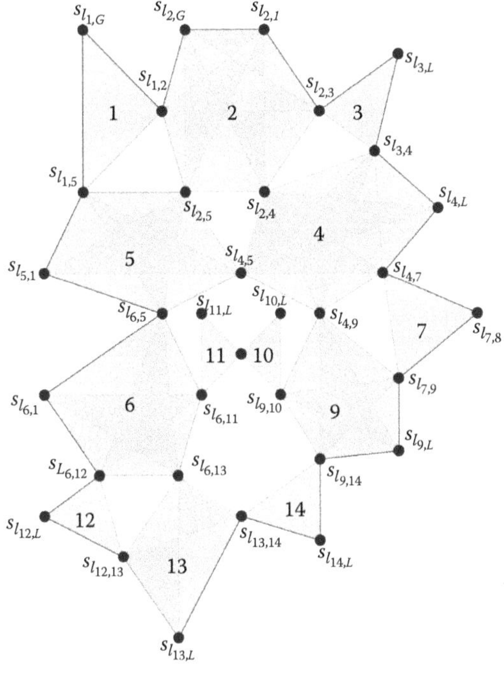

Figure 5.10 Simplicial complex of full sensor coverage in IEEE 14-bus system.

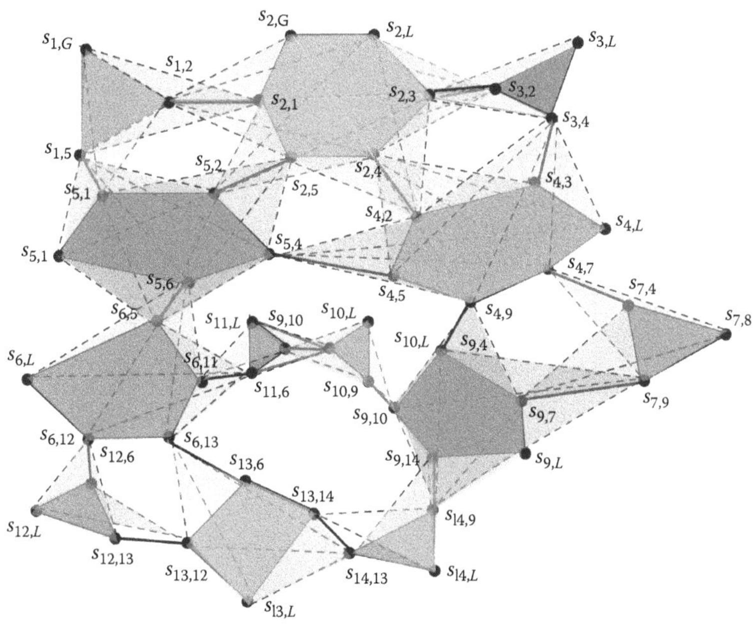

Figure 5.11 Extended simplicial complex of full sensor coverage in IEEE-14 bus system.

5.5.2 Reasoning Algorithm for Sensor Failure Detection and Data Restoration

In order to detect sensor failures, the SMS will use a reasoning algorithm, which will utilize the sensors redundancy to identify the faulty sensors and restore their data. The agreement of sensors is tested by checking for satisfaction of the constraint equation. We say that a team of sensors agrees if their measured quantities satisfy the given constraint equation, in this case one of the Kirchhoff's laws.

Example 5.8: Using Chains from Extended Complex

Figure 5.12 shows hypothetical sensor failures in the complex from Figure 5.9. When one sensor fails, the chain composed of the single link $\{F_1, F_2\}$ still provides redundancy. When another sensor fails, both $s_{i,j}$ and $s_{i,b}$ can be computed by substitution. Denote the measured quantity at sensor $s_{i,j}$ by $s^*_{i,j}$:

$$F_1 : s_{i,b} = s^*_{i,a} + s_{i,j} \tag{5.3}$$

$$F_2 : s_{i,j} = s^*_{j,c} + s^*_{j,d} \tag{5.4}$$

$$\{F_1, F_2\} : s_{i,b} = s^*_{i,a} + s^*_{j,c} + s^*_{j,d} \tag{5.5}$$

If they all agree, the algorithm declares that all sensors in the team are healthy, while if there is a team who's sensors do not agree with each other, the algorithm will identify the sensors that do not agree and they will be declared faulty. After a sensor is declared faulty, a process of service restoration will begin. If a failure occurs in one sensor, the SMS will accommodate the fault by isolating the faulty sensor and reconstructing their data from the remaining healthy ones.

5.5.2.1 Simulation Results

To validate our approach for the contingency test, the IEEE-14 bus system is simulated in a MATLAB®/Simulink® environment with three different scenarios for sensor faults: for failures in the sensors $s_{1,2}$ at the time interval [0.1, 0.3s], $s_{2,1}$ at the time interval [0.4, 0.6s], and $s_{2,5}$ at the time interval [0.7, 0.9s].

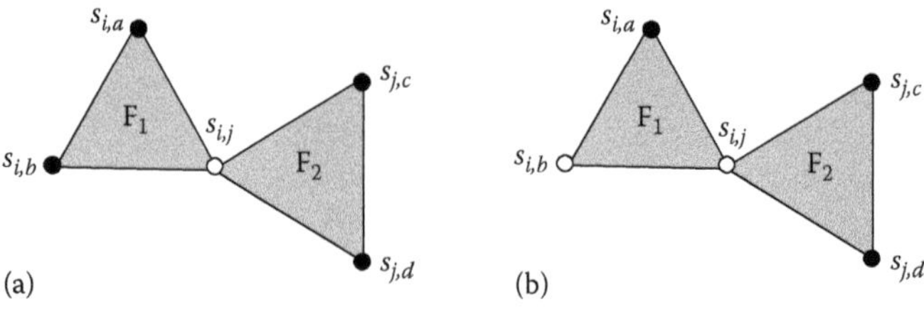

Figure 5.12 (a) First sensor fails, but the chain still provides redundancy; (b) second sensor fails, the chain no longer provides redundancy, but the values of both sensors can still be reconstructed.

Consider the following teams of sensors :

1. Composite: $T_1 = \{s_{1,2}, s_{2,1}, s_{2,3}, s_{2,4}, s_{2,5}, s_{2,L}, s_{2,G}\}$
2. Indirect: $T_2 = \{s_{2,1}, s_{2,3}, s_{2,4}, s_{2,5}, s_{2,L}, s_{2,G}\}$
3. Direct: $T_3 = \{s_{1,2}, s_{2,1}\}$

To test the argument between sensors in a team T_i, we consider the following analytic relations:

$$\mathcal{A}(T_1) : s^*_{1,2} - s^*_{2,1} - s^*_{2,3} - s^*_{2,4} - s^*_{2,5} - s^*_{2,L} - s^*_{2,G} = 0$$
$$\mathcal{A}(T_2) : s^*_{2,1} - s^*2,3 - s^*_{2,4} - s^*_{2,5} - s^*_{2,L} - s^*_{2,G} = 0$$
$$\mathcal{A}(T_3) : s^*_{1,5} - s^*_{1,G} - s^*_{1,2} = 0$$

where $\mathcal{A}(T_i)$ represent the agreement relation. The team T_i is in agreement when the relation $\mathcal{A}(T_i)$ is satisfied, otherwise T_i is in disagreement. If a team of sensors disagrees, then one or more sensors in the team is faulty. Figure 5.13 shows the residues of the agreement test conducted in teams T_1, T_2, and T_3. For the three fault scenarios, the agreement relations $\mathcal{A}(T_1)$, $\mathcal{A}(T_2)$, and $\mathcal{A}(T_3)$ are not satisfied and have non-zero residues when failures occur (Figure 5.13). However, only team T_1 is able to detect the maximum number of failures. In order to identify which sensor

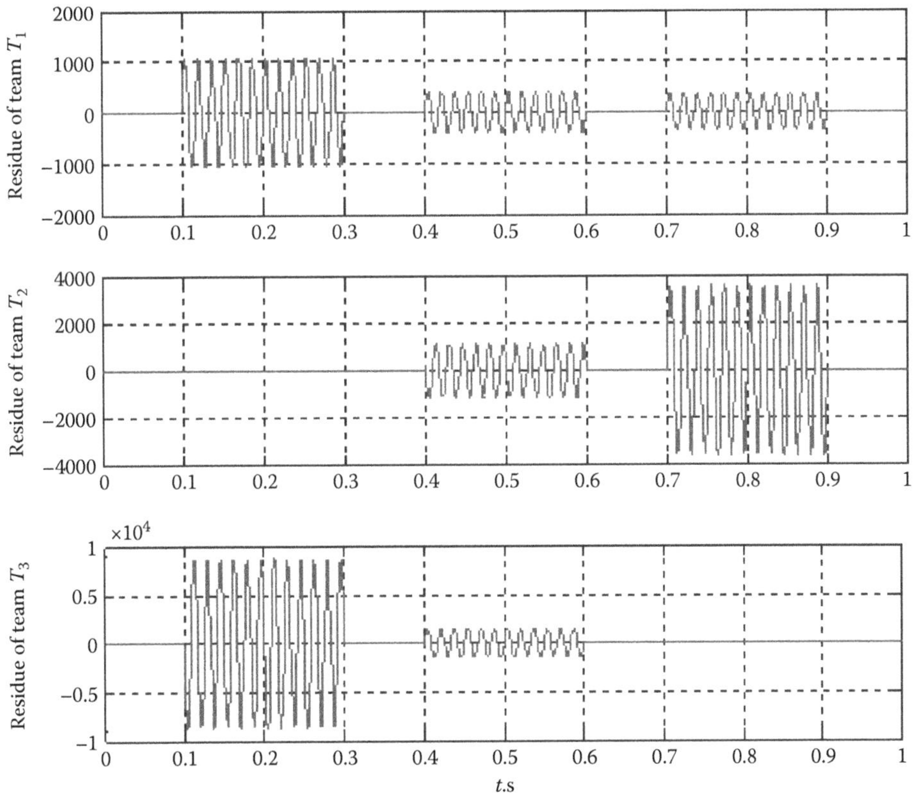

Figure 5.13 Agreement test.

has failed, we compare the agreement of the composite team against the agreement of the direct and the indirect redundant teams:

- At the time interval [0.1, 0.3s], team T_2 is in agreement, while teams T_1 and T_3 are not. This means that the faulty sensor is an element of T_1 and T_3 and not an element of T_2.
- At the time interval [0.4, 0.6s], none of the teams T_1, T_2, and T_3 are in agreement. So the faulty sensor must be an element of T_1, T_2, and T_3.
- At the time interval [0.7, 0.9s], team T_3 agrees, while teams T_1 and T_2 are not in agreement. This means that the faulty sensor is an element of T_1 and T_2 and not an element of T_3.

This comparison allows us to identify the sensor failures when it occurs in power grids.

5.5.2.2 Future Application: Using Analogies in the Reasoning Algorithm for Sensor Team Selection

When not quickly and accurately detected, faults in a power network may lead to a cascade of other failures, for example, a transmission line tripping can cause a transient, overloading lines in other areas. Cascading failures in power systems are the major cause of large blackouts. Even simple events such as line sag close to nearby trees can be the primary source of a cascading failure.

As power systems are generally large-scale distributed systems, redundancy checking can be computationally expensive.* To reduce this burden, the SMS should be able to select only specific teams to test the redundancy when a fault event occurs, instead of checking every redundancy. To achieve this objective, analogies can be a good candidate solution for the team selection. This solution will be based on the strategies of chains that have been used when some specific fault events occurred.

In a power system, due to similar components or topologies, certain fault events can be fairly common and repeat themselves in different areas of the grid. By using analogies to identify these similar events, it may be possible to get a better understanding of the spatial and temporal distribution of similar faults. The knowledge of this distribution can be used by the SMS to select only the teams that are useful to detect the given fault type.

5.6 Conclusion

The integration of a new SMS with the capability of learning, updating knowledge, and making decisions without human intervention can answer the control and monitoring needs of complex engineered systems. The objective of this research is to use a CAT-based KR to develop reasoning algorithms for a distributed intelligent SMS, which will be integrated as a software layer in the control and monitoring system.

To achieve this objective, we developed a set of topological tools and a theory for KR and reasoning. The motivation behind developing an analogical reasoning process using this model is to allow an autonomous intelligent system to act according to structural comparisons among its beliefs about the world as opposed to just what is explicitly representable with logical rules.

We use ASCs as a tool to model knowledge that will serve as a basis for reasoning operations. ASCs have been chosen in this research because of their rich structure, which allows the encoding

* The presented method relies on matrix multiplication and so has complexity $O(n^3)$.

of a large range of knowledge. These developed tools will have an impact beyond the field of sensing and measurement systems, as the generality of our approach will yield a theory that can be used in numerous branches of control, such as system health management, robotics, and unmanned space exploration. The development of technologies (e.g., actuators, sensors, etc.) presents an opportunity to develop a new intelligent control paradigm based on reasoning by analogy with the goal of improving the capability of systems. Furthermore, some of the biggest obstacles in large-scale complex systems are (1) uncertainty, (2) controllability, and (3) unpredictability because of the system's dynamic nature. The proposed tools for using analogies for problem solving in real-time will go a long way in addressing these concerns.

References

1. A. Abur and A. Gómez-Expósito. *Power System State Estimation Theory and Implementations.* Marcel Dekker, New York, NY, 2004.
2. R.H. Atkin. From cohomology in physics to q-connectivity in social science. *International Journal of Man–Machine Studies*, 4(2):139–167, 1972.
3. A. Collins, E.H. Warnock, N. Aiello, and M.L. Miller. Reasoning from incomplete knowledge. Technical Report, Bolt Beranek and Newman Inc., Cambridge, MA, March 1975.
4. C.H. Dowker. Homology groups of relations. *The Annals of Mathematics, 2nd Series*, 56(1):84–95, 1952.
5. J.-L. Giavitto and E. Valencia. Algebraic topology for knowledge representation in analogy solving. In *ECAI98: European Conference on Artificial Intelligence*, Brighton, England, 1998.
6. S. Faridi. The facet ideal of a simplicial complex. *Manuscripta Mathematica*, 109:159–174, 2002. 10.1007/s00229-002-0293-9.
7. D. Gentner and B. Bowdle. *Metaphor as Structure Mapping*, New York: Cambridge University Press, 2008, pp. 109–128.
8. D. Gentner and K. Forbus. Computational models of analogy. *WIREs Cognitive Science*, 2:266–276, 2011.
9. D. Gentner, K.J. Holyoak, and B.N. Kokinov. *The Analogical Mind: Perspectives from Cognitive Science.* The MIT Press, Cambridge, MA, March 2001.
10. J. Hadamard. *The Psychology of Invention in the Mathematical Field.* Mineola, NY: Dover Publications, 1954.
11. A. Hatcher. *Algebraic Topology*, Vol. 227. Cambridge University Press, Cambridge, UK, 2002.
12. R. Manger. A new path algebra for finding paths in graphs. In *26th International Conference on Information Technology Interfaces*, Cavtat, Croatia, 2004.
13. A. Monticelli. Electric power system state estimation. *Proceedings of the IEEE*, 88(2):262–282, 2000.
14. J. Munkres. *Elements of Algebraic Topology.* New York: Perseus Books, 1993.
15. D.P. Nedic, I. Dobson, D.S. Kirschen, B.A. Carreras, and V.E. Lynch. Criticality in a cascading failure blackout model. *International Journal of Electrical Power & Energy Systems*, 28(9):627–633, 2006.
16. W. Qiao, G.K. Venayagamoorthy, and R.G. Harley. Missing-sensor-fault-tolerant control for SSSC facts device with real-time implementation. *IEEE Transactions on Power Delivery*, 24(2):740–750, 2009.
17. M.A. Rodriguez and P. Neubauer. A path algebra for multi-relational graphs. In *IEEE* 27th *International Conference–Data Engineering Workshops*, Hannover, Germany, 2011.

Chapter 6

Intelligent Sensor-to-Mission Assignment

Hosam Rowaihy

Contents

6.1 Introduction

Wireless sensor networks consist of a large number of small sensor devices that have the capability to take various measurements of their environment. These measurements include seismic, acoustic, magnetic, IR and video information. These devices are highly resource-constrained, equipped with a small processor and wireless communication antenna, and battery powered. To be used, sensors are scattered around a sensing field to collect information about their surroundings. For example, sensors can be used in a battlefield to gather information about enemy troops, detect events such as explosions, and track and localize targets. Upon deployment in a field, they form a wireless ad hoc network and communicate with each other and with data processing centers.

A sensor network is typically expected to perform multiple tasks or *missions*. A mission, in this context, is any job that requires some amount of sensing resources to be accomplished such as video monitoring a field, tracking a target, or localizing an event. Missions can be divided into multiple sub-missions. For example, monitoring a large field can be divided into monitoring multiple smaller areas. Each mission can be modeled with a *demand*, which measures its need for sensing resources, and a *profit*, which represents its importance. In heterogeneous networks, the requirement of a mission is specified by its need for different types of sensors. For example, a mission may require video imaging and seismic data to identify an object. In such cases, sensors must be bundled together then assigned to missions. The *utility* (or amount/quality of information) that a sensor can provide to a mission depends on several factors. These include the type of the sensor, its sensing range, its geographic location relative to the mission, and its current operational status, such as its remaining energy.

Due to the limited number of sensors and the potentially large number of missions, competition will arise. In such cases, it might not be possible to satisfy the requirements of all missions using available sensors. Given all currently available information, the network should intelligently choose the "best" assignment of the available sensors to the missions to maximize the utility of the network. In this chapter, we discuss this *sensor-mission assignment problem*.

Although certain types of sensors, such as seismic or acoustic sensors, can receive data from their local surroundings as a whole, other types of sensors need to be directed to a certain location such as those used for imaging purposes. In these cases, the direction of each sensor, and thus the mission it serves, must be chosen appropriately since it may only benefit one mission. The focus here will be on directional sensors as they typically pose more challenging problems. Almost all the problems considered here are NP-hard, and hence we discuss some heuristics that were proposed in the literature to solve such problems.

6.1.1 Problem Variants

There are two broad settings in which sensors need to be assigned to missions: static and dynamic. In the *static* setting, all the missions need to be satisfied simultaneously. Hence, the information about all missions, including their profits and demands, is available at once when making the assignment decisions. This setting is useful when the system has a set of long-lived missions such as perimeter monitoring applications. It can also be useful in systems that are not expected to have fast response, in which case missions arriving at different times may be batched together and start at a single point in time. In the *dynamic* setting, missions start at different times and have different durations. This is a more practical setting in which a sensor network is expected to operate. In such a setting, the system is expected to have a prompt response to incoming missions and should be able to adapt quickly to changes.

There are different constraints for the sensor-mission assignment problem that will be considered. In some environments, missions may have *budget* constraints to limit the number of sensors they can use; in this case, sensors will have associated *cost*. In other environments, *lifetime* of sensors constrain the amount of utility they can provide.

Solutions we discuss in this chapter can be divided into two main categories: centralized and distributed. In a *centralized* solution, all assignment decisions are made by a single node (typically a base station that is deployed in the field) that has all the information about missions. To do this the base station collects all the required information about each sensor. Then, it runs a local algorithm to decide on the assignments and sends these assignments to the respective sensors. Due to its global view of the field, this approach can provide high quality solutions, but can be expensive in terms of communication overhead and introduces a single point of failure. In a *distributed* solution, sensors make these decisions on their own. In such an approach, mission information is disseminated to the network and individual nodes decide on the assignments. This is a more fault-tolerant approach and can be more efficient in terms of communication overhead, but might not provide as good a solution as the centralized approach.

6.1.2 Related Work

There has been some work in defining frameworks for sensor-mission assignment problems. For example, [3] defines a framework for the assignment problem in which the goal is to maximize the utility while staying under a predefined budget. However, the authors do not consider the case of competing missions. The general problem of sensor selection to achieve an objective has also received sizable attention lately. For example, in [10,13] the authors solve the coverage problem, which is a related problem, using the least number of sensors to conserve energy. Another related problem is to efficiently locate and track targets such as in [7,8,16]. A survey of the different sensor selection and assignment algorithms including theoretical models of the problem can be found in [11].

In this chapter, we discuss several sensor-mission assignment problems in the presence of different constraints. Both the static and dynamic settings are considered. We discuss different intelligent solutions to solve these problems based on centralized and distributed approaches. The main goal of these solutions is to intelligently assign sensing resources to the competing missions. Mainly, we focus on the model and solutions proposed in [12] and [6].

6.1.3 Network Model

In the network model we consider in this chapter, it is assumed that a set of static sensors are predeployed in a field. Missions can arrive and depart over time. A mission is a primitive sensing task that requires information of a certain type, which may be contributed by one or more sensors. Each mission is defined by a specific geographic location. An example of a mission is video monitoring an area of interest. General missions that cover large areas, such as perimeter monitoring, can be divided into multiple missions each having its own location. The deployed sensors are directional in nature and hence each of them can be assigned to a single mission (i.e., directed to one location). The direction of a sensor can be changed when the assignment is changed. A video camera is a good example of such sensors.

6.1.4 Roadmap

The rest of this chapter is organized as follows: Section 6.2 introduces the sensor-mission assignment problem in the presence of competition and discusses different solutions. Section 6.3 extends the problem by considering extra budget constraint in the static setting while Section 6.4 considers lifetime constraints in dynamic environments. Finally, Section 6.5 concludes this chapter by looking at future research directions in this area.

6.2 Basic Sensor-Mission Assignment

As mentioned previously, missions may vary in both importance (*profit*) and difficulty (*demand*), and these properties need not be correlated. An ongoing surveillance mission, for example, may be expensive but of minor importance, whereas an urgent mission for information about one particular spot may be of low-demand but very important. In many applications, partial satisfaction will be no better than zero satisfaction. If the goal of a given mission is to reconstruct the 3D shape of an object, for example, then this may be accomplished with images from two cameras, but an image from just one camera will be useless. Indeed, accepting the single image could actually be harmful since the drain on the sensor's battery could preclude a future mission that might otherwise have been satisfiable. The model in [12] considers two profit functions: (1) only receives profit from missions whose demands are fully met and (2) considers profits from ones that reached a preset *threshold*. Hence the problem is to choose the "best" assignment of sensors to missions, in the sense that profits from satisfied missions are maximized.

In some networks, there may simply be a static set of long-term missions, in which case the aspect of time may be eliminated. In other settings, mission arrivals and departures may be infrequent, so that for each block of time, sensor assignment can be solved as a static problem. Even in this static setting, this problem is computationally hard to solve optimally. Thus approximation algorithms and heuristics are used.

A centralized approach to sensor assignment will collect all the relevant information at a central location for decision-making and then distribute assignments. Such an approach can be expensive in terms of communication overhead, however. Another approach is to have nodes make these assignment decisions locally, in a distributed manner, using mission information that is disseminated into the network. While this should decrease communication costs, a centralized algorithm may be able to guarantee a better solution.

Since the problem (in its most general form) is NP-hard, we consider constrained version for which approximation algorithms exist. We consider a geometric constraint; only sensors within a bounded sensing range from a mission can be assigned to that mission, which is a reasonable assumption in realistic settings. The problem is generalized further by allowing a mission to be successful even if its demand is not fully met. This is done by setting a *threshold* that specifies the minimum fraction of the demand to be met for a mission to succeed. In this case, the mission will not be awarded the full profit, but rather a fraction based on its satisfaction level.

In this section, we consider the problem of assigning directional sensors to missions in wireless sensor networks. We study both centralized and distributed approaches to solving the dynamic problem. As energy is a critical resource in wireless sensor networks, we discuss an energy-aware extension to the distributed algorithm that extends network lifetime. This section is based on [12].

6.2.1 Problem Definition

Here, we present the formal definition of the core sensor-mission assignment problem and extend it to the dynamic setting.

6.2.1.1 Static Problem

The core problem, which is called *Semi-Matching with Demand* (SMD) [1], is modeled as a weighted bipartite graph whose vertex sets consist of sensors $S = \{S_1, \ldots, S_n\}$ and missions $\{M_1, \ldots, M_m\}$ (see Figure 6.1). A sensor S_i may be able, depending on its type and location, to provide mission M_j with some data. A positively weighted edge (S_i, M_j) means that S_i is applicable to M_j. The weight of the edge (S_i, M_j) is denoted by e_{ij} and indicates the utility (or quality of information) that S_i could contribute to M_j if this assignment were chosen. The utility may vary depending on the sensor's type, location, or other properties. Also given is a positive-valued demand d_j associated with

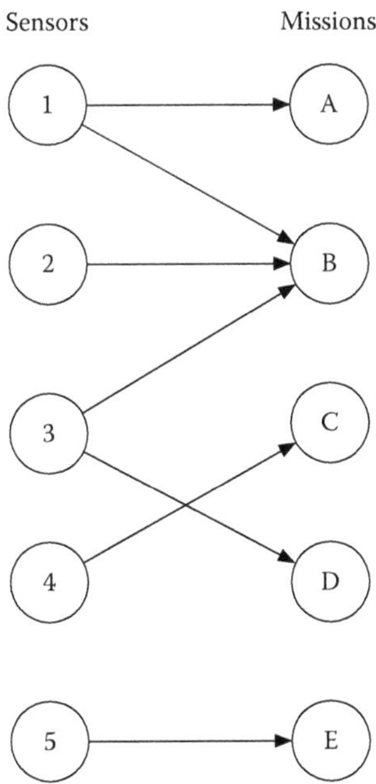

Figure 6.1 Modeling the problem as a bipartite graph.

each mission M_j indicating the total utility the mission requires. The problem can be simplified by assuming that the utility amounts received by a mission are additive. That is, the total utility received by a mission is equal to the sum of the utilities provided by sensors assigned to it. While this may be realistic in some settings such as sensing applications in which high-quality measurements can be obtained by, for example, either taking a single high-quality reading or averaging together several lower-quality readings, in others it is not; for the purpose of the discussion this assumption is sufficient.

A solution to the problem would seek a *semi-matching* of sensors to missions, so that, ideally, each mission demand is satisfied. That is a sensor may be assigned to at most one of the missions to which it is applicable, but a mission can accept utility from multiple sensors. Of course, satisfying all missions may not be feasible; in general, the goal is to maximize a weighted sum of the *satisfied missions*. Since there is a profit p_j associated with achieving mission M_j, the goal becomes to maximize the total satisfied profits.

The problem can be generalized by introducing the concept of a threshold. In this case, missions can be partially successful if they reached a minimum threshold of utility. The demand may now be interpreted as the total utility the mission desires. Profit for mission M_j indicates the importance of the mission and is awarded based on the percentage of satisfied demand, but only if this percentage reaches a satisfaction threshold T; p_j is the maximum profit receivable for mission M_j. The goal is to maximize total profits.

The problem instance and goal are defined as follows:

Instance: A global threshold $T \in [0, 1]$ and a weighted bipartite graph $G = (S, M, P, D, E)$, where $S = \{S_1, \ldots, S_n\}$ is a collection of sensors and $M = \{M_1, \ldots, M_m\}$ is a collection of missions; each mission M_j is associated with a profit $\{p_j\}$ and a demand $\{d_i\}$; each edge in $S \times M$ has an edge-weight e_{ij} indicating utility.

Goal: Find a semi-matching $F \subseteq S \times M$ (no two chosen edges share the same *sensor*), in which $\sum_j p_j(u_j)$ is maximized, where u_j is the total utility received by mission M_j divided by demand d_j. The profit functions are defined as follows:

$$p_j(u_j) = \begin{cases} p_j, & \text{if } u_j \geq 1 \\ p_j \cdot u_j, & \text{if } T \leq u_j < 1 \\ 0, & \text{if } u_j < T \end{cases}$$

The problem can be formulated as an integer program (IP). The following IP employs the decision variable x_{ij} indicating whether sensor S_i is assigned to mission M_j. Finding a solution can be seen as a two-step process: decide which missions to satisfy and then decide how to satisfy them. Each mission M_j has a constraint requiring that the sum of utility received by M_j be at least the value T, which is a user-defined variable. Here is the IP:

$$\begin{aligned} \text{Maximize: } & \textstyle\sum_j p_j(u_j) \\ \text{Such that: } & \textstyle\sum_{i=1}^{n} x_{ij}e_{ij} \geq T, \text{ for each mission } M_j, \\ & \textstyle\sum_{j=1}^{m} x_{ij} \leq 1, \text{ for each sensor } S_i, \text{ and} \\ & x_{ij} \in \{0, 1\}, \text{ for each variable } x_{ij} \text{ and } u_j \in [0, 1], \text{ for each variable } u_j \end{aligned}$$

6.2.1.2 Dynamic Problem

An orthogonal generalization of the original problem, in terms of time, is used to model more realistic scenarios in which missions arrive and depart over time. The problem statement is the same, except that now each mission is associated with a start time and an end time. A mission's demand and maximum profit are constant over time. Awarded profit for a mission is computed at each discrete timestep, based on the satisfaction level at that instant. Total profit for a mission is simply the sum of the instantaneous profits. It is not required that a mission's demand be met over its entire lifetime in order to receive profit. The profit model is in this sense fractional *in terms of time*. The dynamic version is thus given essentially by the same in Section 6.2.1.1 mathematical program (MP) given in Section 6.2.1.1 except that each variable now has an additional time index.

As a generalization of the static problem, previous hardness results apply also to the dynamic version. Indeed, a natural strategy for the dynamic problem is to solve the static problem at each timestep.

6.2.2 Centralized Algorithm

A greedy algorithm to solve this problem will repeatedly satisfy the most *currently profitable* mission, that is, the mission that can be satisfied with the greatest profit, using the currently available sensors. If $S' \subset S$ is the set of not-yet-assigned sensors (initially $S' = S$) and $u_j = \sum_{S_i \in S'} e_{ij}$, then the profit currently achievable by mission M_j is $p_j(u_j)$. Of course, it may be that not all sensors are needed to achieve this profit; conversely, if the demand threshold is not met, this profit is 0. The algorithm repeatedly select a mission M_j of maximum current profitability, and then satisfies it with available sensors (which are removed from S'), in order of decreasing contribution value e_{ij}, until either M_j is *fully* satisfied or all sensors with nonzero offers to M_j have been used. When there are no remaining missions with nonzero current profitability, the algorithm completes. The running time of the algorithm as written is $O(n(m + \log n))$, but it is easy to improve this to $O(mn \log n)$ by updating the u_j values over time rather than computing them from scratch. The details of the algorithm are as follows:

6.2.3 Distributed Algorithms

Although centralized algorithms such as the greedy algorithm discussed in the previous section may provide better solutions to the sensor-mission assignment problem due to their global view of

Algorithm 6.1 Greedy algorithm

```
INPUT: S, M, e_ij, ∀(S_i, M_j) ∈ S × M  and p_j, d_j ∀M_j ∈ M
while true
    for each available mission M_j
        u_j ← Σ_{S_i unused} e_ij
    j ← arg max_j p_j(u_j)
    if p_j(u_j) = 0 then break
    u_j ← 0
    for each unused S_i in decreasing order of e_ij
        if u_j ≥ d_j or e_ij = 0 then break
        assign S_i to M_j
        u_j = u_j + e_ij
OUTPUT: sensor assignment
```

the field, they can be expensive in terms of communication cost. Because a centralized algorithm requires global information about all sensors in the network such as their locations, utilities to the different missions and other stats, the number of messages required to be sent to the base station can become very large, especially in dense networks. This communication cost becomes even higher for dynamic environments in which missions arrive and depart at different points in time, requiring the base station to continually gather information about sensors in the field.

To avoid this cost, distributed algorithms are developed to solve the problem. In such an approach, a *mission leader* is selected for each mission. This should be a sensor that is close to the mission's location. Finding the leader can be done using geographic-based routing techniques such as [2] or [9]. The leaders are informed about the missions' demands, profits, and locations by the base station. Then they run a local protocol to match nearby sensors to their respective missions. It is assumed that the contribution a sensor can provide to a mission is a function of the geographic distance between them and hence only nearby sensors are considered. In the following we discuss a multi-round proposal algorithm (MRPA) that works on static settings, which is then adapt to dynamic cases. We also discuss an energy-aware extension to the dynamic algorithm that helps in prolonging the network lifetime [12].

6.2.3.1 Bidding Algorithm

In this algorithm, each mission leader advertises its mission information (demand, profit, and location) to the nearby sensors by means of broadcast. If the advertisement message needs to be sent over multiple hops then neighboring sensors rebroadcast the message so their neighbors can hear it. The number of hops over which the advertisement message is sent depends on the relation between the communication range, which is the maximum distance over which two sensors can communicate, and the sensing range. If the sensing range is larger than the communication range then sensors that are further away should be notified.

When nearby nodes receive mission advertisements from one or more missions, they decide on which mission(s) to bid. To achieve the highest possible profit, the bidding price (B_{ij}) that sensor S_i sends to mission M_j is set to the product of the sensor-mission contribution and the mission's profit. Using the notation of Section 6.2.1, $B_{ij} = e_{ij}/d_j \times p_j$. The sensor sorts the bidding prices in decreasing order and sends bids to the first N (decided by a protocol parameter).

Mission leaders wait for some time to receive all bids; then they select the best sensors for their needs and send them assignment messages. A sensor is assigned to the first mission it receives an assignment message from. In this algorithm, missions compete for sensors. Once a mission leader selects a sensor, other mission leaders competing for that sensor are notified that it is no longer available. This last requirement makes this algorithm impractical in real systems.

6.2.3.2 Multi-Round Proposal Algorithm

In this algorithm, each mission leader advertises its mission information (demand, profit, and location) to the nearby sensors similar to the previous two algorithms. When a nearby sensor hears such an advertisement message for one or more missions (the set of advertising missions is denoted with Q), it sends a single proposal to the mission it perceives to be its best match. The ranking of missions is based on the profit of a mission weighted by the fraction of the mission's demand that the sensor can satisfy. Using the notation introduced in Section 6.2.1, sensor S_i ranks mission M_j according to B_{ij}, where $B_{ij} = e_{ij}/d_j \times p_j$. The leader, on the other hand, selects for the set of proposing sensors G in a greedy fashion according to their contribution to the mission. If the leader of a mission does not select a proposing sensor, then in the next round the sensor proposes to the next mission on its list. This algorithm consists of a series of proposal-reply rounds. The more rounds, the better the assignment may be. However, as the number of rounds increases, the communication cost grows and only diminishing returns can be obtained. Hence, there is a trade-off between solution quality and communication cost.

Since the aim is to achieve the highest profit from *successful* missions, a mechanism to prevent missions that will never be fully successful from holding up sensors that can help other missions is used. In each round, mission leaders assess the satisfaction level of their missions. If the level is not greater than an increasing threshold ($\alpha(k)$ for round k) then the mission is assumed to be unattainable and all its sensors are released. The threshold is initialized to a fixed value (e.g., 10% satisfaction) and incremented each round (e.g., by 10%). After a sufficient number of rounds, it will reach T, the preset value of the success threshold, at which time all missions that are not yet successful release their sensors. The rising threshold therefore yields two benefits: increasing the chance that the most satisfied missions will become fully satisfied and preventing sensors from spending their energy on missions that will not reach the minimum success threshold (for which no profit is received). Algorithm 6.2 summarizes the steps taken by both the mission leaders and surrounding sensors.

Algorithm 6.2 Multi-round proposal

For leader of mission M_j:

    INPUT: Set of proposing sensors G, $e_{ij} \forall S_i \in G$, d_j, $\alpha(.)$ and number of rounds
    $u_j \leftarrow 0$
    for each round k
        send advertisement
        wait for proposing sensors
        sort proposing sensors in decreasing order of e_{ij}
        while $u_j < d_j$
            assign the next S_i (in sorted order) to M_j
            $u_j = u_j + e_{ij}$
        if M_j's satisfaction level $< \alpha(k)$ **then**
            release all sensors assigned to M_j
    OUTPUT: sensor assignment

For sensor S_i:

    INPUT: Set of advertising missions Q, $e_{ij}, p_j, d_j \forall M_j \in Q$ and number of rounds
    for each round k
        if S_i is *unassigned* **then**
            receive advertisement from all nearby missions
            ignore any mission already considered
            $j \leftarrow \arg\max_j B_j = (e_{ij}/d_j) \times p_j$
            send proposal to M_j
        else break
    OUTPUT: mission M_j to which sensor S_i assigned

We now discuss both the runtime complexity and message complexity of Algorithm 6.2, assuming that the number of sensors is n, number of missions is m, and number of rounds is k. The running time for sensor S_i is $O(km)$ as in each round a sensor may consider up to m missions. The mission leader's complexity is $O(n \log n)$ as in each round the mission leader has to sort up to n proposing sensors and select the best ones, but over all the rounds each sensor proposes to a mission at most one time. Assuming that advertisement messages are only broadcast to immediate neighbors, the message complexity of the algorithm including both sides, the sensor, and the mission, is $O(m + kn)$, as there are m advertisement messages, $O(kn)$ proposals by sensors, and $O(n)$ reply messages from mission leaders.

6.2.3.3 Dynamic Proposal Algorithm

MRPA is designed to work in the cases in which we have prior knowledge about the missions. The multiple rounds allow it to work well even if there are several missions that compete for the same sensing resources. Since it requires complete knowledge about missions it can also work in an environment that does not need fast response to new missions. An example would be a system that batches together a number of missions and runs the assignment algorithm periodically, for example, every few hours. However, in a fully dynamic setting, the network needs to have a fast response to incoming and outgoing missions. By handling missions as they arrive, it is expected to encounter less competition for sensing resources, and so a lighter-weight algorithm that does not need multiple rounds to complete can be used. This algorithm is called the *Dynamic Proposal Algorithm* or DPA [12].

As in MRPA, each mission has a leader that advertises mission information to nearby sensors. A sensor that hears this announcement can be in one of two states: (1) *not assigned*, in which case it proposes to the mission with its utility or (2) *assigned to a mission*, in which case the sensor calculates its effective profit for both missions (which is the B_{ij} value found above) and chooses either to stay with the current mission or to propose to the new mission, depending on which value is higher. So, this algorithm allows a mission to preempt an ongoing mission to increase the overall profit of the network.

After the mission leader collects the proposals, it tries first to satisfy the mission demands with sensors in the *not assigned* state by greedily picking sensors with highest utility. If these sensors are not sufficient, it tries to steal sensors from other ongoing missions, that is, it chooses from sensors in the *assigned* state. If the collected utility at that time is at least T, then the mission leader sends assignment messages to the respective sensors which start collecting information to support the mission. If a sensor is selected which preempts an existing mission, the following procedure is followed.

Let us say that a new mission M_j with leader L_j started in an area close to an ongoing mission M_k with leader L_k. If a sensor S_i that is currently assigned to M_k decides that its contribution will generate better profit if it is assigned to M_j, it notifies L_k of its intention. L_k then tries to find one or more sensors to replace S_i. If no such sensor(s) are found, the leader will agree on the reassignment as long as its current satisfaction level does not drop below T, which will cause the mission to fail. If the release of S_i will bring the allocated utility to lower than T, then the reassignment is temporarily denied. If sensor S_i is *critical* to the new mission M_j, that is, without it the mission will fail, a second test is performed. If the current profit value of M_j with S_i assigned to it is greater than that of M_k, the leader of M_k will release its hold on S_i and agree on the reassignment even if it will cause its own mission to fail. The reassignment becomes final once S_i is selected by M_j. Only at that time are the replacement sensor(s) activated. Algorithm 6.3 summarizes the steps taken by a mission leader and sensor S_i's response to the reception of an advertisement message.

Algorithm 6.3 Dynamic proposal

For leader of mission M_j:

INPUT: Set of proposing sensors G, $e_{ij} \forall S_i \in G$ and d_j
send advertisement of mission M_j
sort proposing sensors in decreasing order of e_{ij}
while $u_j < d_j$
  assign the next S_i (in sorted order) to M_j
  $u_j = u_j + e_{ij}$
OUTPUT: sensor assignment

For sensor S_i:

INPUT: Set of advertising missions Q and $e_{ij}, p_j, d_j \forall M_j \in Q$
receive advertisement of mission M_j
if S_i is not active **then**
  propose to M_j with offer e_{ij}
else if S_i is assigned to M_k with $\frac{e_{ik}}{d_k} \times p_k < \frac{e_{ij}}{d_j} \times p_j$ **then**
  ask current leader L_k for reassignment
  propose to M_j with offer e_{ij} only if current leader agrees
  if selected for mission M_j **then**
    notify leader of M_k to assign replacement(s)
  else
    continue operation on M_k
OUTPUT: mission M_j to which sensor S_i assigned

To reduce both the interruption of ongoing missions and the communication overhead, preemption is limited to one level. That is, if mission M_j preempted mission M_k, M_k will try to satisfy its demand with only available sensors and will not try to steal sensors that are already assigned. When a mission ends, the leader sends out a message to announce that the mission has ended and all assigned sensors are released.

Because the system is dynamic, missions that are not fully satisfied after the first assignment process will retry to obtain more sensors after some time. However, they only retry if there will be more available sensors. This can happen in the case when a nearby mission terminates and has its sensors released. This information can be learned either from the base station or by overhearing the message announcing the end of a mission.

We now discuss the runtime complexity and message complexity of Algorithm 6.3. Assume that the number of sensors is n and the number of missions is m, then the running time for sensor S_i is $O(m)$ as it may consider up to m missions. The reassignment takes constant time for the sensor. The mission leader's complexity, on the other hand, is $O(n \log n)$ as the mission leader has to sort up to n proposing sensors and select the best ones. Again, assuming that mission advertisement messages are only broadcast to immediate neighbors, the message complexity of the algorithm including both sides, the sensor and the mission, is $O(m + n)$ as there are m advertisement messages, $O(n)$ proposals by sensors, and $O(n)$ replies from mission leaders. As the algorithm does not need several rounds to complete, a saving of factor k in the number of messages sent by the sensors is realized over MRPA.

6.2.3.4 Energy-Aware Dynamic Proposal Algorithm

A drawback of DPA is that it does not consider the remaining energy in sensors when making assignment decisions. It selects a sensor based only on the utility it provides. However, the energy level of such a sensor may have been depleted over time and using it for sensing will consume its remaining energy leading to its death. This can happen while other sensors around it that provide lower utility may still have full energy. With this observation, DPA can be extended to make it energy-aware [called *energy-aware dynamic proposal algorithm* (EDPA)]. EDPA uses information about the proposing sensors current remaining energy level to make better assignment decisions that would ultimately lead to a longer lifetime.

Instead of using the utility of a sensor to the mission alone to make assignment decision, EDPA uses a function (f) of utility (U) and fraction of remaining energy (E). We define

$$f(U,E) = U \times E^{\beta}$$

where β is a design parameter. If β is zero, EDPA becomes DPA and hence only the utility is considered. A higher value of β gives more preference to sensors with more remaining energy.

To consume energy more evenly among sensors, after the initial assignment, possible sensor candidates for a mission send periodic updates to the leader including their current energy levels. The leader then checks if it has consumed energy unevenly among all sensors that can contribute to the mission. At that time, the leader may choose to change the assignments of sensors by reapplying the decision function f. This rotation of active sensors is similar to the technique used in LEACH to rotate cluster heads [5].

The periodic updates increase the communication overhead, but as is shown in the following section, this increase is not very high. As expected, this algorithm works better in a dense network in which there are many sensors that apply to a mission and hence more choices are available to the leader.

6.2.4 Performance Evaluation

The algorithms were evaluated using simulator built using Java and tested them using randomly generated problem instances [12]. Two sets of experiments were performed for the static and dynamic settings. For both cases the performance of the centralized greedy algorithm and the distributed algorithms were tested and compared. In the static setting, it is assumed that the entire problem instance, including all sensors and all missions, is given simultaneously. In the dynamic setting, missions arrive over time and depart after spending a certain amount of time being active. For the dynamic case, EDPA is shown to improve network lifetime by using information about remaining energy of sensors to make better selection decisions.

6.2.4.1 Assumptions

Each mission has a demand, an abstract value of the amount of sensing resources it requires, which is exponentially distributed with an average of 2 and a maximum of 6. Also associated with each mission is a profit value, which measures its importance. The profit is also exponentially distributed, but with an average of 1. This simulates common scenarios in which many missions demand few sensing resources and a smaller number demand more resources. The same applies to profit. The profit obtained from a successful mission M_j is equal to $p_j(u_j)$ as defined in Section 6.2.1. A mission is considered successful if it receives at least 50% of its demanded utility from allocated sensors (i.e., $T = 0.5$). Each sensor can only be assigned to a single mission.

The utility that sensor S_i provides to mission M_j is defined as a function of the distance D_{ij} between them. Many types of sensors exhibit some kind of quality deterioration or signal attenuation based on distance. In order to evaluate their utilities to missions, it is assumed that all sensors know their geographical locations. Formally, the potential utility contribution is

$$e_{ij} = \begin{cases} \frac{1}{1+D_{ij}^2/c}, & \text{if } D_{ij} \leq R_s \\ 0, & \text{otherwise} \end{cases} \tag{6.1}$$

where $R_s = 30$ m is the global sensing range. This models typical signal attenuation, which is an inverse function of the distance squared. Note that this utility function is only used for testing and is not meant to model the exact behavior of any sensor. In the following experiments $c = 60$.

Sensors are deployed in uniformly random locations in a 400 m × 400 m field (the base station is located in the center of the left edge). Missions also are created in uniformly random locations in the field. The communication range of sensors is set to 40 m.

As an upper bound the profit results for the optimal fractional solution are included. This is the optimal solution for the relaxed fractional problem in which sensors may divide their utility between multiple missions and all fractional profits are counted, regardless of whether the success threshold is reached or not. The algorithms' performance may be judged in comparison to this upper bound.

6.2.4.2 Static Scenarios

Setup: In this experiment all missions occur simultaneously, with the same start and end times. The number of sensors in the field is fixed to 500 and the number of missions varies from 10 to 100. The results show the average of 10 runs.

For the centralized approaches, results for the greedy algorithm are shown. For the distributed approach, results for MRPA are shown with one round, three rounds, and six rounds. These results illustrate the trade-off between solution quality and communication overhead. The growing threshold α for a mission to release sensors in MRPA is set to 10% in the first round and is increased by 10% for each subsequent round until it reaches T, or 50%. Advertisement messages are sent from mission leaders to all sensors within two hops.

Results: The first set of results (Figure 6.2) shows the fraction of the maximum mission profits achieved by the different algorithms compared to the optimal fractional profits. The maximum profit is the sum of all missions profits. The greedy centralized solution performs best followed by the six-round proposal. However, its advantage lessens as the density of the sensors increases. For a same-sized field with 1000 sensors deployed (not shown in the figures), all the curves except the one-round proposal become nearly aligned. This is expected since there are more sensors that can be assigned to the different missions. Note that the improvement in MRPA when going from a single round to three rounds is very pronounced. However, the improvement gained when jumping to six rounds is less apparent and may not justify the necessary communications overhead.

Figure 6.3 shows the communication overhead of the different algorithms. As expected, the centralized algorithm has the highest overhead. With MRPA, as the number of rounds increase more messages are exchanged. The savings in the number of exchanged messages become more evident if a dynamic system, in which sensor-mission utility values can change over time, is considered.

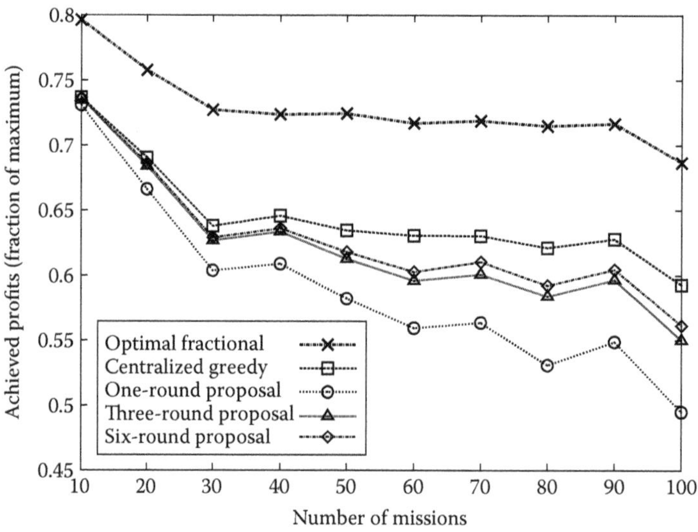

Figure 6.2 Achieved profits.

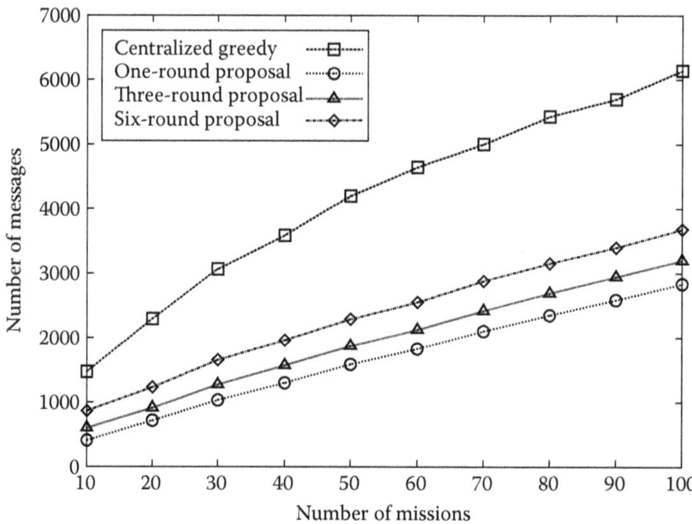

Figure 6.3 Number of messages.

In this case, the centralized algorithm needs to collect information about current sensor utility values before running an assignment process for each new mission that arrives. Distributed algorithms, on the other hand, require the exchange of fewer messages since information about utility values only needs to be sent to the leader of the new mission, which is just a few hops away.

As the number of sensors and missions increase, the distributed algorithms encounter greater overlap in the areas of local assignment, which leads to more exchanged messages. This is true because as densities of sensors and missions increase more sensors can contribute to each mission and at the same time each sensor can contribute to more missions.

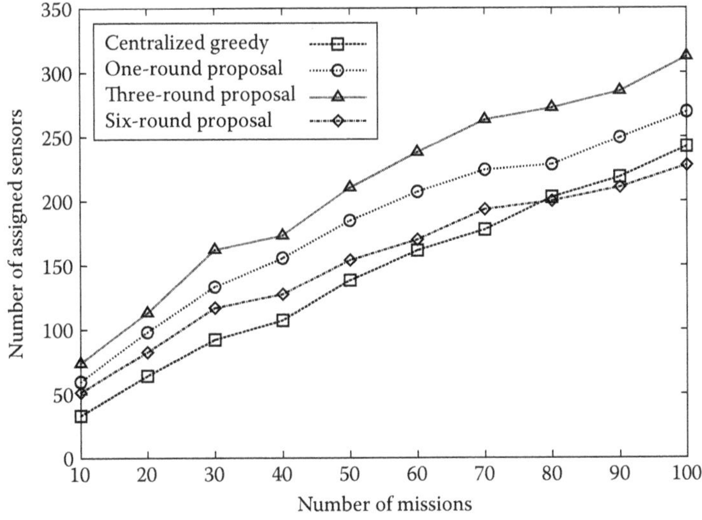

Figure 6.4 Number of assigned sensors.

The number of sensors assigned can be used as a proxy for the amount of energy used (shown in Figure 6.4). The number of sensors used by the centralized algorithm is very close to that of the six-round proposal algorithm. For MRPA, few sensors are assigned if one round is used. This is because this setting does not allow sensors that were rejected in this round to repropose to other missions. With three rounds, the mission leaders release sensors that are not useful which allows these sensors to repropose to other missions and hence the number of assigned sensors becomes higher. With six rounds most of the unused sensors are released which brings the number of the assigned sensors down to about that of the centralized algorithm. Note that the algorithms do not use all the available sensors even when the number of missions is large. This is because some sensors are not within the sensing range of any mission and hence remain idle. When considering the results in this figure we should take into consideration the achieved profits in Figure 6.2. For example, even though the centralized greedy algorithm assigns close to 250 sensors when 100 missions are present, it achieves less than 60% of the possible profits.

Finally, Figure 6.5 shows the fraction of satisfied missions for the different algorithms. Note that the goal is not to maximize this number, but rather to achieve the highest profit. The centralized algorithm is successful in achieving the highest profit values, but not always the largest fraction of satisfied missions. The six-round proposal achieves higher fraction when the number of missions is large. This happens because the greedy centralized algorithm assigns sensors to missions in order of profit and hence may stop satisfying missions after a certain point because sensors are no longer available. So, even though the fraction of satisfied missions is less than that of the six-round proposal, the amount of profits is higher as more profitable missions were picked. Because of its lack of global view, the six-round proposal algorithm is satisfying more missions, but with lower profits. Between the three multi-round algorithms tested, there is a significant increase in the fraction of satisfied missions between one round and three rounds and less improvement between three rounds and six rounds.

From the results obtained, we can see that the distributed algorithms perform well. The difference in achieved profit values compared to the centralized algorithm is less than 8%. At the same time, it saves as much as 50% of the transmitted messages.

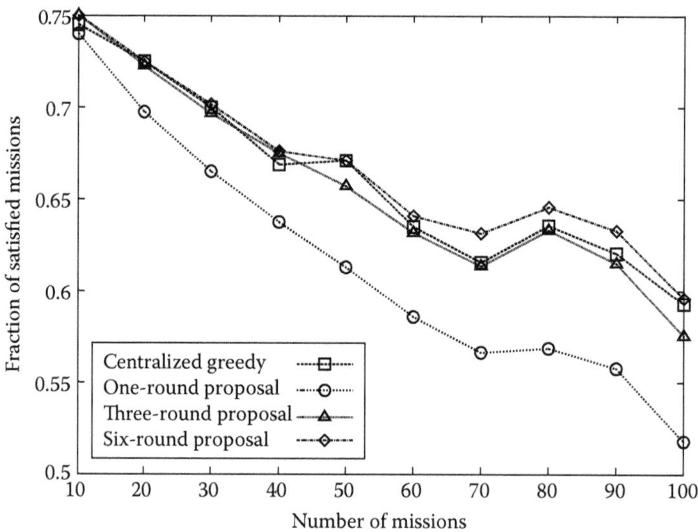

Figure 6.5 Fraction of satisfied missions.

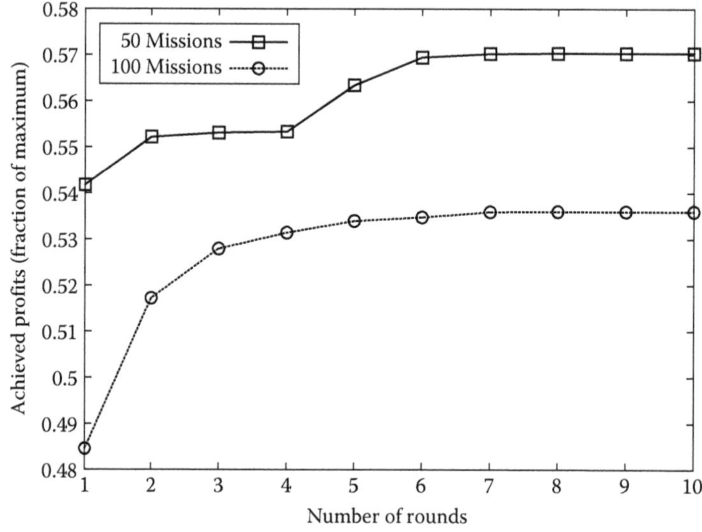

Figure 6.6 Achieved profits vs. rounds.

Figure 6.6 shows the relation between achieved profits and number of rounds. The number of sensors is fixed to 500 and missions to 50 or 100 missions (depending on the experiment). As can be expected, achieved profits initially increase with the number of rounds. However, the additional gains beyond eight rounds are small since by that time all attainable missions have reached their success threshold. Figure 6.7 shows the communication overhead that increases linearly with the number of rounds.

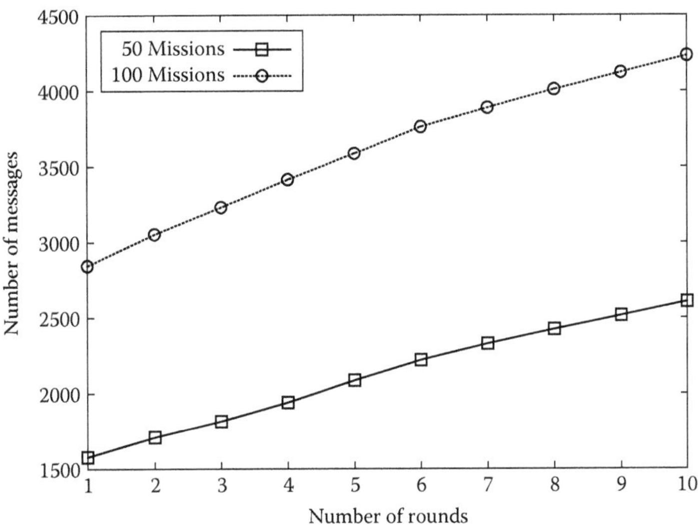

Figure 6.7 Number of messages vs. rounds.

6.2.4.3 Dynamic Scenarios

Setup: Now we discuss the performance of the distributed proposal algorithm in a dynamic setting in which missions arrive over time. The same aforementioned assumptions apply here. Moreover, it is assumed that missions arrive according to a Poisson distribution. The mission lifetimes are selected according to an exponential distribution with an average lifetime of 1 h and a maximum of 4 h. The exponential distribution is heavy-tailed that models realistic scenarios in which there are many short-lived missions and few long-lived ones. Although a centralized algorithm is impractical in dynamic scenarios, due to high communication cost, its results are included to measure the performance of the distributed algorithm. The centralized algorithm, in this case, is rerun for each mission arrival and departure.

Results: The performance of DPA is compared to results achieved by the greedy centralized algorithm and the optimal fractional solution. Figures 6.8 and 6.9 show a trace of the achieved network profits during a period of 12 h for arrival rate, $\lambda = 3$ missions/h and 6 missions/h respectively, with a network of 500 sensors. The simulation is started at time zero and the trace is collected after 10 h to allow the network to reach steady state. As can be seen from the figures, the performance of DPA, which utilizes local information about missions, is very close to that of the centralized algorithm.

Figures 6.10 and 6.11 show the average performance over a period of 50 h (averaged over 10 runs) for a network with 500 sensors and 1000 sensors. Figure 6.10 shows the average achieved profits per unit of time (fraction of maximum) as the mission arrival rate is varied. We see that both the centralized and DPA perform almost equally. Note that, as expected, with a larger number of sensors the network can achieve higher profits.

The communication overhead of DPA is shown in Figure 6.11. The number of messages grows linearly as the number of missions in the network increases. The number of messages for the centralized algorithm is not shown as this value will be very large compared to DPA. For each mission arrival and departure, the base station needs to collect information from all sensors that can contribute to the arriving mission to get status updates. The average number of messages exchanged

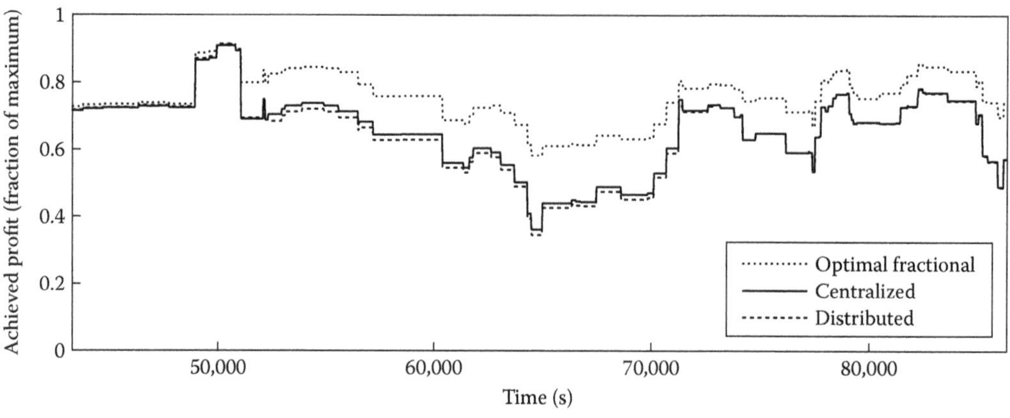

Figure 6.8 Trace of network performance (λ = 3 missions/h).

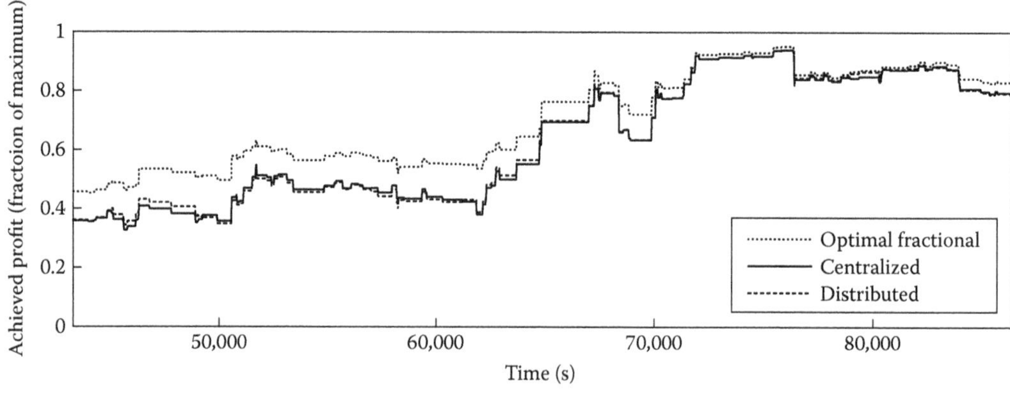

Figure 6.9 Trace of network performance (λ = 6 missions/h).

per mission is around 40 for 500 sensors and 80 for 1000 sensors. This includes all the messages needed to advertise the mission and makes all the assignment decisions including reassignments.

Network lifetime: Figures 6.12 through 6.14 show the results for EDPA in a network of 500 sensors and 1000 sensors (average of 10 runs). The mission arrival rate is set to 6 missions/h. All sensors start with energy to support 10 h of continuous sensing. Only energy consumed for sensing is considered. Sensor reassignment is performed every 20 min to balance energy consumption. Choosing a smaller period may yield a more uniform assignment, but will have a larger communication overhead.

The network lifetime is defined here as the time until the first sensor dies. Figure 6.12 shows the lifetime of the network for different values of β, a parameter used to control dependence on remaining energy. Recall that when $\beta = 0$, EDPA becomes DPA. The results show that when EDPA is used, network lifetime increases by 50% and 70%, for networks with 500 sensors and 1000 sensors, respectively. The increase is notable when β goes from 0 to 1, that is when energy is taken into account. After that point, the increase in lifetime is not very pronounced. The denser the network, the more options the assignment algorithm has and hence it is able to achieve longer lifetime.

Figure 6.13 shows the total achieved profits (sum of profit in every second during lifetime) for the different values of β. This can be thought of as the area under the curves for the two algorithms

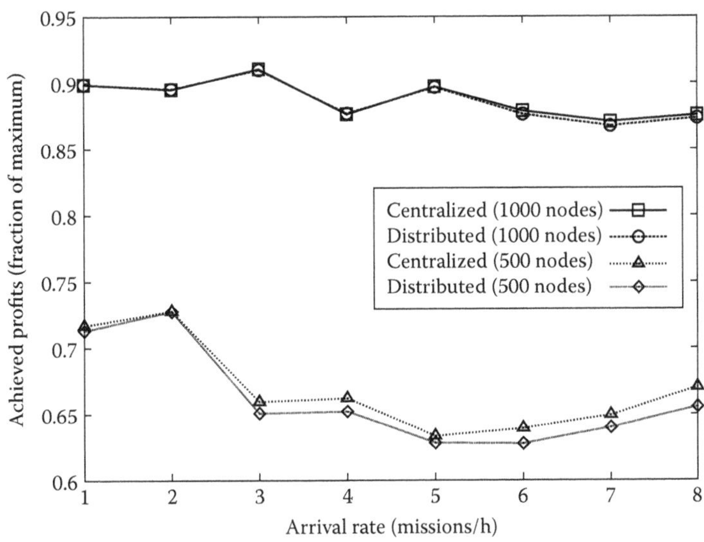

Figure 6.10 Average achieved profits (per unit of time) in period of 50 h.

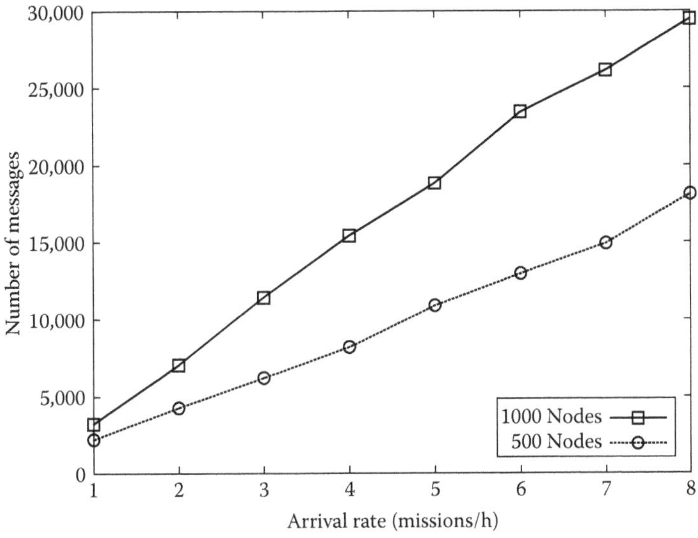

Figure 6.11 Number of exchanged messages in period of 50 h.

shown in Figures 6.8 and 6.9. As expected, the profits increase when lifetime increases. But when there are more sensors, the profit increase in more prominent due to the longer lifetime and the fact that more sensors allow for more satisfaction for missions.

Due to periodic updates, the communication overhead increases when EDPA is used ($\beta > 0$). Figure 6.14 shows the average number of messages per each attempted mission. Note that there is an average increase of around 50% for both network sizes. Although the percentage increase may seem high, the actual number of exchanged messages per mission (around 60 for 500 sensors and

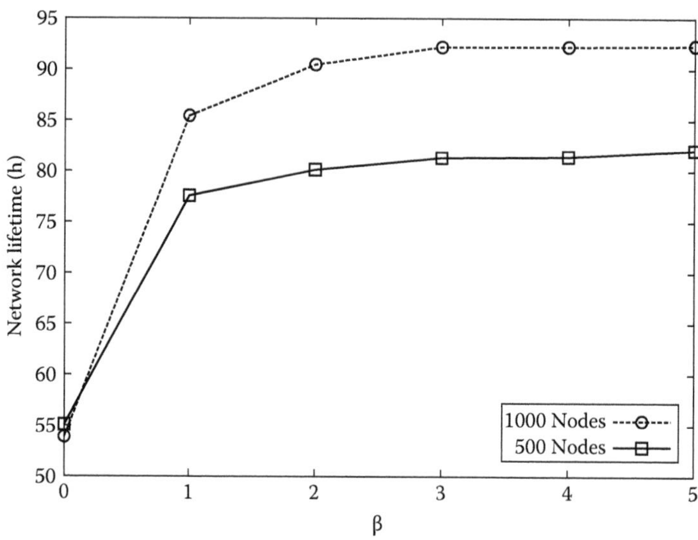

Figure 6.12 Network lifetime (in hours) using EDPA.

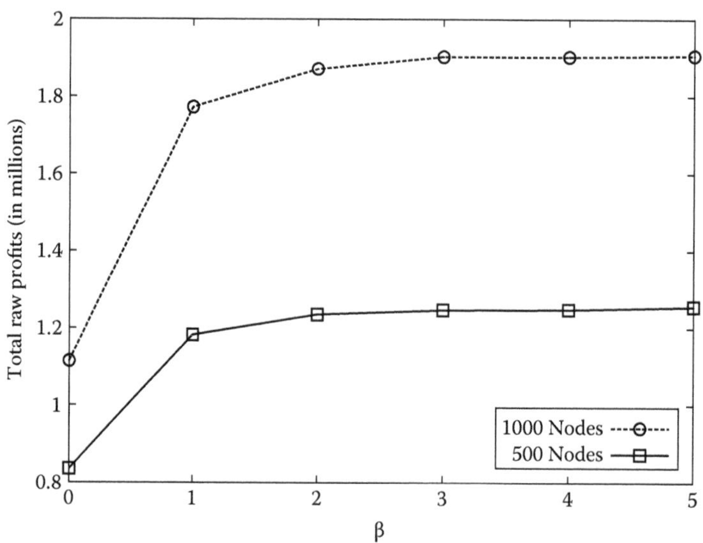

Figure 6.13 Total raw profits using EDPA.

120 for 1000 sensors) is relatively small, especially considering that this number includes all the messages exchanged to setup a mission and is amortized over its lifetime.

6.3 Sensor Assignment in Budget-Constrained Environments

In this section and the next, we examine other variants of sensor-assignment problems motivated by conservation of resources [6]. Again, we consider two broad classes of environments: static and dynamic (see Section 6.4). The *static* setting is motivated by situations in which different users are

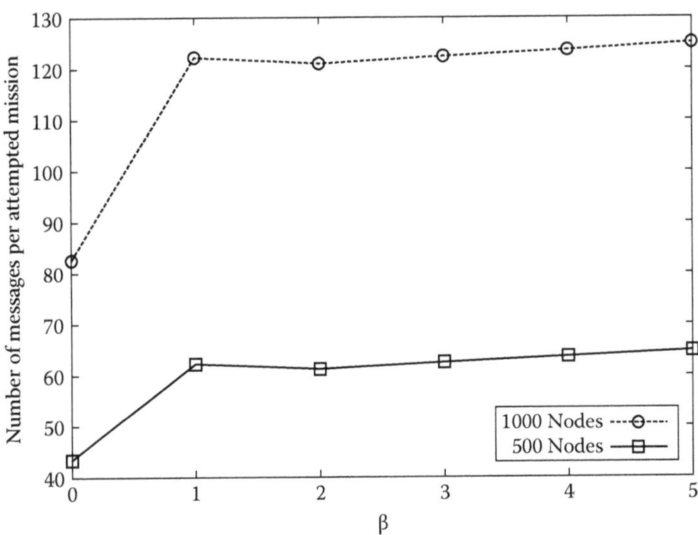

Figure 6.14 Messages per mission using EDPA.

granted control over the sensor network at different times. During each time period, the current user may have many simultaneous missions. While the current user will want to satisfy as many of these as possible, sensing resources may be limited and expensive, both in terms of equipment and operational cost. In some environments, replacing batteries may be difficult, expensive, or dangerous. Furthermore, a sensor operating in active mode (i.e., assigned to a mission) may be more visible than a dormant sensor, and so is in greater danger of tampering or damage. Therefore, we give each mission in the static problem a budget so that no single user may overtax the network and deprive future users of resources. This budget serves as a constraint in terms of the amount of resources that can be allocated to a mission regardless of profit.

In this section, we discuss an efficient greedy algorithm and an MRPA whose subroutine solves a generalized assignment problem (GAP). The performance evaluation results show that in dense networks both algorithms perform well, with the GAP-based algorithm slightly outperforming the greedy algorithm. This section is based on [6].

6.3.1 Problem Definition

With multiple sensors and multiple missions, sensors should be assigned in an intelligent way. This goal is shared by all the problem settings we consider. There are a number of attributes, however, that characterize the nature and difficulty of the problem.

The static setting is similar to the one presented in the Section 6.2. Given is a set of sensors $S_1, \ldots, S_n$ and a set of missions $M_1, \ldots, M_m$. Each mission is associated with a utility demand d_j, indicating the amount of sensing resources needed, and a profit p_j, indicating the importance of the mission. Each sensor–mission pair is associated with a utility value e_{ij} that mission j will receive if sensor i is assigned to it. This can be a measure of the quality of information that a sensor can provide to a particular mission. To simplify the problem, it is again assumed that the utility values e_{ij} received by a mission j are additive (similar to Section 6.2). Finally, a budgetary restriction is given in some form, either constraining the entire problem solution or constraining individual missions as follows: each mission has a budget b_j, and each potential sensor assignment has cost c_{ij}. All the aforementioned values are positive reals, except for costs and utility, which could be zero. The most general problem is defined by the following MP **P**:

$$\begin{aligned}
\textit{Maximize:} \quad & \textstyle\sum_{j=1}^{m} p_j(y_j) \\
\textit{Such that:} \quad & \textstyle\sum_{i=1}^{n} x_{ij}e_{ij} \geq d_j y_j, \text{ for each } M_j, \\
& \textstyle\sum_{i=1}^{n} x_{ij}c_{ij} \leq b_j, \text{ for each } M_j, \\
& \textstyle\sum_{j=1}^{m} x_{ij} \leq 1, \text{ for each } S_i, \\
& x_{ij} \in \{0, 1\} \ \forall x_{ij} \text{ and} \\
& y_j \in [0, 1] \ \forall y_j
\end{aligned}$$

A sensor can be assigned ($x_{ij} = 1$) at most once. Profits are received per mission based on its satisfaction level y_j. Note that y_j corresponds to u_j/d_j within the range [0,1] where $u_j = \sum_{i=1}^{n} x_{ij}e_{ij}$. With *strict* profits, a mission receives exactly profit p_j iff $u_j \geq d_j$. With *fractional* profits, a mission receives a fraction of p_j proportional to its satisfaction level y_j and at most p_j. More generally, similar to the problem introduced in Section 6.2.1, profits can be awarded fractionally, after reaching a fractional satisfaction threshold T:

$$p_j(u_j) = \begin{cases} p_j, & \text{if } u_j \geq d_j \\ p_j \cdot u_j/d_j, & \text{if } T \leq u_j/d_j \\ 0, & \text{otherwise} \end{cases}$$

When $T = 1$, program **P** is an IP; when $T = 0$, it is a mixed IP with the decision variables x_{ij} still integral.

The edge values e_{ij} may be arbitrary non-negative values, or may have additional structure. If sensors and missions lie in a metric space, such as the line or plane, then edge values may be based in some way on the distance D_{ij} between sensor i and mission j. In the *binary sensing* model, e_{ij} is equal to 1 if distance D_{ij} is at most the sensing range R_s, and 0 otherwise. In another geometric setting, e_{ij} may vary smoothly based on distance, according to a function such as $1/(1 + D_{ij})$.

Similarly, the cost values c_{ij} could be arbitrary or could exhibit some structure: the cost could depend on the sensor involved, or could, for example, correlate directly with distance D_{ij} to represent the difficulty of moving a sensor to a certain position. It could also be unit, in which case the budget would simply constrain the number of sensors.

Even if profits are unit, demands are integers, edge values are 0/1, and budgets are infinite, then this problem is NP-hard and as hard to approximate as maximum independent set as shown in [12].

6.3.2 Algorithms

In this section, let us describe two algorithms to solve the static-assignment problem: *greedy* and *multi-round generalized assignment problem* (*MRGAP*). The former requires global knowledge of all missions to run and hence is considered centralized, whereas the latter can be implemented in both centralized and distributed environments, a benefit in the sensor network domain.

6.3.2.1 Greedy

The first algorithm we consider (Algorithm 6.4) is a greedy algorithm that repeatedly attempts the highest-potential-profit untried mission. Because fractional profits are awarded only beyond the threshold percentage T, this need not be the mission with maximum p_j. For each such mission, sensors are assigned to it, as long as the mission budget is not yet violated, in decreasing order of cost-effectiveness, that is, the ratio of edge utility for that mission and the sensor cost. The running time of the algorithm is $O(n(m + \log n))$. No approximation factor is given for this efficiency-motivated algorithm since, even for the first mission selected, there is no guarantee that its feasible solution will be found. This by itself is, after all, an NP-hard 0/1 Knapsack problem.

6.3.2.2 Multi-Round GAP

The idea of the second algorithm (shown in Algorithm 6.5) is to treat the missions as knapsacks that together form an instance of the GAP. The strategy of this algorithm is to find a good solution for the problem instance *when treated as GAP*, and then to do postprocessing to enforce the lower bound constraint of the profit threshold, by removing missions whose satisfaction percentage is too low. Releasing these sensors may make it possible to satisfy other missions, which suggest a series of rounds. In effect, missions not making good progress toward satisfying their demands are precluded from competing for sensors in later rounds.

Cohen et al. [4] give an approximation algorithm for GAP that uses a knapsack algorithm as a subroutine. If the knapsack subroutine has approximation guarantee $\alpha \geq 1$, then the Cohen GAP algorithm offers an approximation guarantee of $1 + \alpha$. The standard knapsack FPTAS [15] is used, which yields a GAP approximation guarantee of $2 + \epsilon$. The post-processing step is used to enforce lower bounds on profits for the individual knapsacks. This is an essential feature of the sensor-assignment problem that is not considered by GAP.

Algorithm 6.4 Greedy algorithm for budget-constrained SMD

INPUT: S, M, $e_{ij}, c_{ij} \forall (S_i, M_j) \in S \times M$ and $p_j, d_j, b_j \forall M_j \in M$
while true **do**
  for each available mission M_j **do**
    $u_j \leftarrow \sum_{S_i(unused)} e_{ij}$
  $j \leftarrow \arg\max_j p_j(u_j)$
  if $p_j(u_j) = 0$ **then break**
  $u_j \leftarrow 0$
  $c_j \leftarrow 0$
  for each unused S_i in decreasing order of e_{ij}/c_{ij} **do**
    if $u_j \geq d_j$ or $e_{ij} = 0$ **then break**
    if $c_j + c_{ij} \leq b_j$
      assign S_i to M_j
      $u_j \leftarrow u_j + e_{ij}$
      $c_j \leftarrow c_j + c_{ij}$
OUTPUT: sensor assignment

Algorithm 6.5 Multi-round GAP algorithm for budget-constrained SMD

INPUT: $S,\ M,\ e_{ij}, c_{ij} \forall (S_i, M_j) \in S \times M,\ p_j, d_j, b_j \forall M_j \in M$, and T
while true **do**
  initialize set of remaining missions $M \leftarrow \{M_1 \ldots M_m\}$
  for $t = 0$ to T step 0.05 **do**
    run the GAP subroutine on M and the *unassigned sensors*
    in the resulting solution, release any superfluous sensors
    if M_j's satisfaction level is $< T$, for any j **then**
      release all sensors assigned to M_j
      $M \leftarrow M - \{M_j\}$
    if M_j is completely satisfied OR has no remaining budget, for any j
      $M \leftarrow M - \{M_j\}$
OUTPUT: sensor assignment

The algorithm works as follows: The threshold is initialized to a small value, for example, 5%. In each round, we run the GAP algorithm of [4] as a subroutine and find the solution based on the remaining sensors and missions. After each round, missions not meeting the threshold are removed, and their sensors are released. Any sensors assigned to a mission that has greater than 100% satisfaction, and which can be released without reducing the percentage below 100%, are released. (Such sensors are *superfluous.*) Sensors assigned to missions meeting the threshold remain assigned to those missions. These sensors will not be considered in the next round, in which the new demands and budgets of each mission will become the remaining demand and the remaining budget of each one of them. Finally, the threshold is incremented, with rounds continuing until all sensors are used or all missions have succeeded or been removed.

The GAP instance solved at each round is defined by the following linear program:

$$\begin{aligned}
\textit{Maximize:}\quad & \textstyle\sum_j \sum_i p_{ij} x_{ij} \text{ (with } p_{ij} = p_j e_{ij} / \hat{d}_j) \\
\textit{Such that:}\quad & \textstyle\sum_{S_i \text{unused}} x_{ij} c_{ij} \le \hat{b}_j, \text{ for each remaining } M_j, \\
& \textstyle\sum_{M_j \text{remaining}} x_{ij} \le 1, \text{ for each unused } S_i, \text{ and} \\
& x_{ij} \in \{0, 1\}\ \forall x_{ij}
\end{aligned}$$

Here $\hat{d}_j$ is the remaining demand of M_j, that is, the demand minus utility received from sensors assigned to it during previous rounds. Similarly, $\hat{b}_j$ is the remaining budget of M_j. The concepts of demand and profit are encoded in the gap model as $p_{ij} = p_j \cdot e_{ij} / \hat{d}_j$. This parameter represents the fraction of demand satisfied by the sensor, scaled by the priority of the mission. In each GAP computation, an assignment of sensors that maximizes the total benefit brought to the demands of the remaining mission is sought.

One advantage of MRGAP is that it can be implemented in a distributed fashion. For each mission there can be a sensor, close to the location of the mission, that is responsible for running the assignment algorithm. Missions that do not contend for the same sensors can run the knapsack algorithm simultaneously. If two or more missions contend for the same sensors, that is, they are within distance $2R_s$ of one other, then synchronization of rounds is required to prevent them from running the knapsack algorithm at the same time. To do this, one of the missions (e.g., the one with the lowest id) can be responsible for broadcasting a synchronization message at the beginning of each new round. However, since R_s is typically small compared to the size of the field, it can be expected that many missions will be able to do their computations simultaneously.

The total running time of the algorithm depends on the threshold T and the step value chosen, as well as on the density of the problem instance, which will determine to what degree the knapsack computations in each round can be parallelized.

6.3.3 Performance Evaluation

In this section, we discuss the performance evaluation results that were generated using a custom-built Java simulator tested with randomly generated problem instances [6].

6.3.3.1 Simulation Setup

In this experiment, all missions occur simultaneously. It is assumed that mission demands are chosen from an exponential distribution with an average of 2 and a minimum of 0.5. Profits for the different missions are also exponentially distributed with an average of 10 and a maximum of 100. This simulates realistic scenarios in which many missions demand few sensing resources and a smaller number demand more resources. The same applies to profit. The simulator filters out any mission that is not individually satisfiable, that is, satisfiable in the absence of all other missions. For a sufficiently dense network, however, it can be expected that there will be few such impossible missions. Nodes are deployed in uniformly random locations in a 400 m × 400 m field. Missions are created in uniformly random locations in the field. The communication range of all sensors is 40 m.

The utility of a sensor S_i to a mission M_j is defined as a function of the distance D_{ij} between them. In order for sensors to evaluate their utilities to missions, it is assumed that all sensors know their geographical locations. Formally the utility is

$$e_{ij} = \begin{cases} \frac{1}{1+D_{ij}^2/c}, & \text{if } D_{ij} \leq R_s \\ 0, & \text{otherwise} \end{cases}$$

where R_s is the global sensing range. This follows typical signal attenuation models in which signal strength decays inversely with distance squared. In the experiments, $c = 60$ and $R_s = 30$ m.

The number of sensors in the field is fixed and the number of missions is varied from 10 to 100. Each sensor has a cost, chosen uniformly at random from $[0, 1]$, which does not depend on the mission it is assigned to. This can represent the sensor's actual cost in real money or, for example, a value indicating the risk of discovery if the sensor is activated. Each mission has a budget drawn from a uniform distribution with an average of 3 in the first experiment and varying from 1 to 10 in the second.

6.3.3.2 Results

The first series of results show the fraction of the maximum mission profits (i.e., the sum of all missions profits) achieved by the different algorithms. We see profit results for the greedy algorithm, MRGAP, and an upper bound on the optimal value running on two classes of sensor networks, sparse (250 nodes) and dense (500 nodes) (Figures 6.15 and 6.16, respectively). The upper bound

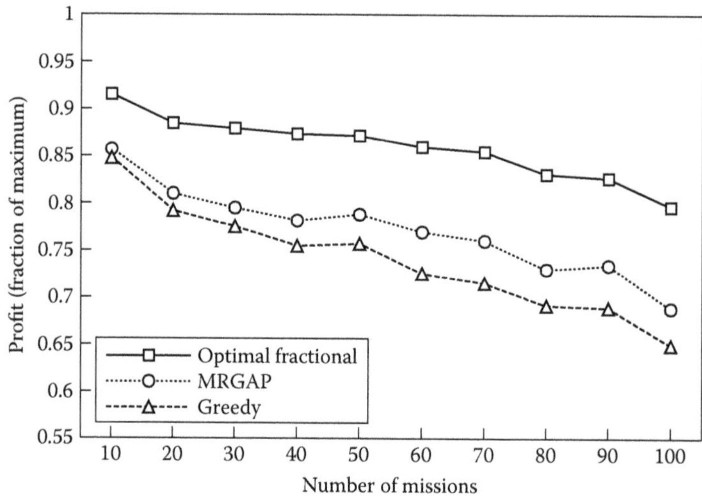

Figure 6.15 Fraction of maximum profit achieved (250 nodes)

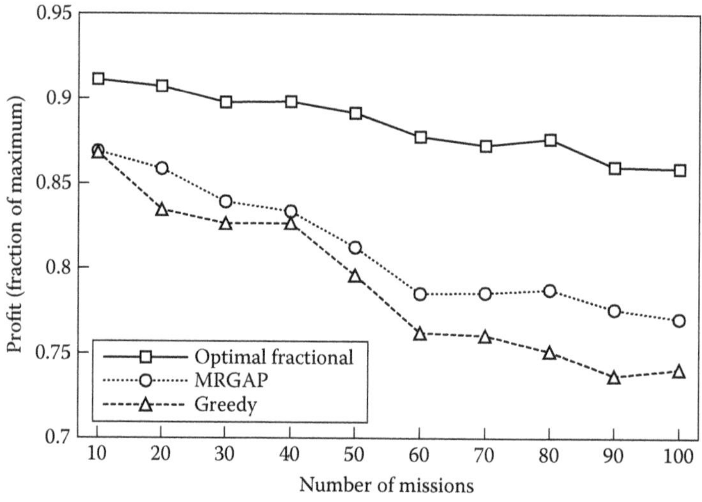

Figure 6.16 Fraction of maximum profit achieved (500 nodes)

on the optimal is obtained by solving the LP relaxation of program **P**, in which all decision variables are allowed to take on fractional values in the range [0, 1], and the profit is simply fractional based on satisfaction fraction, that is, $p_j y_y$ for mission M_j with no attention paid to the threshold T. The actual optimal value will be lower than the fractional one.

The MRGAP algorithm, which recall can be implemented in a distributed fashion, achieves higher profits in all cases than does the greedy algorithm, which is strictly centralized (because missions have to be ordered in terms of profit). The difference, however, is not very large. It can be noted that with 500 nodes the network is able to achieve higher profits which is expected.

Figure 6.17 shows the fraction of the total budget each algorithm spent to acquire the sensing resources it did in a network with 250 nodes. The MRGAP algorithm achieves more profit than the

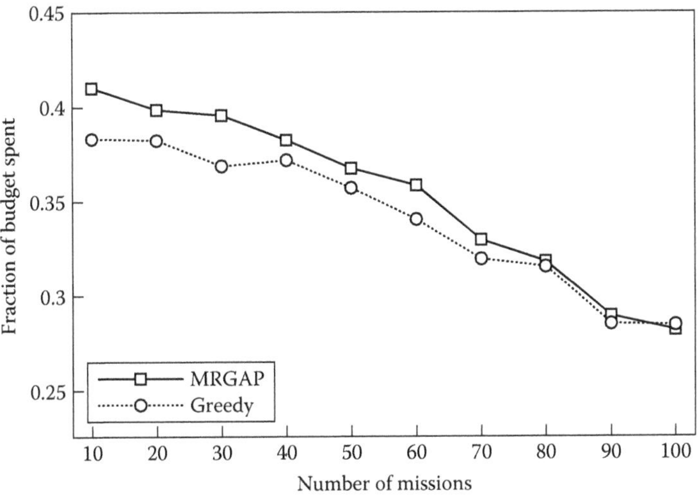

Figure 6.17 Fraction of spent budget (250 nodes).

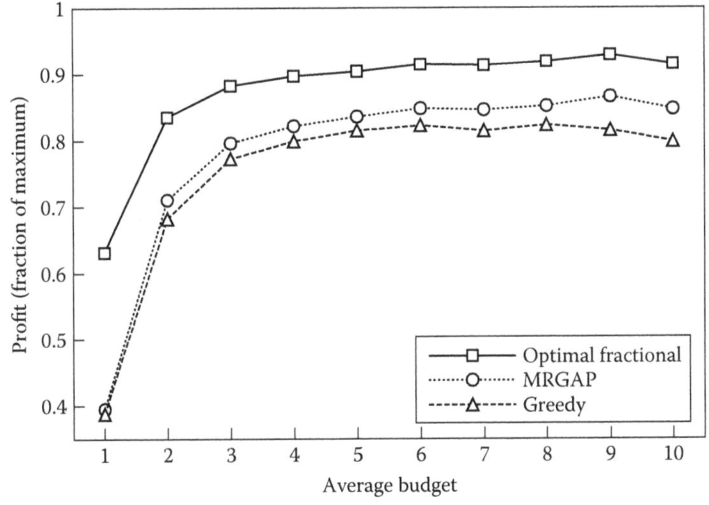

Figure 6.18 Varying the average budget (250 nodes).

greedy algorithm and spends a modest amount of additional resources. The fraction of remaining budget is significant (more than 60% in almost all cases), which suggests either that successful missions had higher budgets than they could spend on available sensors or that unsuccessful missions had lower budgets than necessary and hence they were not able to reach the success threshold and so their budgets were not spent. When the number of missions is large, this can be attributed to the fact that there simply were not enough sensors due to high competition between missions.

Another set of experiments, in which the number of missions was fixed at 50 and average budget given to missions was varied from 1 to 10, are also performed. Figure 6.18 shows the results for a network with 250 nodes. We observe that the achieved profit initially increases rapidly with the budget size but slows as animating influence shifts from budget limitations to competition between missions. We observe the same pattern: MRGAP achieves highest profits followed closely by the greedy algorithm. Figures 6.19 and 6.20 show the same results when 500 nodes are deployed in the same area.

From these results we can see that in dense networks both algorithms perform well, with the GAP-based algorithm slightly outperforming the greedy.

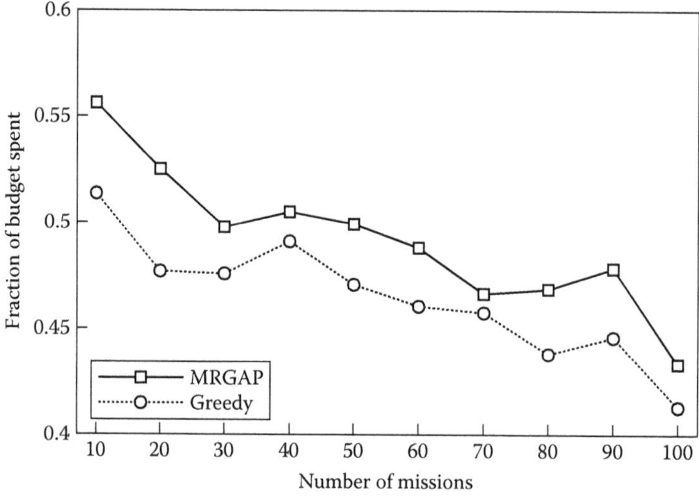

Figure 6.19 Fraction of spent budget (500 nodes).

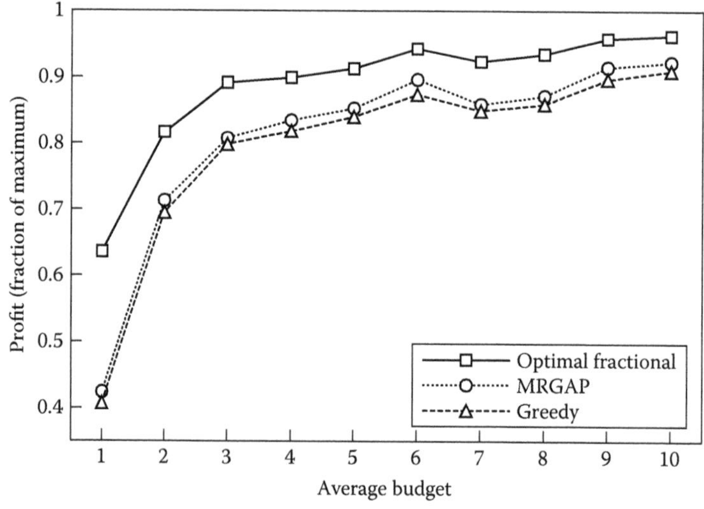

Figure 6.20 Varying the average budget (500 nodes).

6.4 Sensor Assignment in Lifetime-Constrained Environments

In this section, we focus on lifetime-constrained environments. In this *dynamic* setting, missions may start at different times and have different durations. In these, cases, explicit budgets, like the ones discussed in Section 6.3, may be too restrictive because the network must react to new missions given the current operating environment, that is, the condition of the sensors will change over time. Instead, battery lifetime is used as a means of discouraging excessive assignment of sensors to any one mission, evaluating trade-offs between the relative value of a given assignment and the expected profit earned with each sensor (given the mission statistics).

There are two cases in this dynamic setting. In the first case, no advanced knowledge of the target network lifetime is assumed, that is, we do not know for how long the network will be required to operate. This is called the *general dynamic setting.* In the second case, it is assumed that the system operator has knowledge of the target network lifetime, that is, the network is needed for a finite duration. This is called the *dynamic setting with a time horizon.* In the following algorithms, the aggressiveness with which sensors accept new missions are adjusted based on trade-offs between target network lifetime, remaining sensor energy, and mission profit, rather than using hard budgets.

In the following we discuss distributed algorithms that adjust sensors' eagerness to participate in new missions based on their current operational status and the target network lifetime (if known). This section is based on [6].

6.4.1 Problem Definition

In this dynamic setting, we no longer consider explicit budgets for missions and explicit costs for sensor assignments which were used in the Section 6.3. We keep, however, the same network model and the basic structure of the problem. In this case, what constraints the assignment problem is the limited energy that sensors have. The other essential change is the introduction of a time dimension. In this setting, each sensor has a battery size B, which means that it may only be used for at most B timeslots over the entire time horizon. Missions may arrive at any point in time and may last for any duration.

If a sensor network is deployed with no predetermined target lifetime, then the goal may be to maximize the profit achieved by each sensor during its own lifetime. However, if there is a finite target lifetime for the network, then the goal is to earn the maximum total profits over the entire time horizon. The profit for a mission that lasts for multiple timeslots is the sum of the profits earned over all timeslots during the mission's lifetime.

The danger in any particular sensor assignment is then that the sensor in question might somehow be better used at a later time. Therefore, the challenge is to find a solution that competes with an algorithm that knows the characteristics of all future missions before they arrive. The general dynamic problem is specified by the following MP $\mathbf{P}'$:

$$
\begin{aligned}
\textit{Maximize:}\quad & \textstyle\sum_t \sum_{j=1}^m p_j(y_{jt}) \\
\textit{Such that:}\quad & \textstyle\sum_{i=1}^n x_{ijt} e_{ij} \geq d_j y_{jt}, \text{ for each } M_j \text{ and } t, \\
& \textstyle\sum_{j=1}^m x_{ijt} \leq 1, \text{ for each } S_i \text{ and time } t, \\
& \textstyle\sum_t \sum_{j=1}^m x_{ijt} \leq B, \text{ for each } S_i, \\
& x_{ijt} \in \{0,1\}\ \forall x_{ijt} \text{ and} \\
& y_{jt} \in [0,1]\ \forall y_{jt}
\end{aligned}
$$

If *preemption* is allowed, that is, a new mission is allowed to preempt an ongoing mission and grab some of its sensors, then in each timeslot sensors that are assigned to other missions can be freely reassigned based on the arrival of new missions, without reassignment costs. In this case, a long mission can be thought of as a series of unit-time missions, and so the sensors and missions at each timeslot form an instance of the NP-hard static problem. If preemption is forbidden, then the situation for the online algorithm is in a way simplified. If we assume without loss of generality that no two missions will arrive at exactly the same time, then the online algorithm can focus on one mission at a time. Nonetheless, the dynamic problem remains as hard as the static problem, since a reduction can be given in which the static missions are identified with dynamic missions of unit length, each starting ϵ after the previous one. In fact, we can give a stronger result covering settings both with and without preemption.

6.4.2 Algorithms

In this section, we discuss heuristic-based algorithms that intelligently and dynamically assign sensors to missions. These heuristics are similar in operation to the dynamic proposal algorithm discussed in Section 6.2, but with a new focus. Rather than maximizing profit by trying to satisfy all available missions, the focus here is on maximizing the profit over network lifetime by allowing the sensors to refuse participation in missions they deem not worthwhile.

Missions are dealt with as they arrive. A *mission leader*, a node that is close to the mission's location, is selected for each mission. The mission leaders are informed about their missions' demands and profits by a base station. They then run a local protocol to match nearby sensors to their respective missions. Since the utility a sensor can provide to a mission is limited by sensing range, only nearby nodes are considered. The leader advertises its mission information (demand and profit) to the nearby nodes (e.g., two-hop neighbors).

When a nearby sensor hears such an advertisement message, it makes a decision either to propose to the mission and become eligible for selection by the leader or to ignore the advertisement. The decision is based on the current state of the sensor (and the network if known) and on potential contribution to mission profit that the sensor would be providing. Knowledge of the (independent) distributions of the various mission properties (namely, demand, profit, and lifetime) is needed to make proper assignment decisions. Such information can be learned from historical data. To determine whether a mission is worthwhile, a sensor considers a number of factors:

- Mission's profit, relative to the maximum profit
- Sensor's utility to the mission, relative to the mission's demand
- Sensor's remaining battery level
- Remaining target network lifetime, if known

After gathering proposals from nearby sensors, the leader selects sensors based on their utility offers until it is fully satisfied or there are no more sensor offers. The mission (partially) succeeds if it reaches the success threshold; if not, it releases all sensors.

Since it is assumed that all distributions are known, the share of mission profit potentially contributed by the sensor (i.e., if its proposal is accepted) can be compared to the expectation of this value. Based on previous samples, the expected mission profit $E[p]$ and demand $E[d]$ can be estimated. Also, knowing the relationship between sensor–mission distance and edge utility, and assuming a uniform distribution on the locations of sensors and missions, the expected utility contribution $E[u]$ that a sensor can make to a typical mission *in its sensing range* can be computed.

The following expression is used to characterize the expected partial profit a sensor provides to a typical mission:

$$E\left[\frac{u}{d}\right] \times \frac{E[p]}{P} \tag{6.2}$$

We consider two scenarios. In the first, the target network lifetime is unknown, that is, we do not know for how long will the network be needed. In this case, sensors choose missions that provide higher profit than the expected value and hence try to last longer in anticipation of future high profit missions. In the second, the target network lifetime is known, that is, we know the duration for which the network will be required. In this case, sensors take the remaining target network lifetime into account along with their expected lifetime when deciding whether to propose to a mission or not. In the following we describe solutions to these two settings.

6.4.2.1 Energy-Aware Algorithm

In this algorithm, the target lifetime of the sensor network is unknown. For a particular sensor and mission, the situation is characterized by the actual values of mission profit (p) and demand (d) and by the utility offer (u), as well as the fraction of the sensor's remaining energy (f). For the current mission, a sensor computes this value:

$$\frac{u}{d} \times \frac{p}{P} \times f \tag{6.3}$$

Each time a sensor becomes aware of a mission, it evaluates expression (6.3). It makes an offer to the mission only if the value computed is greater than expression (6.2). By weighting the actual profit of a sensor in (6.3) by the fraction of its remaining battery value, the sensors start out eager to propose to missions, but become increasingly selective and cautious over time, as their battery levels decrease. The lower a sensor's battery gets, the higher relative profit it will require before proposing to a mission. Since different sensors' batteries will fall at different rates, in a dense network it is expected that most feasible missions will still receive enough proposals to succeed.

6.4.2.2 Energy and Lifetime-Aware Algorithm

If the target lifetime of the network is known, then sensors can take it into account when making their proposal decisions. To do this, a sensor needs to compute the *expected occupancy time* t_α, the amount of time a sensor expects to be assigned to a mission during the remaining target network lifetime. To find this value, it is required to determine how many missions a given sensor is expected to see. Using the distribution of mission locations, we can compute the probability that a random mission lies within a given sensor's range. Combining this with the remaining target network lifetime and arrival rate of missions, we can find the expected number of missions to which a given sensor will have the opportunity to propose. Thus if the arrival rate and the (*independent*) distributions of the various mission properties are known, we can compute t_α as follows:

$$t_\alpha = \tau \times \lambda \times g \times \gamma \times E[l]$$

where

- τ is the remaining target network lifetime, that is, the initial target network lifetime minus current elapsed time
- λ is the mission arrival rate
- $g = \pi R_s^2/A$ is the probability that a given mission location (chosen uniformly at random) lies within sensing range, R_s is the sensing range, and A is the area of the deployment field
- $E[l]$ is the expected mission lifetime
- γ is the probability that a sensor's offer is accepted*

For each possible mission, the sensor now evaluates an expression which is modified from (6.2). The sensor considers the ratio between its remaining lifetime and its expected occupancy time. If t_b is the amount of time a sensor can be actively sensing, given its current energy level, the expression then becomes

$$\frac{u}{d} \times \frac{p}{P} \times \frac{t_b}{t_\alpha} \tag{6.4}$$

If the value of expression (6.4) is greater than that of expression (6.2), then the sensor proposes to the mission. Moreover, if the sensor's remaining target lifetime is greater than its expected occupancy time, the sensor proposes to *any* mission since in this case it expects to survive until the end of the target time. The effect on the sensor's decision of weighting the mission profit by the ratio (t_b/t_α) is similar to the effect of weighting the fraction of remaining energy (f) had in expression (6.3); all things being equal, less remaining energy makes a sensor more reluctant to propose to a mission. As the network approaches the end of its target lifetime, however, this ratio will actually increase, making a sensor more willing to choose missions with profits less than what it "deserves" in expectation. After all, there is no profit at all for energy conserved past the target lifetime.

6.4.3 Performance Evaluation

The algorithms were tested using a simulator similar to the one discussed in Section 6.3. In the dynamic problem, however, missions arrive without warning over time, and the sensors used to satisfy them have limited battery lives. The goal is to maximize the total profit achieved by missions over the entire duration of the network. In this section, the dynamic heuristic algorithms are tested on randomly generated *sensor network histories* in order to gauge the algorithms' real-world performance.

The energy-aware (*E-aware*) and the energy and lifetime-aware (*E/L-aware*) algorithms are compared with a basic algorithm (*Basic*) that does not take energy or network lifetime into account when making the decision on to which mission to propose (i.e., sensors propose to any mission in their range). For comparison purposes, the performance of a network infinite-energy batteries (i.e., $B \geq$ simulation time) is also shown.

Since finding the true optimal performance value is NP-hard, an exact comparison is not possible. A standard strategy in this kind of situation is to compare with an LP relaxation of the offline IP, that is, of program **P′** of Section 6.4.1. This is a relaxation in the sense of allowing fractional values for the decision variables, as well as allowing full preemption. Its solution value necessarily upper bound the true offline optimal value, just as the LP relaxation of program **P**

* *Computing* this value would imply a circular dependency; in the simulations it was chosen a priori to be 0.25.

did for the experiments in the static setting presented in Section 6.3. Although solving such an LP is theoretically tractable, the size of the LP program is significant, with a decision variable for each (S_i, M_j, t) triple. Even with a problem instance as small as 100 sensors and 50 missions, the LP solver used [14] performs the computation with a gigabyte of memory. Clearly solving more significant problem sizes would be impractical by this method. Instead, a *further* relaxation, program $\mathbf{P}''$, is used which is also upper bounding the optimal:

$$\begin{aligned} &\textit{Maximize:} && \textstyle\sum_{j=1}^{m} p_j l_j y_j \\ &\textit{Such that:} && \textstyle\sum_{i=1}^{n} x_{ij} e_{ij} \geq d_j l_j y_j, \text{ for each mission } M_j, \\ &&& \textstyle\sum_{j=1}^{m} x_{ij} \leq b, \text{ for each sensor } S_i, \text{ and} \\ &&& x_{ij} \in [0, 1]\ \forall x_{ij} \text{ and } y_j \in [0, 1]\ \forall y_j \end{aligned}$$

This formulation condenses the entire history into a single timeslot. The profits and demands are multiplied by the duration of the mission l_j. Since time is elided, the sensor constraint now asserts only that at sensor be used (fractionally) at b times, over the entire history. The optimal solution of this LP will upper bound the optimal solution achievable in the original problem. Note that the solution value provided is the total profits over the entire history, not a time-series. This value is indicated in the plots by a straight line drawn at the average profit corresponding to this total. In the experiments, $\mathbf{P}''$ was solved with a software package [14].

6.4.3.1 Simulation Setup

The simulators use assumptions similar to the ones made in Section 6.3.3. In addition to those assumptions, here it is also assumed that missions arrivals are generated by a Poisson process, with an average arrival rate of 4 missions/h or 8 missions/h depending, on the experiment. Each sensor starts with a battery that will last for 2 h of continuous sensing (i.e., $B = 7200$ s). It is assumed that the battery is used solely for sensing purposes, not for communication. Mission lifetimes are exponentially distributed, with an average of 1 h, a minimum of 5 min and a maximum of 4 h. The sensor network comprises 500 nodes. In order to reward cumulative and broad-based success, the simulator continues to award profit to a mission only so long as total achieved profit per-timestep is at least 50% of the optimal.

6.4.3.2 Results

In both the E-aware and the E/L-aware algorithms, and in order to determine whether the sensor should propose to the mission, the expected profit of a mission (expression (6.2)) needs to be computed. Because the demand and profit value distributions are capped, the actual averages are not equal to the a priori averages of the distributions. It was found, empirically, that the average demand of $d' = 1.2$ and an average profit of $p' = 10.9$. The empirical average duration, which is used to evaluate expression (6.3) was found to be 3827.8 s (roughly 1 h).

Figure 6.21 shows the achieved profit (as a fraction of maximum available) per timestep. The results are the average of 200 runs. The target network lifetime is 3 days (shown as a fine vertical line), but the simulations continue for one full week. Knowledge of the target network lifetime is used by the E/L-aware and in the LP. The other algorithms assume potentially infinite duration. The 50% minimum profit is shown with a fine horizontal line.

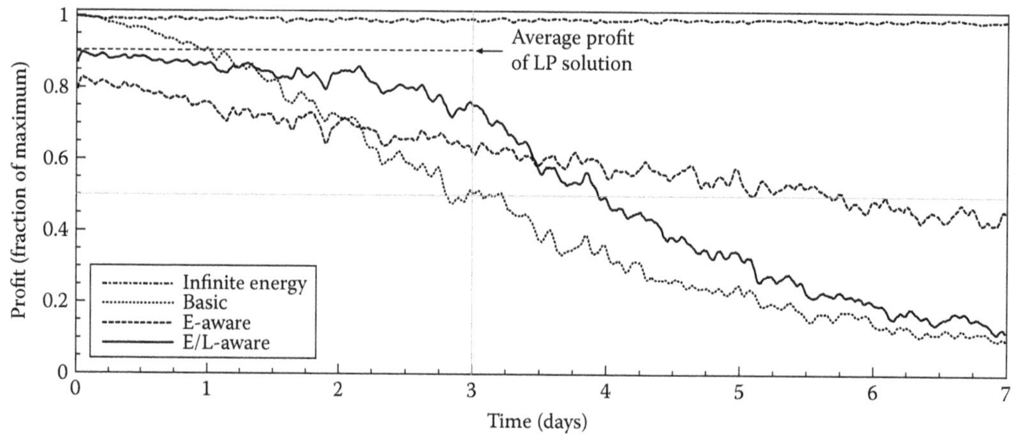

Figure 6.21 Fraction of achieved profits (arrival rate = 4 missions/h).

From Figure 6.21 we see that the profits of all algorithms stay above the 50% threshold for the target lifetime. Basic achieves most of its profits in the beginning and then profits go down (almost linearly) as time progresses. The E-aware algorithm tries to conserve its resources for high profit missions. Because it ignores the fact that we care more about the first 3 days than anytime after that, it becomes overly conservative and ignores many missions. Such an algorithm is better suited to the case when there is no known target lifetime for the network and we want the network to last as long as possible. We see that the profit for E-aware does not fall below the 50% threshold until the end of the 6th day.

In the E/L-aware algorithm, nodes will initially be aggressive in accepting missions that might not provide their expected value, but become more cautious as their energy is used. However, unlike E-aware, as their remaining expected occupancy time approaches their remaining lifetime, sensors will again accept missions with lower and lower profits. The curves for E-aware and E/L-aware cross by the middle of the 4th day, after which point E-aware dominates. When compared to the average LP solution, we see that E/L-aware does very well, within a few percentage points of the optimal (on average). In terms of total *target* lifetime profit (i.e., the area under the curve for the first 3 days), the E/L-aware was found to achieve about 84% of the profits compared to 72% for the E-aware. This means that E/L-aware achieves close to 17% higher profits. If the sensor's battery lifetime is increased from 2 to 3 h, the percentage increase becomes about 22%.

The fraction of extant sensors (i.e., those sensors still *alive*, whose batteries are not yet exhausted) over time is shown in Figure 6.22. Because sensors propose to any mission within its range in Basic, no matter how low the profit is, nodes start dying rapidly. By the end of the third day, only half the nodes can be used for sensing, and by the end of the seventh day this falls below 15%. In E-aware, nodes become very cautious as their batteries run low, which helps the network to last longer without significant sacrifice of achieved profits per timeslot. By the end of the 7 days, about 72% of the nodes remain living. For E/L-aware, sensors accept more missions, and hence are used at a higher rate, as the target lifetime of the network approaches. In the figure, we can see this happening by the 2nd day, when the curve of E/L-aware diverges from that of E-aware. By the end of the 7th day, E/L-aware has used nearly as much energy as Basic.

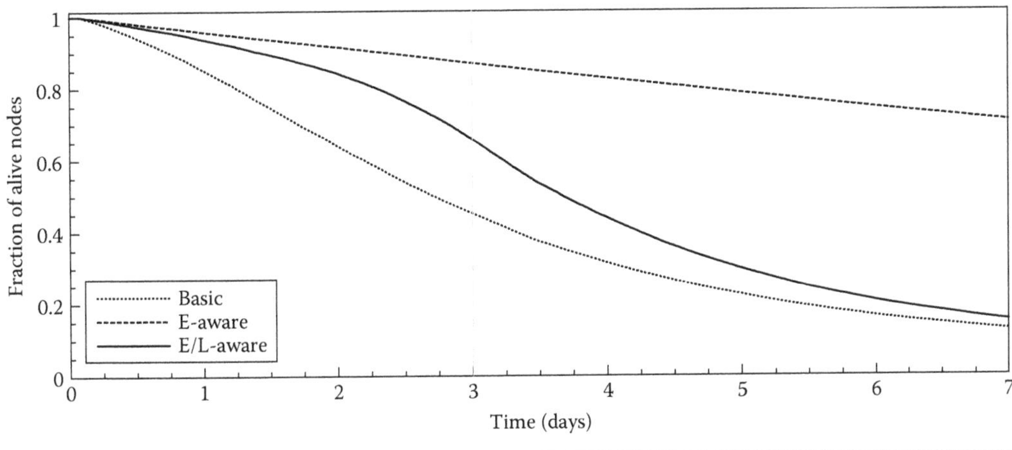

Figure 6.22 Fraction of extant nodes (arrival rate = 4 missions/h).

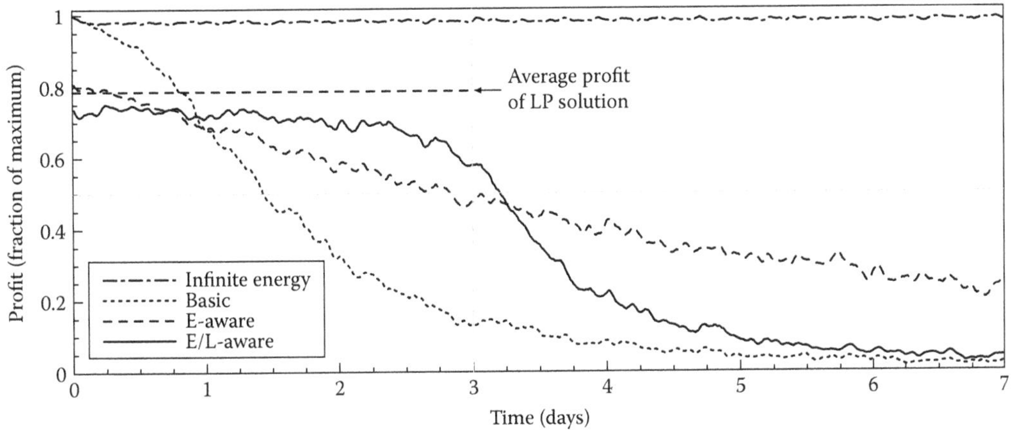

Figure 6.23 Fraction of achieved profits (arrival rate = 8 missions/h).

One thing to note is that E/L-aware acts like Basic once the target lifetime of the network has passed, that is, under it sensors propose to all available missions. If this behavior changed to emulate E-aware's, we could expect the energy usage, and the exhaustion of sensor batteries, to slow down. With more nodes remaining alive for longer times, the decrease in profit following the target network lifetime point would be less dramatic.

Figures 6.23 and 6.24, respectively, show the fraction of achieved profit and fraction of extant nodes over time, with twice the previous arrival rate. Due to the increased number of missions, sensors are used more rapidly and hence both the profit and fraction of extant nodes decrease quickly. Basic passes the 50% profit line by the middle of the 2nd day and both E-aware and E/L-aware pass that point in the beginning of the 4th day. But by that point, E/L-aware achieves significantly higher profits than E-aware. Similar effects are seen on the fraction of extant nodes.

Finally, Figure 6.25 shows the effect of the initial battery lifetime on the performance of both the E-aware and E/L-aware in terms of achieved profits. Here, an arrival rate of 4 missions/h is

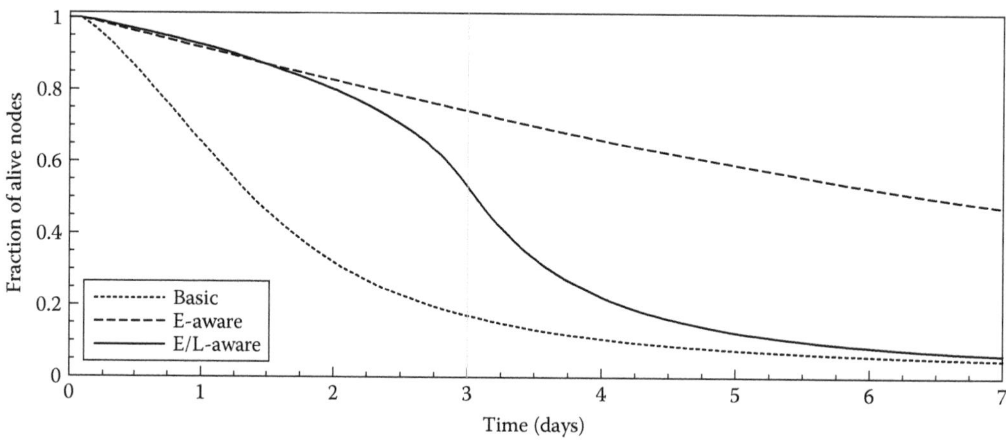

Figure 6.24 Fraction of alive nodes (arrival rate = 8 missions/h).

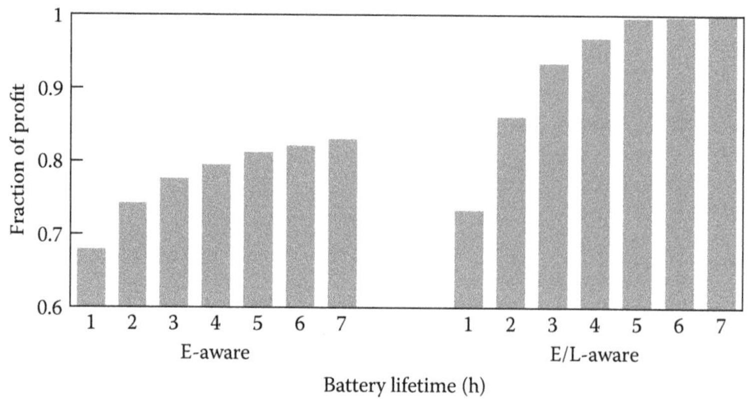

Figure 6.25 Effect of initial battery time on profit (arrival rate = 4 missions/h).

assumed and the fraction of the total profits achieved in the first 3 days is counted. Each of these fractions can be thought of as the ratio of the left rectangle (the first 3 days) in Figure 6.21 and the area under the curve of the corresponding algorithm. The initial energy lifetime of sensors is varied from 1 to 7 h. The effect of increasing the battery lifetime is most pronounced in the beginning. After all, the closer the amount of stored energy gets to the expected occupancy time, the greater the likelihood that energy will be left unspent. We also note that E/L-aware uses the increased battery lifetime more effectively since it takes both the battery and occupancy time into account. The energy-aware would only consider the fraction of used energy, which declines in influence as battery life increases.

The performance evaluation results show that knowledge of energy level and network lifetime is an important advantage: the algorithm given both of these values significantly outperforms the algorithm using only the energy level and the algorithm that uses neither. Given knowledge of both energy and target lifetime, the algorithm can achieve profits 17%–22% higher than if only energy is known.

6.5 Conclusion and Research Directions

In this chapter, we discussed different aspects of the sensor-mission assignment problem in which the objective is to maximize the overall utility of the network. We defined the problem and considered it under different constraints; for each variation we discussed different algorithms to solve the problem. The limitations of sensor devices, such as energy and budget, were key considerations in the design process of the solutions we considered. Another consideration was the ad hoc nature of wireless sensor networks that makes centralized solutions hard to deploy; hence we considered distributed algorithms as well.

There are several issues that remain open for research. In this chapter, we only considered directional sensors for which each sensor can only be assigned to a single mission. Directional sensors, however, can also serve multiple missions through time-sharing. For example, a mission may require an image of a target every 30 s. In this case, multiple missions can use the same camera by rotating the camera to different directions as long as it can meet the requirements of the different missions.

Some sensors are omnidirectional, that is, they can sense from multiple directions. For example, the information provided by a single sensor that measures the ambient temperature can be used to support multiple missions given that they all lie within its sensing range. Although using only directional sensors can be more challenging because the utility from one sensor can only benefit a single mission, having omnidirectional sensors in the network can create a problem instance that is different from what we have considered in this dissertation. Such a problem is typically easier to solve in terms of finding a feasible solution that satisfies all missions. What may be challenging, however, is to optimize the sensor assignment so that we can find the smallest set of sensors that can satisfy the requirements of all missions, which is desirable to conserve resources.

Another venue for research is studying the effects of mobility on the solutions and how we can design better solutions to utilize it. We can see two types of mobility: controlled and uncontrolled. Controlled mobility is the ability to move all or some of the sensing resources in order to achieve better assignments to missions. In uncontrolled mobility the sensing resources might be mounted on people (e.g., helmets of solders in a battlefield) in which case the system has no control on where the sensors are to be moved.

References

1. A. Bar-Noy, T. Brown, M. P. Johnson, T. La Porta, O. Liu, and H. Rowaihy. Assigning sensors to missions with demands. In *ALGOSENSORS 2007*, Wroclaw, Poland, July 2007.
2. P. Bose, P. Morin, I. Stojmenovic, and J. Urrutia. Routing with guaranteed delivery in ad hoc wireless networks. *Wireless Networks*, 7(6):609–616, 2001.
3. J. Byers and G. Nasser. Utility-based decision-making in wireless sensor networks. Technical Report CS-TR-2000-014, Boston University, June 2000.
4. R. Cohen, L. Katzir, and D. Raz. An efficient approximation for the generalized assignment problem. *Information Processing Letters*, 100(4):162–166, 2006.
5. W. Heinzelman, A. Chandrakasan, and H. Balakrishnan. Energy-efficient communication protocols for wireless microsensor networks. In *Proceedings of 33rd Annual Hawaii International Conference on System Sciences (HICSS-33)*, 4–7 January, 2000, Maui, Hawaii. *IEEE Computer Society*, 2000.
6. M. Johnson, H. Rowaihy, D. Pizzocaro, A. Bar-Noy, S. Chalmers, T. La Porta, and A. Preece. Sensor-mission assignment in constrained environments. *IEEE Transaction on Parallel Distributed Systems*, 21(11):1692–1705, 2010.

7. L. Kaplan. Global node selection for localization in a distributed sensor network. *IEEE Transactions on Aerospace and Electronic Systems*, 42(1):113–135, January 2006.
8. L. Kaplan. Local node selection for localization in a distributed sensor network. *IEEE Transactions on Aerospace and Electronic Systems*, 42(1):136–146, January 2006.
9. B. Karp and H. Kung. Greedy perimeter stateless routing for wireless networks. In *Proceedings of the sixth annual international conference on Mobile computing and networking (MOBICOM)*, Boston, MA, USA, August 6–11, 2000. ACM 2000.
10. M. Perillo and W. Heinzelman. Optimal sensor management under energy and reliability constraints. In *Proceedings of the IEEE Conference on Wireless Communications and Networking*, New Orleans, Luisiana, March 2003.
11. H. Rowaihy, S. Eswaran, M. P. Johnson, D. Verma, A. Bar-Noy, T. Brown, and T. La Porta. A survey of sensor selection schemes in wireless sensor networks. In *SPIE Defense and Security Symposium*, Orlando, Florida, April 2007.
12. H. Rowaihy, M. Johnson, O. Liu, A. Bar-Noy, T. Brown, and T. La Porta. Sensor-mission assignment in wireless sensor networks. *ACM Transactions on Sensor Networks*, 6(4):1–33, July 2010.
13. K. Shih, Y. Chen, C. Chiang, and B. Liu. A distributed active sensor selection scheme for wireless sensor networks. In *Proceedings of the IEEE Symposium on Computers and Communications*, Sardinia, Italy, June 2006.
14. LP solver version 5.5. http://lpsolve.sourceforge.net/5.5/
15. V.V. Vazirani. *Approximation Algorithms*. Springer, Berlin, Germany, 2001.
16. F. Zhao, J. Shin, and J. Reich. Information-driven dynamic sensor collaboration. *IEEE Signal Processing Magazine*, 19(2):61–72, March 2002.

Chapter 7

Prediction-Based Data Collection in Wireless Sensor Networks

Yann-Aël Le Borgne and Gianluca Bontempi

Contents

Most sensor network applications aim at monitoring the spatiotemporal evolution of physical quantities, such as temperature, light, or chemicals, in an environment. In these applications, low-resources sensor nodes are deployed and programmed to collect measurements at a predefined sampling frequency. Measurements are then routed out of the network to a network node with higher resources, commonly referred to as base station or sink, where the spatiotemporal evolution of the quantities of interest can be monitored.

In many cases, the collected measurements exhibit high spatiotemporal correlations and follow predictable trends or patterns. An efficient way to optimize the data collection process in these settings is to rely on machine learning techniques, which can be used to model and predict the spatiotemporal evolution of the monitored phenomenon.

This chapter presents a survey of the learning approaches that have been recently investigated for reducing the amount of communication in sensor networks by means of learning techniques. We have classified the approaches based on learning into three groups, namely, model-driven data acquisition, replicated models (RM), and aggregative approaches.

In model-driven approaches, the network is partitioned into two subsets, one of which is used to predict the measurements of the other. The subset selection process is carried out at the base station, together with the computation of the models. Thanks to the centralization of the procedure, these approaches provide opportunities to produce both spatial and temporal models. Model-driven techniques can provide high energy savings as part of the network can remain in an idle mode. Their efficiency in terms of accuracy is, however, tightly dependent on the adequacy of the model to the sensor data. We present these approaches in Section 7.2.

RM encompass a set of approaches where identical prediction models are run in the network and at the base station. The models are used at the base station to get the measurements of sensor nodes, and in the network to check that the predictions of the models are correct within some user defined ϵ. A key advantage of these techniques is to guarantee that the approximations provided by the models are within a strict error threshold ϵ of the true measurements. We review these techniques in Section 7.3.

Aggregation approaches allow to reduce the amount of communication by combining data within the network, and provide to a certain extent a mixture of the characteristics of model-driven and RM approaches. They rely on the ability of the network routing structure to aggregate information of interest on the fly, as the data are routed to the base station. As a result, the base station receives aggregated data that summarize in a compact way information about sensor measurements. The way data are aggregated depends on the model designed at the base station, and these approaches are therefore in this sense model driven. The resulting aggregates may, however, be communicated to all sensors in the network, allowing them to check the approximations against their actual measurements, as in RM approaches. We discuss aggregative approaches in Section 7.4.

7.1 Problem Statement

The main application considered in this chapter is the *periodic data collection*, in which sensors take measurements at regular time intervals and forward them to the base station [35]. Its main purpose is to provide the end-user, or observer, with periodic measurements of the whole sensor field.

This task is particularly useful in environmental monitoring applications where the aim is to follow both over space and time the evolution of physical phenomena [29,35].

7.1.1 Definitions and Notations

The set of sensor nodes is denoted by $\mathcal{S} = \{1, 2, \ldots, S\}$, where S is the number of nodes. The location of sensor node i in space is represented by a vector of coordinates $c_i \in \mathbb{R}^d$, where d is the dimension of the space, typically 2 or 3. Besides the sensors and the battery, a typical sensor node is assumed to have a CPU of a few mega hertz, a memory of a few tens of kilobytes, and a radio with a throughput of a few hundreds of kilobits per second [35].

The entity of interest that is being sensed is called the *phenomenon*. The vector of the measurements taken in the sensor field at time t is denoted by $s[t] = (s_1[t], s_2[t], \ldots, s_S[t]) \in \mathbb{R}^S$. The variable $s[t]$ may be univariate or multivariate. Whenever confusion is possible, the notation $s_i[t]$ is used to specify the sensor node concerned, with $i \in \mathcal{S}$. The time domain is the set of natural numbers $\mathcal{T} = \mathbb{N}$, and the unit of time is called an *epoch*, whose length is the time interval between two measurements.

Most of the prediction models considered in this chapter aim at approximating or predicting sensor measurements. Denoting by $\hat{s}_i[t]$ the approximation of $s_i[t]$ at time t, the prediction models will typically be of the form

$$h_\theta : \mathcal{X} \to \mathbb{R}$$

$$x \mapsto \hat{s}_i[t] = h_\theta(x)$$

The input x will in most cases be composed of the sensor measurements, sensor coordinates, and/or time.

Example 7.1:

In many cases, there exist spatiotemporal dependencies between sensor measurements. A model may, therefore, be used to mathematically describe the relationship existing between one sensor and a set of other sensors. For example, a model:

$$h_\theta : \mathbb{R}^2 \to \mathbb{R}$$

$$x = (s_j[t], s_k[t]) \mapsto \hat{s}_i[t] = \theta_1 s_j[t] + \theta_2 s_k[t]$$

can be used to approximate the measurement of sensor i on the basis of a linear combination of the measurements of sensors j and k, with $i, j, k \in \mathcal{S}$, and the parameters $\theta = (\theta_1, \theta_2) \in \mathbb{R}^2$. The use of such a model may allow to avoid collecting the measurement from sensor i if the model is assumed to be sufficiently accurate.

Example 7.2:

A model may aim at approximating the scalar field of the measurements. The input domain is the space and time domains $\mathbb{R}^d \times \mathbb{R}$, and the output domain is the set of real numbers, representing an approximation of the physical quantity at one point of space and time. For example, a linear

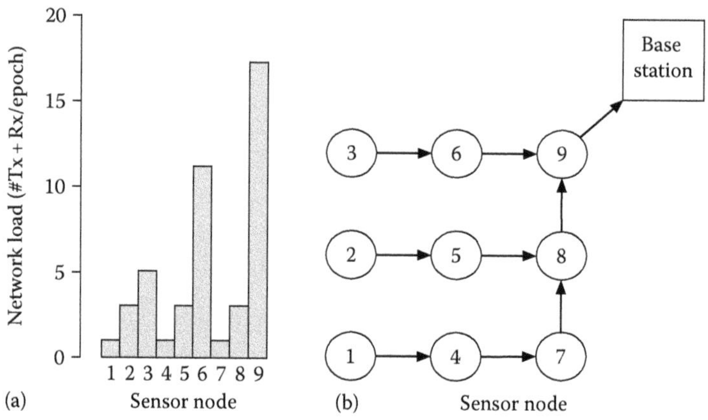

Figure 7.1 Network load sustained by each sensor node in a network of nine sensors, for two different routing trees. The bar plot (left) gives the total number of packets processed (receptions and transmissions) by each node in the network (right). (a) First configuration and (b) Second configuration.

combination of time and spatial coordinates $c_i \in \mathbb{R}^2$ in a two-dimensional space domain can be used to create the following model:

$$\begin{aligned} h_\theta &: (\mathbb{R}^2 \times \mathbb{R}) \to \mathbb{R} \\ x &= (c_i, t) \mapsto \hat{s}_i[t] = x\theta^T \end{aligned}$$

where $\theta = (\theta_1, \theta_2, \theta_3) \in \mathbb{R}^3$ are the weights of the linear combination which form the parameter vector θ. If the weights are known or can be estimated on the basis of past data, the model can be used to get approximated measurements of the sensors without actually collecting data from the network. The model can, moreover, be used to predict the measurements at future time instants or at locations where sensors are not present.

Let us assume that the data are retrieved from the network by means of a routing tree as shown in Figure 7.1. The *observer* specifies the sampling frequency at which the measurements must be retrieved, using, for example, an aggregation service such as TAG or Dozer [3,24].

7.1.2 Network Load and Sensor Lifetime

The purpose of prediction models is to trade data accuracy for communication costs. Prediction models are estimated by learning techniques, which use past observations to represent the relationships between measurements by means of parametric functions. The use of prediction models allows to provide the observer with approximations $\hat{s}[t]$ of the true set of measurements $s[t]$ and to reduce the amount of communication by either subsampling or aggregating data.

Given that the radioactivity is the main source of energy consumption, the reduction of the use of radio is the main way to extend the lifetime of a sensor network. In qualitative terms, the lifetime is the time span from the deployment to the instant when the network can no longer perform the task [34]. The lifetime is application specific. It can be, for example, the instant when a certain fraction of sensors die, loss of coverage occurs (i.e., a certain portion of the desired area can no longer be monitored by any sensor), or loss of connectivity occurs (i.e., sensors can no longer communicate with the base station). For periodic data collection, the lifetime can be more specifically defined as *the number of data collection rounds until* α *percent of sensors die, where* α *is specified by the system designer* [31]. This definition makes the lifetime independent of the sampling rate at which data are collected from the network. Depending on α, the lifetime is, therefore, somewhere in between the number of rounds until the first sensor runs out of energy and the number of rounds until the last sensor runs out of energy.

The communication costs related to the radio of a node i is quantified by **the network load** L_i, which is **the sum of the number of received and transmitted packets during an epoch**. Denoting by Rx_i and Tx_i the number of packet receptions and packet transmissions for node i during the epoch of interest, we have

$$L_i = \mathrm{Rx}_i + \mathrm{Tx}_i$$

A network packet is assumed to contain one piece of information. A sensor transmitting its measurements and forwarding the measurement of another sensor therefore processes three packets during an epoch, that is, one reception and two transmissions.

Figure 7.1 illustrates how the network load is distributed among sensors during an epoch. The network loads are reported for each sensor on the left of Figure 7.1, for two different routing trees built such that the number of hops between the sensor nodes and the base station is minimized. Leaf nodes sustain the lowest load (only one transmission per epoch), whereas the highest load is sustained in both cases by the root node (8 receptions and 9 transmissions, totalizing a network load of 17 packets per epoch).

In data collection, it is important to remark that the nodes that are likely to have the lowest lifetime are the nodes close to the base station, as their radioactivity is increased by the forwarding of data. The lifetime of these nodes is therefore closely related to the network lifetime; since once these nodes have run out of energy, the rest of the network gets out of communication range of the base station. A particular attention therefore is, given in this chapter to the number of packets received and transmitted by the root node. More generally, we will often aim at quantifying the upper bound

$$L_{\max} = \max_i L_i \tag{7.1}$$

of the distribution of the network loads in the network. Most of the methods and techniques discussed in this chapter aim at reducing this quantity, which will be referred to as **highest network load**. In order to consider the effects of collisions, interference or radio malfunctioning, we will use orders of magnitudes instead of precise counting of the packets. Without the use of learning strategies, the order of magnitude of the highest network load is in

$$L_{\max} \sim O(S)$$

where S is the number of nodes.

7.1.3 Data Accuracy

The quantification of the error implied by a prediction model is an important practical issue, as in many cases the observer needs to know how far the predictions obtained at the base station are from the true measurements. Three levels of accuracy are encountered in this chapter.

First, *probabilistic bounded approximation errors* will refer to approximations where

$$P(|s_i[t] - \hat{s}_i[t]| > \epsilon) = 1 - \delta, \ \forall i \in \mathcal{S}, t \in \mathcal{T} \tag{7.2}$$

which guarantees that, with probability $1 - \delta$, approximations do not differ by more than ϵ from the true measurements. The observer can set the error threshold ϵ and the probability guarantee δ.

Second, *bounded approximation errors* will refer to approximations where

$$|s_i[t] - \hat{s}_i[t]| < \epsilon, \ \forall i \in \mathcal{S}, t \in \mathcal{T} \tag{7.3}$$

which ensures the observer that all approximations $\hat{s}_i$ obtained at the base station are within $\pm\epsilon$ of the true measurement $s_i[t]$. This level of accuracy is the highest, as it allows the observer to precisely define the tolerated error threshold.

Finally, *unbounded errors* will refer to modeling schemes where there is no bound between the approximations $\hat{s}_i[t]$ obtained at the base station and the true measurement $s_i[t]$ taken by sensor i.

7.2 Model-Driven Acquisition

In model-driven data acquisition [6,7,15,21,22,37], a model of the sensor measurements is estimated at the base station and used to optimize the acquisition of sensor readings. The rationale of the approach is to acquire data from sensors only if the model is not sufficiently rich to provide the observer with the requested information. An overview of the approach is presented in Figure 7.2. In the first stage, measurements are collected over N epochs from the whole network at the base station, and stored in a matrix X of dimension $N \times S$, where column j contains the measurements from sensor j over the N epochs, and row i contains the measurements from the whole network during the ith epoch. The data set X is then used to estimate a model able to answer the users' queries without collecting data from the whole network. More precisely, the model-driven system aims at finding a subset of sensors $\mathcal{S}_q \subseteq \mathcal{S}$ from which the measurements of the other sensors $\mathcal{S}_p = \mathcal{S} \backslash \mathcal{S}_q$ can be predicted. The subscripts p and q refer to the *queried* and *predicted* subsets. Once the subsets $\mathcal{S}_q$ and $\mathcal{S}_p$ have been identified, measurements are only collected from the sensors in $\mathcal{S}_q$.

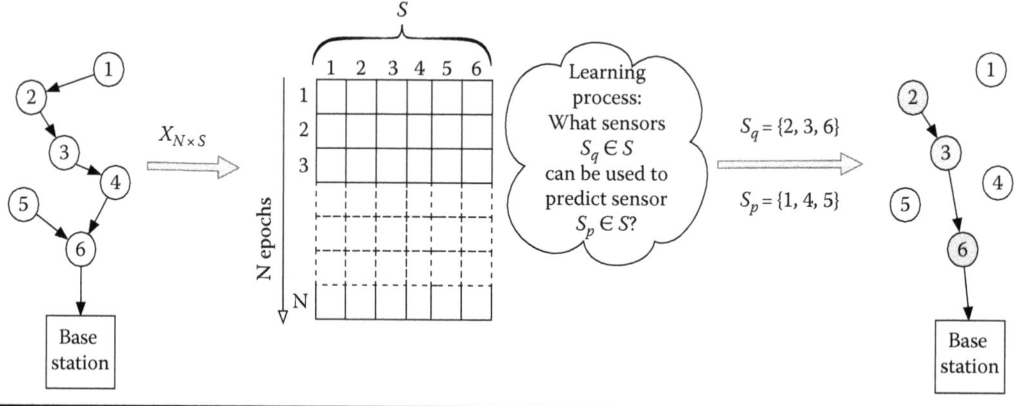

Figure 7.2 Model-driven approach: The learning process takes place at the base station. It aims at finding a subset of sensors from which the measurements of the other sensors can be predicted.

In practice: Model-driven approaches are particularly appropriate for scenarios where groups of sensor nodes have correlated measurements. One node of the group sends its measurements to the base station, which uses them to predict the measurements of other nodes in the group. Thanks to the fact that the base station has a global view of the measurements collected, model-driven approaches allow to detect correlation between sensor nodes that may be far away in the network. A priori information on the periodicity or stationarity of the measurements can be used to determine how many observations N should be collected before using the model. For example, a network collecting outdoor temperature measurements is likely to exhibit diurnal patterns. If the patterns are consistent over days, observations can be taken over a one day period and used to model the measurements of the following days.

7.2.1 Optimization Problem

The model used at the base station aims at predicting a vector of measurements $\hat{s}_p[t]$ for sensors in $\mathcal{S}_p$ with a prediction model

$$h_\theta : \mathbb{R}^{|\mathcal{S}_q|} \to \mathbb{R}^{|\mathcal{S}_p|}$$
$$\hat{s}_q[t] \mapsto \hat{s}_p[t] \tag{7.4}$$

where the input $\hat{s}_q[t]$ is the vector of measurements collected from sensors in $\mathcal{S}_q$ at time t and θ is a vector of parameters. The model-driven approach allows to trade energy for accuracy by carefully choosing the subsets $|\mathcal{S}_q|$ and $|\mathcal{S}_p|$.

Cost: The cost associated to the query of a subset of sensors $\mathcal{S}_q$ is denoted by $C(\mathcal{S}_q)$, and aims at quantifying the energy required to collect the measurements from $C(\mathcal{S}_q)$. The cost is divided into *acquisition* and *transmission* costs in [6,7]. Acquisition refers to the energy required to collect a measurement, and transmission to the energy required for sending the measurement to the base station. The transmission costs are in practice difficult to estimate because of multi-hop routing and packet loss issues. A simple and qualitative metric is to define the cost as the number of sensors in $\mathcal{S}_q$.

Accuracy: Let $s_{p_i}[t]$ and $\hat{s}_{p_i}[t]$ be the true measurement and the by prediction for the ith sensor in $\mathcal{S}_p$ at time t. The accuracy associated to $\hat{s}_{p_i}$ is denoted by $R(\mathcal{S}_{p_i})$, and the accuracy associated to the vector of prediction $\hat{s}_p$ is denoted by $R(\mathcal{S}_p)$. Different choices are possible to define how accuracy is quantified. In [6,7], authors suggest using

$$R(\mathcal{S}_{p_i}) = P(s_{p_i}[t] \in [\hat{s}_{p_i}[t] - \epsilon, \hat{s}_{p_i}[t] + \epsilon]) \tag{7.5}$$

where $i \in \mathcal{S}_p$ and ϵ is a user-defined error threshold. This accuracy metric quantifies the probability that the true measurement $s_{p_i}[t]$ is within $\pm\epsilon$ of the prediction $s_{p_i}[t]$. The overall accuracy of the vector of prediction $\hat{s}_p[t]$ is defined as the minimum of $\hat{s}_{p_i}[t]$, that is,

$$R(\mathcal{S}_p) = \min_i R(\mathcal{S}_{p_i}) \tag{7.6}$$

Optimization loop: The goal of the optimization problem is to find the subset $\mathcal{S}_q$ that minimizes $C(\mathcal{S}_q)$, such that $R(\mathcal{S}_p) > 1 - \delta$, where δ is a user-defined confidence level. An exhaustive search among the set of partitions $\{\mathcal{S}_q, \mathcal{S}_p\}$ can be computationally expensive. There exists 2^S combinations for the set of predicted sensors, and S can be large. In order to speed up this process, an incremental search procedure similar to the forward selection algorithm can be used. The search is initialized with an empty set of sensors $\mathcal{S}_q = \emptyset$. At each iteration, for each $i \in \mathcal{S}_p$, the costs $C(\mathcal{S}_q \cup i)$ and accuracy $R(\mathcal{S}_p \backslash i)$ are computed. If one sensor i can be found such that $R(\mathcal{S}_p \backslash i) > 1 - \delta$, the procedure returns $\mathcal{S}_q \cup i$ as the set of sensors to query and $\mathcal{S}_p \backslash i$ as the set of sensors to predict. Otherwise, the sensor node that provided the *best* trade–off is added to the subset $\mathcal{S}_q$ and removed from $\mathcal{S}_p$. The *best* sensor node is the node that maximizes the ratio $R(\mathcal{S}_p \backslash i)/C(\mathcal{S}_q \cup i)$ [6,7].

7.2.2 Multivariate Gaussians

Different learning procedures can be used to compute the model h_θ in (7.4). Authors in [6,7] suggest the use of a multivariate Gaussian to represent the set of measurements, which allows to compute predictions and confidence bounds using computationally efficient matrix products. Denoting by $\mathbf{s} = (\mathbf{s}_1, \mathbf{s}_2, \ldots, \mathbf{s}_S) \in \mathbb{R}^S$ the random vector of the measurements, where the ith value represents the measurement of the ith sensor, the Gaussian probability density function (pdf) of $\mathbf{s}$ is expressed as

$$p(\mathbf{s} = s) = \frac{1}{\sqrt{(2\pi)^S |\Sigma|}} \exp^{\left(-\frac{1}{2}(s-\mu)^T \Sigma^{-1} (s-\mu)\right)}$$

where μ and Σ are the mean and covariance matrix of the random vector $\mathbf{s}$ (Figure 7.2). Figure 7.3 illustrates a Gaussian over two correlated variables $\mathbf{s}_1$ and $\mathbf{s}_2$. For a Gaussian, the mean is the point at the center of the distribution, and the covariance matrix Σ characterizes the spread of the distribution. More precisely, the ith element along the diagonal of Σ, $\sigma_{(i,i)}$ is the variance of the ith variable, and off-diagonal elements $\sigma_{(i,j)}$ characterize the covariances between the pairs (i, j) of variables. A high covariance between two variables means that their measurements are correlated, such as variables $\mathbf{s}_1$ and $\mathbf{s}_2$ in Figure 7.3.

When sensor measurements are correlated, information on some measurements constrains the values of other measurements to narrow probability bands. Let $s_q[t] \in \mathcal{S}_q$ and $s_p[t] \in \mathcal{S}_p$ be the vectors of measurements of sensors in $\mathcal{S}_q$ and $\mathcal{S}_p$ at time t, with $\mathcal{S}_p = \mathcal{S} \backslash \mathcal{S}_q$. The Gaussian pdf

$$p(s_p | s_q[t]) = \frac{1}{\sqrt{(2\pi)^{|\mathcal{S}_p|} |\Sigma|}} \exp^{\left(-\frac{1}{2}(s_p-\mu_{p|q})^T \Sigma_{p|q}^{-1} (s_p-\mu_{p|q})\right)}$$

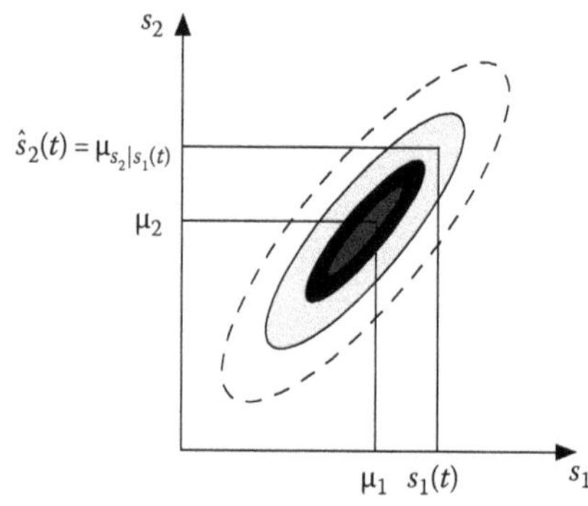

Figure 7.3 Gaussian model of two correlated variables. Knowledge of the outcome of s_1 allows to better estimate the outcome of s_2, thanks to conditioning.

of the random variable $\mathbf{s}_p$ can be computed using the mean μ and covariance matrix Σ of the model $p(s)$. This computation, called conditioning, gives [26]

$$\mu_{p|q} = \mu_p + \Sigma_{pq}\Sigma_{qq}^{-1}(s_q[t] - \mu_q)$$
$$\Sigma_{p|q} = \Sigma_{pp} - \Sigma_{pq}\Sigma_{qq}\Sigma_{qp}$$

where Σ_{pq} denotes the matrix formed by selecting the rows $\mathcal{S}_p$ and the columns $\mathcal{S}_q$ from the matrix Σ. After conditioning, the best approximations $\hat{s}_p[t]$ to sensors in $\mathcal{S}_p$ are given by the mean vector

$$\hat{s}_p[t] = \mu_{p|q} \tag{7.7}$$

The probability

$$P\left(s_i[t] \in \left[\hat{s}_i[t] - \epsilon, \hat{s}_i[t] + \epsilon\right]\right) \tag{7.8}$$

depends on the variance of the measurements of the sensors in $\mathcal{S}_p$ after the conditioning. These variances are actually known as they are the diagonal elements of the covariance matrix $\Sigma_{p|q}$, and allow to estimate the quantity (7.8) by referring to a student's t table [19].

7.2.3 Discussion

With model-driven data acquisition, the communication savings are obtained by reducing the number of sensors involved in the data collection task. These approaches can provide important communication and energy savings, by leaving the set of sensors in $\mathcal{S}_p$ in an idle mode. For continuous queries, however, the constant solicitation of the same subset of sensors leads to an unequal energy consumption among nodes. This issue can be addressed by recomputing from time to time the subset of sensors $\mathcal{S}_p$ in such a way that sensors whose remaining energy is high are favored [16,37].

In terms of communication savings, the network load is reduced by a factor that depends on the ratio of the total number of sensors over the number of queried sensors. The highest load is in

$$L_{\max} \sim O(|\mathcal{S}_q|)$$

and is sustained by the root node of the tree connecting the sensors in $\mathcal{S}_q$.

The main issue of the model-driven approach probably lies in the assumption that the model is correct, which is a necessary condition for the scheme to be of interest in practice. The model parameters μ and Σ must be estimated from data, and any changes in the data distribution will lead to unbounded errors. Also, the presence of outliers in the queried measurements may potentially strongly affect the quality of the predictions. Model-driven data acquisition therefore allows potentially high communication savings, but is little robust to unexpected measurement variations and nonstationary signals.

7.3 Replicated Models

RM form a large class of approaches allowing sensor nodes to remove temporal or spatial redundancies between sensor measurements in order to decrease the network load. They were introduced in the field of sensor networks by Olston et al. in 2001 [27]. Their rationale

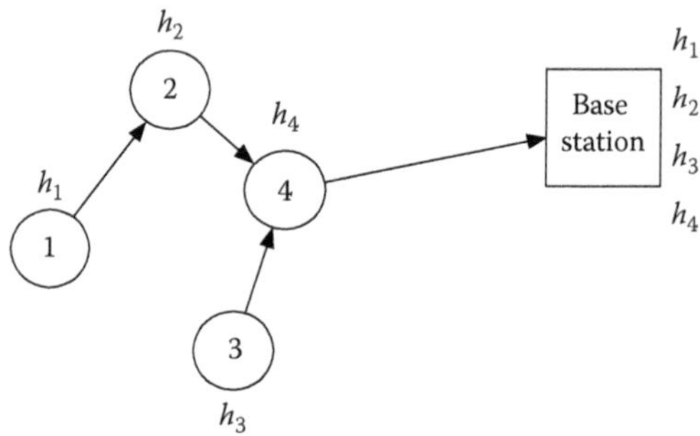

Figure 7.4 Replicated models: Only the models are communicated to the base station. As long as the model is correct, no communication is necessary.

consists in having identical prediction models running both within the network and at the base station. Other names have been given to this approach in the literature, such as *approximate caching* [27], *dual Kalman filters* [11], *replicated dynamic probabilistic models* [5], *dual prediction scheme* [18,32], and *model-aided approach* [36]. We will adopt here the *RM* denomination since it is the one that better expresses in our sense the common rationale of these approaches.

Figure 7.4 gives an illustration of the RM approach. Four models, h_i, are used to predict the sensor measurements, one for each sensor i. The base station and the sensors use the same models. At the base station, the model is used to predict the measurements. On the sensor nodes, the model is used to compute the same prediction as the base station, and to compare it with the true measurement. If the prediction is more than a user-defined ϵ away from the true measurement, a new model is communicated. RM approaches, therefore, guarantee the observer with bounded approximation errors. RM may be used to predict measurement both over the temporal and spatial domains. We first cover the approaches where models only take into account temporal variations, and then present the strategies that have been investigated to extend RM to the modeling of spatial variations.

In practice: In many environmental monitoring applications, the observer can tolerate approximations for the collected measurements. For example, in plant growth studies, ecologists reported that it is sufficient to have accuracy of $\pm 0.5°$C and 2% for temperature and humidity measurements, respectively [7]. RM provide an appropriate mechanism to deal with these scenarios.

7.3.1 Temporal Modeling

Temporal modeling with RM is the simplest approach, as sensors independently produce prediction models for their own measurements [11,27,32]. Models are of the kind

$$\hat{s}_i[t] = h_i(x, \theta)$$

where x is a vector of inputs that consists, for example, in the past measurements of sensor i, and θ is a vector of parameters. There is one model for each sensor node i, whose task is to produce predictions for the upcoming measurements of the sensor node. Once such a model is produced, it is communicated to the base station, which uses it to infer the actual measurements taken by sensor i.

Sensor nodes use their own copy of the model to compare the model predictions with the sensor measurements. The model is assumed to be correct as long as the predicted and the true measurements do not differ by more than a user, that is defined error threshold ϵ,

$$|s_i[t] - \hat{s}_i[t]| < \epsilon$$

Once the prediction is more than ϵ away from the true measurement, an update is sent to the base station to notify that the error threshold is not satisfied. The update may consist of the measurement, or of the parameters of a new model built on the basis of the more recent measurements. This way, RM guarantee the observer that all measurements provided at the base station are within $\pm\epsilon$ of the true measurements.

RM allow to reduce communication since, as long as the model predictions are within $\pm\epsilon$ of the true measurement, no updates occur between the sensor nodes and the base station. The inputs of the model, if they depend on the sensor measurements, are inferred by the base station using the measurements predicted by the model. The sensor nodes also use past predicted measurements as inputs to their models. This way, the sensor nodes and the base station apply exactly the same procedure. Sensor measurements are sent only when a model update is needed.

The pseudocode for running RM on a sensor node is given by Algorithm 7.1. The subscript i is dropped for the sake of clarity. On the base station, the scheme simply consists in using the most recently received model to infer the sensor's measurements. Four types of techniques have been proposed to run temporal RM [12,14,27,32,37] that are presented in detail in the following. The following summary gives an overview of their differences:

Algorithm 7.1 RM—Replicated model algorithm

Input:
ϵ: Error threshold.
h: Model.

Output:
Packets containing model updates sent to the base station (BS).

```
t ← 1
θ[t] ← init(h)
θ^last ← θ[t]
sendNewModel(h, θ^last) to BS
while True do
    t ← t + 1
    s[t] ← getNewReading()
    ŝ[t] ← getPrediction(h, θ^last)
    θ[t] ← update(h, θ[t − 1], s[t])
    if |ŝ[t] − s[t]| > ε then
        θ^last ← θ[t]
        sendNewModel(h, θ^last) to BS
    end if
end while
```

- Constant model [14,27]: The most simple model, nonetheless often efficient. It does not rely on any learning procedure, and as a result cannot represent complex variations.
- Kalman filter [12]: The technique provides a way to filter the noise in the measurements.
- Autoregressive model [37]: It allows to predict more complex variations than the constant model, but requires a learning stage.
- Least mean square (LMS) filter [32]: The filter is based on an autoregressive model that adapts its coefficients over time.

7.3.1.1 Constant Model

In Refs. [14,27], RM are implemented with constant prediction models

$$\hat{s}_i[t] = s_i[t-1]$$

Although simple, the constant model is well suited for slowly varying time series. Also, it has the advantage of not depending on any parameters. This keeps the size of an update to a minimum, which consists only in the new measurement $s_i[t]$. This makes the constant model bound to reduce the communication between the sensor and the base station. Figure 7.5 illustrates how a constant model represents a temperature–time series. The time series was obtained from the ULB Greenhouse dataset [15] on August 16, 2006. Data were taken every 5 min, for a one day period, giving a set of 288 measurements. The measurements are reported with the red dashed lines, and the approximations obtained by a constant model with an error threshold of $\epsilon = 1°C$ are reported with the black solid line. Updates are marked with black dots at the bottom of the figure.

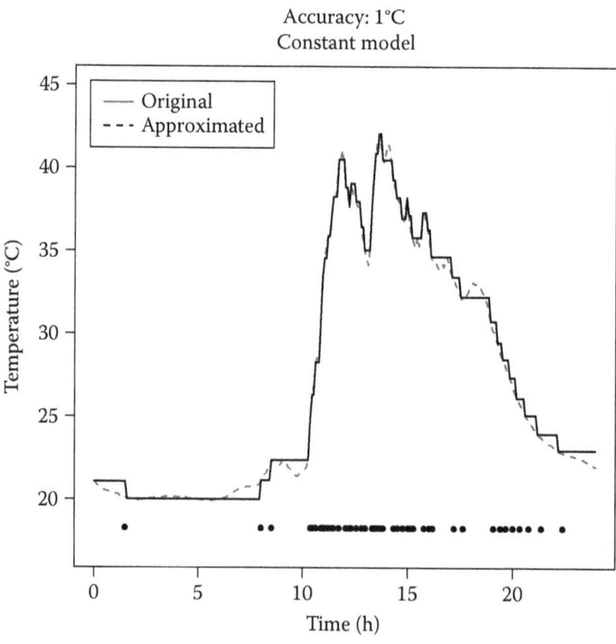

Figure 7.5 A constant model acting on a temperature–time series from the Ulb Greenhouse [15] (August 16, 2006) with a constant model and an error threshold set to $\epsilon = 1°C$.

Using RM, the constant model allows to reduce the number of measurements transmitted to 43, resulting in about 85% of communication savings.

7.3.1.2 Kalman Filter

In Ref. [12], authors suggested to use Kalman filters as a generic approach to modeling sensor measurements for RM. A Kalman filter is a stochastic, recursive data filtering algorithm introduced in 1960 by Kalman [13]. It can be used to estimate the dynamic state of a system at a given time t by using noisy measurements issued at time $t_0 : t-1$.

The main goal of a Kalman filter is to provide good estimations of the internal state of a system by using a priori information on the dynamic of the system and on the noise affecting the measurements. The system model is represented in the form of the following equations

$$\mathbf{x}[t] = F\mathbf{x}[t-1] + \nu_p[t] \tag{7.9}$$

$$\mathbf{s}[t] = H\mathbf{x}[t-1] + \nu_m[t] \tag{7.10}$$

where $\mathbf{x}[t]$ is the internal state of the process monitored, which is unknown, and $\mathbf{s}[t]$ is the measurement obtained by a sensor at time instant t. The matrix F is the state transition matrix, which relates the system states between two consecutive time instants. The matrix H relates the system state to the observed measurement. Finally, $\nu_p[t]$ and $\nu_m[t]$ are the process noise and measurement noise, respectively.

The state of the system is estimated in two stages. First, a *prediction/estimation* stage is used to propagate the internal state of the system by means of Equation (7.10). Second, a *correction stage* fine-tunes the prediction step by incorporating the actual measurement in such a way that the error covariance matrix between the measurements and the predictions is minimized. Eventually, a prediction $\hat{s}[t]$ is obtained, expected to be closer to the true state of the system than the actual measurement $s[t]$. The details of the mathematical steps can be found in [12].

7.3.1.3 Autoregressive Models

In [37], authors suggest the use of autoregressive models, a well-known family of models in time series prediction [2,4]. An autoregressive model is a function of the form

$$\hat{s}[t] = \theta_1 s[t-1] + \theta_2 s[t-2] + \ldots + \theta_p s[t-p] \tag{7.11}$$

that aims at predicting the measurement at time t by means of a linear combination of the measurements collected at the previous p time instants. The vector of parameters has p elements $\theta = (\theta_1, \theta_2, \ldots, \theta_p)^T$, and p is called the order of the model. Using the notations $x[t] = (s[t-1], s[t-2], \ldots, s[t-p])^T$ to denote the vector of inputs of the model at time instant k, the relationship can be written as

$$\hat{s}[t] = x[t]^T\theta \tag{7.12}$$

The estimation of the vector of parameters θ is obtained by first collecting N measurements, and by applying the standard procedure of regression θ [9] on the sensor node. The set of parameters is then communicated to the sink, and used until the prediction error becomes higher than the user predefined threshold ϵ. A new model is then sent to the base station. Authors in [37] also considered the fact that a measurement not correctly predicted by the model may be an outlier, and suggested to use statistical tests to determine whether or not to send an update.

7.3.1.4 Least Mean Square Filter

In Ref. [32], authors argue that the adaptive filter theory [1] offers an alternative solution for performing predictions, without requiring a priori information about the statistical properties of the phenomenon of interest. Among the variety of techniques, they chose the LMS algorithm, arguing that it is known to provide good performances in a wide spectrum of applications. As in Equation (7.12), the LMS is an autoregressive filter, where the output is a linear combination of previous measurements $\hat{s}[t] = x[t]^T \theta[t]$. In contrast to the proposed approach by [37], the vector of parameters $\theta[t]$ is updated over time as new measurements are taken. The LMS theory gives the following set of three equations for updating the parameters:

$$\hat{s}[t] = x[t]^T \theta[t]$$

$$e[t] = s[t] - \hat{s}[t]$$

$$\theta[t+1] = \theta[t] + \mu x[t]^T e[t]$$

where $e[t]$ is the error made by the filter at epoch t and μ is a parameter called the step size, which regulates the speed of convergence of the filter. The choice of the order p of the filter and of the step size μ are the only parameters that must be defined a priori. Authors of [32] suggested on the basis of their experiments that orders from 4 to 10 provided good results. Concerning the choice for the parameter μ, they suggest to estimate it by setting $\mu = 10^{-2}\frac{1}{E}$, where $E = \frac{1}{T}\sum_{t=1}^{T} |s[t]|^2$ is the mean input power of the signal.

7.3.2 Spatial Modeling

Two different approaches were investigated to model spatial dependencies with RM. In [33], RM are used on each edge of a routing tree that connects the nodes to the base station (edge monitoring). In Ref. [5], the network is partitioned in groups of nodes, with one model for each group (clique models).

7.3.2.1 Edge Monitoring

In this approach [33], it is assumed that nodes are connected to the base station by means of a routing tree. As with temporal modeling, there is one model h_i for each sensor i. Additionally, there is also one model for each edge connecting neighboring nodes in the routing tree. More specifically, assuming that there is an edge connecting node i to j, a model h_{j-i} is also produced by node j to represent the difference in value between the measurement $s_j[t]$ and the approximation $\hat{s}_i[t]$. Denoting this difference by $\Delta_{j-i}[t] = s_j[t] - \hat{s}_i[t]$, the model has the form

$$\hat{\Delta}_{j-i}[t] = h_{j-i}(x, \theta)$$

where x is a vector of inputs that consists, for example, of the past difference between measurements of sensor i and j, and θ is a vector of parameters. The most simple model is the constant model,

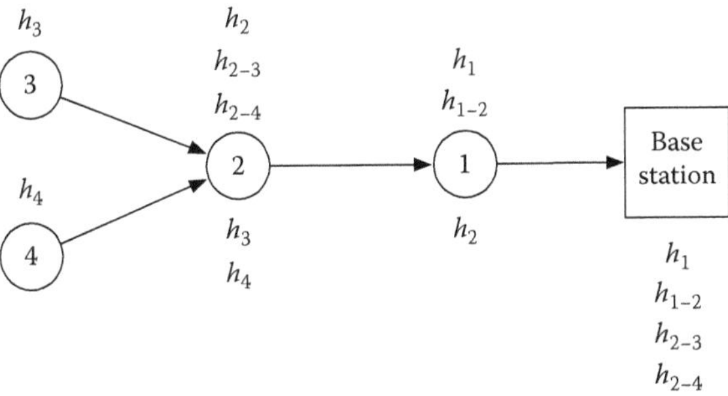

Figure 7.6 Replicated models with edge monitoring: Models h_i are used to infer sensor measurements, while models h_{j-i} are used to monitor the differences between the measurements of two adjacent nodes.

which was the only one considered in [33], which is defined by

$$\hat{\Delta}_{j-i}[t] = \Delta_{j-i}[t-1]$$

Note that more complex models such as autoregressive models could also be used.

As in temporal modeling, each sensor node i has its own model h_i, which is updated when the prediction is above the user-defined error threshold ϵ. The copy of the model is, however, not maintained by the base station, but by the parent node in the routing tree. Each node j that has children maintains a model h_{j-i} for each of its children i. A copy of these models is maintained by the base station. A second user-defined error threshold ϵ_Δ is used to determine when to update these models. The way models are distributed in a network is illustrated in Figure 7.6 for a network of four nodes and a routing tree of depth three. Models produced by nodes are listed above each node, and models used to get predictions are listed below each node.

The measurement of a node i is obtained by the base station by summing the prediction for the root node measurement, and the predictions for the differences between all pairs of nodes that separate node i from the base station. For example, in the network illustrated in Figure 7.6, a prediction for sensor 4 is obtained by summing

$$\hat{s}_4[t] = \hat{s}_1[t] + \hat{\Delta}_{1-2}[t] + \hat{\Delta}_{2-4}[t]$$

Since $\Delta_{j-i}[t] = s_j[t] - \hat{s}_i[t]$ and that the RM scheme ensures $|\hat{s}_i[t] - s_i[t]| < \epsilon$, $|\hat{\Delta}_{j-i}[t] - \Delta_{j-i}[t]| < \epsilon_\Delta$, it follows that

- A prediction for the root node can be obtained with an accuracy $\pm\epsilon$.
- A prediction for a node l hops away from the root node and can be obtained with an accuracy $\pm(\epsilon + l(\epsilon + \epsilon_\Delta))$.

The accuracy therefore depends on the depth of a node in the routing tree, and lower ϵ values must therefore be used to achieve the same accuracy as in temporal modeling.

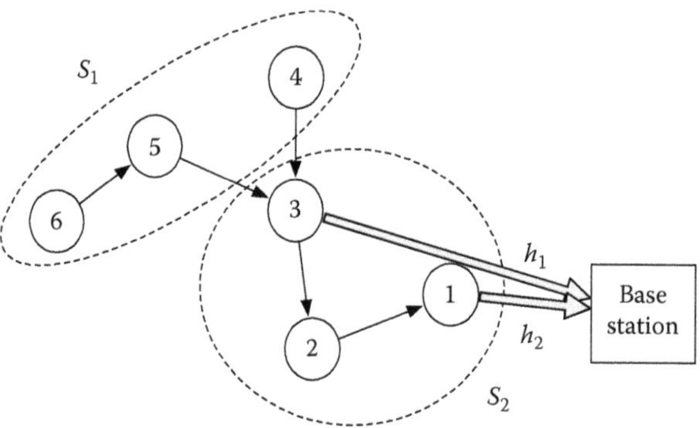

Figure 7.7 Clique models: The network here is partitioned into two cliques $\mathcal{S}_1 = \{4, 5, 6\}$ and $\mathcal{S}_2 = \{1, 2, 3\}$, with replicated models h_1 and h_2 for each clique. The clique root of $\mathcal{S}_1$ is sensor node 3 and the clique root of $\mathcal{S}_2$ is sensor node 1.

7.3.2.2 Clique Models

The rationale of clique models, investigated in [5], is to partition the network into groups of sensors, called *cliques*, with one RM for each clique. Note that the term "clique" here refers to group of nodes and is not related to the notion of clique in graph theory. The measurements of sensors in a clique are gathered at a common sensor node called clique root. The clique root may not be part of the clique. Let $\{\mathcal{S}_k\}_{1 \le k \le K}$ be a partitioning of $\mathcal{S}$ in K cliques, that is, a set of K cliques such that $\mathcal{S} = \cup_{1 \le k \le K} \mathcal{S}_k$. Let $i_k^{root} \in \mathcal{S}_k$ be the sensor node that takes the role of the root of clique k.

The measurements of the set of sensors $i \in \mathcal{S}_k$ are gathered at the clique root i_k^{root}, where data are modeled by a model $h_k(x, \theta)$. Inputs x may take values in the set of measurements of the sensors in the clique. The clique root then transmits to the sink the minimal subset of parameters/data such that the measurements $s_i[t]$ and model counterpart $\hat{s}_i[t]$ do not differ by more than ϵ. An illustration of a clique partition is given in Figure 7.7. Note that updating the model may require a clique root to rely on multi-hop routing. For example, sensor node 3 may not be in communication range of the base station and may require some nodes of $\mathcal{S}_2$ to relay its data.

The choice of the model and the partitioning of cliques are clearly central pieces of the system. Inspired by the work of [6] concerning model-driven data acquisition, authors in [5] suggest to use Gaussian models. The clique root collects data from all the sensors in the clique for N epochs, compute the mean vector and covariance matrix, and communicate these data to the base station. Then, at every epoch, the clique root determines what measurements must be sent so that the base station can infer the missing measurements with a user-defined ϵ error threshold.

The goal of the approach is to reduce the communication costs. These are divided into intra-source and source-base station costs. The former is the cost incurred in the process of collecting data by the clique root to check if the predictions are correct. The latter is the cost incurred while sending the set of measurements to the sink. Authors show that the problem of finding optimal cliques is NP-hard, and propose the use of a centralized greedy algorithm to solve the problem. The heuristic starts by considering a set of S cliques, that is, one for each sensor, and then assesses the reduction of communication obtained by fusing all combinations of cliques. The algorithm stops once fusion leads to higher communication costs.

7.3.3 Discussion

RM have a high potential in reducing communications, and their main advantage is to guarantee bounded approximation errors. Temporal modeling is easy to implement, and the different proposed approaches do not require much computation, which makes them suitable for the limited computational resources of sensor nodes. In terms of communications savings, the network load of sensors is reduced by a factor proportional to the number of updates required to maintain synchronized models between the sensors and the base station. The highest network load is sustained by the base station and depends on the number of updates sent during an epoch. If all the sensor measurements can be predicted, then no update is sent and the load is therefore null for all sensors. At the other extreme, if all sensor nodes send an update, the load distribution is similar to that of collecting all the measurements. Therefore, RM give

$$L_{\max} \sim O(S)$$

The modeling of spatial dependencies is attractive as spatial correlations are very often observed in sensor network data. Also, it is likely that all temporal models have to update their parameters at about the same time, that is, when an unexpected event occurs in the environment, for example. The modeling of spatial dependencies, however, raises a number of concerns. In the edge-monitoring approach, the error tolerance ϵ must be reduced in order to provide the same accuracy guarantee as in temporal modeling approaches [33]. In clique models, the partitioning of the network is a computationally intensive task that, even if undertaken by the base station, raises scalability issues [5]. In terms of communication savings, the network load of sensors is reduced, as for temporal modeling, by a factor that depends on the average number of updates. The savings can, however, be much higher, particularly, in situations where all the measurements increase by the same amount, for example.

An issue common to all the approaches based on RM is packet losses. In the absence of notification from a sensor node, the base station deems the prediction given by the shared model to fall within the ϵ error tolerance. Additional checking procedures must, therefore, be considered for this scheme to be completely reliable. To that end, a "watchdog" regularly checking the sensor activity and the number of sent packets can be set up, as discussed in [32], for example. By keeping the recent history of sent updates in the sensor node memory, these can be communicated to the sink at checking intervals if the number of sent packets differ from the number of received packets. Node failure is detected by the absence of acknowledgment from the sensor node to the watchdog request. Finally, the choice of the model is also an important factor in the efficiency of RM. Techniques based on model selection can be used to tackle this problem [15,18].

7.4 Aggregative Approaches

Aggregative approaches are based on aggregation services, which allow to aggregate sensor data in a time- and energy-efficient manner. A well-known example of aggregation service is the TAG system, developed at the University of Berkeley, California [24,25]. TAG stands for Tiny AGgregation and is an aggregation service for sensor networks that has been implemented in TinyOS, an operating system with a low memory footprint specifically designed for wireless sensors [20]. In TAG, an epoch is divided into time slots so that sensors' activities are synchronized according to their depth in the routing tree. Any algorithm can be relied on to create the routing tree, as long as it allows data to flow in both directions of the tree and does not send duplicates [24].

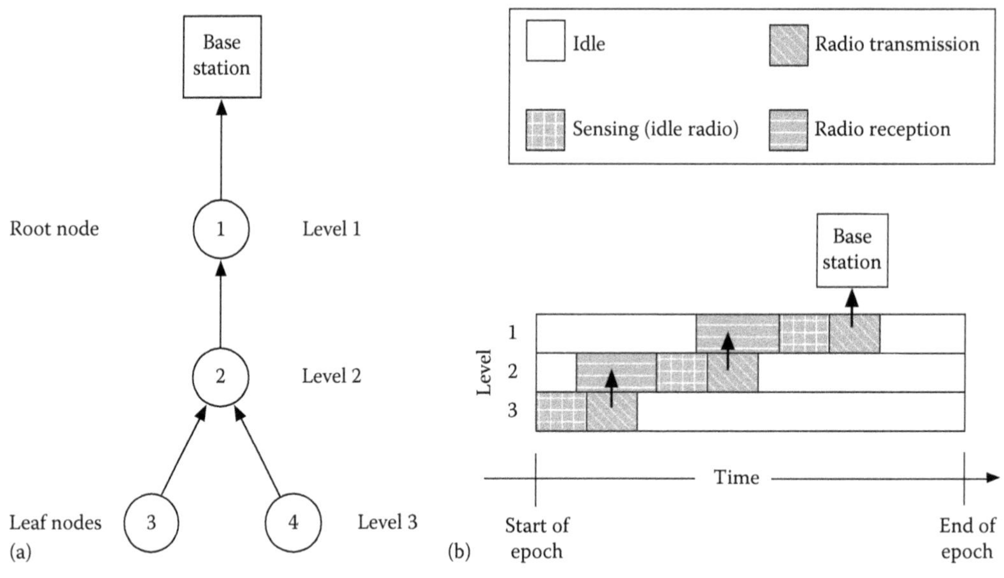

Figure 7.8 Multi-hop routing along a routing tree, and node synchronization for an efficient use of energy resources. (a) Routing tree of depth three. (b) Activities carried out by sensors depending on their level in the routing tree. (Adapted from Madden, S. et al., TAG: a tiny aggregation service for ad-hoc sensor networks. In *Proceedings of the 5th ACM Symposium on Operating Design and Implementation (OSDI),* Vol. 36, pp. 131–146. ACM Press, 2002. With permission.)

The TAG service focuses on low-rate data collection task, which permit loose synchronization of the sensor nodes. The overhead implied by the synchronization is therefore assumed to be low. The goal of synchronization is to minimize the amount of time spent by sensors in powering their different components and to maximize the time spent in the idle mode, in which all electronic components are off except the clock. Since the energy consumption is several orders of magnitude lower in the idle mode than when the CPU or the radio is active, synchronization significantly extends the wireless sensors' lifetime. An illustration of the sensors' activities during an epoch is given in Figure 7.8 for a network of four nodes with a routing tree of depth three. Note that the synchronization is maintained at the transport layer of the network stack and does not require precise synchronization constraints. Aggregation services such as TAG allow to reduce energy consumption both by carefully scheduling sensor node's activity and by allowing the data to be aggregated as they are routed to the base station.

Using the terminology of [24,25], an aggregate of data is called a *partial state record* and is denoted by $\langle . \rangle$. It can be any data structure, such as a scalar, a vector, or a matrix, for example. Partial state records are initialized locally on all nodes, and then communicated and merged in the network. When the partial state record is eventually delivered by the root node to the base station, its elements may be recombined in order to provide the observer with the final output. Methods based on aggregation require the definition of three primitives [24,25]:

- An initializer *init* that creates a partial state record
- An aggregation operator f, that merges partial state records

- An evaluator e that returns, on the basis of the partial state record finally delivered to the base station, the result required by the application

The partial state records are merged from the leaf nodes to the root, along a synchronized routing tree such as in Figure 7.8

In practice: Aggregation services were first shown [24,25] to be able to compute simple operations like the minimum, the maximum, the sum, or the average of a set of measurements. For example, for computing the sum, the following primitives can be used:

$$\begin{cases} init(s_i[t]) = \langle s_i[t] \rangle \\ f(\langle S1 \rangle, \langle S2 \rangle) = \langle S1 + S2 \rangle \\ e(\langle S \rangle) = S \end{cases}$$

Measurements are simply added as they are forwarded along the routing tree. The resulting aggregate obtained at the base station is the overall sum of the measurements. The main advantage of aggregation is that the result of an operator is computed without collecting all the measurements at the base station. This can considerably reduce the amount of communication.

In data modeling, aggregation services can be used to compute the parameters of models. For example, let us consider a sensor network monitoring the temperature in a room where an air conditioning system is set at 20°C. Most of the time, the temperature measurements of sensor nodes are similar, and can be approximated by their average measurements. The *average model* [5,10,24] was one of the first proposed, as its implementation is fairly straightforward. The interest of such a model can be to detect, for example, when a window or a door is opened. To do so, the average of the measurements is first computed by means of an aggregation service and retrieved at the base station. The result is then transmitted back to the sensor nodes, which can compare their measurements with average measurement. If the difference is higher than some user-defined threshold, a sensor node can notify the base station that its local measurement is not in agreement with the average measurement.

In [8], authors showed that the coefficient of a linear model can be computed using aggregation services. Their approach, called *distributed regression*, allows to use more complex models for representing sensor measurements as a function of spatial coordinates. The average model is first presented in the following, and is followed by the presentation of the distributed regression algorithm.

7.4.1 Average Model

The average model is a simple model that well illustrates the rationale of aggregation. It consists in modeling all the sensor measurements by their average value, that is,

$$\hat{s}_i[t] = \mu[t]$$

where $\mu[t] = \frac{\sum_{i=1}^{S} s_i[t]}{S}$ is the average of all the sensor measurements at epoch t [5,10,24]. The average is obtained by the following primitives:

$$\begin{cases} init(s_i[t]) = \langle 1, s_i[t] \rangle \\ f(\langle C1, S1 \rangle, \langle C2, S2 \rangle) = \langle C1 + C2, S1 + S2 \rangle \\ e(\langle C, S \rangle) = \frac{S}{C} \end{cases}$$

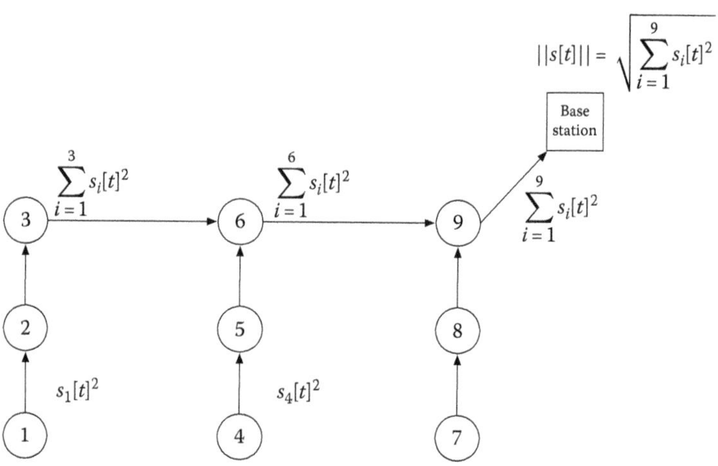

Figure 7.9 Aggregation service at work for computing the average $\mu[t] = \frac{\sum_{i=1}^{9} s_i[t]}{\sum_{i=1}^{9} 1}$ of the measurements taken by sensors at epoch t.

The partial states record $\langle C, S\rangle$ is a vector of two elements consisting of the count and the sum of sensor measurements. The aggregation process is illustrated in Figure 7.9. This way, only two pieces of data are transmitted by all sensor nodes. The main advantage of aggregation is that all nodes send exactly the same amount of data, which is independent of the number of sensors. It, therefore, provides a scalable way to extract information from a network. Also, it may dramatically decrease the highest network load, typically sustained by the root node. In the network of nine sensors represented in Figure 7.9, the transmission of all measurements to the base station would cause the root node to send nine measurements at every epoch. This number is reduced to two thanks to the aggregation process.

Once obtained at the base station, aggregates can be transmitted from the base station to sensor nodes [8]. This allows sensor nodes to compare their measurement to the average measurement, and to provide the base station with their measurement if

$$|s_i[t] - \mu| > \epsilon$$

where ϵ is a user-defined error threshold.

This is illustrated in Figure 7.10, where nodes 1 and 8 actually send their true measurement after receiving the feedback $\mu[t]$ from the base station. Such a strategy allows to bound the approximation errors of the average model. It, however, implies additional communication rounds between the base station and the sensor nodes, which are intuitively expensive in terms of communication.

7.4.2 Distributed Regression

Similar to the average model, an aggregative approach can be used to compute the regression coefficients of a linear model. The approach was investigated by Guestrin et al. in [8], who relied on basis functions to approximate sensor network measurements [9]. The basis functions may be defined over space and time, allowing in some cases to compactly represent the overall spatiotemporal variations by a small set of coefficients.

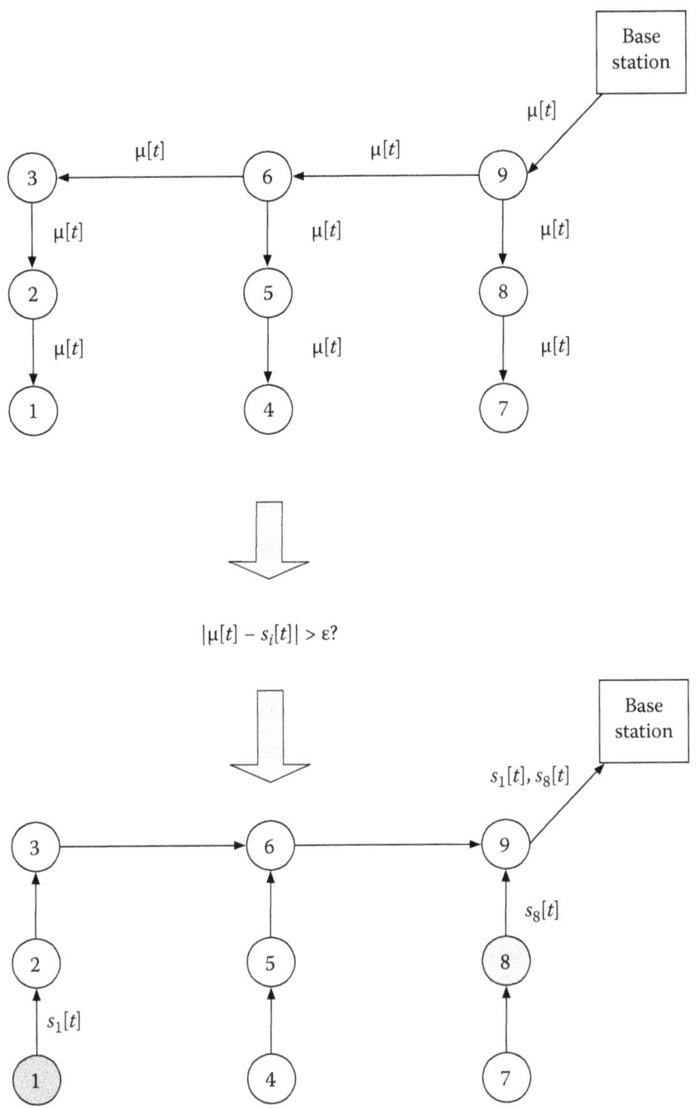

Figure 7.10 The aggregate $\mu[t]$ can be communicated to all sensors allowing them to compute locally $|s_i[t] - \mu[t]|$ and to send their true measurement $s_i[t]$ if the difference $|s_i[t] - \mu[t]|$ is higher than a user-defined error threshold ϵ. In this example, sensors 1 and 8 update their measurements.

More precisely, let $\mathcal{H} = \{\pi_1, \ldots, \pi_p\}$ be a set of p basis functions which are used to represent the sensor measurements. This set must be defined by the observer, prior to running the algorithm. The inputs of these basis functions can be functions of the time t or of the sensor coordinates $c = (c_1, c_2, c_3)$ (assuming 3D coordinates), for example. Let p be the overall number of basis functions and let θ_j be the coefficient of the jth basis function. The approximation to a sensor measurement at time t and location c is given by

$$\hat{s}(c,t) = \sum_{j=1}^{p} \theta_j \pi_j(c,t)$$

Example 7.3:

A quadratic regression model over time $\hat{s}(c,t) = \theta_1 t + \theta_2 t^2$ is defined by two basis functions $\pi_1(c,t) = t$ and $\pi_2(c,t) = t^2$. The addition of $\pi_3(c,t) = c_1$ and $\pi_4(c,t) = c_2$ gives a model that captures correlations over space. An intercept can be added with $\pi_5(c,t) = 1$.

Using the notations defined earlier, we have a model h that represents the overall variations in the sensor field by

$$h_\theta : \mathbb{R}^p \rightarrow \mathbb{R}$$

$$x \mapsto \hat{s}(c,t) = \sum_{j=1}^{p} \theta_j \pi_j(c,t)$$

where the inputs $x = (\pi_1(c,t), \ldots, \pi_p(c,t))$ are functions of time and coordinates, and the parameters $\theta = (\theta_1, \ldots, \theta_p)$ are the coefficients of the linear model. Approximations $\hat{s}_i[t]$ for sensor i are given by specifying the coordinates c_i of sensor in the model, that is, $\hat{s}_i[t] = \sum_{j=1}^{p} \theta_j \pi_j(c_i, t)$. An interesting feature of this model is that it not only provides approximations to sensor measurements but also allows to provide predictions for all locations in the field.

Assuming that N_i measurements $s_i[t]$ have been taken at locations c_i for N_i epochs t, let $N = \sum_{i=1}^{S} N_i$ be the overall number of measurements taken by sensor nodes. The coefficients θ_j can be identified by minimizing the mean squared error between the actual and approximated measurements.

For this, let Y be the $N \times 1$ matrix that contains these measurements, and let θ be the column vector of length p that contains the coefficients θ_j. Finally, let X be the $N \times p$ matrix whose columns contain the values of the basis functions for each observation in Y. Using this notation, the optimization problem can be stated as

$$\theta^* = \arg\min_{\theta} ||X\theta - Y||^2$$

which is the standard optimization problem in regression [9]. The optimal coefficients are found by setting the gradient of this quadratic objective function to zero, which implies

$$(X^T X)\theta^* = X^T Y \tag{7.13}$$

Let $A = X^T X$ and $b = X^T Y$. A is referred to as the scalar product matrix and b as the projected measurement vector. In distributed regression, the measurements $s_i[t]$ do not need to be transmitted to the base station. Instead, the matrix A and vector b are computed by the aggregation service. Once aggregated, the coefficients θ can be computed at the base station by solving Equation 7.13.

Let X_i be the $N_i \times p$ matrix containing the values of the basis functions for sensor i, and let Y_i be the $N_i \times 1$ matrix that contains the measurements taken by sensor i. Both X_i and Y_i are available at sensor i, which can therefore compute locally $A_i = X_i^T X_i$ and $b_i = X_i^T Y_i$.

The matrix A and the vector b are actually sums of $A_i = X_i^T X_i$ and $b_i = X_i^T Y_i$. Using this fact, A and b can be computed by an aggregation service by merging along the routing tree the contributions A_i and b_i of each sensor.

Indeed, assuming S sensors, each of which collects a measurement $s_i[t]$, we have

$$a_{j,j'} = \sum_{i=1}^{S} \pi_j(c_i, t)\pi_{j'}(c_i, t) \tag{7.14}$$

where $a_{j,j'}$ is the entry at the jth row, j'th column of the scalar product matrix A. Similarly, we have

$$b_j = \sum_{i=1}^{S} \pi_j(c_i, t)s_i[t] \tag{7.15}$$

where b_j is the jth element of the projected measurement vector. All the elements $a_{j,j'}$ and b_j can be computed by means of an aggregation service, using the following primitives:

- For elements $a_{j,j'}$:

$$\begin{cases} init(\mathrm{i}) = \langle \pi_j(c_i, t)\pi_{j'}(c_i, t)\rangle \\ f(\langle S1\rangle, \langle S2\rangle) = \langle S1 + S2\rangle \\ e(\langle S\rangle) = S \end{cases}$$

- For elements b_j:

$$\begin{cases} init(\mathrm{i}) = \langle \pi_j'(c_i, t)s_i[t]\rangle \\ f(\langle S1\rangle, \langle S2\rangle) = \langle S1 + S2\rangle \\ e(\langle S\rangle) = S \end{cases}$$

The computation of the matrix A requires the aggregation of p^2 elements, while the aggregation of the vector b requires the aggregation of p elements. Once all the elements are retrieved at the base station, the set of coefficients θ can be computed by solving the system

$$A\theta = b$$

The resulting model allows to get approximations to sensor measurements. Depending on the model, approximations may also be obtained for other spatial coordinates or at future time instants (cf. example 3). As with the average model, the parameters θ may be communicated to all nodes, allowing each sensor to locally compute the approximation obtained at the base station. This makes it possible to check that approximations are within an error threshold ϵ defined by the observer. All sensors whose approximations differ by more than $\pm\epsilon$ may notify their true measurements to the base station.

7.4.3 Discussion

The main advantage of aggregative approaches is that they allow to represent the variations of sensor measurements by means of models whose number of coefficients is independent of the number of sensor nodes in the network. This makes these approaches scalable to large networks. Furthermore, they allow to evenly distribute the number of radio transmissions among sensor nodes. Compared to model-driven and RM approaches, the network load of leaf nodes is increased, whereas the load of nodes close to the base station is reduced. Considering that the highest network load primarily determines the network lifetime, aggregative approaches are particularly attractive for reducing the load of sensor nodes close to the base station.

The highest network load depends on whether bounded approximation errors are required. Retrieving the model coefficients causes a highest network load of

$$L_{\max} \sim O(p^2)$$

where p is the number of parameters of the model. If bounded approximation errors are required, the p coefficients must be communicated to all nodes, causing p transmissions. Depending on the number of sensors for which approximations are more than ϵ away from their true measurements, an additional number of updates of up to S may be sent. Denoting by $L_{\max}^{\text{check}}$ the highest network load when approximations are checked against the true measurements, we have

$$L_{\max}^{\text{check}} \sim O(p^2 + S)$$

The upper bound is higher for data collection where all measurements are collected, and for which we had $L_{\max} = O(S)$ (cf. Section 7.1). The distributed regression with bounded errors, therefore, may lead to higher communication costs if the model does not properly reflect the variations of sensor measurements.

The choice of the model is thus an important issue, particularly when there is no a priori information on the type of variations. In practice, a solution may be to collect data from the whole network in order to get an overview of the types of measurement patterns. On the basis of this initial stage, different models may be tried and assessed at the base station, in order to select a model that properly fits the data. In this respect, it is worth mentioning that aggregative approaches can also be applied for dimensionality reduction purposes, using the principal component analysis [17] and the compressed sensing frameworks [23,30,38].

Finally, it is noteworthy that different optimizations can be brought to these approaches. In particular, the elements of the matrix A do not depend on sensor measurements. In the case of spatial models, they only depend on the spatial coordinates of the sensors. If these coordinates are known by the base station, the matrix A may be computed straightaway at the base station, thus saving $0(p^2)$ transmissions. In the same way, the base station may also infer the entries of A when time is involved. The load can therefore be reduced to $O(p)$ if no error threshold is set.

7.5 Conclusion

This chapter provided a state of the art on the use of learning techniques for reducing the amount of communication in sensor networks. Classifying these approaches in three main types, namely model driven, RM, and aggregative approaches, we outlined for each of them their strengths and their limits. Table 7.1 gives a summary of the different learning schemes in terms of error type and highest network load.

- Approaches based on *model-driven acquisition* reduce the highest network load to $|S_q|$, that is., the number of sensors whose measurements are effectively collected. The main characteristic of these approaches is that part of the network can remain in the idle mode. Model-driven data-acquisition, therefore, not only reduces the highest network load, but also allows to reduce to a negligible level the energy consumption of the sensor nodes not queried. In the idle mode, the energy consumption is about four orders of magnitude lower than in the active mode (e.g., 5 μA in the idle mode against 19.5 mA with MCU active for the Telos node [28]).

The subset of sensor nodes whose measurements are collected can be changed over time to distribute the energy consumption. Indeed, there exists in most cases different pairs of set of queried and predicted sensors for which the observer's accuracy requirements can be satisfied [16].

The error type entailed in the obtained predictions depends on whether the model can be trusted. If the model is correct, predictions can be bounded with an error threshold ϵ and a confidence level $1 - \delta$. However, the drawback of model-driven approaches is that unexpected events in the monitored phenomenon may not be detected if they concern locations where measurements are predicted. The use of multiple pairs of sets of queried and predicted sensors can be used to address this issue. Indeed, assuming that all sensor nodes will at some point be part of the set of queried sensors, an unexpected event will be detected when the sensor nodes monitoring the location of the event are queried. The time elapsed before the event is detected depends on the frequency at which subsets of sensors are changed. This time may be long, and therefore model-driven approaches are not well-suited to event detection tasks.

- The main characteristic of *RM* is to guarantee ϵ-bounded prediction errors, even in the case of unexpected events. These approaches, however, require the sensor nodes to take their measurements at every epoch, so that they can be compared with the predicted measurements. The energy savings depend on the frequency of updates of the model. In the optimal case, the predictions obtained by the model are always within $\pm\epsilon$ of the true measurements, and therefore no communication is needed. The energy savings are in this case around one order of magnitude (1.8 mA in the idle mode against 19.5 mA with MCU active for the Telos node [28]).

  In the worst case, updates are needed at every epoch. The Highest Network Load (HNL - cf. Section 7.1.2) is in $O(S)$ for this worst scenario, and has therefore the same order of magnitude as the default data collection scheme. It is therefore worth noting that depending on the number of parameters sent in each model update, the exact amount of communication can even be higher with RM than for the default data collection scheme.

Table 7.1 Comparison of Performances of the Different Modeling Approaches.

Learning Scheme	*Error Type*	*Highest Network Load*
Model-driven acquisition	Probabilistic bounded or unbounded	$L_{max}^{MD} \sim O(\lvert S_q \rvert)$
Replicated models	ϵ-bounded	$L_{max}^{RM} \sim O(S)$
Aggregative approaches	Unbounded or ϵ-bounded	$L_{max}^{DR} \sim O(p^2)$ $L_{max}^{DR_{check}} \sim O(p^2 + S)$

- Finally, *aggregative approaches* lead to either unbounded or ϵ-bounded error. The type of error depends on whether aggregates obtained at the base station are communicated to sensor nodes. If no check is made against the true measurement, the highest network load is reduced to the number of aggregates collected, that is, $O(p^2)$, where p is the number of parameters in a model, and optimization techniques can be used to reduce the HNL in $O(p)$. The main characteristic of aggregative approaches with unbounded errors is that the HNL does not depend on the number of sensors. This makes these approaches scalable to large networks.

  The model coefficients computed at the base station can be communicated to sensor nodes. This allows sensor nodes to compute locally the predictions obtained at the base station, enabling aggregative approaches to deal with event detection. The use of this feature can, however, be expensive in terms of communications, as an additional network load in $O(p + S)$ can be reached in the worst case.

In summary, modeling techniques can greatly reduce energy consumption, up to several orders of magnitude with model-driven approaches. The use of models, however, implies some approximations of the sensor measurements. For an observer, it is often important that the approximation errors are bounded, that is, within $\pm\epsilon$ of the true measurements. This guarantee is only possible with RM and aggregative approaches, which allow to compare the model predictions with the true measurements.

References

1. S.T. Alexander. *Adaptive Signal Processing: Theory and Applications*. Springer-Verlag, New York, 1986.
2. P.J. Brockwell and R.A. Davis. *Introduction to Time Series and Forecasting*. Springer, Berlin, 2002.
3. N. Burri and R. Wattenhofer. Dozer: Ultra-low power data gathering in sensor networks. In *Proceedings of the Sixth International Conference on Information Processing in Sensor Networks*, pp. 450–459. ACM, New York, NY, USA, 2007.
4. C. Chatfield. *Time-Series Forecasting*. Chapman & Hall/CRC, Boca Raton, FL, USA, 2001.
5. D. Chu, A. Deshpande, J.M. Hellerstein, and W. Hong. Approximate data collection in sensor networks using probabilistic models. In *International Conference on Data Engineering (IEEE)*, Piscataway, NJ, USA, 2006.
6. A. Deshpande, C. Guestrin, S. Madden, J. Hellerstein, and W. Hong. Model-driven data acquisition in sensor networks. In *Proceedings of the 30th Very Large Data Base Conference (VLDB'04)*, Toronto, Canada, 2004.
7. A. Deshpande, C. Guestrin, S.R. Madden, J.M. Hellerstein, and W. Hong. Model-based approximate querying in sensor networks. *The VLDB Journal, The International Journal on Very Large Data Bases*, 14(4):417–443, 2005.
8. C. Guestrin, P. Bodi, R. Thibau, M. Paski, and S. Madden. Distributed regression: An efficient framework for modeling sensor network data. *Proceedings of the Third International Symposium on Information Processing in Sensor Networks*, pp. 1–10, ACM, New York, NY, USA, 2004.
9. T. Hastie, R. Tibshirani, and J.H. Friedman. *The Elements of Statistical Learning*. Springer, Berlin, 2009.
10. C. Intanagonwiwat, R. Govindan, and D. Estrin. Directed diffusion: A scalable and robust communication paradigm for sensor networks. In *Proceedings of the ACM/IEEE International Conference on Mobile Computing and Networking*, pp. 56–67, ACM, New York, NY, USA, 2000.
11. A. Jain and E.Y. Chang. Adaptive sampling for sensor networks. *ACM International Conference Proceeding Series*, pp. 10–16, ACM, New York, NY, USA, 2004.

12. A. Jain, E.Y. Chang, and Y.-F. Wang. Adaptive stream resource management using Kalman filters. In *Proceedings of the ACM SIGMOD International Conference on Management of Data (SIGMOD '04)*, pp. 11–22, ACM, New York, NY, USA, 2004.
13. R.E. Kalman. A new approach to linear filtering and prediction problems. *Journal of Basic Engineering*, 82(1):35–45, 1960.
14. I. Lazaridis and S. Mehrotra. Capturing sensor-generated time series with quality guarantee. In *Proceedings of the 19th International Conference on Data Engineering (ICDE'03)*, Bangalore, India, March 2003.
15. Y. Le Borgne. *Learning in Wireless Sensor Networks for Energy-Efficient Environmental Monitoring*. PhD thesis, Université Libre de Bruxelles, Brussels, Belgium, 2009.
16. Y. Le Borgne and G. Bontempi. Round robin cycle for predictions in wireless sensor networks. In M. Palaniswami, ed. *Proceedings of the Second International Conference on Intelligent Sensors, Sensor Networks and Information Processing (ISSNIP)*, pp. 253–258, Piscataway, NJ, IEEE Press, 2005.
17. Y. Le Borgne, S. Raybaud, and G. Bontempi. Distributed principal component analysis for wireless sensor networks. *Sensors Journal*, 8(8):4821–4850, 2008.
18. Y. Le Borgne, S. Santini, and G. Bontempi. Adaptive model selection for time series prediction in wireless sensor networks. *Journal of Signal Processing*, 87(12):3010–3020, 2007.
19. M. Lefebvre. *Applied Stochastic Processes Universitext*. Springer, Berlin, 2007.
20. P. Levis, S. Madden, J. Polastre, R. Szewczyk, K. Whitehouse, A. Woo, D. Gay, J. Hill, M. Welsh, E. Brewer et al. TinyOS: An operating system for sensor networks. *Ambient Intelligence*, volume 35, 115–148, Springer, Berlin, 2005.
21. Z.Y. Li and R.C. Wang. Secure coverage-preserving node scheduling scheme using energy prediction for wireless sensor networks. *The Journal of China Universities of Posts and Telecommunications*, 17(5):100–108, 2010.
22. J.C. Lim and C.J. Bleakley. Extending the lifetime of sensor networks using prediction and scheduling. In *International Conference on Intelligent Sensors, Sensor Networks and Information Processing, 2008. ISSNIP 2008*, pp. 563–568. IEEE, 2008.
23. C. Luo, F. Wu, J. Sun, and C.W. Chen. Efficient measurement generation and pervasive sparsity for compressive data gathering. *IEEE Transactions on Wireless Communications*, 9(12):3728–3738, 2010.
24. S. Madden, M.J. Franklin, J.M. Hellerstein, and W. Hong. TAG: A Tiny AGgregation service for ad-hoc sensor networks. In *Proceedings of the Fifth ACM Symposium on Operating System Design and Implementation (OSDI)*, Vol. 36, pp. 131–146. ACM Press, New York, NY, USA, 2002.
25. S.R. Madden, M.J. Franklin, J.M. Hellerstein, and W. Hong. TinyDB: An acquisitional query processing system for sensor networks. *ACM Transactions on Database Systems*, 30(1):122–173, 2005.
26. T.M. Mitchell. *Machine Learning*. McGraw-Hill, New York, NY, USA, 1997.
27. C. Olston, B.T. Loo, and J. Widom. Adaptive precision setting for cached approximate values. *ACM SIGMOD Record*, 30(2):355–366, 2001.
28. J. Polastre, R. Szewczyk, and D. Culler. Telos: Enabling ultra-low power wireless research. In *Proceedings of the Fourth International Symposium on Information Processing in Sensor Networks*, pp. 364–369, ACM, New York, NY, USA, 2005.
29. J. Porter, P. Arzberger, H.W. Braun, P. Bryant, S. Gage, T. Hansen, P. Hansen, C.C. Lin, F.P. Lin, T. Kratz, W. Michener, S. Shapiro, and T. Williams. Wireless sensor networks for ecology. *BioScience*, 55(7):561–572, 2005.
30. G. Quer, R. Masiero, D. Munaretto, M. Rossi, J. Widmer, and M. Zorzi. On the interplay between routing and signal representation for compressive sensing in wireless sensor networks. In *Information Theory and Applications Workshop, 2009*, pp. 206–215. IEEE, 2009.
31. R. Rajagopalan and P.K. Varshney. Data aggregation techniques in sensor networks: A survey. *IEEE Communications Surveys and Tutorials*, 8(4):48–63, 2006.

32. S. Santini and K. Römer. An adaptive strategy for quality-based data reduction in wireless sensor networks. In *Proceedings of the Third International Conference on Networked Sensing Systems (INSS 2006)*, Chicago, IL, June 2006.
33. A. Silberstein, R. Braynard, G. Filpus, G. Puggioni, A. Gelfand, K. Munagal, and J. Yang. Data-driven processing in sensor networks. *Ambient Intelligence*, pp. 115–148, 2005.
34. A. Swami, Q. Zhao, Y.-W. Hong, and L. Tong. *Wireless Sensor Networks: Signal Processing and Communications*. John Wiley & Sons, New York, NY, USA, 2007.
35. S. Tilak, N.B. Abu-Ghazaleh, and W. Heinzelman. A taxonomy of wireless micro-sensor network models. *ACM SIGMOBILE Mobile Computing and Communications Review*, 6(2):28–36, 2002.
36. C. Zhang, M. Li, M. Wu, and W. Zhang. Monitoring wireless sensor networks using a model-aided approach. *Lecture Notes in Computer Science*, 3976:1210, 2006.
37. D. Tulone and S. Madden. PAQ: Time series forecasting for approximate query answering in sensor networks. In *Proceedings of the Third European Workshop on Wireless Sensor Networks*, pp. 21–37. Springer, Berlin, 2006.
38. L. Xiang, J. Luo, and A. Vasilakos. Compressed data aggregation for energy efficient wireless sensor networks. In *Eighth Annual IEEE Communications Society Conference on Sensor, Mesh and Ad Hoc Communications and Networks (SECON), 2011*, pp. 46–54. IEEE, 2011.

Chapter 8

Neuro-Disorder Patient Monitoring via Gait Sensor Networks

Toward an Intelligent, Context-Oriented Signal Processing

Fei Hu, Qingquan Sun, and Qi Hao

Contents

8.1 Introduction

Based on World Health Organization (WHO) report [1], over 1 billion people worldwide suffer from neuro-disorder diseases (NDDs), ranging from epilepsy to Parkinson's disease. Every year over 7 million people die as a result of NDDs [1]. In the United States alone, NDDs cost over $148 billion per year, and the annual cost of care for each NDD patient is over $64,000 per year [2]. Before determining the medication or surgery treatments for many different types of NDDs (>50 [3]), a neurology doctor needs to analyze the patient's concrete NDD symptoms, which include abnormal gaits (in daytime walking) or motor disorders (mostly occurring in nighttime [4–6]).

Despite NDDs' extremely high medical cost, up to this point we still rely on *labor-intensive observations* to determine neuro-disorder symptoms. For example, in the initial observation phase of epilepsy, a patient often needs to stay in the hospital for at least a few days (each day of in-hospital cost is around $1500 today in the U.S. hospitals [7]). Today a NDD physician either asks the patient to orally report the symptoms, or uses recorded video to search each symptom [4]. The oral report is unreliable since the patient cannot clearly recall what occurred. The video-based analysis still costs significant time and effort from a NDD physician. A programmable, PTZ-capable camera could cost more than $1000 each. Even a low-resolution video sensor could cost >$300 each.

Therefore, it is critical to design a *gait recognition* system for accurate capture of NDD symptoms. Such an automatic gait monitoring system has to be low-cost, and uses highly motion-sensitive sensors and accurate gait pattern recognition algorithms. Pyroelectric sensors have many advantages: they are inexpensive, only $3 each. They are very small, only the size of semiconductor tubes. A pyroelectric sensor generates very small sensing data (a few bits for each event). No complex calculation is needed. Moreover, the low-cost sensor can detect human's radiation (8–14 μm) [8] with a high sensitivity. They can also accurately capture angular velocities of a thermal source (0.1–3 rad/s).

We have successfully built an *Intelligent Compressive Multi-Walker Recognition and Tracking (iSMART)*[9–14]. The neighboring sensors combine their readings to achieve a global view of the walkers' paths through the recognition and trace of their gaits. It is challenging to distinguish among different people's gait sensing signals when their signals are mixed together. Binary signal separation algorithms are needed to associate binary sequences to different individuals. iSMART control needs the knowledge from multiple disciplines: machine learning, pattern recognition, signal processing, computer networking, embedded systems (sensors), and computer programming [11].

Figure 8.1 shows the basic principle of iSMART system. A PSN uses multi-hop wireless communications to send data to a base station (for centralized data processing) or to perform localized data

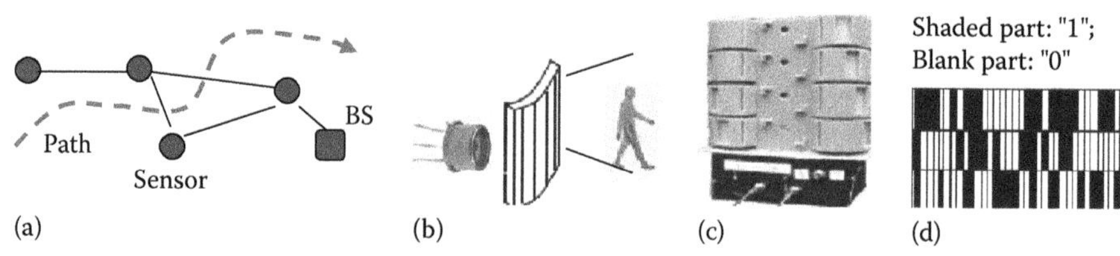

Figure 8.1 (a) PSN for tracking (b) Pyroelectric sensor with Fresnel lens (c) Fresnel lens (d) Binary data.

processing without sending to base station (called distributed processing). When a patient passes through the sensor network, the system is able to recognize and track the patient (Figure 8.1a). To generate rich visibility modulation from human thermal sources, we put a Fresnel lens before the pyroelectric sensor (see Figure 8.1b). The lens has different hole patterns that filter thermal signals in certain mode. We have tried dozens of different lens patterns, and found the pattern shown in Figure 8.1c is especially sensitive to different gaits, that is, different patients generate more discriminative sensor data. The lens is made of plastic materials and could be very inexpensive after batch production. Note that Figure 8.1c shows that we use one lens to perform signal modulation for multiple sensors simultaneously. Figure 8.1d is the observation data. The shaded part means "1" and blanked part means "0." That is, the observed data is binary format (later on we will explain how we get such binary data). Note that Figure 8.1d shows a matrix of binary data. This is because we use each row to represent the data from the same sensor. Each column means data in the same time (from different sensors).

On "binary" observation data: Although all computer data are in binary format, here "binary observation data" do not refer to computer digital representation. Instead, we mean that a patient is "detected" (use "1" to represent it) or "not detected" (use "0") in a certain time from the perspective of a sensor. The binary generation procedure is shown in Figure 8.2. In Figure 8.2a, we can see that when an analog signal from pyroelectric sensor passes through a Sine filter, we get most energy for that signal. Then compare its amplitude to a threshold, if it is higher than the threshold, we interpret it as "1"; otherwise, it is "0." Figure 8.2b shows the signal waveform change from original oscillating signals to interpreted binary data.

In this chapter, we will report our recent research results on a challenging issue in iSMART: *context feature extraction for crowded scene patient gait identification*. Context information can help us quickly locate an object. For instance, when looking for specific objects in complex and cluttered scenes, human and other intelligent mammals use *visual context* information to facilitate the search, by directing their attention to *context-relevant* regions (e.g., searching for cars in the street, looking for a plate on the table).

In Figure 8.3, the input to the context filter is the binary observations from multiple sensors. We divide the stream into different observation windows. Each window is a matrix. We then process such a matrix via context filter, which runs signal projection algorithms (such as principal component analysis) to extract dominant features. Based on these features, we then determine whether or not this window of data has RoI. Such a RoI could mean any scenarios such as A and B

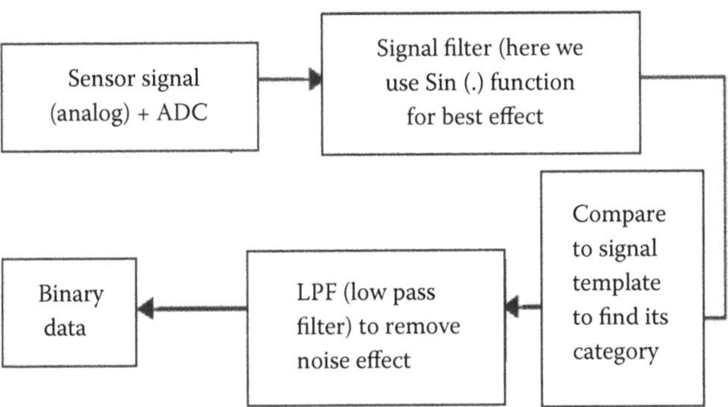

Figure 8.2 Generate binary observation data. (a) Binary data generation via threshold comparison. (b) Waveform (from raw signals to binary).

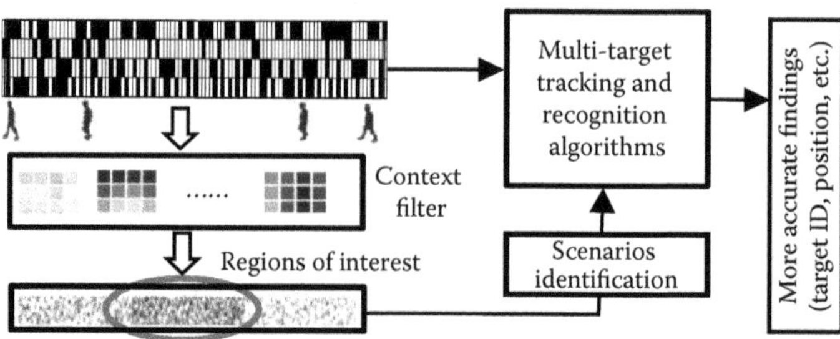

Figure 8.3 Functional diagram of context awareness.

are passing by. As shown in Figure 8.3, besides passing through the context filter, the binary data will go through Bayesian-based patient tracking and recognition algorithms. By combining with the identified scenarios (i.e., RoIs), we will be able to obtain more accurate information such as what person is passing through what location.

The rest of this chapter will be organized as follows. Section 8.2 provides an overview of related work. The big picture of gait context awareness is stated in Section 8.3. Next, in the main body of this paper (Section 8.4), we elaborate the use of NMF to extract patient gait context. Sections 8.5 and 8.6 briefly introduce some basic algorithms to be used in performance comparisons between different gait context extraction methods. Section 8.7 details our experiment results. Section 8.8 concludes this paper and some future works are mentioned.

8.2 Related Works

The work close to ours is gait context awareness in video-based gait recognition and human tracking system. For instance, in Refs. [15,16] the concept of region of interest (RoI) has been defined to refer to a target or special scene to be searched in a complex, large-scale picture pixels. They utilize humans' RoI capture behaviors: typically humans look at a scene in a top-down approach. That is, we first take a glance at the entire scene without carefully looking at each detail. If we find an interesting profile, we then look at details to ensure this is what we are looking for (i.e., RoI). In Refs. [17,18] the context concept is enhanced by the definition of saliency, which means how different a local image profile is different from background image. Bayesian framework is used to deduce the saliency and context values.

This top-down context capture approach [17–19] can also be used for our case. When we receive a window of pyroelectric sensor data, we may use fast statistical methods (such as energy functions) to see whether this window has different statistics. If so, more accurate context extraction algorithms (such as signal projection methods) can be used to extract the signal bases and the corresponding basis weights (i.e., coefficients). However, because the binary pyroelectric data does not have visual interpretation (except some 0, 1 binary values), the saliency-based context extraction used in traditional video systems cannot be used to capture human gait features. Therefore, we propose to use binary-oriented NMF and PCA algorithms to extract the most dominant sensing signal features.

8.3 Principle of Gait Context Extraction

Our iSMART system uses a gait sensor cluster architecture as shown in Figure 8.4. Instead of evenly distributing sensors everywhere, we deploy sensors into "clusters." This can fully utilize the sensitive, wide-angle thermal detection capability of pyroelectric sensors and thus reduce repeated sensor

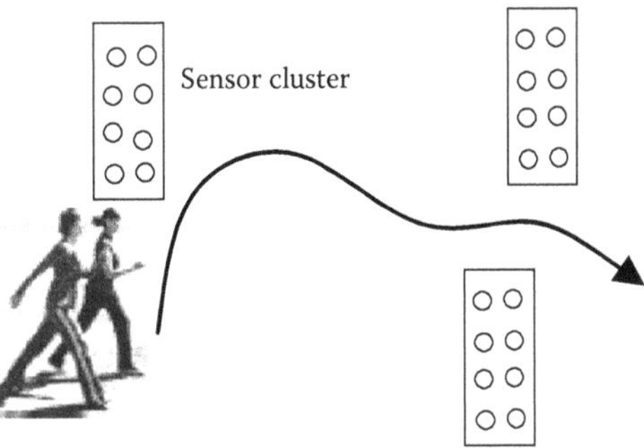

Figure 8.4 Multicluster gait sensing.

measurements. Moreover, since a microcontroller has multiple ADC (analog-to-digital converter) interfaces, by grouping multiple sensors in one cluster, we can use one RF communication board to send out multiple sensors' data, which reduces the hardware cost. Through careful control of each sensor's facing direction, we could well capture a 360° view of a neighborhood around a cluster.

Assume a cluster has N sensors. For such N-dimension data stream, we will use the principle shown in Figure 8.5 to identify a new gait *context* (called *context extraction*). As shown in Figure 8.5, for such a N-dimension data, first we need to segment it into different windows. The window size depends on how much data a sensor can handle in real-time. Here we use an 8×16 window size to form a binary matrix, called observation data $\mathbf{X}$. Each value is either 1 (means "detected") or 0 (means "not detected"). The context extraction system includes two phases:

1. *Training phase*: It is important to identify some common aspects to be compared between different scenarios. For instance, in traditional video systems, to identify human faces, we typically use eye size, nose length, distance between eyes, etc. to serve as comparison "bases." Likewise, we need to identify some gait "bases" in our PSN, although each basis may not have

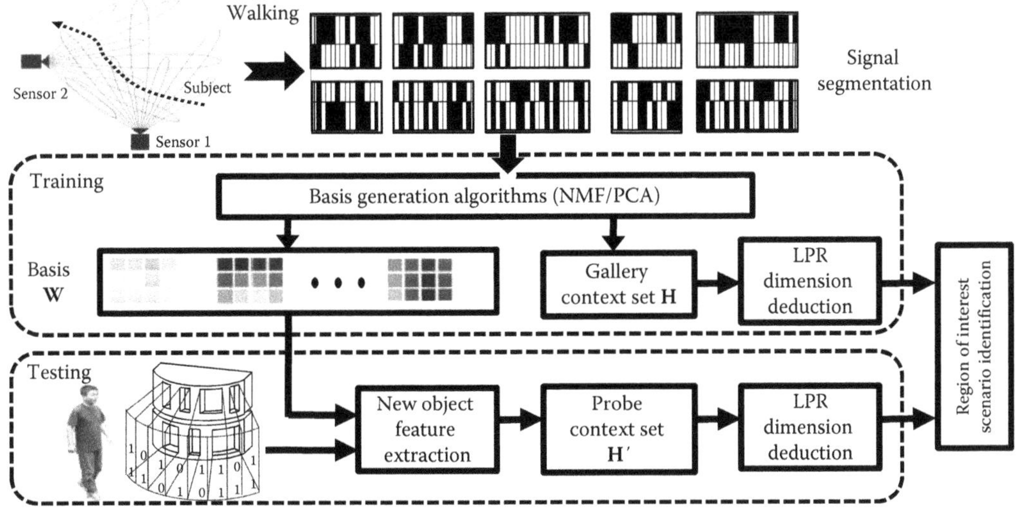

Figure 8.5 Gait context detection system.

physical interpretation of gaits as clear as human faces. Depending on which signal projection algorithm we use (such as PCA, NMF, etc.), we could obtain a set of bases for all pyroelectric sensor data to be trained. We select windows of binary data for almost all different contexts. For each of those contexts, we obtain the corresponding coefficients (called weights) for each basis. All contexts' bases and weights are stored in context template database for testing purpose.

2. *Testing phase*: When a new window of data comes, to extract the context information from this window, we project the data into the bases prestored into the context database and calculate the corresponding basis coefficients (weights). We then calculate the similarity level between the new calculated weights to the ones in the database. The closest match indicates a found "context." In Figure 8.5, we use **H** to represent the context weights prestored in the database, and use $\mathbf{H}'$ to represent new tested context weights. In order to visualize the context features, we utilize the *linear principal regression* (LPR) to project the multidimension vectors (**H** or $\mathbf{H}'$) to a two-dimension space (Section 8.6 will have algorithms).

8.4 Context Awareness Model: Parts-Based Approach

8.4.1 Hidden Context Pattern for Binary Pyroelectric Sensing Data

We format the gait context identification problem as the issue of identifying the hidden context patterns (HCPs) for a given observed sensing data (OSD), which is high-dimensional *binary* data. Particularly, we attempt to answer two questions: (1) Since HCPs (matrix **H**) describe context features of each patient walking scenario, it should be extracted from data **X**. As a matter of fact, we can regard HCPs as the intrinsic sensing characteristics of each cluster of sensors (Figure 8.4). Then how do we identify the real-valued HCPs (Matrix **H**) from each binary-valued OSD **X**? (2) Each OSD can be seen as the mixture of different HCPs (Figure 8.6). Then how do we determine the mixture coefficients (also called "weights") that form matrix **W**?

To answer the aforementioned two questions, we first model the individual gait sensor value (either 0 or 1) as a Poisson-distributed random variable. Suppose there are N sensors in each cluster. Also assume each observation window has *M times of measurement*. Thus OSD is a $N \times M$

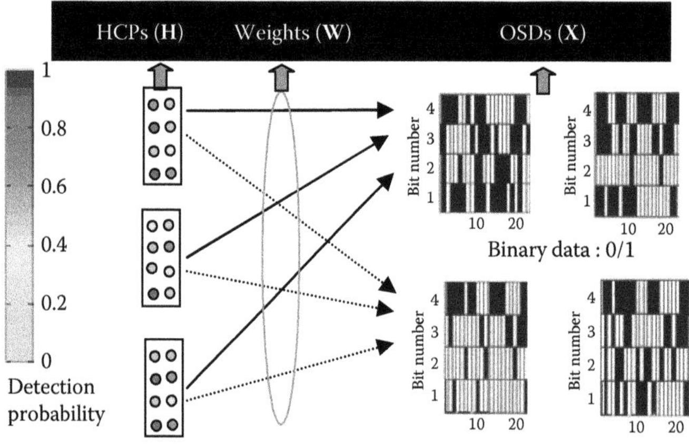

Figure 8.6 Gait context pattern models.

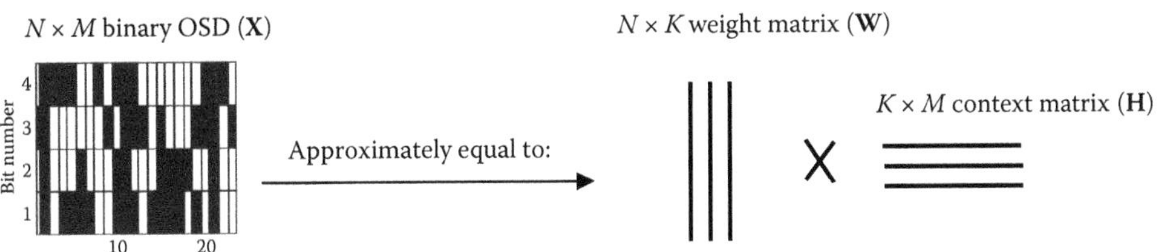

Figure 8.7 Interpretation of OSD and HCP.

matrix (**X**). In each position (i, j) of matrix **X**, the change of its value X_{ij} can be seen as a Poisson-distributed random variable with parameter λ, that is,

$$X_{ij} = \begin{cases} 1 & \text{with probability } e^{-\lambda_{ij}} \text{ (means “detected by sensors,” i.e., shaded part, Figure 8.1d)} \\ 0 & \text{with probability } 1 - e^{-\lambda_{ij}} \text{ (“not detected by sensors,” i.e., blank part, Figure 8.1d)} \end{cases} \quad (8.1)$$

To reflect the nature of HCPs and weight in Figure 8.6, we define a M-dimension gait context pattern that is determined by cluster k as a row vector in context matrix

$$\mathbf{H} \cdot \mathbf{H}_k = (H_{k1}, \ldots, H_{kM}) \geq 0, \; k = 1, 2, 3, \ldots, K.$$

Here we assume there are totally K clusters. Note that the number of clusters K is a small number. It is usually chosen such that $(N + M) \cdot K \leq N \cdot M$. The reason of doing this is to eventually express X in a *compressed* format, that is: $\mathbf{X} \approx \mathbf{WH}$ (Figure 8.7). If we assume that cluster k is the only factor that causes a value “1” in **X**, H_{kj} can then be seen as the *probability* (a real value) of the observation $X_{ij} = 1$. In other words, H_{kj} expresses the probability that observation X_{ij} is either 0 or 1. Since it is a probability, it could be a noninteger number between 0 and 1. Therefore, if cluster k is the only event that causes a “1” in **X**, $H_{ka} < H_{kb}$ could be interpreted as follows: it is more likely to detect the object in time b than time a from cluster k' sensors viewpoint.

Now, suppose each cluster $K_{k^*} (k = 1, 2, \ldots, k)$ contributes the detection of an object with different weights, denoted as W_{ik}, then the overall detection probability from those K clusters is

$$P(X_{ij} = 1 | W_{i1}, \ldots, W_{iK}, H_{1j}, \ldots, H_{Kj}) = \prod_{k=1}^{K} e^{-W_{ik} \cdot H_{kj}} = e^{-[\mathbf{WH}]_{ij}} \quad (8.2)$$

Likewise, the overall of missing (i.e., not detected) probability should be

$$P(X_{ij} = 0 | W_{i1}, \ldots, W_{iK}, H_{1j}, \ldots, H_{Kj}) = 1 - \prod_{k=1}^{K} e^{-W_{ik} \cdot H_{kj}} = 1 - e^{-[\mathbf{WH}]_{ij}} \quad (8.3)$$

For convenience of representation, we summarize the aforementioned conclusion as follows:

For any given OSD X_{ij}, and K clusters of pyroelectric sensors, we have the following detection and missing probability:

$$\begin{cases} \text{Detected}: & P(X_{ij} = 1 | \mathbf{W}, \mathbf{H}) = e^{-[\mathbf{WH}]_{ij}} \\ \text{Missing}: & P(X_{ij} = 0 | \mathbf{W}, \mathbf{H}) = 1 - e^{-[\mathbf{WH}]_{ij}} \end{cases} \quad (8.4)$$

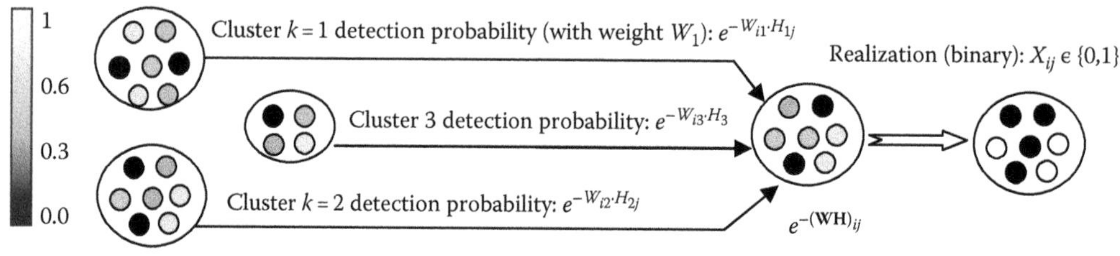

Figure 8.8 Binary pattern detection.

Note that although we could have noninteger numbers in weight matrix **W** and context matrix **H**, the observed data X_{ij} is always a binary number (0 or 1). Figure 8.8 illustrates such a concept.

Now our task is to seek for the context matrix **H** and weight matrix **W**. Their selection should maximize the *likelihood* of modeling the binary observed data matrix **X**. For a single observation X_{ij}, since it is a binary value, a natural choice is to use Bernoulli distribution to describe this single-value (X_{ij},) likelihood as given in the following:

$$P(X_{ij}|\mathbf{W},\mathbf{H}) = e^{-[\mathbf{WH}]_{ij}}\left(1 - e^{-[\mathbf{WH}]_{ij}}\right) \tag{8.5}$$

Therefore, the overall likelihood for all vales in data matrix **X** should be (take "log") (called Bernoulli Log-Likelihood):

$$\begin{aligned} L = \ln(P([\mathbf{X}|\mathbf{W},\mathbf{H}) &= \ln\left(\prod_i \prod_j \left(e^{-[\mathbf{WH}]_{ij}}\right)^{X_{ij}} \left(1 - e^{-[\mathbf{WH}]_{ij}}\right)^{1-X_{ij}}\right) \\ &= \sum_{i=1}^{N}\sum_{j=1}^{M} ((1 - X_{ij})\ln(1 - \exp(-[\mathbf{WH}]_{ij})) - X_{ij}[\mathbf{WH}]_{ij}) \end{aligned} \tag{8.6}$$

Obviously, we should seek non-negative matrixes **W**, **H** to maximize L, that is,

$$\text{Seek } \mathbf{W},\mathbf{H} \text{ to } \arg\max(L(\mathbf{W},\mathbf{H})) \text{ subject to } \mathbf{W},\mathbf{H} > 0$$

This approach could obtain the solution of **W** and **H** by using alternating least square (ALS) algorithm. That is, each element of matrix **W** and **H** could be solved as follows:

$$H_{kj}^{\text{New}} \leftarrow H_{kj}^{\text{Old}} + \lambda_H \frac{\partial L}{\partial H_{kj}}, \quad W_{ik}^{\text{New}} \leftarrow W_{ik}^{\text{Old}} + \lambda_W \frac{\partial L}{\partial W_{ik}} \tag{8.7}$$

The important idea is to keep one matrix fixed while updating the other one. Due to the constraint of $\mathbf{W},\mathbf{H} > 0$, the step size parameters $\lambda_{\mathbf{W}}$ and $\lambda_{\mathbf{H}}$ should be chosen carefully. However, ALS can only be guaranteed to find a local maximum for sufficiently small $\lambda_{\mathbf{W}}$ and $\lambda_{\mathbf{H}}$.

8.4.2 Probabilistic NMF Model

Based on the aforementioned analysis, our task is to actually seek a matrix factorization solution for $\mathbf{X} = \mathbf{W}\,\mathbf{H}$, where **X** is the data matrix, **W** is the weight matrix, and **H** is the context matrix.

However, we have the constraints that **X** is a binary matrix and **W**, **H** are non-negative matrix. This reminds us of the NMF (Non-negative Matrix Factorization) [21]. Given a target's signal (i.e., a binary matrix **V**), NMF intends to find the basis matrix **W** and feature set matrix **H** as follows:

$$\mathbf{V} \approx \mathbf{WH}, \quad \mathbf{W} \in R^{n\times r}, \quad \mathbf{H} \in R^{r\times m} \tag{8.8}$$

We then apply expectation-maximization (EM)-like algorithm to recursively update basis matrix and feature matrix as follows:

$$W_{ia}^{\text{new}} \leftarrow W_{ia}^{\text{old}} \sum_k \frac{V_{ik}}{(WH)_{ik}} H_{ak} \quad H_{ak}^{\text{new}} \leftarrow H_{ak}^{\text{old}} \sum_i W_{ia}^{\text{old}} \frac{V_{ik}}{(WH)_{ik}} \tag{8.9}$$

where $i = 1, 2, \ldots, m$, $a = 1, 2, \ldots, d$, and $k = 1, 2, \ldots, n$.

The NMF model mentioned earlier has a shortcoming: it cannot reflect the random nature of the error: $\mathbf{E} = \mathbf{X} - \mathbf{WH}$. If we assume **E** is a Poisson distribution, we first write down general Poisson distribution as follows:

$$f(X|\theta) = \frac{\theta^X e^{-\theta}}{X!} \tag{8.10}$$

where

θ is Poisson distribution parameter (its original meaning is the event inter-arrival rate)

X is the random variable

For our case, if we assume that

$$X = WH + E, \quad \text{where error } E_{ij} \overset{i.i.d.}{\sim} \text{Poisson}(\theta) \tag{8.11}$$

we can then determine the Poisson Likelihood as follows:

$$L_{\text{Poisson}}(\Theta) = P(X|\mathbf{W},\mathbf{H}) = \prod_i \prod_j \frac{[WH]_{ij}^{X_{ij}} e^{-[WH]_{ij}}}{X_{ij}!} \tag{8.12}$$

For calculation convenience, we take Natural Log as follows:

$$\begin{aligned} L_{\text{Poisson}}(\Theta) &= \ln(P(X|\mathbf{W},\mathbf{H})) = \ln\left(\prod_i \prod_j \frac{[WH]_{ij}^{X_{ij}} e^{-[WH]_{ij}}}{X_{ij}!}\right) \\ &= \sum_i \sum_j \left(X_{ij} \ln[WH]_{ij} - [WH]_{ij} - \ln(X_{ij}!)\right) \end{aligned} \tag{8.13}$$

We use Stirling's formula [22].

$$\ln(X_{ij}!) \approx X_{ij}\ln(X_{ij}) - X_{ij} \tag{8.14}$$

Then the aforementioned Log-Likelihood becomes

$$L_{\text{Poisson}}(\Theta) = \ln(P(X|\mathbf{W},\mathbf{H})) = -\sum_i \sum_j \left(X_{ij}\ln\frac{X_{ij}}{[WH]_{ij}} + [WH]_{ij} - X_{ij}\right) \tag{8.15}$$

It is very interesting to see that this expression is actually the negative of Kullback–Leibler (KL) Divergence between variable **X** and **WH**, that is,

$$L_{\text{Poisson}}(\Theta) = \ln(P(X|\mathbf{W},\mathbf{H})) = -D_{KL}(X, WH) \tag{8.16}$$

As we know, KL Divergence can measure the information entropy difference between two random variables with different probability distributions. The less the KL Divergence, the more similar two distributions are. From the aforementioned equation we can see that our goal is to maximize the Log-likelihood, i.e., to minimize the KL Divergence.

$$\text{Seek W, H to Minimize } D_{KL}\ (\mathbf{X},\mathbf{WH})$$

We can now understand why many NMF applications use the KL Divergence as the cost function:

$$D_{KL}(\mathbf{X},\mathbf{WH}) = \sum_i \sum_j \left(X_{ij} \ln \frac{X_{ij}}{[WH]_{ij}} + [WH]_{ij} - X_{ij} \right) \tag{8.17}$$

Such a KL cost function can give us a better solution of **W**, **H** than *conventional NMF cost function* which does not lead to unique solutions.

$$D_{NMF}(\mathbf{X},\mathbf{WH}) = \frac{1}{2} \sum_i \sum_j (X_{ij} - [WH]_{ij})^2 \tag{8.18}$$

As a matter of fact, the aforementioned cost function comes from the assumption that the error **E** = **X** − **WH** obeys Normal (Gaussian) distribution, that is

$$\mathbf{X} = \mathbf{WH} + \mathbf{E}, \quad \text{where error } E_{ij} \overset{i.i.d.}{\sim} \text{Gaussian } (0, \sigma) \tag{8.19}$$

Thus we obtain the Gaussian Likelihood as follows:

$$L_{\text{Gaussian}}(\Theta) = P(X|\mathbf{W},\mathbf{H}) = \left(\frac{1}{\sqrt{2\pi}\sigma} \right)^{NM} \prod_i \prod_j e^{-\frac{1}{2}\left(\frac{(X_{ij}-[WH]_{ij})}{\sigma} \right)^2} \tag{8.20}$$

If we take Natural Log, we have the following:

$$L_{\text{Gaussian}}(\Theta) = \ln(P(X|\mathbf{W},\mathbf{H})) = -NM \ln(\sqrt{2\pi}\sigma) - \frac{1}{2\sigma^2} \sum_i \sum_j (X_{ij} - [WH]_{ij})^2 \tag{8.21}$$

As we can see, to maximize the likelihood, we can just minimize the second item, which is conventional NMF cost function:

$$\text{Maximize} = \text{Minimize } D_{\text{NMF}}(\mathbf{X},\mathbf{WH}) = \frac{1}{2} \sum_i \sum_j (X_{ij} - [WH]_{ij})^2 \tag{8.22}$$

8.4.3 Seek W, H through Their Prior Distributions

Although ML (Maximize Likelihood) could give us the solution of **W**, **H** as long as we have enough **X** sample values, however, in practical Pyroelectric sensor network applications, we may not be able to obtain large samples of **X**. Moreover, to achieve real-time NMF operation, it is not realistic to wait for the collection of large **X** samples to calculate **W** and **H**. Large **X** matrix causes long calculation time.

To solve the dilemma mentioned earlier, i.e., seek matrix factorization $\mathbf{X}=\mathbf{WH}$ under small **X** matrix, we resort to MAP (Maximize a priori), which is defined in Bayesian theorem:

$$\text{Posterior probability} = \frac{\text{Likelihood} \times \text{Prior}}{\text{Normalization factor}} \tag{8.23}$$

where

Likelihood is the aforementioned likelihood function for **X** matrix

Prior is the preassumed probability distribution for the parameters of **X** distribution

Normalization factor can ensure the integral of the right side is 1 (thus making the *posterior probability* fit the scope of a probability: 0–1)

If we use $P(\Theta)$ to denote the prior belief (distribution) of the parameters **W** and **H**, and assume data **X** is generated by a model with parameters $\Theta = \{\mathbf{W}, \mathbf{H}\}$, then we can use the following way to represent Bayes' formula (Figure 8.9):

Therefore, our goal is to maximize $P(\Theta|\mathbf{X})$. If we take Log-posterior, we have the following:

$$\begin{aligned}\text{Maximize } \ln P(\Theta|\mathbf{X}) &= \text{Maximize } \ln P(\mathbf{W},\mathbf{H}|\mathbf{X}) = \text{Maximize } (\ln P(\mathbf{X}|\mathbf{W},\mathbf{H}) \\ &\quad + \ln(P(W) + \ln(P(H))\end{aligned}$$

Without the loss of generality, we assume the sensors' event capture $P(\mathbf{X}|\mathbf{W},\mathbf{H})$ follows a Gaussian Likelihood. We also assume that **W** and **H** are independent, that is,

$$P(\mathbf{W},\mathbf{H}) = P(\mathbf{W})P(\mathbf{H}) \tag{8.24}$$

Now we need to select the proper prior distributions for **W** and **H**. For each element of **H** matrix, we can assume its prior distribution is an even distribution as long as its value does not go beyond an upper threshold (which is the case for context matrix):

$$P(H_{kj}) = \begin{cases} \text{const.}, & 0 \le H_{kj} \le H_{\max} \\ 0, & \text{otherwise} \end{cases} \tag{8.25}$$

Since the posterior distribution belongs to exponential family, we choose exponent distribution for the prior distribution of W matrix.

$$P(\mathbf{W}) = \prod_i \prod_k \alpha e^{-\alpha W_{ik}}, \quad \alpha > 0 \tag{8.26}$$

$$\underbrace{P(\mathbf{X},\Theta)}_{\text{Joint distribution}} = \underbrace{P(\mathbf{X}|\Theta)}_{\text{Likelihood}} \underbrace{P(\Theta)}_{\text{Prior}} = \underbrace{P(\Theta|\mathbf{X})}_{\text{Posterior}} \underbrace{P(\mathbf{X})}_{\text{X distribution (evidence; normalization constant)}}$$

Figure 8.9 Bayesian theorem.

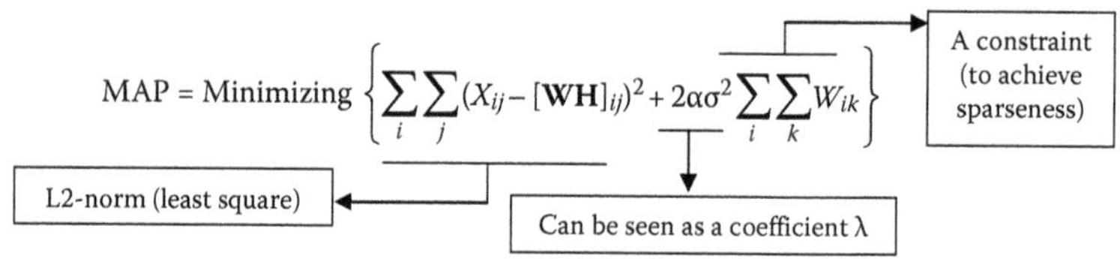

Figure 8.10 Maximum a priori (MAP).

Based on Equations 8.21, the equation (8.23) will be (in Log-Posterior) as follows:

$$\ln P(\mathbf{W},\mathbf{H}|\mathbf{X}) \propto \ln P(\mathbf{X}|\mathbf{W},\mathbf{H}) + \ln(P(\mathbf{W}) + \ln P(\mathbf{H})$$

$$\propto \left\{-NM\ln(\sqrt{2\pi}\sigma) - \frac{1}{2\sigma^2}\sum_i\sum_j(X_{ij}-[\mathbf{WH}]_{ij})^2\right\} + \left\{-\alpha\sum_i\sum_k W_{ik}\right\} + \text{Const}$$

$$\propto -\left\{\sum_i\sum_j(X_{ij}-[\mathbf{WH}]_{ij})^2 + 2\alpha\sigma^2\sum_i\sum_k W_{ik}\right\} + \text{Const} \tag{8.27}$$

Therefore, MAP (i.e., maximizing $\ln P(\mathbf{W},\mathbf{H}|\mathbf{X})$) can be obtained by minimizing the following L2-norm with a constraint item (Figure 8.10):

Based on Lagrange Multipliers, we can concert the aforementioned problem to a special NMF model:

Given a limited **X** sample, seek weight matrix W and context matrix H, to achieve:

$$\text{Minimize}\sum_i\sum_j(X_{ij}-[\mathbf{WH}]_{ij})^2,\ \text{given constraints: (1)}\mathbf{W},\mathbf{H} > 0;\ (2)\sum_i\sum_k W_{ik} < \text{const}$$

Due to the importance of NMF with sparseness constraint, we will provide strict models next.

8.4.4 Context Pattern Seeking under Sparseness Constraint

One of NMF advantages is its basis sparseness. For example, if we extract some common features (to form "basis") from humans' faces, such as eyes, nose, mouth, and so on, we could use different ways (NMF, PCA, Wavelet, etc.) to search for those bases. It was found that NMF gave us the sparsest bases [23]. In Section 8.7 (experiments), we will illustrate this point. The sparseness is important to the reduction of memory storage and calculation complexity. More importantly, it makes NMF more like "parts-based" feature extraction, i.e., we can easily recognize an object by looking at its few features.

The measurement of sparseness can be regarded as a mapping from $\Re^n$ to $\Re$ which to quantify how much energy of a vector is packed into a few components. Without the loss of generality, we adopt a sparseness definition used in Ref. [23] that considers an observation vector **X** with n elements $(x_1, x_2, \ldots, x_n)$:

$$\text{Sparseness}\,(\mathbf{X}) = \frac{\sqrt{n} - \left(\sum |x_i|\right) / \sqrt{\sum x_i^2}}{\sqrt{n} - 1} \tag{8.28}$$

As we can see, the scope of the aforementioned sparseness function is [0,1]: when all **X**'s elements are equal, we have the least sparseness of 0; when **X** has only one single non-zero element, we get the maximum sparseness of 1.

Then our goal is to use NMF or the earlier discussed methods to decompose **X** into a weight matrix **W** and a context matrix **H** *with desired degrees of sparseness*. Then our question is: should **W** or **H** be made sparse? This depends on different applications. For instance, if a doctor wants to find the cancer patterns from large sample cases, it will be reasonable to assume that cancers are rare among all cases. Assume each column of **X** represents one patient and each row of **X** represents the symptoms of a specific disease. Then we expect the occurrence of a disease (a cancer) is rare (i.e., sparse), that is, the weight matrix **W** should be sparse. However, the doctor wishes that the symptoms of each disease should be detailed as possible, that is, the context matrix **H** should NOT be sparse.

In our case, we wish to identify the walker's gait patterns. We wish to use "sparse" features (represented by context matrix **H**) to identify gaits. On the other hand, we do not expect to have many sensor "clusters" in order to save cost. In other words, the weight matrix **W** should also be sparse. Therefore, both **W** and **H** should be sparse.

Definition—sparse pyroelectric sensing: For a pyroelectric sensor network with K clusters and each cluster has N sensors, if we sense $N \times M$ data **X**, the goal of "sparse pyroelectric sensing" can be formulated into a matrix factorization procedure: we seek non-negative weight matrix **W** and context matrix **H**, such that the least square meets, that is, Minimize $||\mathbf{X} - \mathbf{W}\,\mathbf{H}||^2$. In the meantime, **W** and **H** should meet two sparseness constraints: (1) Sparseness $(w_i) = S_w$, for any ith row of **W**. S_w is the desired **W** sparseness (Preset by user; Range: [0,1]). (2) Sparseness $(h_i) = S_h$, for any ith column of **H**. S_h is the desired **H** sparseness (Preset by user; Range: [0,1]).

We can use the projected gradient descent NMF algorithm (such as the one used in [20]) to seek sparse **W** and **H** in the aforementioned definition. The basic procedure is to take a step in the direction of the negative gradient, and then project onto the constraint space. The taken step should be small enough to reduce the $||\mathbf{X} - \mathbf{W}\,\mathbf{H}||^2$ in each step. The critical operation is projection operator that updates **W** and **H** when sparseness does not meet in step i

$$\begin{cases} W_i^{\text{Next}} = W_i \otimes (XH_i^T) \diamond (W_i H_i H_i^T) \\ H_i^{\text{Next}} = H_i \otimes (W_i^T X) \diamond (W_i^T W_i H_i) \end{cases} \tag{8.29}$$

where

$\otimes$ means multiplication for matrix operation,

$\diamond$ is elementwise division.

8.4.5 Smoothing Context Features

Although sparseness can make NMF more like "parts-based" feature recognition, too sparse matrix representation could not accurately describe an object since most elements of the context matrix **H** will be zero (or very small values). Therefore in some applications, we could control the "richness" of context matrix **H** by adding smoothness constraints to **H**. Smoothness tries to reduce the big differences among elements to make values have "smooth" differences. Here we define a smoothing function S as follows [24]:

$$S = (1 - \theta)\mathbf{I} + \frac{\theta}{K}11^T \tag{8.30}$$

where
- **I** is the identity matrix
- **1** is a vector of ones [1 1 1 ...]

The parameter θ is important. Its range is [0, 1]. Assume $\mathbf{H}' = \mathbf{SH}$. The larger θ is, the more smooth effect we can get. This can be seen from following fact: If $\theta = 0$, $\mathbf{H}' = \mathbf{H}$, and no smoothing on **H** has occurred. However, when θ get approaching to 1, that is, $\theta \rightarrow 1$, $\mathbf{H}'$ tends to be a constant matrix with all elements almost equal to the average of the elements of **H**. Therefore, parameter θ determines the extent of smoothness of the context matrix.

The solution to a NMF with smoothness constraints is straightforward. We can just revise the following NMF iteration equation from $\mathbf{W} \rightarrow \mathbf{WS}$ and $\mathbf{H} \rightarrow \mathbf{SH}$:

$$W_{ia}^{\text{new}} \leftarrow W_{ia}^{\text{old}} \sum_k \frac{V_{ik}}{(WH)_{ik}} H_{ak}, \quad H_{ak}^{\text{new}} \leftarrow H_{ak}^{\text{old}} \sum_i W_{ia}^{\text{old}} \frac{V_{ik}}{(WH)_{ik}} \tag{8.31}$$

where $i = 1, 2, \ldots, m$, $a = 1, 2, \ldots, d$, and $k = 1, 2, \ldots, n$.

8.5 Linear Principal Regression for Feature Visibility

In order to *visualize the context features* of different scenarios, we project the multidimension vectors to a two-dimension space. Here we utilize the linear principal regression (LPR) to accomplish the data dimensionality reduction.

The singular value decomposition of a binary signal matrix could be represented by

$$S_{m\times n} = U_{m\times m} \sum\nolimits_{m\times n} V_{n\times n}^T \tag{8.32}$$

where
- U and V are orthogonal matrixes
- $\sum$ is a diagonal matrix with eigenvalues
- m is the number of samples
- n is the dimension of samples

Then S could be approximated by its first k eigenvalues as follows:

$$S \approx TP^T \quad \text{where } T_{m\times k} = \tilde{U}_{m\times k} \sum\nolimits_{k\times k},\ P = \tilde{V}_{n\times k} \tag{8.33}$$

To find the regression vector in k dimensions, we seek the least square solution for the equation

$$I_{m\times 1} = T_{m\times k} f_{k\times 1} \quad \text{where } f_{k\times 1} = (T^T T)^{-1} T^T I \tag{8.34}$$

The regression vector could be solved as

$$F_{n\times 1} = P_{n\times k} f_{k\times 1} \tag{8.35}$$

8.6 Similarity Score for Context Understanding

During gait context understanding (testing phase), we use K-means cluster and vector distance to compute the similarity score between H and H'.

$$\arg\min \sum_{i=1}^{k} \sum_{h_j \in C_i} \left\| h_j - \mu_i \right\|^2 \tag{8.36}$$

where

h_j is a context feature vector
μ_i is the mean of cluster i
$C_1, C_2, \ldots, C_k$ are k clusters

Scenario identification decision: For a context feature h_j, we will have $k + 1$ hypothesis to test for k registered scenarios, i.e., $\{H_0, \ldots, H_k\}$. The hypothesis H_0 represents "none of the above." The decision rule is then

$$h_j \in \begin{cases} H_0, & \text{if } \max_i p(h_j|H_i) < \gamma \\ H_i : i = \arg\max_i p(h_j|H_i), & \text{otherwise} \end{cases} \tag{8.37}$$

where

$p(h_j|H_i)$ is the likelihood of h_j associating with the ith hypothesis
γ is a selected acceptance/rejection threshold

8.7 Experiment Results

In the following discussion, we provide part of our experimental results. For more detailed descriptions on our results, please refer to our other publications [9–14].

8.7.1 Gait Context Identification via Traditional PCA Signal Projection

By using real application data from our iSMART platform, we first test the use of traditional, nonbinary PCA algorithm for signal projection in order to find out hidden gait context patterns. Here we aim to find four different gait contexts: (1) 1-patient scenario, (2) 2-patient scenario (two patients' gaits signals are mixed together), (3) 3-patient scenario, and (4) 4-patient scenario. As we know, PCA has both pattern recognition and dimension reduction functions. Since we use general PCA that is not optimized for binary data, the context identification accuracy and calculation complexity would be high for our binary data matrix. Figure 8.11 shows the bases we found.

As we can see from Figure 8.11, those four contexts share four common bases. Based on the weights (coefficients) for each basis, we are able to detect different contexts. In order to better see the differences among different contexts, we use PCA to project the context weights to a 2D coordinate. As shown in Figure 8.12, PCA can clearly map each context to different clusters.

Following general pattern recognition methodology, we use *receiver operating characteristics (ROC)* graph to illustrate context identification accuracy. Figure 8.13 shows the similarity score distributions for self-testing and crosstesting when using PCA. As we can see, self-testing can yield true positive data, and crosstesting can yield false positive data, those two distributions can be almost completely separated from each other. This indicates PCA can efficiently recognize different contexts.

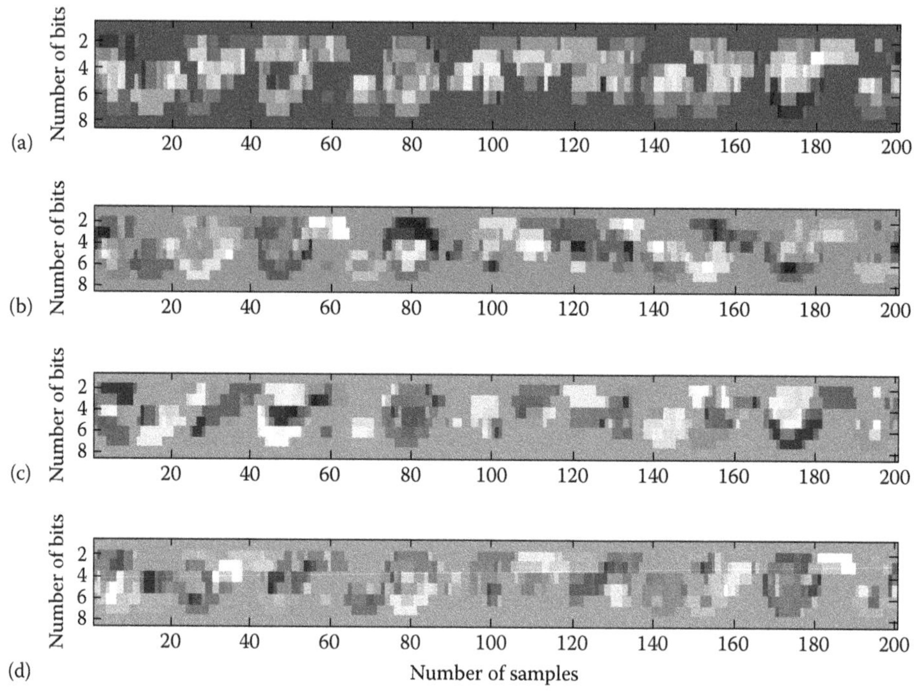

Figure 8.11 Contextual bases achieved by principal components analysis. (a) Eigen base 1. (b) Eigen base 2. (c) Eigen base 3. (d) Eigen base 4.

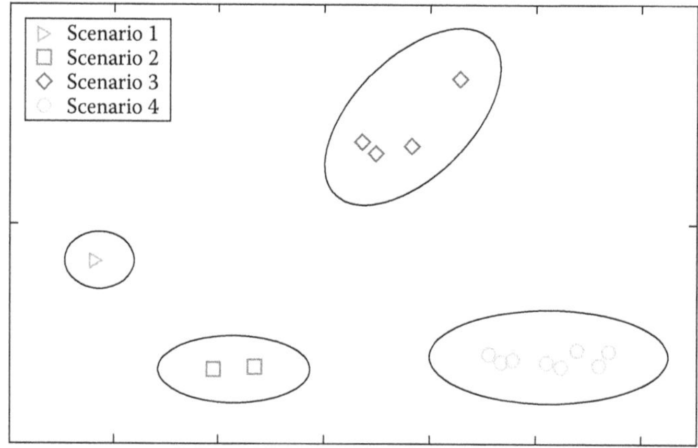

Figure 8.12 Feature clusters of four scenarios using principal components analysis.

8.7.2 Gait Context Identification Using Probabilistic Matrix Factorization Model

We then use the probabilistic NMF to identify those four contexts. As discussed in Section 8.4, NMF is parts-based signal projection. It can use simpler bases (than PCA) to identify a context. Our experiment verifies this point. Figure 8.14 clearly shows that the bases generated from NMF

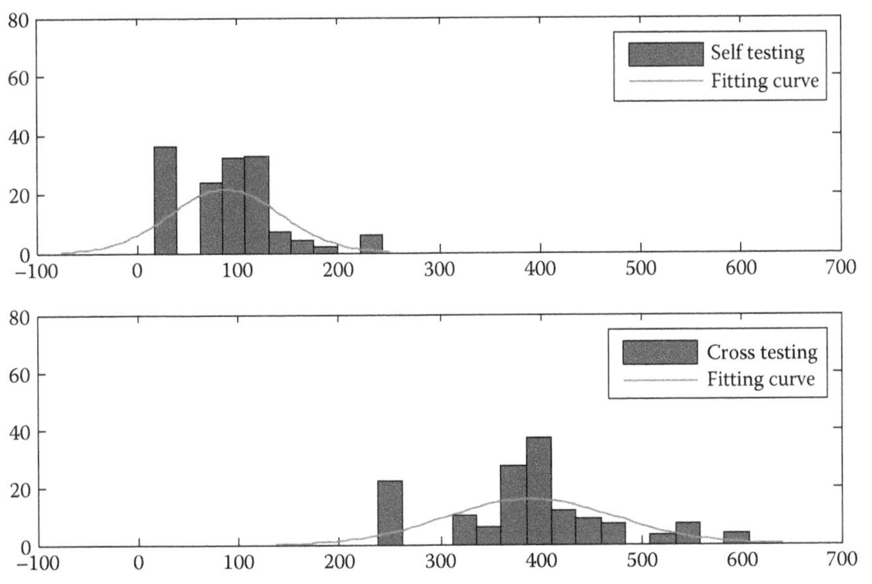

Figure 8.13 Similarity score distributions of self-testing and crosstesting using PCA.

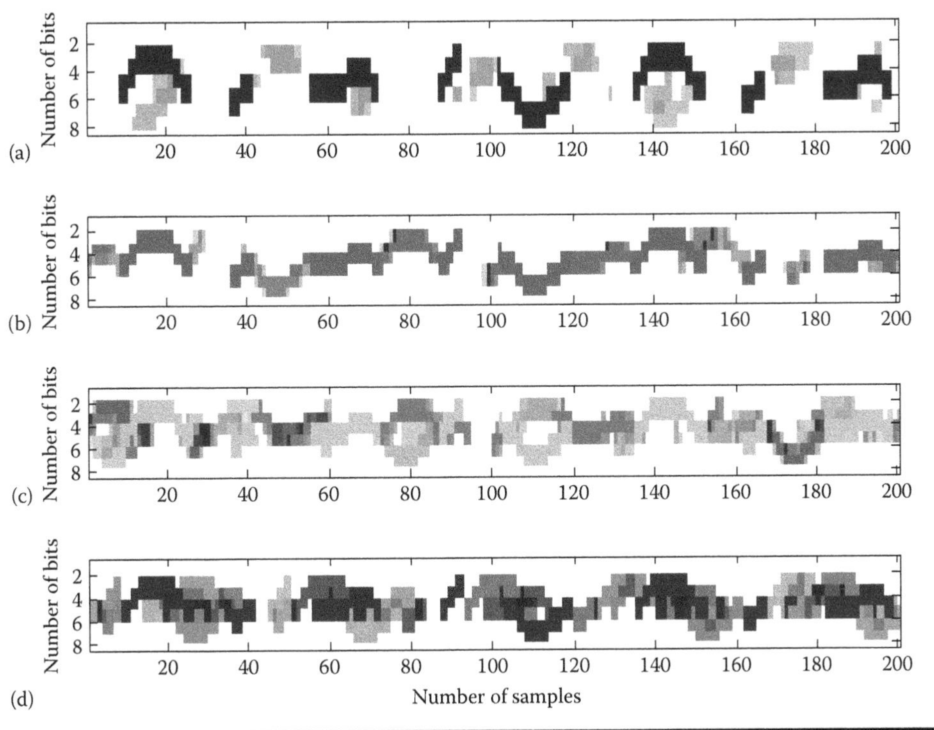

Figure 8.14 Region of interest of NMF. (a) Base 1. (b) Base 2. (c) Base 3. (d) Base 4.

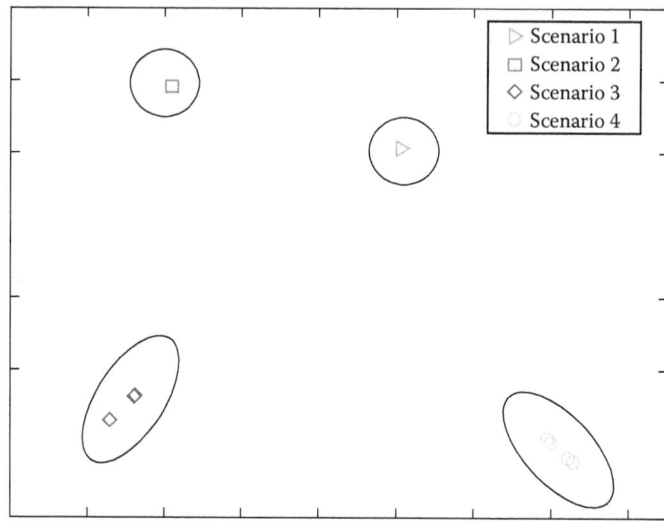

Figure 8.15 Feature clusters of four scenarios using non-negative matrix factorization.

are sparser than PCA case (Figure 8.11). This is beneficial to the savings of memory storage and the reduction of weights calculation time. As a matter of fact, because NMF uses non-negative bases and weights, while PCA allows negative weights, NMF can interpret signal features in an intuitive way.

We again use NMF to project pyroelectric sensor data to cluster architecture. As shown in Figure 8.15, NMF can completely separate different contexts. Moreover, compared to PCA case (Figure 8.12), NMF can make contexts falling into the same category converge into a smaller cluster, which indicates the advantage of NMF over PCA from context identification viewpoint.

We then tested the similarity scores (explained in Section 8.6) in NMF case. As shown in Figure 8.16, the two distributions (self-testing and cross-testing) can be completely separated from each other. Moreover, compared to PCA case (Figure 8.13), the self-testing scores occupy only a small region and crosstesting scores cross a larger region. These results show that NMF can detect context more accurately than PCA.

Then we compare the ROC graphs between PCA and NMF. As we can see in Figure 8.17, NMF is always 1 no matter what value the false alarm rate is. PCA is <1 in certain rates.

8.7.3 NMF with Smoothness or Sparseness Constraints

To test the performance of constrained NMF (see Sections 8.4.4 and 8.4.5 on sparseness and smoothness), we have used real pyroelectric sensor data to perform ROC test. The result is shown in Figure 8.18. It shows that NMF under smoothness constraints have the best performance. This could be due to two reasons:

1. *Scenario-dependent context prefers smoothness constraints*: Two kinds of hidden context patterns can be extracted via NMF algorithm: One is scenario-dependent context (i.e., how many walkers are in the scenario simultaneously); the other is path-dependent context (i.e., the same walker changes paths each time he/she walks through the sensors). The former (scenario-dependent) generates holistic context patterns (i.e., all NMF weights tend to be more evenly

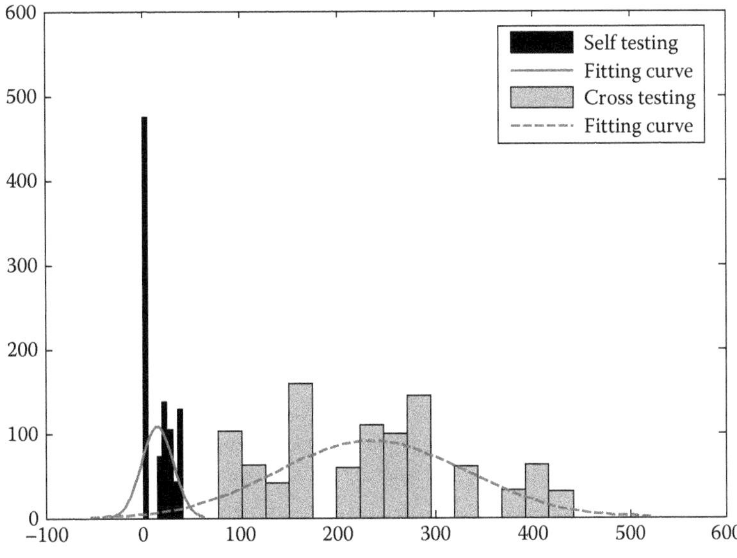

Figure 8.16 Similarity score distributions of self-testing and crosstesting using non-negative matrix factorization.

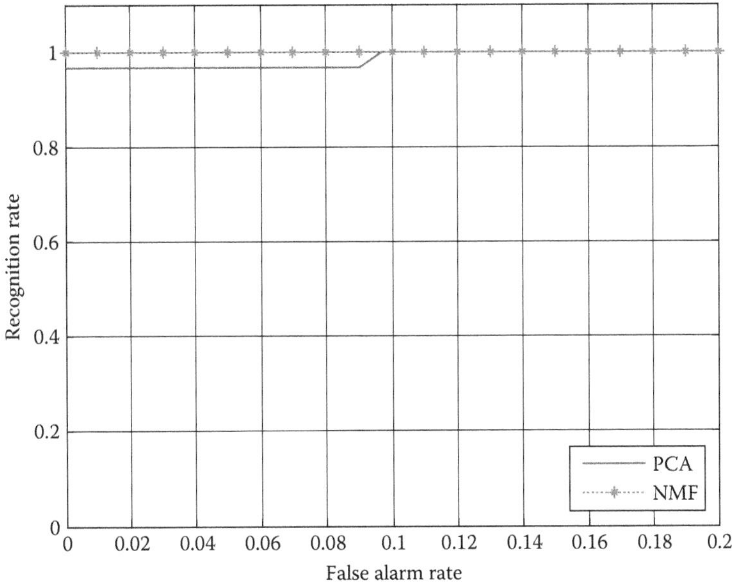

Figure 8.17 Receiver operating characteristics of PCA and NMF.

distributed), while the latter (path-dependent) generates local context characteristics (i.e., NMF weights tend to be distributed in two extremities). Our experiments have chosen the former (scenario-dependent) as the context identification objective. Therefore, adding smoothness constraint makes the NMF weights look more holistic (i.e., all weights become more evenly distributed), which makes context extraction more convenient.

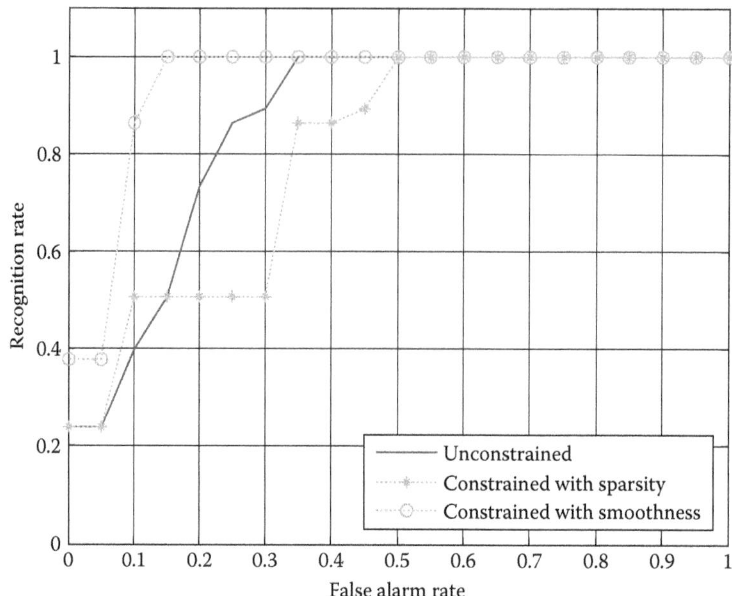

Figure 8.18 ROC of NMF with the constraints of sparseness and smoothness.

2. *Smoothness constraint fits our K-means cluster algorithm*: We have adopted k-means cluster and vector distance in the calculation of similarity score function. Due to the various value positions in the feature vectors (i.e., NMF weights) of the same cluster, k-means algorithm tends to smooth the feature vectors. In other words, K-means tends to reduce the sparseness degree of NMF weights, which is against the sparseness constraint operation. Therefore, adding sparseness constraint to NMF weight matrix would get a worse performance. On the contrary, adding smoothness constrain is consistent with the function of k-means cluster algorithm, which brings a better context identification performance.

8.7.4 Pseudo-Random Field of Visibility Modulation

We have also investigated the use of *pseudo-random modulation of the field of view* (explained in Section 8.7) to enhance context identification performance. All the earlier experiments were implemented via regular visibility modulation. In this section, we use *Hadamard* code-based pseudo-random visibility modulation, see Table 8.1. *Hadamard matrix* encodes visibility mask by replacing the "−1" with "0" in a Hadamard matrix.

Figure 8.19 shows the similarity score comparison. As we can see, those two distributions are completely separate from each other, which indicates the perfect context identification rate.

Figure 8.20 shows the ROC comparisons. We can see that the context identification rate is always 100% for *pseudo-random visibility* case. However, in *regular visibility* case, the ROC is <100% when the false alarm rate is <0.3. Therefore, the pseudo-random visibility modulation can better capture data patterns and improve the context identification performance.

Table 8.1 Coding Scheme for Visibility Modulation

Sensor Number	*Regular Visibility*	*Pseudo-Random Visibility*
1	[1 1 1 0 0 0 0 0 0 0 0 0 0 0]	[1 0 1 0 1 0 1 0 1 0 1 0 1 0]
2	[0 1 1 1 1 0 0 0 0 0 0 0 0 0]	[1 1 0 0 1 1 0 0 1 1 0 0 1 1]
3	[0 0 0 1 1 1 1 0 0 0 0 0 0 0]	[1 0 0 1 1 0 0 1 1 0 0 1 1 0]
4	[0 0 0 0 0 1 1 1 1 0 0 0 0 0]	[1 1 1 1 0 0 0 0 1 1 1 1 0 0]
5	[0 0 0 0 0 0 0 1 1 1 1 0 0 0]	[1 0 1 0 0 1 0 1 1 0 1 0 0 1]
6	[0 0 0 0 0 0 0 0 0 1 1 1 1 0]	[1 1 0 0 0 0 1 1 1 1 0 0 0 0]
7	[0 0 0 0 0 0 0 0 0 0 1 1 1 1]	[1 0 0 1 0 1 1 0 1 0 0 1 0 1]
8	[0 0 0 0 0 0 0 0 0 0 0 1 1 1]	[1 1 1 1 1 1 1 1 0 0 0 0 0 0]

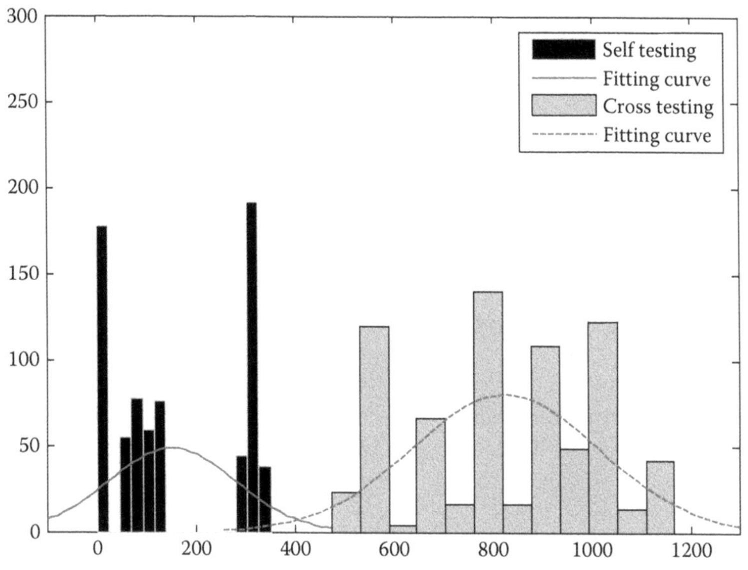

Figure 8.19 Similarity score distributions using regular visibility and pseudo-random visibility modulations.

8.8 Concluding Marks and Future Work

In this chapter, we have presented our recent research results on gait context identification in mobile patient tracking scenarios, which are monitored through extremely low cost sensors—pyroelectric sensors. Those sensors form a multicluster network for binary data collection ("1" means detected; "0" means not detected.) The next step research will focus on two aspects: (1) First, we will determine a good window size when we segment the incoming binary data. The window size should not sacrifice real-time context identification performance; (2) We will investigate the distributed

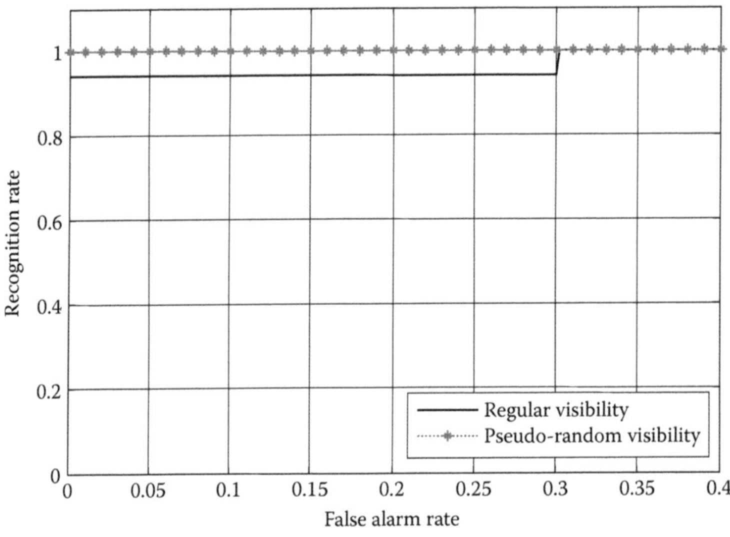

Figure 8.20 ROC from regular visibility and pseudo-random visibility modulations.

NMF implementation in sensors instead of sending all data to the base-station for centralized processing. Such distributed context identification can greatly reduce network processing overhead and delay.

Acknowledgments

This research has been supported by the U.S. National Science Foundation (NSF) IIS # 0915862 and a faculty research support grant from the University of Alabama. All results presented in this paper do not necessarily reflect NSF's opinions. The authors also acknowledge the communications and help from other group members (Xin Zhou, Qi Wei, Dong Zhang, Jiang Lu).

References

1. Brussels/Geneva, Neurological disorders affect millions globally: WHO report, Feb 27, 2007. Neuro-disorder data (2007) from World Health Organization (WHO) report (visited in November 2011): http://www.who.int/mediacentre/news/releases/2007/pr04/en/index.html
2. The Silver Book, The economic burden of neurological disorder: http://www.silverbook.org/browse.php?id=52
3. WiKi's definition on the types of neuro-disorders symptoms: http:/en.wikipedia.org/wiki/List_of_neurological_disorders
4. B. Toro, C. Nester, P. Farren, A review of observational gait assessment in clinical practice, *Physiother Theory Pract*, 19(3): 137–150, 2003.
5. L. Sudarsky, Gait disorders: Prevalence, morbidity, and etiology, *Adv Neurol*, 87: 111–117, 2001.
6. J. Jankovic, J. G. Nutt, L. Sudarsky, Classification, diagnosis and etiology of gait disorders, *Adv Neurol*, 87: 119–133, 2001.

7. AHRQ (Agency for Healthcare Research and Quality), Healthcare cost for hospital stays of stroke and other cerebrovascular diseases, 2005 statistics: http://www.hcup-us.ahrq.gov/reports/statbriefs/sb51.pdf (accessed: October 2011).
8. S. B. Lang, Pyroelectricity: From ancient curiosity to modern imaging tool, *Phys Today*, 58(8): 31–35, August 2005.
9. Q. Hao, F. Hu, Y. Xiao, Multiple human tracking and identification with wireless distributed pyroelectric sensors, *IEEE Syst J* (special issue on *Biometrics*), 3(4): 428–439, December 2009.
10. X. Zhou, Q. Hao, F. Hu, 1-bit walker recognition with distributed binary pyroelectric sensors, *IEEE International Conference on Multisensor Fusion and Integration*, Salt Lake City, UT, September 5–7, 2010, pp. 168–173.
11. Q. Sun, F. Hu, Q. Hao, Context awareness emergence for distributed binary pyroelectric sensors, *IEEE International Conference on Multisensor Fusion and Integration*, Salt Lake City, UT, September 5–7, 2010, pp. 162–167.
12. Q. Hao, F. Hu, J. Lu, Distributed multiple human tracking with wireless binary pyroelectric infrared (PIR) sensor networks, *IEEE Conference on Sensors*, Hawaii, November 2010, pp. 946–950 .
13. F. Hu, Q. Sun, Q. Hao, Mobile targets region-of-interest via distributed pyroelectric sensor network: Towards a robust, real-time context reasoning, *IEEE Conf Sens*, Hawaii, November 2010, pp. 1832–1836.
14. Q. Hao, F. Hu, Y. Xiao, Multiple human tracking and recognition with wireless distributed pyroelectric sensor systems, *IEEE Syst J*, 3(4): 428–439, December 2009.
15. S. E. Palmer, The effects of contextual scenes on the identification of objects, *Memory Cognit*, 3: 519–526, 1975.
16. A. Torralba, Modeling global scene factors in attention, *J Opt Soc Am A*, 20(7): 1407–1418, July 2003.
17. A. Torralba, P. Sinha, Statistical context priming for object detection: Scale selection and focus of attention, in *Proceedings of the International Conference on Computer Vision*, IEEE Computer Society Press, Los Alamitos, CA, 2001, Vol. 1, pp. 763–770.
18. A. Torralba, Contextual modulation of target saliency, in *Advances in Neural Information Processing Systems*, T. G. Dietterich, S. Becker, Z. Ghahramani, eds., MIT Press, Cambridge, MA, 2002, Vol. 14, pp. 1303–1310.
19. M. M. Chun, Y. Jiang, Contextual cueing: Implicit learning and memory of visual context guides spatial attention, *Cogn Psychol*, 36: 28–71, 1998.
20. R. Schachtner, G. Pöppel, E. W. Lang, 2009. Binary nonnegative matrix factorization applied to semi-conductor wafer test sets (both paper and slides), in *Proceedings of the 8th International Conference on independent Component Analysis and Signal Separation*, T. Adali, C. Jutten, J. M. Romano, and A. K. Barros, eds., Lecture Notes in Computer Science, Paraty, Brazil, March 15–18, 2009, Springer-Verlag, Berlin, Germany, Vol. 5441, pp. 710–717.
21. D. D. Lee, H. S. Seung, Learning the parts of objects by non-negative matrix factorization, *Nature*, 401: 788–791, 1999.
22. Stirling's Formula: http://www.sosmath.com/calculus/sequence/stirling/stirling.html
23. P. O. Hoyer, Non-negative matrix factorization with sparseness constraints, *J Mach Learn Res*, 5: 1457–1469, December 2004.
24. A. Pascual-Montano, J. M. Carazo, K. Kochi, D. Lehmann, R. D. Pascual-Marqui, Nonsmooth nonnegative matrix factorization (NsNMF), *IEEE Trans Pattern Anal Mach Intell*, 28(3), March 2006, pp. 403–415.

Chapter 9

Cognitive Wireless Sensor Networks

Sumit Kumar, Deepti Singhal, and Rama Murthy Garimella

Contents

9.1 Introduction to Wireless Sensor Networks

A wireless sensor network (WSN) consists of a set of sensor nodes that are deployed in a field and interconnected with a wireless communication network. In general, they have short-range communication capability. These nodes cooperatively monitor physical or environmental conditions, such as vibration, motion, temperature, sound, etc. Each of these scattered sensor nodes has the capability to collect data and route the data back to the sink/base station [1,2].

Each sensor node comprises of a sensing unit, data processing unit, communication unit, and power unit. There may be additional components such as the localization unit, energy producer, position changer, etc. depending on the application. The general architecture of a sensor node is shown in Figure 9.1

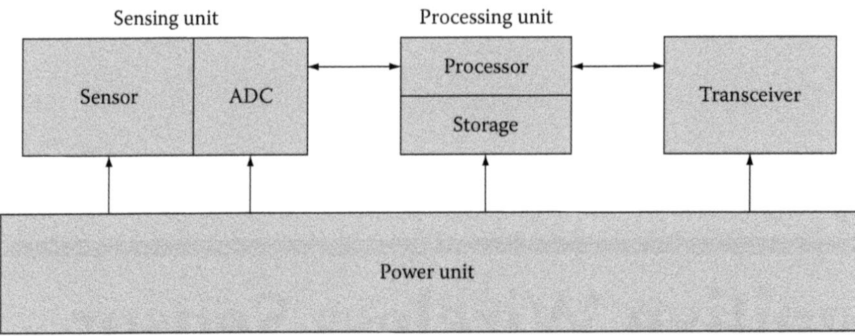

Figure 9.1 Basic units of a typical wireless sensor node.

The strength of WSNs lies in their flexibility and scalability. The nodes are capable of forming self-organized networks using multihop communications. In many applications, it is impractical to recharge nodes after they are deployed, as WSN nodes do not receive personal human interaction/care and usually get deployed in the field at random unknown locations. Thus, sensor nodes show strong dependence on battery life. In WSN, each node plays the dual role of data collector and data router. Hence, malfunctioning of a few nodes can cause significant topological changes. It might require rerouting of packets and reorganization of the complete network. Therefore, energy efficient communication is a very significant constraint in WSN, and thus computational operations of nodes and communication protocols must be made as energy efficient as possible.

Traditionally, almost all WSNs operate in unlicensed frequency bands, which are also used by other wireless applications, such as Wi-Fi, Bluetooth WiMAX, and ZigBee. Other applications such as microwave ovens and cordless phones also operate in those bands. This makes unlicensed bands overcrowded, which creates scarcity of spectrum and is also one of the significant problems for WSNs. With the rapid growth in ubiquitous low cost wireless hardware applications utilizing the unlicensed spectrum, demand and competitiveness for spectrum increases. Hence, network-wide performance degradation of traditional WSNs is expected. The condition may be more severe in the populated urban areas.

The coexistence issues in unlicensed bands have been the subject of extensive research [3,4]. In particular, it has been shown that IEEE 802.11 networks [5] can significantly degrade the performance of ZigBee/802.15.4 networks when operating in overlapping frequency bands [4].

Also, currently WSNs lack the capability of fine-tuning their radio configuration parameters dynamically to meet the challenges of a dynamic operating environment, for instance, floods, seasonal changes in vegetation, spectrum congestion, etc. Wireless channel properties also keep changing randomly. Other wireless devices operating in the same environment add up to the changing wireless environment for WSNs. This may result in degradation in radio link performance and unreliable network connectivity [6].

9.1.1 Common Design Issues in WSN

Routing: Routing is the process of selecting paths in a network along which network traffic is to be sent. In WSN, topology changes dynamically either because sensor nodes can move in the field or because a node exhausts its available energy. Moreover, sometimes the available channel may go into deep fade, which will then ask for change in the route.

Fusion of information with fault tolerance: Information fusion deals with the combination of information from the same source or different sources to obtain improved fused estimate with greater quality or greater relevance.

Security in WSN: Information based on sensed data can be used in agriculture and livestock, assisted driving, or even in providing security at home or in public places. A key requirement from both the technological and commercial points of view is to provide adequate security capabilities.

Scalability: It may happen that some nodes die or some more nodes join the existing wireless sensor network. The sensor network should adapt to the changes in the network size, node density, and topology.

Deployment of network: Deployment means setting up an operational sensor network in a real world environment. Sensor nodes can be deployed either by placing one after another in a sensor field in a deterministic manner or by dropping them from a plane randomly. Various deployment issues need to be taken care of because they directly affect routing and localization.

Spectrum scarcity: Rapid growth of wireless applications that work in the unlicensed band had created spectrum scarcity as the WSNs also operate in that band. Too much interference or unavailability of channel at proper time may lead to a disastrous situation.

Localization of sensor nodes: Sensing data without knowing the sensor location are meaningless; hence, localization deals with determining the locations of wireless devices (sensor nodes) in a WSN. The challenge comes here due to the fact that either some GPS-based mechanism or some local positioning–based mechanism is deployed.

9.2 Cognitive Radio

Traditionally, wireless networks run with fixed spectrum assignment policy regulated by government agencies. A spectrum is assigned to service providers on a long-term basis for large geographical regions. These spectrums can only be allowed to be used by licensed users, but Federal Communications Commission (FCC) measurements have indicated that many licensed frequency bands remain unused 90% of the time [7]. In order to better utilize the licensed spectrum, the FCC has launched a secondary market initiative [8], the goal of which is to remove regulatory barriers and facilitate the development of secondary markets in spectrum usage rights among the wireless radio services. This introduces the concept of dynamic spectrum allocation. Dynamic spectrum allocation is a technology that senses open channels and allows devices to communicate in underused parts of the spectrum. These underused parts of the spectrum, also called *spectrum holes*, can be visualized as in Figure 9.2. Dynamic spectrum allocation implicitly requires the use of cognitive radios to improve spectral efficiency. The inefficient usage of the existing spectrum can be improved through opportunistic access to the licensed bands without interfering with the

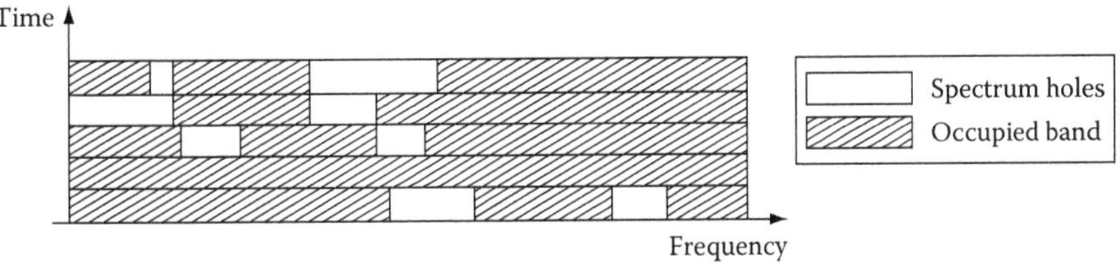

Figure 9.2 A typical spectrum occupancy chart.

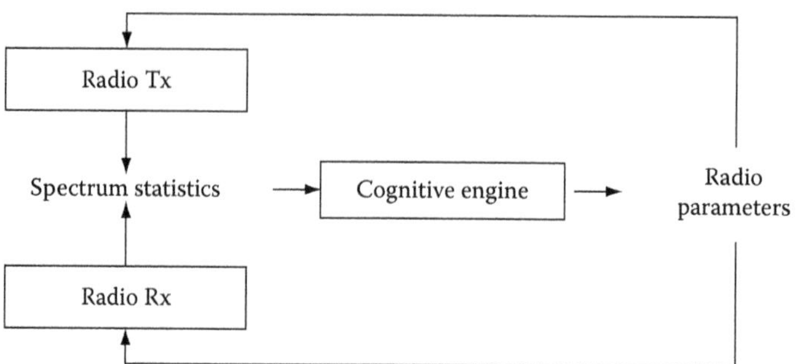

Figure 9.3 Basic units of a typical cognitive radio.

primary users (PUs). Cognitive radio (CR) is an intelligent wireless communication system that is aware of its surrounding environment. The idea is to use intelligent signal processing and decision making to enable the radio not just to utilize the spectrum efficiently but also to manage/adapt the other wireless parameters. These radios would dynamically reconfigure center frequency, waveform design, time diversity, and spatial diversity options. CR dynamically adapts the transmission or reception parameters of either a network or a node to achieve efficient communication without interfering with PUs. In simple words, a CR network consists of primary and secondary users (SUs). The PUs are the licensed users and hence have exclusive right to access the radio spectrum, whereas the SUs/cognitive users are the unlicensed users that can opportunistically access the free spectrum bands, without causing harmful interference to PUs.

Thus, radios that are capable of these intelligent decisions are called *cognitive radios*. These radios observe their wireless environment and then use their intelligent algorithms and computational learning methods and take actions accordingly to optimize the behavior of the network. A simple model of a CR where the cognitive engine interacts with the radio is shown in Figure 9.3.

9.3 Cognitive Radio–Based WSN

In both CR and WSN, sensing tasks are performed to collect information from the operating environment about spectrum occupancy and environmental parameters, respectively, and then appropriate actions are taken accordingly. CR-based WSN speaks about application of CR in WSN not only in the PHY and MAC layer. It enables cognition by taking advantage in all the layers and thus employs a cross-layered approach. It is promising and also challenging for WSNs to adopt the CR technology. One of the biggest advantages is that it enables the WSN to sense spectrum hole and utilize the vacant frequencies to improve spectrum utilization. It is also capable of increasing its own quality of service and throughput by adaptively and cognitively changing the various transmission and reception parameters such as transmitted power, operating frequency, modulation, pulse shape, symbol rate, coding technique, and constellation size. A wide variety of data rates and Quality of Service (QoS) can be achieved that improve the power consumption and network lifetime in a WSN.

CR technology can provide access not only to new spectrum bands but also to spectrum bands with better propagation characteristics. Generally, the lower frequencies have better propagation characteristics than the higher ones. The operation of WSNs at lower frequency bands allows range extension and higher energy efficiency. This helps in getting simpler topology as well as fewer sensor nodes to cover a given area as with lower frequencies the transmission range of the same node with same transmit power gets increased.

Some of the advantages of using low frequency for transmission are

- Higher transmission range
- Fewer sensor nodes required to cover a specific area
- Lower energy consumption
- Less number of hops to the destination
- Lowered end-to-end delay

Higher transmission range improves several important factors in WSNs including network connectivity, lifetime, and end-to-end delay.

Another advantage of using CR technology in wireless sensor networks is that data from various sensor nodes which are not spatially correlated or having low redundancy can be transmitted to the sink simultaneously in different channels which are free using the CR technology. This reduces the delay and enables the sink to monitor a large number of nonspatially correlated information in a real-time manner.

9.4 CWSN Architecture

Cognitive radio-based WSN is also like WSN in the sense that it consists of several tiny sensor nodes with all the constraints which a normal WSN has, especially the "limited battery energy." They differ in their transceiver hardware architecture and states. In a CWSN, the hardware also consists of a cognitive module that is responsible for spectrum sensing and adaptively changing the transmission parameters in a reconfigurable transceiver (reconfigurable transceiver is another advantage of the CWSN). Apart from these, there is also a cognitive engine in the cognitive module, which is responsible for learning and achieving cognition in order to make the changes automatically without human intervention. This cognitive engine is also responsible for controlling the changes which should take place in transmitter and receiver.

Cognitive engine works on the concept of cognition cycle (CC) [9]. States of CC are shown in Figure 9.4. Cognitive engine is comprised of six main states: observe, orient, act, decide, plan, and learn. It enables the nodes to achieve context awareness and intelligence so that it can be aware of its operating environment in order to sense for the white spaces, and use them in an intelligent and efficient manner. With regard to WSN, this cognitive engine may also assist in intelligent localization, routing, and scheduling in WSN.

CWSN nodes have an additional state called sensing state where they keep sensing their environmental parameters. Spectrum sensing state consumes a lot of energy because it is directly related to transmission and reception, which is the most energy consuming activity of a CWSN/WSN node (Figure 9.5) [10].

Sensing the spectrum can be done either in a distributed fashion or in a centralized fashion. A separate discussion about spectrum sensing in cognitive wireless sensor networks has been explained in next section.

Spectrum sensing can be done in distributed as well as centralized fashion, as shown in Figure 9.6. In the distributed scheme, all the nodes keep sensing the spectrum environment by their own and compete with other sensor nodes to grab the unoccupied spectrum. But the problem in this architecture is that all the nodes essentially need a spectrum sensing module, which may not be economically feasible.

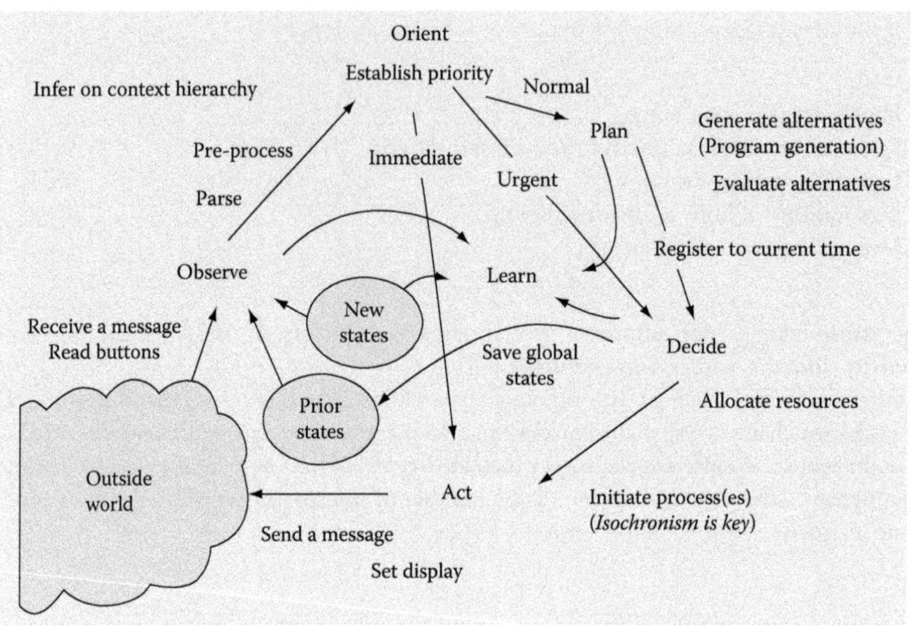

Figure 9.4 Cognitive cycle. (From Howitt, I.; Gutierrez, J.A.; , "IEEE 802.15.4 low rate - wireless personal area network coexistence issues," Wireless Communications and Networking, 2003. WCNC 2003. 2003 IEEE , vol. 3, no., New Orleans, Louisiana, USA, pp.1481–1486 vol. 3, 16–20 March 2003.)

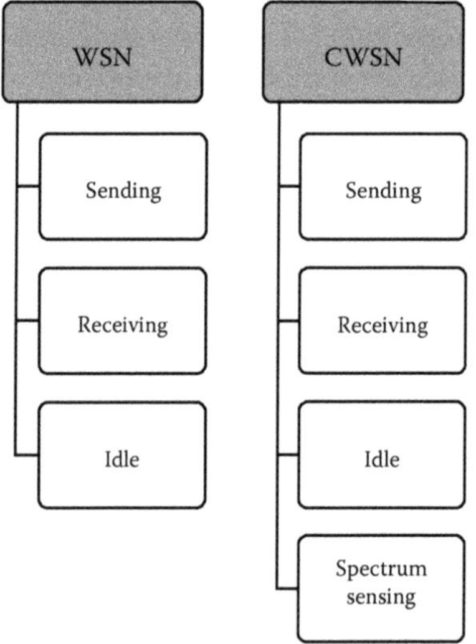

Figure 9.5 Simplified diagrammatic comparison of WSN and CWSN. (From Mitola III, J. and Maguire, G.Q., Cognitive radio: Making software radios more personal, Personal Communications, IEEE, 1999.)

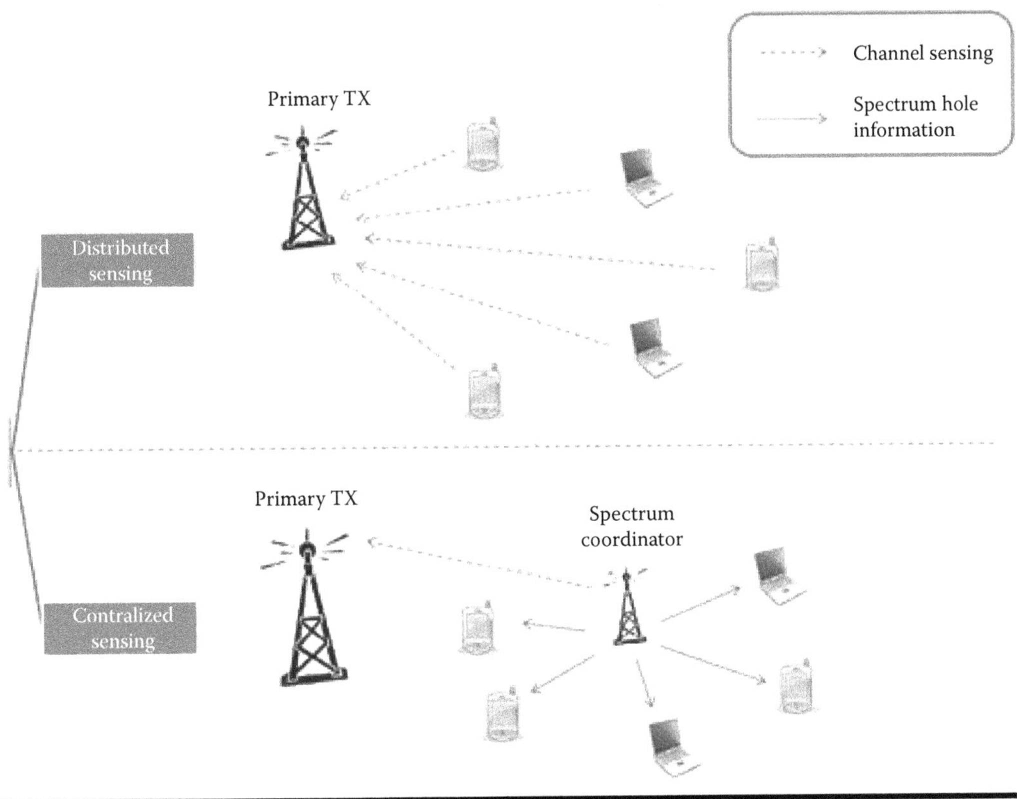

Figure 9.6 Distributed and centralized spectrum sensing. (From Cavalcanti, D. et al., Cognitive radio based wireless sensor networks[C]//, *Proceedings of the 17th International Conference on Computer Communications and Network [S. l.]*, IEEE Press, Washington, DC, pp. 1–6, 2008.)

In the centralized scheme, there is one network coordinator which has the responsibility of spectrum sensing as well as spectrum scheduling, that is, allotting the free channels to the needy nodes as per their need in preferential order.

But in a centralized control there is a difficulty. In centralized control, there has to be a separate control channel. On this control channel, there is broadcast of *channel switch command* according to which the sensor nodes change their Tx/Rx frequencies. The control channel can be from licensed or Industrial, Scientific, Medical (ISM) band. But there is always a fear that if the control channel somehow gets faded, and, in such case, the entire network will be in chaos.

In a common WSN scenario where there are a large number of sensor nodes, it may not be feasible to have a spectrum sensing module on each node. The feasible architecture will be to make only a few specialized nodes capable to do the spectrum sensing. The network coordinator or the cluster head can do this task in realistic deployment. If we take a deterministic deployment scenario like home or office, then easily a few specialized nodes capable of spectrum sensing can be deployed especially for the spectrum sensing [12]. Since spectrum sensing is a repetitive process, which would consume extra energy from battery powered sensors, implementing spectrum sensing in all nodes in a WSN may not be efficient in terms of energy consumption.

Sensing the target frequency/frequencies can be fixed or vary depending on the changing environmental and mobility condition of the nodes. Here, the cognitive engine can play an important role. After collecting spectrum sensing data over a long period of time, the cognitive engine will be able to know at what time which channel or specific band has to be sensed, instead of sensing all the channels every time. The channel occupancy pattern also varies spatially. Hence, the cognitive engine can assist the spectrum sensing nodes to decide which channel/channels to sense at a particular period of time and in a particular geographical region.

Sensing duration depends on the accuracy of sensing and also the required probability of detection. The more the sensing duration, the more complex algorithm for it, and thus the more energy it will consume. There has to be a trade-off between sensing duration and sensing accuracy. We will also discuss about a mechanism which uses appropriate energy budget for spectrum sensing according to the requirement/priority. During sensing, there has to be network wide quiet period, that is, nodes have to suspend their transmission following some schedule which would have been prebroadcasted well ahead of time by the coordinator to avoid any overlap, otherwise false alarms may occur indicating that some channel is occupied because of its own transmission or by some of the member sensor nodes.

Only during network-wide quiet periods, spectrum sensing is performed by the network coordinators. This is done usually to detect incumbents at low IDT (incumbent detection threshold) values and avoid false alarms. All nodes are given a fixed schedule of quiet periods that they have to follow essentially. These quiet periods can be scheduled well ahead of time by a broadcast through the network coordinator so that all nodes can adjust their transmission to avoid overlaps with scheduled quite periods [11]. This will require a very tight and proper coordination among the nodes in WSN Quite Period [13].

QPs may be a critical issue for high-throughput networks, but in WSN the traffic load is typically much lower and it will not be that much of an overhead since most of the time the sensors would be in a "stand by" or "sleep" mode [4].

Now let us talk about some jargons related to spectrum sensing. There have been some limits set by the FCC for IDT, probability of detection (PD), probability of false alarm, maximum probability of false alarm (MPFA), channel move time (CMT), and channel closing transmission time (CCTT) for CR networks. Same set of limits would be applicable to a CWSN. But some parameter values can be relaxed for a WSN due to much lower transmit power used by the WSN transmitters compared to the 802.22 devices for which these protocols and regulations are intended to [11].

Modifying the existing 802.15.4 protocol, that is, ZigBee, to suit for the CWSN physical and MAC layer will be a very good approach. The 802.15.4 standard defines 16 channels, each of 2 MHz bandwidth, in the 2.4 GHz band [14], among which only four are nonoverlapping with 802.11 channels (of 22 MHz bandwidth) in the same band [11].

Whenever an incumbent signal is detected well above IDT, WSN has to switch to a backup channel within the CMT to avoid the interference. This requires the coordinator to broadcast the channel switch command. Channel switch command also consists of the scheduled switching time for the nodes. This command indicates the WSN nodes to switch their channels to the free available channel as specified by the coordinator. In case free channel is not available, they have to utilize the backup channel. This switching is held responsible by the coordinator, which decides how to allocate channels, that is, scheduling of free channels among the needy nodes. The coordinator can use its own spectrum sensing results (centralized spectrum sensing) or may use the spectrum sensing reports (distributed spectrum sensing) from other specialized spectrum sensing nodes.

In practice, the coordinator has to make a list of backup channels available as there may be good probability that many backup channels are available at a particular time and geographical region.

Generally, in WSN, very fast incumbent recovery mechanism is not required. Once the PUs come into picture, the nodes have to vacate the channel very fast. But in some WSNs where there is very tight delay requirement there has to be provision of backup channels. This backup channel can be either from the licensed band shared with the PUs or from the unlicensed band shared with other SUs. Sensing of backup channel has to be done regularly in order to make sure that the backup channel is readily available and clean.

9.5 Spectrum Sensing Schemes for CWSN

Energy efficient spectrum sensing techniques are required to meet the power constraints of the CWSN. Spectrum sensing can be done in both time domain and frequency domain.

A question is raised that how much energy should be spent on channel sensing. Although high energy budgets will result in accurate sensing outcomes, but it may not be needed all the time. Because, in some cases, the interfering signal may be sporadic or may be perceived with very high power and thus is easy to detect. In such cases there is no need for high budget spectrum sensing algorithms. For cognition, there should be provision to tailor the energy budget according to the signal strength. Sensing energy budget should also be tailored according to the size of the packet to be transmitted. In WSN, there are packets of several types and sizes, and these packets have different priorities. Sensor packets can be as small as single to few tens of bytes; therefore, selecting the right amount of energy that has to be devoted to spectrum sensing might significantly improve energy efficiency. Loss of long packet may cause retransmission of the packet, which may be even more costly. Hence, in such cases, high energy budgets can be applied to sense the spectrum. But when it is needed to transmit the low size packets, then low energy budget sensing algorithms can be applied. In this way overall lower energy consumption can be achieved [15].

Now we will have an overview of the spectrum sensing techniques available for CWSN. In [15], a good survey of such spectrum sensing techniques that can be used in CWSN has been provided. Spectrum sensing can be done individually or can be done in a cooperative manner to identify the spectrum holes frequently. The cognitive-transceiver devices have two important functionalities: spectrum sensing and adaptation. The spectrum sensing hardware of cognitive transceiver keeps sensing the spectrum over a wide frequency band. This information is then passed to the SUs (in our case the SUs are WSN nodes).When such a spectrum hole is found, the SUs adapts its transmission power, center frequency, and modulation in order to transmit efficiently as well as minimize the interference to the incumbents [16]. An implied assumption here is that all the nodes have reconfigurable hardware.

Also while the transmission is in process, the cluster head or network coordinator, whichever is doing the task of spectrum sensing, should have the ability to detect the appearance of incumbents so that the SUs are able to change or give off the channel when the PUs starts transmission in that channel.

Spectrum sensing for CWSN can be categorized as blind sensing and signal-specific sensing.

9.5.1 Blind Sensing Techniques

Sensing techniques that do not rely on any special signal features are called blind sensing techniques. There are variants of blind sensing techniques available. We are giving a brief overview of these techniques. More details can be found in Refs. [17,18].

Energy detectors: The energy (power) detector estimates the signal power in the channel and compares that estimate to a threshold to determine whether any incumbent is present or not [17].

Eigenvalue-based sensing: Another blind sensing technique uses eigenvalues of the correlation matrix of the received signal containing both signal and noise [18].

9.5.2 Signal Specific Sensing Techniques

Sensing techniques that utilize specific signal features are mentioned as follows. More details can be found in Refs. [19–24].

Signature sensing for ATSC signal identification: The specific signature found in Advanced Television Systems signals are used to detect the primary transmission [19].

FFT-based carrier sensing: This sensing technique involves estimating the power spectral density in the received signal and detecting the availability of a carrier [20].

Higher order statistics sensing (HOSs): This sensing technique works with the assumption that the noise is Gaussian. These HOSs can be used to clearly estimate how well the distribution of the test statistic meets a Gaussian distribution [21].

PLL-based ATSC pilot sensing: This sensing technique consists of two frequency tracking blocks each attempting to track the ATSC pilot frequency. Multiple methods can be used to implement the frequency tracking block. This method suggests using a digital phase lock loop (PLL) using a version of the Costas loop [22].

Wireless microphone covariance sensing: This method is similar to the eigenvalue-based sensing technique described earlier and it calculates the sample covariance matrix [23].

Correlation sensing of the spectrum: Here the estimate of the Power Spectral Density of the signal is calculated using FFT. This PSD estimate is then correlated with a prestored PSD for the signal of interest. But this technique is not suitable for WSN because of the limited memory available for data storage [24].

9.5.3 Cooperative Sensing

While sensing the spectrum, the two major sources of getting degraded signals are multipath and shadowing for a given frequency. Here the cooperative spectrum sensing can help a lot. Presence of multiple radios helps in reducing the effects of severe multipath because of the achieved diversity. It reduces the probability that all users see deep fades at a time [25].

There are several spectrum sensing techniques. Each one has its own pros and cons. Some are fast with less accuracy and some are slow with more accuracy. Also the energy budget of all the sensing techniques varies. So it is required to optimize so that an overall less energy consumption and reasonable accuracy is achieved.

In Ref. [26], we have suggested a novel architecture for spectrum sensing specially for CWSN. We have come up with a new idea of doubly cognitive architecture where the cognition is achieved not only with respect to time and space but also with respect to traffic.

9.6 Learning in CWSN

In this section we will have a brief overview of the control/learning mechanisms which can be used in CWSN. There can be several approaches as follows:

Centralized: All the nodes are dependent on a centralized spectrum server for all of their spectrum requirements. Spectrum server determines and returns a suitable configuration to the nodes.

Distributed: All the nodes perform direct negotiation with their neighbors and decide for a suitable spectrum allocation.

Local algorithm based: Every node in the network follows some set of rules locally instead of having a centralized controller or distributed negotiation.

In centralized architectures, there is demand for a control channel which is available to every node in the deployment, that is, the channel used to exchange control messages has to be known by every node and unoccupied by other stations in that area. But some situation may occur when this control channel goes into deep fade or gets occupied (in case when it is a shared channel with the PUs). Moreover, there is an issue of system trapped in congestion when more nodes participate in the network. Distributed control is better suited for large environments where the same control channel cannot be maintained throughout the large sensor field. But this is more computationally intensive and the performance may go suboptimal. In Ref. [27], several CR control algorithms have been discussed. All these control mechanisms are some or other form of controlling with coordination between the nodes.

Let us have a brief look on each of them.

Rule-based reasoning: It is a particular type of reasoning that uses "if-then-else" rule statements. Rules are simply patterns, and an inference engine searches for patterns in the rules that match patterns in the data. It has been used in reconfiguring parameters and applications in order to accommodate the changes in the environment [28]. In Ref. [29], an idea of fuzzy rules has been used, where qualitative rules without hard decision boundaries can be employed.

Responsive surface methodology (RSM)/Design of experiment (DoE): RSM is a means by which the relationship between several explanatory variables and one or more response variables are explored. In RSM, sequences of designed experiments are used to obtain an optimal response. This model is easy to estimate and apply, even when little is known about the process. But the estimated optimum may not be the real optimum. In Refs. [30,31], RSM has been used to characterize and learn rules which can be used in adaptation and configuration of the CR parameters.

Game theory: Game theoretic approaches are getting very common nowadays for the control of cognitive networks. Some examples can be taken from Refs. [32–36].

Genetic algorithms: It is a search heuristic which mimics the process of natural evolution. It is then used to generate useful solutions to optimization and search problems. To use genetic algorithm, one needs to represent a solution to the problem as a genome (or chromosome). The genetic algorithm then creates a population of solutions and applies genetic operators such as mutation and crossover to evolve the solutions in order to find the best one. It can be used in order to analyze a predetermined fitness function to optimize an optimal CR configuration. Some researchers have implemented such optimization in systems using genetic algorithms in hardware which can be found in Refs. [37,38].

Linear programming: In linear programming, a mathematical model is developed to determine a way to achieve the best outcome in a given mathematical model for some list of requirements as linear relationships. In Refs. [39–41], the application of linear programming has been discussed, where they have used it in spectrum allocation, dynamic spectrum access for CDMA system, and optimal link activation schedule in CR networks, respectively.

Swarm algorithm: Swarm intelligence is a collective behavior of decentralized, self-organized systems. In such algorithms, the nodes have to follow some very simple rules and there is no centralized control. Here, only the interaction between the nodes will lead to the evolution of intelligent global behavior. Ant colonies, fish schooling, etc. are natural examples. In Ref. [26], this concept is used for local independent control of CR networks, which includes interference avoidance, coordination in the network, and network synchronization. The feasibility of this approach has been shown by its hardware implementation.

9.7 Challenges in Designing CWSN

Lifetime maximization or energy efficiency: In many cases, it is impractical to recharge nodes after they are deployed, as WSN nodes do not receive personal human intervention/care and usually get deployed in field at random unknown locations [42–50]. The same thing applies to CWSN. It inherits the problem of lifetime maximization and energy efficiency from the WSN. In fact, the situation gets deteriorated because apart from doing data sensing the nodes in CWSN are also involved spectrum sensing. Also they have to tune their RF parameters according to the requirement to achieve spectrum efficiency. Various modulation schemes, data rates, and coding schemes also directly influence the power consumption at each node.

PU detection and localization: Sensor nodes are seen as the SUs in CWSN. These nodes share the licensed band with the PUs, but they must avoid the interference to the PUs. In order to avoid interference to PUs, sensor nodes must be aware of the PUs and their location within the region of interest. Hence, CWSN requires localizing the presence of PU within the vicinity of the network.

Fusion: Information fusion deals with the combination of information from same source or different sources to obtain improved fused estimate with greater quality or greater relevance. In CWSN scenario, apart from data sensing, the nodes are also involved in spectrum sensing. Most of the time, this spectrum sensing is done in a cooperative manner where the spectrum sensing nodes share the spectrum sensing information with each other. This makes the fusion task even more challenging. As larger amounts of sensors are deployed in harsher environment, it is important that sensor fusion techniques should be robust and fault tolerant, so that they can handle uncertainty and faulty sensor readouts.

Routing: In CWSN, topology keeps changing dynamically because sensor nodes can adjust their transmission parameters, and CWSN can turn its transceiver on or off based on the presence of PU. If a node exhausts its available energy, it ceases to function. Moreover, in the dynamic environment of CWSN, spectrum is not always available for data transfer to all the sensor nodes. This makes the routes to and from the base station to the nodes very dynamic. This adds new challenges to routing in CWSN.

Resource allocation problem: Resource allocation, that is, spectrum scheduling in CWSN, should allocate spectrum fairly among all the sensor nodes and at the same time it should increase the spectrum utilization.

Power allocation: Power allocation is another challenging problem to consider the co-channel interference when multiple new users decide to use the same frequency. Hence, some sort of control mechanism distributed/centralized is needed in this case to manage the co-channel interference.

Optimization of the radio module: By adapting the modulation type and constellation size and channel coding rate, different data rates can be achieved which will directly influence the power consumption of each node, and in turn will affect the lifetime of the whole network. There may be several trade-offs between parameters to optimize the radio module. The selection of these radio parameters and changing these parameters on the fly is another challenging problem in CWSN.

Spectrum sensing: Spectrum sensing should reduce the sensing duration time as during this interval all traffic is suspended and spectrum sensing is performed by the node. The spectrum sensing duration is also a challenge. Proper choice of quiet periods is also very important as it directly affects the throughput of the network. Apart from this, choosing proper spectrum sensing algorithm according to the size/priority of data packets/signal strength also matters a lot in saving the energy of the spectrum sensing nodes.

Representation of network architecture of CWSN: Network architecture helps in designing and understanding various functions of network and nodes since CWSN has dynamic environment and sometimes it may be mobile. Several responsibilities may be given to network coordinator or may be given to every node in the network and it translates to different architectures. Hence, it is very difficult to find general network architecture.

9.8 Summary

Application of CR and cognitive network concepts in WSN is an emerging technology. The primary difficulty with wireless sensor nodes is their battery life. But another significant problem of spectrum scarcity has emerged these days as the unlicensed spectrum bands are getting overcrowded. At the same time, the licensed spectrum bands are seen to be unoccupied most of the time, which can be used opportunistically. Using CR concept in WSN leads to observing, learning, and adapting in order to achieve spectrum efficiency, good throughput, and better QoS. Several architectures are proposed for CWSN. But it is difficult to say which one is better, as it depends on the environment where the network is deployed as well as the network size. Spectrum sensing is the most important task in any CR-based technology. Several spectrum sensing algorithms have been developed for mobile applications, but they are not suitable for WSN scenario due to limited battery life as well as less complex hardware. The spectrum sensing algorithms need to be dynamically tailored according to signal strength as well as the data packet size/priority. Managing/controlling the entire cognitive network is also challenging and a lot of research is going on in this area. Here also distributed, centralized, and local independent control mechanisms have been proposed for the same. Apart from the limitations which CWSN inherits from WSN, there are some particular challenges like PU detection, PU localization, sensing information fusion, resource (channel and power) allocation, optimization of radio module, spectrum sensing algorithm, and a generalized architecture for such a network. The challenges are big but the technology is very promising.

References

1. I. F. Akyildiz, W. Su, Y. S. Subramaniam, and E. Cayirci, A survey on sensor networks. *Communications Magazine*, IEEE, 40(8):102–114, August 2002.
2. I. F. Akyildiz, W. Su, Y. S. Subramaniam, and E. Cayirci, Wireless sensor networks: A survey. *Computer Networks*, 38:393–422, 2002.
3. I. Howitt and J. Gutierrez, IEEE 802.15.4 low rate—Wireless personal area network coexistence issues. *Proceedings of IEEE WNCN*, Vol. 3, New Orleans, Louisiana, USA, pp. 1481–1486 Vol. 3, 16–20 March 2003.
4. D. Cavalcanti, R. Schmitt, and A. Soomro, Achieving energy efficiency and QoS for low-rate applications with 802.11e, in the *IEEE WCNC 2007*, Hong Kong, China, pp. 2143–2148, 11–15 March 2007.
5. IEEE 802.11 Standard, Wireless LAN medium access control (MAC) and physical layer (PHY) specifications, 1999 (Reaff 2003) Edition.
6. M. Ringwald and K. Romer, Deployment of sensor networks: Problems and passive inspection. The *Fifth IEEE Workshop on Intelligent Solutions in Embedded Systems*, Leganes, Spain, pp. 179–192, June 2007.

7. FCC Spectrum Policy Task Force, FCC report of the spectrum efficiency working group. November 2002.
8. Secondary markets initiative, FCC, 2005, http://wireless.fcc.gov/licensing/secondarymarkets/
9. J. Mitola III and G. Q. Maguire, Cognitive radio: Making software radios more personal, Personal Communications, IEEE, 1999.
10. A. S. Zahmati, S. Hussain, X. Fernando, and A. Grami, Cognitive wireless sensor networks: Emerging topics and recent challenges. *Science and Technology for Humanity (TIC-STH), 2009 IEEE Toronto International Conference*, Toronto, Ontario, Canada, pp. 593–596, September 26–27, 2009.
11. D. Cavalcanti, S. Das, W. Jianfeng et al., Cognitive radio based wireless sensor networks[C]//. *Proceedings of the 17th International Conference on Computer Communications and Network* [S. l.]. IEEE Press, Washington, DC, St. Thomas U.S. Virgin Islands, pp. 1–6, 2008.
12. S. Shankar, C. Cordeiro, and K. Challapali, Spectrum agile radios: Utilization and sensing architecture. *IEEE DySPAN*, pp. 160–169, 8–11, November 2005.
13. IEEE 802.22 draft Standard, IEEE P802.22TM/D0.4 draft standard for wireless regional area networks. http://www.ieee802.org/22/, November 2007.
14. IEEE 802.15.4 Standard, Wireless medium access control (MAC) and physical layer (PHY) specifications for low-rate wireless personal area networks (LR-WPANs), 2003 Edition.
15. L. Stabellini and J. Zander, Energy-aware spectrum sensing in cognitive wireless sensor networks: A cross layer approach. *Wireless Communications and Networking Conference (WCNC), 2010 IEEE*, Sydney, New South Wales, Australia, pp.1–6, 18–21, April 2010.
16. S. Haykin, Cognitive radio brain-empowered wireless communications. *IEEE Journal on Selected Areas in Communications*, 23(2):201–220, February 2005.
17. H. Urkowitz, Energy detection of unknown deterministic signals. *Proceedings of IEEE*, 55:523–531, April 1967.
18. Y. Zeng, Y.-C. Liang, Eigenvalue-based spectrum sensing algorithms for cognitive radio. IEEE Transactions on Communication, 57(6):1784–1793, June 2009.
19. Advanced Television Systems Committee, ATSC Digital Television Standard (A/53, Part 2)–RF/Transmission System Characteristics, January 2007.
20. T. Tomioka, T. Tomizawa, and T. Kobayashi, High-sensitivity carrier sensing using overlapped FFT for cognitive radio transceivers, April 2009.
21. J. M. Mendel, Tutorial on higher order statistics (spectra) in signal processing and systems theory: Theoretical results and some applications. *Proceedings of IEEE*, 79:278–305, March 1991.
22. G. Chouinard, 802.22 presentation to the ECSG on white space, IEEE P802.22 Wireless RANs, March 2009.
23. Y. Zeng, Y.-C. Liang, November 2006, grouper.ieee.org/.../22-06-0187-01-0000_I2R-sensing-2.ppt
24. J.-K. Lee, J.-H. Yoon, and J.-U. Kim, A new spectral correlation approach to spectrum sensing for 802.22 WRAN system, October 2007.
25. S. M. Mishra, A. Sahai, and R. W. Brodersen, Cooperative sensing among cognitive radios.
26. S. Kumar, D. Singhal, and G. Rama Murthy. Doubly cognitive architecture based cognitive wireless sensor networks. *International Journal of Wireless Networks and Broadband Technologies (IJWNBT)*, 1(2):30–35, accessed January 2, 2012.
27. C. Doerr, D. Grunwald, and D. C. Sicker, Local independent control of cognitive radio networks. *3rd International Conference on Cognitive Radio Oriented Wireless Networks and Communications, 2008. CrownCom 2008*, Singapore, pp. 1–9, May 15–17, 2008.
28. M. Bandholz, J. Riihijärvi, and P. Mähönen, Unified layer-2 triggers and application-aware notifications. *IWCMC'06. Proceeding of the 2006 International Conference on Communications and Mobile Computing*, ACM Press, New York, pp. 1447–1452, 2006.

29. N. Baldo and M. Zorzi, Fuzzy logic for cross-layer optimization in cognitive radio networks. *Workshop on Digital Rights Management Impact on Consumer Communications (CCNC)*, Las Vegas, NV, 46(4):64–71, April 2008.
30. K. K. Vadde and V. R. Syrotiuk, Factor interaction on service delivery in mobile ad hoc networks. *IEEE Journal on Selected Areas in Communications*, 22:1335–1346, September 2004.
31. T. Weingart, G. V. Yee, D. Sicker, and D. Grunwald, A dynamic cognitive radio configuration algorithm, *IEEE Communications*, 2006.
32. J. Neel, R. Buehrer, B. Reed, and R. Gilles, Game theoretic analysis of a network of cognitive radios. *The 2002 45th Midwest Symposium on Circuits and Systems (MWSCAS)*, Vol. 3, Tulsa, Oklahoma, pp. III-409- III-412 Vol. 3, 4–7, August 2002.
33. J. Neel, J. H. Reed, and R. P. Gilles, Convergence of cognitive radio networks. *IEEE Wireless Communications and Networking Conference (WCNC)*, Vol. 4, no., Atlanta, Georgia USA, pp. 2250–2255 Vol. 4, 21–25, March 2004.
34. J. O. Neel and J. H. Reed, Performance of distributed dynamic frequency selection schemes for interference reducing networks. *Military Communications Conference MILCOM*, Washington, DC, pp. 1–7, 2006.
35. A. B. MacKenzie and S. B. Wicker, Game theory in communications: Motivation, explanation, and application to power control. *IEEE Globecome*, San Antonio, TX, pp. 821–826, 2001.
36. R. Menon, A. MacKenzie, R. Buehrer, and J. Reed, Game theory and interference avoidance in decentralized networks, in *SDR Forum Technical Conference*, Phoenix, Arizona, 2004.
37. T. W. Rondeau, B. Le, C. J. Rieser, and C. W. Bostian, cognitive radios with genetic algorithms: Intelligent control of software defined radios, in *SDR Forum Technical Conference*, Atlanta, Georgia, USA, 2004.
38. C. J. Rieser, T. W. Rondeau, C. W. Bostian, and T. M. Gallagher, Cognitive radio testbed: Further details and testing of a distributed genetic algorithm based cognitive engine for programmable radios. *Proceedings of the IEEE Military Communications Conference*, Monterey, California, Vol. 3, pp. 1437–1443, 31 October-3 November 2004.
39. O. Ileri, D. Samardzija, and N. Mandayam, Demand responsive pricing and competitive spectrum allocation via a spectrum serve. *First IEEE International Symposium on New Frontiers in Dynamic Spectrum Access Networks*, Baltimore, MD, pp. 194–202, 2005.
40. M. Buddhikot and K. Ryan, Spectrum management in coordinated dynamic spectrum access based cellular networks. *First IEEE International Symposium on New Frontiers in Dynamic Spectrum Access Networks*, Baltimore, MD, pp. 299–307, 2005.
41. C. Raman, R. Yates, and N. Mandayam, Scheduling variable rate links via a spectrum server. *First IEEE International Symposium on New Frontiers in Dynamic Spectrum Access Networks*, Baltimore, MD, pp. 110–118, 2005.
42. B. Fette, *Cognitive Radio Technology*. Newnes Publication, Boston, MA, 2006.
43. D. Cavalcanti, S. Das, J. Wang, and K. Challapali, Cognitive radio based wireless sensor networks. *Proceedings of 17th International Conference on Computer Communications and Networks*, 2008. ICCCN'08, pp. 1–6, 2008.
44. S. Gao, L. Qian, and D. R. Vaman, Distributed energy efficient spectrum access in wireless cognitive radio sensor networks. *Wireless Communications and Networking Conference 2008, WCNC'08*, Las Vegas, NV, pp. 1442–1447, March 2008.
45. A. Mirza and R. Garimella, PASCAL: Power aware sectoring based clustering algorithm for wireless sensor networks. *International Conference on Information Networking, ICOIN 2009*, Chiang Mai, Thailand, pp. 1–6, 2009.
46. K. Akkaya and M. Younis, A survey on routing protocols for wireless sensor networks. *Ad Hoc Networks*, 3:325–349, 2005.

47. W. Heinzelman, A. Chandrakasan, and H. Balakrishnan, Energy-efficient communication protocol for wireless microsensor networks. *Proceedings of the 33rd Annual Hawaii International Conference on System Sciences*, Maui, Hawaii, Vol. 2, p. 10, 2000.
48. R. C. Luo, C.-C. Yih, and K. L. Su, Multisensor fusion and integration: Approaches, applications, and future research directions. *Sensors Journal, IEEE*, 2(2):107–119, April 2002.
49. E. F. Nakamura, A. A. F. Loureiro, and A. C. Frery, Information fusion for wireless sensor networks: Methods, models, and classifications. *ACM Computing Surveys*, 39, pp. 1–7, September 2007.
50. L. Liu, Z. Li, and C. Zhou, Back propagation-based cooperative localization of primary user for avoiding hidden-node problem in cognitive networks. *International Journal of Digital Multimedia Broadcasting*, 9, 2010, http://www.informatik.uni-trier.de/~ley/db/journals/ijdmbc/ijdmbc2010.html#LiuLZ10

INTELLIGENT SENSOR NETWORKS: SIGNAL PROCESSING

Chapter 10

Routing for Signal Processing

Wanzhi Qiu and Efstratios Skafidas

Contents

10.1 Introduction

Signal processing (SP) is the process of information extraction or decision making by analyzing relevant observational or experimental data. SP such as estimation and detection is fundamental to numerous applications in wireless sensor networks (WSNs) such as environmental monitoring, industrial control, and military surveillance. WSNs designed for such applications normally consist of a large number of nodes densely scattered over an area of interest where multihop transmission is the only practical way to move data across the network [1]. Networking, therefore, is a vital process of multihop WSNs, and is carried out by routing protocols that are responsible for establishing routes between source and destination nodes.

One of the major constraints of WSNs is the limited energy supply. Since radio communication is often the most expensive operation a node performs in terms of energy usage, it is crucial for maximizing the lifetime of WSN that data packets are routed to the destination in an energy-efficient manner [2]. Indeed, intensive research has addressed energy-efficient routing for WSN (see Ref. [3] and the references therein). Common route metrics include the number of hops [4–6] and energy expenditure [7–11]. Hierarchical protocols [12,13] group nodes into clusters where cluster heads are responsible for intracluster data aggregation and intercluster communication in order to save energy. Location-based protocols utilize information on node locations to increase energy efficiency by relaying data only to certain desired regions [14]. These generic routing algorithms establish routes without considering the performance of signal processing (PoSP) that is achievable from the data being forwarded along the routes.

Protocols that incorporate application performance or data quality into routing are also available [15–17]. In particular, in data-centric routing, the node desiring certain types of information sends queries to certain regions and waits for data from the nodes located in the selected regions [15,16]. Alternatively, reference [17] introduces information-directed routing to minimize communication cost while maximizing information gain. Reference [18] proposes a link metric that considers packets delay as well as network lifetime.

SP in WSN has also been extensively studied. Traditionally, SP is carried out in a centralized manner where measurements from the nodes are collected and processed at a central location. In contrast, distributed methods [19–28] spread the computation across participating nodes to reduce communication cost and computational burden on any particular point of the system. For example, Ref. [26] proposes a hybrid energy-driven scheme where each sensor node sends out its 1 bit decision if that decision exceeds a predetermined detection accuracy threshold, and sends out all its observations otherwise. Two multihop fusion schemes are proposed in Ref. [27]. In the first scheme, each sensor transmits the histogram of the observations of its descendants and itself. In the second scheme, the normalized log-likelihood ratio (LLR) values for subsets of nodes are computed and propagated along the minimum spanning tree to the fusion center (FC) which decides the hypothesis based on the acquired LLR values. An energy-efficient distributed source localization algorithm is proposed in Ref. [28], where the intermediate estimates are progressively processed by nodes along the routes and the refined results are further relayed to the fusion centre.

Note that routing is not addressed in these distributed SP schemes and, therefore, a requirement for applying them is that routes between source and destination nodes are preestablished. One work related to joint optimization of routing and detection is Ref. [29]. It develops a serial fusion method for signal detection and proposes a routing scheme which is essentially a depth-first traversal technique enhanced with knowledge of locations of nodes. The intermediate detection decision is passed along the route which tries to traverse the network with as few hops as possible until a final detection decision can be made. However, the role of each node is considered identical; no quality of measurement of individual nodes is considered. Research on routing for signal processing (RfSP) protocols can be best represented by [30,31]. The protocols there facilitate joint optimization of PoSP and energy efficiency via metrics which directly connect PoSP with energy consumption associated with sensing and data transmission of each link along the routes. In particular, the problem of routing for the detection of a correlated random signal field is studied in Ref. [30]. A new link metric using the Chernoff information is proposed which characterizes detection performance based on the number and locations of nodes along the route. This novel metric captures the contribution of a given link to the decay rate of error probability of detection, and the route is determined using the shortest path framework though a centralized optimization algorithm. The problem of energy-efficient routing for detection subject to Neyman–Pearson criterion is studied in Ref. [31]. There, routing metrics that connect the gain in signal-to-noise ratio and energy cost of each link are proposed, and combinatorial optimization programs are developed to solve for the best route in terms of a detection-probability-to-energy ratio.

Protocols in Refs. [30,31] are both centralized RfSP schemes, where the routes are computed centrally requiring complex optimization algorithms and global information such as locations and observation coefficients of all nodes in the network. Reference [32] proposes a distributed RfSP scheme for the same problem as in Ref. [31] where the routing decision is made at each node autonomously based on locally available information only. Clearly, for large-scale networks, or networks with dynamically changing topologies, distributed routing schemes which require neither global information nor centralized optimization would be more practical due to their superior flexibility and scalability.

In this chapter we address issues of building link metrics and the associated data aggregation schemes that realize RfSP, review RfSP protocols, and discuss the potential of and challenges facing RfSP. We also discuss in greater detail on how distributed RfSP schemes can be designed. Finally, we conduct numerical simulations to reveal the SP performance and energy efficiency of both centralized and distributed RfSP schemes as compared with generic routing protocols.

This chapter is organized as follows. Section 10.2 describes the problem concerning SP and RfSP in WSN. Section 10.3 reviews centralized RfSP schemes. Section 10.4 introduces distributed RfSP strategies. Section 10.5 provides the simulation results and Section 10.6 draws the conclusion.

10.2 Signal Processing and RfSP

In this section we give a general introduction to RfSP and discuss issues of building link metrics and the associated data aggregation schemes that one has to address in developing RfSP protocols.

10.2.1 Signal Processing Problem

Consider the scenario where sensor nodes are deployed over an area to collect measurements on a particular event or phenomenon. These measurements are then fed to a SP algorithm for the purpose of, for example, detecting the occurrence of the event or determining some information about the phenomenon. The data observed at node N_k can be described by

$$y_k = g_k s_k + w_k \tag{10.1}$$

where

s_k is the unknown signal sampled by N_k

w_k is the observation noise

g_k is the observation coefficient which is in general dependent on the characteristics of node N_k and its physical location

In a centralized SP scheme, all raw measurements $\{y_k\}$ are transmitted to a central location, that is, the FC, where the processing outcome D is achieved through a centralized computation:

$$D = F(\{y_k\}) \tag{10.2}$$

where $F(\cdot)$ denotes the fusion rule applied to the data.

One major issue of centralized SP schemes is the energy inefficiency associated with radio transmissions since no data compression is made when the packets are being relayed by intermediate nodes. As stated in Refs. [30,31], the energy consumption at each node N_k can be modeled by

$$e_{k,j} = \delta\Delta_{k,j}^{\gamma} + e_0, \tag{10.3}$$

where

e_0 accounts for energy consumption for sensing and processing

$\delta\Delta_{k,j}^{\gamma}$ is the energy of one data transmission over the link from N_k to its next-hop N_j, which depends on the distance $\Delta_{k,j}$ between N_k and N_j

Consider a route of M equally spaced nodes, $\Omega = \{N_1, N_2, \ldots, N_M\}$, where the M-th node N_M is acting as the FC. Let Δ denote the node spacing, it can be easily verified that the total energy expenditure for the transmission along the route is. $(1 + 2 + \cdots (M - 1))(\delta\,\Delta^{\gamma} + e_0)$

10.2.2 Distributed Signal Processing

In a distributed SP scheme, nodes collectively achieve a global objective by each performing some processing based on locally available information and relaying the intermediate results. In particular, apart from the first node, the k-th node on the route receives the intermediate result u_{k-1} from the $(k - 1)$-th node and combines it with its local observation y_k to generate its own intermediate result u_k, that is,

$$u_k = \begin{cases} F_k(u_{k-1}, y_k) & k \neq 1 \\ F_k(y_k) & k = 1 \end{cases} \tag{10.4}$$

which is relayed to the $(k + 1)$-th node. Note that u_k may consist of multiple values. For example, it can include an estimate of a parameter and the associated confidence level. The last node on the route (i.e., the FC) generates the final result

$$D = u_M = F_M(u_{M-1}, y_M) \tag{10.5}$$

The strength of distributed SP is its potential of reducing energy consumption in transmission. Suppose u_k can be sent with one transmission, the total energy expenditure for the transmission along the route of M equally spaced nodes would be $(M - 1)(\delta\Delta^{\gamma} + e_0)$. In addition, in a distributed SP scheme, there is no special requirement on energy or computational power imposed on the FC. Here, any node can potentially act as a FC and the fusion (processing) process can stop at any node as soon as a preset performance measure (e.g., confidence level associated with the estimation) is achieved.

Despite all of its advantages, distributed SP algorithms are generally nontrivial to design. In other words, for a centralized fusion rule $F(\cdot)$ of (10.2), it is a challenging task to produce equivalent local fusion rules $\{F_k(\cdot)\}$ that collectively and incrementally achieve the objective and performance of $F(\cdot)$. This is a problem-specific practice and sometimes simplifying assumptions and approximations are necessary [18,24,28].

10.2.3 Routing for Signal Processing

RfSP protocols further exploit the potential of distributed SP schemes. In addition to energy-efficient data delivery via distributed processing, the issue of where to collect these data is addressed in RfSP. The goal of RfSP is the joint optimization of PoSP and energy efficiency via routing metrics which directly connect SP performance and energy consumption. The advantages of RfSP over generic routing protocols are the ability to achieve optimum PoSP-to-energy-consumption ratio and the controlled trade-offs between PoSP and energy efficiency. The solutions to RfSP are highly dependent on the specific SP problems to solve, and usually involve the following three critical processes:

1. Design local fusion rules $\{F_k(\cdot)\}$ to be applied by nodes progressively and associated intermediate results to be forwarded further downstream along the route.
2. Develop a link metric $q_{k,j}$ that characterizes the contribution of a link to the performance of SP. It measures the gain of the link from node N_k to its next-hop N_j by means of a quantity directly related to PoSP, for example, estimation variance or detection probability, and therefore is in general dependent on the data quality (as measured by the observation coefficient) of N_j and the physical locations of both the local and next-hop nodes.
3. Build up a routing metric $\lambda(\cdot)$ that relates the total PoSP contribution $(Q(\Omega))$ and energy expenditure $(E(\Omega) = \sum_k e_{k,(k+1)})$ of a route Ω, and obtain the desired route Ω^* through the following constrained optimization:

$$\Omega^* = \arg\max_{\Omega \in \prod} \lambda(Q(\Omega), E(\Omega)) \tag{10.6}$$

   where $\prod$ is the set of routes which satisfy certain constraints. Typical constraints are
   C1: $E(\Omega) \leq E_{\text{lim}}$—route energy constraint
   C2: $Q(\Omega) \geq Q_{\text{lim}}$—guaranteed performance

For example, the routing metric $\lambda(\cdot) = Q(\Omega)/E(\Omega)$ under C1 leads to the route that has the best performance-to-energy ratio among all routes with energy expenditure up to E_{lim}. Note that the maximization in (10.6) could be replaced by minimization depending on how the metric $\lambda(\cdot)$ is formulated.

The task of developing RfSP protocols is, in general, challenging. First of all, as mentioned previously, the "splitting" of a central SP algorithm $F(\cdot)$ into local rules $\{F_k(\cdot)\}$ is nontrivial. Secondly, the development of the link metric which captures the gain in performance of individual links can be very complex. In addition, in order to facilitate solving the optimization in (10.6), it is important for a link metric to be (1) independent—contribution of a link is independent of past and future links; and (2) additive $-Q(\Omega) = \sum_k q_{k,(k+1)}$. As will be demonstrated in Section 10.3, depending on the specific SP problem, approximations are often required in deriving a link metric of desired forms.

Note that the optimization in the form of (10.6) is carried out centrally, demanding global information such as locations and observation coefficients of all nodes in the network. For such centralized RfSP schemes, although the SP is performed in a distributed fashion, the routes have to be precomputed and nodes programmed for each source-destination pair. Therefore, distributed RfSP (D-RfSP) schemes which require neither global information nor centralized optimization are more practical for large-scale networks, or networks with dynamically changing topologies. In a D-RfSP protocol, each node makes its routing decision autonomously based on locally available information only. However, since D-RfSP protocols only utilize local information, they will be suboptimum as compared to their centralized counterparts. Details on how to design D-RfSP protocols will be provided in Section 10.4.

10.3 Centralized RfSP Schemes

In this section we highlight the core development process of two innovative RfSP schemes [30,31] to illustrate how joint optimization of SP performance and energy efficiency can be achieved for specific SP problems.

10.3.1 Chernoff Routing

The detection of a correlated Gaussian signal field is considered in Ref. [30]. Let hypothesis H_1 denote the presence of the phenomenon within the sensor network and H_0 its absence, and assume all sensors have identical observation coefficients ($g_k = 1$), the observations at each node N_k along the route $\Omega = \{N_1, N_2, \ldots, N_M\}$ under each hypothesis are given by

$$H_0{:}\, y_k = w_k, \quad k = 1, \ldots, M$$
$$H_1{:}\, y_k = s_k + w_k, \quad k = 1, \ldots, M$$

where the s_k's are correlated Gaussian samples of the signal with $s_k \sim N(0, \sigma_s^2)$ and a nondiagonal covariance matrix. The noise samples w_k are i.i.d. Gaussian with $w_k \sim N(0, \sigma_w^2)$. The centralized decision rule is to decide H_1 if

$$\ln \frac{p_1(y)}{p_0(y)} \geq \ln \frac{\pi_0}{\pi_1} \tag{10.7}$$

where

$y = [y_1, \ldots, y_M]$

π_j and $p_j(y)$, $j = 0, 1$, are the prior probability of H_j and probability density function (PDF) of the joint Gaussian random variables under H_j, respectively

The development of the link metric for RfSP starts with using the Chernoff information as a tractable metric which captures the probability of detection error of a given route. Then, Schweppe's recursive representation of the likelihood function is used to express the Chernoff information in an additive form. Finally a link metric which is independent from link to link is created by assuming a Gauss–Markov signal correlation model.

As stated in Ref. [33], the average error probability of the rule (10.7) is bounded by

$$P_\varepsilon \leq \pi_0^{1-s} \pi_1^s e^{\mu(s)}, \quad 0 \leq s \leq 1$$

where $\mu(s)$ is the cumulant generating function of the LLR under H_0, that is,

$$\mu(s) = \ln E_0 \left\{ \exp\left(s \ln \frac{p_1(y)}{p_0(y)} \right) \right\}, \quad 0 \leq s \leq 1$$

with $E_j(\cdot), j = 0, 1$, denoting the mathematical expectation under H_0 and H_1, respectively. The Chernoff information [34] is defined as the $\mu(s)$ evaluated at the maximizing value of s, *that is*,

$$Q = \arg \max_{0 \leq s \leq 1} \{-\mu(s)\} \tag{10.8}$$

In order to reveal the contribution of each link to P_ε, Schweppe's recursive representation of the likelihood function [35] is utilized and it is shown that (10.8) can be approximated by $Q = \sum_{k=1}^{M} q_k$, with

$$q_k = \frac{1}{2}\ln\left(1 + \frac{P_{k/k-1}}{\sigma_W^2}\right) \tag{10.9}$$

where $P_{k/k-1} = E_1\{s_k - E_1\{s_k/y_1, \ldots, y_{k-1}\}\}^2$ is interpreted as the power of the signal innovation. The proposed link metric (10.9) is monotonic in k. It indicates the amount of uncertainty resolved by collecting a sample from node N_k, and the optimal route as the one that provides the maximum reduction in uncertainty.

Since the link metric (10.9) is in general not independent from link to link, in order to make the optimization of finding the optimal route tractable, the signal field is assumed to be a Gauss–Markov process. For this special case, after some approximations, the link metric can be expressed as a function of the link length Δ_k,

$$q_k = \frac{1}{2}\ln\left(\text{SNR} + 1 - (\text{SNR} - 1)e^{-2A\Delta_k}\right) \tag{10.10}$$

where A is a parameter in the signal model, and $\text{SNR} = \sigma_s^2/\sigma_w^2$. The Gauss–Markov model also allows the detection be carried out in a distributed fashion using the Kalman aggregation, where each node, based on data from previous-hop and local measurement, calculates and sends to its next-hop three quantities: the accumulated likelihood function, the variance of innovation, and the predicted measurement.

With this link metric which characterizes the additive and independent contribution of each link to detection performance, the Chernoff routing [30] finds the optimal route using the shortest path framework. In shortest path routing [36], each link in the network is assigned a link cost $\gamma_{k,j}$ which quantifies the consumed resources of the link N_k to N_j, and the "least cost" route, where the cost of a route is simply the sum of its link costs, is sought. A constant link cost, that is, $\gamma_{k,j} = \varepsilon$, leads to the minimum-hop routing. Alternatively, setting $\gamma_{k,j} = e_{k,j}$, where $e_{k,j}$ is the link energy consumption given by (10.3), results in the minimum-energy routing. The link cost of the Chernoff routing introduces a weighting factor $\alpha \geq 0$ to control trade-offs between detection performance and energy:

$$\gamma_{k,j} = (e_{k,j} - \alpha q_j)_\varepsilon^+$$

where

$$(x)_\varepsilon^+ = \begin{cases} x & x > 0 \\ \varepsilon & x \leq 0 \end{cases}$$

It is shown in Ref. [30] that with proper values of the design parameters α and ε, the Chernoff routing is able to achieve better detection performance for the same energy consumption than the minimum-hop and minimum-energy protocols.

10.3.2 Combinatorial Optimization Routing

The problem of target detection is considered in Ref. [31]. The measurement y_k at node N_k along the M-nodes route Ω depends on which of the two hypotheses, the noise-only hypothesis H_0 and target-present hypothesis H_1, is true. That is,

$$H_0 : y_k = w_k, \quad k = 1, 2, \ldots, M \tag{10.11}$$

$$H_1 : y_k = Bg_k + w_k, \quad k = 1, 2, \ldots, M \tag{10.12}$$

where g_k is the observation coefficient of N_k, $B > 0$ represents the reflection strength of the target in response to illuminating signals and is assumed unknown. The noise w_k is white Gaussian with zero mean and variance σ^2. The PDFs under each hypothesis are, therefore, given by

$$p_0(y) = \frac{1}{(2\pi\sigma^2)^{M/2}} \exp\left(-\frac{1}{2\sigma^2} \sum_{k=1}^{M} y_k^2\right) \tag{10.13}$$

$$p_1(y) = \frac{1}{(2\pi\sigma^2)^{M/2}} \exp\left(-\frac{1}{2\sigma^2} \sum_{k=1}^{M} \left(y_k - Bg_k\right)^2\right) \tag{10.14}$$

According to the Neyman–Pearson criterion that maximizes the probability of detection P_D for a given probability of false alarm P_{FA} [37], H_1 is decided if the likelihood ratio $L(x) = p_1(y)/p_0(y)$ is greater than a threshold. By utilizing (10.13) and (10.14), this criterion can be realized by the test

$$T(y) = \sum_{k=1}^{M} g_k y_k > \gamma \tag{10.15}$$

where γ is determined via

$$P_{FA} = \int_{\{y:L(y)>y\}} p_0(T(y))d_y \tag{10.16}$$

The test (10.15) immediately suggests a distributed processing scheme where each nondestination node $N_I(I < M)$ computes the locally accumulated test statistic $\sum_{k=1}^{I} g_k y_k$ and forward it to the next-hop, and the detection decision is made at the destination node based on the accumulated $T(y)$.

The link metric for RfSP is derived by examining the performance of this detector as follows. Since the noise samples are uncorrelated and the test statistic $T(y)$ is Gaussian for both hypotheses, it can be shown that

$$p_0(T(y)) \sim N(0, \sigma^2\varepsilon)$$

$$p_1(T(y)) \sim N(B\varepsilon, \sigma^2\varepsilon)$$

where $\varepsilon = \sum_{k=1}^{M} g_k^2$.

The threshold γ can then be determined according to (10.16) without knowing the value of B, and the probability of detection can be found to be

$$P_D = \int_{\gamma}^{\infty} p_1(y)d_x = Q_f\left(Q_f^{-1}(P_{FA}) - \sqrt{\kappa^2}\right) \tag{10.17}$$

where

$$Q_f(x) = \int_x^{\infty} \frac{1}{\sqrt{2\pi}} \exp\left(-\frac{1}{2}t^2\right) d_t$$
$$\kappa^2 = B^2 \varepsilon / \sigma^2$$

It can be seen from (10.17) that the detection performance is totally characterized by ε. Since $Q_f(x)$ is a monotonically decreasing function, P_D is monotonically increasing with increasing ε which is defined as the performance contribution of the route:

$$Q(\Omega) \stackrel{\Delta}{=} \varepsilon = \sum_{k=1}^{M} g_k^2 \tag{10.18}$$

The corresponding link metric g_k^2 captures the performance contribution of each node on the route and, as demonstrated in (10.18), has the properties of additivity and independence. To facilitate joint optimization of detection performance and energy efficiency, the mean detection-probability-to-energy is introduced in Ref. [31]:

$$\lambda(\Omega) = \frac{Q(\Omega)}{E(\Omega)} \tag{10.19}$$

and the optimum route is obtained through the following optimization:

$$\Omega^* = \arg\max_{\Omega} \lambda(\Omega) \tag{10.20}$$

Two variants of (10.20) are also proposed in Ref. [31] where constraints on $Q(\Omega)$ and $E(\Omega)$ are introduced, respectively, to the optimization so that solutions with guaranteed performance or constrained energy can be sought. These routing metrics are shown in Ref. [31] to achieve superior detection performance and energy consumption trade-offs over the traditional minimum-hop and minimum-energy routing protocols.

10.4 Distributed RfSP Protocols

The centralized RfSP schemes demand global information such as locations and observation coefficients $\{g_k\}$ of all nodes in the network be available at a central location where Ω^* is computed in a static manner. As discussed in Section 10.1, in practice not only the network topology can change, $\{g_k\}$ can also change in situations such as target movements. Such dynamics would make recalculating and updating the centrally optimized routes unrealistic. In this section we present strategies for distributed RfSP (D-RfSP) where the next-hop is calculated on the fly using only locally available information. That is, each node selects the next-hop autonomously with the goal of maximizing the SP performance associated with unit energy expenditure. One protocol of such nature is proposed in Ref. [32] for a signal detection problem. In a D-RfSP scheme, a local node N_k evaluates the performance gain ($q_{k,j}$) each neighbor N_j can potentially provide and the associated energy cost ($e_{k,j}$), and selects the one which yields the largest performance-gain-to-energy ratio as the next-hop, that is,

$$N_j^* = \arg\max_{N_j \in \Theta} \frac{q_{k,j}}{e_{k,j}} \tag{10.21}$$

where Θ denotes the set of N_k's neighbors which meet certain conditions. The neighbors of a node are the nodes that physically positioned within its radio range such that direct communications are possible. Apart from designing a distributed signal fusion scheme and the link metric $q_{k,j}$, two unique issues need to be addressed in developing a D-RfSP scheme. Firstly, energy constraining applied locally at each node is essential for the route to converge. Otherwise the route will attempt to traverse the whole network. Secondly, revisits should be avoided unless absolutely necessary since a revisited node does not provide any new contribution to the SP performance. The solutions to these matters are reflected in the design of the rules for Θ. In this section we first address these two maters and then present in detail a D-RfSP strategy that is based on logical relationships of nodes.

10.4.1 Meeting Energy Constrain

D-RfSP protocols rely on information of neighboring nodes. This information is readily available for most of WSN systems. In fact, in order to maintain connection to the network, a node needs to keep some information such as addresses of its neighbors. This is usually done through a local record termed as the neighbor table. The neighbor table is generally built up during a node's join process when it scans its neighborhood in order to discover its neighbors and find a potential parent node to join [38,39]. The ZigBee standard requires each node to keep the neighbor table up-to-date. This can be achieved, for example, by periodically scanning and/or monitoring the neighborhood.

If the locations of neighbors and the destination are available, to control energy expenditure a local node N_k can regard the neighbor N_j as a next-hop candidate (i.e., a member of Θ), only if it satisfies the following distance condition:

$$\delta\Delta_{k,j}^{\gamma} + e_0 + \delta\Delta_{j,M}^{\gamma} + e_0 < E_{k\_\lim} \tag{10.22}$$

where $E_{k\_\lim}$ is the energy constraint of the route from N_k to the destination N_M. Since N_k does not know the hops beyond N_j, here the energy consumption from the neighbor N_j to N_M is only an approximation. Therefore (10.22) is a so-called soft constraint; it can be slightly dishonored if necessary. In Section 10.4, we will show that if the number of hops from N_j to the destination can be determined, the corresponding energy cost can be estimated by the average energy per hop.

10.4.2 Avoiding Unnecessary Revisits

When the next-hop is determined locally, visiting a node more than once by the same route can happen, although this would rarely occur for densely deployed sensor networks. Therefore, a mechanism should be built into a D-RfSP protocol to avoid unnecessary revisits while the route traverses to the destination. This can be accomplished by the visited node with observation coefficient $g_j > 0$ informing all of its neighbors that it will no longer contribute to the SP performance and should now be considered as having $g_j = 0$. One way to implement this messaging is through dedicated application messages with the associated additional cost in energy and bandwidth.

Another way to achieve this notification is to exploit the services of CSMA/CA-based MAC (medium access control). By using a reserved bit in the RTS (request-to-send) and CTS (clear-to-send) packets, the (re)visit status of a node can be probed and conveyed. This mechanism has been proposed in Ref. [40] for the purpose of informing neighbors of the type of packets to be transmitted.

The third way of carrying out this messaging is to take advantage of the broadcast nature of wireless transmission, where the radio transmission will be heard by all neighbors within the transmission range, not just the intended recipient. In the ZigBee protocol, the network layer is responsible for discarding overheard packets by checking the receiver address field in the received packets. Overhearing has been employed in WSN for network time synchronization [41,42], reducing redundant transmissions [43], and propagating the location of the mobile sink to the source node [44]. As proposed in the distributed RfSP protocol in Ref. [32], when a node with $g_j > 0$ is visited for the first time, it uses the energy which is enough to reach its farthest neighbor when transmitting to the next-hop. Upon hearing the transmission, both the intended recipient and overhearing nodes mark the transmitting node as *VISITED* and treat it as having $g_j = 0$ in subsequent routing. This mark is cleared at the end of current routing session through, say, a time-out at each node. The extra energy in transmission (to reach the farthest neighbor) is the cost one has to pay for lacking of centralized coordination and optimization.

10.4.3 Logical Topology and Tree Routing

This section discusses the logical structure expressed by nodes' addresses upon which the D-RfSP strategy in the next section will be built.

Sensor networks are normally constructed in a spanning tree manner by starting with a root node and growing as new nodes join the existing nodes as child nodes. Each node has one and only one parent while a parent can have multiple children. The resultant logical relationship (LR) is a simple tree structure although the network's physical topology can be quite complex. One parameter associated with each node is its network depth d. The root node has $d = 0$ and a nonroot node has a nonzero depth which equals its parent's $d + 1$. The depth indicates the minimum number of hops a transmitted frame must travel, using only parent–child links, to reach the root node. When a packet from a node N_j travels upward along the tree to the root, it will reach all its ancestors. Similarly, a node-to-root route from another node N_d will cover all ancestors of N_d. The node where the two node-to-root routes merge is termed as the joint node (JN) of the two starting nodes. Note that if the two nodes are the same, the JN is the node itself and, if one node is an ancestor of the other, their JN is the ancestor node.

To make peer-to-peer communication possible, each node must have a unique identity that is typically in the form of its address. Due to the ad-hoc nature of the network topology, an address assignment scheme has to be in place to assign addresses to nodes when they join the network. We are particularly interested in those schemes where, instead of using randomly generated numbers, each parent node autonomously generates addresses for joining nodes according to specific rules. Various such structured address assignment schemes have been reported [38,45–52]. The most notable of them is the one specified in the ZigBee standard [38].

For such structured addresses it has been shown [38] that the LR of nodes *on the same branch* can be determined from their addresses. In particular, from the address of a node, the addresses of upstream nodes (i.e., its parent and all other ancestors) and downstream nodes (i.e., its children and all other descendants) can be determined. This vertical relationship has led to the so-called tree routing (TR) protocol. The TR is a distributed algorithm where the routing decision is made locally and the route is restricted to parent–child links only. In particular, when a packet is received by an intermediate node, it is forwarded either downward or upward along the logical tree. If it is determined that the destination is a descendant node, the data is sent down along the tree to either the destination node if it is a child or, otherwise, the ancestor of the destination node. Otherwise the parent node is chosen as the next-hop.

10.4.4 LR-Based RfSP Protocols

The TR protocol can be very inefficient because of the restrictions on next-hop candidates. In fact, relationship richer than that utilized in TR can be identified in networks with structured address assignment schemes and, therefore, more efficient LR-based routing is possible. It is shown in Ref. [53] that for such networks the logical distance between any two nodes, in terms of the number of hops via parent–child links, can be determined by their addresses. In particular, for any source-destination pair (N_d, N_j), the hop-count from N_j to N_d via parent–child links is given by

$$H_j = d_j + d_d - 2d_{jd}, \tag{10.23}$$

where d_j, d_d, and d_{jd} are the depths of N_j, N_d, and the join-node of them, respectively. All these three depths can be readily determined from addresses of N_d and N_j. Note that this is a conservative prediction of the hop-count. That is, it is guaranteed that at most H_j hops are required for a packet to travel from N_j to N_d; if a cross-branch jump happens at a downstream node along the route, the actual hops will be less.

This property can be used to evaluate all neighbors, not just parent and children, for finding the shortest possible route to the destination. As a result, the route cannot only traverse vertically along parent–child links, but also "jump" horizontally to other branches of the tree, generating more efficient routes. We now present a LR-based D-RfSP protocol which utilizes this routing potential. We assume that homogenous nodes are densely deployed and a structured address assignment scheme is in place to allow determination of (10.24). Each node is required to record locally the addresses and observation coefficients of its neighbors in its neighbor table. In addition, the distances to each neighbor are needed; their absolute locations are not required.

The routing starts at the source node N_1 which is assumed to be the only node with the knowledge of the route energy constraint E_{lim}. It transforms E_{lim} to the maximum number of hops of the route based on the average transmission energy to communicate with all of its neighbors:

$$H_{\max} = \frac{E_{\text{lim}}}{\frac{1}{K}\sum_{j=1}^{K}(\delta\,\Delta_{1,j}^{\gamma} + e_0)} \tag{10.24}$$

where K is the number of neighbors of N_1. This conversion is valid due to the dense deployment and homogeneous nature of the network. Node N_1 then groups all neighbors which satisfy the following condition into the set Θ and selects the next-hop according to (10.21):

$$H_j + 1 \leq H_{\max} \tag{10.25}$$

where H_j is the number of hops N_j takes to reach the destination. The value of $H_{\max}$ is transmitted along with locally derived intermediate SP results to the chosen next-hop, which uses $(H_{\max} - 1)$ as the maximum number of hops in selecting its next-hop. It is assumed that the address of the destination node is known to all nodes. Otherwise, the destination address could be conveyed from the source node to subsequent nodes in the same way $H_{\max}$ is propagated. This process continues until the destination is reached.

In essence, the next-hop is the one with maximum performance-gain-to-energy ratio among the neighbors that have guaranteed number of hops to the destination. If all these neighbors have zero observation coefficients (i.e., $g_j = 0$ for all the nodes in Θ), then the minimum number of hops to the destination is obtained as follows:

$$H_{\min} = \arg\min_{N_j \in \Theta} H_j \tag{10.26}$$

and the node with $H_{\min}$ and is physically closest to the local node N_k is chosen as the next-hop, that is,

$$N_j^* = \arg\min_{N_j \in \Theta} \Delta_{k,j} \quad \text{subject to } H_j = H_{\min} \tag{10.27}$$

This D-RfSP strategy is readily implementable in ZigBee networks, where a structured address assignment scheme is in place and the overhearing based revisit avoidance can be employed. In the next section, we evaluate the performance of this D-RfSP protocol as applied in a signal estimation problem.

10.5 Performance Evaluation

We conduct simulations to reveal the SP performance and energy efficiency of the LR-based D-RfSP scheme in comparison with that of its centralized counterpart (C-RfSP), the TR, minimum hop (min-hop), and minimum-energy (min-energy) protocols.

Consider the following data observation model at the k-th node of route $\Omega = \{N_1, N_2, \ldots, N_M\}$:

$$y_k = g_k\theta + w_k \quad k = 1, 2, \ldots, M \tag{10.28}$$

where

$g_k \geq 0$ is the observation coefficient
w_k is the white Gaussian noise with zero mean and variance σ^2
θ is the unknown signal to be estimated

It is shown in Ref. [54] that the best linear unbiased estimate and its variance are given by

$$\hat{\theta} = \frac{\sum_{k=1}^{M} g_k y_k}{Q(\Omega)} \tag{10.29}$$

$$\text{var}(\hat{\theta}) = \frac{\sigma^2}{Q(\Omega)} \tag{10.30}$$

where $Q(\Omega) = \sum_{k=1}^{M} g_k^2$. The estimation (10.29) can be carried out in a distributed manner where each intermediate node $N_I (I < M)$ sends locally computed results $\sum_{k=1}^{I} g_k y_k$ and $\sum_{k=1}^{I} g_k^2$ to the next-hop. It can be seen from (10.30) that $q_{k,j} = g_j^2$ captures the performance contribution of each neighbor N_j and is thus chosen as the link metric for RfSP schemes. Accordingly, the D-RfSP protocol implements the LR-based scheme introduced in Section 10.4 (may incur extra transmission energy as per overhearing based revisit avoidance), and the C-RfSP protocol obtains the optimum route through the following optimization:

$$\Omega^* = \arg\max_{\Omega} \lambda(\Omega) \quad \text{subject to } E(\Omega) \leq E_{\text{lim}} \tag{10.31}$$

where $E(\Omega)$ is the route energy consumption and $\lambda(\Omega) = Q(\Omega)/E(\Omega)$ is the estimation-gain-to-energy ratio.

An event-driven simulator developed in MATLAB® is employed. In all the simulation scenarios, the ZigBee network parameter is set as $(C_M, R_M, L_M) = (5, 5, 6)$, where C_M, R_M, and

L_M denote the maximum number of children a parent can have, maximum number of router children a parent can have, and maximum network depth, respectively. Two types of nodes are specified in ZigBee networks, namely routers that can accommodate other nodes as children and end devices that cannot have children of their own. The network setup process including address assignment scheme of the ZigBee standard is followed. In particular, after the nodes are deployed with every node having acquired a physical location, the root node is powered on to start the network. All the other nodes then power on and search the neighborhood for a potential parent which is by definition a router already joined the network and having capacity to accommodate more children. The joining node and potential parent then exchange join request and response messages to complete the join process by which the joining node is assigned an address. The network is established after all nodes have joined the network. The target is then deployed and each node builds up its neighbor table which contains the observation coefficients of and distances to its neighbors.

For the C-RfSP optimization (10.31), all routes satisfying the energy constraint are sought [55] and the one with the largest estimation-gain-to-energy ratio is selected [31]. The routes by the min-hop and min-energy protocols are found using the algorithms in Ref. [55] as well. All these operations require centralized optimization based on the deployment information—nodes' locations and observation coefficients. The solutions of these schemes are, therefore, globally optimal in terms of the corresponding metrics. For the D-RfSP scheme, the routes are determined progressively at each node as the routes traverse to the destination. The TR protocol also determines the route locally but, as the min-hop and min-energy protocols, does so without considering the estimation performance. An event is defined as the transmission of a network layer packet from the source node to the destination node along the route determined by a routing protocol under study. The events happen sequentially, that is, an event starts after the previous one finishes. As a result, there is no packet collision or channel contention during packet transmission. This allows us to focus on the routing protocols under study by examining the performance of the routes generated.

We consider the scenarios where 100 nodes each with a transmission range of 200 m are randomly deployed in a square region with size 1000 m by 1000 m. The destination node is the root node which is located at the centre of the region (500 m, 500 m) and the signal source (i.e., the target) is located at (100 m, 100 m). The observation coefficient of a node within sensing range T_s is inversely proportional to its distance (Δ_k) to the target, that is,

$$g_k = \left\{ \begin{array}{cc} 1/\Delta_k & \text{if} \quad \Delta_k \leq T_s \\ 0 & \text{otherwise} \end{array} \right\} \tag{10.32}$$

For each simulation scenario, 200 instances of the sensor network are randomly generated. For each instance, a source node is randomly chosen among the nodes within 100 m away from the target and the routes to the destination node established by the routing protocols are studied, that is, their estimation variances and estimation-gain-to-energy ratios $\lambda(\Omega)$ are determined. The results over all the network instances are then averaged to produce the final measure. The following parameters of the data model (10.28) and energy consumption model (10.3) are chosen: $\theta = 1, \delta = 10^{-2}$, $\nu = 2, e_0 = 0.5\,\mu$J and the noise variance is chosen to be $\sigma^2 = 3.98 \times 10^{-5}$ which amounts to a 4-dB SNR at a node 100 m away from the target.

Figure 10.1 shows one instance of the network and the routes determined by the routing protocols when sensing range $T_s = 200$ m and energy constraint $E_{\text{lim}} = 5\ \mu$J, where the star

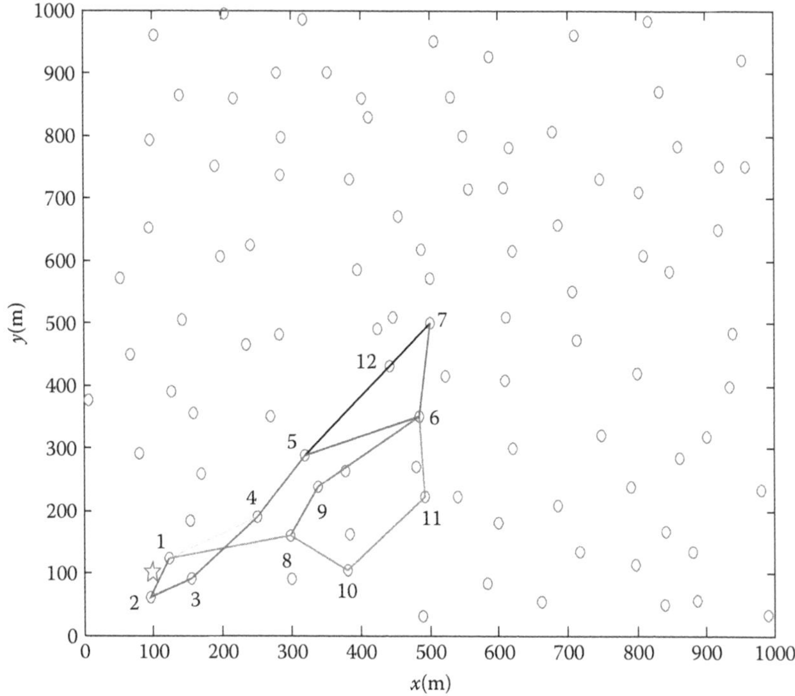

Figure 10.1 One instance of the network and routes.

Table 10.1 Routes in Figure 10.1 and Their Performance Metrics

Protocol	Ω	$E(\Omega)$ *in* μ*J*	$Q(\Omega)$	$\lambda(\Omega)$ *in* 1/μ*J*	$var(\hat{\theta})$
Min-hop	$1 \to 8 \to 9 \to 6 \to 7$	3.4691	0.9050e−3	0.2609e−6	0.0440
Min-energy	$1 \to 4 \to 5 \to 12 \to 7$	3.2878	0.9375e−3	0.2852e−6	0.0425
TR	$1 \to 8 \to 10 \to 11 \to 6 \to 7$	4.0759	0.9050e−3	0.2220e−6	0.0440
C-RfSP	$1 \to 2 \to 3 \to 4 \to 5 \to 12 \to 7$	4.3613	1.8972e−3	0.4350e−6	0.0210
D-RfSP	$1 \to 2 \to 3 \to 4 \to 5 \to 6 \to 7$	5.2261	1.8972e−3	0.3630e−6	0.0210

denotes the target and small circles denote sensor nodes. Table 10.1 lists the routes and associated performance metrics.

It can be seen from Table 10.1 that the min-energy route, as expected, has the least energy consumption. Since the min-hop route is not unique and the earliest identified one is chosen, it does not have the same performance metrics as the min-energy route even though they have the same number of hops in this particular case. The TR route has comparable performance both in estimation performance and energy efficiency as the other two generic protocols since in essence it is a hop-count-based scheme. Both the C-RfSP and D-RfSP routes traverse away from the destination initially in order to cover nodes with more information (i.e., the ones close to the target.)

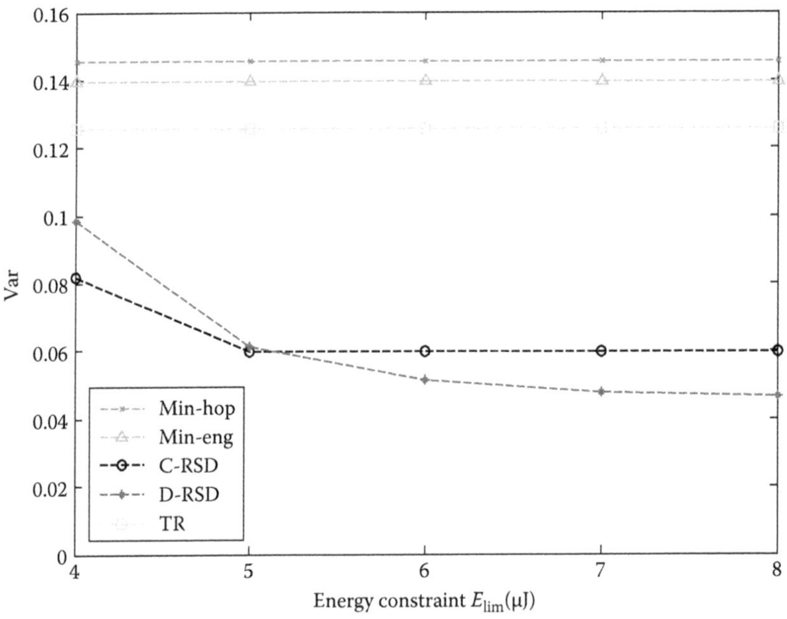

Figure 10.2 Estimation variances with changing energy constraint.

The energy efficiency $\lambda(\Omega)$ of the D-RfSP route is less than that of the C-RfSP route which is the optimum solution in this regard. This inferior performance of the D-RfSP is due to the fact that a node only knows its neighbors and makes a routing decision based on local information only. Both RfSP routes are shown to have more accurate estimates and higher energy efficiency than that of the three generic routing routes.

Figures 10.2 and 10.3 show the var($\hat{\theta}$) and $\lambda(\Omega)$, respectively, of the routes when $T_s = 200$ m and E_{lim} changes from 4 to 8μJ. The estimation performance and energy efficiency of the TR, min-hop, and min-energy routes remain constant since they are independent of energy constraints. Both centralized and distributed RfSP protocols have increased estimation accuracy when the energy constraint is increased. For the D-RfSP protocol, it appears that when E_{lim} is tight (<5 μJ) there is not much room to maneuver, and when E_{lim} is loose (>7 μJ) the route traversals a larger number of nodes to increase the estimation gain $Q(\Omega)$, resulting in reduced energy efficiency in both cases. The two RfSP schemes are shown to yield better estimates and energy efficiency than the three generic routing protocols.

Figures 10.4 and 10.5 show the var($\hat{\theta}$) and $\lambda(\Omega)$, respectively, of the routes when $E_{\text{lim}} = 5$ μJ and T_s changes from 100 to 500 m. When T_s increases, all the five routes are shown to produce better estimates since more nodes are included in the sensing range. These additional nodes, however, have relatively small observation coefficients g_k because they are further away from the target (see (10.32)). This leads to reduced energy efficiency of D-RfSP routes since they try to capture more estimation again by visiting increased number of nodes. A possible modification to the D-RfSP protocol is to incorporate a threshold τ and treat all neighbors with $g_j < \tau$ as having $g_j = 0$. Once again, the two RfSP protocols outperform the three generic protocols in both estimation performance and energy efficiency.

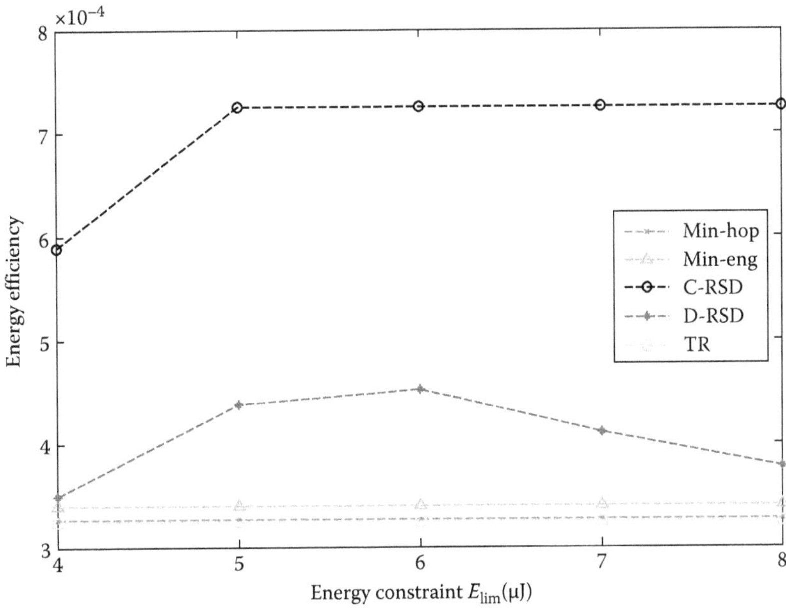

Figure 10.3 Energy efficiency with changing energy constraint.

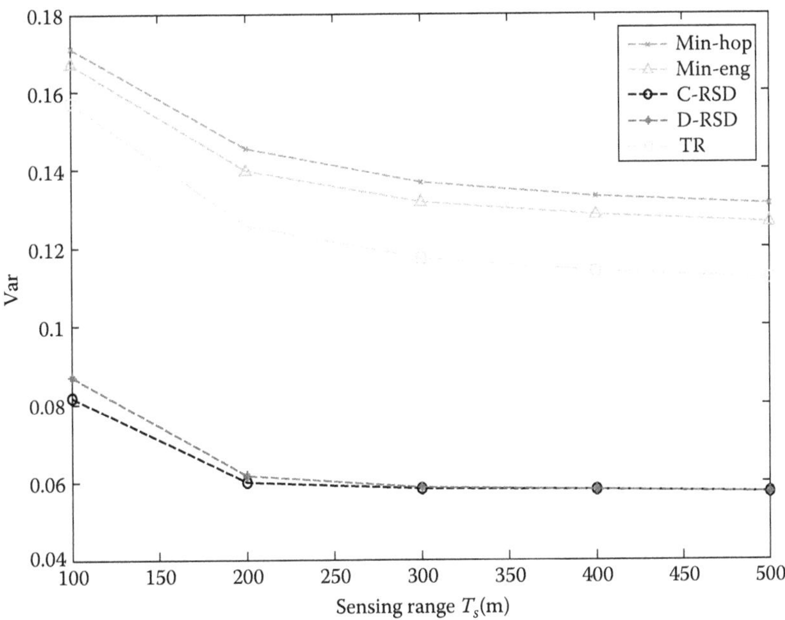

Figure 10.4 Estimation variances with changing sensing range.

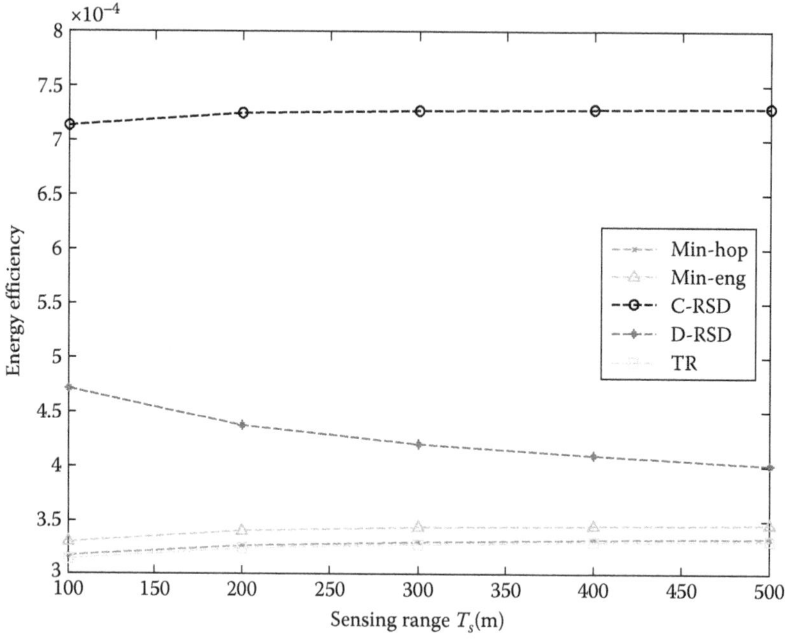

Figure 10.5 Energy efficiency with changing sensing range.

10.6 Concluding Remarks

Routing protocols that jointly optimize SP performance and energy efficiency have great potential in achieving energy-efficient routing. Although being inferior to the global optimum solution in terms of energy efficiency, distributed RfSP protocols do not require centralized optimization and, therefore, are viable alternatives to centralized schemes for large-scale networks due to their flexibility and scalability. Simulation results have shown that both centralized and distributed RfSP protocols are able to achieve significant improvement in both SP performance and energy efficiency over generic routing schemes.

Developing RfSP schemes is challenging and problem specific. Future work in this area could involve extending link metrics for Gauss–Markov signals to more general models and dealing with colored or non-Gaussian noise. Another interesting work is to develop RfSP schemes for other WSN SP algorithms such as beam-forming [56], channel identification [57], and principal component analysis [58].

References

1. F. Zhao and L. Guibas, *Wireless Sensor Networks: An Information Processing Approach*, Boston, MA: Elsevier-Morgan Kaufmann, 2004.
2. C. Chong and S.P. Kumar, Sensor networks: Evolution, opportunities, and challenges, *Proceedings of the IEEE*, 91, 1247–1256, 2003.
3. K. Akkaya and M. Younis, A survey on routing protocols for wireless sensor networks, *Ad Hoc Networks*, 3, 325–349, 2005.

4. C. Perkins and P. Bhagwat, Highly dynamic destination-sequenced distance-vector routing (DSDV) for mobile computers, in *Proceedings of the ACM SIGCOMM*, London, UK, pp. 1–11, 1994.
5. D.B. Johnson and D.A. Maltz, Dynamic source routing in ad hoc wireless networks, in *Mobile Computing*, (eds.) Tomasz Imielinski and Henry F. Korth, pp. 153–181, Boston, MA: Kluwer Academic Publishers, March 1996.
6. C.E. Perkins and E.M. Royer, Ad hoc on-demand distance vector routing, in *Proceedings of Second IEEE Workshop Mobile Computing Systems and Applications*, New Orleans, Louisiana, USA, pp. 90–100, 1999.
7. J.H. Chang and L. Tassiulas, Maximum lifetime routing in wireless sensor networks, *IEEE/ACM Transactions on Networking*, 12, 609–619, August 2004.
8. M. Zimmerling, W. Dargie, and J.M. Reason, Energy-efficient routing in linear wireless sensor networks, in *IEEE International Conference on Mobile Ad hoc and Sensor Systems*, Pisa, Italy, pp. 1–3, 2007.
9. F. Shao, X. Shen, and L. Cai, Energy efficient reliable routing in wireless sensor networks, in *First International Conference Communications and Networking*, Beijing, China, pp. 1–5, 2006.
10. H. Chang and L. Tassiulas, Energy conserving routing in wireless ad hoc networks, in *Proceedings of IEEE INFOCOM 2000*, Tel-Aviv, Israel, pp. 22–31, March 2000.
11. S. Singh, M. Woo, and C. Raghavendra, Power-aware routing in mobile ad hoc networks, in *Proceedings 4th ACM/IEEE Conference on Mobile Computing and Networking (MobiCom)*, Seattle, Washington, USA, pp. 181–190, October 1998.
12. W. Heinzelman, A. Chandrakasan, and H. Balakrishnan, Energy-efficient communication protocol for wireless sensor networks, in *Proceeding of the Hawaii International Conference System Sciences*, Honolulu, HI, Vol. 8, pp. 1–10, 2000.
13. S. Lindsey and C.S. Raghavendra, PEGASIS: Power efficient gathering in sensor information systems, in *Proceedings of the IEEE Aerospace Conference*, Big Sky, MT, Vol. 3, pp. 1125–1130, 2002.
14. K. Sohrabi et al., Protocols for self-organization of a wireless sensor network, *IEEE Personal Communications*, 7(5), 16–27, 2000.
15. C. Intanagonwiwat, R. Govindan, and D. Estrin, Directed diffusion: A scalable and robust communication paradigm for sensor networks, in *Proceedings of the 6th Annual International Conference on Mobile Computing and Networks (MobiCom 2000)*, ACM Press, Boston, MA, pp. 56–67, 2000.
16. D. Braginsky and D. Estrin, Rumor routing algorithm for sensor networks, in *Proceedings of the 1st ACM International Workshop on Wireless Sensor Networks and Applications*, Atlanta, GA, pp. 22–31, 2002.
17. J. Liu, F. Zhao, and D. Petrovic, Information-directed routing in ad hoc sensor networks, *IEEE Journal on Selected Areas Communications*, 23, 851–861, April 2005.
18. T. Girici and A. Ephremides, Joint routing and scheduling metrics for ad hoc wireless networks, in *Proceedings of the 36th Asilomar Conference on Signals, Systems and Computers*, College Park, MD, pp. 1155–1159, November 2002.
19. R. Viswanathan and P.K. Varshney, Distributed detection with multiple sensors: Part I—Fundamentals, *Proceedings of the IEEE*, 85, 54–63, January 1997.
20. R.S. Blum, S.A. Kassam, and H.V. Poor, Distributed detection with multiple sensors: Part II—Advanced topics, *Proceedings of the IEEE*, 85, 64–79, January 1997.
21. J.-F. Chamberland and V.V. Veeravalli, Decentralized detection in sensor networks, *IEEE Transactions on Signal Processing*, 51(2), 407–416, February 2003.
22. B. Chen, L. Tong, and P.K. Varshney, Channel aware distributed detection in wireless sensor networks, *IEEE Signal Processing Magazine*, 23, 16–26, July 2006.
23. Y. Huang and Y. Hua, Multi-hop progressive decentralized estimation of deterministic vector in wireless sensor networks, in *Proceedings of Forty-First Asilomar Conference on Signals*, Systems and Computers, Pacific Grove, CA, USA, pp. 2145–2149, 2007.

24. M.G. Rabbat and R.D. Nowak, Decentralized source localization and tracking, in *ICASSP'04*, Montreal, Quebec, Canada, Vol. 3, pp. 17–21, May 2004.
25. A. Bertrand and M. Moonen, Distributed adaptive estimation of node-specific signals in wireless sensor networks with a tree topology, *IEEE Transactions on Signal Processing*, 59(5), 2196–2210, 2011.
26. L. Yu, L. Yuan, G. Qu, and A. Ephremides, Energy-driven detection scheme with guaranteed accuracy, in *Proceedings of the 5th International Conference Information Processing Sensor Networks (IPSN)*, New York, pp. 284–291, April 2006.
27. W. Li and H. Dai, Energy efficient distributed detection via multi-hop transmission in sensor networks, in *Proceedings of the Military Communications Conference (MILCOM)*, Washington, DC, pp. 1–7, October 2006.
28. W. Qiu and E. Skafidas, Distributed source localization based on TOA measurements in wireless sensor networks, *Journal of Electrical and Computer Engineering*, Vol. 8, pp. 1–5, January 2009.
29. S. Patil, S.R. Das, and A. Nasipuri, Serial data fusion using space filling curves in wireless sensor networks, in *Proceedings of the IEEE Communications Society Conference on Sensor Ad Hoc Communications and Networks (SECON)*, Santa Clara, CA, pp. 182–190, October 2004.
30. Y. Sung, S. Misra, L. Tong, and A. Ephremides, Cooperative routing for distributed detection in large sensor networks, *IEEE Journal on Selected Areas in Communications*, 25, 471–483, February 2007.
31. Y. Yang, R.S. Blum, and B.M. Sadler, Energy-efficient routing for signal detection in wireless sensor networks, *IEEE Transactions on Signal Processing*, 57(6), 2050–2063, June 2009.
32. W. Qiu and E. Skafidas, Distributed routing for signal detection in wireless sensor networks, July 2011, submitted.
33. H.V. Poor, *An Introduction to Signal Detection and Estimation*, New York: Springer-Verlag, 1994.
34. A. Dembo and O. Zeitouni, *Large Deviations Techniques and Applications*. New York: Springer, 1998.
35. F.C. Schweppe, Evaluation of likelihood functions for Gaussian signals, *IEEE Transactions on Information Theory*, IT-1, 61–70, 1965.
36. D. Bertsekas and R. Gallager, *Data Networks*, 2nd edn., New York: Springer, 1991.
37. S.M. Kay, *Fundamentals of Statistical Signal Processing: Detection Theory*, Englewood Cliffs, NJ: PTR Prentice-Hall, 1998.
38. ZigBee Specification Version 1.0, ZigBee Alliance. 2005.
39. Institute of Electrical and Electronics Engineers, Inc., IEEE Std. 802.15.4–2003, IEEE Standard for Information Technology—Telecommunications and Information Exchange between Systems—Local and Metropolitan Area Networks—Specific Requirements—Part 15.4: Wireless Medium Access Control (MAC) and Physical Layer (PHY) Specifications for Low Rate Wireless Personal Area Networks (WPANs), New York: IEEE Press, 2003.
40. C.M. Vuran and I.F. Akyildiz, Spatial correlation-based collaborative medium access control in wireless sensor networks, *IEEE/ACM Transactions on Networking*, 14(2), 316–329, April 2006.
41. K. Cheng, K. Lui, Y. Wu, and V. Tam, A distributed multihop time synchronization protocol for wireless sensor networks using pairwise broadcast synchronization, *IEEE Transactions on Wireless Communications*, 8(4), 1764–1772, 2009.
42. K. Nohk, E. Serpendine, and K. Qaraqe, A new approach for time synchronization in wireless sensor networks: Pairwise broadcast synchronization, *IEEE Transactions on Wireless Communications*, 7(9), 3318–3322, 2008.
43. H. Le, H. Guyennet, and V. Felea, OBMAC: An overhearing based MAC protocol for wireless sensor networks, in *Proceedings of the International Conference on Sensor Technologies and Applications*, Washington, DC, pp. 547–553, 2007.
44. F. Yu, S. Park, E. Lee, and S.H. Kim, Elastic routing: A novel geographic routing for mobile sinks in wireless sensor networks, *IET Communications*, 4(6), 716–727, 2010.

45. M. Mohsin and R. Prakash, IP address assignment in a mobile ad hoc network, *Proceedings of Milicom 2002*, 2, 856–861, 2002.
46. G. Bhatti and G. Yue, A structured addressing scheme for wireless multi-hop networks, Technical Report TR2005–149, Mitsubishi Electronic Research Laboratories, June 2006, http://www.merl.com/reports/docs/TR2005–149.pdf
47. H.J. Ju, D. Ko, and S. An, Enhanced distributed address assignment scheme for ZigBee-based wireless sensor networks, in *Proceedings of 2009 International Conference on Intelligent Networking and Collaborative Systems*, Washington, DC, pp. 237–240, 2009.
48. M.S. Pan, H.W. Fang, Y.C. Liu, and Y.C. Tseng, Address assignment and routing schemes for ZigBee-based long-thin wireless sensor networks, in *IEEE Vehicular Technology Conference*, Marina Bay, Singapore, VTC Spring 2008, pp. 173–177, 2008.
49. Y. Li, H. Shi, and B. Tang, Address assignment and routing protocol for large-scale uneven wireless sensor networks, in *International Symposium on Computer Network and Multimedia Technology, CNMT2009*, Wuhan, China, pp. 1–4, 2009.
50. J. Jobin, S.V. Krishnamurthy, and S.K. Tripathi, A scheme for the assignment of unique addresses to support self-organization in wireless sensor networks, in *IEEE 60th Vehicular Technology Conference, VTC2004-Fall*, Los Angeles, CA, USA, Vol. 6, pp. 4578–4582, 2004.
51. M.S. Pan, C.H. Tsai, and Y.C. Tseng, Theorphan problem in ZigBee wireless networks, *IEEE Transactions on Mobile Computing*, 8(11), 1573–1584, 2009.
52. L.H. Yen and W.T. Tsai, Flexible address configurations for tree-based ZigBee/IEEE 802.15.4 wireless networks, in *22nd International Conference on Advanced Information Networking and Applications*, Okinawa, Japan, pp. 395–402, 2008.
53. W. Qiu, E. Skafidas, and P. Hao, Enhanced tree routing for wireless sensor networks, *Ad Hoc Networks*, 7, 638–650, 2009.
54. S.M. Kay, *Fundamentals of Statistical Signal Processing: Estimation Theory*, Englewood Cliffs, NJ: PTR Prentice-Hall, 1993.
55. V.M. Jiménez and A. Marzal, Computing the K shortest paths: A new algorithm and an experimental comparison, in *Lecture Notes in Computer Science*, Berlin, Germany: Springer-Verlag, Vol. 1668, pp. 15–29, 1999.
56. M.F.A. Ahmed and S.A. Vorobyov, Collaborative beam forming for wireless sensor networks with Gaussian distributed sensor nodes, *IEEE Transactions on Wireless Communications*, 8(2), 638–643, 2009.
57. X. Li, Blind channel estimation and equalization in wireless sensor networks based on correlations among sensors, *IEEE Transactions on Signal Processing*, 53(4), 1511–1519, 2005.
58. S. Yang and J.N. Daigle, APCA-based vehicle classification system in wireless sensor networks, in *Proceedings of IEEE Wireless Communications and Networking Conference*, Las Vegas, NV, USA, 2006, pp. 2193–2198.

Chapter 11

On-Board Image Processing in Wireless Multimedia Sensor Networks

A Parking Space Monitoring Solution for Intelligent Transportation Systems

Claudio Salvadori, Matteo Petracca, Marco Ghibaudi, and Paolo Pagano

Contents

11.1 Introduction

Wireless sensor networks (WSNs) have been experiencing a rapid growth in recent years due to the joint efforts of both academia and industry in developing this technology. If on one hand the academia is going ahead in developing new innovative solutions looking at enabling sensor network pervasiveness, on the other hand the industry has started to push on standardization activities and real implementations concerning reliable WSN-based systems [1,2]. WSNs are nowadays envisioned to be adopted in a wide range of applications as an effective solution able to replace old wired and wireless systems that are more expensive and hard to set up because of their necessity of power and connection cables. A reduced set of WSN applications include climatic monitoring [3,4], structural monitoring of buildings [5,6], human tracking [7], military surveillance [8], and, more recently, multimedia-related applications [9–11].

The wireless multimedia sensor networks (WMSNs) development has been mainly fostered by a new generation of low-power and high performance microcontrollers able to speed up the processing capabilities of a single wireless node, as well as the development of new microcameras and microphones imported from the mobile phones industry. Along with classical multimedia streaming applications in which voice and images can be sent through the network, pervasive WMSNs, consisting in large deployments of camera equipped devices, may support new vision-based services. By collecting and analyzing images from the scene, anomalous and potentially dangerous events can be detected [12], advanced applications based on human activities [13] can be enabled and intelligent services, such as WMSNs-based intelligent transportation systems (ITS) [14], can be provided.

A successful design and development of vision-based applications in WMSNs cannot be achieved without adopting feasible solutions of the involved computer vision techniques. In such a context, state-of-the-art computer vision algorithms cannot be directly applied [15] due to reduced capabilities, in terms of memory availability, computational power, and CMOS resolution, of the camera nodes. In fact, since WMSNs usually require a large number of sensors, possibly deployed over a large area, the unit cost of each device should be as small as possible to make the technology affordable. As a consequence of the strong limitations in hardware capabilities, low-complexity computer vision algorithms must be adopted, while reaching a right trade-off between algorithms performance and resource constraints.

In this chapter, we tackle with the problem of developing low-complexity computer vision algorithms targeted to WMSNs devices for enabling pervasive ITS. To this end, we present a parking space monitoring algorithm able to detect the occupancy status of a parking space while filtering spurious transitions in the scene. The algorithm has been developed by adopting only basic computer vision techniques and its performance evaluated in terms of sensitivity, specificity, execution time, and memory occupancy by means of a real implementation in a WMSN device.

11.2 Smart Cameras for Wireless Multimedia Sensor Networks

The performance of vision-based algorithms targeted at WMSN devices mainly depends on the computational capabilities of the whole smart camera system. In this section, we provide an overview of the most popular embedded vision platforms in the WMSNs domain, as well as a description of their main hardware and image processing characteristics. A final comparison overview among the described platforms is reported in Table 11.1, while in Figure 11.1 some pictures of the devices have been reported.

Table 11.1 Platforms Characteristics Comparison

Platform	*Sensor*	*CPU*	*Application*
WiCa [16]	2 Color CMOS 640×480	Xetal IC3D	Local processing, collaborative reasoning
Cyclops [17]	Color CMOS 352×288	ATMega 128	Collaborative object tracking
MeshEye [18]	Color CMOS 640×480	ARM7	Distributed surveillance
CMUcam3 [19]	Color CMOS 352×288	ARM7	Local image analysis
CITRIC [20]	Color CMOS 1280×1024	XScale PXA270	Compression, tracking, localization
Vision Mesh [21]	Color CMOS 640×480	ARM9	Image-based water analysis

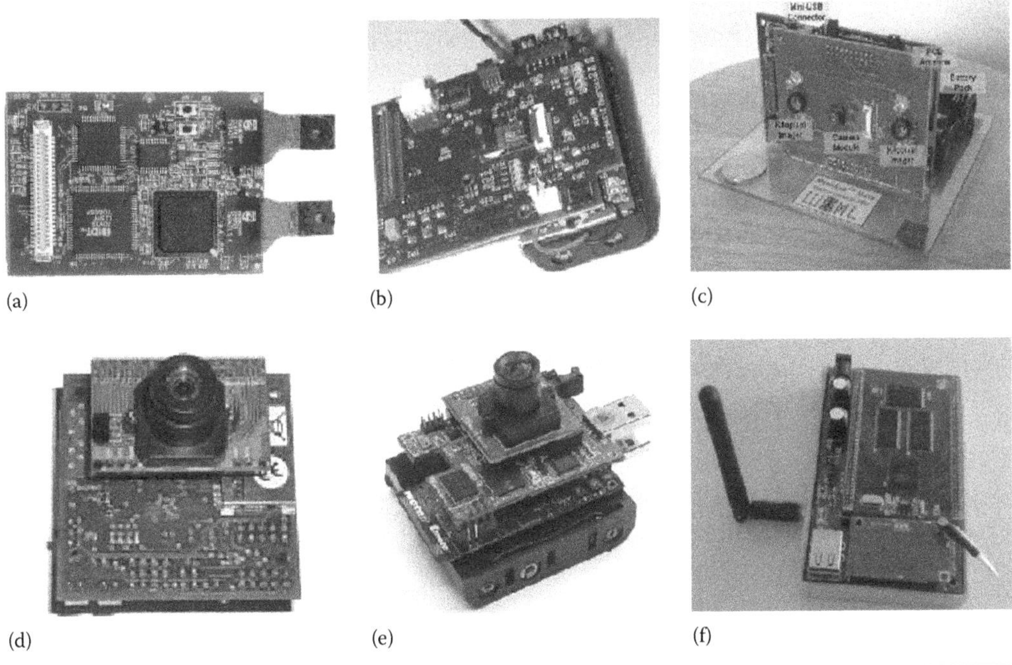

Figure 11.1 Smart cameras for wireless multimedia sensor networks. (a) WiCa. (b) Cyclops. (c) Mesheye. (d) CMUcam3. (e) CITRIC. (f) Vision Mesh.

In the last years, several research initiatives produced prototypes of smart cameras able to perform an onboard image processing. Among the first developed devices must be cited the WiCa [16] camera, developed by NXP Semicondutcors Research. The platform is equipped with NXP Xetal IC3D processor based on an SIMD architecture with 320 processing elements and can host up to two CMOS cameras at VGA resolution (640 × 480). The communication standard adopted to send data through a wireless network is the IEEE802.15.4. WiCa has been adopted for image-based local processing and collaborative reasoning applications.

The Cyclops [17] project is another research initiative aimed at developing an embedded vision platform for WMSNs. The device is equipped with a low performance ATMega128 8 bit RISC microcontroller with 128 kB of FLASH program memory and only 4 kB of SRAM data memory. The CMOS sensor supports three image formats of 8 bit monochrome, 24 bit RGB color, and 16 bit YCbCr color at CIF resolution (352 × 288). The board is not equipped with wireless transceivers; however, wireless communications can be enabled in IEEE802.15.4 networks via MicaZ mote. In the Cyclops board, the camera module enables for a whole image processing pipeline for performing demosaicing, image size scaling, color correction, tone correction, and color space conversion. The Cyclops effectiveness has been demonstrated in collaborative object tracking applications.

In the MeshEye [18] project, an energy-efficient smart camera mote architecture based on the ARM7 processor was designed, mainly targeted at intelligent surveillance application. MeshEye mote has an interesting special vision system based on a stereo configuration of two low-resolution, low-power cameras, coupled with a high-resolution color camera. In particular, the stereo vision system continuously determines position, range, and size of moving objects entering its fields of view. This information triggers the color camera to acquire the high-resolution image subwindow containing the object of interest, which can then be efficiently processed. To communicate among peer devices, MeshEye is equipped with an IEEE802.15.4 compliant transceiver.

Another interesting example of low cost embedded vision system is represented by the CMU-cam3 [19] developed at the Carnegie Mellon University. More precisely, the CMUcam3 is the third generation of the CMUcam series, which has been specially designed to provide an open source, flexible, and easy development platform targeted to robotics and surveillance applications. The hardware platform is more powerful with respect to previous CMUcam boards and may be used to equip low cost embedded systems with vision capabilities. The hardware platform is constituted by a CMOS camera, an ARM7 processor, and a slot for MMC cards. Wireless transceivers are not provided on board and communications can be enabled through mote systems (e.g., IEEE802.15.4-based communications via FireFly mote).

More recently, the CITRIC [20] platform integrates in one device a camera sensor, an XScale PXA270 CPU (with frequency scalable up to 624 MHz), a 16 MB FLASH memory, and a 64 MB RAM. Such a device, once equipped with a standard wireless transceiver, is suitable for the development of WMSNs. The design of the CITRIC system allows to perform moderate image processing tasks in network nodes. In this way, there are less stringent issues regarding transmission bandwidth with respect to simple centralized solutions. CITRIC capabilities have been illustrated by three sample applications: image compression, object tracking by means of background subtraction(BS), and self-localization of the camera nodes in the network.

Finally we cite the Vision Mesh [21] platform. The device integrates an Atmel 9261 ARM9 CPU, 128 MB NandFlash, 64MB SDRAM, and a CMOS camera at VGA resolution (640×480). The high computational capabilities of the embedded CPU permit to compress acquired images in JPEG format as well as to perform advanced computer vision technique targeted to water conservancy engineering applications. In-network processing of the acquired visual information may be enabled by means of IEEE802.15.4-based communications.

All reported smart camera devices represent an effective solution for enabling vision-based applications in WMSNs. The general trend in developing such devices is to increase computational capabilities without taking into account power consumption issues, the lowest experienced power consumption among the presented smart cameras devices is bigger than 650 mW [22], still a prohibitive figure for autonomous set ups in pervasive contexts.

11.3 Image Processing for Parking Space Occupancy Detection

WMSNs are considered a key technology in enabling pervasive ITS [23]. The use of low-cost smart cameras to detect parking spaces occupancy levels provides an effective and cheaper alternative to state-of-the-art magnetic field sensors installed under the asphalt [24]. In this section, we present a low-complexity computer vision solution aiming at detecting the occupancy status of a single parking space. The algorithm can be easily instantiated in multiple instances in real smart camera devices dedicated to monitor a set of parking spaces, until reaching the full coverage of the parking lot.

11.3.1 Background Subtraction Approach in WMSNs

Classical computer vision approaches for monitoring applications usually consist on BS-based algorithms [25]. As a function of the required constraints in terms of frame rate and image size, as well as of the adopted technique to model the background, (e.g., mixture of Gaussians, kernel density estimation, etc.) a BS approach can respond to a variety of performance and complexity levels.

A BS-based approach, with a frame differencing (FD) enforcement, is at the basis of the presented algorithm in which a low-complexity objective has been followed. As already discussed in Section 11.1, state-of-the-art computer vision algorithms cannot be directly applied in a WMSNs scenario due to several smart camera constraints: memory size, computational power, CMOS resolution, and energy consumption. The reduced amount of memory and CMOS capabilities have a strong impact on image frame resolutions and color depths: feasible resolution values are 160×120, 320×240, and 640×480 pixels, usually in gray scale. The energy consumption constraint is directly related to the maximum allowable frame rate, feasible values are lower than 2 fps (state-of-the-art computer vision algorithms usually work at 25, 30 fps), due to the necessity of increasing the device idle time during which all the peripherals are turned off. Regarding the limited computational capabilities, these require the development of simple background modeling techniques, strongly related to the developed application. In our approach a custom background modeling technique is defined with the aim to react to permanent changes in the scene (e.g., luminosity variation) and filter spurious transitions (e.g., once-off variations) while guaranteeing a real-time response.

11.3.2 Parking Space Status Analysis

In order to effectively model the background of the monitored parking space scene, a behavioral analysis regarding the possible status of a parking space must be performed. Considering a parking space identified by a region of interest (ROI), as depicted in Figure 11.2, three main possible states are possible: *full*, *partially full/empty*, and *empty*. While the full and empty states do not require further investigations, the partially full/empty one must be better detailed. The car that is parking or leaving the monitored space usually requires several video frames to complete all the maneuvering, thus giving the possibility to slowly move from full to empty or vice versa. In a better explicative way, it is possible to call the partially full/empty state as *transition* state. Although the transition state can model all the car maneuvering, it can also be used to model possible errors due to car and people passing through the ROI and causing false empty to full transitions.

The aforementioned full parking and leaving process can be modeled by means of the three state Markov chain depicted in Figure 11.3. In fact, the probability to be in a state at time $i + 1$ depends only by the state at time i. This observation can be expressed in mathematical terms as

Figure 11.2 Parking spaces identified by their own ROIs.

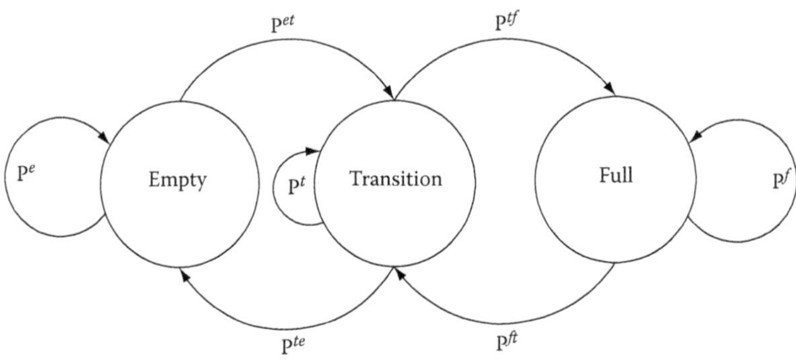

Figure 11.3 Markov chain–based parking space model.

$$P\{s_{i+1} = x_{i+1}|s_0 = x_0, s_1 = x_1, \ldots, s_i = x_i\} = P\{s_{i+1} = x_{i+1}|s_i = x_i\} \tag{11.1}$$

where

s_i is the status at time i

$x_{i+1}, x_i, \ldots, x_1, x_0 \in \{empty, transition, full\}$

Regarding the transition probabilities of the Markov chain, these represent the usage trends of the parking space and can be experimentally evaluated in time windows. The effective transition frequencies can be obtained by means of a ground-truth human analysis, and will be in turn used to measure the performance of a given detection algorithm. Better algorithms give transition values closer to the human ground-truth.

11.3.3 Background Modeling

In BS-based algorithms, the reference background of the scene requires to be updated at runtime to react to luminosity variations while filtering once-off changes. In order to create a background modeling technique feasible to be implemented in a camera network device, simple computer vision techniques have been implemented and applied following the Markov chain behavioral model discussed in Section 11.3.2. A background modeling technique specifically designed for the final application can guarantee good performance while keeping low the computational complexity.

In the proposed parking space monitoring application, the developed background modeling technique aims at compensating the luminosity variations and once-off changes effects in a predictable way starting from the system state knowledge, thus guaranteeing the background model consistency with respect to the real system state. The background luminosity variation compensation is performed by adopting an exponential forgetting algorithm [26]. Each background pixel is updated according to the following equation:

$$B_{i,j}(t_n) = (1 - \alpha)B_{i,j}(t_{n-1}) + \alpha I_{i,j}(t_n) \tag{11.2}$$

where

$B(t_{n-1})$ is the old background
$I(t_n)$ is the last acquired frame
α is the learning rate ($\alpha \in (0, 1)$)

The reported background update process for luminosity variation is performed only in the stable states of the system, empty and full, while it is avoided in the transition state, to fully control the once-off changes filtering procedure. In case of a change in the scene is detected, ROI partially occluded due to maneuvering or passing cars, and system move from empty/full to transition the exponential forgetting is not applied until the system moves into a stable state. When a transition in one of the two stable states is considered complete, the last acquired image is set as background and the exponential forgetting is enabled again. The background update policy as a function of the system state has been depicted in Figure 11.4a through c where the current state is identified by a colored area, light gray for the states in which the exponential forgetting is performed and dark gray otherwise, while a transition from a previous state is identified by a light gray arrow.

11.3.4 Status Change Detection

In the three states-based Markov chain adopted to describe the parking space behavior, and used to define the background modeling logic, a transition from one state to the other is achieved when a change in the scene is detected. In the proposed work, the change detection is based on a joint BS and FD approach.

In all possible states of the system, both BS and FD are performed. The BS procedure is performed subtracting the last acquired frame to the background image and counting the difference image pixels (n_{BS}) bigger than a TH_D threshold. In case n_{BS} is bigger than a TH_{BS} threshold, a possible change in the system status is detected. The FD is based on the same logic, the last acquired frame is subtracted to the previous image frame and the number of the difference image pixels (n_{FD}) is evaluated against a TH_{FD} threshold.

When the system is in one of the two stable states (i.e., empty and full), the condition $n_{BS} < TH_{BS}$ confirms that the system is in a stable state and the background can be updated by Equation 11.2. In this case, an FD can be used to cross-check the results retrieved by BS: the condition $n_{FD} < TH_{FD}$ confirms a lack of dynamics in the scene. If $n_{BS} > TH_{BS}$ and $n_{FD} > TH_{FD}$ a change in the system status is detected and the state of the system moves from empty/full to transition. The BS output can be seen as a system trigger, enforced by FD, moving from a stable state to the transition one.

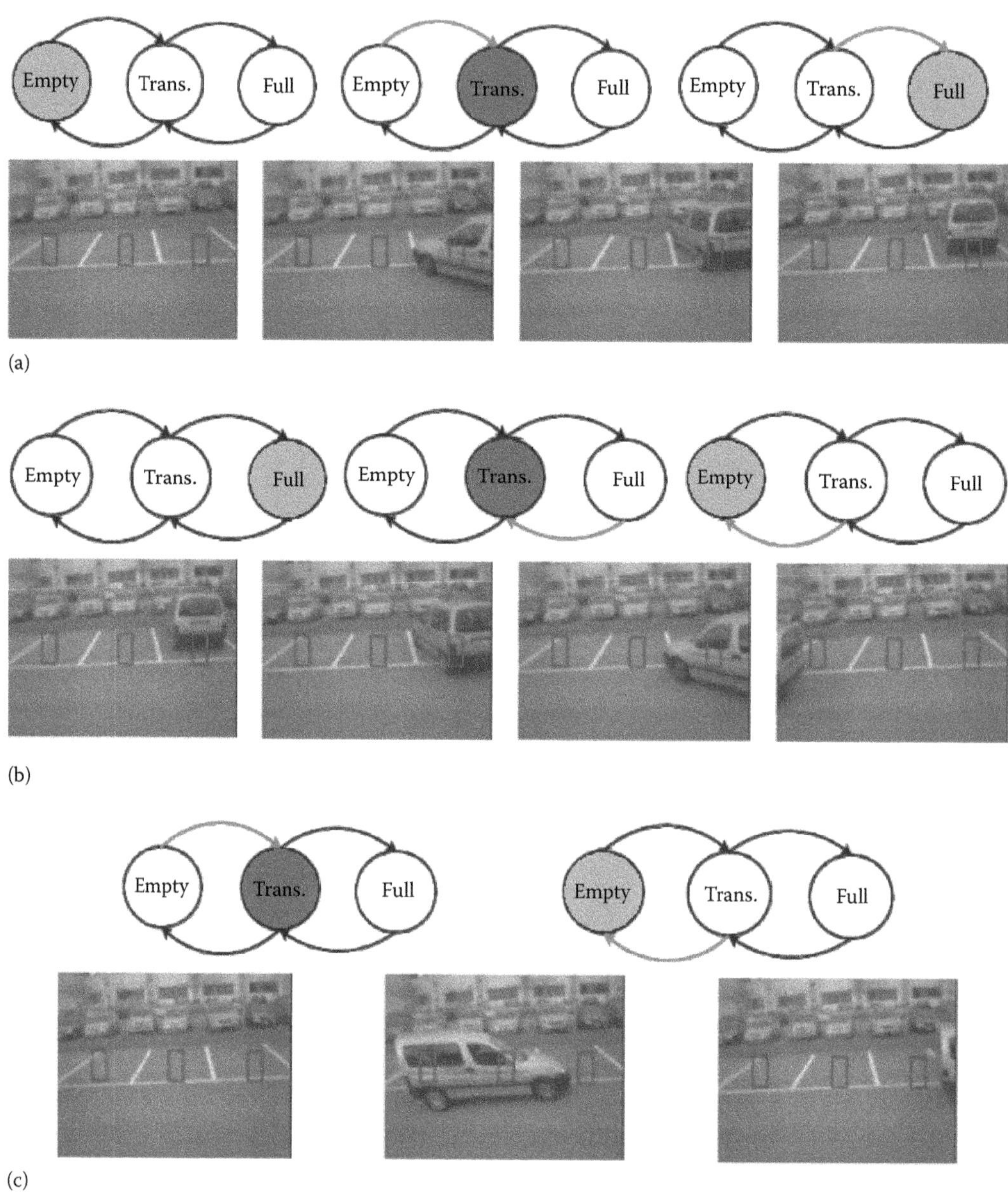

Figure 11.4 Transitions due to car parking activities and spurious events. (a) From empty to full transitions, (b) from full to empty transition, and (c) spurious event transitions.

Once the system is in the transition state, the FD is used as a main metric to move to a stable one. When $n_{FD} > TH_{FD}$, moving objects are still detected (e.g., car maneuvering, people going down from the car, etc.) and no possible changes to stable states are considered. When $n_{FD} < TH_{FD}$ is found for a number of frame bigger than a given TH_N threshold, the system moves in a stable state decided according to n_{BS}. If $n_{BS} < TH_{BS}$ the new state is the same hold before the transition, otherwise a status change has occurred. The FD output can be seen as a system trigger able to move

Figure 11.5 Car parking dynamics at 1 fps.

from the transition state to the stable ones: to this end the n_{FD} variable used to detect the event stability must be kept as low as possible to avoid inefficiencies.

The FD enforcement adopted in the change detection logic avoids the use of computational expensive recognition techniques, even if it imposes the use of frame rates relatively high with respect to a car parking dynamic (e.g., 1 or 2 fps instead of one image every minute). The use of frame rates lower than 1 or 2 fps could result in a wrong synchronization between state and background thus giving a wrong output regarding the parking space occupancy status. This situation could happen when a black car exits from the parking space and a white car enters in the same. In case of an excessively high sample time, the change will be interpreted as a change in the system state, from full to empty (BS above threshold and FD lower than the threshold in the next frame) even if the parking spaces are still full. This is depicted in Figure 11.5; an acquisition time equal to 1 fps is enough to understand the car parking dynamics.

11.3.5 Confidence Index

In this section, we describe the logic for deciding whether the parking space is empty or full. The adopted metric is a *confidence index* (CI) since it describes the probability of a parking space to be full. The CI is evaluated as a function of the time and it is retrieved by the parking space occupancy algorithm. The index is evaluated at the end of each change detection evaluation and then quantized on a 8 bit value in order to reduce the packet payload in a wireless communication. In an application scenario in which several parking spaces are monitored, several CI must be sent together with other possible acquired data (e.g., temperature, light, and CO_2 level), the use of a tiny amount of bytes for each status notification allows to reduce the transceiver usage, thus saving energy.

Due to the necessity of describing the parking space status according to the three states of the Markov chain, the CI values range has been divided in three main parts, as depicted in Figure 11.6, and each of them mapped in a possible state of the system. The range from the empty state goes from 0 to T_e, while the full from T_f to 255, where T_e and T_f are close to 0 and 255, respectively.

Moving from empty to full, the CI increases as a broken line from 0 to 255 (as shown in Figure 11.7), following two different behaviors in the transition state. The change detection procedure splits the transition zone in two parts: a *transition unstable zone* and a *transition stable zone*. The transition unstable zone is close to the previous stable state and represents the period of

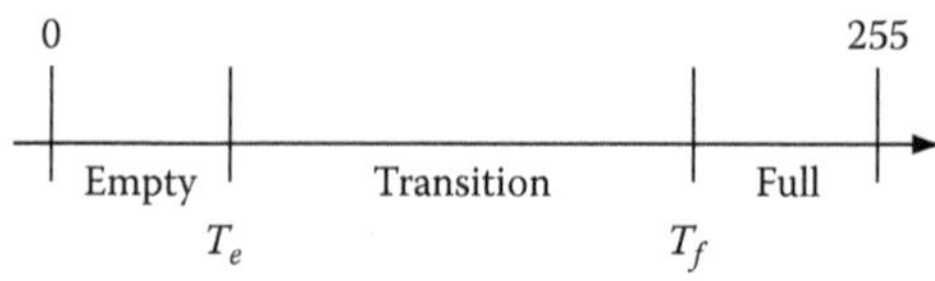

Figure 11.6 Confidence index values range and system states.

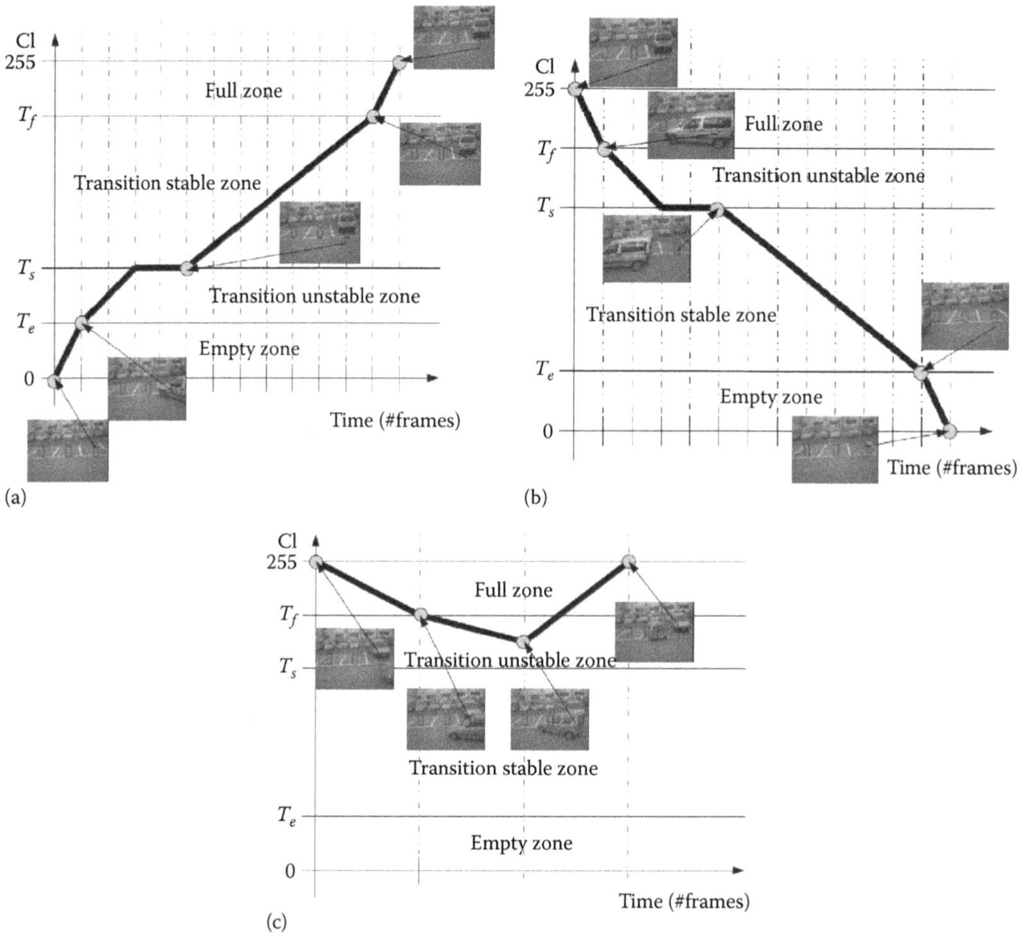

Figure 11.7 Effect of different types of transitions on CI values. (a) From empty to full transition, (b) from full to empty transition, and (c) spurious event transitions.

time dedicated to enter or leave the parking space. The transition stable zone, instead, represents the period of time between the end of the car maneuvering and the transition to the stable state.

11.3.6 Parking Space Occupancy Detection Algorithm Pseudocode

In this section, we report the algorithm pseudocode while explaining its components with respect to the logic described in the previous sections. The pseudocode reported in Algorithm 11.1 is applied to a single ROI covering a parking space in the scene.

Algorithm 11.1 Parking space monitoring algorithm pseudocode

```
/*Initialization step*/
roi = get_ROI();
prev=roi;
bgnd=roi;
state = init_state;
p_state = init_state;
cont = 0;
while 1 do
   /*Get a new image and perform BS and FD*/
   roi = get_ROI();
   n_bs = n_diff_over_th(roi, bgnd);
   n_fd = n_diff_over_th(roi, prev);
   if (state = FULL) or (state = EMPTY) then
      /*Full/empty state analysis*/
      if (n_bs >TH_BS) and (n_fd >TH_FD) then
         p_state = state;
         state = TRANSITION;
      else
         p_state = state;
         state = state;
         bgnd = update_bgnd(roi, bgnd);
      end if
   else if (state = TRANSITION) then
      /*Transition state analysis*/
      if (n_bs >TH_BS) then
         /*Real transition*/
         if (n_fd >TH_FD) then
            /*Transition unstable zone*/
            cont = 1;
         else
            /*Transition stable zone*/
            if (cont >TH_N) then
               cont = 0;
               if (p_state = EMPTY) then
                  p_state = TRANSITION;
                  state = FULL;
                  bgnd = roi_copy(roi);
               else
                  p_state = TRANSITION;
                  state = EMPTY;
                  bgnd = roi_copy(roi);
               end if
            end if
         end if
      else
         /*Spurious event*/
         state = p_state;
         p_state = TRANSITION;
         cont = 0;
      end if
   end if
   ci = compute_ci(state);
   prev = roi_copy(roi);
end while
```

The first step in the algorithm is an initialization procedure in which the last acquired ROI is used as background and previous frame. At this stage the state of the system must be known and imposed to the algorithm. This initialization step corresponds to a situation in which the device is installed and manually configured for the utilization. While the algorithm is running, for each acquired ROI the BS and FD procedures are performed. When the previous state is a stable state, and the conditions $n_{BS} > TH_{BS}$ and $n_{FD} > TH_{FD}$ occur, the state changes from stable to transition. In all the other cases, the state does not change and the background is updated by the exponential forgetting algorithm reported in Equation 11.2.

When the state is equal to transition, the condition $n_{BS} < TH_{BS}$ means that a spurious event has happened and the state is changed with the last stable one, instead $n_{BS} > TH_{BS}$ confirms a possible status change. In this last case, FD is used to evaluate whether the system enters the transition stable zone: when this happens, the new state is set and the background updated.

11.4 Algorithm Thresholds Tuning and Performance Evaluation

The parking space status detection algorithm described in Section 11.3 permits to decide whether a parking space is full or empty as a function of several thresholds used for both BS and FD algorithms. In this section, we first discuss how to tune the thresholds to reduce possible incorrect decisions, then we show the algorithm performance in terms of sensitivity, specificity, execution time, and memory occupancy by means of a real implementation in a WMSN device.

11.4.1 Algorithm Thresholds Tuning

The effectiveness of the proposed algorithm can be seen as its ability in reflecting the real behavioral trend of monitored parking spaces. In terms of performance, the algorithm detection capabilities can be measured with respect to real ground-truth values evaluated by means of a human-based analysis, considering that better algorithm performance means detection outputs much more similar to the reference ground-truth. As a consequence, an effective algorithm thresholds tuning process must therefore select the best thresholds to reach detection performance consistent to the human ground-truth. To this end, starting from real images belonging to the IPERDS [27] dataset collected within the IPERMOB [28] project, we first evaluated the real ground-truth of tuning image sequences with a human-based frame-by-frame process, then we tuned all the algorithm thresholds to make it able to follow the real trend.

As previously introduced, the image dataset adopted in the tuning process is the IPERDS, which is basically a collection of gray scale images acquired with a resolution of 160×120 pixels at 1 fps and related to traffic and parking spaces conditions. All the images composing the dataset have been collected by using a real WMSN device equipped with a low-cost camera, hence they have all the necessary characteristics to prototype video streaming and computer vision algorithms targeted to low-end devices. Among all the IPERDS traces related to parking space monitoring we selected one characterized by heavy shadows effects. The selected trace, in fact, can be considered the most challenging for the developed algorithm, because false change transitions can be detected in case of shadows in the selected ROI, causing in turn a wrong synchronization between the real status and the algorithm output.

The real ground-truth of the IPERDS trace adopted to tune algorithm thresholds has been evaluated by a human operator with a frame-by-frame analysis. In order to have a human output in the same range of the algorithm (CI output range) the empty status has been notified with the value 0, the transition with 127, and the full with 255. Two main rules have been imposed in evaluating the ground-truth: only parking cars can trigger status transitions, thus filtering moving people, and a transition ends when people inside the car leave the monitored ROI. A snapshot of the adopted tuning trace with the considered parking spaces is reported in Figure 11.8, while the time behavior ground-truth for each parking space is depicted in Figure 11.9. As it is possible to see from the plots, all the four parking spaces are characterized by status changes in the considered window time (more than 15 min) with shadows on neighboring parking spaces. Although the time-related ground-truth is enough to evaluate the algorithm thresholds, a secondary outcome of the performed analysis is the parking space usage trend model. In fact, considering the frequencies of each event it is possible to evaluate all the probabilities of the Markov chain introduced in Section 11.3.2. Table 11.2 reports the parking spaces probabilities for the selected tuning trace.

Starting from a human-based ground-truth' it is possible to tune all the algorithm thresholds by means of a comparison with the algorithm output. Although the thresholds introduced in Section 11.3.4 are four, only two of them must be properly tuned: TH_D and TH_N. The two

Figure 11.8 Parking spaces considered in the tuning trace.

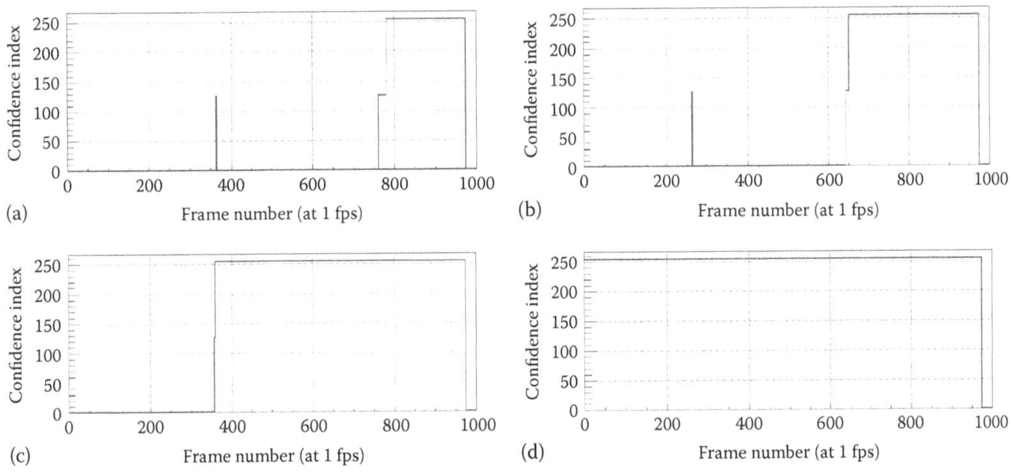

Figure 11.9 Ground-truth confidence index trend for the tuning trace. (a) P11, (b) P12, (c) P13, and (d) P14.

Table 11.2 Ground-Truth Transition Probabilities for the Tuning Trace

Parking Space ID	p_e	p_t	p_f	p_{et}	p_{ft}	p_{te}	p_{tf}
P11	0.779	0.019	0.198	0.002	0.000	0.001	0.001
P12	0.657	0.008	0.331	0.002	0.000	0.001	0.001
P13	0.367	0.001	0.630	0.001	0.000	0.000	0.001
P14	0.000	0.000	1.000	0.000	0.000	0.000	0.000

remaining thresholds, TH_{BS} and TH_{FD}, are dependent on TH_D, hence it is possible to set them equal to a portion of the ROI area while tuning TH_D appropriately. The TH_{BS} threshold has been imposed equal to 1/4 of the ROI area due to its requirements in detecting changes in the scene to trigger state transitions, while TH_{FD} has been imposed equal to 1/8 of the ROI area due to its requirements in guarantee event stability. The TH_D and TH_N have been jointly varied in the range from 50 to 60 and from 1 to 15, respectively, while evaluating the ground-truth similarity trend. In mathematical terms, the tuning procedure consists in finding from a set $\Im$ of possible TH_D and TH_N thresholds combinations the pair $TH = (TH_D, TH_N) \in \Im$, which minimizes the difference of the algorithm output from the human-based ground-truth. As similarity measure between algorithm output and ground-truth, we adopted the relation reported in the following:

$$S = \sqrt{\sum_{k=1}^{N} \frac{(G_{gt}[k] - G[k])^2}{N}} \qquad (11.3)$$

where

G_{gt} is the ground-truth value
G is the algorithm output with a specific $TH = (TH_D, TH_N)$ pair
N is the total number of image frames

S is an averaged Euclidean distance among CI outputs where lower values indicate a better similarity between the considered outputs. To thresholds tuning purposes, the similarity S has been calculated for all the four parking spaces selected in the tuning trace and then averaged among them in order to have an overall comparison value among $TH = (TH_D, TH_N)$ combinations. A graphical representation of the performed analysis is reported in Figure 11.10, where for three TH_D values the similarity S as a function of TH_N is shown.

The performed similarity analysis shows that TH_D must be set larger than 55. Adopting the lowest selected value, 50, the problem pointed out at the beginning of this section occurs, so that the parking space P14 loses the state/background synchronization due to luminosity variations caused by shadows (Figure 11.11). This behavior is confirmed by higher values of S for TH_D equal to 50 (Figure 11.10a). Regarding TH_N, a suitable value coming from the realized analysis is bigger than 1, even if bigger values can be adopted due to a possible increase in spurious transition filtering capabilities with no sensitive differences in similarity. As a consequence of the performed analysis results, we selected TH_D and TH_N equal to 60 and 5, respectively, in order to better filter spurious transition and guarantee a correct state stabilization. Algorithm CI outputs with $TH_D = 60$ and

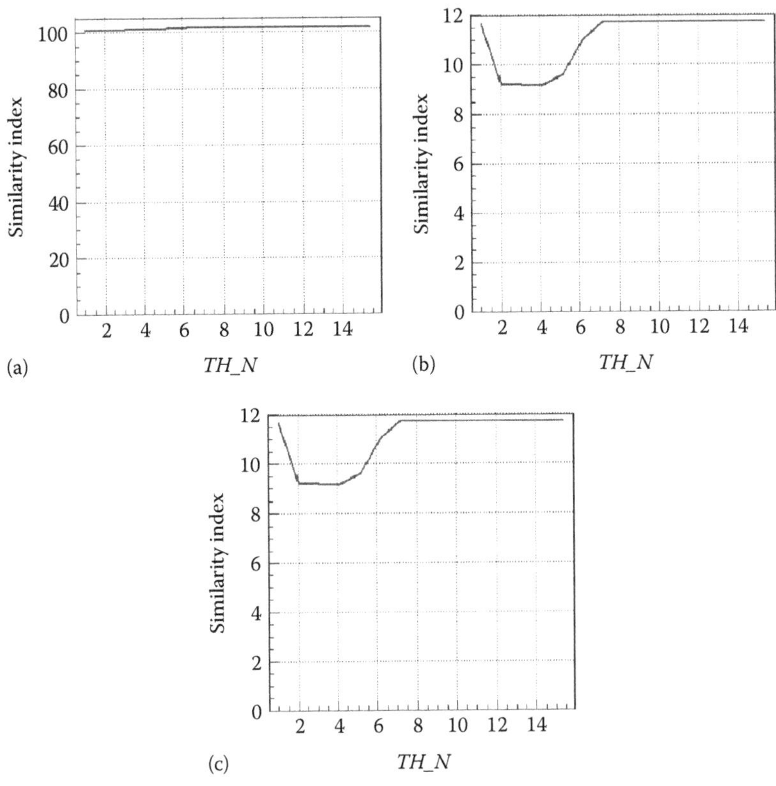

Figure 11.10 Similarity trend analysis. (a) $TH_D = 50$, (b) $TH_D = 55$, and (c) $TH_D = 60$.

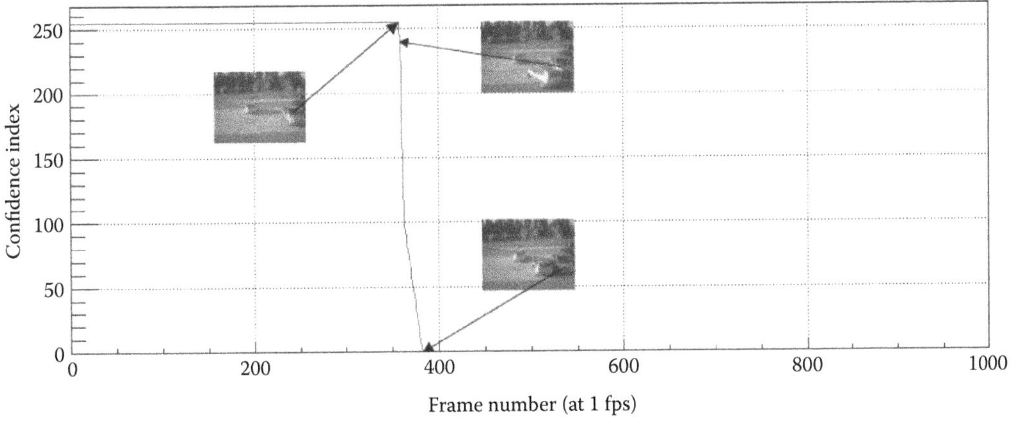

Figure 11.11 Wrong state/background synchronization behavior in P14.

$TH_N = 5$ are depicted in Figure 11.12 for all considered parking spaces; to be noticed is the strong similarity with the human-based ground-truth shown in Figure 11.9.

A validation process regarding the chosen thresholds values can be easily performed by evaluating the Markov chain transition probabilities coming out from the algorithm and comparing them with the one obtained by the human ground-truth analysis. The whole algorithm transition probabilities

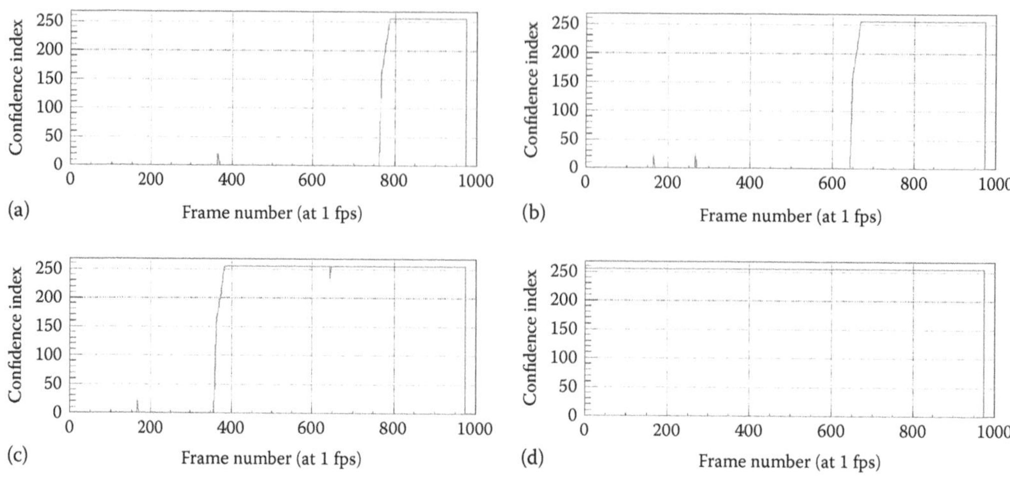

Figure 11.12 Algorithm confidence index trend for the tuning trace ($TH_D = 60$, $TH_N = 5$). (a) P11, (b) P12, (c) P13, and (d) P14.

Table 11.3 Algorithm Transition Probabilities for the Tuning Trace ($TH_D = 60$, $TH_N = 5$)

Parking Space ID	P_e	P_t	P_f	P_{et}	P_{ft}	P_{te}	P_{tf}
P11	0.780	0.005	0.211	0.002	0.000	0.001	0.001
P12	0.656	0.005	0.333	0.003	0.000	0.002	0.001
P13	0.366	0.005	0.623	0.002	0.001	0.001	0.002
P14	0.000	0.000	1.000	0.000	0.000	0.000	0.000

with the adopted thresholds values are reported in Table 11.3. Comparing such results with the one reported in Table 11.2 by means of the overall Euclidean distance between the vector of ground-truth probabilities and the one of the algorithm probabilities, the distances for the considered parking spaces are minimum: 0.019 for P11, 0.004 for P12, 0.007 for P13, and 0.000 for P14. Moreover, it must to be noticed that the differences among the probabilities in Tables 11.2 and 11.3 are minimum for the stable states (P_e and P_f), while the biggest differences are reached in the transition state (Pt) where the human ground-truth is substantially different from the algorithm output.

11.4.2 Algorithm Occupancy Status Detection Performance

The detection performance of the developed algorithm have been evaluated by means of simulations using an algorithm implementation suitable to run in real embedded devices. By adopting the threshold values selected in Section 11.4.1, the algorithm sensitivity and specificity [29] have been evaluated using IPERDS traces characterized by high movements in the scene with luminosity variations and regular shadows.

Figure 11.13 Parking spaces considered in a testing trace.

The sensitivity of the algorithm has been evaluated as the number of true positive events over the sum of true positive and false negative events, as reported in Equation 11.4, and indicates the ability of the algorithm of correctly detecting full state events. Regarding the specificity, this performance parameter has been evaluated as the number of true negative events over the sum of true negative and false positive events, see Equation 11.4, and indicates the ability of the algorithm of correctly detecting empty state events. The performance of the developed algorithm with respect to these two metrics are 99.92% for the sensitivity and 95.59% for the specificity. As it is possible to see from the reported results, the proposed algorithm with a properly tuning process can correctly detect the status of a parking space both in full and empty conditions:

$$Sensitivity = \frac{TP}{TP + FN}, \quad Specificity = \frac{TN}{TN + FP} \tag{11.4}$$

Considering a testing trace in which four parking spaces are monitored (Figure 11.13), a graphical comparison analysis between the human-based ground-truth and algorithm output (Figures 11.14 versus 11.15) confirms the algorithm capabilities in detecting the effective parking spaces occupancy status.

11.4.3 Algorithm Performance in the SEED-EYE Camera Network

The algorithm performance in terms of execution time and memory occupancy has been evaluated by means of a real implementation in the SEED-EYE [30] camera network device. The SEED-EYE board, depicted in Figure 11.16, is an innovative camera network device developed by Scuola Superiore Sant'Anna and Evidence within the IPERMOB [28] project. The board is equipped with a Microchip PIC32 microcontroller working at a frequency of 80 MHz and embedding 512 kB of Flash and 128 kB of RAM. The device mounts a CMOS camera that can be programmed to acquire images at various resolutions (up to 640×480) and frame rates (up to 30 fps). As network interfaces, an IEEE802.15.4 compliant transceiver and an IEEE802.3 module have been installed in order to enable wireless communications among peer devices and possible connections to backhauling networks. The SEED-EYE board has been specifically designed to support high-demanding multimedia applications while requiring low power consumption during image acquisition and processing. Performance evaluation executed in laboratory has shown that the board can acquire and process 160×120 images at 30 fps while experiencing a maximum power

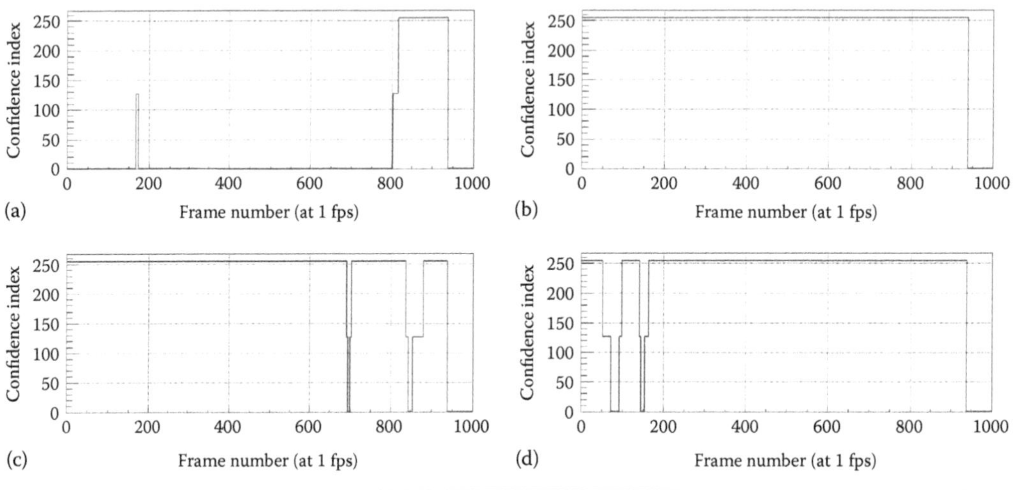

Figure 11.14 Ground-truth confidence index trend for a testing trace. (a) P21, (b) P22, (c) P23, and (d) P24.

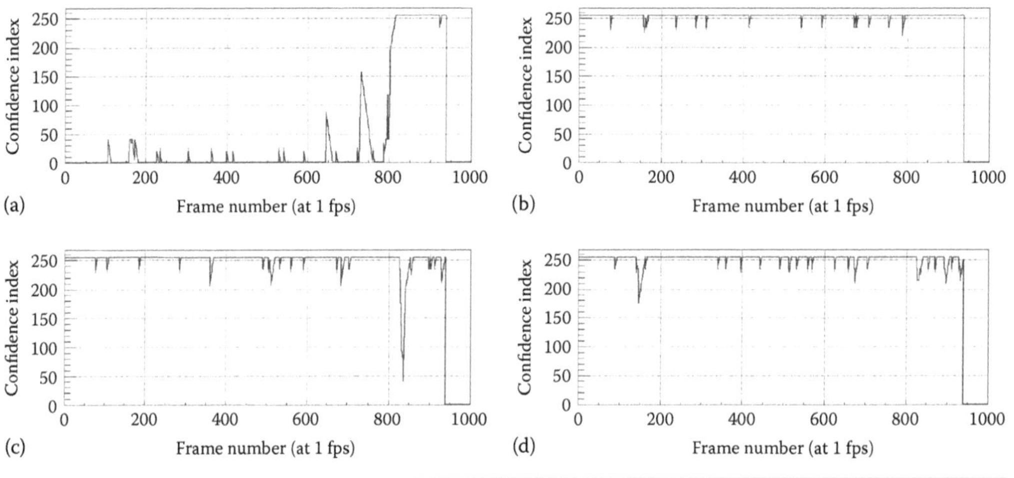

Figure 11.15 Algorithm confidence index trend for a testing trace ($TH_D = 60$, $TH_N = 5$). (a) P21, (b) P22, (c) P23, and (d) P24.

consumption equal to 450 mW when all the peripherals are activated. Lower power consumption values can be achieved reducing the image acquisition frame rates.

To evaluate the algorithm execution time in the SEED-EYE camera network device, the algorithm implementation adopted in Section 11.4.2 has been ported as a custom application on the top of the Erika Enterprise (EE) [31,32] OS, an innovative real-time operating system for small microcontrollers that provides an easy and effective way for managing tasks. More in detail, Erika is a multiprocessor real-time operating system kernel, implementing a collection of application programming interfaces similar to those provided by the OSEK/VDX standard for automotive embedded controllers. The algorithm execution time on top of EE OS has been measured performing several execution runs while considering a single parking space. The whole performance evaluation results are presented in Figure 11.17 in terms of execution time distribution.

Figure 11.16 The SEED-EYE board developed within the IPERMOB project.

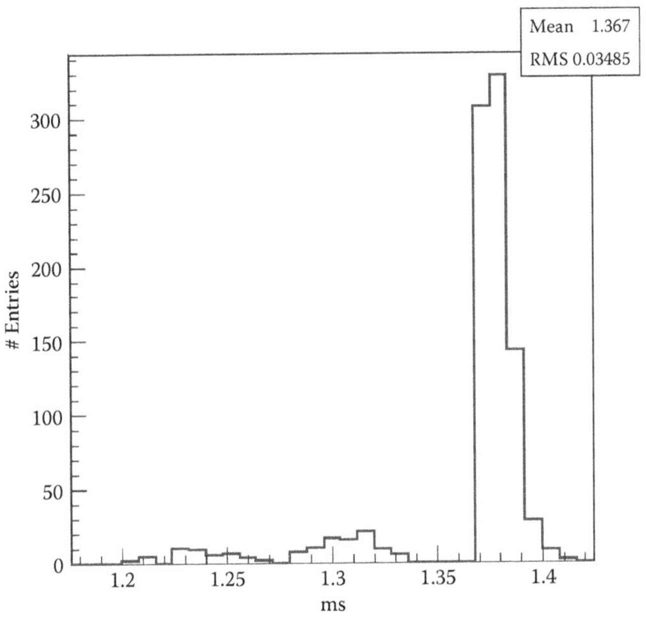

Figure 11.17 Execution time distribution.

As it is possible to see from the plot, the algorithm execution time is not constant due to the priority-based scheduling policies adopted in EE OS giving higher priorities to basic operating system tasks. As overall result, the algorithm shows an average execution time of 1.37 ms with a standard deviation of 0.05 ms.

Regarding the memory occupancy on both Flash and RAM, these values has been obtained by Microchip tools and are equal to 80.5 kB and 26.7 kB, respectively, for Flash and RAM. The percentage of total required memory is equal to 16.75%.

11.5 Conclusions

In this chapter, we focus on the development of onboard image processing techniques for detecting the occupancy status of parking spaces. The developed techniques are presented as an effective solution for vehicle parking lot monitoring applications in the domain of ITS. Starting from the adoption of classical BS techniques, we propose a modeling background process specially designed for the considered application in order to follow a low-complexity approach. Moreover, in the chapter the process for appropriately tuning all the algorithm parameters is exhaustively presented starting from a human-based ground-truth behavioral comparison. In a real implementation on a camera network device, the developed algorithm can reach 99.92% sensitivity and 95.59% specificity in detecting the parking spaces occupancy status, while showing an average execution time of 1.37 ms with a memory occupancy of 80.5 kB in Flash and 26.7 kB in RAM.

References

1. IEEE Computer Society, Wireless Medium Access Control (MAC) and Physical Layer (PHY) Specifications for Low-Rate Wireless Personal Area Networks (LR-WPAN), The Institute of Electrical and Electronics Engineers, Inc., October 2003.
2. J. Song, S. Han, A.K. Mok, D. Chen, M. Lucas, M. Nixon, and W. Pratt, WirelessHART: Applying wireless technology in real-time industrial process control, in *Proceedings of Real-Time and Embedded Technology and Applications Symposium*, April 2008, pp. 377–386.
3. K. Martinez, R. Ong, and J. Hart, Glacsweb: A sensor network for hostile environments, in *Proceedings of the IEEE Sensor and Ad Hoc Communications and Networks Conference*, October 2004, pp. 81–87.
4. I. Talzi, A. Hasler, S. Gruber, and C. Tschudin, PermaSense: Investigating permafrost with a WSN in the Swiss Alps, in *Proceedings of the Fourth Workshop on Embedded Networked Sensors*, June 2007, pp. 8–12.
5. R. Lee, K. Chen, S. Chiang, C. Lai, H. Liu, and M.S. Wei, A backup routing with wireless sensor network for bridge monitoring system, in *Proceedings of the Communication Networks and Services Research Conference*, May 2006, pp. 161–165.
6. M. Ceriotti, L. Mottola, G.P. Picco, A.L. Murphy, S. Guna, M. Corra, M. Pozzi, D. Zonta, and P. Zanon, Monitoring heritage buildings with wireless sensor networks: The torre aquila deployment, in *Proceedings of the International Conference on Information Processing in Sensor Networks*, April 2009, pp. 277–288.
7. S.D. Feller, Y. Zheng, E. Cull, and D.J. Brady, Tracking and imaging humans on heterogeneous infrared sensor arrays for law enforcement applications, in *Proceedings of SPIE Aerosense*, 2002, pp. 212–221.
8. I. Bekmezci and F. Alagoz, New TDMA based sensor network for military monitoring (MIL-MON), in *Proceedings of IEEE Military Communications Conference*, October 2005, pp. 2238–2243.
9. I.F. Akyildiz, T. Melodia, and K.R. Chowdury, A survey on wireless multimedia sensor networks, *Computer Networks (Elsevier)*, 51(4), 921–960, 2007.
10. M. Petracca, G. Litovsky, A. Rinotti, M. Tacca, J.C. De Martin, and A. Fumagalli, Perceptual based voice multi-hop transmission over wireless sensor networks, in *Proceedings of IEEE Symposium on Computers and Communications*, July 2009, pp. 19–24.
11. M. Petracca, M. Ghibaudi, C. Salvadori, P. Pagano, and D. Munaretto, Performance evaluation of FEC techniques based on BCH codes in video streaming over wireless sensor networks, in *Proceedings of IEEE Symposium on Computers and Communications*, July 2011, pp. 43–48.

12. A. Adam, E. Rivlin, I. Shimshoni, and D. Reinitz, Robust real-time unusual event detection using multiple fixed-location monitors, *IEEE on Transactions Pattern Analysis and Machine Intelligence*, 30(3), 555–560, 2008.
13. G. Srivastava, H. Iwaki, J. Park, and A.C. Kak, Distributed and lightweight multi-camera human activity classification, in *Proceedings of International Conference on Distributed Smart Cameras*, September 2009, pp. 1–8.
14. M. Chitnis, C. Salvadori, M. Petracca, P. Pagano, G. Lipari, and L. Santinelli, Distributed visual surveillance with resource constrained embedded systems, in *Visual Information Processing in Wireless Sensor Networks: Technology, Trends and Applications*, chapter 13, pp. 272–292. IGI Global, 2011.
15. P. Pagano, F. Piga, G. Lipari, and Y. Liang, Visual tracking using sensor networks, in *Proceedings of International Conference on Simulation Tools and Techniques for Communications, Networks and Systems*, 2009, pp. 1–10.
16. A.A. Abbo and R.P. Kleihorst, A programmable smart-camera architecture, in *Proceedings of Advanced Concepts for Intelligent Vision Systems*, 2002.
17. M.H. Rahimi, R. Baer, O.I. Iroezi, J.C. García, J. Warrior, D. Estrin, and M.B. Srivastava, Cyclops: In situ image sensing and interpretation in wireless sensor networks, in *Proceedings of Conference on Embedded Networked Sensor Systems*, 2005, pp. 192–204.
18. S. Hengstler, D. Prashanth, S. Fong, and H. Aghajan, Mesheye: a hybrid-resolution smart camera mote for applications in distributed intelligent surveillance, in *Proceedings of International Symposium on Information Processing in Sensor Networks*, 2007, pp. 360–369.
19. A. Rowe, A.G. Goode, D. Goel, and I. Nourbakhsh, CMUcam3: An open programmable embedded vision sensor, CMU-RI-TR-07-13, Robotics Institute, Technical Report, May 2007.
20. P. Chen, P. Ahammad, C. Boyer, S. Huang, L. Lin, E. Lobaton, M. Meingast, O. Songhwai, S. Wang, Y. Posu, A.Y. Yang, C. Yeo, L. Chang, J.D. Tygar, and S.S. Sastry, Citric: A low-bandwidth wireless camera network platform, in *Proceedings of International Conference on Distributed Smart Cameras*, September 2008, pp. 1–10.
21. M. Zhang and W. Cai, Vision mesh: A novel video sensor networks platform for water conservancy engineering, in *Proceedings of International Conference on Computer Science and Information Technology*, July 2010, pp. 106–109.
22. Z. Zivkovic and R. Kleihorst, Smart cameras for wireless camera networks: Architecture overview, in *Multi-Camera Networks: Principles and Applications*, chapter 21, pp. 497–510. Elsevier, 2009.
23. P. Pagano, M. Petracca, D. Alessandrelli, and C. Nastasi, Enabling technologies and reference architecture for a eu-wide distributed intelligent transport system, in *Proceedings of European ITS Congress*, June 2011.
24. J. Chinrungrueng, S. Dumnin, and R. Pongthornseri, iParking: A parking management framework, in *Proceedings of International Conference on ITS Telecommunications*, August 2011, pp. 63–68.
25. M. Piccardi, Background subtraction techniques: A review, in *Proceedings of International Conference on Systems, Man and Cybernetics*, October 2004, pp. 3099–3104.
26. X. J. Tan, J. Li, and C. Liu, A video-based real-time vehicle detection method by classified background learning, *World Transactions on Engineering and Technology Education*, 6(1), 189–192, 2007.
27. IPERDS, http://rtn.sssup.it/index.php/projects/prjipermob/ipermobdataset, September 2011.
28. IPERMOB: A pervasive and heterogeneous infrastructure to control urban mobility in real-time, http://www.ipermob.org, July 2009.
29. N. Lazarevic-McManus, J.R. Renno, D. Makris, and G.A. Jones, An object-based comparative methodology for motion detection based on the f-measure, *Computer Vision and Image Understanding*, 111(1), 74–85, 2008.

30. B. Dal Seno, M. Ghibaudi, and C. Scordino, Embedded boards for traffic monitoring, in *Poster Session of the IPERMOB Project Final Workshop*, http://www.ipermob.org/files/DemoSAT/afternoon/2011-05-18_poster_hw_oo3.pdf, May 2011.
31. P. Gai, E. Bini, G. Lipari, M. Di Natale, and L. Abeni, Architecture for a portable open source real time kernel environment, in *Real-Time Linux Workshop and Hand's on Real-Time Linux Tutorial*, November 2000.
32. The Erika Enterprise Real-time Operating System, http://erika.tuxfamily.org

Chapter 12

Signal Processing for Sensing and Monitoring of Civil Infrastructure Systems

Mustafa Gul and F. Necati Catbas

Contents

12.1 Introduction

12.1.1 Current Condition of Civil Infrastructure Systems

It is widely accepted and acknowledged that the civil infrastructure systems (CIS) all around the world are aging and becoming more vulnerable to damage and deterioration. As an example, many of the 600,000 highway bridges existing today in the United States were constructed from 1950 to 1970 for the interstate system. Having an approximately 50 year design life, most of these bridges are either approaching or have surpassed their intended design life (Figure 12.1). Highway agencies are struggling to keep up with the increasing demands on their highways, and deteriorating bridges are becoming severe choke points in the transportation network. It is estimated that more than 25% of the bridges in the United States (~150,000) are either structurally deficient or functionally obsolete and that it will cost $1.6 trillion to eliminate all bridge deficiencies in the United States. In addition to highways and bridges, similar problems exist for other CIS such as buildings, energy systems, dams, levees, and water systems. Degradations, accidents, and failures indicate that there is an urgent need for complementary and effective methods for current assessment and evaluation of the CIS. At this point, structural health monitoring (SHM) offers a very promising tool for tracking and evaluating the condition and performance of different structures and systems by means of sensing and analysis of objective measurement data.

12.1.2 Structural Health Monitoring of CIS

SHM is the research area focusing on condition assessment of different types of structures including aerospace, mechanical, and civil structures. Though the earliest SHM applications were in aerospace engineering, mechanical and civil applications have gained momentum in the last few decades.

Different definitions of SHM can be found in the engineering literature. For example, Aktan et al. (2000) defined SHM as follows: SHM is the measurement of the operating and loading environment and the critical responses of a structure to track and evaluate the symptoms of operational incidents, anomalies, and/or deterioration or damage indicators that may affect operation, serviceability, or safety and reliability. Another definition was given by Farrar et al. (1999) and Sohn et al. (2001), where the researchers stated that SHM is a statistical pattern recognition process to implement a damage detection strategy for aerospace, civil, and mechanical engineering

(a)

(b)

Figure 12.1 Seymour Bridge in Cincinnati, Ohio, was constructed in 1953 and was decommissioned approximately after 50 years of service. General view (a) and the condition of the deck at the time of decommissioning (b).

(a)

(b)

Figure 12.2 Sunrise Bridge, a bascule-type movable bridge in Ft. Lauderdale, Florida, was instrumented with more than 200 sensors for monitoring of structural, mechanical, and electrical components: bridge when opened (a) and instrumentation of the structural elements (b).

infrastructure and it is composed of four portions: (1) operational evaluation; (2) data acquisition, fusion, and cleansing; (3) feature extraction; and (4) statistical model development.

The starting point of an SHM system may be considered as the sensing and data acquisition step. The properties of the data acquisition system and the sensor network are usually application specific. The number and types of the sensors have a direct effect on the accuracy and the reliability of the monitoring process. With the recent technological developments in reduced cost sensing technologies, large amounts of data can be acquired easily with different types of sensors. The data collected during an SHM process generally include the response of the structure at different locations and information about the environmental and operational conditions. The measurements related to the structural response may include strain, acceleration, velocity, displacement, rotation, and others (Figure 12.2). On the other hand, data related to environmental and operational conditions may include temperature, humidity, wind speed, weigh in motion systems, and others.

After collection, SHM data can be analyzed by means of various methodologies to obtain useful information about the structure and its performance. Unless effective data analysis methodologies are implemented to an SHM system, problems related to data management will be inevitable. This may not only cause an overwhelming situation for handling large amounts of data effectively but also cause missing critical information. In addition to the analysis of experimental data, interpretation might require modeling and simulation where the analytical and numerical results may be combined or compared with experimental findings. Finally, information extracted from the data is used for decision-making about the safety, reliability, maintenance, operation, and future performance of the structure.

12.2 Data Analysis and Processing for SHM of CIS

Although sometimes SHM is used (rather incorrectly) as a synonym to damage detection, it actually refers to a much broader research area that can be employed for different purposes such as validation of the properties of a new structure, long-term monitoring of an existing structure, structural control, and many others. Brownjohn et al. (2004) presents a good review of civil infrastructure SHM applications. On the other hand, it should also be emphasized that damage

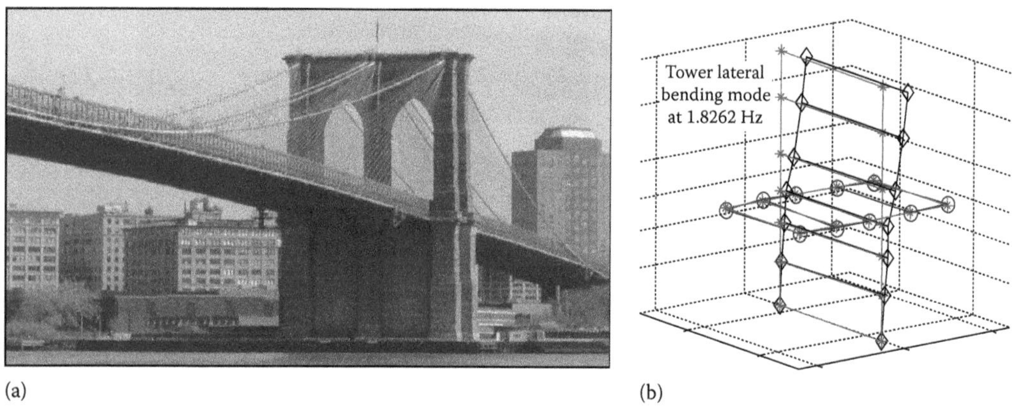

Figure 12.3 Brooklyn Bridge in New York City (a) was tested with 43 accelerometers to obtain the dynamic properties of the tower and deck: lateral bending mode of tower at 1.8262 Hz (b).

detection is a very critical component of SHM. Identifying the presence of the damage might be considered as the first step to take preventive actions and to start the process toward understanding the root causes of the problem.

Various methodologies have been proposed for detecting damage using SHM data. For global condition assessment, most of these methodologies employ vibration data by using one or a combination of different time domain and/or frequency domain algorithms (Figure 12.3). The aim is to extract features that will be sensitive to the changes occurring in the structure and relatively insensitive to other interfering effects (e.g., noise and operational and environmental effects). Some of these methodologies can be found in literature and the references therein (Hogue et al. 1991; Toksoy and Aktan 1994; Doebling et al. 1996; Worden et al. 2000; Sohn and Farrar 2001; Bernal 2002; Chang et al. 2003; Kao and Hung 2003; Sohn et al. 2003; Giraldo and Dyke 2004; Lynch et al. 2004; Alvandi and Cremona 2005; Nair and Kiremidjian 2005; Catbas et al. 2006; Sanayei et al. 2006; Carden and Brownjohn 2008; Gul and Catbas 2008, 2009; Gul and Catbas 2011).

12.2.1 Parametric Data Analysis Using Modal Models

A considerable number of damage detection efforts focus on parametric methods that generally assume that the a priori model related to the physical characteristics of the system is known. The aim of such methods is usually to compute the unknown parameters of this model. These parameters are mostly related to physical quantities such as mass, damping, and stiffness of the system, and the change in these parameters is used for damage detection.

Although a variety of parametric methods exist for damage detection applications, modal parameter estimation is one of the most commonly used parametric system identification approaches where the aim is to identify the unknown parameters of a modal model (modal frequencies, damping ratios, mode shape vectors, and modal scalings) of the system from given input–output or output-only data sets. One of the advantages of using the modal parameters is that they can easily be related to the physical characteristics of the structure. Therefore, a large body of research effort has been conducted investigating the modal parameter-based damage indices for SHM.

A very comprehensive and highly referenced review by Doebling et al. (1996) discusses and summarizes different methodologies to identify the damage by using the modal parameters and modal parameter-based damage indices.

Modal parameter estimation research has yielded different methods and approaches especially for mechanical and aerospace engineering applications in the last four to five decades. More recently, research for civil infrastructure applications using different forms of input–output or output-only dynamic tests has contributed new and/or revised algorithms, methods, and methodologies. These can be mainly grouped into two categories according to their working domain, i.e., time and frequency. There are several insightful review studies in the literature that would provide more detailed information about some of these methods (Maia and Silva 1997; Allemang and Brown 1998; Ewins 2000; Fu and He 2001; Peeters and Ventura 2003; Alvandi and Cremona 2005).

12.2.1.1 Modal Parameter Identification in Time Domain

Time domain modal parameter identification methods, as the name implies, extract the modal information from the time history data. These methods are generally developed from control theory concepts. The starting point of these methodologies is usually the free response or the pulse response of the system. However, most of these methodologies can be used with ambient vibration data after some preprocessing of the raw data to obtain an estimation of the free decay time response data.

Although these methodologies are usually numerically stable and give satisfactory results, their application to heavily damped systems is limited since they require a large amount of time domain data. Some of the widely used methods include complex exponential algorithm (CEA), polyreference time domain (PTD) method, Ibrahim time domain (ITD) method, and Eigensystem realization algorithm (ERA). Detailed discussions about the time domain methodologies can be found in the literature including Maia and Silva (1997), Allemang and Brown (1998), Allemang (1999), and Peeters and Ventura (2003). Since ERA is one of the widely used time domain methods and it provides a generic framework for CEA and ITD, the following discussions give more details about this technique. ERA was developed by Juang and Pappa (1985) and is based on the minimal realizations to obtain a state space system with minimum orders to represent a given set of input output relations.

A discrete time nth-order state space system with r inputs and m outputs can be written as

$$\begin{aligned} x(k+1) &= A_D x(k) + B_D u(k) \\ y(k) &= C_D x(k) + D_D u(k) \end{aligned} \tag{12.1}$$

where

$y_{m\times 1}(k)$ is the output vector
$u_{r\times 1}(k)$ is the input vector
$x_{n\times r}(k)$ is the state vector
$A_{D\,n\times n}$, $B_{D\,n\times r}$, $C_{D\,m\times n}$, and $D_{D\,m\times r}$ are the time-invariant system matrices

If the system is assumed to be excited with a unit impulse function and if the initial conditions of the system are zero as shown in Equation 12.2, then the response of the system can be calculated as written in Equation 12.3:

$$
\begin{aligned}
&u(k) = 1,\ x(k) = 0 \quad \text{for } k = 0 \\
&u(k) = 0,\ x(k) \neq 0 \quad \text{for } k \neq 0
\end{aligned}
\tag{12.2}
$$

$$
y(0) = D_D,\ y(1) = C_D B_D,\ y(2) = C_D A_D B_D,\ \ldots,\ y(k) = C_D A_D^{k-1} B_D \tag{12.3}
$$

The parameters shown in Equation 12.3 are known as Markov parameters. These parameters are collected in a so-called Hankel matrix, denoted with Π, as in Equation 12.4:

$$
\begin{aligned}
\Pi(0) &= \begin{bmatrix} y(1) & y(2) & \cdots & y(i) \\ y(2) & y(3) & \cdots & y(i+1) \\ \vdots & \vdots & \ddots & \vdots \\ y(j) & y(j+1) & \cdots & y(i+j-1) \end{bmatrix} \\
&= \begin{bmatrix} C_D B_D & C_D A_D B_D & \cdots & C_D A_D^{i-1} B_D \\ C_D A_D B_D & C_D A_D^2 B_D & \cdots & C_D A_D^{i} B_D \\ \vdots & \vdots & \ddots & \vdots \\ C_D A_D^{j-1} B_D & C_D A_D^{j} B_D & \cdots & C_D A_D^{j+i-2} B_D \end{bmatrix}
\end{aligned}
\tag{12.4}
$$

where i and j are the number of the columns and rows in the Hankel matrix, respectively. After building the Hankel matrix, the system matrices are retrieved by using singular value decomposition (SVD) of the Hankel matrix:

$$
\Pi(0) = USV^T = [U_1\ U_2] \begin{bmatrix} S_1 & 0 \\ 0 & 0 \end{bmatrix} \begin{bmatrix} V_1^T \\ V_2^T \end{bmatrix} \tag{12.5}
$$

where
 U and V are unitary matrices
 S is a square diagonal matrix

The system matrices can be obtained by using the following:

$$
A_D = S_1^{-1/2} U_1^T \Pi(1) V_1 S^{-1/2} \tag{12.6}
$$

$$
B_D = S_1^{1/2} V_1^T E_r \tag{12.7}
$$

$$
C_D = E_m U_1 S_1^{1/2} \tag{12.8}
$$

where $E_r = [I_{r\times r}\ 0\ \ldots\ 0]$ and $E_m = [I_{m\times m}\ 0\ \ldots\ 0]$. The modal parameters can be extracted by using the system matrices obtained with earlier equations. The natural frequencies can be obtained directly from the eigenvalues of the system matrix A_D. The mode shapes can be computed by multiplying the corresponding eigenvectors of A_D with the output matrix C_D.

12.2.1.2 Modal Parameter Identification in Frequency Domain

Frequency domain methods transform the time histories to frequency domain and extract the modal parameters in the frequency domain. These methodologies use the frequency response functions (FRFs) to compute the modal parameters. One of the main advantages of the frequency domain methodologies is that less computational modes (noise modes) are obtained in comparison with time domain algorithms. Some of the disadvantages of these methodologies are due to the restrictions of the fast Fourier transform (FFT). For example, leakage is one of the commonly encountered problems because FFT assumes that the signal is periodic within the observation time. The effect of leakage can be eliminated by using windowing functions but it cannot be avoided completely.

One of the simplest frequency domain methods is peak picking method where the modes are selected from the peaks of the FRF plots. If the system is lightly damped and if the modes are well separated, the natural frequencies (eigenfrequencies) can be estimated from the FRF plots. The damping ratios can be determined by using the half-power method. Rational fraction polynomial is a high-order frequency domain methodology where the following formulation is used to identify the modal parameters. The coefficients of these polynomials can be estimated from the FRF measurements by using a linear least squares solutions. Then the modal parameters are computed by using the polynomial coefficients:

$$H(j\omega) = \frac{\left[(j\omega)^p \beta_p + (j\omega)^{p-1}\beta_{p-1} + \cdots + \beta_0\right]}{\left[(j\omega)^p I + (j\omega)^{p-1}\alpha_{p-1} + \cdots + \alpha_0\right]} \tag{12.9}$$

where

H is the FRF

ω is the frequency in radians

α, β are the polynomial coefficients

Another method called complex mode indicator function (CMIF) where the modal parameters are identified by using the SVD of the output spectrum matrix is described in detail in the following since it is the methodology used for the examples presented in this text. Shih et al. (1988a) initially introduced CMIF as a mode indicator function for MIMO data to determine the number of modes for modal parameter estimation. Then, CMIF was successfully used as a parameter estimation technique to identify the frequencies and unscaled mode shapes of idealized test specimens (Shih et al. 1988b; Fladung et al. 1997). Catbas et al. (1997, 2004) modified and further extended CMIF to identify all of the modal parameters including the modal scaling factors from MIMO test data. In these studies, it was shown that CMIF is able to identify physically meaningful modal parameters from the test data, even if some level of nonlinearity and time variance were observed. Figure 12.4 shows the basic steps of the methodology.

The first step of the CMIF method is to compute the SVD of FRF matrix, which is given in Equation 12.10:

$$[H(\omega_i)]_{(N_o\times N_i)} = [U]_{(N_o\times N_i)}\,[S]_{(N_i\times N_i)}\,[V]^H_{(N_i\times N_i)} \tag{12.10}$$

where

$[S]$ is singular value matrix

$[U]$ and $[V]$ are left and right singular vectors, respectively

$[V^H]$ indicates the conjugate transpose of $[V]$

The earlier equation shows that the columns of the FRF matrix $[H(\omega_i)]$ are linear combination of the left singular vectors and, similarly, the rows of the FRF matrix are linear combination of the right singular vectors. Since the left and right singular vectors are unitary matrices, the amplitude

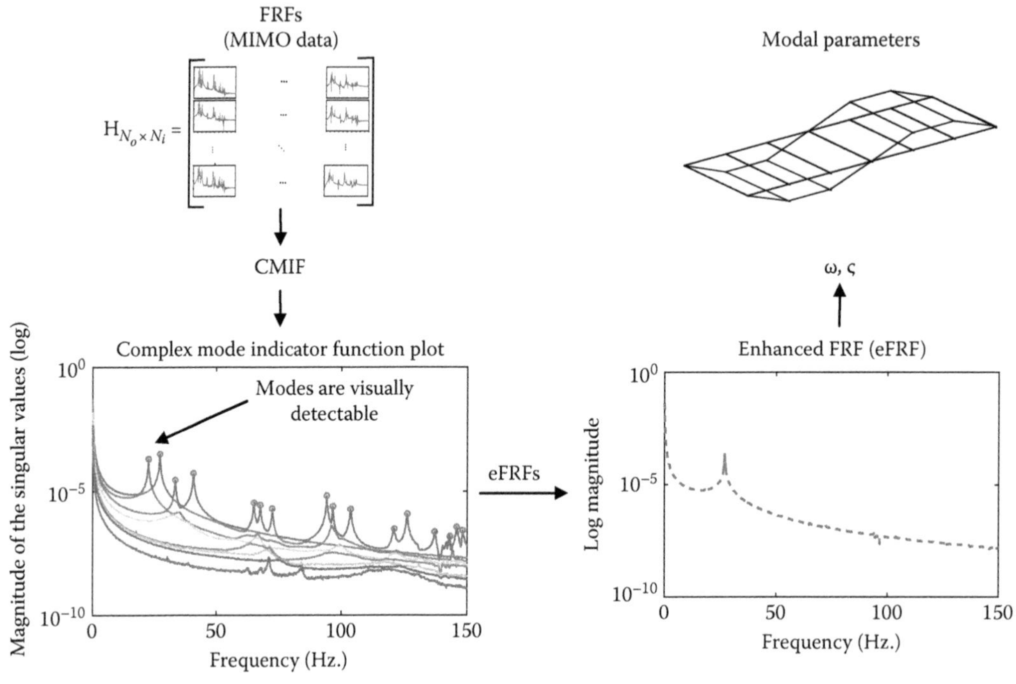

Figure 12.4 Summary of CMIF method.

information of the FRF matrix is carried within the singular value matrix. The CMIF plot shows the singular values as a function of frequency. The number of singular values at a spectral line and therefore the number of lines in the CMIF plot depends on the number of excitation points, N_i, in Equation 12.10, assuming that N_i is smaller than N_o. For this reason, MIMO data CMIF plots indicate multiple lines enabling tracking of the actual physical modes of the structure. The FRF matrix can be expressed in a way different from Equation 12.10 in terms of modal expansion using individual real or complex modes as given in Equation 12.11:

$$[H(\omega_i)]_{(N_o \times N_i)} = [\Psi]_{(N_o \times 2N)} \left[\frac{1}{j\omega_i - \lambda_r} \right]_{(2N \times 2N)} [L]^T_{(2N \times N_i)} \tag{12.11}$$

where

- ω is frequency
- λ_r is rth complex eigenvalue or system pole
- $[\Psi]$ is the mode shapes
- $[L]$ is the participation vectors

Note that while Equation 12.10 is a numerical decomposition, Equation 12.11 incorporates the physical characteristics, such as mode shapes and frequencies. As mentioned before, the left and right singular vector matrices $[U]$ and $[V]$ are unitary matrices in SVD formulation. Furthermore, $[\Psi]$ and $[L]$ are constant for a particular mode. The system pole and the driving frequency are closer along the frequency line near resonance, which results in a local maximum in CMIF plot. Therefore, there is a very high possibility that the peak singular values in CMIF plot are the pole locations of the system. In addition, since the left singular vector is the response at the resonance, it is a good approximation of the modal vector at that frequency. Then these modal vectors are used as modal filters to condense the measured response to as many single-degree-of-freedom (SDOF) systems as the number of selected peaks. This, in fact, is a transformation of the response from physical coordinates to the modal coordinates. After the transformation is completed, enhanced frequency response functions (eFRFs) are calculated for each SDOF system. Then the entire system is uncoupled to a vector of SDOF system for mode m with Equation 12.12:

$$eH(\omega_i)_m = \{U\}_m^H \left[[\Psi] \left[\frac{1}{j\omega_i - \lambda_r} \right] [L]^T \right] \{V\}_m \tag{12.12}$$

The level of the enhancement depends on the inner product of the left singular vector and the modal vector $[\Psi]$. If the modal vectors are mutually orthogonal, then the eFRF will be completely uncoupled, showing a single-mode FRF with a strong peak. However, if some of the modes are non-orthogonal, then those modes will have some contribution to the eFRF, which will cause another peak or peaks to appear.

After obtaining the set of SDOF systems, the second part of the method is about determining the modal frequency, damping, and modal scaling for each separate mode. Since the system is now transformed to a set of SDOF systems using the eFRFs, the following equation can be written in the frequency domain to compute the system poles:

$$\left[(j\omega_i)^2 \alpha_2 + (j\omega_i)\alpha_1 + \alpha_0 \right] \{eH(\omega_i)\} = \left[(j\omega_i)^2 \beta_2 + (j\omega_i)\beta_1 + \beta_0 \right] \{RH(\omega_i)\} \tag{12.13}$$

In Equation 12.13, $\{R(\omega)\}$ is the index vector showing the coordinates of the forcing locations and α and β are unknown coefficients. Since the eFRF matrix is generated in the first phase, $eH(\omega)$ and $(j\omega)$ are now known quantities. If there is no noise or residual terms in the data sets, just β_0 is sufficient, but to handle the noise β_1 and β_2 can be added to the right-hand side of the equation to enhance the results (Catbas et al. 1997; Fladung et al. 1997). If either α_0 or α_2 is assumed to be unity, a least-square solution can be applied and then the eigenvalue problem can be formulated and solved for the poles of the SDOF system. The poles $(\lambda_r = \sigma_r + j\omega_r)$ of the system are determined on a mode-by-mode basis.

12.2.1.3 Ambient Vibration Data Analysis

As mentioned earlier, the ideal case for modal parameter identification is one where both the input and output data are available. However, for most of the real-life applications for CIS, such experimental setups cannot be implemented since it is generally neither feasible nor possible to excite the large constructed structures with a known input especially if the structure is tested during usual operation particularly in the case of existing bridges. Therefore, identification of modal parameters from ambient vibration test data has attracted attention in the last decades, and several studies are available in the literature (Beck et al. 1994; Brincker et al. 2000; Peeters and De Roeck 2001; Brownjohn 2003; Caicedo et al. 2004; Yang et al. 2004; Catbas et al. 2007; Gul and Catbas 2008).

Several methods have been proposed to identify the modal parameters of the structures with output-only data. These methods are usually based on the methods discussed in previous sections of this text where data are preprocessed with different methodologies to obtain output spectra, unscaled free responses, or unscaled FRFs. For example, peak picking method, which was mentioned in the previous sections, can be used for ambient data analysis. However, for output-only analysis, the auto- and cross-power spectral densities of the ambient outputs are used instead of FRFs (Ren et al. 2004). Another method called frequency domain decomposition has also been used for ambient analysis by using SVD of the output spectrum matrix (Brincker et al. 2000; Peeters and De Roeck 2001). This method is also referred as CMIF (Catbas et al. 1997; Peeters and De Roeck 2001).

There are a number of different approaches for ambient vibration data analysis. One example of time domain methods for ambient data analysis is the ITD used in conjunction with random decrement (RD) (Huang et al. 1999). In another approach, Caicedo et al. (2004) combined the natural excitation technique with ERA to identify the modal and stiffness parameters. Stochastic subspace identification (SSI) is another commonly used method (Van Overschee and DeMoor 1996; Peeters and De Roeck 2001; Ren et al. 2004), which is based on writing the first-order state space equations for a system by using two random terms, i.e., process noise and measurement noise, which are assumed to be zero mean and white. After writing the first-order equations, the state space matrices are identified by using SVD. Then, the modal parameters are extracted from the state space matrices (Peeters and De Roeck 2001). Details and examples for ambient vibration data analysis are not presented in this text for the sake of brevity.

12.2.1.4 Modal Parameters and Damage Detection

After obtaining the modal parameters, these parameters or their derivatives can be used as damage-sensitive metrics. Some of the common modal-parameter-based features may be summarized as the natural frequencies, mode shapes and their derivatives, modal flexibility matrix, modal curvature, and others. It has been shown that natural frequencies are sensitive to environmental conditions, especially to temperature changes, yet they are not sufficiently sensitive to damage. In addition, since damage is a local phenomenon most of the time, the lower-frequency modes are usually not affected by the damage. The higher-frequency modes may indicate the existence of the damage because they generally represent local behavior but it is more difficult to identify those modes compared to identification of lower-frequency modes.

Unlike natural frequencies, which do not usually provide any spatial information, mode shapes provide such information and thus they are generally a better indicator of damage than natural frequencies. In theory, mode shapes would indicate the location of the damage; however, a dense array of sensors may be needed to capture those modes. Modal assurance criterion is one of the commonly used modal vector comparison tool (Allemang and Brown 1982). Modal curvature, which is usually obtained by taking derivative of the mode shapes, has also been used for damage detection purposes. Modal flexibility is another damage indicator, which can be obtained by using the frequencies and mass-normalized mode shapes. A review of these damage features was given by Doebling et al. (1996) and Carden and Fanning (2004). For the examples presented here, modal flexibility and modal curvature are obtained from the MIMO data sets with CMIF as summarized in Figure 12.5. The details and formulations of these damage features are explained later.

First developed by Maxwell (1864), the flexibility is a displacement influence coefficient of which the inverse is stiffness. Flexibility is a significant index as it characterizes input–output relationship for a structure. It has been shown to be a robust and conceptual condition index for constructed facilities. To find the modal flexibility of a structure, one can use modal parameters

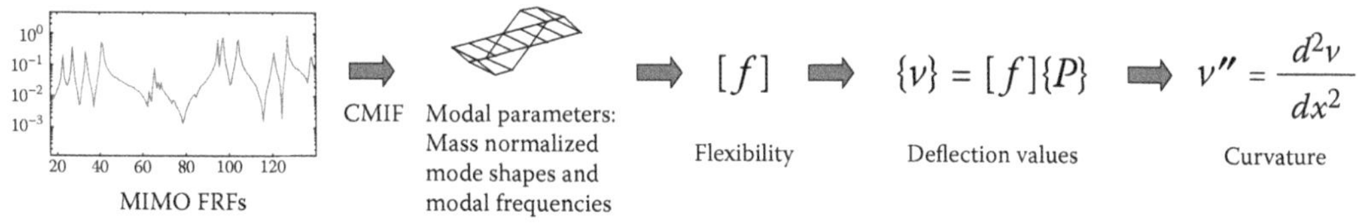

Figure 12.5 Obtaining the modal flexibility and modal curvature.

from dynamic testing. Flexibility has been proposed and shown as a reliable signature reflecting the existing condition of a bridge. For this reason, flexibility-based methods in bridge health monitoring are promising. Flexibility has been extracted and used in a number of different ways, and further studies can be found in Toksoy and Aktan (1994), Catbas and Aktan (2002), Bernal (2002), Bernal and Gunes (2004), Alvandi and Cremona (2005), Huth et al. (2005), Catbas et al. (2004, 2006, 2008a), and Gao and Spencer (2006). If an approximation to real structural flexibility is needed, the input force must be known in order to obtain the scaling of the matrix. In addition, it is always an approximate index since not all the modes can ever be included in the calculation of the flexibility matrix (Catbas et al. 2006).

The derivation of the modal flexibility can be better understood when looked at the FRF between points p and q written in partial fraction form as in Equation 12.14:

$$H_{pq}(\omega) = \sum_{r=1}^{m} \left[\frac{(A_{pq})_r}{j\omega - \lambda_r} + \frac{(A^*_{pq})_r}{j\omega - \lambda^*_r} \right] \tag{12.14}$$

where

$H_{pq}(\omega)$ is FRF at point p due to input at point q
ω is frequency
λ_r is rth complex eigenvalue or system pole
$(A_{pq})_r$ is residue for mode r
$(*)$ indicates the complex conjugate

Equation 12.14 can be rewritten by using the modal parameters as in Equation 12.15:

$$H_{pq}(\omega) = \sum_{r=1}^{m} \left[\frac{\psi_{pr}\psi_{qr}}{M_{Ar}(j\omega - \lambda_r)} + \frac{\psi^*_{pr}\psi^*_{qr}}{M^*_{Ar}(j\omega - \lambda^*_r)} \right] \tag{12.15}$$

where

ψ_{pq} is the mode shape coefficient between point p and q for the rth mode
M_{Ar} is the modal scaling for the rth mode

Finally, the modal flexibility matrix can be computed by evaluating $H_{pq}(\omega)$ at $\omega = 0$ as in Equation 12.16:

$$H_{pq}(\omega = 0) = \sum_{r=1}^{m} \left[\frac{\psi_{pr}\psi_{qr}}{M_{Ar}(-\lambda_r)} + \frac{\psi^*_{pr}\psi^*_{qr}}{M^*_{Ar}(-\lambda^*_r)} \right] \tag{12.16}$$

The flexibility formulation is an approximation to actual flexibility matrix because only a finite number of modes can be included in the calculations. The number of modes, m, is to be determined such that sufficient modes are selected (i.e., temporal truncation is minimized) and a good approximation to actual flexibility is achieved. Catbas et al. (2006) used a modal convergence criterion to determine the number of the modes necessary to construct a reliable flexibility matrix. In the examples presented here, around 15 modes are used to construct the flexibility matrices for each case. The modal flexibility matrix can also be written as in Equation 12.17. After obtaining the scaled modal flexibilities, the deflections under static loading can be calculated easily for any given loading vector $\{P\}$, which is shown in Equation 12.18:

$$[H] = \begin{bmatrix} H_{11}\,(\omega = 0) & H_{12}\,(\omega = 0) & \cdots & H_{1N}\,(\omega = 0) \\ H_{21}\,(\omega = 0) & \cdots & \cdots & \cdots \\ \cdots & \cdots & \cdots & \cdots \\ H_{N1}\,(\omega = 0) & \cdots & \cdots & H_{NN}\,(\omega = 0) \end{bmatrix} \tag{12.17}$$

$$[\text{deflection}] = \{v\} = [H]\,\{P\} \tag{12.18}$$

As another damage index, curvatures of modal vectors have been presented in the literature (Pandey et al. 1991; Maeck and De Roeck 1999, 2003). In this text, the deflections and curvatures are created from the modal flexibility. These indices are implemented here in terms of displacement vectors resulting from uniform loads applied to the modal flexibility matrices. However, the limitations of the curvature method are to be recognized. First, the spatial resolution (i.e., number of sensors) should be sufficient to describe a deflection pattern along a girder line. In order to obtain a good approximation to actual flexibility, both dynamic inputs and outputs are to be measured. In addition, modal truncation is to be minimized since modal flexibility has to approximate actual flexibility. Finally, taking derivatives of the data that include random noise and experimental errors might create numerical errors (Chapra and Canale 2002). However, the derivation presented in this text is based on the combination of all modes and associated deflections. Therefore, the random numerical errors are averaged out and may have less effect than taking the derivative of a single-mode shape.

As given in mechanics theory, curvature and deflection are related for a beam type by Equation 12.19:

$$v'' = \frac{d^2v}{dx^2} = \frac{M}{EI} \tag{12.19}$$

where

- v'' is the curvature at a section
- M is the bending moment
- E is the modulus of elasticity
- I is the moment of inertia

Since curvature is a function of stiffness, any reduction in stiffness due to damage should be observed by an increase in curvature at a particular location. The basic assumption for applying this relation to bridges is that the deformation is a beam-type deformation along the measurement line. This assumption can yield reasonably good approximation for bridges with girder lines. However, if it is to be used for damage identification for a two-way plate-type structure, the curvature formulation should be modified to take two directions into account. To calculate the curvature of the displacement vectors, the central difference approximation is used for numerical derivation as in Equation 12.20:

$$v''_{q,i} = \frac{v_{q-1,i} - 2v_{q,i} + v_{q+1,i}}{\Delta x^2} \tag{12.20}$$

where

q represents the elements of the ith displacement vector
Δx is the length between measured displacement points

12.2.2 Nonparametric Data Analysis Using Time Series Analysis

Use of statistical pattern recognition methods offers promise for handling large amounts of data. Most of the studies focusing on statistical pattern recognition applications on SHM use a combination of time series modeling with a statistical novelty detection methodology (e.g., outlier detection). One of the main advantages of such methodologies is that they require only the data from the undamaged structure in the training phase (i.e., unsupervised learning) as opposed to supervised learning where data from both undamaged and damaged conditions are required to train the model. The premise of the statistical pattern recognition approach is that as the model is trained for the baseline case, new data from the damaged structure will likely be classified as outliers in the data.

Most of these statistical models are used to identify the novelty in the data by analyzing the feature vectors, which include the damage-sensitive features. For example, Sohn et al. (2000) used a statistical process control technique for damage detection. Coefficients of auto-regressive (AR) models were used as damage-sensitive features and they were analyzed by using X-bar control charts. Different levels of damage in a concrete column were identified by using the methodology. Worden et al. (2000) and Sohn et al. (2001) used Mahalanobis distance-based outlier detection for identifying structural changes in numerical models and in different structures. Worden et al. (2000) used transmissibility function as damage-sensitive features whereas Sohn et al. (2001) used the coefficients of the AR models. Manson et al. (2003) also used similar methodologies to analyze data from different test specimens including aerospace structures.

In another study, Omenzetter and Brownjohn (2006) used auto-regressive integrated moving average (ARIMA) models to analyze the static strain data from a bridge during its construction and when the bridge was in service. The authors were able to detect different structural changes by using the methodology. They also mentioned the limitations of the methodology, for example, it was unable to detect the nature, severity, and location of the structural change. Nair et al. (2006) used an auto-regressive moving average (ARMA) model and used the first three AR components as the damage-sensitive features. The mean values of the damage-sensitive features were tested using a hypothesis test involving the t-test. Furthermore, the authors introduced two damage localization indices using the AR coefficients. They tested the methodology using numerical and experimental results of the ASCE benchmark structure. It was shown that the methodology was able to detect and locate even different types of damage scenarios for numerical case. However, it was concluded by the authors that more investigations were needed for analysis of experimental data.

Another methodology was proposed by Zhang (2007), where the author used a combination of AR and ARX (auto-regressive model with eXogenous output) models for damage detection and localization. The standard deviation of the residuals of the ARX model was used as damage-sensitive feature. Although the methodology was verified by using a numerical model, the author indicated that further studies should be conducted to make the methodology applicable in practice. In a recent study, Carden and Brownjohn (2008) used ARMA models and a statistical pattern classifier, which uses the sum of the squares of the residuals of the ARMA model. The authors stated that the algorithm was generally successful in identifying the damage and separating different damage cases from each other. However, the authors noted that the vibration data were coming from forced excitation tests and the methodology may not be applicable for structures with only ambient dynamic excitation.

12.2.2.1 Review and Formulations of Time Series Modeling

Time series modeling (or analysis) is a statistical modeling of a sequence of data points that are observed in time. It has been used in many different fields including structural dynamics and system identification. In the following sections, a brief discussion about time series modeling is given. A more detailed discussion about the theory of the time series modeling is beyond the scope of this text and can be found in the literature (Pandit and Wu 1993; Box et al. 1994; Ljung 1999).

A linear time series model representing the relationship of the input, output, and the error terms of a system can be written with the difference equation shown in Equation 12.21 (Ljung 1999). A compact form of this equation is shown in Equation 12.22:

$$\begin{aligned} &y(t) + a_1 y(t-1) + \cdots + a_{n_a} y(t-n_a) \\ &\quad = b_1 u(t-1) + \cdots + b_{n_b} u(t-n_b) + e(t) + d_1 e(t-1) + \cdots + d_{n_d} e(t-n_d) \end{aligned} \tag{12.21}$$

$$A(q)y(t) = B(q)u(t) + D(q)e(t) \tag{12.22}$$

where

$y(t)$ is the output of the model
$u(t)$ is the input to the model
$e(t)$ is the error term

The unknown parameters of the model are shown with a_i, b_i, and d_i and the model orders are shown with n_a, n_b, and n_d. $A(q)$, $B(q)$, and $D(q)$ in Equation 12.22 are polynomials in the delay operator q^{-1} as shown later in Equation 12.23. The model shown in Equation 12.22 can also be referred as an ARMAX model (auto-regressive moving average model with eXogenous input), and a block diagram of an ARMAX model can be shown as in Figure 12.6.

$$\begin{aligned} A(q) &= 1 + a_1 q^{-1} + a_2 q^{-2} + \cdots + a_{n_a} q^{-n_a} \\ B(q) &= b_1 q^{-1} + b_2 q^{-2} + \cdots + b_{n_b} q^{-n_b} \\ D(q) &= 1 + d_1 q^{-1} + d_2 q^{-2} + \cdots + d_{n_d} q^{-n_d} \end{aligned} \tag{12.23}$$

By changing the model orders of an ARMAX model, different types of time series models can be created. For example, if $n_b = n_d = 0$, the model is referred as an AR model, whereas an ARMA model is obtained by setting n_b to zero. The structure of an AR model is shown in Equation 12.24 whereas the block diagram of the model is shown in Figure 12.7.

$$A(q)y(t) = B(q)u(t) + e(t) \tag{12.24}$$

Estimating the unknown parameters of a time series model from the input–output data set (i.e., system identification) is of importance since the identified model can be used for many different

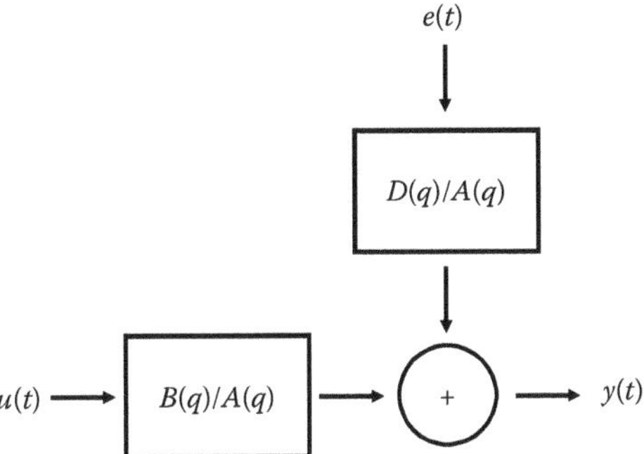

Figure 12.6 Block diagram of an ARMAX model. (Adapted from Ljung, L., *System Identification: Theory for the User*, Prentice-Hall, Upper Saddle River, NJ, 1999.)

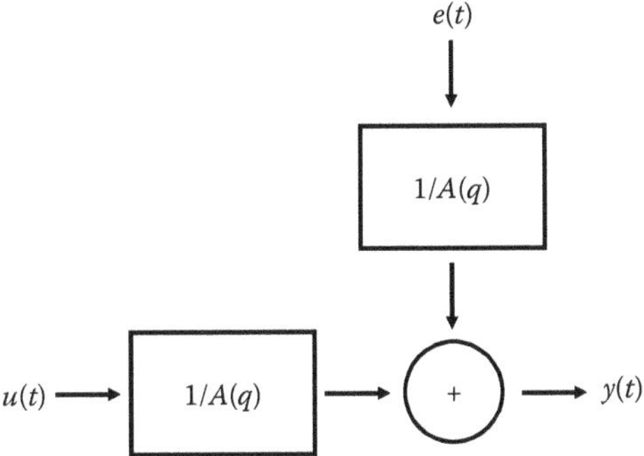

Figure 12.7 Block diagram of an AR model. (Adapted from Ljung, L., *System Identification: Theory for the User*, Prentice-Hall, Upper Saddle River, NJ, 1999.)

purposes including prediction and novelty detection. To estimate the unknown parameters, the difference equation of an ARX model can be written as in Equation 12.25:

$$y(t) = -a_1 y(t-1) - \cdots - a_{n_a} y(t-n_a) + b_1 u(t-1) + \cdots + b_{n_b} u(t-n_b) + e(t) \tag{12.25}$$

Equation 12.25 can be written for the previous time step as in Equation 12.26:

$$\begin{aligned} y(t) = {} & -a_1 y(t-1) - \cdots - a_{n_a} y(t-n_a-1) \\ & + b_1 u(t-2) + \cdots + b_{n_b} u(t-n_b-1) + e(t-1) \end{aligned} \tag{12.26}$$

Considering that these equations can be written for each time step, the equations can be put in a matrix form as in Equation 12.27:

$$Y = X\theta + E \tag{12.27}$$

where

$$Y = \left[y(t)\; y(t-1)\; \cdots y(t-n+1)\right]^T \tag{12.28}$$

$$X = \begin{bmatrix} -y(t-1) & \cdots & -y(t-n_a) & -u(t-1) & \cdots & -u(t-n_b) \\ -y(t-2) & \cdots & -y(t-n_a-1) & -u(t-2) & \cdots & -u(t-n_b-1) \\ \vdots & \vdots & \vdots & \vdots & \vdots & \vdots \\ -y(t-n) & \cdots & -y(t-n_a-n) & -u(t-n) & \cdots & -u(t-n_b-n) \end{bmatrix} \tag{12.29}$$

$$\theta = \begin{bmatrix} a_1 & \cdots & a_{n_a} & b_1 & \cdots & b_{n_b} \end{bmatrix}^T \tag{12.30}$$

$$E = \begin{bmatrix} e(t) & e(t-1) & \cdots & e(t-n+1) \end{bmatrix}^T \tag{12.31}$$

where n is the number of the equations. It is observed that Equation 12.27 is a linear matrix equation and the vector θ containing the unknown parameters can be estimated by using linear regression as shown in Equation 12.32. This solution also guarantees that the error vector E is minimized.

$$\theta \left(X^T X\right)^{-1} X^T Y \tag{12.32}$$

12.2.2.2 Time Series Modeling in Conjunction with Novelty Detection

This section demonstrates the implementation of the time series modeling for novelty detection for CIS. The RD method (not presented here, details can be found in Cole [1968], Asmussen [1997], and Gul and Catbas [2009]) is used to normalize the data and obtain the pseudo free responses from the ambient data. By doing so, the effect of the operational loadings on the data is minimized. Therefore, different data sets from different operational conditions can be compared more reliably. The methodology is illustrated in Figure 12.8.

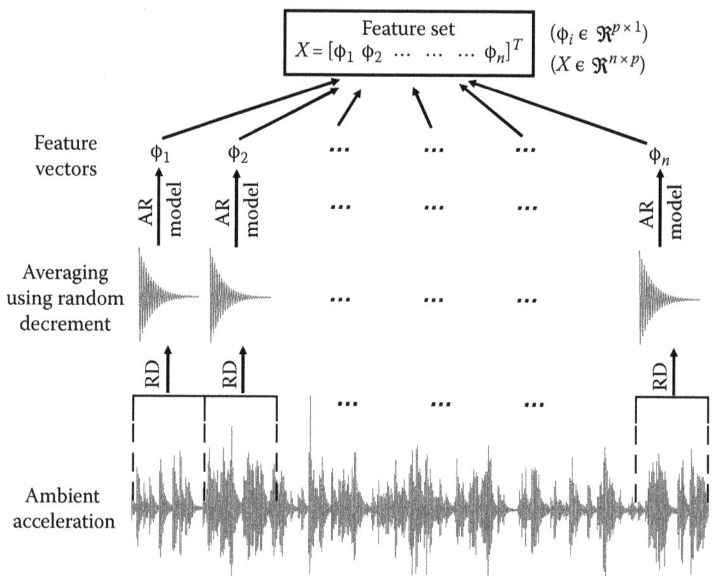

Figure 12.8 Implementing RD to ambient acceleration data before obtaining the feature sets using time series modeling.

After obtaining the pseudo free response functions, AR models of these free responses are created. Although a more detailed discussion about time series modeling was given in the previous sections, a brief discussion about AR models is given here. An AR model estimates the value of a function at time t based on a linear combination of its prior values. The model order (generally shown with p) determines the number of past values used to estimate the value at t (Box et al. 1994). The basic formulation of a pth-order AR model is defined in Equation 12.33:

$$x(t) = \sum_{j=1}^{p} \varphi_j x\left(t - j\Delta t\right) + e(t) \tag{12.33}$$

where

$x(t)$ is the time signal
φ is the model coefficients
$e(t)$ is the error term

After obtaining the coefficients of the AR models, they are fed to the outlier detection algorithm, where the Mahalanobis distance between the two different data sets is calculated.

After obtaining the AR model coefficients for different data sets, these coefficients are now used for outlier detection. Outlier detection can be considered as the detection of clusters, which deviate from other clusters so that they are assumed to be generated by another system or mechanism. Outlier detection is one of the most common pattern recognition concepts among those applied to SHM problems. In this section, Mahalanobis distance-based outlier detection is used to detect the novelty in the data.

The outlier detection problem for univariate (1D) data is relatively straightforward (e.g., the outliers can be identified from the tails of the distribution). There are several discordance tests, but one of the most common is based on deviation statistics and it is given by Equation 12.34:

$$z_i = \frac{d_i - \bar{d}}{\sigma} \tag{12.34}$$

where

z_i is the outlier index for univariate data
d_i is the potential outlier
$\bar{d}$ and σ are the sample mean and standard deviation, respectively

The multivariate equivalent of this discordance test for $n \times p$ (n is the number of the feature vectors and p is the dimension of each vector) data set is known as the Mahalanobis squared distance (Mahalanobis 1936). The Mahalanobis squared distance will be referred as Mahalanobis distance after this point and it is given by Equation 12.35:

$$Z_i = (x_i - \bar{x})^T \sum^{-1} (x_i - \bar{x}) \tag{12.35}$$

where

Z_i is the outlier index for multivariate data
x_i is the potential outlier vector
$\bar{x}$ is the sample mean vector
Σ is the sample covariance matrix

By using the earlier equations, the outliers can be detected if the Mahalanobis distance of a data vector is higher than a preset threshold level. Determining this threshold value is very critical, and different frameworks such as the one presented by Worden et al. (2000) can be used.

12.2.2.3 Damage Detection with Time Series Modeling Using Sensor Clustering

In this section, a modified time series modeling for damage detection and localization will be described. As the starting point, the equation of motion for an N degrees of freedom (DOF) linear dynamic system can be written as in Equation 12.36:

$$M\ddot{x}(t) + C\dot{x}(t) + Kx(t) = f(t) \tag{12.36}$$

where
$M \in \Re^{N\times N}$ is the mass matrix
$C \in \Re^{N\times N}$ is the damping matrix
$K \in \Re^{N\times N}$ is the stiffness matrix

The vectors $\ddot{x}(t), \dot{x}(t)$, and $x(t)$ are acceleration, velocity, and displacement, respectively. The external forcing function on the system is denoted with $f(t)$. The same equation can be written in matrix form as shown in Equation 12.37 (t for time is omitted):

$$\begin{bmatrix} m_{11} & \cdots & m_{1N} \\ \vdots & \ddots & \vdots \\ m_{N1} & \cdots & m_{NN} \end{bmatrix} \begin{Bmatrix} \ddot{x}_1 \\ \vdots \\ \ddot{x}_N \end{Bmatrix} + \begin{bmatrix} c_{11} & \cdots & c_{1N} \\ \vdots & \ddots & \vdots \\ c_{N1} & \cdots & c_{NN} \end{bmatrix} \begin{Bmatrix} \dot{x}_1 \\ \vdots \\ \dot{x}_N \end{Bmatrix} + \begin{bmatrix} k_{11} & \cdots & k_{1N} \\ \vdots & \ddots & \vdots \\ k_{N1} & \cdots & k_{NN} \end{bmatrix} \begin{Bmatrix} x_1 \\ \vdots \\ x_N \end{Bmatrix} = \begin{Bmatrix} f_1 \\ \vdots \\ f_N \end{Bmatrix} \tag{12.37}$$

The equality in Equation 12.38 is obtained if the first row of Equation 12.37 is written separately. By rearranging Equation 12.38, it is seen in Equation 12.39 that the output of the first DOF can be written in terms of the excitation force on first DOF, the physical parameters of the structure, and the outputs of the other DOFs (including itself). Furthermore, in case of free response, the force term can be eliminated and the relation is written as shown by Equation 12.40:

$$(m_{11}\ddot{x}_1 + \cdots + m_{1N}\ddot{x}_N) + (c_{11}\dot{x}_1 + \cdots + c_{1N}\dot{x}_N) + (k_{11}x_1 + \cdots + k_{1N}x_N) = f_1 \tag{12.38}$$

$$\ddot{x}_1 = \frac{f_1 - (m_{12}\ddot{x}_2 + \cdots + m_{1N}\ddot{x}_N) - (c_{11}\dot{x}_1 + \cdots + c_{1N}\dot{x}_N) - (k_{11}x_1 + \cdots + k_{1N}x_N)}{m_{11}} \tag{12.39}$$

$$\ddot{x}_1 = -\frac{(m_{12}\ddot{x}_2 + \cdots + m_{1N}\ddot{x}_N) + (c_{11}\dot{x}_1 + \cdots + c_{1N}\dot{x}_N) + (k_{11}x_1 + \cdots + k_{1N}x_N)}{m_{11}} \tag{12.40}$$

It is seen from Equation 12.40 that if a model is created to predict the output of the first DOF by using the DOFs connected to it (neighbor DOFs), the change in this model can reveal important information about the change in the properties of that part of the system. Obviously, similar equalities can be written for each row of Equation 12.37, and different models can be created for each equation. Each row of Equation 12.37 can be considered as a sensor cluster with a reference DOF and its neighbor DOFs. The reference DOF for Equation 12.40, for example, is the first DOF, and neighbor DOFs are the DOFs that are directly connected to the first DOF. Therefore, it is observed that different linear time series models can be created to establish different models for each sensor cluster, and changes in these models can point the existence, location, and severity of the damage. The details of the methodology are explained in the following sections.

As explained in the previous chapter, a general form of a time series model can be written as in Equation 12.41, and an ARX model is shown in Equation 12.42:

$$A(q)y(t) = B(q)u(t) + D(q)e(t) \tag{12.41}$$

$$A(q)y(t) = B(q)u(t) + e(t) \tag{12.42}$$

The core of the methodology presented in this part is to create different ARX models for different sensor clusters and then extract damage-sensitive features from these models to detect the damage. In these ARX models, the $y(t)$ term is the acceleration response of the reference channel of a sensor cluster, the $u(t)$ term is defined with the acceleration responses of all the DOFs in the same cluster, while $e(t)$ is the error term. Equation 12.43 shows an example ARX model to estimate the first DOF's output by using the other DOFs' outputs for a sensor cluster with k sensors:

$$A(q)\ddot{x}_1(t) = B(q)\begin{bmatrix}\ddot{x}_1(t) & \ddot{x}_2(t) & \cdots & \ddot{x}_k(t)\end{bmatrix}^T + e(t). \tag{12.43}$$

To explain the methodology schematically, a simple three-DOF model is used as an example. Figure 12.9 shows the first sensor cluster for the first reference channel. The cluster includes first and second DOFs since the reference channel is connected only to the second DOF. The input vector u of the ARX model contains the acceleration outputs of first and second DOFs. The output of the first DOF is used as the output of the ARX model as shown in the figure. When the second channel is the reference channel, Figure 12.10, the sensor cluster includes all three DOFs since they are all connected to the second DOF. The outputs of the first, second, and third DOFs are used as the input to the ARX model and then the output of the second DOF is used as the output of this model. Likewise, for the reference channel three, Figure 12.11, the inputs to the ARX model are

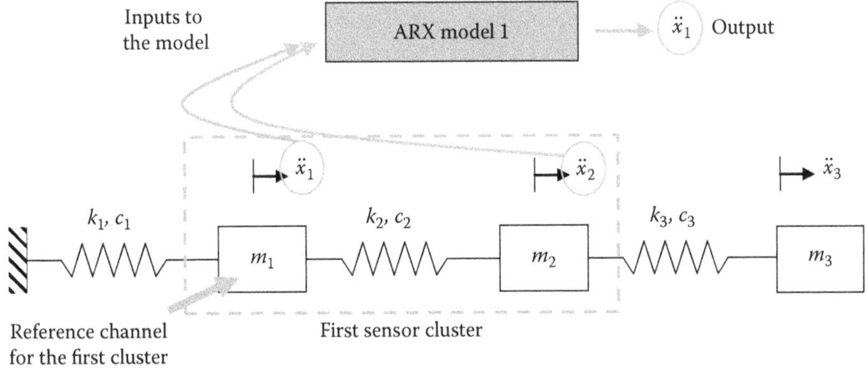

Figure 12.9 Creating different ARX models for each sensor cluster (first sensor cluster).

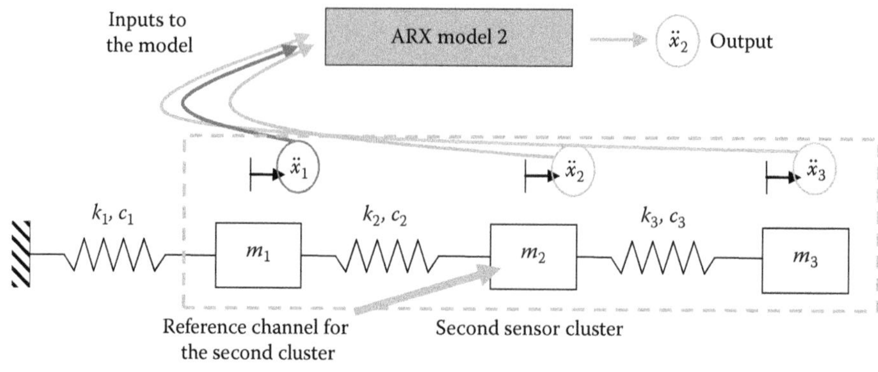

Figure 12.10 Creating different ARX models for each sensor cluster (second sensor cluster).

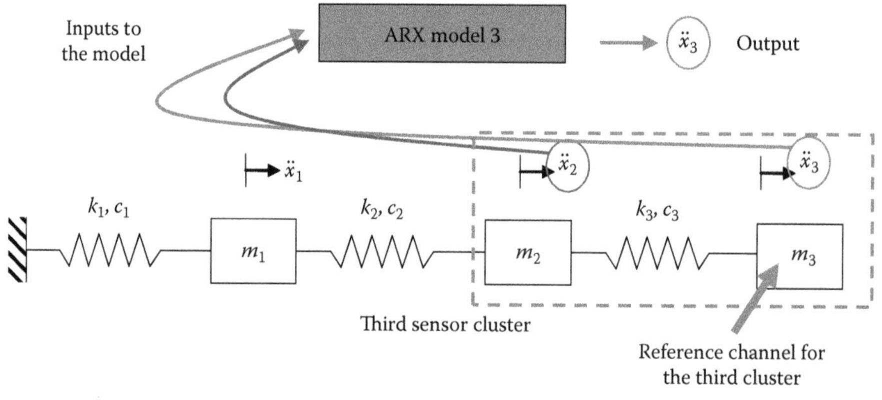

Figure 12.11 Creating different ARX models for each sensor cluster (third sensor cluster).

the output of the second and third channels and the output of the model is the third channel itself. The equations of the ARX models created for the example system are shown in Equations 12.44 through 12.46:

$$A_1(q)\ddot{x}_1(t) = B_1(q)\left[\begin{array}{cc}\ddot{x}_1(t) & \ddot{x}_2(t)\end{array}\right]^T + e_1(t) \tag{12.44}$$

$$A_2(q)\ddot{x}_2(t) = B_2(q)\left[\begin{array}{ccc}\ddot{x}_1(t) & \ddot{x}_2(t) & \ddot{x}_3(t)\end{array}\right]^T + e_2(t) \tag{12.45}$$

$$A_3(q)\ddot{x}_3(t) = B_3(q)\left[\begin{array}{cc}\ddot{x}_2(t) & \ddot{x}_3(t)\end{array}\right]^T + e_3(t) \tag{12.46}$$

After creating the ARX models for the baseline condition, different approaches may be implemented for detecting damage. For example, comparison of the coefficients of the ARX models for each sensor cluster before and after damage can give information about the existence, location, and severity of the damage. For the approach adapted here, the fit ratios (FRs) of the baseline ARX model when used with new data are employed as a damage-sensitive feature. The difference between the FRs of the models is used as the damaged feature (DF) (Gul and Catbas 2011). The FR of an ARX model is calculated as given in Equation 12.47:

$$\text{Fit Ratio } (FR) = \left(1 - \frac{\|\{y\} - \{\hat{y}\}\|}{\|\{y\} - \{\bar{y}\}\|}\right) \times 100 \tag{12.47}$$

where

$\{y\}$ is the measured output
$\{\hat{y}\}$ is the predicted output
$\{\bar{y}\}$ is the mean of $\{y\}$
$\|\{y\} - \{\hat{y}\}\|$ is the norm of $\{y\} - \{\hat{y}\}$

The DF is calculated by using the difference between the FRs for healthy and damaged cases as given in Equation 12.48:

$$\text{Damage Feature } (DF) = \frac{FR_{healthy} - FR_{damaged}}{FR_{healthy}} \times 100 \tag{12.48}$$

12.3 Laboratory Studies

12.3.1 Steel Grid Structure

Before the routine applications of SHM systems to real-life structures, the methodologies should be verified on analytical and physical models. Although analytical studies are necessary in the first phases of verification, laboratory studies with complex structures are also essential. Laboratory studies with large physical models are a vital link between the theoretical work and field applications if these models are designed to represent real structures where various types and levels of uncertainties can be incorporated.

For this section, data from a steel grid structure is employed for the experimental verifications of the methods discussed in this text. This model is a multipurpose specimen enabling researchers to try different technologies, sensors, algorithms, and methodologies before real-life applications. The physical model has two clear spans with continuous beams across the middle supports. It has two 18 ft girders (S3 × 5.7 steel section) in the longitudinal direction. The 3 ft transverse beam members are used for lateral stability. The grid is supported by 42 in. tall columns (W12 × 26 steel section). The grid is shown in Figure 12.12 and more information about the specimen can be found in Catbas et al. (2008b).

A very important characteristic of the grid structure is that it can be easily modified for different test setups. For example, with specially designed connections (Figure 12.13), various damage cases can be simulated. In addition, several different boundary conditions and damage cases (e.g., pin supports, rollers, fixed support, and semi-fixed support) can be simulated by using the adjustable connections.

The grid structure can be instrumented with a number of sensors for dynamic and static tests. For the dynamic tests that are in the scope of this text, the grid is instrumented with 12 accelerometers in vertical direction at each node (all the nodes except N7–N14 in Figure 12.14). The accelerometers used for the experiments are ICP/seismic-type accelerometers (Figure 12.15) with a 1000 mV/g sensitivity, 0.01–1200 Hz frequency range, and ∓2.5 g of measurement range.

Figure 12.12 Steel grid model used for the experimental studies.

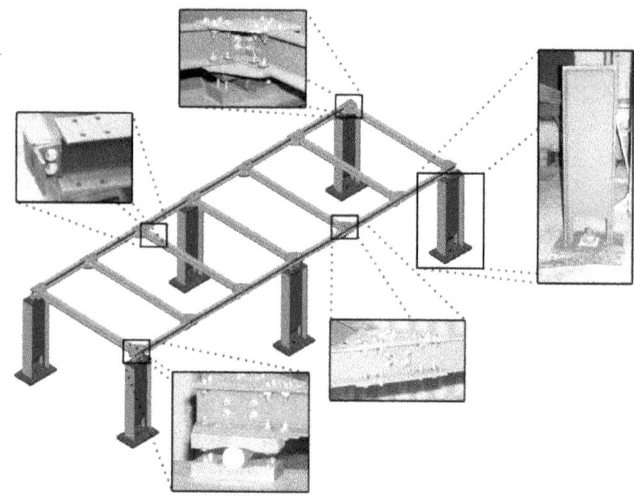

Figure 12.13 CAD drawing and representative pictures showing the details of the grid.

To record the dynamic response, an acquisition system from VXI and Agilent Technologies is used. The MTS-Test software package was used for acquisition control of the impact tests.

For the impact tests, the grid was excited at nodes N2, N5, N6, and N12 and five averages were used to obtain the FRFs as it is suggested in the literature. The sampling frequency is 400 Hz. For the impact tests, an exponential window is applied to both input and output data sets whereas a force window is applied only to the input set. Both time history and FRF data from MTS software were recorded. The ambient vibration was created by random tapping of two researchers with fingertips simultaneously. The researchers were continuously moving around the structure to make the excitation as random as possible. The ambient data were recorded by using VXI DAQ Express software.

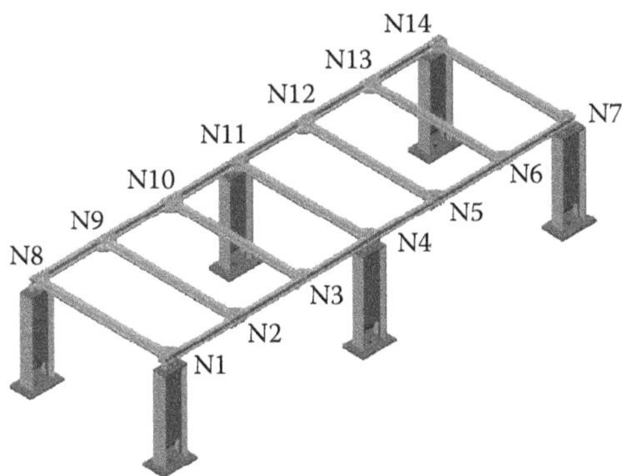

Figure 12.14 Node numbers for the steel grid.

Figure 12.15 Accelerometers used in the experiments.

12.3.1.1 Damage Simulations

A number of different damage scenarios were applied to the grid. These damage cases are simulated to represent some of the problems commonly observed by bridge engineers and Department of Transportation officials (Burkett 2005). Two different damage cases investigated here are summarized in the following text. One of these damage cases is devoted to local stiffness loss whereas the other case is simulated for boundary condition change.

Baseline case (BC0): Before applying any damage, the structure is tested to generate the baseline data so that the data from the unknown state can be compared to the baseline data for damage detection.

Damage case 1 (DC1): Moment release and plate removal at N3: DC1 simulates a local stiffness loss. The bottom and top gusset plates at node N3 are removed in addition to all bolts at the connection (Figure 12.16). This is an important damage case especially for CIS applications since gusset plates are very critical parts of steel structures. Furthermore, it has been argued that inadequate

Figure 12.16 Plate removal at N3 for DC1.

gusset plates might have contributed to the failure of the I-35W Mississippi River Bridge (Holt and Hartmann 2008).

Damage case 2 (DC2): Boundary restraint at N7 and N14: DC2 is created to simulate some unintended rigidity at a support caused by different reasons such as corrosion. The oversized through-bolts were used at N7 and N14 to introduce fixity at these two supports (Figure 12.17). Although these bolts can create considerable fixity at the supports, it should be noted that these bolts cannot guarantee a perfect fixity.

12.3.2 Damage Detection Results Using Parametric Methods

Generally, the first step in damage detection is to define the baseline state. Therefore, the damage features are first evaluated for the healthy case. The modal parameters of the healthy structure are identified by using CMIF that was outlined in the previous sections. Sample data, FRFs and the CMIF plot for BC0 are shown in Figures 12.18 through 12.20. There are 17 vertical modes identified for this case and the first 10 vertical modes are shown in Figure 12.21. After the modal parameters were identified, the modal flexibility was calculated. The deflection profile obtained with modal flexibility is shown in Figure 12.22. The deflected shape is obtained by applying a 100 lb uniform load to the measurement locations (i.e., 100 lb at each node).

After obtaining the deflections, the curvature is obtained by using the deflected shapes as shown in Equation 12.20. The modal curvature of the baseline case is shown in Figure 12.23. It should be noted that the spatial resolution of the sensors would affect the quality of the curvature data considerably. A denser sensor array would further improve the results. However, the sensor spatial resolution in this study is defined so that it represents a feasible sensor distribution for real-life applications on short- and medium-span bridges. Another consideration for modal curvature calculation is that there is no curvature value at the beginning and end measurement locations due

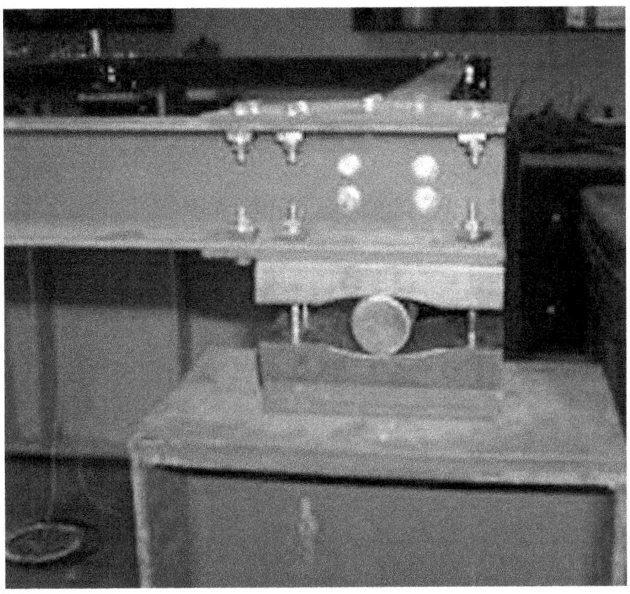

Figure 12.17 Boundary fixity at N7 and N14 for DC2.

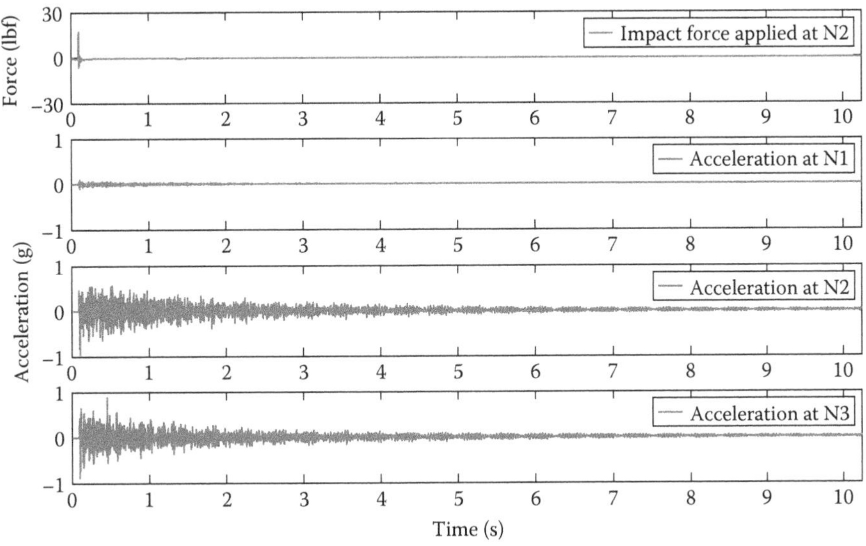

Figure 12.18 Sample data from impact testing for BC0.

to the numerical approximation of the central difference formula. The curvatures at the supports are assumed to be zero since the roller supports cannot resist moment thus indicating that there cannot be any curvature at these points. Finally, the curvature plot is very similar to that of a moment diagram (*M*/*EI*) for a girder under uniform load. As such, the interpretation and evaluation of the curvature plot become very intuitive for structural engineers as in the case of deflected shapes.

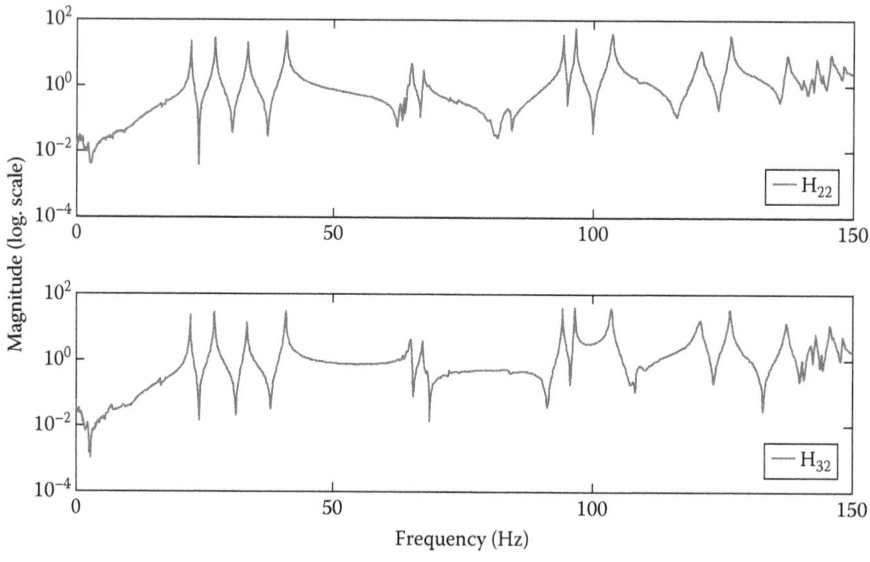

Figure 12.19 Sample FRF data obtained from impact testing for BC0.

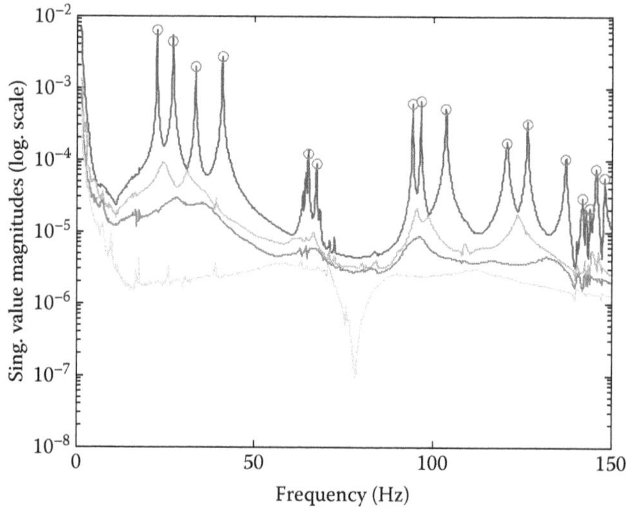

Figure 12.20 CMIF plot for BC0 obtained with impact testing.

Damage case 1 (DC1): Moment release and plate removal at N3: When the same procedure is repeated for DC1, it is observed in Figure 12.24 that the maximum deflection change obtained is at the damage location (N3) and is around 2.8%. The changes at the other nodes are around 1%–2%. Although this change may possibly be considered as an indicator of the damage occurred at N3 for this laboratory case, it should also be noted that a 3% change in the flexibility coefficients might not be a clear indicator of the damage, especially for real-life applications. Looking at the curvature comparisons, Figure 12.25 shows that the maximum curvature change is around 10.7%

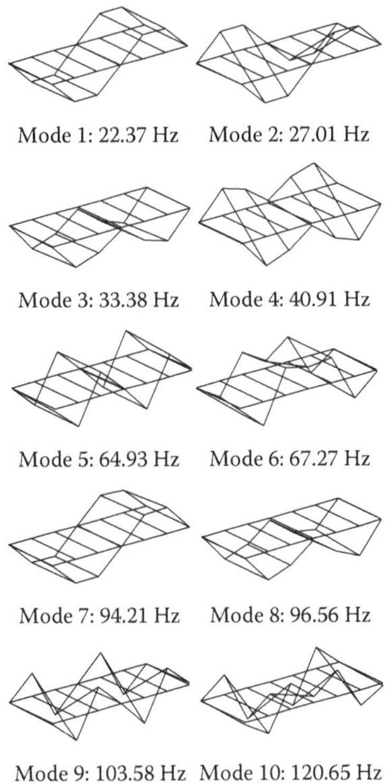

Figure 12.21 First 10 vertical modes for BC0 obtained with impact testing.

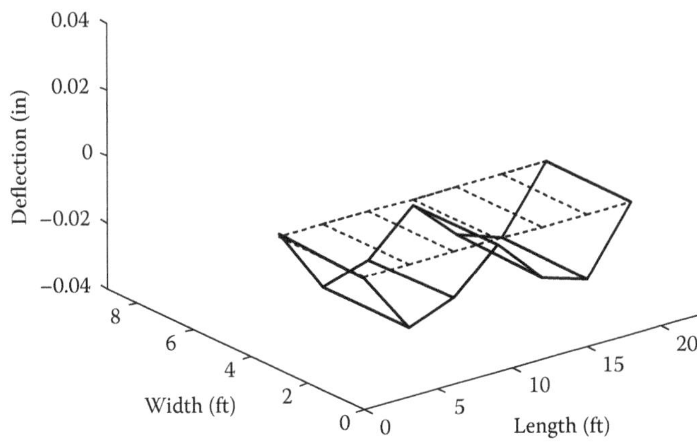

Figure 12.22 Deflection profile obtained using the modal flexibility for BC0.

and is obtained at the damage location. The changes in the curvature for other points are less than 1%–3% except at N12, where 4.7% increase in the curvature is computed.

Damage case 2 (DC2): Boundary restraint at N7 and N14: For DC2, the damage can be visually observed from the deflection patterns as seen in Figure 12.26. It should be noted that this damage case can be considered as a symmetric damage case (both N7 and N14 are restrained) and thus a

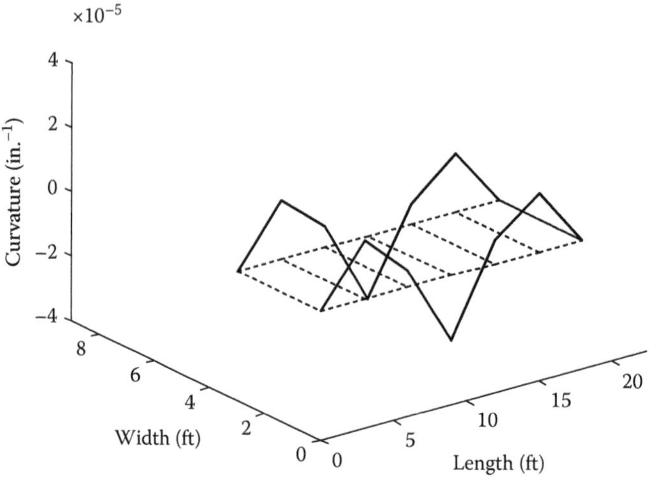

Figure 12.23 Curvature profile obtained using the modal curvature for BC0.

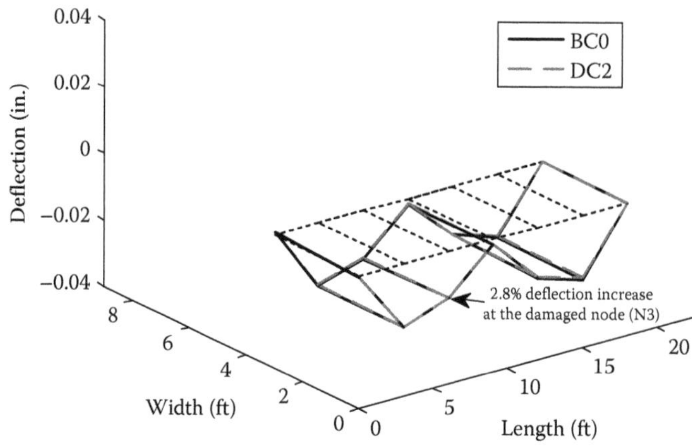

Figure 12.24 Deflection comparison for BC0 and DC1.

clear change in the deflected shapes is observable for both of the girders. The deflection reduced by about 30%–50% at the span where joint restraint damage scenario was implemented.

Curvature from the deflected shapes was determined subsequently for DC2. As was mentioned before, the curvature at the roller supports was assumed zero for pin-roller boundary conditions. For the restrained case, however, this assumption is not correct for N7 and N14 since the moment at these fixed supports is nonzero. However, for visualization purposes, the curvatures at N7 and N14 are still assumed as zero. It is seen from Figure 12.27 that a clearly observable 30% decrease in the curvature exists near the damage location. Here, we see a decrease in the curvature because the structure with restrained support is stiffer than the baseline with roller supports. It is clear that a finer resolution of sensors, especially around the end supports, would yield more accurate results in terms of curvature.

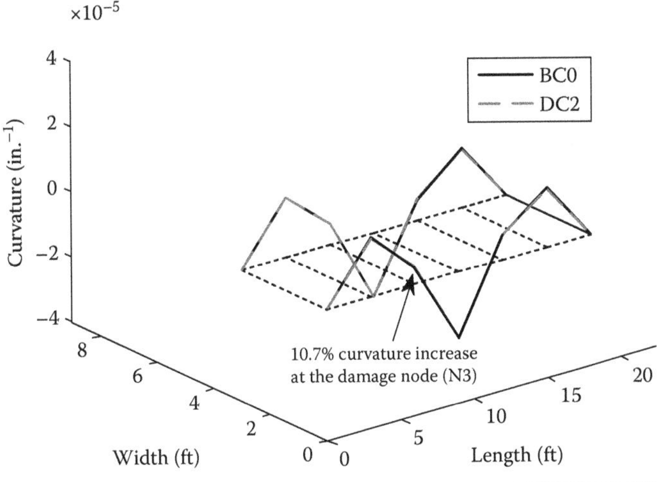

Figure 12.25 Curvature comparison for BC0 and DC1.

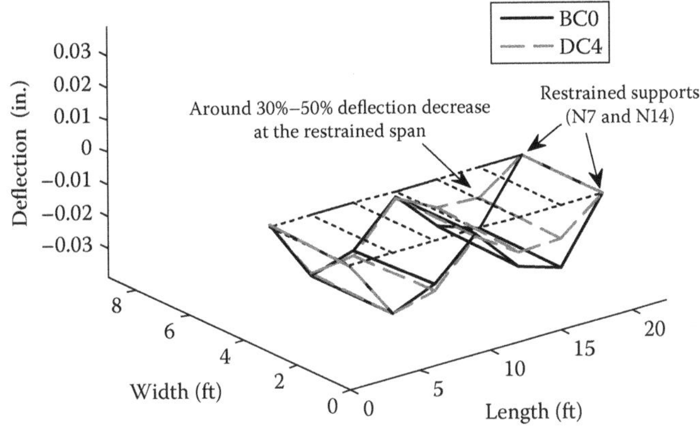

Figure 12.26 Deflection comparison for BC0 and DC2.

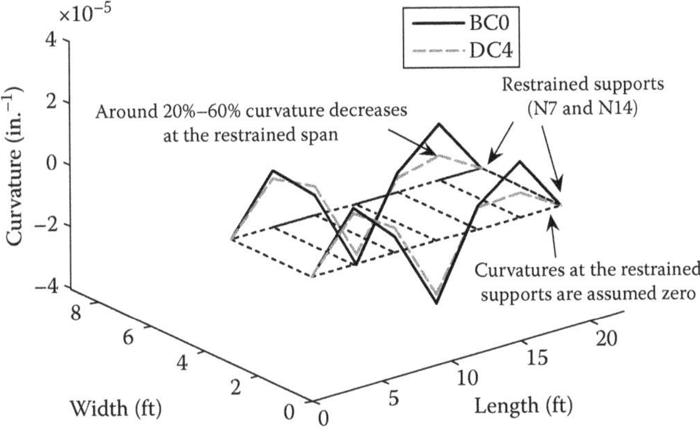

Figure 12.27 Curvature comparison for BC0 and DC2.

12.3.3 Damage Detection Results Using Nonparametric Methods

12.3.3.1 AR in Conjunction with Novelty Detection

For this section, ambient vibration data from the test grid created by random hand tapping of two researchers simultaneously are used. Sampling rate is 400 Hz for the experiments. There are 23 data blocks for each case. The acceleration data are averaged by using RD where the reference channel for RD process is node 2 (the location of node 2 can be seen in Figure 12.28). The model order for the AR models has been determined to be 10. The threshold is calculated as 180.

Baseline case (BC0): Before the analysis of the data from the damage cases, it is investigated whether the data from the baseline (healthy) grid structure are under the determined threshold value or not. Figure 12.28 shows analysis results of the baseline data acquired on the same day (first and second half of one data set). It is seen from the figure that all the values are under the threshold value (no false positives). This indicates that the numerically evaluated threshold value is consistent with the experimental results.

Damage case 1 (DC1): Moment release and plate removal at N3: Figure 12.29 shows the same plots for DC1. For this case, it is observed that the features for the second data set are clearly separated from the features from the baseline case. This shows that the damage applied at N3 can be identified by using the methodology. However, there is no clear information about the location of the damage since approximately same amount of separation is obtained at all nodes.

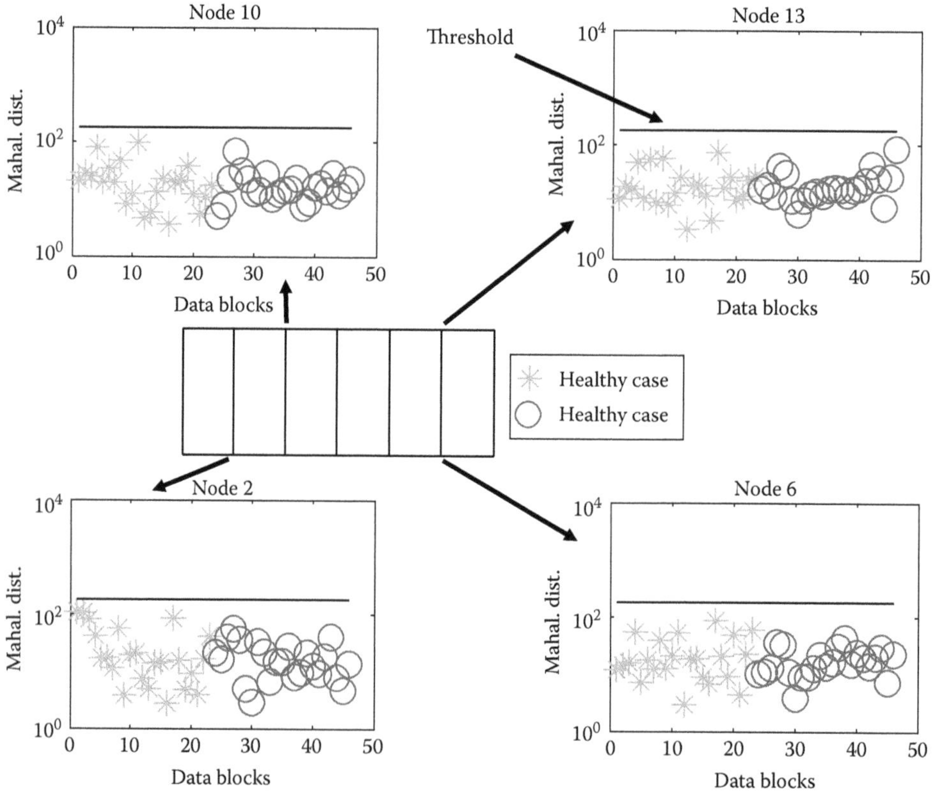

Figure 12.28 Verification of the threshold value with experimental data.

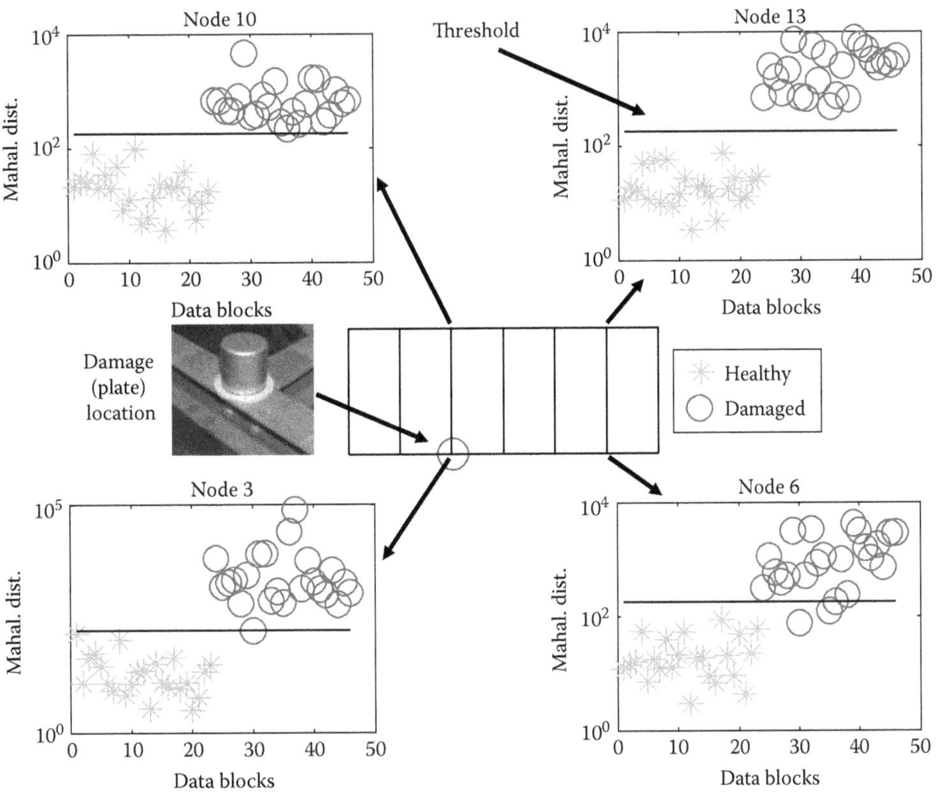

Figure 12.29 Mahalanobis distance plots for DC1.

Damage case 2 (DC2): Boundary restraint at N7 and N14: As for DC2 (Figure 12.30), the majority of the values are still above the threshold value; however, some false negatives are also observed. These results are somewhat surprising since the severe damage at the boundaries should be detected with a smaller number of false negatives.

The results presented in the preceding sections show that the methodology is capable of detecting changes in the test structures for most of the cases. However, the methodology does not provide enough information to locate the damage. It should also be noted that there are a number of issues to be solved before the methodology can be successfully applied to real-life structures in an automated SHM system.

For example, it was noted that determining the right threshold value is one of the important issues to solve. The threshold value depends on both the length of the feature vector and the number of the features in the vector. Therefore, a different threshold value might be obtained when a different model order is used since the AR model order determines the length of the feature vector. It was also noted that determining this threshold is not a trivial problem, and it might require some trial and error process during the monitoring process. If the threshold is set too low, most of the healthy data can be identified as outliers, increasing the number of the false positives. On the contrary, if the threshold value is set too high, data from damaged structure can be classified as inliers, which is not a desired situation, either. Further investigation is needed for demonstration of the threshold value for an automated SHM application. Second, determining the order of the AR

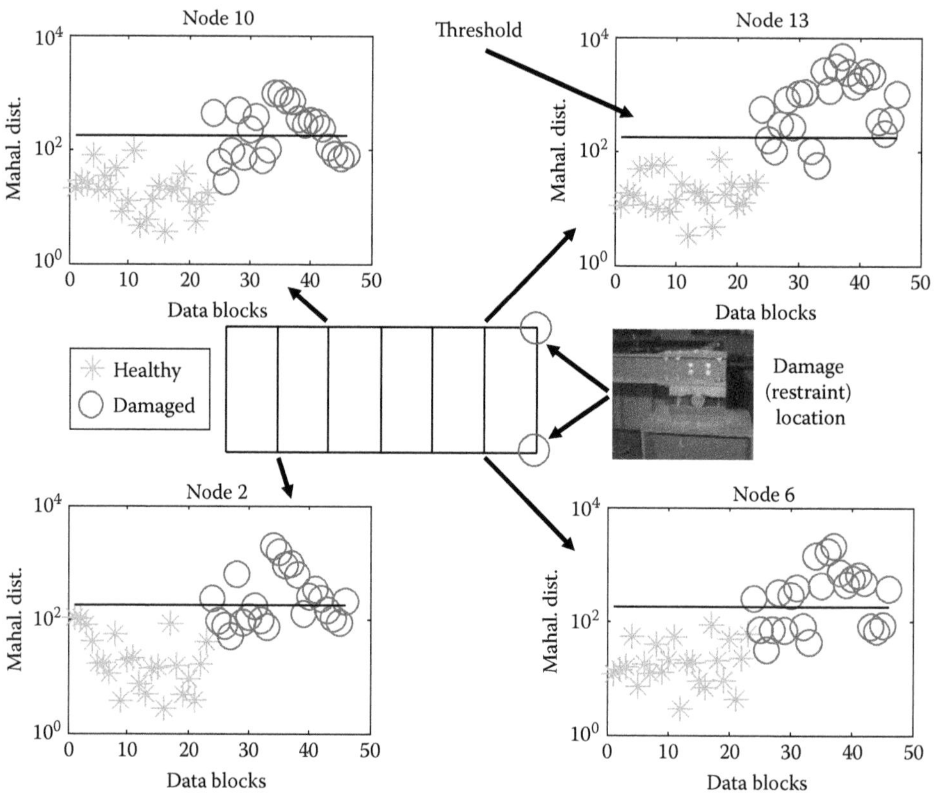

Figure 12.30 Mahalanobis distance plots for DC2.

model in an automated SHM system might be difficult. A high model number might be required for complex structures (and therefore a longer feature vector), and this can make it more difficult to identify the outliers because of "the curse of the dimensionality." For example, it is shown that the model order p for the grid structure is 10, and this number might be quite high for a real-life structure such as a long span bridge.

12.3.3.2 ARX with Sensor Clustering

For this part, the free responses that are obtained from the impact tests are used. The first 100 points (out of 4096 data points), which cover the duration of the impulse, of each data set are removed so that the impact data can be treated as free responses. Five impact data sets were used for each case.

Damage case 1 (DC1): Moment release and plate removal at N3: For DC1, Figure 12.31 shows that the DFs for N3 are considerably higher than the threshold and other nodes. This fact is due to the plate removal at this node. It is also observed from the figure that the DFs for N2 are also relatively high since N2 is the closest neighbor of N3. Finally, the secondary effect of the damage on N5 and N10 is also seen. Therefore, the methodology was very successful at detecting and locating the damage for this experimental damage case. Finally note that the DFs for other nodes are around the threshold showing us that these nodes are not affected significantly from the localized damage.

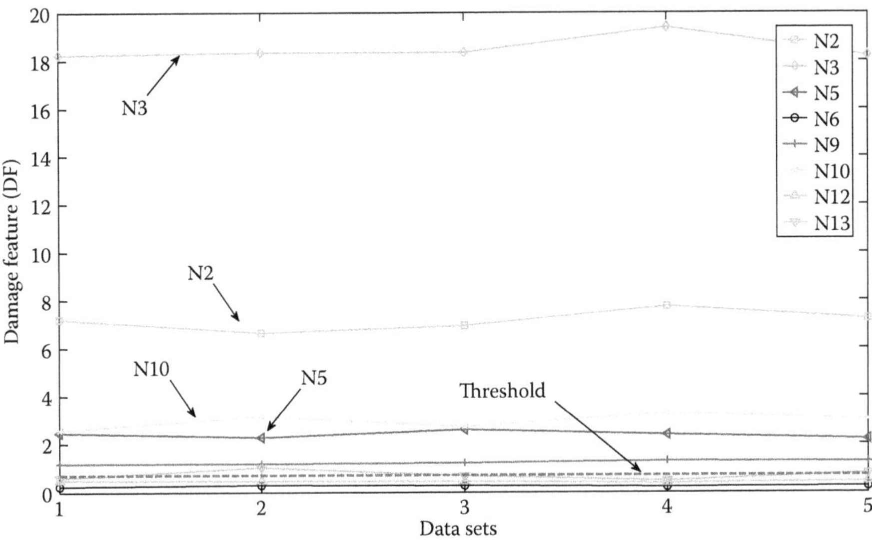

Figure 12.31 DFs for DC1.

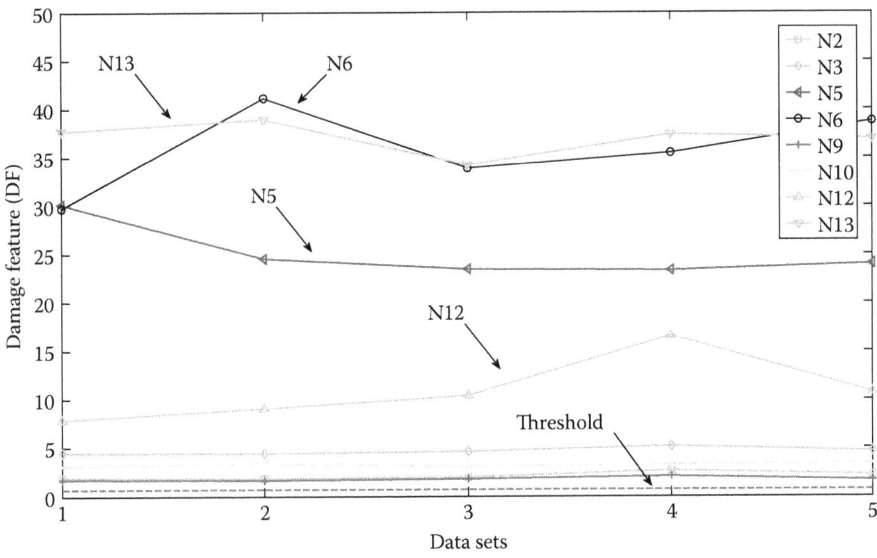

Figure 12.32 DFs for DC2.

Damage case 2 (DC2): Boundary restraint at N7 and N14: The results for the second damage case, DC2, are presented in Figure 12.32. It is seen that the DFs for N6 and N13 are considerably higher than the other nodes. This is because of the fact that they are the closest nodes to the restrained supports (N7 and N14). The DFs for N5 and N12 are also high since they are also affected by the damage. The DFs for the remaining nodes are also slightly higher than the threshold since the structure is changed globally for DC2.

12.4 Summary and Concluding Remarks

The main objective of this chapter is to provide a brief review of different signal processing techniques for damage detection in the context of SHM with applications to CIS. Presented discussions can mainly be summarized in three parts: (1) the evaluation of parametric methods and damage features, (2) investigation of nonparametric methods and features for damage detection, and (3) demonstration of the effectiveness of the methodologies with laboratory experiments.

For the parametric damage evaluation approaches, a general outline for different data analysis and feature extraction methods is discussed. Examples of modal parameter-based parametric damage features such as modal flexibility and modal curvature are presented, and formulations for extracting these parameters from vibration data are presented. Afterward, statistical pattern recognition approaches for *nonparametric damage detection* are presented. Time series analysis along with its implementation with outlier detection for damage detection is discussed. These methodologies can be considered as a complement to commonly employed damage detection methods. A sensor cluster-based time series modeling is also discussed as a powerful damage detection methodology.

After discussing the theoretical background, the performances of these damage detection methods and damage features are exemplified by using experimental data from a steel grid for different damage scenarios. The experimental studies show that these methodologies perform successfully for most of the cases. However, it is also noted that the success rates of the techniques may differ for different cases. One important point here is that a particular data analysis method may not be able to answer every SHM problem. Therefore, a combination of different approaches should be adapted for a successful and reliable SHM system.

Based on the authors' experiences, one very critical challenge in SHM research is the effect of the environmental and operational effects on the structure, which also has critical effects on the data. Damage detection process may easily get very complicated if there is a considerable change in the operational and environmental conditions during the data collection process. Therefore, robust methodologies for elimination of these external effects should be developed and combined with damage detection methodologies.

It is also seen from the results that for an automated SHM system, it might be necessary to set certain rules about the number of outliers so that a decision can be made whether damage has occurred in the structure or not. For example, if a certain number of the new data points are determined as outliers as opposed to a single outlier, then it might indicate a possible structural change where further precautions may be necessary. After determining that a structural change has occurred, more rigorous analyses can be conducted by using different methodologies to determine the location, severity, and the nature of the change.

Finally, it is emphasized that laboratory studies with large, complex, and redundant test specimens are a critical link between the theoretical work and field applications. The reliability of the methods should be verified for damage detection for different cases, loadings, and structures. After making sure that it can be used for a variety of (laboratory and real life) structures under different loading and environmental conditions, the methodology can be implemented to different sensor networks. Embedding these algorithms to different sensor networks for an automated data analysis process will facilitate a better management of CIS in terms of safe and cost-effective operations.

References

Aktan, A. E., F. N. Catbas et al. (2000). Issues in infrastructure health monitoring for management. *Journal of Engineering Mechanics* ASCE **126**(7): 711–724.

Allemang, R. J. (1999). *Vibrations: Experimental Modal Analysis*. Course Notes, Structural Dynamics Laboratory, University of Cincinnati, Cincinnati, OH.

Allemang, R. J. and D. L. Brown (1982). Correlation coefficient for modal vector analysis. *1st International Modal Analysis Conference*, Orlando, FL, pp. 110–116.

Allemang, R. J. and D. L. Brown (1998). A unified matrix polynomial approach to modal identification. *Journal of Sound and Vibration* **211**(3): 301–322.

Alvandi, A. and C. Cremona (2005). Assessment of vibration-based damage identification techniques. *Journal of Sound and Vibration* **292**(1–2): 179–202.

Asmussen, J. C. (1997). *Modal Analysis Based on the Random Decrement Technique-Application to Civil Engineering Structures*. Doctoral dissertation, University of Aalborg, Aalborg, Denmark.

Beck, J. L., M. W. May et al. (1994). Determination of modal parameters from ambient vibration data for structural health monitoring. *Proceedings of the 1st Conference on Structural Control*, Pasadena, CA, pp. TA3:3–TA3:12.

Bernal, D. (2002). Load vectors for damage localization. *Journal of Engineering Mechanics* **128**(1): 7–14.

Bernal, D. and B. Gunes (2004). Flexibility based approach for damage characterization: Benchmark application. *Journal of Engineering Mechanics* **130**(1): 61–70.

Box, G. E., G. M. Jenkins et al. (1994). *Time Series Analysis: Forecasting and Control*. Prentice-Hall, Upper Saddle River, NJ.

Brincker, R., L. Zhang et al. (2000). Modal identification from ambient responses using frequency domain decomposition. *Proceedings of the 8th International Modal Analysis Conference (IMAC)*, San Antonio, TX, pp. 625–630.

Brownjohn, J. M. (2003). Ambient vibration studies for system identification of tall buildings. *Earthquake Engineering and Structural Dynamics* **32**(1): 71–95.

Brownjohn, J. M., S.-C. Tjin et al. (2004). A structural health monitoring paradigm for civil infrastructure. *1st FIG International Symposium on Engineering Surveys for Construction Works and Structural Engineering*, Nottingham, U.K., pp. 1–15.

Burkett, J. L. (2005). Benchmark studies for structural health monitoring using analytical and experimental models. MS thesis, University of Central Florida, Orlando, FL.

Caicedo, J. M., S. J. Dyke et al. (2004). Natural excitation technique and eigensystem realization algorithm for phase I of the IASC-ASCE benchmark problem-simulated data. *Journal of Engineering Mechanics* **130**(1): 49–60.

Carden, E. P. and J. M. Brownjohn (2008). ARMA modelled time-series classification for structural health monitoring of civil infrastructure. *Mechanical Systems and Signal Processing* **22**(2): 295–314.

Carden, E. P. and P. Fanning (2004). Vibration based condition monitoring: A review. *Structural Health Monitoring* **3**(4): 355–377.

Catbas, F. N. and A. E. Aktan (2002). Condition and damage assessment: Issues and some promising indices. *Journal of Structural Engineering, ASCE* **128**(8): 1026–1036.

Catbas, F. N., D. L. Brown et al. (2004). Parameter estimation for multiple-input multiple-output modal analysis of large structures. *Journal of Engineering Mechanics ASCE* **130**(8): 921–930.

Catbas, F. N., D. L. Brown et al. (2006). Use of modal flexibility for damage detection and condition assessment: Case studies and demonstrations on large structures. *Journal of Structural Engineering, ASCE* **132**(11): 1699–1712.

Catbas, F. N., S. K. Ciloglu et al. (2007). Limitations in structural identification of large constructed structures. *Journal of Structural Engineering, ASCE* **133**(8): 1051–1066.

Catbas, F. N., M. Gul et al. (2008a). Conceptual damage-sensitive features for structural health monitoring: Laboratory and field demonstrations. *Mechanical Systems and Signal Processing* **22**(7): 1650–1669.

Catbas, F. N., M. Gul et al. (2008b). Damage assessment using flexibility and flexibility-based curvature for structural health monitoring. *Smart Materials and Structures* **17**(1): 015024 (015012).

Catbas, F. N., M. Lenett et al. (1997). Modal analysis of multi-reference impact test data for steel stringer bridges. *15th International Modal Analysis Conference*, Orlando, FL, pp. 381–391.

Chang, P. C., A. Flatau et al. (2003). Review paper: Health monitoring of civil infrastructure. *Structural Health Monitoring* **2**(3): 257–267.

Chapra, S. C. and R. Canale (2002). *Numerical Methods for Engineers with Software Programming Applications*. McGraw Hill, New York.

Cole, H. A. (1968). On-the-line analysis of random vibrations. *American Institute of Aeronautics and Astronautics* **68**(288).

Doebling, S. W., C. R. Farrar et al. (1996). Damage identification in structures and mechanical systems based on changes in their vibration characteristics: A detailed literature survey. Los Alamos National Laboratory Report No. LA-13070-MS, Los Alamos, NM.

Ewins, D. J. (2000). *Modal Testing, Theory, Practice and Application*. Research Studies Press Ltd, Baldock, U.K.

Farrar, C. R., T. A. Duffey et al. (1999). A statistical pattern recognition paradigm for vibration-based structural health monitoring. *2nd International Workshop on Structural Health Monitoring*, Stanford, CA, pp. 764–773.

Fladung, W. A., A. W. Philips et al. (1997). Specialized parameter estimation algorithm for multiple reference testing. *15th International Modal Analysis Conference*, Orlando, FL, pp. 1078–1085.

Fu, Z.-F. and J. He (2001). *Modal Analysis*. Butterworth-Heinemann, Oxford, U.K.

Gao, Y. and B. F. Spencer (2006). Online damage diagnosis for civil infrastructure employing a flexibility-based approach. *Smart Material and Structures* **15**: 9–19.

Giraldo, D. and S. J. Dyke (2004). Damage localization in benchmark structure considering temperature effects. *7th International Conference on Motion and Vibration Control*, St. Louis, MO, pp. 11

Gul, M. and F. N. Catbas (2008). Ambient vibration data analysis for structural identification and global condition assessment. *Journal of Engineering Mechanics* **134**(8): 650–662.

Gul, M. and F. N. Catbas (2009). Statistical pattern recognition for structural health monitoring using time series modeling: Theory and experimental verifications. *Mechanical Systems and Signal Processing* **23**(7): 2192–2204.

Gul, M. and F. N. Catbas (2011). Structural health monitoring and damage assessment using a novel time series analysis methodology with sensor clustering. *Journal of Sound and Vibration* **330**(6): 1196–1210.

Hogue, T. D., A. E. Aktan et al. (1991). Localized identification of constructed facilities. *Journal of Structural Engineering, ASCE* **117**(1): 128–148.

Holt, R. and J. Hartmann (2008). Adequacy of the U10 & L11 gusset plate designs for the Minnesota bridge No. 9340 (I-35W over the Mississippi River)—Interim report, Federal Highway Administration Turner-Fairbank Highway Research Center, McLean, IL.

Huang, C. S., Yang Y. B. et al. (1999). Dynamic testing and system identification of a multi-span highway bridge. *Earthquake Engineering and Structural Dynamics* **28**(8): 857–878.

Huth, O., G. Feltrin et al. (2005). Damage identification using modal data: Experiences on a prestressed concrete bridge. *Journal of Structural Engineering, ASCE* **131**(12): 1898–1910.

Juang, J.-N. and R. S. Pappa (1985). An eigensystem realization algorithm for modal parameter identification and model reduction. *Journal of Guidance Control and Dynamics* **8**(5): 620–627.

Kao, C. Y. and S.-L. Hung (2003). Detection of structural damage via free vibration responses generated by approximating artificial neural networks. *Computers and Structures* **81**: 2631–2644.

Ljung, L. (1999). *System Identification: Theory for the User*. Prentice-Hall, Upper Saddle River, NJ.

Lynch, J. P., A. Sundararajan et al. (2004). Design of a wireless active sensing unit for structural health monitoring. *Proceedings of the SPIE, NDE for Health Monitoring and Diagnostics*, San Diego, CA, pp. 12.

Maeck, J. and G. De Roeck (1999). Dynamic bending and torsion stiffness derivation from modal curvatures and torsion rates. *Journal of Sound and Vibration* **225**(1): 157–170.

Maeck, J. and G. De Roeck (2003). Damage assessment using vibration analysis on the Z24-bridge. *Mechanical Systems and Signal Processing* **17**(1): 133–142.

Mahalanobis, P. C. (1936). On the generalized distance in statistics. *Proceedings of the National Institute of Sciences of India* **2**: 49–55.

Maia, N. M. M. and J. M. M. Silva (1997). *Theoretical and Experimental Modal Analysis*. Research Studies Press Ltd, Somerset, U.K.

Manson, G., K. Worden et al. (2003). Experimental validation of a structural health monitoring methodology. Part II. Novelty detection on a GNAT aircraft. *Journal of Sound and Vibration* **259**(2): 345–363.

Maxwell, J. C. (1864). On the calculation of the equilibrium and stiffness of frames. *Philosophical Magazine* **27**(4): 294–299.

Nair, K. K. and A. S. Kiremidjian (2005). A comparison of local damage detection algorithms based on statistical processing of vibration measurements. *Proceedings of the 2nd International Conference on Structural Health Monitoring and Intelligent Infrastructure (SHMII)*, Shenzhen, Guangdong, China.

Nair, K. K., A. S. Kiremidjian et al. (2006). Time series-based damage detection and localization algorithm with application to the ASCE benchmark structure. *Journal of Sound and Vibration* **291**(1–2): 349–368.

Omenzetter, P. and J. M. Brownjohn (2006). Application of time series analysis for bridge monitoring. *Smart Material and Structures* **15**: 129–138.

Pandey, A. K., M. Biswas et al. (1991). Damage detection from changes in curvature mode shapes. *Journal of Sound and Vibration* **145**(2): 321–332.

Pandit, S. M. and S. M. Wu (1993). *Time Series and System Analysis with Applications*. Krieger Pub. Co., Malabar, FL.

Peeters, B. and G. De Roeck (2001). Stochastic system identification for operational modal analysis: A review. *Journal of Dynamic Systems, Measurements and Control* **123**: 659–667.

Peeters, B. and C. Ventura (2003). Comparative study of modal analysis techniques for bridge dynamic characteristics. *Mechanical Systems and Signal Processing* **17**(5): 965–988.

Ren, W.-X., T. Zhao et al. (2004). Experimental and analytical modal analysis of steel arch bridge. *Journal of Structural Engineering, ASCE* **130**(7): 1022–1031.

Sanayei, M., E. S. Bell et al. (2006). Damage localization and finite-element model updating using multiresponse NDT data. *Journal of Bridge Engineering* **11**(6): 688–698.

Shih, C. Y., Y. G. Tsuei et al. (1988a). A frequency domain global parameter estimation method for multiple reference frequency response measurements. *Mechanical Systems and Signal Processing* **2**(4): 349–365.

Shih, C. Y., Y. G. Tsuei et al. (1988b). Complex mode indication function and its application to spatial domain parameter estimation. *Mechanical Systems and Signal Processing* **2**(4): 367–377.

Sohn, H., J. A. Czarnecki et al. (2000). Structural health monitoring using statistical process control. *Journal of Structural Engineering, ASCE* **126**(11): 1356–1363.

Sohn, H. and C. R. Farrar (2001). Damage diagnosis using time series analysis of vibration signals. *Smart Materials and Structures* **10**: 1–6.

Sohn, H., C. R. Farrar et al. (2001). Structural health monitoring using statistical pattern recognition techniques. *Journal of Dynamic Systems, Measurement, and Control, ASME* **123**: 706–711.

Sohn, H., C. R. Farrar et al. (2003). A review of structural health monitoring literature: 1996–2001, Los Alamos National Laboratory Report, Los Alamos, NM.

Toksoy, T. and A. E. Aktan (1994). Bridge condition assessment by modal flexibility. *Experimental Mechanics* **34**(3): 271–278.

Van Overschee, P. and B. De Moor (1996). *Subspace Identification for Linear Systems: Theory, Implementation and Applications*. Kluwer Academic Publishers, Dordrecht, the Netherlands.

Worden, K., G. Manson et al. (2000). Damage detection using outlier analysis. *Journal of Sound and Vibration* **229**(3): 647–667.

Yang, J. N., Y. Lei et al. (2004). Identification of natural frequencies and dampings of in situ tall buildings using ambient wind vibration data. *Journal of Engineering Mechanics* **130**(5): 570–577.

Zhang, Q. W. (2007). Statistical damage identification for bridges using ambient vibration data. *Computers and Structures* **85**(7–8): 476–485

Chapter 13

Data Cleaning in Low-Powered Wireless Sensor Networks

Qutub Ali Bakhtiar, Niki Pissinou, and Kia Makki

Contents

13.1 Introduction

A wireless sensor network (WSN) [1] is a special kind of a peer-to-peer (P2P) network where the nodes communicate with the sink wirelessly to transmit the sensed information. In contemporary world, a WSN utilizes different technological advancements in low power communications and very large scale integration (VLSI) [2] to support functionalities of sensing, processing, and communications. WSNs have penetrated in all walks of life ranging from health care to environmental monitoring to defense-related applications.

A WSNs is often followed by an application that uses sensed observations to perform a certain task. For example, a WSN may be designed to gather and process data from the environment to regulate the temperature level of a switching station. A fundamental issue in WSN is the way the data are collected, processed, and used. Specifically, in case of data processing, the initial step is to verify whether the data are correct or conform to a specified set of rules. In this context, data cleaning for WSN arises as a discipline that is concerned with the identification and rectification of errors given the nodes inability to handle complex computations and low energy resources. In a nutshell, data cleaning in WSN can be defined as the process to detect and correct the inaccurate sensor data by replacing, modifying, or deleting the erroneous data.

The challenging aspect of data cleaning in WSN is that the cleaning mechanism is online; that is, the cleaning process is carried out for continuous stream of data generated [3]. After the cleaning process is executed, ideally the data stream becomes consistent, free from any errors, suitable for the applications to use, and make any decisions based on it. However, in spite of technological advances in sensor processing, no perfect data cleaning mechanism exists for WSN so far.

WSNs are generally deployed in environments where nodes are exposed to nonideal conditions as a result of which the nodes' ability to accurately record or relay the information is hampered [4,5]. For example, a WSN may be asked to report the temperature or other environmental attributes of a nuclear reactor where nuclear/electromagnetic radiations pose problems for the nodes to precisely record the observations and relay them. Therefore, before deploying a WSN the possibility of imprecision (or even loss depending on the surroundings) of sensor measurements should be considered. Sometimes, even in case of ideal conditions, sensors may not perfectly observe the environment they monitor. This is due to the hardware inaccuracies, imprecision, and imperfection of the sensing mechanism imparted when the device is manufactured. More specifically in this case, the imperfections represent the inability of technology to perfectly manufacture a sensing device.

WSNs also tend to malfunction because of natural/man-made conditions they operate in. For example, a sensing node monitoring the water level of a levee might get washed away due to floods or a node monitoring wind speed might be blown away due to heavy hurricane winds. Sensor nodes under severe conditions have the tendency to ill perform, and the breakdown of a single node might compromise the overall perception and/or the communication capability of the network. The perception ability of a sensing node corresponds to the extent of exposure [6].

WSN also faces limitations due to its position and time of operation of its constituent nodes. The sensing range of a node might not cover the entire region as required by the application. For example, a temperature monitoring node might partially cover the area it is required to observe. Also, a node cannot be active all the time because of its limited energy constraints. As a result, the sensing operation of the node is activated and deactivated based on a sampling rate defined to make sure that no relevant event is missed. For example, a node monitoring the moving activity of an object might not be active when the object passes through its sensing region. Spatial [7] and temporal [8] coverage in WSNs has been explored in different scenarios, from target tracking [9] to node scheduling [10,11] to cover the aforementioned drawback.

Through this chapter, a paradigm for validating extent of spatiotemporal associations among data sources to enhance data cleaning is discussed by means of pairwise similarity verification techniques. The primary work described in this chapter establishes belief [12] on potential sensor nodes of interest to clean data by combining the time of arrival of data at the sink and extent of spatiotemporal associations among the sensor nodes.

13.2 Background

Data cleaning in WSN essentially suppresses the drawbacks and shortcomings mentioned in the Section 13.1. One of the major challenges of data cleaning in WSN is that the cleaning process has to be performed online on the incoming stream of data, depending on the requirements of the application query. The fundamental features of data cleaning in WSN involve outlier detection, adaptation, and estimation.

An outlier is an error observation whose value lies at an abnormal distance from other samples in an expected data set. Subramaniam et al. [13] in their work classified outliers as distance based or local metric based. Distance-based outliers do not require any prior knowledge of the underlying data distribution, but rather use the intuitive explanation that an outlier is an observation that is sufficiently far from most other observations in the data set. Local metric–based outlier is observed when the comparison between the samples with the neighborhood count differs significantly. Local metric–based outliers utilize the spatial associations between the nodes. Jeffery et al. [14] classified outliers for Radio-Frequency Identification (RFID) systems as false positives—a reported observation that indicates an existence of a tag when it does not exist—and false negatives—an absence of reading where a tag exists, but is not reported. The aim of any data cleaning mechanism is to suppress false positives and regenerate false negatives

Many data cleaning approaches have been presented in literature to cleanse the corrupted data from WSN to deal with outliers and replenish the data stream with an appropriate value.

The first independent data cleaning mechanism, Extensible Sensor stream Processing (ESP) for WSN was proposed by Jeffery et al. [15,16] in collaboration with a research group responsible for developing Stanford data stream management system (STREAM) [17–19]. ESP essentially is a framework for building sensor data cleaning infrastructures for use in pervasive applications. ESP's pipeline approach (as shown in Figure 13.1) has been designed in such a way that it can easily be integrated into STREAM. ESP accommodates declarative queries from STREAM, which performs cleaning based on spatial and temporal characteristics of sensor data. Incoming data stream is passed sequentially to different stages of ESP to remove the unreliable data and generate the lost data.

ESP systemizes the cleaning of sensor stream into a cascade of five declarative programmable stages: point, smooth, merge, arbitrate, and virtualize. Each stage operates on a different aspect of data ranging from finest (single readings) to coarsest (readings from multiple sensors and other data sources). The primary objective of the point stage is to filter out the individual values (e.g., distance-based outliers). Smooth performs the functionality of aggregation to output a processed reading, and then advances the streaming window by one input reading. Merge stage uses spatial commonalities in the input stream from a single type of device and clusters the readings in a single group. The arbitrate stage filters out any conflicts such as duplicate readings, between data streams from different spatial clusters. The virtualize stage combines readings from different types of devices and different spatial clusters.

Sarma [20] along with the developers of ESP proposed a quality estimation mechanism while object-detection data streams are cleaned to overcome the drawback of the sequence of different stages in ESP. The work essentially develops a quality check mechanism in parallel to the pipelined data cleaning processes. Every step of the cleaning mechanism is associated with a quality testing mechanism defined by means of two parameters, confidence and coverage.

Confidence is defined as the measure of evidence of an object being present in a logical area; the evidence typically is the data about the phenomena being sensed provided by different sensors. Coverage is a window-level value assigned to a set of readings for a given time period T. It gives the fraction of readings from a stream representing the actual environment in the time period T.

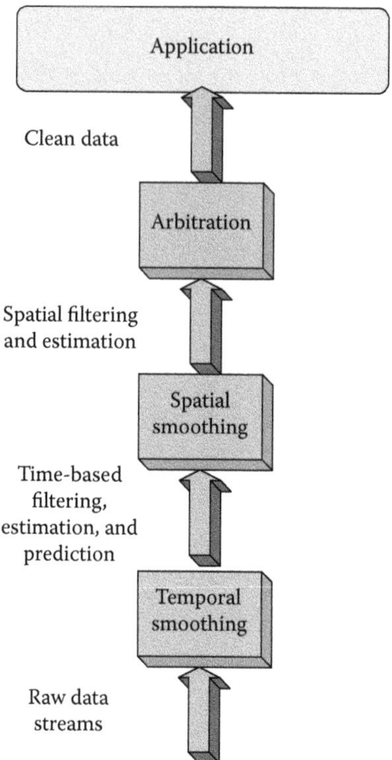

Figure 13.1 Fundamental pipeline stream architecture for data cleaning in wireless sensor networks.

The confidence and coverage parameters are calculated for every instance of the incoming stream based on which the current quality of the cleaning mechanism is determined.

The possibility of incorporating the sampling rate and the variable window size of the data stream into data cleaning process motivated the developers of ESP to propose SMURF [14]. SMURF is an adaptive smoothing filter developed to provide a preamble to ESP's architecture. It models the unreliability of RFID readings by viewing RFID streams as a statistical sample of tags in the physical world, and exploits techniques based on sampling theory to drive its cleaning processes. Through the use of tools such as binomial sampling and estimators, SMURF continuously adapts the smoothing window size in a principled manner to provide accurate RFID data to applications.

SMURF's adaptive algorithm models the RFID data stream [21] as a random sample of the tags in a reader's detection range. It contains two primary cleaning mechanisms aimed at producing accurate data streams for individual tag-ID readings (individual tag cleaning) and providing accurate aggregate estimates over large clusters of multi-tags using Horvitz–Thompson estimators [22]. It focuses more on capturing the data sensed by the network rather than filtering out the anomalies (noise and data losses) in the data. The SUMRF filter adapts and varies the size of the window based on the evaluation of binomial distribution of the observation observed. SMURF is a tool used in ESP to achieve improved efficiency and accuracy while performing data cleaning.

SMURF and ESP have been extended to develop a metaphysical independence layer [23] between the application and the data gathered. The key philosophy behind metaphysical data independence (MDI) is that sensor data are abstracted as data about the physical world; that is, applications interact with a reconstruction of the physical world in the digital world, as if the physical-digital divide does not exist. Essentially, MDI-SMURF is an RFID middleware platform organized as a pipeline of processing stages with an associated uncertainty-tracking shadow pipeline.

MDI-SMURF uses temporal smoothing filters that use statistical framework to correct for dropped readings common in RFID data streams. Additionally, the filter estimates the resulting uncertainty of the cleaned readings. These cleaned readings are then streamed into spatial-SMURF, a module that extends temporal-SMURF's statistical framework to address errors and semantic issues that arise from multiple RFID readers deployed in close proximity.

Zhuang et al. [24] in contrast to ESP presented a smart weighted moving average (WMA) algorithm [25] that collects confident data from sensors according to the weights associated with the nodes. The rationale behind the WMA algorithm is to draw more samples from a particular node that is of great importance to the moving average, and provide a higher confidence weight for this sample, such that this important sample is quickly reflected in the moving average. In order to accomplish the task of extracting the confident data, the sampling rate of the sensors is increased by means of a weighting mechanism dependent on the identification of significant change in the incoming data stream. A change is said to be significant whenever there is enough confidence to prove that the new value exceeds the prediction range. The identified sample, as well as its confidence level, is sent to the sink, and if the sample is finally proved to be in the prediction range, only the proved sample (without attaching the confidence) is sent to the sink.

Elnahrawy and Nath [26–28] along the lines of WMA use Bayesian classifiers [29] to map the problem of learning and utilizing contextual information provided by wireless sensor networks over the problem of learning the parameters of a Bayes classifier. The adapted Bayes classifier is later on used to infer and predict future data expected from WSN. The work proposes a scalable and energy-efficient procedure for online learning of these parameters in network, in a distributed fashion by using learning Bayesian networks. Elnahrawy and Nath use the current readings of the immediate neighbors (spatial) and the last reading of the sensor (temporal) to incorporate the spatial and temporal associations in the data. The authors rightly use Markov models with short dependencies in order to properly model the nonlinear observations provided by the network.

The model initially was not proposed to perform data cleaning. However, a new version of the model was presented by the same authors [28] to identify the uncertainties associated with the data that arise due to random noise, in an online fashion. The approach combines prior knowledge of the true sensor reading, the noise characteristics of sensor, and the observed noise reading in order to obtain a more accurate estimate of the reading. This cleaning step is performed either at the sensor level or at the base station.

Peng et al. [30] combine the concepts of graph theory and business process logic to develop a data cleaning model for RFID systems. The model presented collaboratively sends and receives messages between related nodes, and is capable of detecting and removing false positives and false negatives to clean the data. The work envisages WSN to be a graph and nodes as the vertices of the graph. A small relevant number of vertexes in the network are involved to form a data cleaning cluster. Nodes in the cluster are related by the business processing logic. The P2P-collaborated data cleaning process is divided into three phases: initialization phase, local correction phase, and peer correction phase. In the initialization phase, when one of the tagged items is detected, the information of the detected node will be sent to the previous and next nodes that fall within the path of traversal for the data from the node to the sink. In the local correction phase, the false negatives detected are identified and corrected locally, whereas in the peer correction phase the cleaning is performed using the previous, current, and the next patterns as well as the messages received from other adjacent nodes.

In general, a data cleaning mechanism for WSN attempts to solve the following problem:

Problem definition: Given a set of spatially and temporally related nodes continuously monitoring some distributed attributes. If $x_{n_i}(t)$ is the data received by the sink through node n_i at time instant t then minimize the difference between the actual data representing the environment and the data received at the sink, that is,

$$minimize\|x_{n_i}(t) - \overline{x}_{n_i}(t)\| \tag{13.1}$$

where $\overline{x}_{n_i}(t)$ is the actual data representing the environment.

13.3 Data Cleaning Using TOAD

In Section 13.2, all the data cleaning mechanisms for WSN proposed so far utilize the spatiotemporal associations among the nodes in some way or the other. However, majority of them fail to verify or validate the extent of spatiotemporal associations among them before they are exploited. This deficiency is critical considering factors such as uncertainty in the communicating medium, improbable behavior of the sensing environment, and the low energy resources of the nodes, which are responsible for forcing the data to lose its spatiotemporal associations with other nodes and the past data. Banking on this drawback, Ali et al. [12] presented a data cleaning design which makes sure the validation of spatial and temporal data smoothing techniques in fact facilitates data cleaning.

The main feature of the design is the usage of time of arrival t_k of data. Intuitively, it can be said that if there are delays in data arrival, the conditions are nonconducive for communication and the likelihood of error in the data is high.

Time of arrival identifies the current state of the network and the ability of a node to communicate clean data whereas correlation validation measures the degree of association/disassociation among the nodes, so that appropriate smoothing mechanism is identified and used for cleaning the data.

The data cleaning mechanism referred to as TOAD (Time Of Arrival for Data cleaning)[12] embeds t_k into the spatiotemporal characteristics of data and provides a belief-based mechanism to filter out any anomalies in the data.

What demarcates TOAD from its peers is the presence of a belief mechanism that not only identifies nodes with highest confidence in providing clean data to the sink, but also selects an apt smoothing mechanism based on the confidence exhibited by the incoming stream of data.

13.4 TOAD Architecture

The architecture of TOAD as shown in Figure 13.2 comprises belief component, data smoothing component, rules component, and arbitration component classified according to their functionalities. The rules component is responsible for creating rules and settings that affect the precision of the output of the cleaning mechanism and the methodology adopted for smoothing the data. The data smoothing component has three filters which are responsible to manage the false positives and false negatives [14] that arise in the data stream. Belief component is responsible for selecting the method of cleaning and smoothing the data at a particular instance, spatially, temporally, and spatiotemporally from average (A), temporal (T), and tap-exchange (TE) smoothing filters, respectively. TOAD contains a feedback loop from the three smoothing filters to the belief component

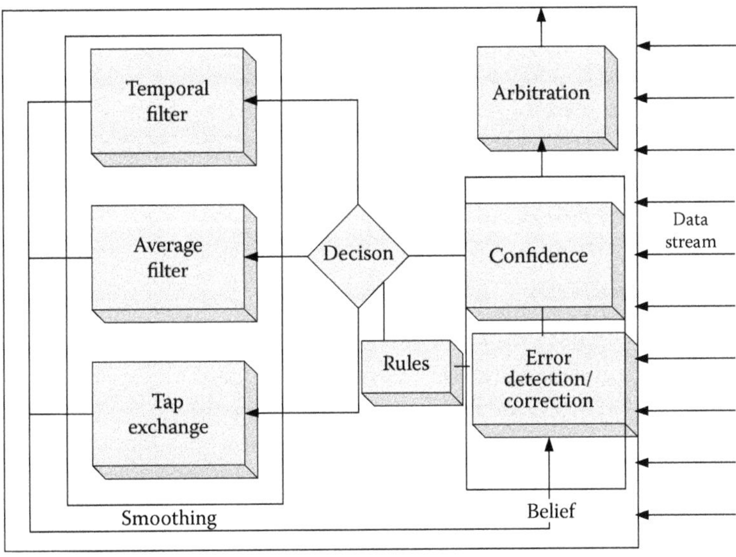

Figure 13.2 TOAD architecture.

in order to compare the predicted/estimated output with the received sample (if available), identify the dirty and lost data, compensate for it and provide a support mechanism in case of data losses. Although there exists a vast literature on spatial and temporal smoothing procedures [16,26,31], to elucidate the ability of TOAD architecture an adaptive filter approach [32,33] similar to the work presented in [34–36] is adopted.

13.5 Preamble

WSN generates a stream of data, wand the cleaning mechanism does not have the liberty of processing the entire data set at a given instance. Techniques based on windowing where a fixed amount of data samples are temporarily stored, processed, and purged after certain time limit are used to process data streams. The size of the window is critical for any data processing, system as there are instances where more samples might be required to achieve better results.

Primarily, there are two techniques for defining the window size:

1. Time based
2. Count based

The time-based window system temporarily stores data for the past few time instances. For example, a window can store all the data samples that reach the sink for the past 5 min. The drawback in this approach is that the space to store the number of samples is indefinite as the availability of them from the past t time instants is not fixed. This poses a major challenge while data processing, as the size of the window dynamically varies. The count-based approach, on the other hand, uses a fixed number of samples in the window at any given instance of time and updates the window for every newly arrived sample. The utilization of time-based and count-based window systems vary from application to application and the sinks storage and processing ability.

13.6 TOAD Components

TOAD uses the count-based window system and sets the window size equal to that of the number of taps of the adaptive filter used for smoothing. Upon the arrival of a new sample from a node, the window slides by deleting the oldest sample in it. Typically, the N-sized window contains the current value and the past $N - 1$ samples of the data stream, and every window is responsible for generating the $(N + 1)$th sample.

13.6.1 Belief Component

The major difference between TOAD and the other data cleaning systems [14,26] is that of verifying the strength of correlation and the measure of the node's ability to send the clean data uniformly. Belief component of TOAD is responsible for incorporating a mechanism that calculates the correlation among the nodes along with their ability to send the data uniformly. The major functionalities of the belief component involve confidence evaluation of a node and error detection (and correction), which are achieved by its respective modules.

13.6.1.1 Confidence Module

The primary objective of the confidence module is to evaluate the degree of belief on each node that contributes for data cleaning, through a variable called belief parameter. Let R_{n_i,n_j} be the correlation coefficient [25] obtained between the previous N samples of the nodes n_i and n_j (N = size of the window and number of taps used in the filter). Let

$$\mathbf{x}_{n_i}(k) = x_{n_i}(k) \quad x_{n_i}(k-1) \quad \ldots \quad x_{n_i}(k-N+1) \tag{13.2}$$

be the previous N samples at node n_i and

$$\mathbf{x}_{n_j}(k) = x_{n_j}(k) \quad x_{n_j}(k-1) \quad \ldots \quad x_{n_j}(k-N+1) \tag{13.3}$$

be the previous N samples at node n_j, then the value of correlation coefficient is given by

$$\begin{aligned} \mathbf{R}(n_i, n_j) &= \frac{\mathbf{cov}(\mathbf{x}_{n_i}, \mathbf{x}_{n_j})}{\sigma_{n_i}\sigma_{n_j}} \\ &= \frac{E(\mathbf{x}_{n_i}, \mathbf{x}_{n_j}) - E(\mathbf{x}_{n_i})E(\mathbf{x}_{n_j})}{\sqrt{E(\mathbf{x}_{n_i}^2) - E^2(\mathbf{x}_{n_i})}\sqrt{E(\mathbf{x}_{n_j}^2) - E^2(\mathbf{x}_{n_j})}} \end{aligned} \tag{13.4}$$

where $\mathbf{cov}(\mathbf{x}_{n_i}, \mathbf{x}_{n_j})$ = covariance between nodes $\mathbf{x}_{n_i}, \mathbf{x}_{n_j}$

$$\begin{aligned} E(\mathbf{x}_{n_i}) &= \frac{\sum \mathbf{x}_{n_i}(k)}{N} \\ E(\mathbf{x}_{n_i}, \mathbf{x}_{n_j}) &= \frac{\sum \mathbf{x}_{n_i}(k)\mathbf{x}_{n_j}(k)^T}{N} \\ \sigma_{n_i}(\text{standard deviation}) &= \sqrt{E(\mathbf{x}_{n_i}^2) - E^2(\mathbf{x}_{n_i})} \end{aligned} \tag{13.5}$$

The value of the correlation coefficient has two properties that represent the degree of similarity between the nodes[25]:

1. Magnitude: The magnitude of the correlation coefficient determines the extent of a relationship between the two entities that are being compared. The maximum value of magnitude is 1 (highly correlated) and the minimum is 0 (uncorrelated).
2. Sign: The direction of relationship is determined by the sign of the correlation coefficient. If the correlation coefficient is positive then the samples in the windows are related in a linearly increasing order, whereas if the correlation coefficient is negative then the samples in the windows are related in a linearly decreasing order.

From Equations 13.4 and 13.5, it is imperative that the value of the correlation coefficient $\mathbf{R}(n_i, n_j)$ tends to a nondeterministic form if the standard deviation in any of the windows containing the samples from the nodes is approximately equal to 0. Low standard deviation implies low variance, which means that the value of samples in the window hardly changes (possess a constant value). If the standard deviation of any window is approximately equal to 0, then TOAD assigns 0 to the correlation coefficient and subsequently 0 to the belief parameter.

In order to accommodate inconsistencies in the time of arrival of data from the nodes, TOAD uses a time consistency variable ρ, which is defined as recurring sum of the ratio of difference between the waiting interval and the sampling interval of the data. The waiting interval here is the time difference of arrival of two consecutive samples. The value of ρ is maintained for every node and is updated whenever a new sample arrives. Let δ be the sampling interval, that is, the time duration after which each sample is sensed and expected to reach the sink, and t_k be time of arrival of the kth sample at any node in a real-time scenario; the time consistency variable of a node can then be defined as

$$\begin{aligned}
\rho_{n_i}(k) &= \rho_{n_i}(k-1) + \frac{abs(|t_k - t_{k-1}| - \delta)}{\delta} \\
\rho_{n_i}(k-1) &= \rho_{n_i}(k-2) + \frac{abs(|t_{k-1} - t_{k-2}| - \delta)}{\delta} \\
&\;\;\vdots \qquad\qquad\qquad \vdots \\
\rho_{n_i}(1) &= 1 + \frac{abs(|t_1 - t_0| - \delta)}{\delta}
\end{aligned} \tag{13.6}$$

where $t_k - t_{k-1}$ is the waiting time spent by the module for the sample to arrive at the kth instance. ρ_{n_i} also depicts the ability of a node to consistently relay the data at a fixed rate because of its recursive nature. Also, ρ provides deviation of actual arrival rate of data from the expected arrival rate which at times can also be used to assess the communication inaccuracies. Larger the value of ρ_{n_i}, less the reliability of the data received by the node and vice versa. Whenever a new sample arrives, the time consistency variable is reset to 1 to inform the system that the sample containing the actual information about the environment has arrived from a specific node.

The belief parameter is defined as the ratio of the correlation coefficient and the time consistency variable ρ_{n_i}. Based on the value of the belief parameter, one can identify the node that is correlated and consistent to provide clean data to the sink:

$$\beta_{n_i,n_j} = \frac{\mathbf{R}(n_i, n_j)}{\rho_{n_j}} \tag{13.7}$$

The belief component uses sample windows and time consistency variables of nodes sensing the same attribute to calculate the degree of belief.

13.6.1.2 Error Detection and Correction Module

The other major functionality of the belief module is to detect and rectify the corrupted data (outliers) that are present in the data stream. Although there exist many formal definitions of outliers, TOAD uses distance-based outliers and local metric–based outliers as described in [13].

If a sample of a data stream lies outside the feasible zone as specified by the application, then it is labeled as a distance-based outlier. The distance-based outliers can be handled well if the end application/user has prior information about the accurate threshold range that the samples of the data stream lie in. For example, the temperature of a steel furnace cannot reach 0 degrees during its period of operation.

A sample of the data stream is termed to be a local metric–based outlier if its value differs from the mean value of samples of similar nodes, significantly. The similarity between two nodes in TOAD is calculated by using the correlation coefficient stated in Equation 13.4. Typically, to rectify these outliers spatial relationships are used.

To identify local metric-based outliers, the belief module calculates the mean of the predicted values of similar nodes from the spatial module and compares with that of the incoming sample. If the difference between them is significant, then the sample is termed to be a local metric-based outlier. These kinds of outliers can be detected if and only if there is more than one highly correlated node. The threshold for the difference can be regulated by the rules component of the system.

13.6.2 Smoothing Component

Depending on the value of the belief parameter, the data stream smoothing process is selected from different smoothing mechanisms (temporal, average, and tap exchange).

Table 13.1 provides the range of the belief parameter and the corresponding smoothing module employed for data cleaning. The range specified in Table 13.1 is calculated on the basis of existing literature on correlation coefficient specified in [25].

Table 13.1 Module Identification

Belief Range	*Module Selected*
$0.8 \leq \lvert\beta\rvert \leq 1$	Spatial
$0.5 \leq \lvert\beta\rvert < 0.8$	Spatiotemporal
$\lvert\beta\rvert < 0.5$	Temporal

13.6.2.1 Temporal Filter

To include the facility of data cleaning temporally, TOAD uses adaptive filters [32]. Each filter contains a window to accommodate N samples of the data stream and each sample is associated with a weight $w_{n_i}(k)$ that is updated whenever the window slides. The sliding window is count based that is, whenever a new sample arrives, the oldest sample in the window is deleted and the position of existing samples is shifted by one place. The adaptive filter as shown in Figure 13.3 uses the current sample $x_{n_i}(k)$ and the previous $N-1$ samples to generate the output $y_{n_i}(k)$ that is then compared to $x_{n_i}(k+1)$ to find the error $e(k)$. The error is then used by the adaptive algorithm to familiarize with the incoming data stream. In case of any data losses, the error is kept constant and the output $y_{n_i}(k)$) is fed back as the input. Let the filter deployed contain $\boldsymbol{N}$ taps:

$$\mathbf{w}_{n_i}(k) = w^1_{n_i}(k) \quad w^2_{n_i}(k) \quad . \quad . \quad . \quad w^N_{n-i}(k), i = 1, \ldots, p \tag{13.8}$$

where

$\mathbf{w}_{n_i}(k)$ is the set of window taps at the k_{th} input sample
n_i is the node identification variable

Let the desired data be denoted by $x_{n_i}(k+1)$ which is the $(k+1)$th instance of the sensed data during adaptation. The error is given by $e(k) = x_{n_i}(k+1) - y_{n_i}(k)$, where $y_{n_i}(k) = \mathbf{w}_{n_i}(k)\mathbf{x}^T_{n_i}(k)$ is the output of the filter. Table 13.2 provides details of some adaptive algorithms* that are typically deployed in machine learning systems to capture temporal relationships among the data. At the physical layer of any communication system, these algorithms are used to minimize interference between the signals [37]. The stability, robustness, and performance of these algorithms have already been proven [32,38] and demonstrated. The algorithms mentioned in Table 13.2 can also be used in TOAD to exploit the temporal relationships between the sensed data by updating window taps $w(k)$ after the arrival of each sample from the nodes.

Each filter temporally evolves by generating weight vectors according to the equations specified in Table 13.2. The weights essentially capture the pattern followed by the past data and vary their

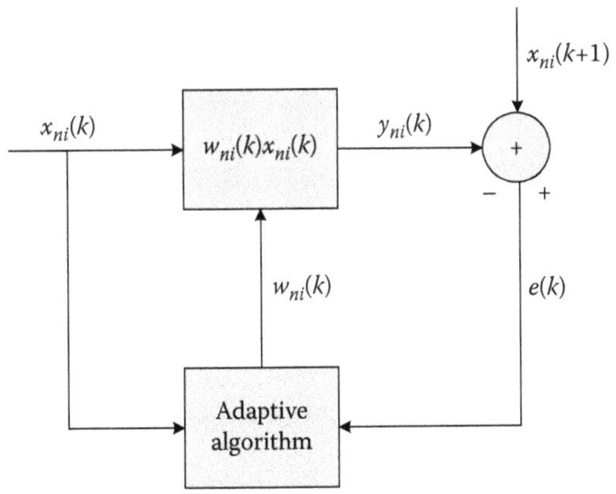

Figure 13.3 Adaptive filter. (From Ali, B.Q. et al., *Wireless Commun. Mobile Comput.*, 12(5), 406, 2012. With permission.)

* The reader is requested to refer the work by Simon Haykin [32,33] for further details about the algorithms mentioned in Table 13.2

Table 13.2 Adaptive Algorithms

Algorithm	*Tap Update Function*
Least mean square	$\mathbf{w}_{n_i}(k+1) = \mathbf{w}_{n_i}(k) + \mu e(k)\mathbf{x}_{n_i}(k)$ $e(k) = [x_{n_i}(k+1) - y_{n_i}(k)]$
Normalized least mean square	$\mathbf{w}_{n_i}(k+1) = \mathbf{w}_{n_i}(k) + \frac{\mu_n}{\|\mathbf{x}_{n_i}\|^2} e(k)\mathbf{x}_{n_i}(k)$ $e(k) = [x_{n_i}(k+1) - y_{n_i}(k)]$
Recursive least square	$\mathbf{K}(k) = \frac{\xi^{-1}\mathbf{P}[k-1]\mathbf{x}(k)}{1+\xi^{-1}\mathbf{x}^H(k)\mathbf{P}[k-1]\mathbf{x}(k)}$ $\mathbf{P}[k] = \xi^{-1}\mathbf{P}[k-1] + \rho^{-1}\mathbf{K}(k)\mathbf{x}(k)$ $\mathbf{w}_{n_i}(k+1) = \mathbf{w}_{n_i}(k) + \mathbf{K}(k)e(k)$ $e(k) = [x_{n_i}(k+1) - y_{n_i}(k)]$

values according to characteristics of the data stream. If the weights vary minimally then the filter is said to have converged. It is at this instance that the error between the estimated and the desired output is approximately equal to 0.

By using adaptive filters, the window of samples used to calculate the belief parameter can also be used by filters to forecast the future sample.

The temporal module is used when the belief parameter is less than 0.5. If a false positive is detected then the corrupted sample at that instance is replaced by the output of the filter. During false negatives, the filter performs prediction by replacing the null values by the output of the filter. The error $e(k)$ is kept constant as the sample $x_{n_i}(k+1)$ is unavailable. Also, during prediction $y_{n_i}(k)$ is fed back to the filter as the input data are unavailable. However, the algorithms alone are not in a position to exploit the spatial relationships among the data independently.

13.6.2.2 Average Filter

When the belief parameter is greater than 0.8, averaging is used for cleaning the data in TOAD. Averaging aggregates data in the space dimension utilizing readings from highly correlated sensors monitoring the same logical area and environment. The false positives and false negatives are replaced by the sample from the node which has the highest belief parameter. The correlation coefficient gives the strength and direction of linear relationships between two different nodes. The higher the value of the belief parameter, the higher the correlation between the nodes.

If a false negative or a positive is needed to be replaced by the value generated by the average filter, then the mean of the difference between the windows of samples is evaluated. The mean is then added to the sample of the corresponding node to get the required sample. The high-level description of the methodology is specified in Algorithm 13.1.

Algorithm 13.1 Spatial module process of smoothing the data stream

Calculate $\beta_{n_i,n_j}\ \forall j = 1..p$
Sort $\beta_{n_i,n_j}\ \forall j = 1..p$
Calculate $\phi = \text{mean}(x_{n_i}(k) - x_{n_j}(k))$ where $k = p...(p+N)$
$x_{n_i}(k+1) = x_{n_j}(k+1) + \phi$

13.6.2.3 Tap-Exchange

Not all burst of samples in the data stream are related spatially and temporally. There are situations when the spatial and temporal relationships among the data have to be used simultaneously for data estimation. The tap-exchange module of TOAD provides such a functionality when the belief parameter lies in between 0.5 and 0.8 (from Table 13.1). TOAD exploits the temporal associations by linear adaptive filters, and introduces the spatial association in the design by exchanging the weights associated with each sample within the window of the temporal filter as shown in Figure 13.4. Essentially, the weights represent changes in the environment that is being sensed. The high-level description of the algorithm is stated in Algorithm 13.2.

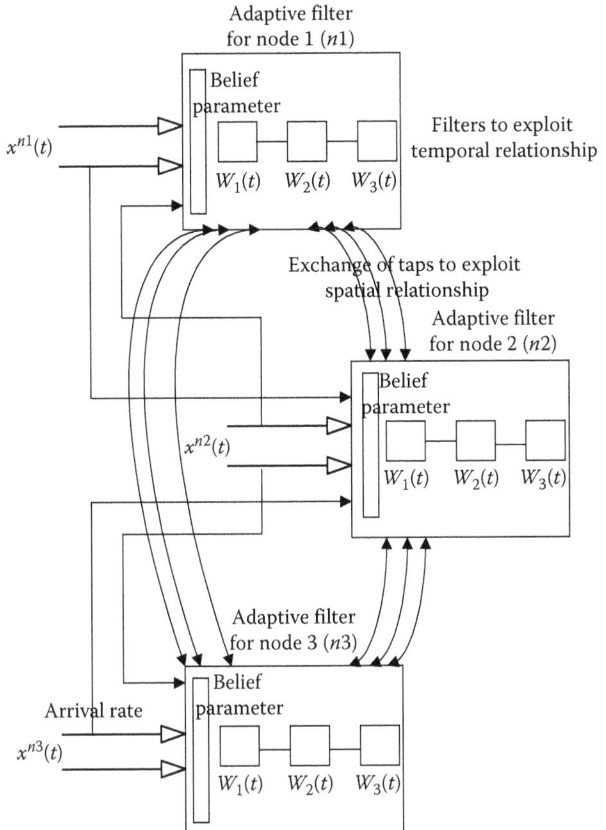

Figure 13.4 Tap-exchange smoothing process.

Algorithm 13.2 Spatiotemporal module

1: Identify n_j with highest β_{n_i,n_j} $\forall j = 1...p$
2: Replace $\mathbf{w}^{n_i}(k)$ with $\mathbf{w}^{n_j}(k)$
3: Estimated output: $y_{n_i}(k) = \mathbf{w}_{n_i}(k)\mathbf{x}^{n_i}(k-1-l)$

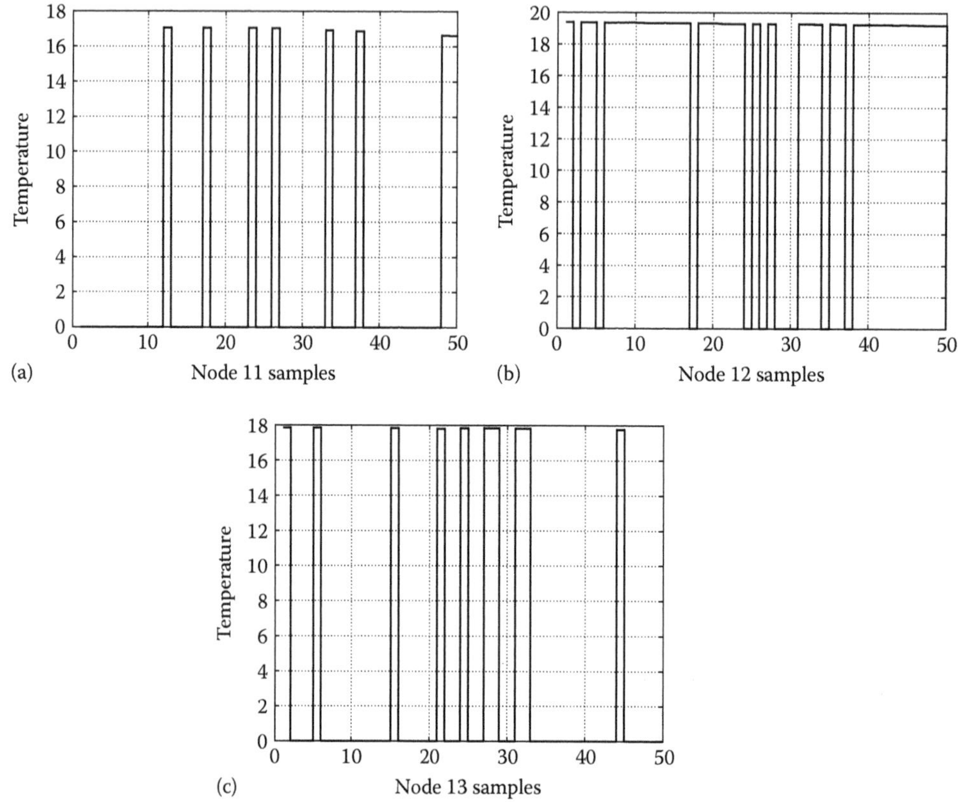

Figure 13.5 Predominant data loss in the sensed data. (a) Node 11 data loss, (b) node 12 data loss, and (c) node 13 data loss. (From Ali, B.Q. et al., *Wireless Commun. Mobile Comput.*, 12(5), 406, 2012. With permission.)

13.6.3 Arbitration

Arbitration component avoids the possible conflicts between the readings from different sensing nodes that are physically close to one another and sensing the same attribute. Any possible aggregation functionality in the request by the application is taken care by this component. Also, arbitration component filters out contradictions, such as duplicate readings, between data streams from different spatially related nodes.

13.7 Case Study

In this section, data cleaning mechanism TOAD is tested by means of simulation on real-time data sets provided by Intel Labs at MIT [39].

Nodes 11, 12, and 13 that measure temperature of the same room (Figure 13.6) for a duration of 3 h and 13 min from 1:12 AM to 4:25 AM on February 28, 2004, are taken into consideration. From Figure 13.7, it is evident that data from the nodes 11, 12, and 13 although spatially correlated do not change linearly with each other and any assumption made to utilize this spatial relation results in estimating an incorrect output.

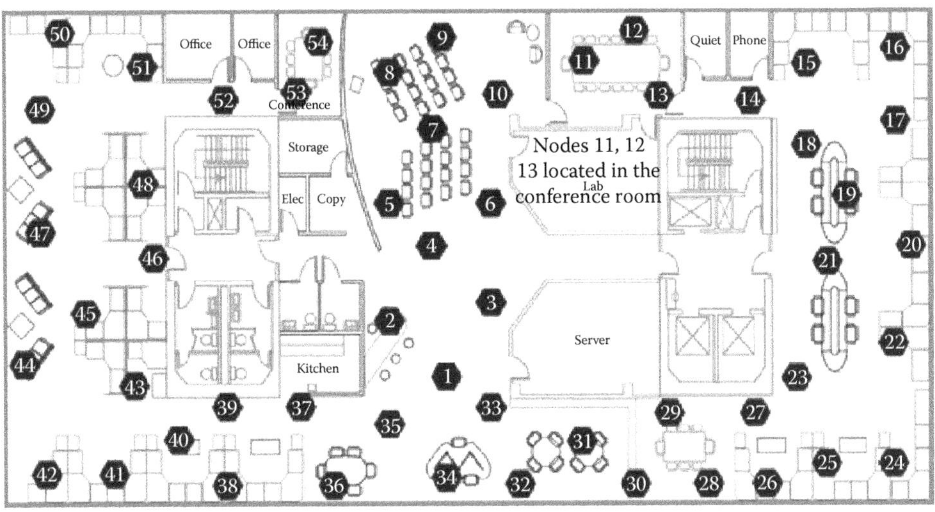

Figure 13.6 MIT sensor test bed. (From Ali, B.Q. et al., *Wireless Commun. Mobile Comput.*, 12(5), 406, 2012. With permission.)

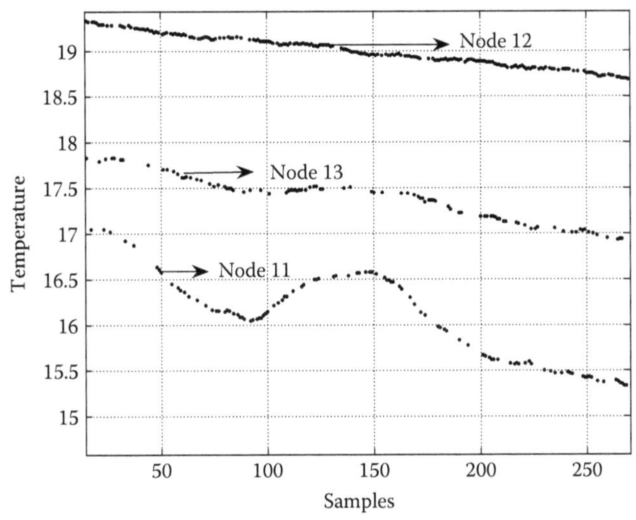

Figure 13.7 Real-time data from nodes 11, 12, and 13 from MIT Lab. (From Ali, B.Q. et al., *Wireless Commun. Mobile Comput.*, 12(5), 406, 2012. With permission.)

Table 13.3 illustrates the calculation process of the belief parameter and selection of the smoothing process used to clean the data. The rules component (shown in Figure 13.2) is used to set the value of the belief parameter based on which decisions regarding the selection of the smoothing process are made.

The correlation coefficient between nodes 11 and 12, and 11 and 13 is represented by $R_{11,12}$ and $R_{11,13}$ whereas ρ_{12} and ρ_{13} represent the time consistency variables (Equation 13.6). The belief parameter defined as the ratio of the correlation coefficient R and the time consistency variable

Table 13.3 Selection Process of the Cleaning Methodology

$x_{n_{11}}$	$R_{11,12}$	ρ_{12}	$R_{11,13}$	ρ_{13}	$\beta_{11,12}$	$\beta_{11,13}$	*A*	*TE*	*T*
15.4706	−0.3483	2.1000	0.7258	4.0000	−0.1659	0.1814			√
15.5082	−0.2288	1.0000	0.6830	1.0000	−0.2288	0.6830		√	
15.6222	0.2459	2.0000	0.9983	2.0000	0.1230	0.4991			√
15.4706	0.5706	1.0000	0.9970	3.0000	0.5706	0.3323		√	
15.3540	−0.0421	2.0000	0.9182	4.0000	−0.0210	0.2295			√
15.7504	0.3120	1.0000	0.9489	1.0000	0.3120	0.9489	√		
15.4804	0.2362	1.0333	0.9512	2.0000	0.2285	0.4756			√
15.4706	0.1319	1.0000	0.9718	1.0000	0.1319	0.9718	√		
15.4608	0.1611	1.0000	0.9534	2.0000	0.1611	0.4767			√
15.5127	0.3412	2.0000	0.9854	1.0000	0.1706	0.9854	√		
15.4097	0.7949	1.0000	0.8334	1.0000	0.7949	0.8334	√		

Note: A, averaging; TE, tap exchange; T, temporal.

ρ is calculated using Equation 13.7. According to measurements made at the Intel Lab [39], only 42% of the estimated data reached the sink, although the environment was indoor and controlled, unreliable wireless communication and low energy resources resulted in huge loss of data. The aforementioned information can be corroborated by examining Figures 13.7 and 13.5.

Figures 13.8 and 13.9 illustrate the behavior of correlation coefficient, variation in time consistency, and fluctuation of belief parameter observed when TOAD is used to clean the data from node 11 using nodes 12 and 13.

Correlation coefficient as shown in Figures 13.8a and 13.9a varies between ± 1 with respect to the incoming data streams from nodes 11, 12, and 13. Equation 13.4 reaches a nondeterministic form when the variance in the participating windows is very low. In such cases, TOAD assigns the value of the correlation coefficient and subsequently of the belief parameter to be 0; that is, it selects the temporal smoothing mechanism for cleaning (from Table 13.1). Essentially low variance implies that there is a minimal change between successive values of the window implying that the samples are near constant within the window. Under these circumstances, temporal smoothing is the best-suited method for generating the corrupted data as the error between the previous sample and the present sample is very low.

Figures 13.9b, 13.11b, and 13.12b display the value of time consistency variable ρ for nodes 11, 12, and 13. The value of ρ increases when the time interval between the successive arrival of samples increases. The maximum value of ρ attained is for node 11 when there is a gap of 32 samples suggesting that the communication is not reliable from it and the chances that the data are corrupted or lost are more. As soon as a new sample is arrived, the time consistency variable is reset to 1 (from Equation 13.6), imparting some confidence to the node in providing clean data. The plot of ρ for node 12 (Figure 13.12b) is more dense around value 1 making its data more reliable to be considered for spatial cleaning, whereas the data from the other nodes are scattered

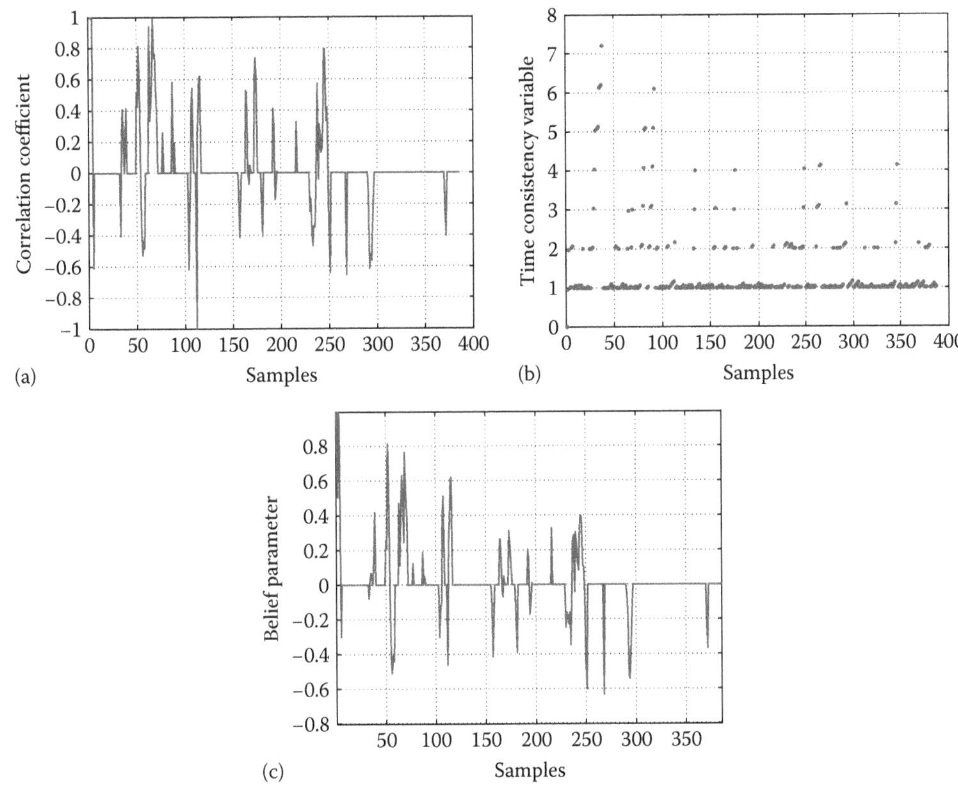

Figure 13.8 Evolution of belief parameter from correlation coefficient and time consistency variable for node 11 with respect to node 12. (a) Node 11 correlation with node 12, (b) time consistency variable of node 12, and (c) belief parameter. (From Ali, B.Q. et al., *Wireless Commun. Mobile Comput.*, 12(5), 406, 2012. With permission.)

throughout, conveying the likelihood of unreliability to be more. Through time consistency variable one can infer that there is some problem during communication but not the cause of the problem.

Figures 13.8c and 13.9c depict the behavior of belief parameters for node 11 with respect to nodes 12 and 13. The plots convey the confidence of nodes 12 and 13 upon 11 to be considered for different smoothing mechanisms used for cleaning. The value of the belief parameter lies between ± 1 similar to that of the correlation coefficient.

Figure 13.10 displays that TOAD suppresses the false negatives in the data from node 11 by dynamically selecting either of the smoothing mechanisms discussed in Section 13.4. Data losses in Figure 13.10 can be inferred by locating the absence of less dense "*" marker in the curve.

The correlation coefficient, time consistency variable, and the belief parameters for node 13 with respect to nodes 11 and 12 are shown in Figures 13.11 and 13.12. From Figure 13.13, similar to that of node 11, it is evident that the false negatives are suppressed by TOAD using an appropriate smoothing mechanism.

An interesting observation can be made by examining Figures 13.9a and 13.11a, depicting the behavior of correlation coefficient, which are similar due to the commutative property of covariance;

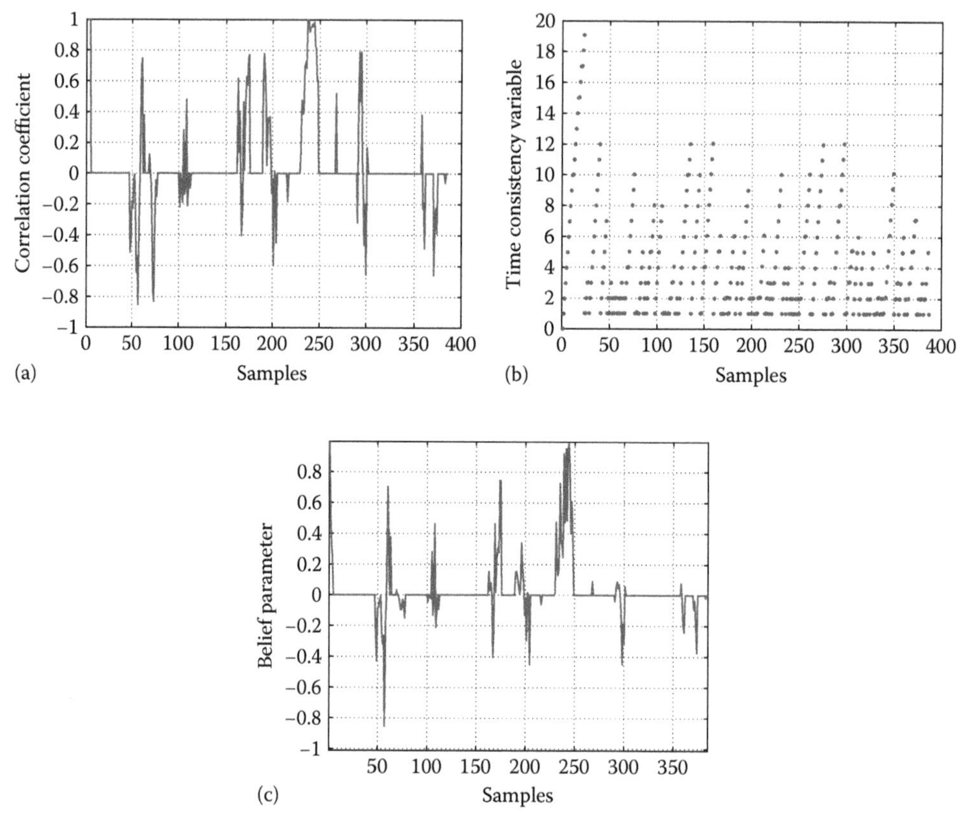

Figure 13.9 Evolution of belief parameter from correlation coefficient and time consistency variable for node 11 with respect to node 13. (a) Node 11 correlation with node 13, (b) time consistency variable of node 13, and (c) Belief parameter.

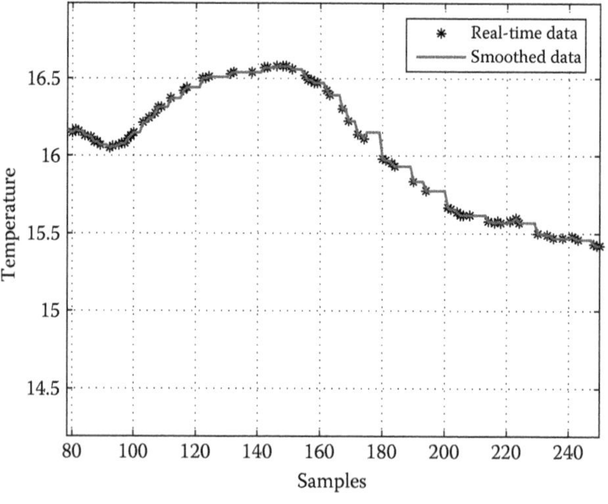

Figure 13.10 Clean data after smoothing for node 11.

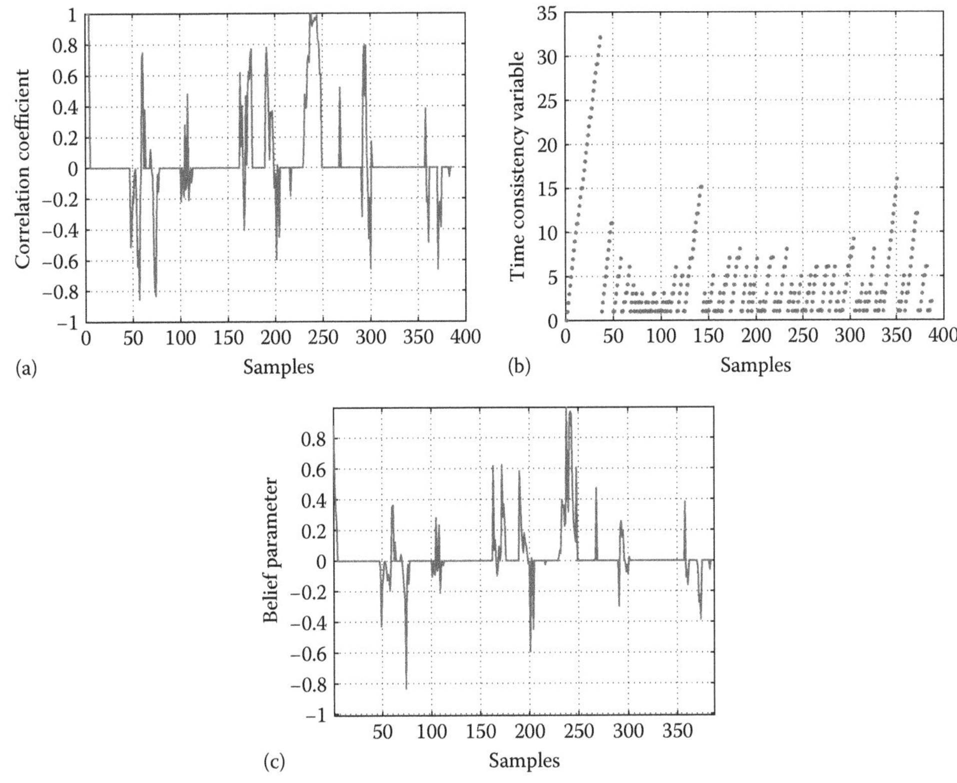

Figure 13.11 Evolution of belief parameter from correlation coefficient and time consistency variable for node 13 with respect to node 11. (a) Node 13 correlation with node 11, (b) time consistency variable of node 11, and (c) belief parameter.

however, the belief parameters of nodes 11 and 13 vary with respect to each other because of the different time consistency variables.

The previous simulations illustrated the effectiveness of TOAD in handling false negatives (i.e., data losses). However, TOAD has the facility to ensure that the false positives or corrupted data (distance-based as well as the metric-based outliers [13]) are suppressed. A sample is termed to be a distance-based outlier if it falls outside the range that is specified by the rules component. It is not always possible to have the exact outlier range available, hence distance-based outlier identification works well if the application has the prior information about the approximate behavior of the input data stream.

To identify and rectify the metric-based outliers, TOAD compares the estimated output of the spatial module with that of the received sample whenever the belief parameter is within the acceptable range. If the error after comparison exceeds the threshold specified by the rules component, then the incoming sample is termed as an outlier and the estimated data from the spatial module are used to reinstate the data stream.

To further test TOAD incoming data stream of node 11 randomly corrupted, a range of 10°–20° is specified to detect the distance-based outliers, and the belief parameter is set to 0.8–1 in order to detect the metric-based outliers. Figure 13.14 depicts that the incoming data stream from node 11 is corrupted and the samples are scattered all over the plot, whereas TOAD cleans

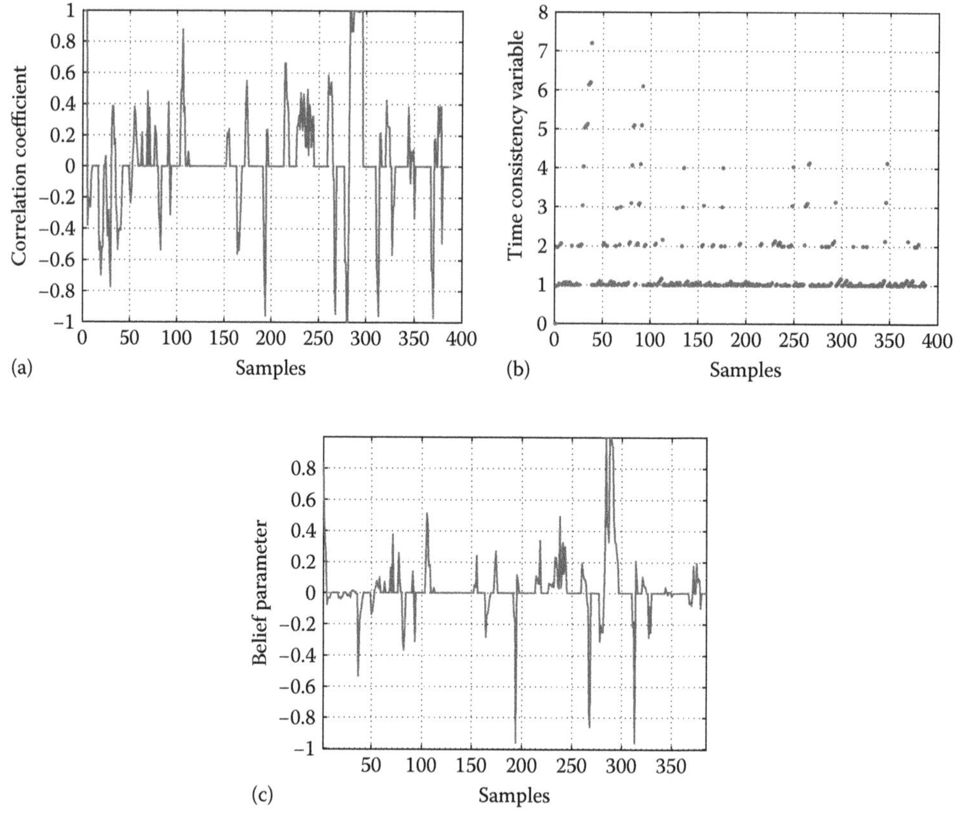

Figure 13.12 Evolution of belief parameter from correlation coefficient and time consistency variable for node 13 with respect to node 12. (a) Node 13 correlation with node 12, (b) time consistency variable of node 12, and (c) belief parameter.

and smoothes the data using its design and architecture. However, there are some instances such as the outlier falling within the range specified, when TOAD is unable to suppress the outliers. When Figures 13.8c, 13.9c and 13.14 are compared, we observe the zones where the belief parameter range is high, the data cleaning is achieved in a better fashion than the zones where the belief parameter is low. If the belief parameter attains high values, then even the outliers that lie within the distance-based range can be identified and suppressed as shown in the Figure 13.14.

13.8 Summary

Due to their pervasive nature and ability to characterize an environment, WSN s have begin to influence the day-to-day activity of human society. As a result, massive amounts of data reflecting the behavior and dynamism of the sensed environment is gathered and processed. The end application utilizes this data and takes an appropriate action. In case the data gathered is corrupted, missed, or inconsistent with the actual reality, the application errs in making the right decision.

Data cleaning in WSN arises as in demand research topic which makes sure that the data gathered from WSN is consistent, full, and clean. Although data cleaning serves a critical role

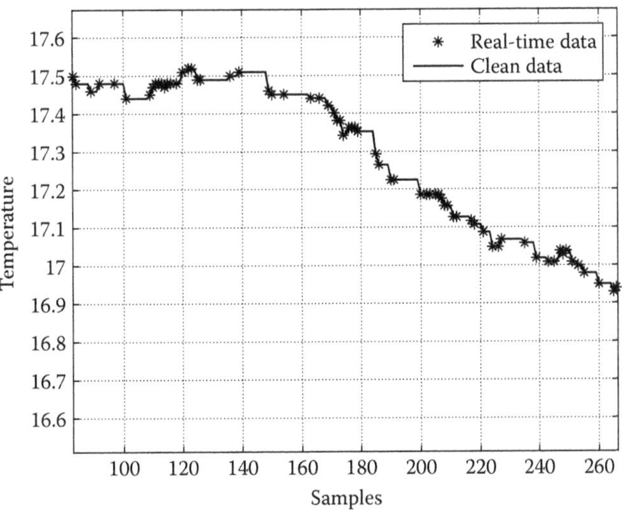

Figure 13.13 Clean data after smoothing for node 13.

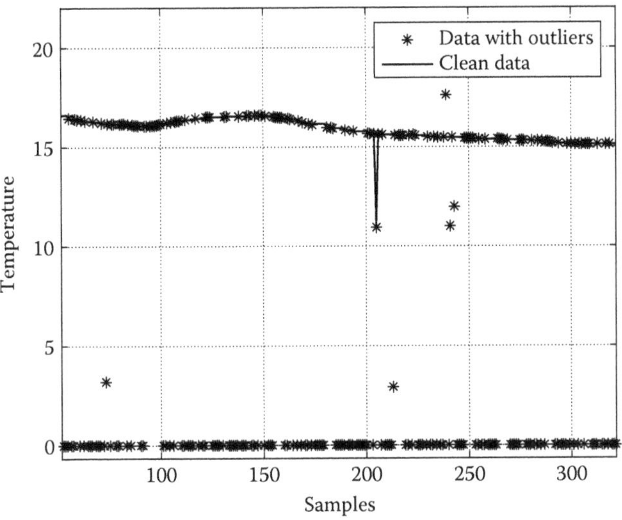

Figure 13.14 Outlier detection and rectification for node 11.

in determining the effectiveness of WSN, until today there is no generic model present in the literature that performs well in the varying and challenging environmental conditions. This chapter is an attempt to summarize different present day data cleaning mechanisms in WSN and discuss its pros and cons. Also an indepth description of a data cleaning mechanism TOAD which uses time of arrival of data in conjunction with validation of spatiotemporal associations within the sensor nodes is presented. The novelty of TOAD when compared to the other data cleaning mechanisms is that it accommodates the changes in the spatiotemporal associations among the nodes of WSN and uses it to clean the gathered data.

References

1. I. F. Akyildiz, S. Weilian, Y. Sankarasubramaniam, and E. Cayirci. A survey on sensor networks. *IEEE Communication Magazine*, 12, 102–114, August 2002.
2. R. Min, M. Bhardwaj, S. H. Cho, and E. S. A. Sinha. Low-power wireless sensor networks. In *In VLSI Design*, pp. 205–210, 2001.
3. J. Gama and M. M. Gaber, Eds., *Learning from Data Streams*. Springer, Berlin, Germany, 2007.
4. I. F. Akyildiz and I. H. Kasimoglu, Wireless sensor and actor networks: Research challenges, *Journal of Ad Hoc Networks*, 2, pp. 351–363, October 2004.
5. I. Akyildiz, Grand challenges for wireless sensor networks. In *MSWiM '05: Proceedings of the Eighth ACM International Symposium on Modeling, Analysis and Simulation of Wireless and Mobile Systems*, New York, pp. 142–142, 2005.
6. S. Megerian, F. Koushanfar, G. Qu, G. Veltri, and M. Potkonjak. Exposure in wireless sensor networks: Theory and practical solutions. *Wireless Networks*, 8(5), 443–454, 2002.
7. S. Meguerdichian, S. Slijepcevic, V. Karayan, and M. Potkonjak. Localized algorithms in wireless ad-hoc networks: Location discovery and sensor exposure. In *MobiHoc '01: Proceedings of the Second ACM International Symposium on Mobile Ad Hoc Networking & Computing*, New York, pp. 106–116, 2001.
8. K. Romer. Temporal message ordering in wireless sensor networks. In *IFIP Mediterranean Workshop on Ad-Hoc Networks*, Tunisia, Mahdia, pp. 131–142, June 2003.
9. X. Yu, K. Niyogi, S. Mehrotra, and N. Venkatasubramanian. Adaptive target tracking in sensor networks. In *Communication Networks and Distributed Systems Modeling and Simulation Conference (CNDS04)*, San Diego, CA, 2004.
10. S. Gandham, M. Dawande, and R. Prakash. Link scheduling in wireless sensor networks: Distributed edge-coloring revisited. *Journal of Parallel Distribution and Computing*, 68(8), 1122–1134, 2008.
11. M. Moges and T. Robertazzi. Wireless sensor networks: Scheduling for measurement and data reporting. *IEEE Transactions on, Aerospace and Electronic Systems*, 42, 327–340, January 2006.
12. B. Q. Ali, N. Pissinou, and K. Makki. Belief based data cleaning for wireless sensor networks. *Wireless Communications and Mobile Computing*, 12(5), 406–409, April 2012.
13. S. Subramaniam, T. Palpanas, D. Papadopoulos, V. Kalogeraki, and D. Gunopulos. Online outlier detection in sensor data using non-parametric models. In *VLDB* Seoul, Korea, 2006.
14. S. R. Jeffery, M. Garofalakis, and M. J. Franklin. Adaptive cleaning for RFID data streams. In *Proceedings of the 32nd international conference on Very large data bases (VLDB '06)*, pp. 163–174, 2006.
15. S. R. Jeffery, G. Alonso, M. J. Franklin, W. Hong, and J. Widom. Declarative support for sensor data cleaning. In *Proceedings of the 4th international conference on Pervasive Computing (PERVASIVE'06). Lecture Notes in Computer Science*, 3968, 83–100, 2006.
16. S. R. Jeffery, G. Alonso, M. J. Franklin, W. Hong, and J. Widom. A pipelined framework for online cleaning of sensor data streams. In *Proceedings of the 22nd International Conference on Data Engineering*, pp. 140–140, April 2006.
17. A. Arasu, B. Babcock, S. Babu, M. Datar, K. Ito, I. Nishizawa, J. Rosenstein, and J. Widom, Stream: The stanford stream data manager (demonstration description). In *SIGMOD '03: Proceedings of the 2003 ACM SIGMOD International Conference on Management of Data*, New York, pp. 665–665, ACM, 2003.
18. R. Motwani, J. Widom, B. B. A. Arasu, M. D. S. Babu, C. O. G. Manku, J. Rosenstein, and R. Varma. Query processing, resource management, and approximation in a data stream management system. In *CIDR*, 2003.
19. B. Babcock, S. Babu, M. Datar, R. Motwani, and J. Widom. Models and issues in data stream systems. In *PODS '02: Proceedings of the 21st ACM SIGMOD-SIGACT-SIGART Symposium on Principles of Database Systems*, New York, pp. 1–16, 2002.

20. A. D. Sarma, S. R. Jeffery, M. J. Franklin, and J. Widom, Estimating data stream quality for object-detection applications. In *Third International ACM SIGMOD Workshop on Information Quality in Information Systems*, Chicago, IL, 2006.
21. Y. Bai, F. Wang, P. Liu, C. Zaniolo, and S. Liu. RFID data processing with a data stream query language. In *IEEE 23rd International Conference on, Data Engineering, 2007. ICDE 2007*, pp. 1184–1193, April 2007.
22. M. M. S. M. Wosten, M. Boeve, W. Gaastra, B. A. M. Van der Zeijst, and van der Reijst, Y.G. Berger Rate of convergence for asymptotic variance of the Horvitz-Thompson estimator. *Journal of Statistical Planning and Inference*, 74, 149–168, October 1 1998.
23. S. R. Jeffery, M. J. Franklin, and M. Garofalakis. An adaptive RFID middleware for supporting metaphysical data independence. *The VLDB Journal*, 17(2), 265–289, 2008.
24. Y. Zhuang, L. Chen, X. S. Wang, and J. Lian, A weighted moving average-based approach for cleaning sensor data. *ICDCS '07. 27th International Conference Distributed Computing Systems, 2007*, pp. 38–38, June 2007.
25. W. G. Cochran and G. W. Snedecor. *Statistical Methods*. Ames, IA: Iowa State University Press, 1989.
26. E. Elnahrawy and B. Nath, Cleaning and querying noisy sensors. In *ACM International Workshop in Wireless Sensor Networks WSNA03*, September 2003.
27. E. Elnahrawy and B. Nath, Poster abstract: Online data cleaning in wireless sensor networks. In *SenSys '03: Proceedings of the First International Conference on Embedded Networked Sensor Systems*, New York, pp. 294–295, 2003.
28. E. Elnahrawy and B. Nath, Context-aware sensors. In *Proceedings of EWSN. Lecture Notes in Computer Science*, 2004, pp. 77–93.
29. O. Pourret, P. Nam, and B. Marcot, *Bayesian Networks: A Practical Guide to Applications (Statistics in Practice)*. John Wiley & Sons, 2008.
30. X. Peng, Z. Ji, Z. Luo, E. Wong, and C. Tan, A p2p collaborative rfid data cleaning model. *The Third International Conference on, Grid and Pervasive Computing Workshops, 2008. GPC Workshops '08*, pp. 304–309, May 2008.
31. A. Jain, E. Chang, and Y. Wang, Adaptive stream resource management using kalman filters. In *ACM SIGMOD/PODS Conference (SIGMOD 04)*, Paris, France, June 2004.
32. S, Haykin, *Adaptive Filter Theory*, 4th edn. Englewood Cliffs, NJ: Prentice Hall, 2002.
33. S. Haykin, *Least Mean Square Adaptive Filters*. New York, Wiley InterScience, 2003.
34. B. Q. Ali, N. Pissinou, and K. Makki, Approximate replication of data using adaptive filters in wireless sensor networks. In *Third International Symposium on Wireless Pervasive Computing, 2008. ISWPC 2008*, Vol. 7–9, pp. 365–369, May 2008.
35. B. Ali, N. Pissinou, and K. Makki. Estimated replication of data in wireless sensor networks. In *Third International Conference on Communication Systems Software and Middleware and Workshops, 2008, COMSWARE 2008*, pp. 107–110, January 2008.
36. B. Ali, N. Pissinou, and K. Makki. Identification and validation of spatio-temporal associations in wireless sensor networks. In *Third International conference on Sensor Technologies and Applications, 2009*, pp. 496–501, June 2009.
37. J. G. Proakis. *Digital Communications*: McGraw-Hill, New York, 2nd edn. 1989.
38. B. Widrow and J. M. E Hoff. Adaptive switching networks. In *IRE WESCON Convention Record*, pp. 96–104, 1960.
39. I. L. Data. http://db.lcs.mit.edu/labdata/labdata.html

Chapter 14

Sensor Stream Reduction

Andre L.L. Aquino, Paulo R.S. Silva Filho, Elizabeth F. Wanner, and Ricardo A. Rabelo

Contents

14.1 Introduction

This chapter introduces a methodology for sensor stream reduction in wireless sensor networks. This methodology allows the integration of techniques and tools in sensor stream reduction applications. The main hypothesis to use the proposed methodology is as follows:

> An appropriate methodology enables an efficient prototype of new wireless sensor networks applications. This prototype considers the project, analysis, implementation, test, and deployment phases. The applications can involve hardware, software, and networking functions aspects.

Considering a sensor stream reduction application, a basic methodology is composed of four phases: characterization, reduction tools, robustness, and conception. These phases are illustrated in Figure 14.1.

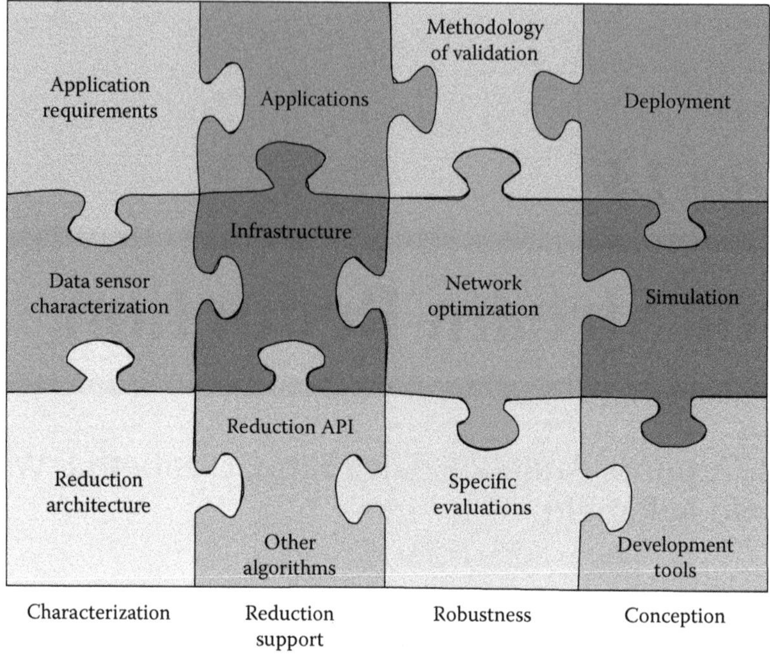

Figure 14.1 Steps to develop a sensor reduction application.

An appropriate methodology must provide (1) a conceptual model to help the reduction application design, that is, the characterization phase; (2) a reduction application programming interface (reduction-API) that can be used in software or hardware, that is, the reduction tools phase; (3) a conceptual model to allow an adequate evaluation of the reduction strategies, that is, the robustness phase; and (4) a friendly user tool to test, develop, and deploy the reduction strategies, that is, the conception phase. A brief description of each phase is presented in the following:

Characterization: This phase provides the requirement list of sensor stream reduction applications and it is applied in the following elements: (1) *reduction architecture* that can be constantly redefined and used in different scenarios; (2) *data sensor characterization* uses models that describe the stream sensed. This characterization is important to identify the data behavior allowing the development of specific and efficient reduction algorithms; and (3) *application requirements* allow to list the application requirements to be considered in the reduction algorithms project, for example, the quality of service (QoS) requirements considered to project a specific reduction algorithms.

Reduction tools: This phase provides support to reduction application development and it is applied in the following elements: (1) *reduction API* contains all reduction algorithms available and it can be combined with network mechanisms or directly in applications; (2) *infrastructure* represents the network mechanisms that can be combined or used with some reduction strategy, for example, clustering based on data sensed; and (3) *applications* that need to execute an explicit reduction strategy, for example, user query applications—in this case, the sensor nodes need to store a reduced metadata used to answer the initial query.

Robustness: This phase provides models that allow the algorithms validation and it is applied in the following elements: (1) *specific evaluations* are used when restricted models are considered to represent a specific data sensed. These evaluations can be applied to obtain a better classification among different algorithms considering a specific application; (2) *network optimization* is necessary when the requirements have to be optimized. In this case, different optimization problems can be investigated to enable the network or algorithm optimization; and (3) *methodology of validation* provides the definition of conceptual models that combine the evaluation of reduction algorithms with optimization models considering the users, infrastructure, or application requirements.

Conception: This phase integrates the simulation and deployment mechanisms and it is applied in the following elements: (1) *development tools* enable an easy and fast application prototypes; (2) *simulation* incorporates the characterized sensor stream, reduction algorithms, and infrastructure mechanisms to the simulators; and (3) *deployment* strategy must be used to allow the solution conception in real environments.

This methodology is an important aspect to be considered during the reduction application development since other applications or solutions do not consider the relation between "data and infrastructure" and neither do they assess adequately the quality of the reduction.

14.2 Sensor Stream Characterization

Wireless sensor network applications, generally, have different configurations, data types, and requirements. Thus, it is important to consider a conceptual model to assist the network managers in the specification of application requirements. In our case, this model will help the sensor stream reduction applications.

The basic element of characterization is the *reduction architecture* proposed by Aquino [1]. This architecture describes a conceptual model that considers general data applications, for example, temperature, pressure or luminosity monitoring. The data reduction process, in this conceptual model, consists of the sensed data characterization, that is, how the data can be collected and represented, and the application specification, that is, what application requirements have to be considered, for example, real-time deadlines, energy consumption, or QoS guarantee.

The other elements related to characterization model of reduction applications are discussed in the following.

Data sensor characterization: To process the data, it is important to know the behavior of phenomenon monitored and how it is performed by the environment samples. Considering this element, the main question is

> *How is the monitored phenomenon characterized and represented in a space-temporal domain?*

Based on this question the following assumptions can be considered:

- The phenomenon is modeled via statistical approaches.
- Using statistical models that describe the monitored data, more efficient and specific algorithms for each phenomenon can be proposed.

A previous knowledge about this question can be seen in Frery et al. [2], which proposes a representation for sensing field used to characterize only one instant of the space-temporal monitored domain.

Application requirements: Other important sets of requirements are determined by reduction application. These requirements—energy consumption, delay, and data quality—allow an efficient data monitoring. Considering this element, the main question is

> *How are the application requirements identified and evaluated to allow an efficient sensor stream reduction?*

Based on this question the following assumptions can be considered:

- The model must provide a tabulation of requirements applications. With this, it is possible to identify "how and where" we can use the reduction algorithms.
- Mathematical models have to be proposed in order to describe the reduction impact considered for each, or for a group, of application requirements.

A previous knowledge about this question can be seen in Aquino et al. [3,4], which shows a tabulation in a real-time application. In this work, a mathematical model to identify the appropriate amount of data to be propagated through the sink is proposed. Moreover, a data-centric routing algorithm is presented to ensure the data delivery at the time.

For instance, consider the general architecture proposed by Aquino [1]. This architecture shows a brief instance of characterization phase applied to wireless sensor network scenario. The way the streams are reduced depends on the moment that the reduction is going to be performed, that is, during sensing or routing streams.

Figure 14.2 [1] shows that the input streams must have been originated by the phenomena (sensing stream) or sent to the sensor node by another node (routing stream). This illustration represents the specific *reduction architecture* element described previously and depicted in Figure 14.1.

The sensing reduction is recommended when the sensor device gets an excessive number of samples, and it cannot be dynamically calibrated to deal with more data than it currently deals with. The routing reduction, in contrast, is performed when the network does not support the

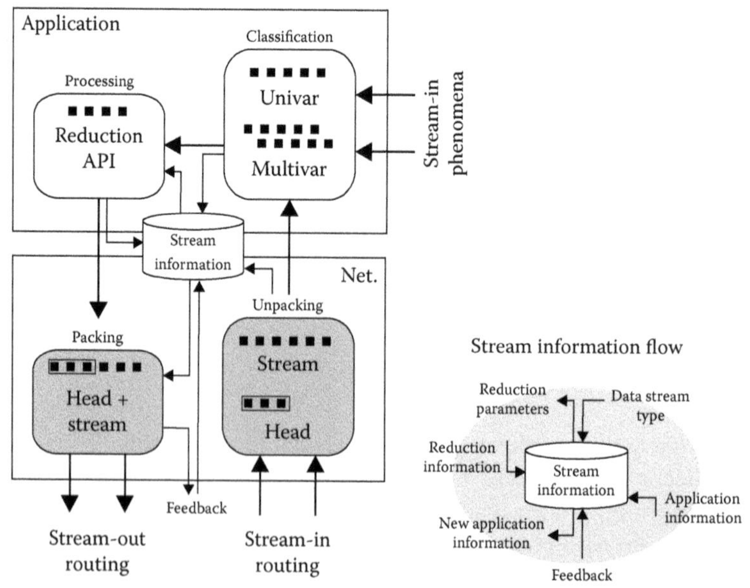

Figure 14.2 General sensor stream architecture.

amount of data being transmitted. This representation defines the *data sensor characterization* element described previously and depicted in Figure 14.1.

For instance, if the application has some requirements regarding the amount of data supported by each sensor node, data stream reduction can avoid uncontrolled data loss while guaranteeing the application requirements. Note that, the sensing stream arrives in the application layer, while the routing stream arrives in the network layer. This identification represents the *application requirements* element described previously and depicted in Figure 14.1.

The data reduction process, represented in Figure 14.2, is described as follows [1]: when data arrive in the network layer, the network packets first need to be unpacked, separating the data stream from the header (that may contain some specific information/restriction of the application). Once it is unpacked, it is sent to the application layer to be processed in the same way as that of the sensing streams. At the same time, stream information is given to a cross-layer (labeled in Figure 14.2 as "stream information") being responsible to make the interface between the application and network layers. The "stream information," highlighted in Figure 14.2, is responsible for choosing which reduction algorithm should be executed and its parameters. The information stored in this cross-layer includes the following:

Feedback: Data received from other sensor nodes in order to perform the reduction calibration in an online fashion.

Application information: Data received from the network layer when the stream is unpacked.

Data stream type: Data received from the application layer after the data stream is classified.

Reduction parameters: Data given to the application layer guiding it to perform an appropriated reduction, considering the "application information" and the "data stream type."

Reduction information: Data received from the application layer after the reduction.

New application information: Data given to the network layer for packing the reduced stream out.

When data streams arrive to the application layer, they first have to be classified according to the number of variables that they monitor. In this context, data streams can be univariate or multivariate. Univariate streams are represented by a set of values read by a unique type of sensor, for example, a sensor node that monitors only environmental temperature. On the other hand, multivariate streams are represented by a set of values coming from different sensors of the same sensor node, for example, a node that monitors temperature, pressure, and humidity simultaneously, or by a set of measurements coming from the same sensor type located in different sensor nodes, for example, a node that processes data from different nodes monitoring only temperature. This classification is important because the data reduction process itself depends directly on the stream type.

After the stream type is known, we have to choose an appropriated stream reduction algorithm to effectively perform the reduction. There are various types of data stream reduction methods, such as online samples, histograms built, and sketches [5,7,9]. The reduction algorithms available in our API will be described in Section 14.3. At the end the reduced data stream is obtained; if the stream was being routed, it is passed back to the network layer, which packs the stream and any information gathered from the cross-layer and sends it to the sink.

In order to illustrate the execution of this specific architecture consider Figure 14.3.

This figure depicts two specific problems: the first one considers a general sensor stream application where a scheduled reduction has to be performed in the source node, that is, it receives data about a phenomena for a certain time and then sends it to the sink; and the second one addresses a real-time application where the reduction is performed during routing.

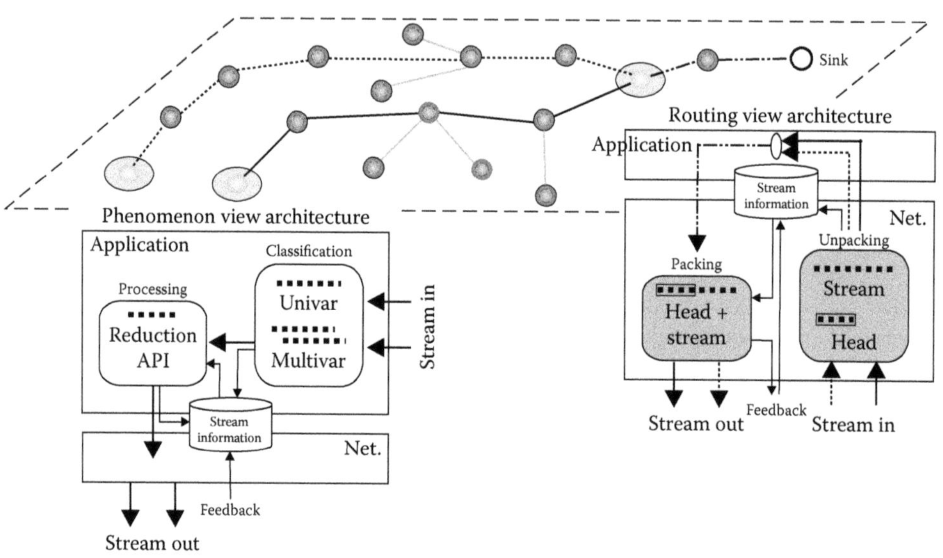

Figure 14.3 General architecture.

14.3 Reduction Tools

A group of reduction solutions must be available to help the application development. Some examples of these solutions are the reduction API, the infrastructure network combined with reduction techniques, and reduction strategies applied in specific applications. These solutions can consider hardware or software implementation.

The basic element of reduction support is the *reduction API* available in the *reduction architecture* proposed by Aquino [1]. The API algorithms consider sampling, sketch, and wavelets stream techniques and they can be applied for univariate or multivariate data stream [6,8,10,11]. Other algorithms can be easily integrated to the API. The general idea is to allow the use of reduction API combined with elements of infrastructure network, for example, routing or density control. In addition, the API is available to applications, for example, query processing, network manager, and software reconfiguration.

The elements related to reduction support are discussed in the following.

Reduction API: Although the reduction API is already available in the methodology, the investigation of new algorithms is always necessary. Considering this element, the main question is

> *Is it possible to improve the data reduction algorithms considering the data representativeness and saving the network resources?*

Based on this question, the following assumptions can be considered:

- By using data reduction algorithms it is possible to save network resources keeping the data representativeness.
- Considering the query processing applications, we can use the data sketched to provide an approximated answer to some query. The data sketched is used instead of sensed data due to the low cost to store the data in the sensor node.

A previous knowledge about this question can be seen in Aquino et al. [3,4,8], which present a set of reduction algorithms based on data stream techniques. The results show that the assumptions described previously can be considered for this algorithm class.

Infrastructure: The network infrastructure uses the data sensed to improve itself. It can be used as the sampled or sketched data. Considering this element, the main question is

Is it possible to use the data reduction to assist the network infrastructure?

Based on this question the following assumption can be considered: The data can be reduced during the data routing based on some network requirements, for example, energy consumption, packet delay, and data quality. In this way, the reduction is performed to achieve these requirements.

A previous knowledge about this question can be seen in Aquino and Nakamura [4], which shows a data-centric routing algorithm for real-time applications.

Applications: As expected the reduction strategies can be used directly in the application layer. Some applications that use the reduction directly are listed as follows:

- *Quality of service:* Among the QoS parameters we can include the data quality parameter. With this, the application has to guarantee a minimum data quality specified by the application. The data are reduced considering a determined data representativeness. Another aspect is the use of reduction to achieve some infrastructure parameters, for example, energy, delay, or packet loss. These parameters are degraded by huge data. The reduction can be used to improve the achievement of these parameters.
- *Real-time applications:* The real-time applications in sensor networks, generally, are soft real-time applications. This occurs because the environment is not controlled and the applications use approximated methods to meet the required deadlines. Thus, we can use the routing combined with the data reduction algorithms to achieve soft deadlines. A previous knowledge about these kinds of applications can be seen in Aquino et al. [3,4,10].

The main aspect of this phase is the *reduction API* element. For instance, the wavelet, sampling, and sketch-based algorithm will be presented.

14.3.1 Wavelet-Based Algorithm

As described in Aquino et al. [8], the wavelet transform of a function $f(x)$ is a two-dimensional function $\gamma(s,\tau)$. The variables s and τ represent the new dimensions, scale, and translation, respectively.

The wavelet transform is composed by functions $\phi(t)$ and $\psi(t)$ in which the admissibility condition $\int \psi(t)\,dt = 0$ holds. This condition states that the wavelet has a compact support, since the biggest part of its value is restricted to a finite interval, which means that it has an exact zero value outside of this interval. Besides, $\psi_{i,k}(t) = 2^{\frac{i}{2}}\psi(2^i - k)$, for the dilated and translated versions of $\psi(t)$. Both these cases characterize the wavelet spacial localization property.

Another property is the wavelet smoothness. Consider the expansion of $\gamma(s,\tau)$, around $\tau = 0$, in a Taylor series of n order,

$$\gamma(s,0) = \frac{1}{\sqrt{s}}\left[\sum_{p=0}^{n} f^{(p)}(0) \int \frac{t^p}{p!}\psi\left(\frac{t}{s}\right)\,dt + O(n+1)\right],$$

where $f^{(p)}$ is the pth derivative of f and $O(n+1)$ represents the rest of the expansion. A wavelet $\psi(t)$ has L null moments if

$$\int_{-\infty}^{\infty} t^k \psi(t)\, dt = 0$$

to $0 \leq k < L$. If

$$M_k = \int t^k \psi(t)\, dt$$

we have

$$\gamma(s,0) = \frac{1}{\sqrt{s}}\left[\frac{f^{(0)}(0)}{0!} M_0 s^1 + \cdots + \frac{f^{(n)}(0)}{n!} M_n s^{n+1} + O(s^{n+2})\right]$$

The vanishing moments are

$$M_n(0,l) = 0 \, for \, l = 0, 1, \ldots, L-1$$

where L is the number of moments of the wavelet. From the admissibility condition, we have that the 0th moment, M_0, is equal to zero. If the other M_n moments are zero, then $\gamma(s,\tau)$ will converge to a smooth function $f(t)$. So, if $f(t)$ is described by a polynomial function of degree up to $(L-1)$, the term $O(s^{n+2})$ will be zero and small values will appear as a linear combination of ψ in the function $\gamma(s,\tau)$.

The regularity degree and decreasing rate of wavelet transforms are related to its number of null moments. This property is important to infer the approximation properties in the multi-resolution space. When a wavelet has various null moments, there will be coefficients with low values. In regions where $f(x)$ is a smooth function, the wavelet coefficients with thin scales in $f(x)$ are null.

The periodic wavelet transforms are applied to the limited interval of functions. In order to apply the periodic transform, it is necessary to consider that the limits of target function are repeated. With the goal of avoiding this condition, the coiflets basis has the null moments property in its scale function. The coiflets basis is an extension of the Daubechies ones having the following properties:

$$\begin{cases} \int \psi(x)dx = 1 \\ \int x^l \psi(x)dx = 0, \quad l = 0, 1, \ldots, L-1 \\ \int x^l \phi(x)dx = 0, \quad l = 0, 1, \ldots, L-1 \end{cases}$$

Due to these properties, ψ and ϕ have null moments. The function ϕ is smoother and more symmetric than the ϕ considered in Daubechies family. Moreover, this function is a better approximation of polynomial functions having interpolation properties. Considering $2L$ the number of coiflets moments and $f(x)$ in $[p,q]$, we have

$$\int_p^q f(x)\phi(x) = \int_p^q f(0) + f'(0) + \cdots$$
$$+ \frac{f^{2L-1}(0)x^{2L-1}}{(2L-1)!} + \cdots \approx f(0)$$

Let $f(x)$ be a polynomial function of order $p \leq 2L - 1$. Then, there is $g(x)$, with the same degree, so that

$$f(x) \approx \sum_{\tau} g(\tau)\phi_{s,\tau}(x)$$

This property states that, during the transformation, some terms can be obtained directly or sampled considering the following approximation error:

$$\left\| f(x) - \sum_{\tau} g(\tau)\phi_{s,\tau}(x) \right\| = O(2^{s.2L}), \tag{14.1}$$

This error has the δ Dirac property $\int f(x)\delta(x)dx = f(0)$ only if it has infinite null moments. Due to this property, the coefficients in $\gamma(s,\tau)$ are sparse representations of $f(x)$ and, consequently, only few coefficients are necessary to approximate $f(x)$. With this, the scale coefficients in $\gamma_{s,\tau}$ can be approximated by sampling of $f(x)$, so that

$$\gamma(s,\tau) = f(2^{-(s)}\tau) + O(2^{s.2L})$$

Considering the Lth order coiflets basis, it is possible to define a fast algorithm using the low error property, when the function $f(t)$ is a smooth function. Each element in $V_j'[t]$ can be approximated by

$$V_j'[t] = 2^{-\frac{i}{2}} f(2^i t)$$

and each approximated wavelet coefficient, $W_t[t]$ at scale 2^i, is computed as

$$W_j[t] = \sum g[2t - n]V_{j+1}'[t]$$

The pseudocode of the wavelet-based algorithm can be seen in Algorithm 14.1 and its analysis of complexity is presented in the following text [8]. In this algorithm, h and g are the discrete forms of ψ and ϕ, respectively; V_j' represents the reduced data, in which j represents the resolution, that is, the sampling rate scale, for example, 2^j, 4^j, and so forth.

If M represents the number of decomposition level, the total number of operations in the data vector with size $|V|$ has a number of operations of order $O(M\,|V|\,log|V|)$. The result of wavelet transform is the signal decomposition in different subspaces, or sub-bands, with different resolutions in time and frequency. Considering the temporal series, they are decomposed in other series that compose the original one.

The main aspects of this sensor stream algorithm can be summarized as follows: (1) the stream item considered is the sensor data buffered V; (2) the online behavior of the algorithm considers one pass in V but the information of the oldest processing is not used; and (3) considering the data representativeness as the most important aspect in our applications, the approximation rate is calculated from Equation 14.1.

14.3.2 Sampling Algorithm

Aquino et al. [6] present a sample-based algorithm that aims to keep the data quality and the sequence of sensor stream. This algorithm provides a solution that allows the balance between best

Algorithm 14.1: Wavelet-based algorithm

Data: V – original data stream
Data: h, g – Coiflets basis filters
Result: V'_j – sampled data stream

Generate filters $h(i)$ and $g(i)$;
for $t \leftarrow 0$ **to** $(|V|/2 - 1)$ **do**
  $u \leftarrow 2t + 1$;
  $V'_{jt} \leftarrow g_1\ V_{u+1}$;
  for $n \leftarrow 1$ **to** $(|h| - 1)$ **do**
    $u \leftarrow u - 1$;
    if $u \leq 0$ **then**
      $u \leftarrow |V| - 1$;
    end
    $V'_{jt} \leftarrow V'_{jt} + g_{n+1}\ V_{u+1}$;
  end
  $V'_{j(t+1)} \leftarrow V'_{jt}$;
end

data quality and network requirements. The sample size can vary, but it must be representative to attend the data similarity requirement.

The sampling algorithm has two versions: random [6] and central [4]. The random one considers the data histogram and chooses the samples randomly in each histogram class. The central one chooses the central elements of each histogram class. The sampling algorithm can be divided into the following steps:

Step 1: Build a histogram of the sensor stream.
Step 2 (*random*): Create a sample based on the histogram obtained in Step 1. To create such sample, we randomly choose the elements of each histogram class, taking into account the sample size and the class frequencies of the histogram. Thus, the resulting sample will be represented by the same histogram.
Step 2 (*central*): Create a sample based on the histogram obtained in Step 1. To create such sample, we choose the central elements of each histogram class, taking into account the sample size and the class frequencies of the histogram. Thus, the resulting sample will be represented by the same histogram.
Step 3: Sort the data sample according to its order in the original data.

These steps are illustrated in Figure 14.4. The original sensor stream is composed of $|V|$ elements. The histogram of the sensor streaming is built in Step 1. A minor histogram is built in Step 2, which has the sample size required, and it keeps the same frequencies of the original histogram. Finally, the minor built histogram is reordered to keep the data sequence in Step 3.

An execution example of sample and central algorithms and their operations is depicted in Figure 14.5 [4]. In the random sampling (a), we have a "stream in V" with 100 elements, $|V'| \rightarrow 50\%$ of V is randomly chosen (this choice is performed in each histogram class), and then a "stream out V'" is generated with $|V'| = 50$. In the central sampling (b), we have again a

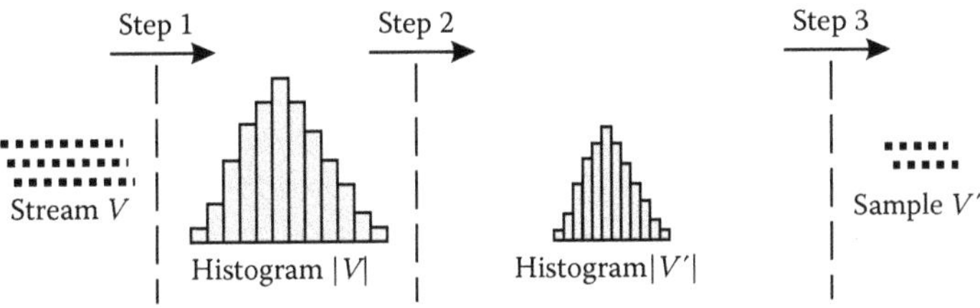

Figure 14.4 Sampling algorithm steps.

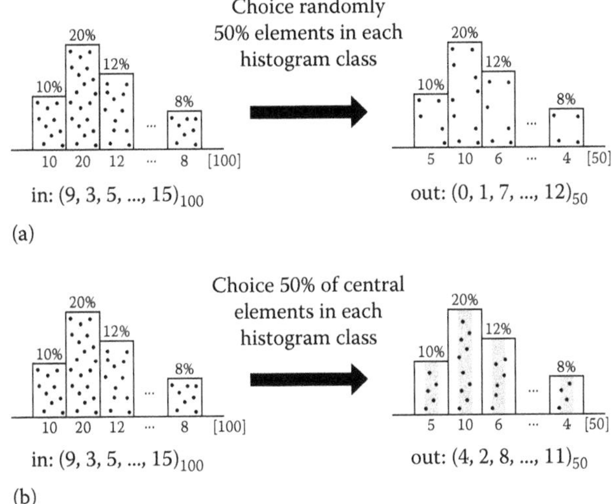

Figure 14.5 Sampling algorithm examples. (a) Random sampling and (b) central sampling.

"stream in V" with 100 elements, $|V'| \rightarrow 50\%$ of V is choice considering the central histogram class elements, and then a "stream out V'" is generated with $|V'| = 50$.

The pseudocode of the sampling algorithm is given in Algorithm 14.2.

In Algorithm 14.2, we have two possibilities for the execution of lines 9 and 13 that represent the choice of samples to compose the sampled data stream V'. In the random version, in both lines we have

$$index \leftarrow Random(pr, pr + n_c'),$$

where the *Random* function returns some integer number between $[pr, pr + n_c']$. In the central version, in line 9 we have

$$index \leftarrow pr + \lceil (n_c - n_c')/2 \rceil$$

and in line 13 we have

$$index \leftarrow index + 1.$$

Algorithm 14.2: Sampling algorithm

```
   Data: V – original data stream
   Data: |V'| – sample size
   Result: V' – sampled data stream
1  Sort(V) {based on data values};
2  lg ← "Histogram classes";
3  pr ← 0 {first index of the first histogram class};
4  n_c ← 0 {number of elements, for each histogram class};
5  k ← 0;
6  for i ← 0 to |V| − 1 do
7      if V[i] > V[pr] + lg or i = |V| − 1 then
8          n'_c ← ⌈n_c |V'|/|V|⌉ {number of elements, for each column in V'};
9          index ← "Index choice following step 2";
10         for j ← 0 to n'_c do
11             V'[k] ← V[index];
12             k ← k + 1;
13             index ← "Index choice following step 2";
14         end
15         n_c ← 0;
16         pr ← i;
17     end
18     n_c ← n_c + 1;
19 end
20 Sort(V') {based on arrival order};
```

Analyzing the Algorithm 14.2 we have

Line 1: Executes in $O(|V| \log |V|)$.
Lines 2–5: Correspond to the initialization of variables.
Lines 10–14: Define the inner loop that determines the number of elements of each histogram class of the resulting sample. Consider H_{cn} as the number of classes of the histograms. The runtime of inner loop is $O(|V'|)$, where in line 11 $n'_c = |V'| \leftrightarrow H_{cn} = 1$, that is, we would have a single class in the histogram of the sample with $|V'|$ elements to be covered.
Lines 6–19: Define the outer loop where the input of data are read and the sample elements are chosen. H_{cn} is the number of histogram classes. Before line 7 is accepted, we execute the outer loop n_c times, which corresponds to the counting of the number of elements of a class of the original histogram (line 18). After the condition in line 7 is accepted, the outer loop is executed n'_c times and the condition in this line is accepted just H_{cn} times. With this, we run $H_{cn}\,(n_c + n'_c)$ for the outer loop. Since $|V| = H_{cn}\,n_c$ and $|V'| = H_{cn}\,n'_c$, we have a runtime for the outer loop of $O(|V| + |V'|)$.
Line 20: Executes in $O(|V'| \log |V'|)$.

Thus, the overall complexity of the sampling algorithm is

$$O(|V| \log |V|) + O(|V| + |V'|) + O(|V'| \log |V'|) = O(|V| \log |V|),$$

since $|V'| \leq |V|$. The space complexity is $O(|V|+|V'|) = O(|V|)$, since we store the input data V and the resulting sample V'. Since each node sends its sample toward the sink, the communication complexity is $O(|V'|D)$, where D is the largest route (in hops) of the network.

14.3.3 Sketch Algorithm

Aquino et al. [6] present a sketch-based algorithm that aims to keep the frequency of the data values without losses, by using a constant packet size. With this information, the data can be generated artificially in the sink node. However, the sketch solution looses the sequence of the sensor stream. The sketch algorithm can be divided into the following steps:

Step 1: Order the data and identify the minimum and maximum values in the sensor stream.
Step 2: Build the data out, only with the histogram frequencies.
Step 3: Mount the sketch stream, with the data out and the information about the histogram.

The execution of algorithm is depicted in Figure 14.6. The original sensor stream is composed of $|V|$ elements. The sensor stream is sorted, and the sketch information is acquired in Step 1. The histogram frequencies are built in Step 2, where $|V'|$ is the number of columns in the histogram. The sketch stream, with the frequencies and sketch information, is created in Step 3.

The pseudocode of the sketch-based algorithm is given in Algorithm 14.3.

Analyzing Algorithm 14.3 we have

Line 1: Executes in $O(|V| \log |V|)$.
Lines 2–6: Correspond to the initialization of variables.
Lines 7–15: Define the loop for the histogram construction and it is executed in $O(|V|)$.

Thus, the overall time complexity is $O(|V| \log |V|) + O(|V|) = O(|V| \log |V|)$. The space complexity is $O(|V|+|V'|) = O(|V|)$ if we store the original data stream and the resulting sketch. Since every source node sends its sketch stream toward the sink, the communication complexity is $O(|V'|D)$, where D is the largest route (in hops) in the network.

14.4 Robustness Evaluations

Once the data are reduced, it is necessary to guarantee its representativeness for each application. Considering the infrastructure operations, some mathematical models are used to provide a good approximation of the solutions presented.

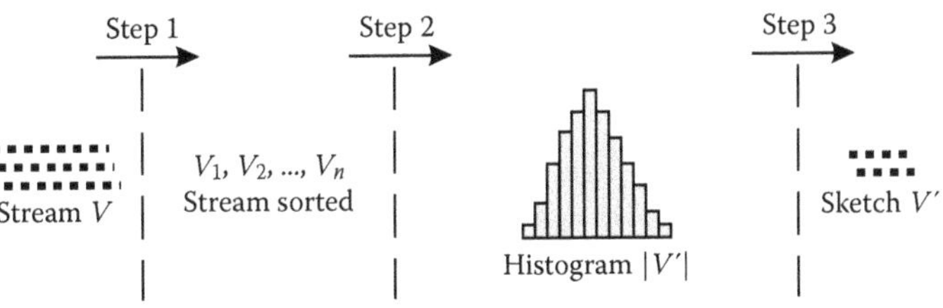

Figure 14.6 Sketch algorithm steps.

Algorithm 14.3: Sketch algorithm

```
Data: V – original data stream
Result: V′ – sketched data stream
Sort(V);
lg ← "Histogram classes";
|V′| ← ⌈(V[|V|] − V[0])/lg⌉;
pr ← V[0] {first index of the first histogram class};
c ← 0 {counter};
index ← 0;
for i ← 0 to |V| − 1 do
    if V[i] > pr + lg or i = |V| − 1 then
        v′[index] ← c;
        index ← index + 1;
        c ← 0;
        pr ← V[i];
    end
    c ← c + 1;
end
```

The basic element of robustness is the *specific evaluations*. For all scenarios considered by Aquino [1], specific evaluations for data quality were performed. These evaluations are required, because when the data are reduced, it is necessary to identify the error generated. Statistical techniques are used to identify the robustness of reduction algorithms. Furthermore, it is important to consider other aspects, for example, mathematical models to optimize some infrastructure or application parameters.

The other elements related to robustness are discussed as follows:

Network optimization: Wireless sensor networks applications always are interested in minimizing the network parameters. When the data reduction is used, the data quality must be maximized. For each network parameter, it is important to determine mathematically the ideal amount of data to be propagated. Considering this element the main question is

How are the reduced data quality and the network parameters economy ensured?

Based on this question the following assumption can be considered: Optimization models can be used to allow the proposition of distributed heuristics for each application.
A previous knowledge about this question, can be seen in Ruela et al. [11,12]. These works consider exact methods, as benders decomposition, and heuristics, as evolutionary algorithms, to tackle the problem. These works do not consider the data reduction as an optimization parameter.

Methodology of validation: Since the data reduction is performed, it is important to identify the error, for each input data and each reduction technique. Considering this element the main question is

Which is the best approach, a specific or a general strategy, to estimate the data quality?

Based on this question the following assumption can be considered: Assuming that we have a good data characterization, the specific strategy is more appropriated. Otherwise, the general strategy is more suitable.

A previous knowledge about this question can be seen in Aquino [1], which presents some strategies based on statistical methods that assess the data representativeness considering the data distribution and data values. Besides, the work proposed by Frery et al. [2] presents a representativeness evaluation based on data reconstruction.

For instance, consider the specific evaluation presented in Aquino [6]. This specific evaluation considers two analyses: (1) the distribution approximation between the original and sampled streams; and (2) the discrepancy of the values in the sampled streams.

The distribution approximation could be identified by using the Kolmogorov–Smirnov test (K-S test) [13,14]. This test evaluates if two samples have similar distributions, and it is not restricted to samples following a normal distribution. The K-S test is described as follows:

1. Build the cumulative distribution Fn of V and V' using the same class.
2. Calculate the differences accumulated for each point and consider the largest ones (D_{max}).
3. Calculate the critical value

$$D_{crit} = y\sqrt{(|V| + |V'|)/|V|\,|V'|}$$

where y is a tabulated value and it represents the significance level.

4. The samples follow the same distribution if

$$D_{max} \leq D_{crit}$$

Consider Figure 14.7 that shows the comparison between the Fn distributions, with $|V| = 256$ and $|V'| = \{\log_2 |V|, |V|/2\}$, where $V' \subset V$. In both cases, through K-S test, V' follows the same distribution of V.

To evaluate the discrepancy of the values in the sampled streams, that is, if they still represent the original stream, the relative absolute error is calculated. The absolute value of the largest distance between the average of the original data and the lower or higher confidence interval values (95%) of the sampled data average is calculated. The average is $\overline{V}$ and the confidence interval values over $\overline{V'}$ is $IC = [v_{inf}; v_{sup}]$. The evaluation steps are described as follows:

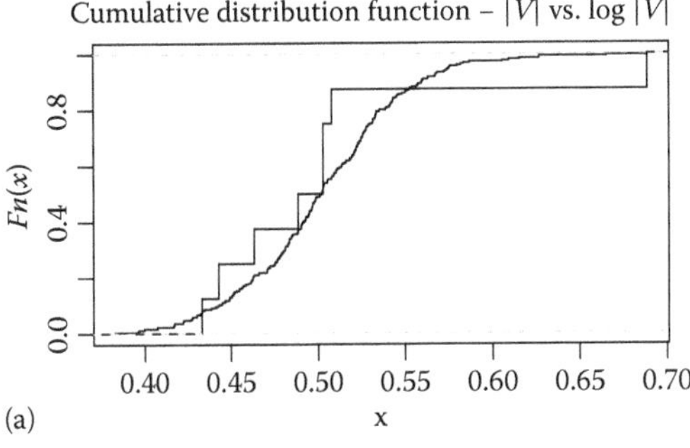

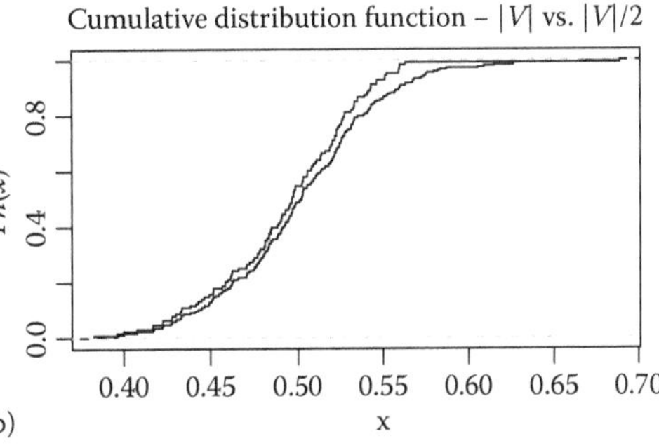

Figure 14.7 Accumulated distribution function. (a) Log of data and (b) half of data.

1. Calculate the average of reduced values and original ones, they are respectively $\overline{V}$ and $\overline{V'}$.
2. Calculate the confidence interval IC with confidence of 95% for $\overline{V'}$.
3. Calculate the relative absolute error

$$\epsilon = \max\{|v_{inf} - \overline{V}|, |v_{sup} - \overline{V}|\}.$$

14.5 Conception by Simulations

The basic element of conception phase is the *development tools*. Lins et al. [15,16] propose a tool for the development of monitoring applications for wireless sensor networks. This tool allows the fast prototype of data reduction applications.

It is necessary some improvements in the simulation tools and the proposition of adequated methodologies to assess the application deployment in real scenarios. The other elements related to robustness are discussed as follows:

Simulation: Several solutions for sensor networks are assessed by simulations, for example, *Network Simulator** and *Sinalgo*.† The reduction solutions must be available in the methodology through these simulators. In this way, it is possible to integrate new solutions and, consequently, to test and to validate them. The simulators must be increased considering, for example, sensed data generators, data traffic generators, and specific simulation traces to allow a better result analysis. Considering only the reduction algorithms, it is important to provide a test tool to assess the data quality separately.

Deployment: Considering the TinyOS operational system,‡ it is possible to develop a basic infrastructure of NesC, allowing an easy NesC-based reduction application. Both, the present solutions and the new ones, can be tested in a real sensor node. Moreover, the development tool proposed by Lins et al. [15,16] could consider the reduction algorithms and solutions allowing the development of reduction applications based on NesC.

The main aspect of this phase is the *simulation* element. Some simulation results and analyses of the wavelet-, sampling-, and sketch-based algorithms are presented in the following text.

14.5.1 Wavelets Algorithm Simulation

Aquino et al. [8] evaluate the impact of the data stream wavelet algorithm Θ to reduce the data sensed V. Considering a previous characterization, two scenarios are considered: with and without an event. Figure 14.8 illustrates these scenarios. To simulate the event, we cause a random perturbation in V so that the values change drastically.

Initially, the wavelet transform with $j = 2$ is considered, using only the V_2' values in which the samples are performed in 2^2 steps, and it is based on coiflets filters and $|V'| = |V|/4$ [8]. In order to compare this strategy, a simple static sampling is used with the same steps, that is, the sampling considers 2^2 static steps instead of the coiflets filters. Figure 14.9 shows the results when there is no external event. The results were sampled of data shown in Figure 14.8a and the proposed strategy

* http://nsnam.isi.edu/nsnam/index.php/Main_Page

† http://dcg.ethz.ch/projects/sinalgo/

‡ http://www.tinyos.net/

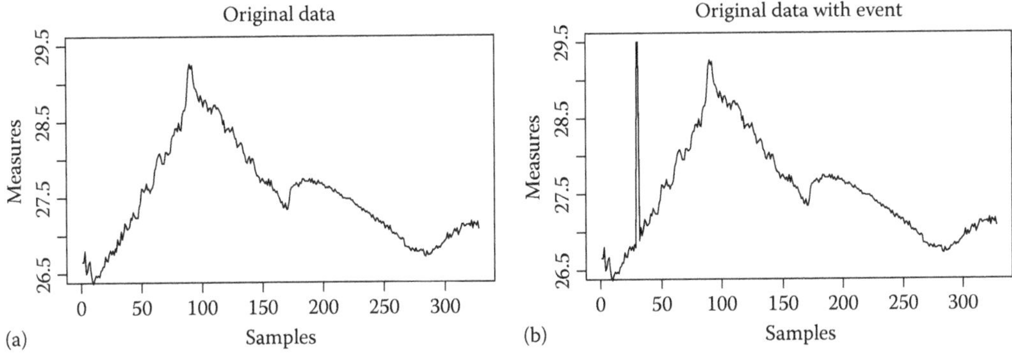

Figure 14.8 Input data for wavelets algorithm. (a) Without events and (b) with events.

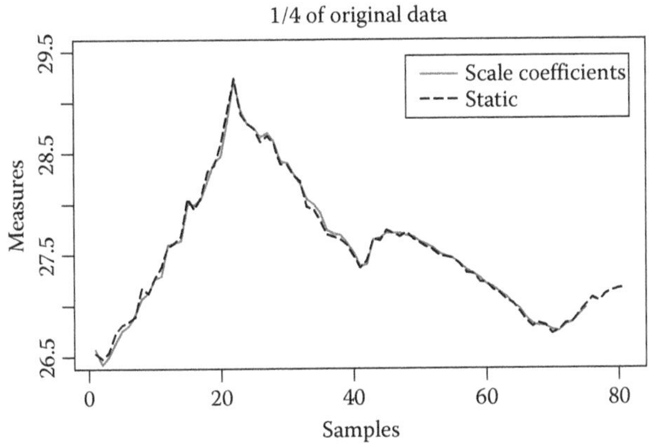

Figure 14.9 Sampling without event presence. The reduction is 1/4 of original data.

is represented by a continuous line that indicates $\left|V_j'\right| \approx \left|2^{i/2} V_i\right|$, where $2^{i/2}$ is the normalization factor at $\psi_{i,k}(t) = 2^{i/2}\psi(2^i - k)$.

In a different way, as showed in Figure 14.10, when we consider the presence of some event (Figure 14.8b), the wavelet-based sampling is able to detect it. This occurs because the coiflets transform is used here to detect the changes in the sampling. It is important to highlight that the error in both cases, Figures 14.9 and 14.10 can be obtained from Equation 14.1.

14.5.2 Sampling and Sketch Algorithms Simulations

Aquino et al. [6] simulate the sampling and sketch algorithms. The simulations are based on the following considerations:

- The evaluation is performed through simulations and uses the NS-2 (Network Simulator 2) version 2.33. Each simulated scenario was executed with 33 random scenes. At the end, for each scenario we plot the average value with 95% of confidence interval.

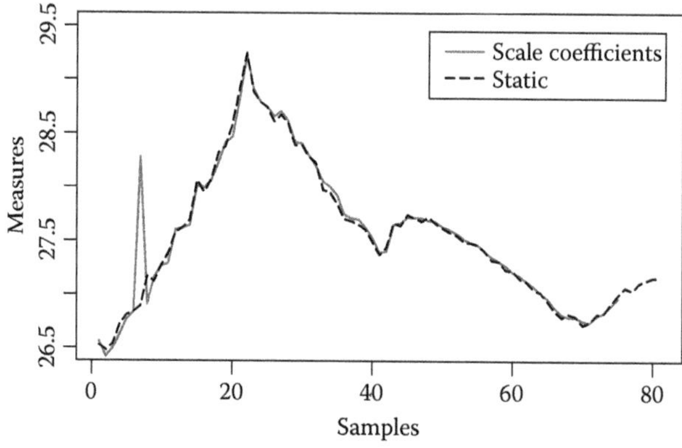

Figure 14.10 Sampling with event presence. The reduction is 1/4 of original data.

- A tree-based routing algorithm called EF-Tree is used [17]. The density is controlled and all nodes have the same hardware configuration. To analyze only the application, the tree is built just once before the traffic starts.
- The data streams used by the nodes are always the same, following a normal distribution, where the values are between [0.0; 1.0], and the periodicity of generation is 60 s. The size of the data packet is 20 bytes. For larger samples, these packets are fragmented by the sources and reassembled at the reception.
- The stream size is varied and the application and the network behavior is analyzed by using sample size of $|V|$, $|V|/2$, and $\log_2 |V|$.

Figure 14.11 [6] shows the energy consumption when the sensor stream varies. We observe, in the sampling solution, that when the sample size is diminished the consumed energy diminishes accordingly. The sketch solution follows the sample $\log n$ result. This occurs because the packet size is constant and close to the sample, that is, a $\log n$ packet size.

The sample of $\log n$ and the sketch have their best performance in all cases, and the energy consumption does not vary when the sample size increases. In the sample of $\log n$, this occurs because the packet size is increased only when we increase the sensor stream size (256, 512, 1024,

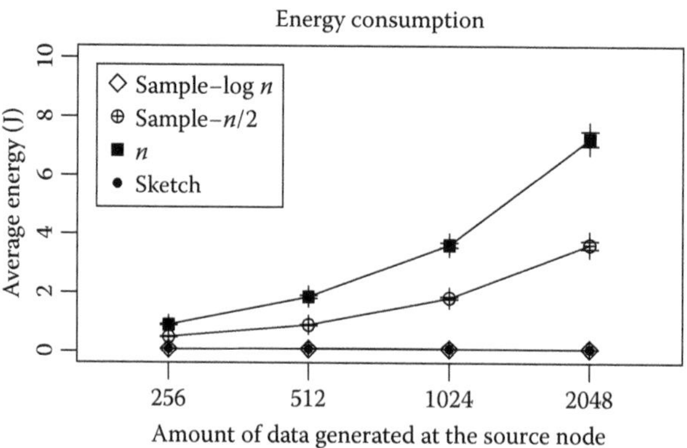

Figure 14.11 Energy network behavior.

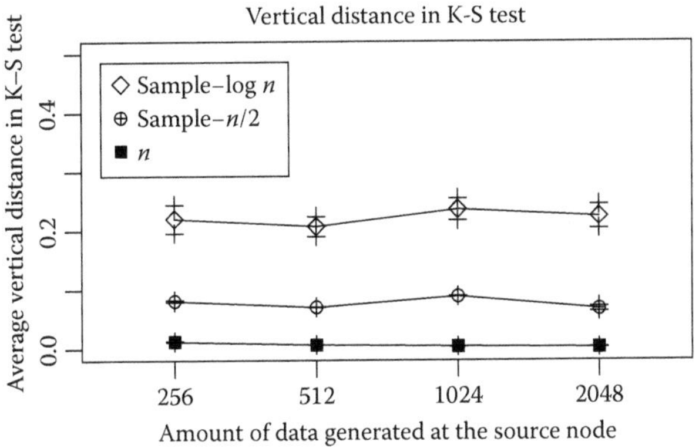

Figure 14.12 K-S data behavior.

2048). In the sketch case the used packet size is always constant. The other results (samples of $n/2$ and n) have a worst performance because the packet size is increased proportionally when the sensor streaming size is increased.

Now, consider the impact of solution by evaluating the data quality [6]. This evaluation is only for the sampling solution, because this solution loses information in its process, and therefore it is important to evaluate its impact on the data quality. In the sketch case, all data can be generated artificially when it arrives in the sink node, and, therefore the losses are not identified when the data tests are applied. The only impact generated by the sketch solution is the loss of the data sequence, which is not evaluated in this work.

Figure 14.12 shows the similarity between the original and sampled stream distributions. The difference between them is called *K-S-diff*.

The results show that when the sample size is decreased, the *K-S-diff* increases. Because the data streams are generated between [0.0; 1.0], *K-S-diff* = 20% for $\log n$ sample sizes, and *K-S-diff* = 10% for $n/2$ sample sizes. In all cases, the error is constant, since the data loss is small. The higher error occurs when we use a minor sample size; however, the data similarity is kept.

14.6 Summary

This chapter addresses some sensor stream reduction issues. A methodology suitable for sensor stream applications is presented. This methodology considers the following phases:

Characterization: This phase provides the requirements list of sensor stream reduction applications. It comprises, the following elements: reduction architecture, data sensor characterization and application requirements.

Reduction tools: This phase provides support to reduction applications development. It comprises the following elements: reduction API, infrastructure, and applications.

Robustness: This phase provides models that allow the algorithms validation. It comprises the following elements: specific evaluations, network optimization, and methodology of validation.

Conception: This phase integrates the simulation and deployment mechanisms. It comprises the following elements: development tools, simulation, and deployment.

References

1. A. L. L. Aquino. A framework for sensor stream reduction in wireless sensor networks. In *The Fifth International Conference on Sensor Technologies and Applications*, pp. 30–35, Nice/Saint Laurent du Var, France, August 2011.
2. A. C. Frery, H. Ramos, J. Alencar-Neto, and E. F. Nakamura. Error estimation in wireless sensor networks. In *23rd ACM Symposium on Applied Computing 2008*, pp. 1923–1927, Fortaleza, Brazil, March 2008. ACM.
3. A. L. L. Aquino, A. A. F. Loureiro, A. O. Fernandes, and R. A. F. Mini. An in-network reduction algorithm for real-time wireless sensor networks applications. In *Workshop on Wireless Multimedia Networking and Performance Modeling*, Vancouver, British Columbia, Canada, October 2008. ACM.
4. A. L. L. Aquino and E. F. Nakamura. Data centric sensor stream reduction for real-time applications in wireless sensor networks. *Sensors Basel*, 9:9666–9688, 2009.
5. S. Muthukrishnan. *Data Streams: Algorithms and Applications.* Now Publishers Inc, Hanover, MA, January 2005.
6. A. L. L. Aquino, C. M. S. Figueiredo, E. F. Nakamura, L. S. Buriol, A. A. F. Loureiro, A. O. Fernandes, and C. N. Coelho Jr. Data stream based algorithms for wireless sensor network applications. In *21st IEEE International Conference on Advanced Information Networking and Applications*, pp. 869–876, Niagara Falls, Canada, May 2007. IEEE Computer Society.
7. A. L. L. Aquino, C. M. S. Figueiredo, E. F. Nakamura, L. S. Buriol, A. A. F. Loureiro, A. O. Fernandes, and C. N. Coelho Jr. A sampling data stream algorithm for wireless sensor networks. In *IEEE International Conference on Communications*, pp. 3207–3212, Glasgow, U.K., June 2007. IEEE Computer Society.
8. A. L. L. Aquino, R. A. R. Oliveira, and E. F. Wanner. A wavelet-based sampling algorithm for wireless sensor networks applications. In *25th ACM Symposium On Applied Computing*, pp. 1604–1608, Sierra, Switzerland, March 2010. ACM.
9. O. Silva Jr., A. L. L. Aquino, R. A. F. Mine, and C. M. S. Figueiredo. Multivariate reduction in wireless sensors networks. In *IEEE Symposium On Computers and Communication,* Sousse, Tunisia, July 2009. IEEE Computer Society.
10. A. L. L. Aquino, C. M. S. Figueiredo, E. F. Nakamura, A. A. F. Loureiro, A. O. Fernandes, and C. N. Coelho Jr. On the use data reduction algorithms for real-time wireless sensor networks. In *IEEE Symposium On Computers and Communications*, pp. 583–588, Aveiro, Potugal, July 2007. IEEE Computer Society.
11. A. S. Ruela, R. da S. Cabral, A. L. L. Aquino, and F. G. Guimaraes. Evolutionary design of wireless sensor networks based on complex networks. In *Fifth International Conference on Intelligent Sensors, Sensor Networks and Information Processing*, Melbourne, Victoria Australia, December 2009.
12. A. S. Ruela, R. da S. Cabral, A. L. L. Aquino, and F. G. Guimaraes. Memetic and evolutionary design of wireless sensor networks based on complex network characteristics. *International Journal of Natural Computing Research*, 1(2):33–53, April–June 2010.
13. E. Reschenhofer. Generalization of the Kolmogorov-Smirnov test. *Computational Statistics and Data Analysis*, 24(4):422–441, June 1997.
14. S. Siegel and John N. Castellan Jr. *Nonparametric Statistics for the Behavioral Sciences.* McGraw-Hill Humanities/Social Sciences/Languages, Columbus, OH N. J. Castellan Jr, 2nd edn., January 1988.
15. A. Lins, E. F. Nakamura, A. A. F. Loureiro, and C. J. N. Coelho Jr. Beanwatcher: A tool to generate multimedia monitoring applications for wireless sensor networks. In A. Marshall and N. Agoulmine, eds., *6th IFIP/IEEE Management of Multimedia Networks and Services*, vol. 2839 of *Lecture Notes in Computer Science*, pp. 128–141, Belfast, Northern Ireland, September 2003. Springer.

16. A. Lins, E. F. Nakamura, A. A. F. Loureiro, and C. J. N. Coelho Jr. Semi-automatic generation of monitoring applications for wireless networks. In *9th IEEE International Conference on Emerging Technologies and Factory Automation*, pp. 506–511, Lisbon, Portugal, September 2003. IEEE Computer Society.
17. E. F. Nakamura, F. G. Nakamura, C. M. S. Figueredo, and A. A. F. Loureiro. Using information fusion to assist data dissemination in wireless sensor networks. *Telecommunication Systems*, 30(1–3):237–254, November 2005.

Chapter 15

Compressive Sensing and Its Application in Wireless Sensor Networks

Jae-Gun Choi, Sang-Jun Park, and Heung-No Lee

Contents

15.1 Introduction

In this chapter, we discuss the application of a new compression technique called compressive sensing (CS) in wireless sensor networks (WSNs). The objective of a WSN we assume in this chapter is to collect information about events occurring in a region of interest. This WSN consists of a large number of wireless sensor nodes and a central fusion center (FC). The sensor nodes are spatially distributed over the said region to acquire physical signals such as sound, temperature, wind speed, pressure, and seismic vibrations. After sensing, they transmit the measured signals to the FC. In this chapter, we focus on the role of the FC, which is to recover the transmitted signals in their original waveforms for further processing. By doing so, the FC can produce a global picture that illustrates the event occurring in the sensed region. Each sensor uses its onboard battery for sensing activities and makes reports to FC via wireless transmissions. Thus, limited power at the sensor nodes is the key problem to be resolved in the said WSN.

CS is a signal acquisition and compression framework recently developed in the field of signal processing and information theory [1,2]. Donoho [1] says that "The Shannon–Nyquist sampling rate may lead to too many samples; probably not all of them are necessary to reconstruct the given signal. Therefore, compression may become necessary prior to storage or transmission." According to Baraniuk [3], CS provides a new method of acquiring compressible signals at a rate significantly below the Nyquist rate. This method employs nonadaptive linear projections that preserve the signal's structure; the compressed signal is then reconstructed from these projections using an optimization process. There are two tutorial articles good for further reading on CS [3,4] published in the *IEEE Signal Processing Magazine* in 2007 and 2008, respectively.

Our aim in this chapter is to determine whether the CS can be used as a useful framework for the aforementioned WSN to compress and acquire signals and save transmittal and computational power at the sensor node. This CS-based signal acquisition and compression are done by a simple linear projection at each sensor node. Then, each sensor transmits the compressed samples to the FC; the FC, which collects the compressed signals from the sensors, jointly reconstructs the signal in polynomial time using a signal recovery algorithm. Illustrating this process in detail throughout this chapter, we check to see if CS can become an effective, efficient strategy to be employed in WSNs, especially for those with low-quality, inexpensive sensors.

In this chapter, as we assume a scenario in which a WSN is used for signal acquisition, we intend to pay some effort in modeling correlation among the signals acquired from the sensors. We discuss a few signal projection methods suggested in the literature that are known to give a good signal recovery performance from the compressed measurements. We also investigate a couple of well-known signal recovery algorithms such as the orthogonal matching pursuit (OMP) (greedy approach) [13,14] and the primal-dual interior point method (PDIP) (gradient-type approach) [5]. Finally, we simulate the considered WSN system and examine how the presence of signal correlation can be exploited in the CS recovery routine and help reduce the amount of signal samples to be transmitted at the sensor node.

15.2 Compressed Sensing: What Is It?

In a conventional communication system, an analog-to-digital converter based on the Shannon–Nyquist sampling theorem is used to convert analog signals to digital signals. The theorem says

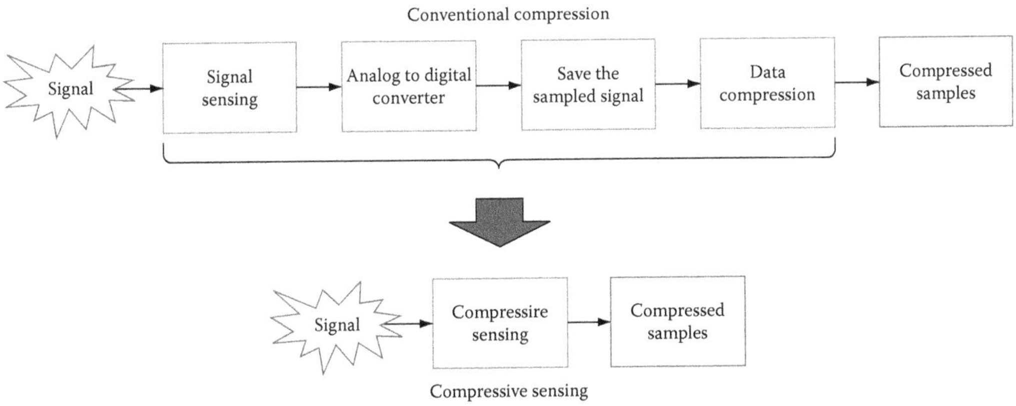

Figure 15.1 Conventional compression and compressive sensing.

that if a signal is sampled at a rate twice, or higher, the maximum frequency of the signal, the original signal can be exactly recovered from the samples. Once the sampled signals are obtained over a fixed duration of time, a conventional compression scheme can be used to compress them. Because the sampled signals often have substantial redundancy, compression is possible. Several compression schemes follow this approach, for example, the MP3 and JPEG formats for audio or image data. However, conventional compression in a digital system is sometimes inefficient because it requires unnecessary signal processing stages, for example, retaining all of the sampled signals in one location before data compression. According to Donoho [1], the CS framework, as shown in Figure 15.1, can bypass these intermediate steps and thus provides a light-weight signal acquisition apparatus that is suitable for those sensor nodes in our WSN.

The CS provides a direct method that acquires compressed samples without going through the intermediate stages of conventional compression. Thus, CS provides a much simpler signal acquisition solution. In addition, the CS provides several recovery routines that the original signal can be regenerated perfectly from the compressed samples.

15.2.1 Background

Let a real-valued column vector **s** be a signal to be acquired. Let it be represented by

$$\mathbf{s} = \Psi \mathbf{x} \tag{15.1}$$

where **x** and $\mathbf{s} \in \mathbf{R}^n$, and **x** is also a real-valued column vector. The matrix $\Psi \in \mathbf{R}^{n \times n}$ is an orthonormal basis, i.e., $\Psi^T \Psi = \Psi \Psi^T = I_n$, the identity matrix of size $\mathbf{R}^{n \times n}$. The signal **s** is called k-sparse if it can be represented as a linear combination of only k columns of Ψ, i.e., only the k components of the vector **x** are nonzero as represented in Equation 15.2:

$$\mathbf{s} = \sum_{i=1}^{n} x_i \psi_i, \quad \text{where } \psi_i \text{ is a column vector of } \Psi \tag{15.2}$$

A signal is called *compressible* if it has only a few significant (large in magnitude) components and a greater number of insignificant (close to zero) components. The compressive measurements **y** (compressed samples) are obtained via linear projections as follows:

$$\mathbf{y} = \Phi\mathbf{s} = \Phi\Psi\mathbf{x} = \mathbf{A}\mathbf{x} \tag{15.3}$$

where the measurement vector is $\mathbf{y} \in \mathbf{R}^m$, with $m < n$, and the measurement matrix $\mathbf{A} \in \mathbf{R}^{m\times n}$. Our goal is to recover $\mathbf{x}$ from the measurement vector $\mathbf{y}$. We note that Equation 15.3 is an underdetermined system because it has fewer equations than unknowns; thus, it does not have a unique solution in general. However, the theory of CS asserts that, if the vector $\mathbf{x}$ is sufficiently sparse, an underdetermined system is guaranteed with high probability to have a unique solution.

In this section, we discuss the basics of CS in more detail.

1. k-sparse signal $\mathbf{x}$ in orthonormal basis

The k-sparse signal, $\mathbf{s}$ in Equation 15.1, has k nonzero components in $\mathbf{x}$. The matrix Ψ is, again, an orthonormal basis, i.e., $\Psi^T\Psi = \Psi\Psi^T = I_n$, the identity matrix of size $\mathbf{R}^{n\times n}$.

2. Measurement vector $\mathbf{y}$ and underdetermined system

The sensing matrices Φ and $\mathbf{A}$ in Equation 15.3, are m by n. Note that m is less than n and thus Equation 15.3 is an underdetermined system of equations. When m is still good enough for signal recovery, a compression effect exists. A good signal recovery with $m < n$ in this underdetermined system of equation is possible with the additional information that the signal is k-sparse. While we address the detail in signal recovery problem in Section 15.2.2, we may consider a simpler case for now. Suppose $\mathbf{x}$ is k-sparse and the locations of the k nonzero elements are known. Then, we can form a simplified equation by deleting all those columns and elements corresponding to the zero-elements, as follows:

$$\mathbf{y} = \mathbf{A}_{\mathsf{K}}\mathbf{x}_{\mathsf{K}} \tag{15.4}$$

where $\mathsf{K} \in \{1, 2, \ldots, n\}$ is the support set, which is a collection of indices corresponding to the nonzero elements of $\mathbf{x}$. Note that the support set K can be any size—k subset of the full index set, $\{1, 2, 3, \ldots, n\}$. Equation 15.4 has the unique solution $\mathbf{x}_{\mathsf{K}}$ if the columns of $\mathbf{A}_{\mathsf{K}}$ are linearly independent. The solution can be found using

$$\mathbf{x}_{\mathsf{K}} = \left(\mathbf{A}_{\mathsf{K}}^T\mathbf{A}_{\mathsf{K}}\right)^{-1}\mathbf{A}_{\mathsf{K}}^T\mathbf{y} \tag{15.5}$$

Thus, if the support set K can be found, the problem is easy to solve provided the columns are linearly independent.

3. Incoherence condition

The incoherence condition is that the rows of Φ should be incoherent to the columns of Ψ. If the rows of Φ are coherent to the columns of Ψ, the matrix $\mathbf{A}$ cannot be a good sensing matrix. In the extreme case, we can show a matrix $\mathbf{A}$ having m rows of Φ that are the first m columns of Ψ.

$$\mathbf{A} = \Phi\Psi = \Psi_{(1:m,:)}^T\Psi = \begin{bmatrix} 1 & 0 & 0 & 0 & 0 & \ldots & 0 \\ 0 & 1 & 0 & 0 & 0 & \ldots & 0 \\ 0 & 0 & 1 & 0 & 0 & \ldots & 0 \\ 0 & 0 & 0 & 1 & 0 & \ldots & 0 \end{bmatrix} \tag{15.6}$$

If **A** of Equation 15.6 is used as sensing matrix, the compressed measurement vector **y** captures only the first m elements of the vector **x**, and the rest of the information contained in **x** is completely lost.

4. Designing a sensing matrix Φ

One choice for designing a sensing matrix Φ is Gaussian. Under this choice, the sensing matrix Φ is designed as a Gaussian, i.e., matrix elements are independent and identically distributed Gaussian samples. This choice is deemed good since a Gaussian sensing matrix satisfies the incoherence condition with high probability for any choice of orthonormal basis Ψ. This randomly generated matrix acts as a random projection operator on the signal vector **x**. Such a random projection matrix need not depend on specific knowledge about the source signals. Moreover, random projections have the following advantages in the application to sensor networks [6]:

1. *Universal incoherence*: Random matrices Φ can be combined with all conventional sparsity basis Ψ, and, with high probability, sparse signals can be recovered by an L_1 minimum algorithms from the measurements **y**.
2. *Data independence*: The construction of a random matrix does not depend on any prior knowledge of the data. Therefore, given an explicit random number generator, only the sensors and the FC are required to agree on a single random seed for generating the same random matrices of any dimension.
3. *Robustness*: Transmission of randomly projected coefficients is robust to packet loss in the network. Even if part of the elements in measurement y is lost, the receiver can still recover the sparse signal, at the cost of lower accuracy.

15.2.2 L_0, L_1, and L_2 Norms

In CS, a core problem is to find a unique solution for an underdetermined equation. This problem is related to the signal reconstruction algorithm, which takes the measurement vector **y** as an input and the k-sparse vector **x** as an output. To solve an underdetermined problem, we consider minimization criteria using different norms such as the L_2, L_1, and L_0 norms. The L_p norm of a vector **x** of length n is defined as

$$\|\mathbf{x}\|_p = \left(\sum_{i=1}^{n} |x_i|^p\right)^{1/p}, \quad p > 0 \tag{15.7}$$

Although we can define the L_2 and L_1 norms as $\|\mathbf{x}\|_2 = \left(\sum_{i=1}^{n} |x_i|^2\right)^{1/2}$ and $\|\mathbf{x}\|_1 = \sum_{i=1}^{n} |x_i|$, respectively, using the definition of L_p norm, L_0 norm cannot be defined this way. The L_0 norm is a pseudo-norm that counts the number of nonzero components in a vector as defined by Donoho and Elad [7]. Using this definition of norms, we will discuss the minimization problem.

1. L_2 norm minimization

$$\begin{aligned}(L_2)\,\hat{\mathbf{x}} &= \arg\min \|\mathbf{x}\|_2 \quad \text{subject to } \mathbf{y} = \mathbf{Ax}, \quad \text{where} \quad \mathbf{A} \in \mathbf{R}^{m\times n},\ rank\,(\mathbf{A}) = m \\ &= \mathbf{A}^T\left(\mathbf{AA}^T\right)^{-1}\mathbf{y}\end{aligned} \tag{15.8}$$

However, this conventional solution yields a non-sparse solution, so it is not appropriate as a solution to the CS problem.

2. L_0 norm minimization

$$(L_0)\ \text{Minimize } \|\mathbf{x}\|_0 \quad \text{subject to } \mathbf{y} = \mathbf{Ax}, \quad \text{where} \quad \mathbf{A} \in \mathbf{R}^{m\times n},\ rank\,(\mathbf{A}) = m \qquad (15.9)$$

The L_0 norm of a vector is, by definition, the number of nonzero elements in the vector. In the CS literature, it is known that the L_0 norm problem can be solved by examining all the possible cases. Since this process involves a combinatorial search for all possible $\binom{n}{k}$ support sets, it is an NP-complete problem. Thus, we cannot solve it within polynomial time. Therefore, we consider L_1 norm minimization as an alternative.

3. L_1 norm minimization

$$(L_1)\ \text{Minimize } \|\mathbf{x}\|_1 \quad \text{subject to } \mathbf{y} = \mathbf{Ax}, \quad \text{where} \quad \mathbf{A} \in \mathbf{R}^{m\times n}, rank\,(\mathbf{A}) = m \qquad (15.10)$$

This L_1 norm minimization can be considered as a relaxed version of the L_0 problem. Fortunately, the L_1 problem is a convex optimization problem and in fact can be recast as a linear programming problem. For example, it can be solved by an interior point method. Many effective algorithms have been developed to solve the minimum L_1 problem, and it will be considered later in this chapter. Here, we aim to study the sufficient conditions under which Equations 15.9 and 15.10 have unique solutions. We provide a theorem related to this issue.

Theorem 15.1 L_0/L_1 equivalence condition Let $\mathbf{A} \in \mathbf{R}^{m\times n}$ be a matrix with a maximum correlation definition μ, $\mu\,(\mathbf{A}) = \max\limits_{i\neq j} \left|\langle \mathbf{a}_i, \mathbf{a}_j\rangle\right|$, where $\mathbf{a}_i$ is the ith column vector of $\mathbf{A}$ with $i = 1, 2, \ldots, n$, and $\mathbf{x}$ is a k-sparse signal. Then, if $k < \frac{1}{2}\,(1 + (1/\mu))$ is satisfied, then the solution of L_1 coincides with that of L_0 [7].

15.3 Wireless Sensor Networks

15.3.1 Network Structure

We consider a WSN consisting of a large number of wireless sensor nodes and one FC (Figure 15.2). The wireless sensor nodes are spatially distributed over a region of interest and observe physical changes such as those in sound, temperature, pressure, or seismic vibrations. If a specific event occurs in a region of distributed sensors, each sensor makes local observations of the physical phenomenon as a result of this event taking place. An example of sensor network applications is area monitoring to detect forest fires. A network of sensor nodes can be installed in a forest to detect when a fire breaks out. The nodes can be equipped with sensors to measure temperature, humidity, and the gases produced by fires in trees or vegetation [8]. Other examples include military and security applications. Military applications vary from monitoring soldiers in the field to tracking vehicles or enemy movement. Sensors attached to soldiers, vehicles, and equipment can gather information about their condition and location to help planning activities on the battlefield. Seismic, acoustic, and video sensors can be deployed to monitor critical terrain and approach routes; reconnaissance of enemy terrain and forces can be carried out [9].

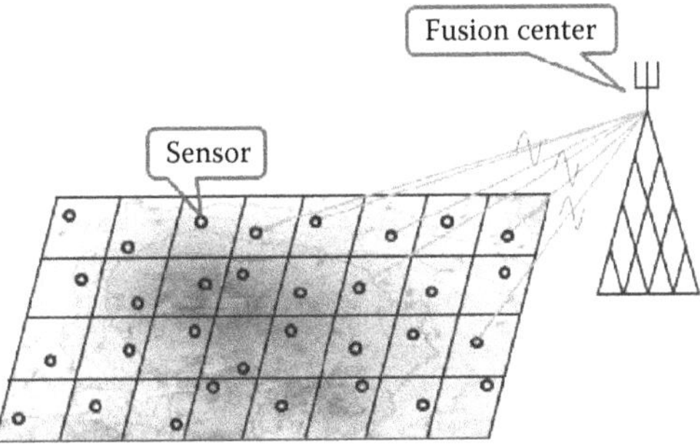

Figure 15.2 Wireless sensor network.

After sensors observe an event taking place in a distributed region, they convert the sensed information into a digital signal and transmit the digitized signal to the FC. Finally, the FC assembles the data transmitted by all the sensors and decodes the original information. The decoded information at the FC provides a global picture of events occurring in the region of interest. Therefore, we assume that the objective of the sensor network is to accurately determine and rapidly reconstruct transmitted information and reconstruct the original signal.

We discuss the resource limitations of WSNs in the next section.

15.3.2 Resource Limitations in WSNs

In this section, we describe the assumptions made in the sensor network we are interested in. We assume that the sensors are distributed and supposed to communicate with the FC through a wireless channel. Because each sensor is an important component of WSN that observes event, they should typically be deployed in a large volume over the region of interest. Therefore, they are usually designed to be inexpensive and small. For that reason, each sensor operates on an onboard battery that is not rechargeable at all; thus, for simplicity, the hardware implementation of sensor nodes can provide only limited computational performance, bandwidth, and transmission power. As a result of limitations on the hardware implementation in sensor nodes, the FC has powerful computational performance and plentiful energy, which naturally performs most of the complex computations.

Under the limited conditions stated earlier for a WSN, CS can substantially reduce the data volume to be transmitted at each sensor node. With the new method, it is possible to compress the original signal using only $O(k \log(n/k))$ samples without going through many complex signal processing steps. These signals can be recovered successfully at the FC. All these are done under the CS framework. As a result, the consumption of power for transmission of signal contents at each sensor can be significantly reduced thanks to decreased data volume. Further, it should be noted that, this data reduction comes without utilizing onboard signal processing units since all the intermediate signal processing steps, shown in Figure 15.1, are not needed. Namely, the sensor nodes can compress the signal while not spending any power for running complex compression algorithms onboard.

15.3.3 Usefulness of CS in WSNs

In this section, we provide a brief comparison of using CS and using the conventional compression in a WSN. This comparison illustrates why CS could be a useful solution for WSNs.

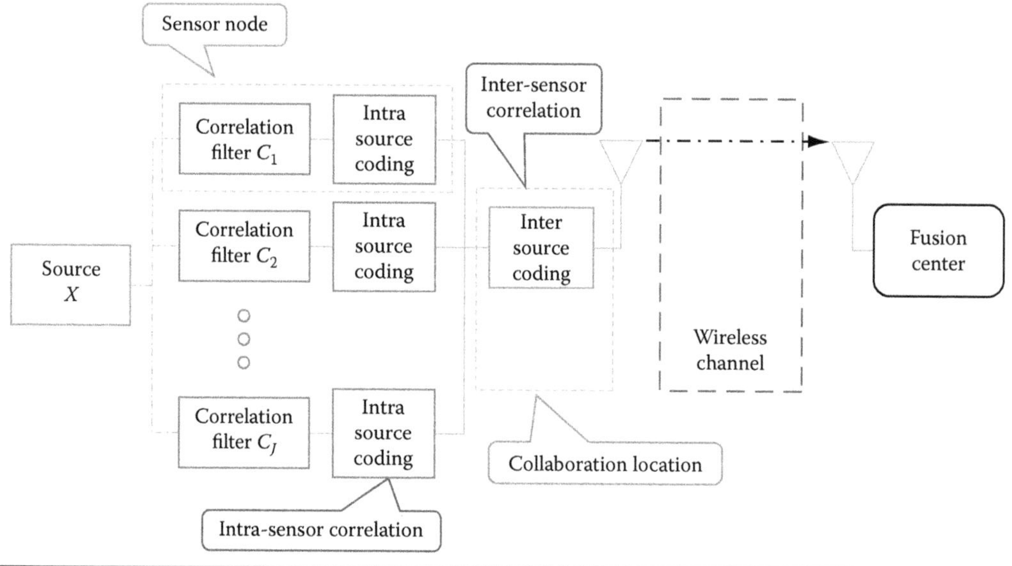

Figure 15.3 Conventional sensor network scheme.

1. Sensor network scheme with conventional compression

For a conventional sensor system, suppose that the system designer has decided to gather all the uncompressed samples at a single location, say one of the sensors, in order to exploit inter-sensor correlation. See diagram shown in Figure 15.3. At the collection point, joint compression can be made and compressed information can be sent to the FC.

This option has a couple drawbacks. First, gathering the samples from all the sensors and jointly compressing them cause a transmission delay. Second, a lot of onboard power should be spent at the collaboration point. Third, each sensor should be collocated so that the transmitted information can be gathered at collaboration location.

Now, we may suppose that the joint compression is not aimed at and each sensor compresses the signal on its own. First, the data reduction effect with this approach will be limited because inter-sensor correlation is not exploited at all. The total volume of the independently compressed data is much larger than that of jointly compressed data. This may produce a large traffic volume in the WSN, and a large amount of transmission power will be wasted from the sensor nodes that transmit essentially the same information to the FC. Thus, this is an inefficient strategy as well.

2. Sensor network scheme with CS

In contrast to the conventional schemes considered in the previous paragraph, the CS method aims to acquire compressed samples directly. If a high-dimensional observation vector x exhibits sparsity in a certain domain (by exploiting intra-sensor correlation), CS provides the *direct method* for signal compression as discussed in Figure 15.1. To compress the high-dimensional signal $\mathbf{x}$ into a low-dimensional signal $\mathbf{y}$, as in Equation 15.3, it uses a simple matrix multiplication with an $m \times n$ projection matrix $\mathbf{A}_j, j \in \{1, 2, \ldots, J\}$, where j is the sensor index, as depicted in Figure 15.4.

In the CS-based sensor network scheme, each sensor compresses the observed signals using a simple linear projection and transmits the compressed samples to the FC. Then, the FC can

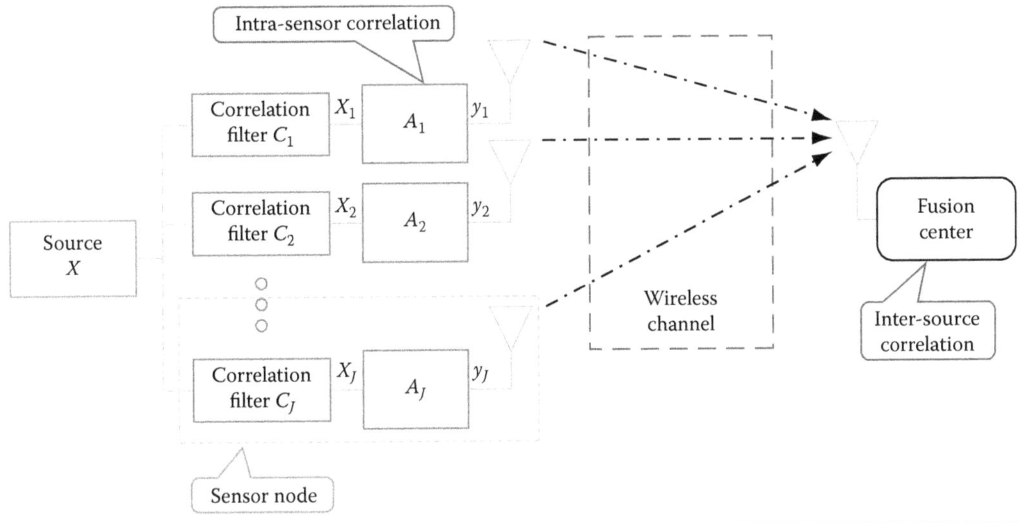

Figure 15.4 CS sensor network scheme.

jointly reconstruct the received signals (by exploiting inter-sensor correlation) using one of the CS algorithms. Therefore, each sensor does not need to communicate with its neighboring sensors for joint compression. Our method is distributed compression without having the sensors to talk to each other; only the joint recovery at the FC is needed. Thus, no intermediate stages are required that are to gather all of the samples at a single location and carry out compression aiming to exploiting inter-sensor correlation. This free of intermediate stages allows us to reduce time delay significantly as well. Therefore, if the original data are compressed by CS, each sensor node produces much smaller traffic volume that can be transmitted to the FC at a much lower transmission power and with a smaller time delay.

15.4 Wireless Sensor Network System Model

15.4.1 Multi-Sensor Systems and Observed Signal Properties

Each sensor can observe only the local part of an entire physical phenomenon, and a certain event of interest is measured by one or more sensors. Therefore, the sensed signals are often partially correlated. These measured signals have two distinct correlations: intra-sensor correlation and inter-sensor correlation. Intra-sensor correlation exists in the signals observed by each sensor. Once a high-dimensional sensed signal has a sparse representation in a certain domain, we can reduce its size by using CS. This process exploits the intra-sensor correlation. By contrast, inter-sensor correlation exists among the signals sensed by different sensors. By exploiting inter-sensor correlation, further reduction in transmitted signals can be made.

These two correlations can be exploited to improve the system performance. As the number of sensors in a region becomes dense, each sensor has a strongly correlated signal that is similar to that of neighboring sensors. By contrast, if we decrease the density of sensors distributed in a given region, the sensed signals will obviously be more weakly correlated with each other. In this section, we discuss two strategies for transmitting signals in a multi-sensor CS-based system. One strategy uses only intra-sensor correlation and the other uses both types of correlation. We illustrate that

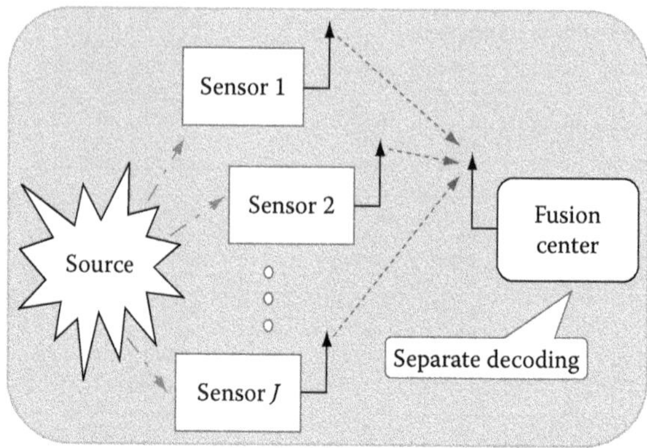

Figure 15.5 Intra-sensor correlation scheme.

CS-based system in WSN exploits the inter-sensor correlation more effectively and simply than that of conventional sensor network.

1. Exploiting only intra-sensor correlation
   In Figure 15.5, each sensor observes the source signal and independently compresses it to a low-dimensional signal. After compression, each sensor transmits the compressed signal to the FC. Without exploiting inter-sensor correlation between transmitted signals, the FC recovers these signals separately. In this case, even if there exists a correlation among the sensed signals, because only intra-sensor correlation is exploited, we cannot gain any advantages from joint recovery. This method has the following characteristics:
   a. Independent compression and transmission at each sensor
   b. Signal recovery by exploiting only intra-sensor correlation at the FC
2. Exploiting both intra- and inter-sensor correlations

Figure 15.6 shows the same process as in situation (1) provided earlier, except that the FC exploits the inter-sensor correlation among sensed signals at signal reconstruction stage. In conventional sensor network system as shown in Figure 15.3, the sensor nodes communicate with their neighboring

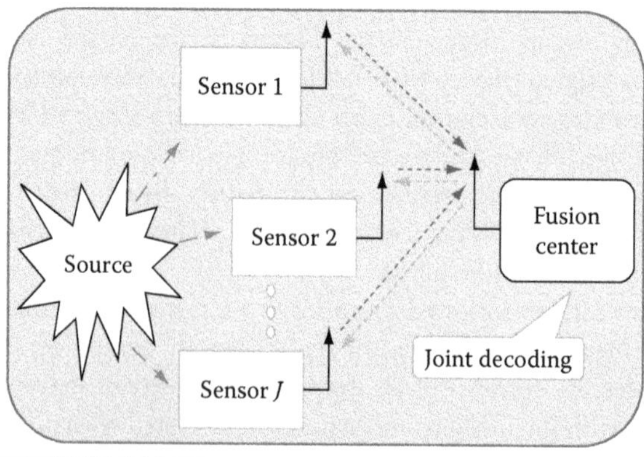

Figure 15.6 Intra/inter-sensor correlation scheme.

Table 15.1 Synthetic Signal Models According to Correlation Degree

	Correlation Characteristics					
	Support Set		Element Value			
Correlation Degree	Common	Innovation	Same	Different	Both	Model Name
Weak	O	O	X	O	X	(Empty)
↕	O	O	X	X	O	*JSM*-1
Strong	O	X	X	O	X	*JSM*-2
	O	X	O	X	X	(Empty)

sensors to take advantage of joint compression by exploiting inter-sensor correlation. However, in the CS-based system, a stage for exploiting inter-sensor correlation is achieved at FC. It means that if inter-sensor correlation exists within the sensed signals, the FC can exploit it. This is done with sensors communicating with the FC but not among the sensors themselves. We refer to this communication strategy as the *distributed compressive sensing* (DCS). Exploitation of inter-sensor correlation should be manifested with the reduction of the measurement size m of matrix $\mathbf{A} \in \mathbf{R}^{m\times n}$, where $\mathbf{y} = \mathbf{Ax}$, required for good single recovery. The characteristics of our DCS sensor network are as follows:

1. Independent compression and transmission at each sensor
2. Exploitation of inter-sensor signal correlation with the joint recovery scheme at the FC
3. Variation of the per sensor CS measurements to manipulate the level of signal correlation

15.4.2 Correlated Signal Models and System Equations

In this section, we introduce how signals with different degrees of correlation can be generated with sparse signal models. Sparse signal is a correlated signal. The degree of sparseness, called the sparsity, is proportional to the amount of correlation. More correlated signal means sparser. In addition, inter-sensor signal correlation can be modeled by (1) the degree of overlaps in the support sets of any two sparse signals and (2) the correlation of nonzero signal values. There are a number of papers that use interesting signal correlation models [6,10–12]. Table 15.1 is useful for identifying the two models we take from these papers and use in the subsequent sections.

Table 15.1 lists the signal models introduced in [10,11]. In those references, the correlation signal is referred to as *JSM*-1 (joint signal model) or *JSM*-2 depending on the correlation type. In *JSM*-1, all of the signals share exactly the same common nonzero components that have the same values, whereas each signal also independently has different nonzero components, which is called innovation. Such a signal is expressed as

$$\mathbf{x}_j = \mathbf{z}_c + \mathbf{z}_j, \qquad j \in \{1, 2, \ldots, J\}, \qquad j \text{ is the index of the sensors} \tag{15.11}$$

where $\|\mathbf{z}_c\|_0 = K$ and $\|\mathbf{z}_j\|_0 = K_j$. Obviously, $\mathbf{z}_c$ appears in all the signals. It can be recognized as the inter-sensor correlation. We note that the intra-sensor correlation is that all of the signals are sparse. The jth sensor transmits $\mathbf{y}_j = \mathbf{A}_j\mathbf{x}_j$ to the FC. After all the sensed signals are transmitted to the FC, the FC aims to recover all the signals. Because inter-sensor correlation exists in the sensed signals, we can obtain several benefits by using the correlated information in the transmitted signals. For ease of explanation, suppose that the WSN contains J sensors and its sensed signal follows *JSM*-1. Then, the FC can exploit both intra- and inter-sensor correlations by solving Equation 15.12 as described in the following:

1. Joint recovery scheme for *JSM*-1

The sensed signals from J sensors can be expressed as follows:

$$\begin{aligned} \mathbf{x}_1 &= \mathbf{z}_c + \mathbf{z}_1 \in \mathbf{R}^n \\ \mathbf{x}_2 &= \mathbf{z}_c + \mathbf{z}_2 \in \mathbf{R}^n \\ &\vdots \\ \mathbf{x}_J &= \mathbf{z}_c + \mathbf{z}_J \in \mathbf{R}^n \end{aligned}$$

where the sparsities of vectors $\mathbf{z}_c$ and $\mathbf{z}_j$ are K and K_j, respectively.

The transmitted signal $\mathbf{y}_j$ can be divided into two parts as follows:

$$\mathbf{y}_j = \mathbf{A}_j(\mathbf{z}_c + \mathbf{z}_j) = \mathbf{A}_j\mathbf{z}_c + \mathbf{A}_j\mathbf{z}_j$$

If the FC received all the signals transmitted from J sensors, it then concatenates the used sensing matrix and received signal using Equation 15.12. Because the common sparsity $\mathbf{z}_c$ appears only once in the equation, the total sparsity is reduced from $J \times (K + K_j)$ to $K + (J \times K_j)$. In the underdetermined problem, low sparsity yields exact reconstruction. We will show the relationship between exact reconstruction and sparsity from simulation results in later section. By solving this equation, the FC can take advantage of exploiting inter-sensor correlation.

$$\begin{bmatrix} \mathbf{y}_1 \\ \mathbf{y}_2 \\ \mathbf{y}_3 \\ \vdots \\ \mathbf{y}_J \end{bmatrix} = \begin{bmatrix} \mathbf{A}_1 & \mathbf{A}_1 & 0 & 0 & \cdots & 0 \\ \mathbf{A}_2 & 0 & \mathbf{A}_2 & 0 & \cdots & 0 \\ \mathbf{A}_3 & 0 & 0 & \mathbf{A}_3 & 0 & \vdots \\ \vdots & \vdots & \vdots & \vdots & \ddots & 0 \\ \mathbf{A}_J & 0 & 0 & 0 & 0 & \mathbf{A}_J \end{bmatrix} \begin{bmatrix} \mathbf{z}_c \\ \mathbf{z}_1 \\ \mathbf{z}_2 \\ \mathbf{z}_3 \\ \vdots \\ \mathbf{z}_J \end{bmatrix} \quad (15.12)$$

However, if the FC recovers the received signals independently without using any correlation information, separate recovery is done. Even if the sensed signals are correlated, separate recovery offers no advantages for signal reconstruction because it does not exploit inter-sensor correlation.

2. Separate recovery scheme for *JSM*-1

Even if a common correlated element exists in the sensed signals, separate recovery does not use that correlation information. Therefore, the received signals are recovered as follows:

$$\begin{bmatrix} \mathbf{y}_1 \\ \mathbf{y}_2 \\ \vdots \\ \mathbf{y}_J \end{bmatrix} = \begin{bmatrix} \mathbf{A}_1 & 0 & 0 & 0 \\ 0 & \mathbf{A}_2 & 0 & \vdots \\ \vdots & \vdots & \ddots & 0 \\ 0 & 0 & 0 & \mathbf{A}_J \end{bmatrix} \begin{bmatrix} \mathbf{x}_1 \\ \mathbf{x}_2 \\ \vdots \\ \mathbf{x}_J \end{bmatrix} \quad (15.13)$$

To solve Equations 15.12 and 15.13, we use the PDIP, which is an L_1 minimization algorithm, and compare the results of the two types of recovery. Using the comparison results in a later section, we can confirm that the measurement size required for perfect reconstruction is smaller for joint recovery than for separate recovery.

Now, we introduce *JSM*-2, which is simpler than *JSM*-1. All the signal coefficients are different, but their indices for nonzero components are the same. Suppose that there exist two signals, $\mathbf{x}_1$ and $\mathbf{x}_2$. The ith coefficient for $\mathbf{x}_1$ is nonzero if and only if the ith coefficient for $\mathbf{x}_2$ is nonzero. This property represents inter-sensor correlation, because if we know the support set for $\mathbf{x}_1$, then we automatically know the support set for $\mathbf{x}_2$.

3. Recovery scheme for *JSM*-2

The prior inter-correlation becomes relevant when the number of sensors is more than two. To reconstruct the transmitted signals of *JSM*-2, we can solve the following equation jointly:

$$\mathbf{y}_j = \mathbf{A}_j\mathbf{x}_j, \quad j \in \{1, 2, \ldots, J\} \tag{15.14}$$

Like the FC in *JSM*-1, the FC in *JSM*-2 can exploit the fact that the support set is shared. By solving Equation 15.14 jointly in *JSM*-2, we obtain several benefits when the FC exploits inter-sensor correlation. If we solve this equation separately, but not jointly, it is separate recovery. As an algorithm for solving the equation of the *JSM*-2 signal, we use a simultaneous OMP modified from an OMP algorithm in order to demonstrate the benefits when the FC exploits inter-sensor correlation. These algorithms are discussed in Section 15.5.

15.5 Recovery Algorithms

In this section, we discuss the recovery algorithms used to solve the underdetermined equation. The recovery algorithms used in CS can be classified as the greedy type and the gradient type.

We introduce representative algorithms from these two types, the orthogonal matching pursuit (OMP) and the primal-dual interior point method (PDIP), respectively.

15.5.1 Orthogonal Matching Pursuit (Greedy-Type Algorithm)

The OMP is a famous greedy-type algorithm [13]. OMP produces a solution within k steps because it adds one index to the sparse set Λ at each iteration. The strategy of OMP is outlined in Tables 15.2 and 15.3.

Table 15.2 Inputs and Outputs of OMP Algorithm

Input	*Output*
A $m \times n$ measurement matrix $\mathbf{A}$ A m − dimensional data vector $\mathbf{y}$ The sparsity level k of the ideal signal	An estimate $\hat{\mathbf{x}}$ in $\mathbf{R}^n$ for the ideal signal. A set Λ_k containing k elements from $\{1, \ldots, n\}$ An m − dimensional approximation $\hat{y}_k$ of the data $\mathbf{y}$ An m − dimensional residual $\mathbf{r}_k = \mathbf{y} - \hat{\mathbf{y}}_k$

Table 15.3 OMP Algorithm

1. Initialize: Let the residual vector be $\mathbf{r}_0 = \mathbf{y}$, the sparse set $\Lambda_0 = \{\}$, and iteration number $t = 1$. 2. Find the index λ_t: $\lambda_t = \arg\max_{i=1,\ldots,n} \left|\langle \mathbf{r}_{t-1}, \mathbf{a}_i \rangle\right|$. The $\mathbf{a}_i$ is the ith column vector of matrix $\mathbf{A}$. 3. Update set: $\Lambda_t = \Lambda_{t-1} \cup \{\lambda_t\}$. 4. Signal estimate: $\mathbf{x}_t(\Lambda_t) = \mathbf{A}^{\dagger}_{\Lambda_t}\mathbf{y}$ and $\mathbf{x}_t\left(\Lambda^C_t\right) = 0$, where $\mathbf{x}_t(\Lambda_t)$ is the set of elements whose indices are corresponding to the sparse set. 5. Get new residual: $\hat{\mathbf{y}}_t = \mathbf{A}_t\mathbf{x}_t$, $\mathbf{r}_t = \mathbf{y} - \hat{\mathbf{y}}_t$. 6. Increment t: Increase iteration number $t = t + 1$, and return to Step 2 if $t < k$.

Let us examine the earlier OMP algorithm. In step 2, OMP selects one index that has a dominant impact on the residual vector $\mathbf{r}$. Then, in step 3, the selected index is added to the sparse set, and the sub matrix $\mathbf{A}_{\Lambda_t}$ is constructed by collecting the column vectors of $\mathbf{A}$ corresponding to the indices of the sparse set Λ_t. OMP estimates the signal components corresponding to the indices of the sparse set and updates the residual vector by removing the estimated signal components in steps 4 and 5, respectively. Finally, OMP finishes its procedures when the cardinality of the sparse set is k.

OMP is a greedy-type algorithm because it selects the one index regarded as the optimal decision at each iteration. Thus, its performance is dominated by its ability to find the sparse set exactly. If the sparse set is not correctly reconstructed, OMP's solution could be wrong. Because OMP is very easy to understand, a couple of modified algorithms based on OMP have been designed and developed. For further information on the OMP algorithm and its modifications, interested readers are referred to two papers [14,15].

We introduce another greedy-type algorithm based on OMP as an example: simultaneous orthogonal matching pursuit (SOMP) [14]. This greedy algorithm has been proposed for treating multiple measurement vectors for *JSM*-2 when the sparse locations of all sensed signals are the same. Namely, SOMP algorithm handles multiple measurements $\mathbf{y}_j$ as an input, when j is the index of distributed sensors, $j \in \{1, 2, \ldots, J\}$. In a later section, we use this algorithm to recover *JSM*-2. The pseudo code for SOMP is shown in Tables 15.4 and 15.5.

15.5.2 Primal-Dual Interior Point Method (Gradient-Type Algorithm)

The L_1 minimization in Equation 15.10 can be recast as linear programming. Here we examine this relationship. Clearly, the L_1 minimization problem in Equation 15.10 is not linear programming because its cost function is not linear. However, by using a new variable, we can transform it to

Table 15.4 Inputs and Outputs of SOMP Algorithm

Input	*Output*
A $m \times n$ measurement matrix $\mathbf{A}_j$ A m-dimensional data vector $\mathbf{y}_j$ The sparsity level k of the ideal signal	An estimate $\hat{\mathbf{x}}_j$ in $\mathbf{R}^n$ for the ideal signal. A set Λ_k containing k elements from $\{1, \ldots, n\}$ An m-dimensional approximation $\hat{\mathbf{y}}_{j,k}$ of the data $\mathbf{y}_j$ An m-dimensional residual $\mathbf{r}_{j,k} = \mathbf{y}_j - \hat{\mathbf{y}}_{j,k}$

Table 15.5 SOMP Algorithm

1. Initialize:
   Let the residual matrix be $\mathbf{r}_{j,0} = y_{j,0}$. The sparse set $\Lambda_0 = \{\}$, and iteration number $t = 1$.
2. Find the index λ_t: $\lambda_t = \arg\max_{i=1,\ldots,n} \sum_{j=1}^{J} \left|\left\langle \mathbf{r}_{j,t-1}, \mathbf{a}_{j,i} \right\rangle\right|$. The $\mathbf{a}_{j,i}$ is the ith column vector of matrix $\mathbf{A}_j$.
3. Update set: $\Lambda_t = \Lambda_{t-1} \cup \{\lambda_t\}$.
4. Signal estimate: $\mathbf{x}_{j,t}(\Lambda_t) = \mathbf{A}^{\dagger}_{j,\Lambda_t}\mathbf{y}_j$ and $\mathbf{x}_{j,t}\left(\Lambda_t^C\right) = 0$, where $\mathbf{x}_{j,t}(\Lambda_t)$ is the set of elements whose indices are corresponding to the sparse set.
5. Get new residual: $\hat{\mathbf{y}}_{j,t} = \mathbf{A}_{j,t}\mathbf{x}_{j,t}, \quad \mathbf{r}_{j,t} = \mathbf{y}_j - \hat{\mathbf{y}}_{j,t}$.
6. **Increment** t: Increase iteration number $t = t + 1$, and return to Step 2 if $t < k$.

linear programming. Thus, the problem that we want to solve is

$$\begin{gathered} \min_{(x,u)} \sum_i u_i \\ \text{subject to} \\ \forall_i \, |x(i)| \le u_i \\ \mathbf{Ax} = \mathbf{b} \end{gathered} \tag{15.15}$$

The solution of the earlier equation is equal to the solution of the L_1 minimization problem. Many approaches to solving Equation 15.15 have been studied and developed. Here, we discuss the PDIP method, which is an example of gradient-type algorithms. First, we have the Lagrangian function of Equation 15.15, as follows:

$$L(\mathbf{t}, \boldsymbol{\lambda}, \mathbf{v}) = \begin{bmatrix} 0_1^T & 1^T \end{bmatrix}\mathbf{t} + \mathbf{v}^T\left(\begin{bmatrix} \mathbf{A} & 0_2 \end{bmatrix}\mathbf{t} - \mathbf{b}\right) + \lambda^T\left(\begin{bmatrix} \mathbf{e} & -\mathbf{e} \\ -\mathbf{e} & -\mathbf{e} \end{bmatrix}\mathbf{t}\right) \tag{15.16}$$

where $\mathbf{e}$ is the $n \times n$ identity matrix, 0_1 is the $n \times 1$ zero vector, 0_2 is the $m \times n$ zero matrix, and 1 is the $n \times 1$ vector whose elements are all one, $\mathbf{t} := \begin{bmatrix} \mathbf{x} \\ \mathbf{u} \end{bmatrix} \in \mathbf{R}^{2n\times 1}$, $\mathbf{v} \in \mathbf{R}^{m\times 1}$, and $\boldsymbol{\lambda} \in \mathbf{R}^{2n\times 1} \ge 0$. From the Lagrangian function, we have several KKT conditions,

$$\begin{aligned} &\begin{bmatrix} 0 \\ 1 \end{bmatrix} + \begin{bmatrix} \mathbf{A}^T \\ 0^T \end{bmatrix}\mathbf{v}^* + \begin{bmatrix} \mathbf{e} & -\mathbf{e} \\ -\mathbf{e} & -\mathbf{e} \end{bmatrix}\boldsymbol{\lambda}^* = 0_3 \\ &\begin{bmatrix} \mathbf{A} & 0_2 \end{bmatrix}\mathbf{t}^* - \mathbf{b} = 0_4 \\ &\begin{bmatrix} \mathbf{e} & -\mathbf{e} \\ -\mathbf{e} & -\mathbf{e} \end{bmatrix}\mathbf{t}^* \le 0_1 \\ &(\boldsymbol{\lambda}^*)^T\begin{bmatrix} \mathbf{e} & -\mathbf{e} \\ -\mathbf{e} & -\mathbf{e} \end{bmatrix}\mathbf{t}^* = 0, \quad \boldsymbol{\lambda}^* \ge 0_3 \end{aligned} \tag{15.17}$$

where 0_3 is the $2n \times 1$ zero vector and 0_4 is the $m \times 1$ zero vector. The main point of the PDIP is to seek the point $(\mathbf{t}^*, \boldsymbol{\lambda}^*, \mathbf{v}^*)$ that satisfies the earlier KKT conditions. This is achieved by defining a mapping function $F(\mathbf{t}, \boldsymbol{\lambda}, \mathbf{v}) : \mathbf{R}^{(2n+m)\times 1} \rightarrow \mathbf{R}^{(2n+m)\times 1}$, which is

$$F(\mathbf{t}, \boldsymbol{\lambda}, \mathbf{v}) = \begin{bmatrix} \begin{bmatrix} 0 \\ 1 \end{bmatrix} + \begin{bmatrix} \mathbf{A}^T \\ 0^T \end{bmatrix} \mathbf{v} + \begin{bmatrix} \mathbf{e} & -\mathbf{e} \\ -\mathbf{e} & -\mathbf{e} \end{bmatrix} \boldsymbol{\lambda} \\ (\boldsymbol{\lambda}^*)^T \begin{bmatrix} \mathbf{e} & -\mathbf{e} \\ -\mathbf{e} & -\mathbf{e} \end{bmatrix} \mathbf{t}^* \\ \begin{bmatrix} \mathbf{A} & 0_2 \end{bmatrix} \mathbf{t} - \mathbf{b} \end{bmatrix} = 0_4 \in \mathbf{R}^{(2n+m)\times 1},$$

$$\begin{bmatrix} \mathbf{e} & -\mathbf{e} \\ -\mathbf{e} & -\mathbf{e} \end{bmatrix} \mathbf{t}^* \leq 0_1, \quad \boldsymbol{\lambda}^* \geq 0_3 \tag{15.18}$$

where 0_4 is the $(2n+1)\times 1$ zero vector. Now, we would like to find the point $(\mathbf{t}^*, \boldsymbol{\lambda}^*, \mathbf{v}^*)$ satisfying $F(\mathbf{t}^*, \boldsymbol{\lambda}^*, \mathbf{v}^*) = 0_4$. Here, we use a linear approximation method. From the Taylor expansions of the function $F(\mathbf{t}, \boldsymbol{\lambda}, \mathbf{v})$, we have

$$F(\mathbf{t} + \Delta\mathbf{t}, \boldsymbol{\lambda} + \Delta\boldsymbol{\lambda}, \mathbf{v} + \Delta\mathbf{v}) \approx F(\mathbf{t}, \boldsymbol{\lambda}, \mathbf{v}) + \nabla_{(\mathbf{t},\boldsymbol{\lambda},\mathbf{v})} F(\mathbf{t}, \boldsymbol{\lambda}, \mathbf{v}) \begin{bmatrix} \Delta\mathbf{t} \\ \Delta\mathbf{v} \\ \Delta\boldsymbol{\lambda} \end{bmatrix} \tag{15.19}$$

Thus, solving the earlier equations yields the direction $(\Delta\mathbf{t}, \Delta\mathbf{v}, \Delta\boldsymbol{\lambda})$. Next, we seek the proper step length along the direction that does not violate $\begin{bmatrix} \mathbf{e} & -\mathbf{e} \\ -\mathbf{e} & -\mathbf{e} \end{bmatrix} \mathbf{t}^* \leq 0_1$ and $\boldsymbol{\lambda}^* \geq 0_3$. The pseudo code for the PDIP algorithm is shown in Table 15.6.

Table 15.6 Primal-Dual Interior Point Method Algorithm

1. Initialize:
   Choose $\mathbf{v}^0 \in \mathbf{R}^{m\times 1}$, $\boldsymbol{\lambda}^0 \geq 0_3$, and $\mathbf{t}^0 = [\mathbf{x}^0 \quad \mathbf{u}^0]^T$, where $\mathbf{x} = \mathbf{A}^\dagger \mathbf{b}$, and $\mathbf{u}^0 = \left|\mathbf{x}^0\right| + \alpha \left|\mathbf{x}^0\right|$ and iteration number k= 1. (The $\mathbf{A}^\dagger = \left(\mathbf{A}^T\mathbf{A}\right)^{-1} \mathbf{A}^T$ is the Moore–Penrose pseudo-inverse of **A** and $\mathbf{A}^T$ denotes the transpose of **A**.)
2. Find the direction vectors $(\Delta\mathbf{t}, \Delta\mathbf{v}, \Delta\boldsymbol{\lambda})$:
   $$\begin{bmatrix} \Delta\mathbf{t} \\ \Delta\mathbf{v} \\ \Delta\boldsymbol{\lambda} \end{bmatrix} = -\left[\nabla_{\left(\mathbf{t}^k,\boldsymbol{\lambda}^k,\mathbf{v}^k\right)} F\left(\mathbf{t}^k, \boldsymbol{\lambda}^k, \mathbf{v}^k\right)\right]^{-1} F\left(\mathbf{t}^k, \boldsymbol{\lambda}^k, \mathbf{v}^k\right).$$
3. Find the proper step length:
   Choose the largest α satisfying $\left\|F\left(\mathbf{t}^k + \alpha, \boldsymbol{\lambda}^k + \alpha, \mathbf{v}^k + \alpha\right)\right\|_2^2 \leq \left\|F\left(\mathbf{t}^k, \boldsymbol{\lambda}^k, \mathbf{v}^k\right)\right\|_2^2$.
4. Update parameters: $\mathbf{t}^{k+1} = \mathbf{t}^k + \alpha\Delta\mathbf{t}$, $\quad \mathbf{v}^{k+1} = \mathbf{v}^k + \alpha\Delta\mathbf{v}$, $\quad \boldsymbol{\lambda}^{k+1} = \boldsymbol{\lambda}^k + \alpha\Delta\boldsymbol{\lambda}$.
5. Update the signal: $\mathbf{x}^{k+1} = \mathbf{x}^k + \mathbf{t}\left[1 : n\right]$.
6. Increment the iteration number k: Increase iteration number $k = k + 1$, and return to Step 2 if $\left\|\mathbf{y} - \mathbf{A}\mathbf{x}^k\right\|_2^2 >$ eps.

15.6 Performance Evaluation

In this section, we investigate the performance of a WSN system that applies CS by using the PDIP or the SOMP. We divide this section into four sections that analyze the relationship among the number of measurements M, sparsity k depending on the number of sensors J, the degree of correlation, and the signal-to-noise ratio (SNR), respectively. To avoid confusion regarding the graphs, we define the notations and metrics used in the experiments in Tables 15.7 and 15.8, respectively.

Table 15.7 Notation Used in Experiments

Notation
N: The length of the signal $\mathbf{x}$ at each sensor
M: the length of measurement $\mathbf{y}$
j: Index of the sensors, $j \in \{1, 2, \ldots, J\}$
$\mathbf{y}$: Signal transmitted from each sensor
$\mathbf{A}$: Sensing matrix, $\mathbf{R}^{M\times N}$, its elements are generated by a Gaussian distribution
$\mathbf{x}$: Sparse signal on the sensor; its elements also have a Gaussian distribution
$\mathbf{n}$: Additive white Gaussian noise (AWGN)
K: Common sparsity number
K_j : Innovation sparsity number

Table 15.8 Metrics Used in Experiments

Metrics
1. SNR (dB): Signal-to-noise ratio, $$10\log_{10}\frac{\|\mathbf{Ax}\|_2^2}{M\sigma_n^2}$$ where σ_n^2 is the variance of noise. 2. MSE: Mean square error, $$\frac{\|\hat{\mathbf{x}}-\mathbf{x}\|_2^2}{\|\mathbf{x}\|_2^2}$$ where $\hat{\mathbf{x}}$ is the recovered signal $\mathbf{x}$ is the original signal

The proposed correlation signals, *JSM*-1 and *JSM*-2, as described in Table 15.1, will also be investigated in terms of various parameters, such as signal length, matrix size, and sparsity number. To recover the *JSM*-1 signal (which includes both common and innovation components, and the common component has the same values for every sensor) from received signal $\mathbf{y}$, we use the PDIP algorithm. However, to recover the *JSM*-2 signal (which includes only a common component that has different values for every sensor), we use SOMP. It is inappropriate to apply SOMP to the *JSM*-1 signals because there exists the innovation component at every sensed signal $\mathbf{x}_j$. Although SOMP can identify the common part exactly, confusion may arise regarding the optimal selection for the innovation component. Because SOMP selects only one index that has the optimal value among the vector elements of length N in every iteration, if the selected index is included in the innovation component of only one sensor node, the solution cannot be correct.

For this reason, we use the SOMP algorithm to recover only the *JSM-2* signal. If we use SOMP to recover the *JSM*-1 signal, we should improve the algorithm for finding the innovation component. From the results of simulations using those two recovery algorithms, we determined the relationship among the sensors, measurement, and amount of correlation in the unknown sensor signals.

15.6.1 Reconstruction Performance as a Function of Sparsity

Figure 15.7 shows the results when the PDIP algorithm was used to reconstruct signals for *JSM*-1. We increased the common sparsity K of each sensor and the number of sensors J while fixing the signal length, the number of measurements, and the innovation sparsity of each sensor at $N = 50$, $M = 20$, and $K_j = 3$, respectively. The FC concatenates the received signals $\mathbf{y}_j, j = \{1, 2, \ldots, J\}$ to

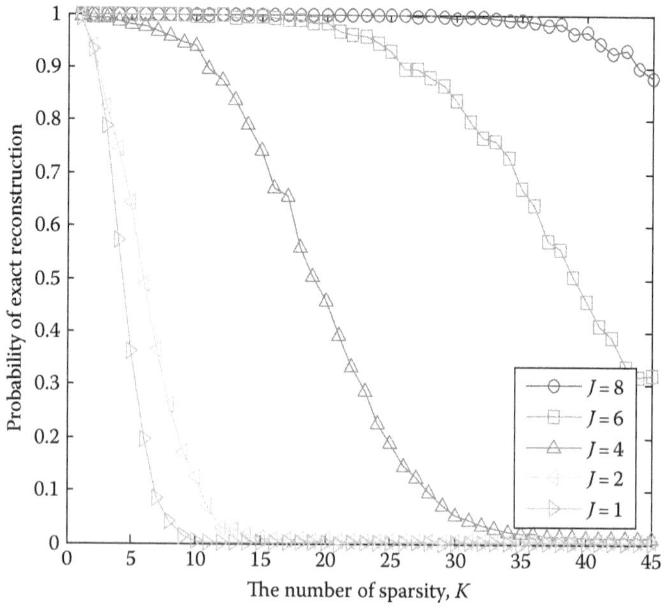

Figure 15.7 Signal reconstructed using PDIP algorithm for *JSM*-1. System parameters are $N = 50$, $M = 20$, and innovation sparsity $K_j = 3$.

$[\mathbf{y}_1, \mathbf{y}_2, \ldots, \mathbf{y}_J]^T$, and puts the sensing matrices to the integrated one as $\mathbf{A}_{PDIP}$ of Equation 15.20. Thus, the equation is $\mathbf{y}_{PDIP} := \mathbf{A}_{PDIP}\mathbf{z}_{PDIP}$, and the number of measurements in this equation is $M_{PDIP} = M \times J$; it then uses PDIP algorithm to get $\mathbf{z}_{PDIP}$ from $\mathbf{y}_{PDIP}$. The recovered signal $[\hat{\mathbf{z}}_c, \hat{\mathbf{z}}_1, \hat{\mathbf{z}}_2, \ldots, \hat{\mathbf{z}}_J]^T$ from $[\mathbf{y}_1, \mathbf{y}_2, \ldots, \mathbf{y}_J]^T$ was compared with the original $[\mathbf{z}_c, \mathbf{z}_1, \mathbf{z}_2, \ldots, \mathbf{z}_J]^T$ in order to calculate the probability of exact reconstruction.

$$\mathbf{y}_{PDIP} := \begin{bmatrix} \mathbf{y}_1 \\ \mathbf{y}_2 \\ \vdots \\ \mathbf{y}_J \end{bmatrix} = \underbrace{\begin{bmatrix} \mathbf{A}_1 & \mathbf{A}_1 & 0 & \cdots & 0 \\ \mathbf{A}_2 & 0 & \mathbf{A}_2 & \cdots & \vdots \\ \vdots & \vdots & \vdots & \ddots & 0 \\ \mathbf{A}_J & 0 & 0 & 0 & \mathbf{A}_J \end{bmatrix}}_{\mathbf{A}_{PDIP}} \underbrace{\begin{bmatrix} \mathbf{z}_c \\ \mathbf{z}_1 \\ \mathbf{z}_2 \\ \vdots \\ \mathbf{z}_J \end{bmatrix}}_{\mathbf{z}_{PDIP}:=} \tag{15.20}$$

In this case, even if the original signal $\mathbf{x}_j = \mathbf{z}_c + \mathbf{z}_j \in \mathbf{R}^n, j \in \{1, 2, \ldots, J\}$ is not sparse, the signals y_j transmitted from sensors can be recovered perfectly at the FC if all the sensors have a small number of innovation component K_j that corresponds to $\mathbf{z}_j$. However, as the number of sensors increases, the integrated matrix also becomes large. Consequently, the computation is complex, and much time is required to obtain the solution.

Figure 15.8 illustrates the use of the SOMP algorithm to recover *JSM*-2. The fixed parameters are the signal length N and measurement size $M$of each sensor. To determine the effect of the number of sensors and sparsity in the WSN, we increased the sparsity K and the number of sensors J. Because the *JSM*-2 signal has the same sparse location for every sensor, the sparse location can be found by using SOMP easily. As the number of sensors increases, the probability of making the optimal decision at each iteration is greater. As a result, exact reconstruction is achieved, as shown in Figure 15.8.

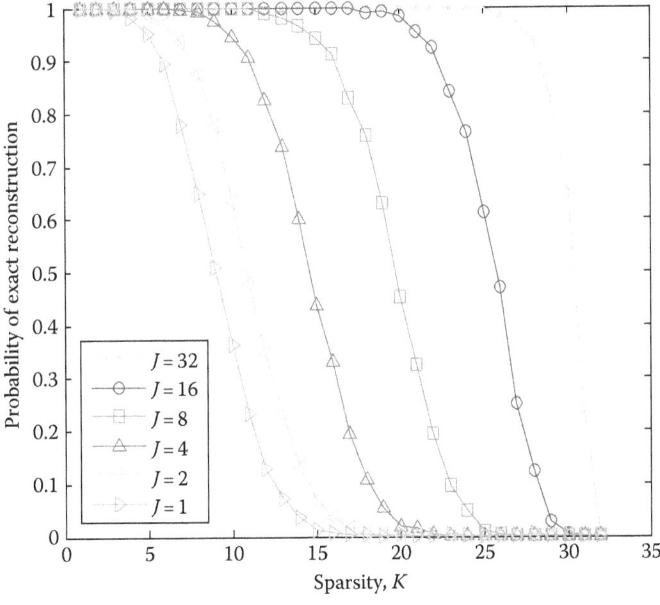

Figure 15.8 Signal reconstructed using SOMP for *JSM*-2 for increasing common sparsity K and number of sensors J. System parameters: $N = 256$ and $M = 32$.

In both cases, we notice that the probability of successful reconstruction increases as the number of sensors increases, because both algorithms use the prior information that the signals are correlated. For example, when we increase only the common sparsity K, we can reconstruct all of the signals by only increasing the number of sensors. Interestingly, the curve of Figure 15.7 in *JSM*-1 experiment does not show convergence as the number of sensors increases. On the other hand, that of Figure 15.8 in *JSM*-2 experiment converges to $M - 1$ as the number of sensors increases. These results are determined from the ratio of the number of measurement to sparsity (M/K) in compressive equation. In the case of Figure 15.7, as the number of sensors increases, the number of measurement M_{PDIP} also increases. Thus, as the number of sensors increases, the ratio is also changed. (In our experiment, we choose $K_j = 3$, where $K_j \ll M$. Therefore, the ratio increases as the number of sensors increases.) In the case of Figure 15.8, there is no change for the ratio regardless of increasing the number of sensors. The varying ratio (M/K) of *JSM*-1 experiment makes the result about no convergence in contrast with that of JSM-2 experiment.

SUMMARY 15.1 RECONSTRUCTION PERFORMANCE AS A FUNCTION OF SPARSITY

We aim at investigating how the increase in sparsity K for signal at each sensor affects reconstruction performance of the joint recovery algorithms, while the signal length N and the number of measurements M are fixed at each sensor. As the common sparsity K of each sensor increases, the probability of exact reconstruction decreases. This is obvious. Equation 15.20 is the result of *JSM*-1 model that can be used to represent both common and innovative elements in each sensor and allows exploitation of inter-sensor correlation. Thus, as the number of sensors increases, the total sparsity and the number of measurements M_{PDIP} also increase as shown in Equation 15.20. In *JSM*-2, the sparsity K and the number of measurements Mper sensor are fixed by the formulation in (15.14), regardless of the number of sensors. The varying ratio between the number of measurement and sparsity makes the results of Figures 15.7 and 15.8, respectively.

15.6.2 Relationship between the Number of Sensors and the Number of Measurements Required for Exact Reconstruction

In Figure 15.9, we show the results when we increased the number of measurements and the number of sensors while fixing the signal length ($N = 50$), common component ($K = 9$), and innovation component ($K_j = 3$). As the number of sensors increased, the number of measurements required for the probability of exact reconstruction to converge to one decreased. Therefore, if we use many sensors to reconstruct the correlated signal, we can reduce the number of measurements, which in turn reduces the transmission power at each sensor. However, as Figures 15.9 and 15.10 show, the decrease in measurement size is limited by the sparsity number $(K + 1)$ in one sensor.

For *JSM*-2 signals, reconstruction is similar to that of *JSM*-1 signals in terms of the effect of increasing the number of sensors when the correlated signal is jointly recovered (Figure 15.10, solid line). However, if signal reconstruction is performed separately, more measurements per sensor are needed as the number of sensors J increases (Figure 15.10, dotted line). Because the transmitted signals from each sensor are reconstructed independently, if the probability p of successful reconstruction is less than or equal to 1, then the total probability of successful reconstruction for all transmitted signals is p^J.

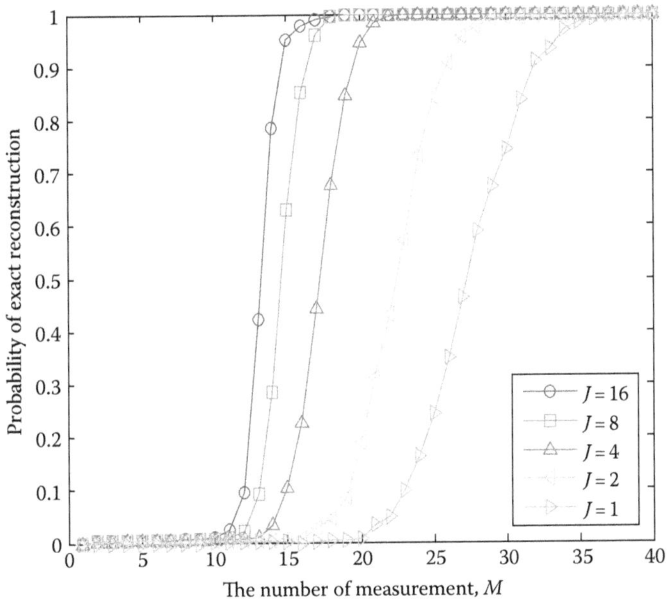

Figure 15.9 Signal reconstructed using PDIP algorithm for *JSM*-1. System parameters: $N = 50$, $K = 9$, and $K_j = 3$.

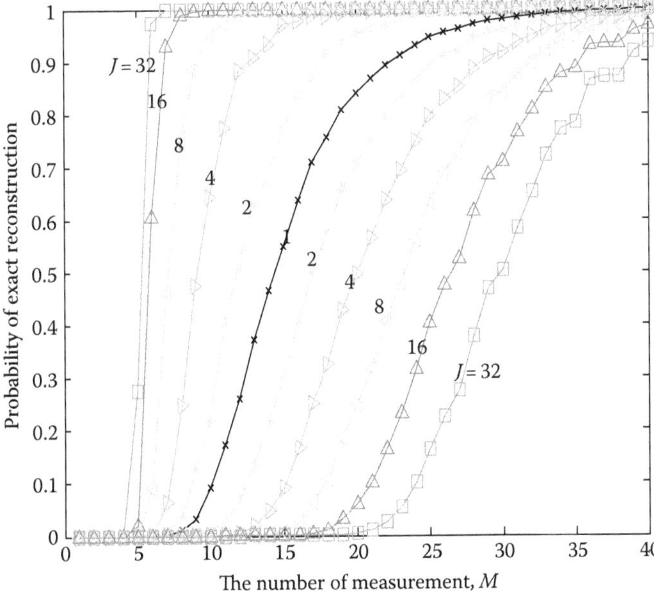

Figure 15.10 Reconstruction using SOMP for *JSM*-2. System parameters: $N = 50$ and $K = 5$. Solid line: joint recovery; dotted line: separate recovery.

SUMMARY 15.2 RELATIONSHIP BETWEEN THE NUMBER OF SENSORS AND THE NUMBER OF MEASUREMENTS REQUIRED FOR EXACT RECONSTRUCTION

We aim at investigating how the probability of exact reconstruction changes with the number of sensors increased. As the number of sensors is increased, the signals FC collects are more inter-sensor correlated and the number of measurements per sensor required for exact reconstruction decreases. Figures 15.9 and 15.10 show that the original signals can be recovered with high probability at the fixed measurement as $J \to \infty$ and the per-sensor measurements required for perfect signal recovery converges to $K + 1$.

15.6.3 Performance as a Function of SNR

In this section, we present the system performance of a WSN that uses CS in an additive white Gaussian noise (AWGN) channel. As in the other experiment, we used a Gaussian distribution to create the sensing matrix $\mathbf{A}_j$, $j \in \{1, 2, \ldots, J\}$, and sparse signal $\mathbf{x}_j$ and then added AWGN n to the measurement $\mathbf{y}_j = \mathbf{A}_j\mathbf{x}_j$. At the FC, the received signal $\tilde{\mathbf{y}}_j = \mathbf{y}_j + n$ was recovered jointly. We increased the number of sensors while fixing the signal length, number of measurements, common sparsity, and innovation sparsity at $N = 50$, $M = 20$, $K = 3$, and $K_j = 2$, respectively. In this experiment, the SNR is set as follows:

$$\text{SNR}\left(\text{signal to noise ratio}\right) = 10\log_{10}\frac{\left\|\mathbf{A}_j\mathbf{x}_j\right\|^2}{M\sigma_n^2}$$

where

$\|\mathbf{A}_j\mathbf{x}_j\|^2$ is the transmitted signal power at sensor j
M is the number of measurements
σ_n^2 is the noise variance

To estimate the reconstruction error between the original signal $\mathbf{x}_j$ and the reconstruction signal $\hat{\mathbf{x}}_j$, we used the mean square error (MSE) as follows:

$$\text{Mean square error} = \frac{\left\|\hat{\mathbf{x}}_j - \mathbf{x}_j\right\|^2}{\left\|\mathbf{x}_j\right\|^2}$$

We applied the PDIP algorithm to solve Equation 15.20 for *JSM*-1 and obtained the solution, $[\hat{\mathbf{z}}_c, \hat{\mathbf{z}}_1, \hat{\mathbf{z}}_2, \ldots, \hat{\mathbf{z}}_J]^T$. Because of the effect of noise, the solution $[\hat{\mathbf{z}}_c, \hat{\mathbf{z}}_1, \hat{\mathbf{z}}_2, \ldots, \hat{\mathbf{z}}_J]^T$ does not have a sparse solution. Therefore, we chose the largest $K + (J \times K_j)$ values from among the elements of the solution. To compare the recovered signal $\hat{\mathbf{x}}_j$ to the original sensed signal $\mathbf{x}_j$, we divided the concatenated solution $[\hat{\mathbf{z}}_c, \hat{\mathbf{z}}_1, \hat{\mathbf{z}}_2, \ldots, \hat{\mathbf{z}}_J]^T$ by each recovered signal $\hat{\mathbf{x}}_j$. The results are shown in Figure 15.11.

To obtain the results in Figure 15.12, we used the SOMP algorithm for *JSM*-2 with the same processing. In contrast to the PDIP algorithm, the SOMP algorithm first searches the support set; therefore, it does not require a step in which the largest K values are chosen from among the elements of the solution. However, if the selected support set is wrong, the reconstruction is also wrong. Both results, Figures 15.11 and 15.12, show that if we increase the number of sensors,

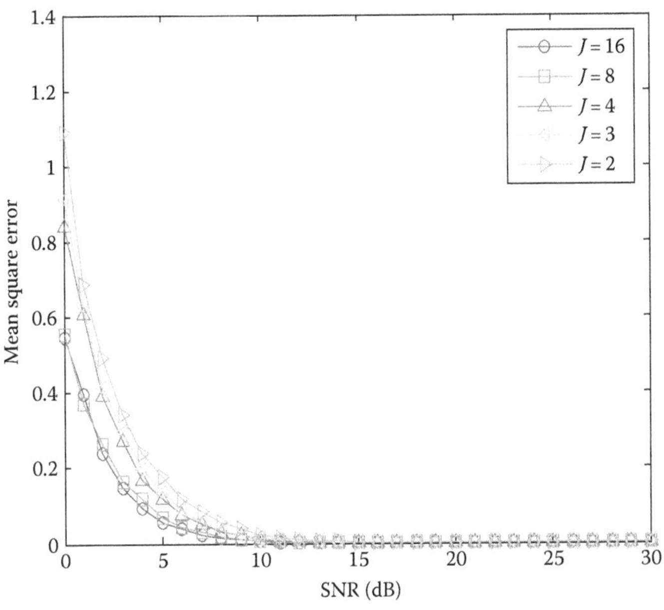

Figure 15.11 Signal reconstructed using PDIP method for *JSM*-1. System parameters: $N = 50$, $M = 20$, $K = 3$, and $K_j = 2$.

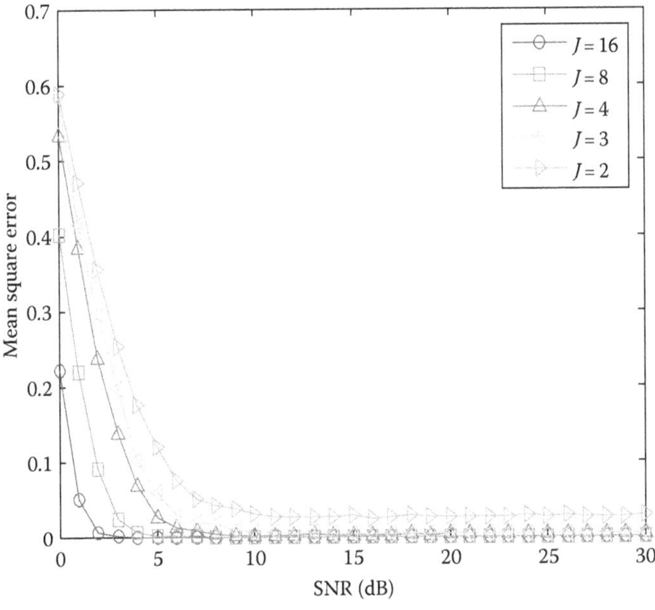

Figure 15.12 Signal reconstructed using SOMP for *JSM*-2. System parameters: $N = 50$, $M = 20$, and $K = 5$.

the MSE is improved and finally converges to zero as the SNR increases. Even if the transmitted signals contain much noise, having a large number of sensors to observe the correlated signal in the sensed region facilitates the search for the exact solution. In Figure 15.12, when the number of sensors is two or three, the MSE does not converge to zero even if the SNR is high. Because the SOMP algorithm uses cross-correlation to find the support set (step 2 of Table 15.5), if the rank of sensing matrix **A** is smaller than the number of columns in **A**, then each column will exhibit significant correlation among themselves. Consequently, the SOMP algorithm selects the wrong support location. However, this problem can be solved by using a large number of sensors.

SUMMARY 15.3 PERFORMANCE AS A FUNCTION OF SNR

We aim to investigate the effect of noise in CS-based WSN. In particular, we experiment how MSE decreases as SNR increases. Figures 15.11 and 15.12 show similar results. As the number of sensors increases, signals are more correlated. This helps signal recovery.

15.6.4 Joint versus Separate Recovery Performance as a Function of Correlation Degree

Now, we compare the results of joint recovery and separate recovery (Figure 15.13). In joint recovery, if a correlation exists between the signals observed from the distributed sensors, the FC can use the correlated information to recover the transmitted signals. In separate recovery, correlated information is not used regardless of whether a correlation pattern exists between the

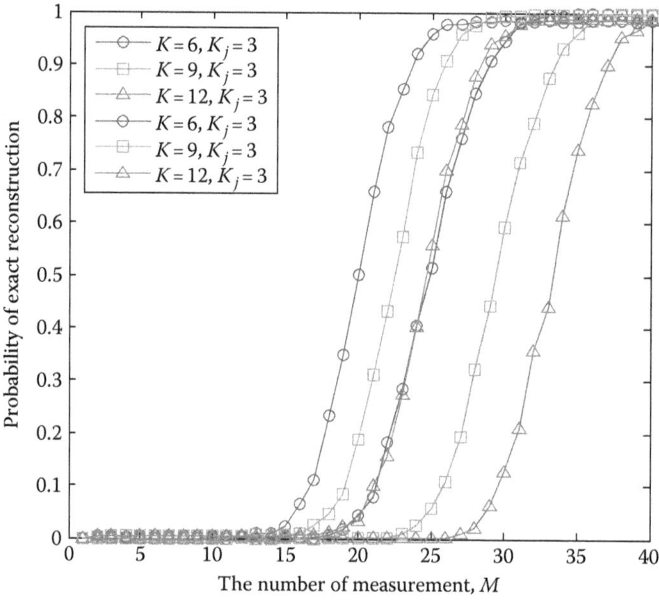

Figure 15.13 Joint (solid line) and separate (dotted line) reconstructions using PDIP algorithm for *JSM*-1. System parameters: $N = 50$ and $J = 2$. The benefits of joint reconstruction depend on the sparsity number K.

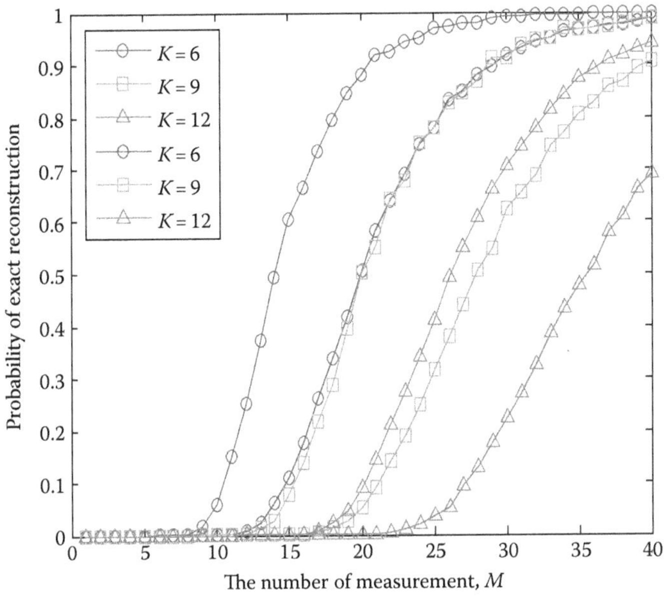

Figure 15.14 Joint (solid line) and separate (dotted line) reconstructions using SOMP for *JSM*-2. System parameters: $N = 50$ and $J = 2$. Joint reconstruction has a higher probability of success than separate reconstruction.

observed signals. In Figure 15.13, solid lines were obtained from joint reconstructions, whereas dotted lines are the results of separate reconstructions.

When we use separate reconstruction, we cannot obtain any benefits from correlated information. However, when we use joint reconstruction, we can reduce the measurement size. For example, in Figure 15.14, the required number of measurements is almost 40 (dashed line and circles, $K = 6$) for perfect reconstruction when we use separate reconstruction. On the other hand, when we use joint reconstruction, it decreases to around 30 (solid line and circles, $K = 6$). Furthermore, as the common sparsity increases, the performance gap increases. For example, when the common sparsity is 9, joint reconstruction has a 90% probability of recovering all the signals at $M = 30$. However, the probability that separate reconstruction can recover all the signals is only 70%. Figure 15.13 also shows that joint reconstruction is superior to separate reconstruction. For example, we need at least 30 measurements for reliable recovery using separate reconstruction. However, we merely need at least 25 measurements for reliable recovery using joint reconstruction.

SUMMARY 15.4 JOINT VERSUS SEPARATE RECOVERY PERFORMANCE AS A FUNCTION OF CORRELATION DEGREE

If a correlation exists between the signals observed from the distributed sensors, and if the FC uses the joint recovery, then it can reduce the measurement size required for exact reconstruction in comparison with that of the separate recovery. As the degree of correlation increases, the gap in the results of two methods (joint recovery and separate recovery) widens as shown in Figures 15.13 and 15.14.

15.7 Summary

In this chapter, we discussed the application of CS for WSNs. We assumed a WSN consisting of spatially distributed sensors and one FC. The sensor nodes take signal samples and pass their acquired signal samples to the FC. When the FC receives the transmitted data from the sensor nodes, it aims to recover the original signal waveforms for later identification of the events possibly occurring in the sensed region (Section 15.3.1).

We discussed that CS is the possible solution that provides simpler signal acquisition and compression. CS is suitable for the WSNs since it allows removal of intermediate stages such as sampling the signal and gathering the sampled signals at one collaboration point, which would usually be the case in a conventional compression scheme. Using CS, the amount of signal samples that need to be transferred to the FC from the sensors can be significantly reduced. This may lead to reduction of power consumption at the sensor nodes, which was discussed in Section 15.3.3. In summary, each sensor with CS can save power by not needing to run complex compression operations onboard and by cutting down signal transmissions.

Distributed sensors usually observe a single globally occurring event and thus the observed signals are often correlated with each other. We considered two types of correlations: intra- and inter-sensor signal correlations. We provided the sparse signal models, which encompass both types of correlation in Sections 15.4.1 and 15.4.2.

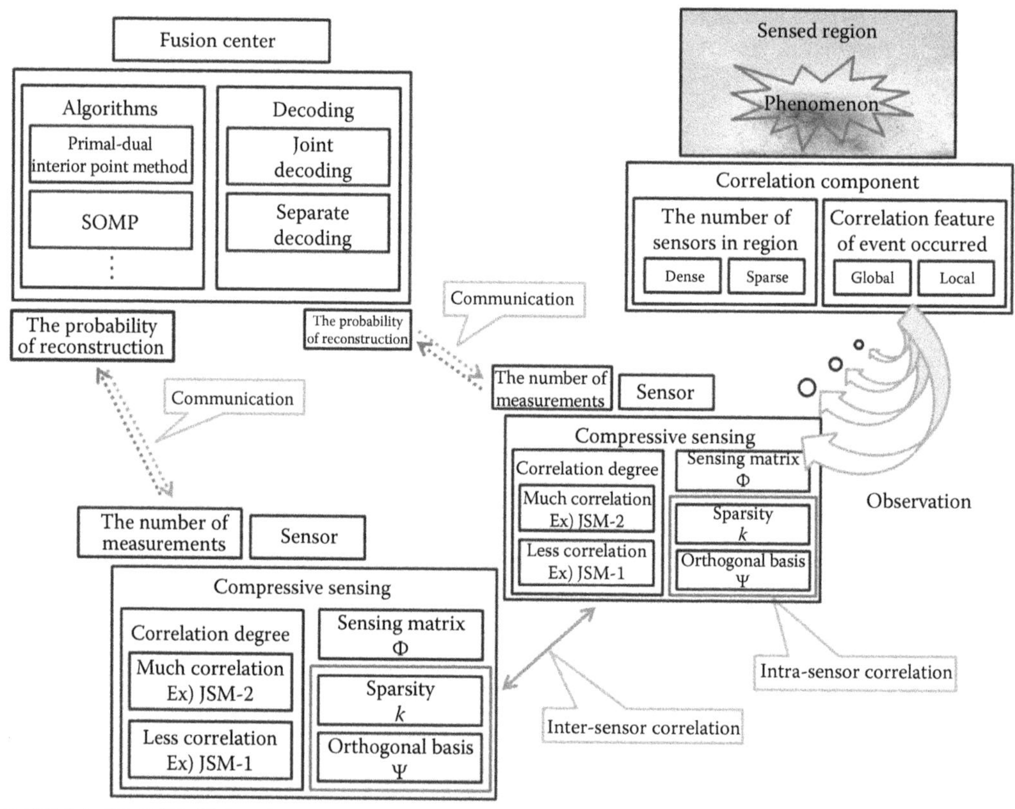

Figure 15.15 Summary of CS application in WSN.

The FC receives the compressed signals from the sensors. The FC then recovers the original signal waveforms from the compressed signals using a CS recovery algorithm. We considered two types of algorithms. One is a greedy type, which includes the OMP and the SOMP algorithms, discussed in Section 15.5.1. The other is a gradient type for which we used the PDIP method, in Section 15.5.2.

Finally, we presented simulations results in which the CS-based WSN system parameters such as the number of measurements, the sparsity, and the signal length were varied. We discussed the use of a joint recovery scheme at the FC. A CS recovery algorithm is referred to as the *joint recovery* scheme when it utilizes inter-sensor signal correlation as well. By contrast, when the inter-sensor signal correlation is not utilized, it is referred to as the *separate recovery* scheme. In the joint recovery scheme, inter-sensor signal correlation information is incorporated in the formation of recovery equation as shown in Equations 15.12 and 15.14. In the separate recovery scheme, a sensor signal recovery is done individually and independently from the recovery of other sensor signals. We compared the results of the joint recovery with those of the separate recovery scheme. We have shown that correlation information can be exploited and the number of measurements needed for exact reconstruction can be significantly reduced as shown in Figure 15.14. It means that the traffic volume transmitted from the sensors to the FC can decrease significantly without degrading the quality of the recovery performance (Section 15.6).

We have shown that the CS is an efficient and effective signal acquisition and sampling framework for WSN (Figure 15.15), which can be used to save transmittal and computational power significantly at the sensor node. This CS-based signal acquisition and compression scheme is *very simple*, so it is suitable for inexpensive sensors. The number of compressed samples required for transmission from each sensor to the FC is *significantly small*, which makes it perfect for sensors whose operational power is drawn from onboard battery. Finally, the joint CS recovery at the FC exploits signal correlation and enables DCS.

Acknowledgement

This work was supported by the National Research Foundation of Korea (NRF) grant funded by the Korean government (MEST) (Do-Yak Research Program, No. 2012-0005656).

References

1. D. L. Donoho, Compressed sensing, *IEEE Transactions on Information Theory*, 52(4), 1289–1306, April 2006.
2. D. L. Donoho and J. Tanner, Precise undersampling theorems, *Proceedings of IEEE*, 98, 913–924, May 2010.
3. R. G. Baraniuk, Lecture notes: Compressed sensing, *IEEE Signal Processing Magazine*, 24(4), 118–121, July 2007.
4. J. Romberg, Imaging via compressive sampling, *IEEE Signal Processing Magazine*, 25(2), 14–20, March 2008.
5. E. Candes and J. Romberg, Caltech, L_1-Magic: Recovery of sparse signals via convex programming, October 2005.
6. A. Y. Yang, M. Gastpar, R. Bajcsy, and S. S. Sastry, Distributed sensor perception via sparse representation, to appear in *Proceedings of IEEE*, 98(6), 1077–1088.

7. D. L. Donoho and M. Elad, Maximal sparsity representation via ℓ_1 minimization, *Proceedings of the National Academy of Sciences of the United States of America*, 100, 2197–2202, March 4, 2003.
8. J. Solobera, Detecting forest fires using wireless sensor networks with waspmote October 1, 2011. http://www.libelium.com/wireless_sensor_networks_to_detec_forest_fires/
9. A. Hac, *Wireless Sensor Network Designs*, John Wiley & Sons, Ltd., New York, 2003.
10. D. Baron, M. F. Duarte, S. Sarvotham, M. B. Wakin, and R. G. Baraniuk, An information theoretic approach to distributed compressed sensing, in *Proceedings of 43rd Allerton Conference on Communication, Control, and Computing*, Allerton, IL, September 2005.
11. M. F. Duarte, S. Sarvotham, D. Baron, M. B. Wakin, and R. G. Baraniuk, Distributed compressed sensing of jointly sparse signals, *Asilomar Conference on Signals*, Systems and Computers, Asilomar, pp. 1537–1541, October 28–November 1, 2005.
12. M. Mishali and Y. C. Eldar, Reduce and boost: Recovering arbitrary sets of jointly sparse vectors, *IEEE Transactions on Signal Processing*, 56(10), 4692–4702, 2008.
13. J. A. Tropp and A. C. Gilbert, Signal recovery from random measurements via orthogonal matching pursuit, *IEEE Transactions on Information Theory*, 53(12), 4655–4666, December 2007.
14. J. A. Tropp, A. C. Gilbert, and M. J. Strauss, Simultaneous sparse approximation via greedy pursuit, in *Proceedings of IEEE International Conference on Acoustics, Speech, and Signal Processing (ICASSP)*, 725, v/72I–v/724, 2005.
15. M. E. Davies and Y. C. Eldar, Rank awareness in joint sparse recovery, Arxiv preprint arXiv: 1004.4529, 2010.

Chapter 16

Compressive Sensing for Wireless Sensor Networks

Mohammadreza Mahmudimanesh, Abdelmajid Khelil, and Neeraj Suri

Contents

Compressive sensing (CS) is a new paradigm in signal processing and sampling theory. In this chapter, we introduce the mathematical foundations of this novel theory and explore its applications in wireless sensor networks (WSNs). CS is an important achievement in sampling theory and signal processing. It is increasingly being implied in many areas like multimedia, machine learning, medical imaging, etc. We focus on the aspects of CS theory that has direct applications in WSNs. We also investigate the most well-known implementations of CS theory for data collection in WSNs.

16.1 Compression of Signals in Sensory Systems

16.1.1 Components of Distributed Sensory Systems

When we look at many sensory systems, we usually observe three different functions.

1. *Sensing*: A sensor records the physical value of interest, like light intensity, sound, pressure, temperature, etc., and an analog-to-digital converter (ADC) quantizes the recorded value and outputs a binary number that reflects the current sensed value.
2. *Transmission*: The sensor recordings are transmitted periodically over a communication channel. Depending on the technology used for transferring data, this process entails different levels of transmission cost. For example, a webcam connected directly to a computer has a fast, accurate, and cheap (low power, inexpensive) communication channel for transferring data. In several applications, the raw data recorded by sensors are transmitted over a noisy wireless channel. For example, in a WSN, a vast amount of data has to be transmitted over a noisy wireless channel. Because of collision, congestion, and inevitable errors, some of the packets are dropped, causing more and more retransmissions. Such a communication channel is too expensive for the sensor nodes (SNs) since each transmission consumes a lot of node battery power. When the communication is too expensive or the amount of raw data is far more than channel capacity, the raw data must be compressed before transmission to utilize the communication line efficiently.
3. *Processing and storage*: The raw data collected from the individual sensors are processed by a fusion center or stored in a persistent storage mechanism for future processing. The data collected from the network can be encoded, encrypted, or compressed. Depending on the application of the sensory systems, a combination of compression, encryption, or other special encodings is applied. In each case, a suitable decoding, decompression, or decryption technique may be applied by the receiver. Note that when the acquired data are too large in raw format, a compression technique must be applied before storage or forwarding.

16.1.2 Compressibility of Signals

Figure 16.1 shows an example of Fourier transform that is performed on signal acquired from a temperature sensor. Although the signal itself in time domain does not appear to have a specific structure, its Fourier transform is quite compressible. We see that only very few Fourier coefficients have a large magnitude and most of other coefficients are quite negligible. One can devise a compression technique by keeping only the largest Fourier coefficients and discarding the others.

To show how efficient transform-coding compression techniques are, we do a simple numerical experiment on the time-domain signal in Figure 16.1. We represent the signal as a vector $f \in \mathbb{R}^N$. Assume $x \in \mathbb{C}^N$ is the Fourier transform of f. We select k largest elements of x and set its other $N - k$ elements to zeros. Let y be the resulting vector. Now, we do an inverse Fourier transform on y and get the vector g, a recovered version of the original signal f. Since the signal is to be recovered from incomplete set of information, reconstruction error is inevitable. We measure the error by calculating mean square error (MSE). Figure 16.2 illustrates that the recovery error rapidly decreases when we increase the number of Fourier measurements. In other words, almost all energy of the time-domain signal in Figure 16.1 is concentrated on the few largest Fourier coefficients of its frequency-domain transform.

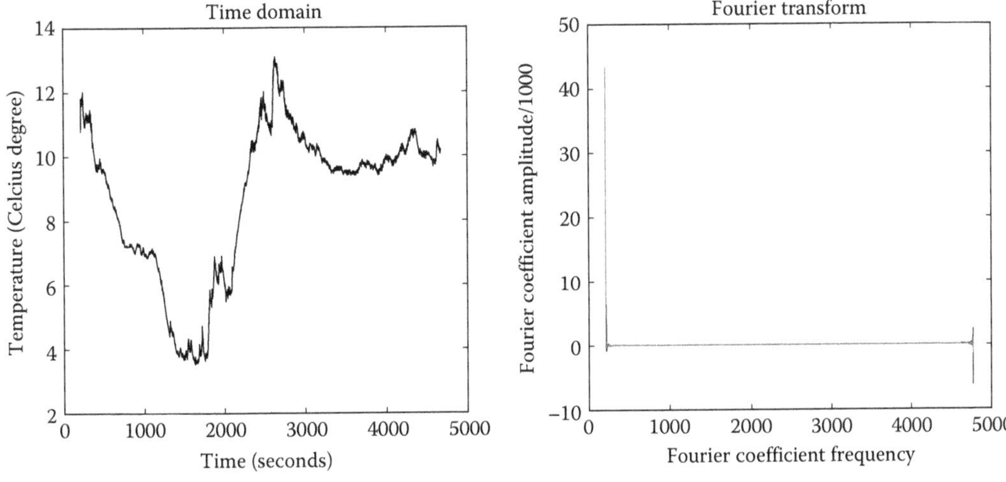

Figure 16.1 Compressibility of a natural signal under Fourier transform.

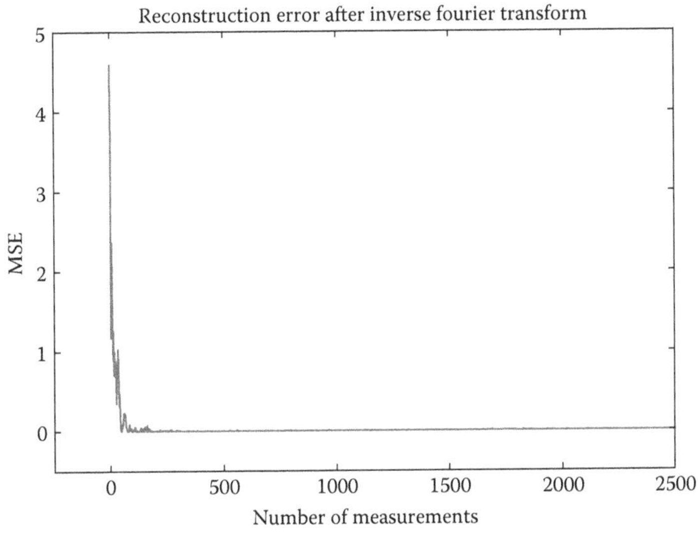

Figure 16.2 Reconstruction error decay for a signal compressible in Fourier domain.

Many other signals acquired from natural phenomena like sound, radiation level, humidity, as well as more complex audiovisual signals recorded by cameras and microphones have a very dense support on a suitably chosen domain like Fourier, wavelet, or discrete cosine transform (DCT). Therefore, we can depict the overall function of a transform coding system in Figure 16.3.

16.1.3 Compression Techniques

In general, we see a variety of compression and decompression requirements in sensory systems, especially distributed sensor networks like WSNs. Compression can be either lossless or lossy.

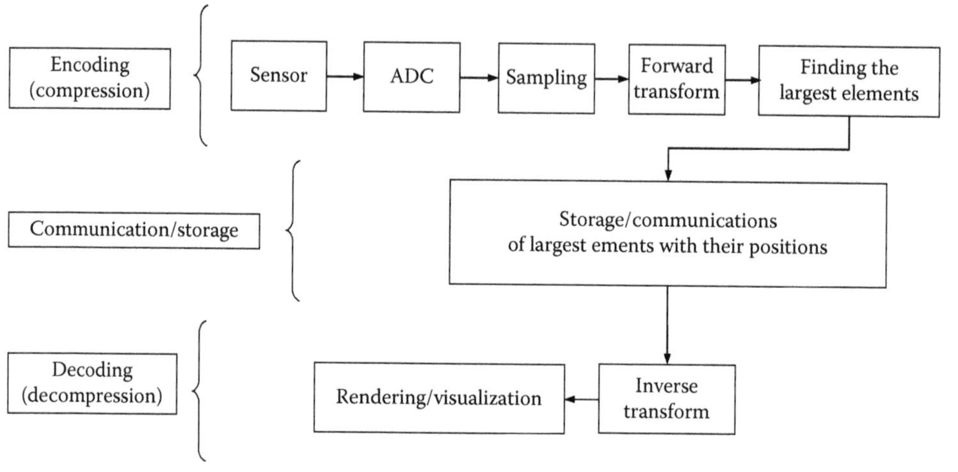

Figure 16.3 Traditional transform coding compression and decompression system.

- *Lossless compression* refers to compression techniques that conserve the whole raw data without losing accuracy. Lossless compression is usually used in a common data compression technique which is widely used in commercial compression applications. Some of the most commonly used lossless compression algorithms are run-length encoding, Lempel–Ziv–Welch (LZW), and Lempel–Ziv–Markov chain algorithm. These algorithms are used in many compression applications and several file formats like graphics interchange format, portable network graphics, ZIP, GNU ZIP, etc.
- *Lossy compression*, on the other hand, allows unimportant data to be lost for the sake of more important data. Most of the lossy compression techniques involve transforming the raw data to some other domain like frequency domain or wavelets domain. Therefore, such techniques are also called *transform coding*. Since decades, it has been known that natural signals like audio, images, video and signals recorded from other physical phenomena like seismic vibrations, radiation level, etc. have a sparse or compressible support in Fourier, DCT, and wavelets. An encoder of such natural signals can transform the raw data using a suitable transformation and throw out negligible values. A decoder can recover the original signal with some small error from fewer number of compressed data items.

The CS theory discusses lossy compression technique that differs from transform coding in measurement acquisition as we explain later in this chapter. First, we study the transform coding compression and decompression technique in more detail and point out its inefficiencies. Moreover, transform coding and CS are strongly connected topics and having a good knowledge of transform coding helps in better understanding of CS and its advantages.

Encoding part of the compression system illustrated in Figure 16.3 is a complex function that is run on a large amount of data. Note that after determining the largest elements of the transformed signal, negligible values are simply thrown away. One can ask why should such a large amount of data be acquired to obtain a small amount of information. This question is answered by compressive sensing theory. It says that under certain conditions, a compressible signal can be recovered from a few number of random linear measurements. The aim of CS is to bring the compression techniques down to the sampling level. As we see later in this chapter, this is especially important in WSN, as we prefer to compress the data as it is being transported over the network. Random measurement makes it easier to implement CS in a distributed manner, hence no centralized decision and control is needed. Linearity of the measurement mechanism simplifies hardware and software design for the sensor nodes. As we describe later in this chapter, acquiring linear measurements from the network involves only multiplication and additions to be done by the nodes and the complexity of the calculations remains to its minimum. Moreover, such operations can be done more efficiently if a cheap low-power hardware is built in the sensor nodes.

16.2 Basics of Compressive Sensing

Throughout this text we may use the terms *signal* and *vector* interchangeably. Both refer to discrete values of a spatiotemporal phenomenon. For example, a WSN consisting of n SNs each of which recording r samples in every T seconds interval, produces a discrete spatiotemporal signal (vector) $\mathbf{f} \in \mathbb{R}^{nr}$. To begin a formal description of the CS theory, we need some mathematical definitions that may apply both to the original signal in time or space domain and its projection on some frequency domain. Therefore, we use the notation $\mathbf{v}$ as a general vector that may refer to a signal or its transformation. The first three definitions are general mathematical definitions that may apply to any vector. Definition 16.4 refers to the CS theory in a more specific manner. This clarification is required to avoid ambiguity in using the mathematical notation for the signal vectors.

Definition 16.1 Vector $\mathbf{v} \in \mathbb{R}^N$ is said to be S-sparse if $\|\mathbf{v}\|_0 = S$, i.e., $\mathbf{v}$ has only S nonzero entries and its all other $N - S$ entries are zero.

Definition 16.2 S-sparse vector $\mathbf{v}_S \in \mathbb{R}^N$ is made from non-sparse vector $\mathbf{v} \in \mathbb{R}^N$ by keeping S largest entries of $\mathbf{v}$ and zeroing its all other $N - S$ entries.

Definition 16.3 Vector $\mathbf{v}$ is said to be compressible if most of its entries are near zero. More formally, $\mathbf{v} \in \mathbb{R}^N$ is compressible when $\|\mathbf{v} - \mathbf{v}_S\|_2$ is negligible for some $S \ll N$.

These definitions may apply to any vector, which can be a signal vector or its projection on some orthonormal basis. In the following sections, we may use Definitions 16.1 through 16.3 both to time-domain or space-domain signals or their projections on orthonormal bases like Fourier, DCT, wavelet, etc. In any case, we can substitute the vector $\mathbf{v}$ in these definitions with the signal vector $\mathbf{f}$ or its compressive projection $\mathbf{x}$, whichever is discussed in the context.

Definition 16.4 Signal $\mathbf{f}$ is compressible under orthonormal basis Ψ when $\mathbf{f} = \Psi\mathbf{x}$ and $\mathbf{x}$ is compressible. The matrix Ψ is a real or complex orthonormal matrix with the basis vectors of Ψ as its rows. We also say that Ψ is a compressive basis for signal $\mathbf{f}$.

Most signals recorded from natural phenomena are compressible under Fourier, DCT, and the family of wavelet transforms. This is the fundamental fact behind every traditional compression technique. Audio signals are compressible under Fourier transform. Images are compressible under DCT or wavelet. WSN also records a distributed spatiotemporal signal from a natural phenomenon, which can be compactly represented under the family of Fourier or wavelet orthonormal bases. CS is distinguished from traditional compression techniques in signal acquisition method as it combines compression into the data acquisition layer and tries to recover the original signal from fewest possible measurements. This is very advantageous in applications where acquiring individual samples is infeasible or too expensive. WSN is an excellent application for CS since acquiring all and every single sample from the whole network leads to a large traffic over the capacity of limited SNs and significantly limits the WSN lifetime.

Definition 16.5 Measurement matrix Φ_m is an $m \times N$ real or complex matrix consisting of $m < N$ basis vectors randomly selected from orthonormal measurement basis Φ that produces an incomplete measurement vector $\mathbf{y} \in \mathbb{C}^m$ such that $\mathbf{y} = \Phi_m \mathbf{f}$.

Definition 16.6 Coherence between the measurement basis Φ and the compressive basis Ψ is denoted by $\mu(\Phi, \Psi)$ and is equal to $\max_{1 \le i,j \le N} |\langle \phi_i, \psi_j \rangle|$ where for each $1 \le i,\ j \le N$, ϕ_i's and ψ_j's are basis vectors of Φ- and Ψ-domain, respectively, and $\langle \cdot \rangle$ is the inner product operation.

Now, we are ready to state the fundamental theorem of CS according to the aforementioned definitions.

Theorem 16.1 [1] Suppose signal $\mathbf{f} \in \mathbb{R}^N$ is S-sparse in Ψ-domain, i.e., $\mathbf{f} = \Psi\mathbf{x}$ and $\mathbf{x}$ is S-sparse. We acquire m linear random measurements by randomly selecting m basis vectors of the measurement basis Φ. Assume that $\mathbf{y} \in \mathbb{C}^m$ represents these incomplete measurements such that $\mathbf{y} = \Phi_m \mathbf{f}$ where Φ_m is the measurement matrix. Then it is possible to recover $\mathbf{f}$ exactly from $\mathbf{y}$ by solving the following convex optimization problem

$$\hat{\mathbf{x}} = \underset{\mathbf{x} \in \mathbb{C}^N}{\operatorname{argmin}} \|\mathbf{x}\|_1 \quad \text{subject to} \quad y_k = \langle \Psi\mathbf{x}, \phi_k \rangle \quad \text{for all} \quad k = 1, \ldots, m. \tag{16.1}$$

where ϕ_k's are the rows of the Φ_m matrix. Recovered signal will be $\hat{\mathbf{f}} = \Psi\hat{\mathbf{x}}$.

In case of real measurement and compressive bases, problem (16.1) can be simplified to a linear program [2]. Noiselet [3] measurement matrices involve complex numbers and thus a linear program cannot solve problem (16.1). Convex optimization problem (16.1) with complex values can be cast to a second-order cone program (SOCP) [4].

Accurate signal recovery is possible when the number of measurements follows

$$m > C \cdot S \cdot \log N \cdot \mu^2(\Phi, \Psi) \tag{16.2}$$

where $C > 1$ is a small real constant [5].

From Equation (16.2), it is clear why sparsity and incoherence are important in CS. To efficiently incorporate the CS theory in a specific sampling scenario, we need the measurement and compressive bases to be incoherent as maximum as possible to decrease parameter μ in Equation 16.2. Moreover, compressive basis must be able to effectively compress the signal $\mathbf{f}$ to decrease S in Equation 16.2. When these two preconditions hold for a certain sampling configuration, it is possible to recover the signal **f** from m measurements, where m can be much smaller than the dimension of the original signal [1].

Haar wavelet and noiselet are two orthonormal bases with perfect incoherence and hence are of special interest for CS theory [6]. Noiselet as measurement basis (Φ) and Haar wavelet as compressive basis (Ψ) have a small constant coherence independent of the dimension of spatial signal [7]. Their perfect incoherence makes them very useful in many CS applications. Although CS theory suggests Haar wavelet and noiselet as the perfect compressive and measurement pair, in practice it may not be plausible to compute noiselet measurements in a WSN unless we embed parts of the noiselet transform matrix inside the SNs. Interestingly, random matrices such as a Gaussian matrix with independent and identically distributed (i.i.d.) entries from a $\mathcal{N}(0, 1)$ have low coherence with any fixed orthonormal basis [5]. The elements of such a random matrix can be calculated on the fly using a pseudorandom number generator, which is common between SNs and the sink. When the normal random generator, at every SN is initialized by the id-number of that SN, the sink can also reproduce the measurement matrix exactly at the sink. Note that in this case there is no need for a centralized control to update the measurement matrix and the values of the measurement matrix are not needed to be stored inside the SN. Therefore, using random measurement matrices gives us more flexibility and requires less memory on the SNs. Instead, because of slightly more coherence between random measurement matrix and the fixed compressive basis, the number of required measurements m will increase according to Equation 16.2.

CS is also very stable against noisy measurement and can also handle signals that are not strictly sparse but compressible. It is a very idealistic condition being able to transform signal **f** to a strictly sparse vector in the Ψ-domain. Instead, **f** is always transformed into a compressible form with many near-zero entries. Candès et al. [8,9] have shown that if $\|\mathbf{x} - \mathbf{x}_S\|_2 < \epsilon$ for some integer $S < N$ and a small real constant ϵ, then the recovery error by solving problem (16.1) will be about $O(\epsilon)$. Similarly, in a noisy environment if the measurement vector is added by an additive white Gaussian noise (AWGN) $\sim \mathcal{N}(0, \sigma^2)$, the recovery error is bounded by $O(\sigma^2)$.

16.3 Implementation of Compressive Sensing for Wireless Sensor Networks

One of the first detailed implementations of CWS is introduced by Luo et al. [10,11] in compressive data gathering (CDG) for large-scale WSN. They made a comprehensive comparison between CS and traditional compression techniques showing that CS leads to a more efficient and stable signal acquisition technique compared to some traditional methods like in-network compression [12] and distributed source coding (DCS) [13]. Using overcomplete dictionaries [14] in solving (16.1), CDG is also able to tackle abnormal sensor readings and cope with unpredicted events in the operational environment of WSN. In their initial attempt, Luo et al. [10] have applied a random matrix as their measurement matrix seeded by a unique number broadcast in the network and a common pseudorandom number generator embedded in all SNs. Random matrices such as a Gaussian matrix with i.i.d. entries from a $\mathcal{N}(0, 1)$ have low coherence with any fixed orthonormal basis [5].

16.3.1 Compressive Wireless Sensing

Bajwa et al. [15] first proposed an implementation of CS for WSNs, which was based on analog data communication. Although the realization of their abstract idea is merely feasible, their model provides a simple technique to acquire compressive measurements from a WSN. Therefore, we begin the study of major CS implementations for WSNs first by introducing CWS. Let us assume Φ is a measurement basis and Φ_m is a measurement matrix that has m rows and N columns. Here, N is the total number of SNs and m is the required number of compressive measurements. m can be determined by Equation 16.2 according to the compressibility level of the spatial signal. The measurement matrix Φ_m is made of m randomly selected basis vectors from the measurement basis Φ. When we are using random Gaussian basis, Φ_m can be simply populated by random real numbers from a Gaussian distribution with zero mean and unit variance. Either using random measurements calculated on the fly, or using embedded measurement vectors, we have to calculate the result of $\Phi_m\mathbf{f}$. Here, $\mathbf{f} \in \mathbb{R}^N$ is the spatial signal with a dimension equal to the number of SNs. The matrix multiplication $\mathbf{y} = \Phi_m\mathbf{f}$ is actually a linear measurement process that must be performed in a distributed manner. The result will be the vector $\mathbf{y} \in \mathbb{R}^m$, which has a lower dimension than the original signal. The CS recovery algorithm then has to recover the original vector $\mathbf{f}$ from $\mathbf{y}$.

To understand how CWS works, we consider the simplest case where all SNs can directly communicate with the sink. We know that Φ_m has N columns. Suppose each SN embeds one column of Φ. When using random measurements, no embedded data are required. The ith column of Φ_m can be generated on the fly. The only requirement here is that the entries of Φ_m can be regenerated at the sink. Therefore, either the columns of Φ are embedded into SNs before deployment of the WSN, or SNs produce their corresponding column using a pseudo-random Gaussian random number generator, which is known to the sink. The same Φ_m can be then reproduced at the sink using the same algorithm and a predetermined seed. When SNs initialize the pseudorandom Gaussian random number generator with their own id, the whole sequences of random entries of the measurement matrix can be regenerated at each sampling period.

All SNs record the value of the intended physical phenomena at the same time. This means that the nodes have synchronized clocks (we discuss later the disadvantages of the need for synchronized clocks). Each SN then multiplies its recorded real value with its own column vector, which is part of the measurement matrix Φ_m. We set $\mathbf{w}_i = f_i\phi_i$ where f_i is the value recorded by that SN and ϕ_i is the ith column of Φ_m. When all SNs have done the calculation of their own vector multiplication, the vector $\mathbf{w}_i$ will be ready to be transmitted from each SN. If all the SNs transmit their own $\mathbf{w}_i$ at the same time, the result of matrix multiplication $\mathbf{y} = \Phi_m\mathbf{f}$ will be accumulated at the sink. Figure 16.4 illustrates this setup.

CWS assumes that the SNs can be perfectly synchronized such that the measurement vector $\mathbf{y}$ can be accurately accumulated at the sink. Bajwa et al. [15] have also considered the effect of AWGN present during the radio communication. As discussed earlier, the recovery error in the presence of a AWGN $\mathcal{N}(0, \sigma^2)$ will be in the order of σ^2.

The accumulation is actually taking place by adding vectors of the same size across the WSN. Therefore, it can be done by any order since the addition is a commutative operation. This property of CS measurement has a very important advantage. Due to the commutative property of addition, accumulation can be done in any order. Therefore, the same approach as discussed earlier can be applied to a star topology, chain topology, or more complex tree topologies. As long as the tree routing is applied, accumulation can be done over many levels. Thus, we can extend the very simple topology in Figure 16.4 to a multi-hop WSN depicted in 16.5.

The CWS model in its very primary form, which is based on synchronized superposition of signals, is infeasible with current hardware of today's typical SNs. However, it has interesting

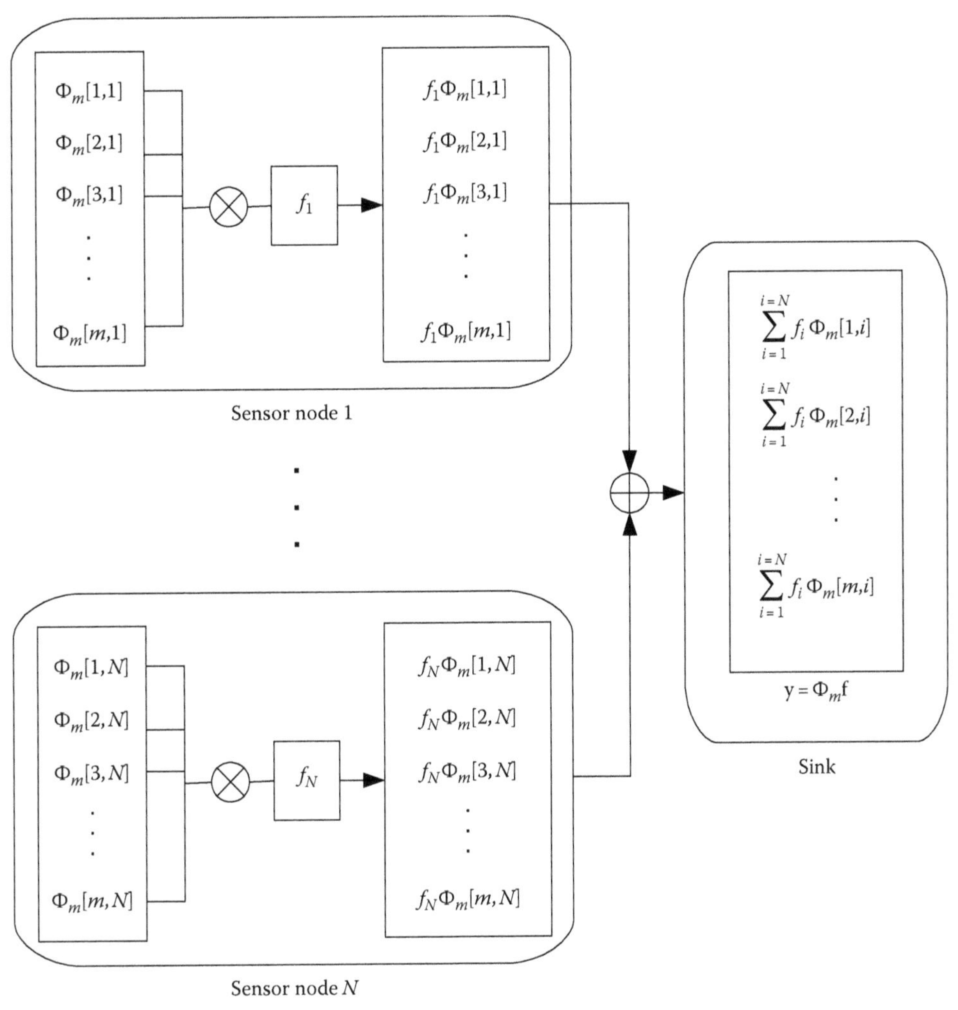

Figure 16.4 CWS in a WSN with star topology.

attributes in its abstract form. Note that the accumulation of the vectors can be done on the air and without further computation by the SNs. In the star topology, the accumulation takes place instantly when all SNs have transmitted their own calculated vector. Today's common wireless medium access control (MAC) protocols for WSNs are all based on detecting and avoiding collision. There has been a large body of continuous work for an efficient MAC protocol that faces minimum collisions. Now, CWS requires a synchronized collision. We believe that implementing a communication layer with perfectly synchronized collisions is even more challenging than making a wireless MAC protocol with minimum collisions. This topic can be considered as an open area in the application of CS in WSN. There can be certain limited applications where analog CWS can be applied instead of using digital communication, which is used in common MAC protocols. The abstract model of the CWS prepares us to get familiar with a more practical implementation of CS for WSNs which we discuss in the next subsection.

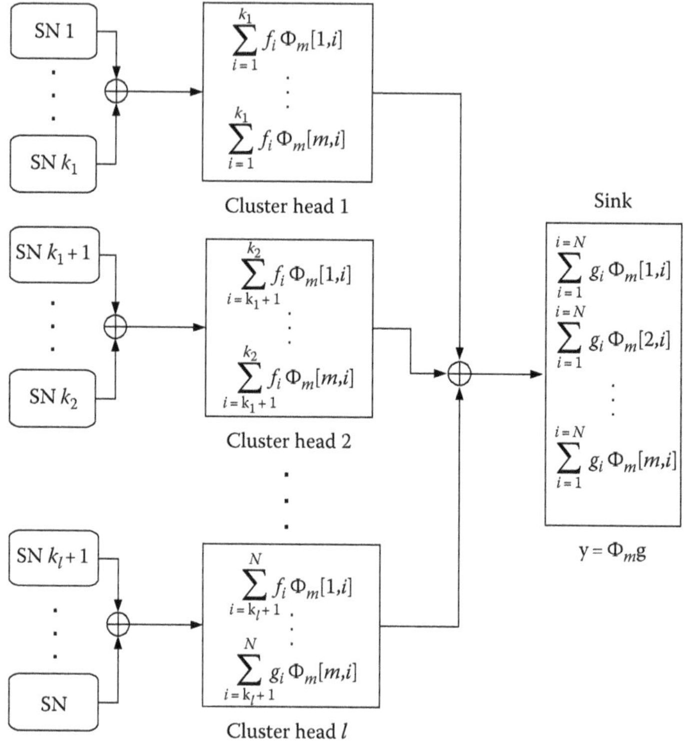

Figure 16.5 CWS in a multi-hop WSN with tree topology.

16.3.2 Compressive Data Gathering

Luo et al. [10,11] introduced a model the abstract for the abstraction of CWS data aggregation as message passing through the WSN. Therefore, we do not discuss the data transfer as analog signals any more. Instead, data transmission is modeled as messages that are transferred over a digital wireless communication channel. CDG in its heart remains very much like CWS. It only considers spatial sampling, and the spatial measurements are acquired periodically. CDG addresses also the problem of abnormal sensor readings. Abnormal readings may render the signal not being compressible on the compressive basis. Figure 16.6 shows an example of the effect of the abnormal samples on the compressibility of a signal. Figure 16.6a shows the original signal, which is sparse in DCT as can bee seen in Figure 16.6b. In Figure 16.6c, we see the same signal with only two abnormal sensor readings. The value of those two samples is either too high or too low compared to the average of the samples. Figure 16.6d shows the DCT transform of the abnormal signal. Obviously, the signal contaminated with abnormal values is not compressible under DCT. This violates the preconditions required for operation of the traditional CS. Fortunately, there is a solution to this problem by using overcomplete dictionaries [14].

CDG proposes the use of overcomplete dictionaries to detect abnormal sensor readings. Assume $\mathbf{d}$ is the spatial signal vector that may have been contaminated with abnormal readings. Remember that we assume that the abnormal readings in the signal are sparse, otherwise, it is not possible to

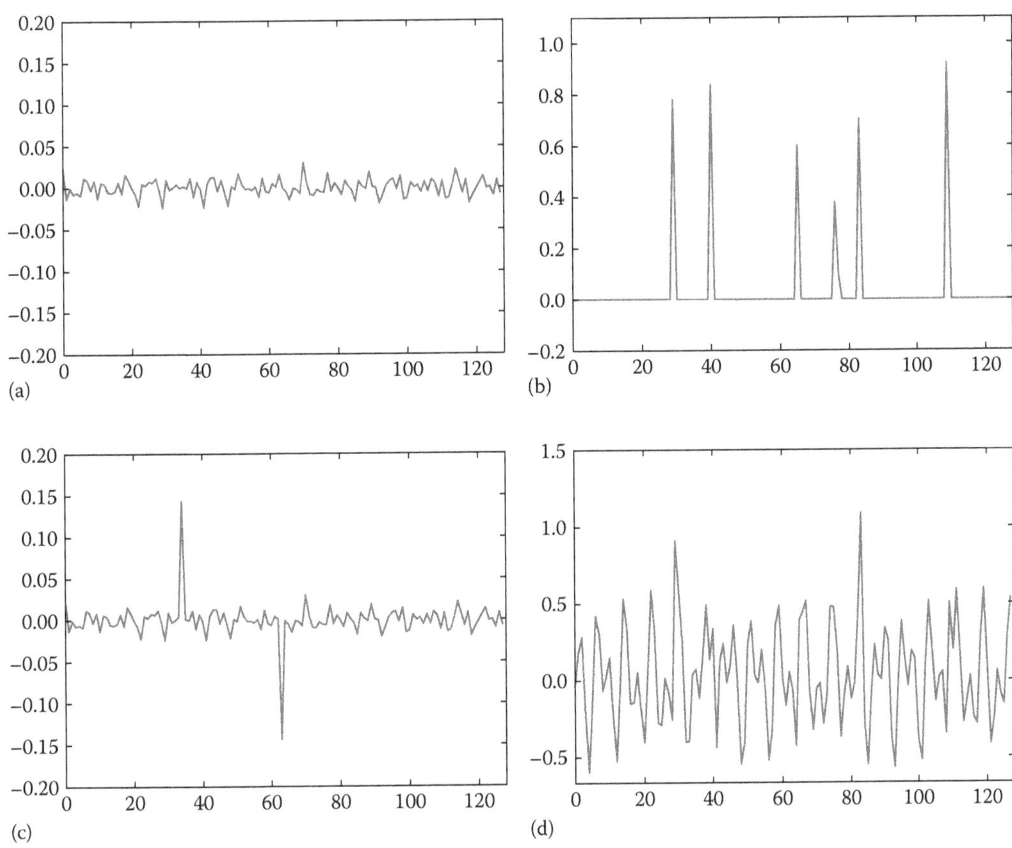

Figure 16.6 Effect of abnormal readings on the DCT projection of a signal. (a) original signal, (b) DCT transform of original signal, (c) original signal contaminated with abnormal readings, and (d) DCT transform of the distorted signal.

recover a heavily distorted signal by any efficient data acquisition mechanism. We can decompose the vector $\mathbf{f}$ into two vectors:

$$\mathbf{f} = \mathbf{f}_0 + \mathbf{d}_s \tag{16.3}$$

where

$\mathbf{f}_0$ is the normal signal, which is sparse or compressible in the compressive basis Ψ

$\mathbf{d}_s$ contains the deviated values of the abnormal readings

The vector $\mathbf{d}_s$ is supposed to be sparse since the abnormal sensor readings are rare and sporadic. Therefore, $\mathbf{d}_s$ is sparse in space domain. The original signal vector $\mathbf{f}$ can be then represented as the linear combination of two sparse signals and we can rewrite Equation 16.3 as

$$\mathbf{f} = \Psi\mathbf{f}_0 + I\mathbf{x}_s \tag{16.4}$$

where $\mathbf{x}_0$ is the sparse projection of the normal component $\mathbf{f}_0$ on the Ψ-domain. Vector $\mathbf{d}_s$ is in fact equal to $\mathbf{x}_s$ since $\mathbf{d}_s$ is already sparse in the space domain. I is the identity matrix and hence

$\mathbf{d}_s = \mathbf{x}_s = I\mathbf{x}_s$. It is possible to project the signal $\mathbf{f}$ on a single overcomplete basis. First, we need to construct an augmented transformation matrix for an overcomplete system named Ψ'. Assume that the matrix Ψ' is made by placing the matrices Ψ and I next to each other. Formally speaking, $\Psi' = [\Psi I]$, which has now N rows and $2N$ columns.

Donoho et al. [14] have shown the possibility of stable recovery of sparse signal under an overcomplete system. They have also proved that their method is effective for measurements contaminated with noise. The recovery takes place using a convex optimization similar to the typical CS recovery method. The recovery error in the presence of noise would be in the order of the additive noise magnitude. Since we are going to search for a solution in an augmented dictionary, our target vector has now the dimension of $2N$ instead of N. Note that $\mathbf{f} \in \mathbb{R}^N$ can be represented by a sparse projection on the $\boldsymbol{\Psi}'$ basis:

$$\mathbf{f} = \Psi'\mathbf{x}, \quad \mathbf{x} = [\mathbf{x}_0{}^T \mathbf{x}_s{}^T]^T \tag{16.5}$$

Here, you see that $\mathbf{x}$ has a dimension of $2N$ and is made by concatenating $\mathbf{x}_s$ and $\mathbf{x}_0$ on top of each other.

Now, we apply the CS Theorem 16.1 for this new pair of signal and measurements. The measurement vector can be calculated using a random measurement matrix as before. The estimated solution vector will be $\hat{\mathbf{x}} = [\hat{\mathbf{x}}_0^T \hat{\mathbf{x}}_s^T]^T$. The original signal can be recovered by calculating $\hat{\mathbf{f}} = \Psi'\hat{\mathbf{x}}$, which is expected to have a limited error compared to $\mathbf{f}$. Again, the recovery error depends on the number of measurements m and the noise that is present in the communication channel. CDG is also able to detect the location of abnormal events in the WSN. First, we need to solve the convex optimization problem (16.1) for the augmented system and find the solution $\hat{\mathbf{x}} = [\hat{\mathbf{x}}_0^T \hat{\mathbf{x}}_s^T]^T$. The nonzero elements in $\hat{\mathbf{x}}_s$ determine the position of the abnormal events.

CDG also provides an analysis of the network capacity when using CS for acquiring the distributed spatial signal. Comparison of baseline data transmission in a WSN with uniformly random distribution of SNs shows that a network capacity gain of N/m is possible using CDG. N is the number of SNs and m is the number of required compressive measurements. m depends on the sparsity or compressibility of the Ψ-transform of signal vector $\mathbf{f}$. For a S-sparse compressible signal, roughly $m = 4S$ measurements is required to recover the signal with an acceptable accuracy [11].

16.3.3 Distributed Compressed Sensing

Wakin et al. in distributed compressed sensing (DCS) [16] give a model of WSN with each node having a direct link to the sink. DCS considers not only the spatial correlation of a distributed signal but also the temporal correlation over a period of time. Therefore, DCS can be regarded as a spatiotemporal sampling technique. The spatiotemporal signal is modeled as a combination of several temporal signals. DCS assumes two components are contributing to the spatiotemporal signal recorded by the WSN. The first component is the common component, which is sensed overall by SNs. For the second component, individual SNs can contribute their own signals, which can be sparse. According to this configuration, DCS defines three joint sparsity models (JSMs), namely JSM-1, JSM-2, and JSM-3 which we briefly explain later. Suppose there are J SNs each of which are producing a temporal signal $\mathbf{f}_j$ with dimension N. This means J SNs are producing N samples in every sampling period. We assume that there is a fixed Ψ-basis for $\mathbb{R}^N$ on which the signals can be projected sparsely. Now, we explain attributes of the three JSM models.

JSM-1: Sparse common component + innovations. In this model, all of the temporal signals share a common component, which is compressible in Ψ-domain while every SN may contribute a sparse innovation component:

$$\mathbf{f}_j = \mathbf{z}_C + \mathbf{z}_j \,, j \in \{1, 2, \ldots, J\} \tag{16.6}$$

such that

$$\mathbf{z}_C = \Psi\theta_C, \quad \|\theta_C\|_0 = K \quad \text{and} \quad \mathbf{z}_j = \Psi\|\theta_j\|_0, \quad \|\theta_j\|_0 = K_j \tag{16.7}$$

In this formula, $\mathbf{z}_C$ is the common component that has the K-sparse representation of θ_C. Every SN has its own innovation $\mathbf{z}_j$, which is K_j-sparse in Ψ-domain. Note that here Ψ is an $N \times N$ orthonormal matrix. This model can describe of wide range WSN applications. Assume a WSN that is deployed in an outdoor environment to record the microclimate data like ambient air temperature. Overall, nearly a constant temperature is observed that is close to the average. There might be some more structures present in the signal values over a geographical area, but the signal is expected to be temporally compressible. In some spots of the operational environment, there is a temperature difference because of local events or conditions like shades, flow of water, etc. Therefore, the signal of this example can be perfectly modeled by JSM-1.

JSM-2: Common sparse supports. In this model, signals recorded by all SNs can be projected sparsely on a single fixed orthonormal basis, but the coefficients may differ for every SN. Formally speaking, this means that

$$\mathbf{f}_j = \Psi\theta_j, \quad j \in \{1, 2, 3, \ldots, J\} \tag{16.8}$$

where each signal vector θ_j can be projected only on the same subset of basis vectors $\Omega \subset \{1, 2, 3, \ldots, N\}$ such that $|\Omega| = K$. Therefore, all temporal signals recorded by SNs have an l_0 sparsity of K, but the amplitude of coefficients may differ for each SN. This model is useful in situations where all signals have the same support in frequency domain, but may experience phase shifts and attenuations due to the signal propagation. In scenarios like acoustic localizations, it is necessary to recover every signal individually.

JSM-3: Non-sparse common + sparse innovations. This model can be regarded as an extension to JSM-1 since it does not require the common signal to be sparse in the Ψ-domain; however, the innovations by individual SNs are still sparse:

$$\mathbf{f}_j = \mathbf{z}_C + \mathbf{z}_j, \quad j \in \{1, 2, \ldots J\} \tag{16.9}$$

$$\mathbf{z}_C = \Psi\theta_C \quad \text{and} \quad \mathbf{z}_j = \Psi\theta_j, \quad \|\theta\|_{l_0} = K_j. \tag{16.10}$$

One application of this model can be in a distributed deployment of cameras where the overall picture can be presented as a reference picture with some small innovations sensed by different cameras. These differences from the base picture depend on the position of each camera.

Wakin et al. [16] have also proposed an efficient novel algorithm especially developed for joint signal recovery. In general, the reconstruction accuracy or recovery probability increases with more measurements. For separate signal recovery, the performance decreases when we add more SNs, that is, when the value of J increases. This is expected according to the traditional theory of the CS. The dimension of the signal increases, but the number of measurements does not increase according to Equation 16.2. On the other hand, according to the numerical experiments done Wakin et al., the accurate reconstruction probability increases when more SNs are added to the network. By increasing J, we will have more correlation between the SNs. Therefore, with fewer measurements the original signal can be recovered accurately. Most interestingly, DCS leads to asymptotically optimal decoding in JSM-2 as the number of SNs increases [16].

16.3.4 Compressive Sensing over ZigBee Networks

ZigBee is a high-level wireless communication specification based on the IEEE 802.15.4 standard [17]. It is widely used in today's SN platforms like MicaZ and Telos. As mentioned earlier, none of the proposed adaptations of the CS to WSN is perfectly suitable for the typical hardware and software platform that is currently being used. CWS requires analog signal superposition and hence perfect time synchronization between nodes. Furthermore, it does not yet consider the effect of multi-path fading and other propagation noises. CDG is basically very similar to CWS and yet does not provide a practical implementation. DCS and JSM models are best suited for star topologies where each node can transmit its data directly to the sink. Again, like CWS, a time-synchronized signal superposition is preferred for DCS, otherwise time division multiple access methods will become very time consuming.

Caione et al. in [18] proposed a simple but very effective improvement technique for the operation of a CS-based signal acquisition for WSNs. In the previous section, we have seen that when using CS, the number of data items to be transmitted by each SN is equal to m and hence the energy consumption is balanced. This means that for a leaf node that needs to send only one value, m transmissions must be performed. We know that most of the nodes in a tree structure belong to the lower levels of the tree (levels near to the leaves). Therefore, a majority of nodes are sending data much more than the useful information that they produce. Caione et al. introduced a *hybrid CS* method for WSNs in which two operations are done by nodes:

- Pack and forward (PF): The SN packs its own recorded data along with the data received from its children SNs.
- CS measurement: The SN accumulates its own data and the data that have been eventually received from its children using CS measurement techniques. It then sends the CS measurement vector to its parent node.

All SNs follow two rules throughout acquiring, packing, forward, and accumulation of data:

- A SN only applies CS measurement accumulation method, when the length of the resulting message is less than that when using PF. In other words, each node decides whether to use PF or CS depending on the length of the outgoing message. This way, the SNs can minimize the energy consumed during radio transmission, which is the major resource bottleneck for SNs.
- When the message received by an SN is a CS measurement, the SN is obligated to apply the CS accumulation process. In other words, when the message changes its type from PF to CS, then it continues to be in the CS form all the way through the network till it reaches the sink.

These two simple rules and operations help to reduce the overall energy consumption of SNs. However, the energy consumption is not balanced any more. In hybrid CS, there is still the jeopardy of network partitioning, since the nodes near the sink may get exhausted and deplete earlier than leaf SNs. Xiang et al. in [19] have tried to optimize the aggregation tree of hybrid CS, though they do not address the unbalanced energy consumption of intermediate level nodes.

16.3.5 Further Improvements to Compressive Sensing in Wireless Sensor Networks

The CS theory has found applications in many scientific and industrial areas like magnetic resonance imaging, multimedia, genetics, and WSNs. In every area, there are customized algorithms and techniques that try to improve the CS performance for a specific data acquisition and recovery technique.

Mahmudimanesh et al. in [20,21] show that it is possible to virtually reorder the samples of a spatial signal in a WSN to get a more compressible view of the signal. When the signal is more compressible, we can recover it with the same quality from fewer number of measurements, or we can recover the signal from a fixed number of measurements while delivering a higher reconstruction quality. The proposed algorithm gets the current state of the environment, that is, the current spatial signal that is distributively recorded by the WSN. The algorithm then produces a permutation of the nodes under which the signal is more compressible when projected on a real fixed compressive basis like DCT of Fourier.

Interestingly, the near-optimal permutation found in one sampling round can be applied to next sampling round and an almost equal performance gain is achievable over next few sampling intervals. This attribute mostly depends on the dynamics of the operational environment. When the spatial signal does not change drastically over a short period of time, it is possible to apply the methods proposed in [21]. In most applications of WSNs, especially when the environment has not very rapid dynamics, these methods result in better performance. They do not discuss the implementation of their method on real-world WSNs. It is a rather general approach that can be applied by CWS, CDG, or DCS with JSM. Hybrid CS that has been discussed in the previous section can be also merged into this reordering technique.

16.4 Summary

In this chapter, we have explained the fundamentals of the emerging theory of CS and its applications in WSNs. We have realized that CS provides a very flexible, tunable, resilient, and yet efficient sampling method for WSNs. We have listed major advantages of CS over similar transform coding or other signal compression methods, which are traditionally used in WSNs. CS can guarantee a balanced energy consumption by all of the SNs. This important property avoids network partitioning and improves overall resource management of the whole network. It is easily implementable into strict hardware or software platform of today's typical SNs. Its resilience against noise and packet loss makes it very suitable for harsh operational environment of WSNs.

We have reviewed the most important variants of CS that are especially devised for WSNs. In recent years, with the progress of the CS theory in WSNs area, more and more realistic applications of CS are presented. CWS was one of the first variants of CS for WSNs. Although it was lacking consideration of established hardware platforms of SNs, it was a novel idea, which opened a new avenue of research in the area of signal acquisition for WSNs. CDG, which was directly based on CWS model, enabled detection of unexpected events. Event detection is a crucial requirement of many critical WSN applications.

DCS with its JSM models has categorized a wide range of applications for CS in WSNs. DCS provided an efficient solution to the problem of joint signal recovery. DCS has many applications in different distributed sampling scenarios, like WSN with star topology, camera arrays, acoustic localization, etc. However, it is not very suitable for multi-hop WSNs. Hybrid CS is a simple and yet effective solution for multi-hop WSNs consisting of very resource-limited SNs. It is one of the most recent improvements to typical CS implementation in WSNs. The only disadvantage of using hybrid CS is the unbalanced energy consumption leading to network partitioning.

CS has proved itself as a very advantageous method for distributed signal acquisition. Multi-hop implementation of CS can be regarded as the next challenge toward realizing an efficient distributed sampling method for WSNs. Moreover, further research is required to examine the effect of faulty nodes on the performance of CWS, DCS, CDG, and hybrid CS. We believe that a fault-tolerant CS-driven sampling and routing method for WSNs can significantly contribute to this research area.

References

1. E. J. Candes and T. Tao, Near-optimal signal recovery from random projections: Universal encoding strategies?, *IEEE Transactions on Information Theory*, 52(12), 5406–5425, 2006.
2. R. Vanderbei, Linear programming: Foundations and extensions, 2nd Edition, ISBN-10 0792373421, ISBN-13 9780792373421 *Springer-Verlag* USA, 2001.
3. R. Coifman, F. Geshwind, and Y. Meyer, Noiselets, *Applied and Computational Harmonic Analysis*, 10(1), 27–44, 2001.
4. S. Winter, H. Sawada, and S. Makino, On real and complex valued l1-norm minimization for overcomplete blind source separation, *IEEE Workshop on Applications of Signal Processing to Audio and Acoustics*, 16–19 October 2005, New York, pp. 86–89, 2005.
5. E. Candes and J. Romberg, Sparsity and incoherence in compressive sampling, *Inverse Problems*, 23(3), 969–985, 2007.
6. T. Tuma and P. Hurley, On the incoherence of noiselet and Haar bases, *International Conference on Sampling Theory and Applications SAMPTA'09*, May 18–22 2009, Marseille, France, pp. 243–246, 2009.
7. E. J. Candes and M. B. Wakin, An introduction to compressive sampling, *IEEE Signal Processing Magazine*, 25(2), 21–30. 2008.
8. E. J. Candès, J. K. Romberg, and T. Tao, Stable signal recovery from incomplete and inaccurate measurements, *Communications on Pure and Applied Mathematics*, 59(8), 1207–1223, 2006.
9. E. Candes, J. Romberg, and T. Tao, Robust uncertainty principles: Exact signal reconstruction from highly incomplete frequency information, *IEEE Transactions on Information Theory*, 52(2), 489–509, 2006.
10. C. Luo, F. Wu, J. Sun, and C. W. Chen, Compressive data gathering for large-scale wireless sensor networks, *Proceedings of the 15th Annual International Conference on Mobile Computing and Networking*, MobiCom 2009, 20–25 September 2009, Beijing, China, pp. 145–156, 2009.
11. C. Luo, F. Wu, J. Sun, and C. Chen, Efficient measurement generation and pervasive sparsity for compressive data gathering, *IEEE Transactions on Wireless Communications*, 9(12), 3728–3738, 2010.
12. A. Ciancio, S. Pattem, A. Ortega, and B. Krishnamachari, Energy-efficient data representation and routing for wireless sensor networks based on a distributed wavelet compression algorithm, *Proceedings of the 5th International Conference on Information Processing in Sensor Networks* - IPSN 2006, April 19–21 2006, Nashville, TN, USA pp. 309–316, ACM, 2006.
13. G. Hua and C. W. Chen, Correlated data gathering in wireless sensor networks based on distributed source coding, *International Journal of Sensor Networks*, 4(1/2), 13–22, 2008.

14. D. L. Donoho, M. Elad, and V. N. Temlyakov, Stable recovery of sparse overcomplete representations in the presence of noise, *IEEE Transactions on Information Theory*, 52(1), 6–18, 2006.
15. W. Bajwa, J. Haupt, A. Sayeed, and R. Nowak, Compressive wireless sensing, in *Proceedings of the 5th International Conference on Information Processing in Sensor Networks*, April 19–21 2006, Nashville, TN, USA, pp. 134–142, 2006.
16. M. B. Wakin, M. F. Duarte, S. Sarvhotam, D. Baron, and R. G. Baraniuk, Recovery of jointly sparse signals from few random projections, *Neural Information Processing Systems*—NIPS, Vancouver, Canada, December 2005.
17. P. Baronti, P. Pillai, V. W. C. Chook, S. Chessa, A. Gotta, and Y. F. Hu, Wireless sensor networks: A survey on the state of the art and the 802.15.4 and ZigBee standards, *Computer Communications*, 30(7), 1655–1695, 26 May 2007. ISSN 0140-3664.
18. C. Caione, D. Brunelli, and L. Benini, Compressive sensing optimization over zigbee networks, *in 2010 International Symposium on Industrial Embedded Systems (SIES)*, Trento, Italy, pp. 36–44, 2010.
19. L. Xiang, J. Luo, and A. Vasilakos, Compressed data aggregation for energy efficient wireless sensor networks, *8th Annual IEEE Communications Society Conference on Sensor, Mesh and Ad Hoc Communications and Networks (SECON)*, 27–30 June 2001, Salt Lake City, Utah, USA, pp. 46–54, 2011.
20. M. Mahmudimanesh, A. Khelil, and N. Yazdani, Map-based compressive sensing model for wireless sensor network architecture, a Starting point, *International Workshop on Wireless Sensor Networks Architectures, Simulation and Programming (WASP)*, 28–29 April 2009, Berlin, Germany, 2009.
21. M. Mahmudimanesh, A. Khelil, and N. Suri, Reordering for better compressibility: Efficient spatial sampling in wireless sensor networks, *IEEE International Conference on Sensor Networks, Ubiquitous, and Trustworthy Computing (SUTC)*, 7–9 June 2010, New Port Beach, CA, USA, pp.50–57, 2010.
22. M. F. Duarte, S. Sarvotham, D. Baron, M. B. Wakin, and R. G. Baraniuk, Distributed compressed sensing of jointly sparse signals, *Conference Record of the Thirty-Ninth Asilomar Conference on Signals, Systems and Computers*, 18–21 Feb. 2005, Pacific Grove, CA, USA, pp. 1537–1541, 2005.

Chapter 17

Framework for Detecting Attacks on Sensors of Water Systems

Kebina Manandhar, Xiaojun Cao, and Fei Hu

Contents

A cyber physical system (CPS) is a system that tightly combines and coordinates its computational/cyber elements together with physical elements. Wireless sensor communications have been extensively used to enhance the intelligence of CPS, whereas sensors gather information from the physical world and convey it to the central cyber system that further processes the sensor data [1]. For example, in [2] the authors have designed a cyber system employing sensor techniques to monitor the algae growth in Lake Tai, China. As stated in the paper, the design comprises of sensors and actuators to monitor the order of severity of the algae bloom and to dispatch salvaging boats. Similarly, the authors in [3] have proposed a CPS approach that navigates users in locations with potential danger, which takes advantage of the interaction between users and sensors to ensure timely safety of the users.

While sensors are widely deployed in the CPS, sensors are often targeted by the attackers. There are many types of physical and cyber attacks possible on a sensor. Physical attacks may range from short circuiting the sensors to damaging the sensors such that they present false data. Physical attacks can generally be mitigated by deploying proper security measures in the area. Cyber attacks generally include software, protocol, algorithm, or network-based attacks, which can be much more complicated and harder to defend. In this chapter, we consider attacks on sensors deployed in a water supply system and present a framework to detect such attacks.

The water supply system is equipped with water level sensors, to measure water level, and drifters, to measure water velocity. Data collected by these sensors are wirelessly transmitted to a central controller that implements a Kalman filter detection system. The Kalman filter generates estimates of the sensor data at next time step based on the data from current time step and other neighbor sensors. These estimated readings are compared with the actual readings to look for discrepancies. Without the loss of generality, the water supply system is modeled using the Saint-Venant models, a nonlinear hyperbolic mathematical model. The model is further refined to yield a linear and discrete mathematical model such that the Kalman filter can be applied.

17.1 Introduction

Recently, the security aspects of the water system drew attentions from the literature. For example, vulnerabilities of supervisory control and data acquisition systems used to monitor and control the modern day irrigation canal systems are investigated in [4]. The paper has linearized the shallow water partial differential equations (PDEs) used to represent water flow in a network of canal pools. As shown in [4], the adversary may use the knowledge of the system dynamics to develop a deception attack scheme based on switching the PDE parameters and proportional (P) boundary control actions, to withdraw water from the pools through offtakes. The authors in [5] have analyzed false data injection attacks on control systems using Kalman filter and Linear Quadratic Gaussian (LQG) controllers. The paper proves that a system with unstable eigenvector can be perfectly attackable and has provided means to derive an attack sequence that can bypass χ^2 detectors of the system.

While being increasingly deployed in CPS such as the water system, sensors are becoming the easy targets for attackers to compromise the security of the system. Since these sensors can be in remote and dangerous areas, deploying human resources to guard these sensors is not a practical approach. Hence, it is important that the cyber system is intelligent enough to detect attacks on the sensors. The water system described here is assumed to have two type of sensors: water level sensor and velocity sensor. The water level sensors are used to measure the water level from ground, and velocity sensors are used to measure the velocity of water at a certain location in the system. The system uses Lagrangian drifters equipped with GPS to calculate the velocity of the water. It is stated that the use of drifters can closely resemble the system that uses fixed sensors to measure velocity of the water flow [6,7]. To study the characteristics of water flow, the Saint-Venant equations have been used to model a wide variety of water flow systems like shallow water flow, one-dimensional flow, two-dimensional flow, etc. [6,7]. For example, a mathematical model using Saint-Venant one-dimensional shallow water equations for water flow problem is discussed in [6].

17.2 Overview of the Framework

Figure 17.1 shows the framework we consider in this work. The framework consists of a typical water system with stationary level sensors and mobile drifters. The drifters sense velocity of the stream and are equipped with GPS. The level sensors sense water level.

The water system is modeled using the Saint-Venant model. The Saint-Venant equations are modeled as hyperbolic nonlinear equations. The Saint-Venant equations are not compatible with the Kalman filter as the filter is designed for linear systems. So, to combine the hyperbolic water system model and the Kalman filter model, the Saint-Venant equations are linearized using Taylor series expansion (as described in Section 17.3.3). Furthermore, Kalman filter requires to break events into discrete events as it estimates the readings from sensor at next time stamp using data from previous time stamps. Hence, we divide time and water channel length into smaller events Δt and Δx, respectively. Lax diffusive scheme is then applied on the linearized Saint-Venant equations to discretize the equations, based on which the Kalman filter can estimate the readings from the sensors. As shown in Figure 17.1, the periodical readings from the wireless sensors and the output generated by the Kalman filter are then fed to a detector. The difference between these two values is compared with a threshold to detect if the sensors have been compromised or are at fault.

17.3 Modeling the Water Supply System

In this section, the Saint-Venant equations are introduced to model the water system. For a steady-state water flow, the hyperbolic and continuous Saint-Venant model is then linearized and discretized for its direct application in the Kalman filter (to be discussed in Section 17.4).

17.3.1 Saint-Venant Model

Saint-Venant equations are derived from the conservation of mass and momentum [8]. These equations are first-order hyperbolic nonlinear PDEs equations and for one-dimensional flow with no lateral inflow, these equations can be written as

$$T\frac{\delta H}{\delta t} + \frac{\delta Q}{\delta x} = 0 \tag{17.1}$$

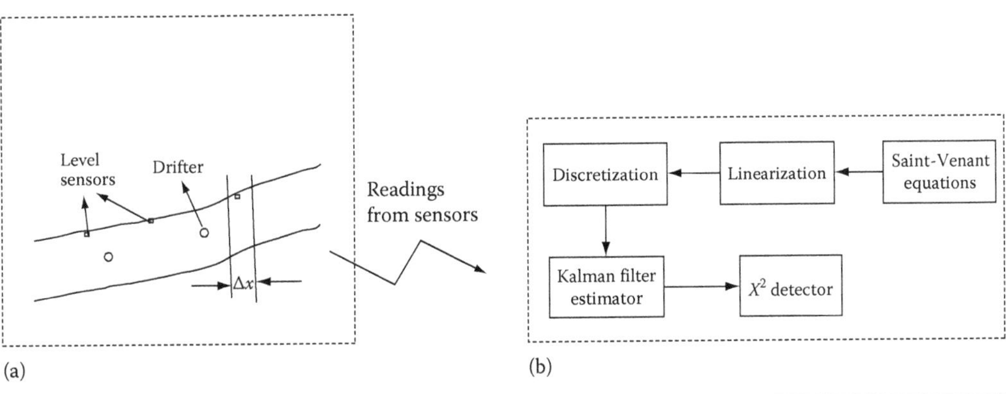

Figure 17.1 The water supply system. (a) physical water system and (b) central system with estimator and detector.

$$\frac{\delta Q}{\delta t}+\frac{\delta}{\delta x}\left(\frac{Q^2}{A}\right)+\frac{\delta}{\delta x}(gh_cA)=gA(S_0-S_f)$$
$$\text{for } (x,t)\in(0,L)X\mathbf{R}^+ \tag{17.2}$$

where
L is the length of the flow (m)
$Q(x,t)=V(x,t)A(x,t)$ is the discharge or flow (m^3/s) across cross section
$A(x,t)=T(x)H(x,t)\cdot V(x,t)$ refers to velocity (m/s)
$H(x,t)$ refers to water depth (m)
$T(x,t)$ refers to the free surface width (m)
$S_f(x,y)$ is the friction slope
S_b is the bed slope
g is the gravitational acceleration (m/s^2)

These equations can be elaborated [6] in terms of water depth and velocity as

$$T\frac{\delta H}{\delta t}+\frac{\delta(THV)}{\delta x}=0 \tag{17.3}$$

$$\frac{\delta V}{\delta t}+V\frac{\delta V}{\delta x}+g\frac{\delta H}{\delta x}=g(S_b-S_f) \tag{17.4}$$

The friction is empirically modeled by the Manning–Stickler's formula:

$$S_f=\frac{m^2V|V|(T+2H)^{\frac{4}{3}}}{(TH)^{\frac{4}{3}}} \tag{17.5}$$

where m is the Manning's roughness coefficient $(s/m^{\frac{1}{3}})$

17.3.2 Steady-State Flow

There exists a steady-state solution of the Saint-Venant equations under constant boundary conditions [7]. We denote the variables corresponding to the steady-state condition by adding suffix 0. By excluding the term containing δt and expanding Equation 17.3, we obtain the following equation:

$$\frac{dV_0(x)}{dx}=-\frac{V_0(x)}{H_0(x)}\frac{dH_0(x)}{dx}-\frac{V_0(x)}{T(x)}\frac{dT(x)}{dx} \tag{17.6}$$

Solving Equations 17.4 and 17.6, we get

$$\frac{dH_0(x)}{dx}=\frac{S_b-S_f}{1-F(x)^2} \tag{17.7}$$

with $F_0=V_0/C_0$, $C_0=\sqrt{gH_0}$. Here, C_0 is the gravity wave celerity, F_0 is the Froude number. We assume the flow to be subcritical, that is, $F_0<1$ [7].

17.3.3 Linearized Saint-Venant Model

The linearized Saint-Venant model can be obtained from the steady-state flow characterized by V_0 and H_0 [7]. Let $v(x, y)$ and $h(x, y)$ denote the first-order perturbations in water velocity and water level. Then,

$$V(x,t) = V_0(x,t) + v(x,t) \tag{17.8}$$

$$H(x,t) = H_0(x,t) + h(x,t) \tag{17.9}$$

The values of H and V are substituted in Equations 17.3 and 17.4 and expanded in Taylor series. We use T_0 in place of T to emphasize that it is uniform. As described in [7], neglecting higher order terms, a given term $f(V, H)$ of Saint-Venant model can be written as $f(V, H) = f(V_0, H_0) + (f_V)_0 v + (f_H)_0 h$ in which $()_0$ indicates steady-state conditions. The linearized Saint-Venant equations can be obtained as follows [6,9]:

$$h_t + H_0(x)v_x + V_0(x)h_x + \alpha(x)v + \beta(x)h = 0 \tag{17.10}$$

$$v_t + V_0(x)v_x + gh_x + \gamma(x)v + \eta(x)h = 0 \tag{17.11}$$

where $\alpha(x), \beta(x), \gamma(x)$, and $\eta(x)$ are given by

$$\alpha(x) = \frac{dH_0}{dx} + \frac{H_0}{T}\frac{dT_0}{dx} \tag{17.12}$$

$$\beta(x) = -\frac{V_0}{H_0}\frac{dH_0(x)}{dx} - \frac{V_0(x)}{T(x)}\frac{dT(x)}{dx} \tag{17.13}$$

$$\gamma(x) = 2gm^2\frac{|V_0|}{H_0^{\frac{4}{3}}} - \frac{V_0}{H_0}\frac{dH_0(x)}{dx} - \frac{V_0(x)}{T(x)}\frac{dT(x)}{dx} \tag{17.14}$$

$$\eta(x) = -\frac{4}{3}gm^2\frac{V_0|V_0|}{H_0^{\frac{7}{3}}} \tag{17.15}$$

17.3.4 Discretization

In order to discretize the linear equations generated in Section 17.3.3, we use the Lax diffusive scheme [6] as follows. The channel is divided into smaller segments of length Δx and a suitable time interval Δt is selected.

$$\frac{\delta v}{\delta t} = \frac{v_i^{k+1} - \frac{1}{2}(v_{i+1}^k + v_{i-1}^k)}{\Delta t} \tag{17.16}$$

$$\frac{\delta v}{\delta x} = \frac{(v_{i+1}^k + v_{i-1}^k)}{2\Delta x} \tag{17.17}$$

$$\frac{\delta h}{\delta t} = \frac{h_i^{k+1} - \frac{1}{2}(h_{i+1}^k + h_{i-1}^k)}{\Delta t} \tag{17.18}$$

$$\frac{\delta h}{\delta x} = \frac{(h_{i+1}^k + h_{i-1}^k)}{2\Delta x} \tag{17.19}$$

Given $(h_i^k, v_i^k)_{i=0}^I$, we want to compute $(h_i^{k+1}, v_i^{k+1})_{i=0}^I$. Here, I is the total number of segments of length Δx. The updated equations for (h_i, v_i) are

$$\begin{aligned} h_i^{k+1} &= \frac{1}{2}(h_{i+1}^k + h_{i-1}^k) \\ &- \frac{\Delta t}{4\Delta x}(H_{0(i+1)} + H_{0(i-1)})(v_{i+1}^k - v_{i-1}^k) \\ &- \frac{\Delta t}{4\Delta x}(V_{0(i+1)} + V_{0(i-1)})(h_{i+1}^k - h_{i-1}^k) \\ &- \frac{\Delta t}{2}\alpha_{i+1}v_{i+1}^k + \alpha_{i-1}v_{i-1}^k \\ &- \frac{\Delta t}{2}\beta_{i+1}h_{i+1}^k + \beta_{i-1}h_{i-1}^k \end{aligned} \tag{17.20}$$

$$\begin{aligned} v_i^{k+1} &= \frac{1}{2}(v_{i+1}^k + v_{i-1}^k) \\ &- \frac{\Delta t}{4\Delta x}(V_{0(i+1)} + V_{0(i-1)})(v_{i+1}^k - v_{i-1}^k) \\ &- \frac{g\Delta t}{2\Delta x}(h_{i+1}^k - h_{i-1}^k) \\ &- \frac{\Delta t}{2}\gamma_{i+1}v_{i+1}^k + \gamma_{i-1}v_{i-1}^k \\ &- \frac{\Delta t}{2}\eta_{i+1}h_{i+1}^k + \eta_{i-1}h_{i-1}^k \end{aligned} \tag{17.21}$$

If assume that Δx is very small, then we can write $h_{i-1} = h_i = h_{i+1}$ and $v_{i-1} = v_i = v_{i+1}$. Equations 17.20 and 17.21 will then become

$$\begin{aligned} h_i^{k+1} &= \left(1 - \frac{\Delta t}{2}\beta_i + \beta_i\right) h_i^k \\ &+ \left(\alpha_i - \frac{\Delta t}{2}\alpha_i\right) v_i^k \end{aligned} \tag{17.22}$$

$$\begin{aligned} v_i^{k+1} &= \left(\eta_i - \frac{\Delta t}{2}\eta_i\right) h_i^k \\ &+ \left(1 - \frac{\Delta t}{2}\gamma_i + \gamma_i\right) v_i^k \end{aligned} \tag{17.23}$$

17.3.5 Discrete Linear State-Space Model

From the discretized equations in Section 17.3.4, state-space model can be formed as follows:

$$x(k+1) = Ax(k) + Bu(k) + w(k) \tag{17.24}$$

where $x(k) = (v_0^k, \ldots, v_I^k, h_0^k, \ldots, h_I^k)^T$ with the applied control $u(k)$ in the form of discharge perturbation at the upstream end v_0^k and the discharge perturbation $w(k)$ at the downstream end v_I^k [8]. Here, w_k, x_0 are independent Gaussian random variables, and $x_0 \sim \mathcal{N}(0, \Sigma)$ and $w_k \sim \mathcal{N}(0, Q)$.

17.4 Detecting Attacks Using Kalman Filter

In this section, we introduce the Kalman filter [10] technique to obtain estimates for the state-space vector $x(k)$ described in Section 17.3.5. Figure 17.2 shows the control system of the Kalman filter with the sensor readings or observations from the water supply system namely, y_i. And x_i denotes the output of the control system that is fed to the controller. The observations (y_i) are forwarded to the central system containing estimator and detector at a regular time interval denoted by Δt. At each time step Δt, the estimator of the system generates estimated readings based on the reading of previous time step. These readings are used by the detector to detect the difference between the newly observed sensor readings and estimated readings.

17.4.1 Kalman Filter

To apply the Kalman filter technique, the observation equation for the preceding system can be written as

$$y_k = Cx_k + v_k \tag{17.25}$$

Here, $y_k = [y_1^k, \ldots, y_m^k]^T \in \mathbf{R^m}$ is measurement vector collected from the sensors and y_i^k is the measurement generated by sensor i at time k. v_k is the measurement noise and assumed to be white Gaussian noise, which is independent of initial conditions and process noise.

Kalman filter can then be applied to compute state estimations $\hat{x}_k$ using observations y_k. Let the mean and covariance of the estimates be defined as follows:

$$\begin{aligned} \hat{x}_{k|k} &= E[x_k, y_0, \ldots, y_k] \\ \hat{x}_{k|k-1} &= E[x_k, y_0, \ldots, y_{k-1}] \\ P_{k|k-1} &= \Sigma_{k|k-1} \\ P_{k|k} &= \Sigma_{k|k-1} \end{aligned} \tag{17.26}$$

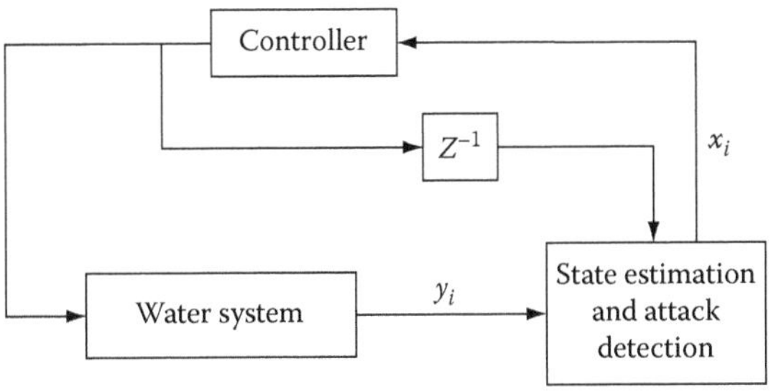

Figure 17.2 Kalman filter.

The iterations of Kalman filter can be written as

$$\begin{aligned}
\hat{x}_{k+1|k} &= A\hat{x}_k + Bu_k \\
P_{k|k-1} &= AP_{k-1}A^T + Q \\
K_k &= P_{k|k-1}C_k^T(C_kP_{k|k-1}C_k^T + R)^{-1} \\
P_{k|k} &= P_{k|k-1} - K_kCP_{k|k-1} \\
\hat{x}_k &= \hat{x}_{k|k-1} + K_k(y_k - C_k\hat{x}_{k|k-1})
\end{aligned} \tag{17.27}$$

Initial conditions being $x_{0|-1} = 0, P_{0|-1} = \Sigma$, and K_k being Kalman gain. We assume that the Kalman gain converges in a few steps and is already in a steady state, then

$$P \triangleq \lim_{k\to\infty} P_{k|k-1}, K = PC^T(CPCT + R)^{-1} \tag{17.28}$$

The Kalman filter equation can be updated as

$$\hat{x}_{k+1} = A\hat{x}_k + Bu_k + K[y_{k+1} - C(A\hat{x}_k + Bu_k)] \tag{17.29}$$

The residue z_{k+1} at time $k+1$ is defined as

$$z_{k+1} \triangleq y_{k+1} - \hat{y}_{k+1|k}$$

Equivalently,

$$z_{k+1} \triangleq y_{k+1} - C(A\hat{x}_k + Bu_k) \tag{17.30}$$

The estimate error e_k is defined as

$$e_k \triangleq x_k - \hat{x}_k \tag{17.31}$$

Equivalently,

$$e_{k+1} \triangleq x_{k+1} - \hat{x}_{k+1} \tag{17.32}$$

Substituting the values x_{k+1} and $\hat{x}_{k+1}$ from Equations 17.24, 17.25, and 17.29, we obtain the following recursive formula for error calculation:

$$e_{k+1} = (A - KCA)e_k + (I - KC)w_k - Kv_k \tag{17.33}$$

17.4.2 Attack/Failure Detection

Since it is assumed that the noises in the system are Gaussian, we use χ^2 detector to compute the difference between the observed value from the sensors and the estimated values from the Kalman filter as the following:

$$g_k = z_k^T \mathcal{P} z_k \tag{17.34}$$

where $\mathcal{P}$ is the covariance matrix of z_k, the residue. The χ^2 detector compares g_k with a certain threshold to detect a failure or attack and triggers the alarm for potential attack or failure [11].

$$g_k > threshold \tag{17.35}$$

where g_k is defined as

$$g_k = g(z_k, y_k, \hat{x}_k, \ldots, z_{k-\tau+1}, y_{k-\tau+1}, \hat{x}_{k-\tau+1}) \tag{17.36}$$

The function g is continuous and $\tau \in \mathbf{N}$ is the window size of the detector [11].

17.5 Summary

The framework designed in this chapter is able to detect attacks or failures by computing the difference between the observed value from the sensors and the estimated values from the Kalman filter. An alarm is triggered if the difference is larger than a given threshold. The threshold also accounts for measurement and system errors and helps avoid the false alarm. Besides easily detecting the attacker naively stealing water, it can effectively detect if the attacker tampers with the sensor readings. However, in [11] the authors have presented an algorithm to attack a system using Kalman filter by generating an attack sequence depending on the statistical property of the system. To defend against such attacks, the users of the water system can precompute all the unstable eigenvectors of the system and recognize the sensors that can be attacked by the attackers. Another approach is that the users can deploy redundant sensors in places where sensors are likely to be compromised, which can help detect the false data injection. On top of these defense schemes, the users can also implement encryption algorithms to further enhance the security of the data transferred from sensors to the central system.

17.A Appendix

Symbols and Abbreviations	*Explanation*
CPS	Cyber physical system
V, H	Water velocity
V_0, H_0	Static velocity and height
v, h	Perturbation in velocity and height
g	Acceleration due to gravity
T	Width of the water flow
A	Cross-section area ($T \times H$)
Q	Flow per unit time ($A \times V$)
S_f	Friction slope

(Continued)

Symbols and Abbreviations	*Explanation*
S_b	Bed slope
m	Manning–Stickler's formula
$C_0 = \sqrt{gH_0}$	Gravity wave celerity
$F_0 = V_0/C_0$	Froude number
$\alpha(x), \beta(x), \gamma(x), \eta(x)$	Coefficients for linearized Saint-Venant equations
n	Dimension of state space
x_k	State space with dimension n at time k
y_k	Measurement vector collected from sensors
P_k	Covariance of the estimates
K_k	Kalman gain
z_k	Residue
e_k	Error estimate

References

1. T. Dillon and E. Chang, Cyber-physical systems as an embodiment of digital ecosystems extended abstract, *Digital Ecosystems and Technologies (DEST), 4th IEEE International Conference*, Dubai, pp. 701, 13–16 Apr. 2010.
2. D. Li, Z. Zhao, L. Cui, H. Zhu, L. Zhang, Z. Zhang, and Y. Wang, A cyber physical networking system for monitoring and cleaning up blue-green algae blooms with agile sensor and actuator control mechanism on Lake Tai, *Computer Communications Workshops (INFOCOM WKSHPS), IEEE Conference*, Shanghai, China, pp. 732–737, 10–15 Apr. 2011.
3. M. Li, Y. Liu, J. Wang, and Z. Yang, Sensor network navigation without locations, *INFOCOM 2009, IEEE*, Rio de Janeiro, Brazil, pp. 2419–2427, 19–25 Apr. 2009.
4. A. Saurabh, L. Xavier, S. Shankar Sastry, and A.M. Bayen, Stealthy deception attacks on water SCADA systems, *In Proceedings of the 13th ACM International Conference on Hybrid Systems: Computation and Control (HSCC '10)*, Stockholm, Sweden, 2010.
5. Y. Mo, False data injection attacks in control systems, *SCS 10 Proceedings of the First Workshop on Secure Control Systems*, Stockholm, Sweden, 2010.
6. M. Rafiee, W. Qingfang, and A.M. Bayen, Kalman filter based estimation of flow states in open channels using Lagrangian sensing, *Decision and Control, Held Jointly with the 28th Chinese Control Conference. CDC/CCC 2009. Proceedings of the 48th IEEE Conference*, Shanghai, China, pp. 8266–8271, 15–18 Dec. 2009.
7. S. Amin, A.M. Bayen, L. El Ghaoui, and S. Sastry, Robust feasibility for control of water flow in a reservoir-canal system, *Decision and Control, 46th IEEE Conference*, New Orleans, LA, pp. 1571–1577, 12–14 Dec. 2007.
8. M. Rafiee, A. Tinka, J. Thai, and A.M. Bayen, Combined state-parameter estimation for shallow water equations, *American Control Conference (ACC)*, San Francisco, CA, pp. 1333–1339, Jun. 29 2011–Jul. 1 2011.

9. X. Litrico and V. Fromion, Infinite dimensional modelling of open-channel hydraulic systems for control purposes, *Decision and Control, Proceedings of the 41st IEEE Conference*, vol. 2, pp. 1681–1686, 10–13 Dec. 2002.
10. R.E. Kalman, A new approach to linear filtering and prediction problems, *Journal of Basic Engineering*, vol. 82, pp. 35–45, 1960.
11. Y. Mo, E. Garone, A. Casavola, and B. Sinopoli, False data injection attacks against state estimation in wireless sensor networks, *Decision and Control (CDC), 49th IEEE Conference*, Atlanta, GA, pp. 5967–5972, 15–17 Dec. 2010.

INTELLIGENT SENSOR NETWORKS: SENSORS AND SENSOR NETWORKS

Chapter 18

Reliable and Energy-Efficient Networking Protocol Design in Wireless Sensor Networks

Ting Zhu and Ping Yi

Contents

Wireless sensor networks have been proposed for use in many challenging long-term applications such as military surveillance, habitat monitoring, infrastructure protection, scientific exploration, participatory urban sensing, and home energy management [1–3]. Compared with other high-end sensing technologies (e.g., satellite remote sensing), sensor networks are acclaimed to be low cost, low profile, and fast to deploy. With rapid advance in fabrication techniques, the constraints of computation and memory of the sensor nodes might not be major issues over time. However, energy will continue to be the victim of Moore's law, that is, more transistors indicates more power consumption. Constrained by the size and cost, existing sensor nodes (e.g., MicaZ, Telos, and mPlatform [4]) are equipped with limited power sources. Yet these nodes are expected to support long-term applications, such as military surveillance, habitat monitoring, and scientific exploration, which require a network life span that can range from a few months to several years.

To bridge the growing gap between lifetime requirement of sensor applications and the slow progress in battery capacity, it is critical to have an energy-efficient communication stack. In this chapter, we introduce reliable and energy-efficient networking protocol designs in wireless sensor networks. Specifically, we present the following three types of representative design: (1) low-duty-cycle network protocol design; (2) exploring link correlation for energy-efficient network protocol design; and (3) energy-efficient opportunistic routing protocol design.

18.1 Low-Duty-Cycle Network Protocol Design

Typically, energy in communication can be optimized through (1) physical-layer transmission rate scaling [5]; (2) link-layer optimization for better connectivity, reliability, and stability [6]; (3) network-layer enhancement for better forwarders and routes [7]; and (4) application-layer improvements for both content-agnostic and content-centric data aggregation and inference [8]. Although these solutions are highly diversified, they all assume a wireless network where nodes are ready to receive packets and focus mainly on the transmission side, a topic of interests for years with hundreds of related publications.

In contrast, wireless networks with intermittent receivers have caught unproportionately little attention, despite the known fact that communication energy is consumed mostly for being ready for potential incoming packets, a problem commonly referred as idle listening. For example, the widely used Chipcon CC2420 radio draws 19.7 mA when receiving or idle listening, which is actually larger than 17.4 mA when transmitting. More importantly, packet transmission time is usually very small (e.g., <1 ms to transmit a TinyOS packet using a CC2420 radio), while the duration of idle listening for reception can be orders of magnitude longer. For example, most environmental applications, such as Great Duck Island [9] and Redwood Forest [10], sample the environment at relatively low rates (on the order of minutes between samples). With a comparable current draw and ~3 to 4 orders of magnitude longer duration waiting for reception, idle listening is a major energy drain that accounts for most energy in communication if it is not optimized.

To reduce the energy lost to idle listening, a low-duty-cycle network is formed by nodes that listen to the channel very briefly and shut down their radios most of the time (e.g., 99% or more). At any given time, this type of network is actually fragmented (partitioned) and network connectivity (topology) becomes intermittent. Uniquely, communication delay in low-duty-cycle networks is dominated by sleep latency—the delay time a sender waits for its receiver to wake up. Although low-duty-cycle networking is an ideal fit for many long-term unattended sensor applications, research has been lacking and predominately focuses on physical-and link-layer designs. To ensure packet reception at low-duty-cycle receivers, several pioneer researchers have proposed B-MAC [11], X-MAC [12], WiseMAC [13], and the 802.15.4 beacon-enable mode, which successfully reduce the amount of idle listening through techniques such as low-power-listening and/or synchronous channel polling. These link-layer designs are effective; however, further improvement becomes difficult without utilizing information about topology and multi-hop connectivity information at the network layer. For example, data reliability is commonly supported by link-layer protocols [11–16] through retransmission to a same receiver if previous transmission fails. In a low-duty-cycle network, without the network layer rerouting capability, these link-layer protocols have to wait for an intended receiver to wake up again, introducing excessive sleep latency in the orders of seconds or possibly minutes.

Motivated by the insufficiency of link-layer designs, researchers proposed to investigate a wide spectrum of low-duty-cycle network configurations to support diversified types of applications.

(1) The static low-duty-cycle wireless network supports mission-driven applications such as military surveillance with a specified network lifetime requirement and a fixed energy budget [17]. For example, a military strategic area shall be covered until a stronghold is established in 2 months. The duty cycle of this type of network is a fix ratio of the battery lifetime and network lifetime. (2) The dynamic low-duty-cycle wireless network supports scientific exploration such as habitat study [18] and structural monitoring [19]. Battery-powered sensors are not desirable in these long-term unattended applications, because replacing thousands of batteries at hard-to-access locations would add significantly to the cost of maintenance. In this type of network, energy harvesting techniques are normally employed [20–27] and the duty-cycle availability changes dynamically with fluctuating ambient energy [28].

In [29], researchers revealed that in low-duty-cycle networks, sleep latency makes traditional routing algorithms [30,31] ineffective, especially when the unreliable nature of wireless links is seriously considered. The key concept developed in [29] is dynamic forwarding (DSF), which utilizes a sequence of receivers to optimally reduce delay in source-to-sink communication. To minimize sleep latency, DSF utilizes multiple potential forwarding nodes at each hop. For a given sink, each node maintains a sequence of forwarding nodes sorted in the order of the wake-up time associated with them. Packet transmission starts with the first node in the sequence. In case of failure, retransmission follows the sequence order until the packet is successfully received by one of forwarding nodes. The key optimization problem in DSF is how to select a subset of forwarding nodes among all eligible nodes. For the first time, researchers in [29] revealed low-duty-cycle networks have fundamental different properties from always-awake networks. Notably, they found temporary routing loops introduced by DSF is actually beneficial in reducing end-to-end delay, a finding that invalidates the deeply rooted belief that network loops are always harmful.

Intended for system-wide dissemination of configurations and code binaries, flooding has been investigated extensively in wireless networks [32–34]. However, only little work has been done for low-duty-cycle wireless sensor networks in which nodes stay sleeping most of the time and wake up asynchronously. In this type of networks, the concept of broadcast has been changed. A broadcasting packet can rarely be received by multiple nodes simultaneously, a constraining feature making the existing solutions unsuitable. In addition, a whole new level of complexity is added when unreliable nature of wireless communication is considered. Researchers try to design an approach to let a node make probabilistic forwarding decisions based on the delay distribution of receiving nodes. Only early packets are forwarded to achieve shorter flooding delays and reduce the level of redundancy.

18.2 Exploring Link Correlation for Energy-Efficient Network Protocol Design

Wireless sensor networks normally work under 2.4 GHz radio spectrum, which is heavily used by lots of other devices, such as microwave ovens, medical diathermy machines, cordless phones, bluetooth devices, NFC devices, and wireless access points. To understand how these devices impact communication performance between wireless sensor devices, extensive research [7,35–43] has been done to measure packet reception quality of *individual links* in realistic environments. These in situ studies have proved that ideal models do not hold well in practice. For example, (1) signal propagation is not a fixed function of distance; (2) RF signal strength does not attenuate identically in all directions; and (3) link quality does not remain constant over time. If designs of protocols are based on simplified assumptions, their performance is very poor in realistic environments. Although empirical studies on link performance are highly diverse, they predominantly focus on the analysis of *individual links*.

On the other hand, little research has been done to investigate the reception correlation among neighboring wireless links, despite the fact that wireless communication essentially occurs in a broadcast medium with concurrent receptions. For the sake of simplicity, nowadays many existing protocols implicitly assume link independence among concurrent receivers, another simplifying assumption that deserves the same level of inspection as the aforementioned ones. As shown in recent work [44], ignoring spatial link correlation would potentially hinder further performance improvement in wireless communication. In [44], researchers demonstrated that if we assume the receptions of a broadcast packet by multiple neighboring nodes are probabilistic independent of each other, it is necessary to have per-node acknowledgment (ACK) from all neighbors to achieve reliable broadcast [45–48] leading to possibly the ACK implosion problem [49] in high-density wireless sensor networks. Similarly, if reception results of all receivers are identical (i.e., strongly correlated), the performance gain of opportunistic routing and collective forwarding [50,51] becomes less significant because strong spatial link correlation reduces multi-receiver diversity gain [52,53] where the probability of successful reception among a cluster of potential receivers is larger than a single receiver.

In [44], researchers conducted empirical study on link correlation. In the experiments, 42 MICAz nodes were used. The experiments were conducted with multiple randomly generated layouts under two scenarios: an open parking lot and an indoor office. In each scenario, the sender was placed in the center of the topology, while the other 41 nodes were randomly deployed as receivers. The sender broadcasted a packet every 200 ms. Each packet was identified by a sequence number. The total number of packets broadcasted was 6000. In both indoor and outdoor experiments, researchers discover that if a packet is received by a sensor node with low packet reception ratio (PRR), most of the time this packet is also received by the high PRR nodes. Figure 18.1 illustrates the first 600 packet receptions of three nodes in indoor experiments. The black bands correspond to the packets received at the nodes. Clearly, there exists a strong correlation of packet receptions among the neighboring nodes. For example, in Figure 18.1, given the two packets (sequence number 282 and 508) received by N_{22}, these two packets were also received by N_{29} and N_{23}. In order to quantify this correlation, researchers define the conditional packet reception probability (CPRP) as *the probability that a node N_h receives a packet M from sender node S, given the condition that the packet M is received by another node N_l*.

In [44], researchers use $P_S(N_h|N_l)$ to denote CPRP, where N_h and N_l are neighboring receivers of the sender S. Clearly, CPRP $P_S(N_h|N_l)$ can be calculated as the percentage of packets received successfully at the node N_h among the packets that have been received by the node N_l. In addition,

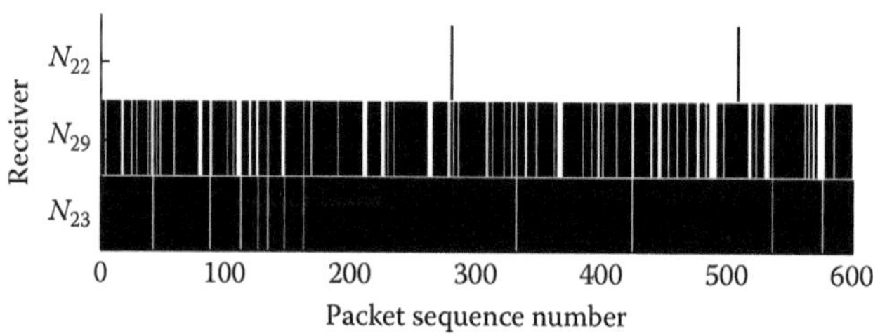

Figure 18.1 Correlation of packet reception among receivers.

packet reception probability $P_S(N_h)$ can also be calculated as the percentage of successful receptions at N_h regardless of the reception result at N_l. If the receptions of wireless links are independent, $P_S(N_h|N_l)$ equals $P_S(N_h)$. To analyze the CPRP among the pairwise receivers more systematically, researchers computed the CPRP for all node pairs with nonzero PRR values. In the indoor experiment, there exists 32 nonzero PRR nodes, which generate $\frac{32\times31}{2} = 496$ combinations of $P_S(N_h|N_l)$. Researchers found that $P_S(N_h|N_l)$ is not equal to $P_S(N_h)$ for about 98.18% of the receiver pairs. This demonstrates the existence of link correlation.

In recent MobiCom work [54], authors demonstrate that link correlation does affect the network performance noticeably, but they did not take further steps to utilize such impact for performance improvement on flooding protocol design as the researchers did in [44,55]. Existing flooding algorithms [45,47,49,56] have demonstrated their effectiveness in achieving communication efficiency and reliability in wireless sensor networks. Further performance improvement, however, has been hampered by the implicit assumption of *link independency* adopted in previous designs. In other words, existing flooding algorithms assume that the receptions of a flooding packet by multiple neighboring nodes are probabilistically independent of each other. Under such an assumption, it is necessary to have an ACK *directly* from all neighboring receivers. This is because a node's ACK cannot be used to confirm the reception at its neighboring receivers if link independency is assumed.

However, direct ACKs per receiver may lead to high collision [57,58], congestion [59], and possibly the ACK storm problem [49] in wireless networks. To address the problem, researchers designed the *Collective Flooding (CF) protocol* [44] that exploits spatial link correlation for performance improvement. The driving idea behind this is *collective ACKs*. Previously, a sender estimated whether a transmission was successful based only on the ACKs from the intended receiver. Instead, the mechanism of collective ACKs allows a sender to infer the success of a transmission to a receiver based on the ACKs from other receivers by utilizing spatial link correlation among them.

Specifically in CF, a node is called a *covered* node if it has already received the flooding packet. Covered nodes are responsible for rebroadcasting the packet to uncovered nodes in the network. The mechanism of collective ACKs allows a node to extract information about the status of its neighboring nodes via receiving or overhearing ACKs from its neighbors. For example, in Figure 18.2, suppose that node S is a covered node while N_1 and N_2 are uncovered. They are within 1-hop communication range from each other. When S broadcasts, if N_1 receives the packet, in traditional flooding protocols without considering the correlation, N_1 only knows that S is covered but still considers N_2 as uncovered (due to unreliable wireless links) until N_1 overhears the rebroadcast from N_2.

The CF protocol takes a different approach. In CF, every node keeps track of its neighbor's coverage probability. As shown in Figure 18.2, from N_1's viewpoint, a packet from S serves two

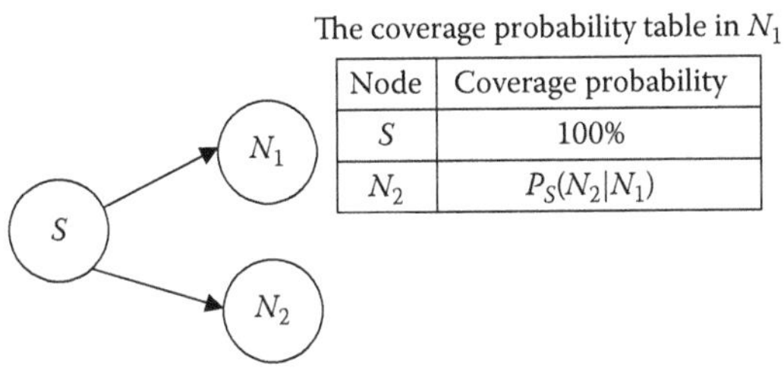

The coverage probability table in N_1

Node	Coverage probability
S	100%
N_2	$P_S(N_2|N_1)$

Figure 18.2 An example of collective ACK.

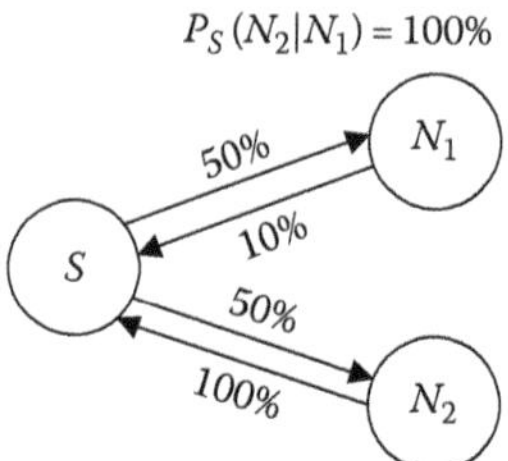

Figure 18.3 Based on N_1's ACK and $P_S(N_1|N_2)$, S knows N_2 is covered.

purposes. First, it is a direct ACK, confirming that S is a covered node. Second, it also serves as a collective ACK to N_1 that N_2 has a reception probability of $P_S(N_2|N_1)$. In traditional designs, overhearing a packet serves only as a direct ACK that the packet sender (e.g., S in Figure 18.2) is covered. In CF, overhearing a packet by N_1 can serve as the partial evidence that the packet receiver (e.g., N_2 in Figure 18.2) is also covered. Such partial evidence can be accumulated in a collective manner to reach a certain coverage probability threshold. Once the threshold is achieved, N_1 considers N_2 covered and hence refrains from redundant rebroadcasting.

Exploring spatial link correlation can greatly reduce the redundant transmission. For the sake of clarity, let us consider a hypothetical example shown in Figure 18.3. The two links from node S to N_1 and N_2 are strongly correlated with 50% PRR. In other words, $P_S(N_2|N_1) = 100\%$, and the link qualities from N_1 and N_2 back to S are 10% and 100%, respectively. In traditional flooding protocols, the sender S treats the receivers packet receptions as independent. To provide reliable broadcasting, S needs to continue transmitting until it receives ACKs or overhears rebroadcast from both N_1 and N_2. Due to the low link quality (10%) from N_2 back to S, S might conduct many unnecessary retransmissions. In contrast, with the knowledge of spatial link correlation, node S can terminate the transmission once it receives ACK from N_1, given the knowledge $P_S(N_2|N_1) = 100\%$. As we can see from the previous simplified example, collective ACKs can improve the efficiency of the reliable flooding protocol by utilizing the spatial link correlation.

Clearly, the previous example can only show potential benefits at the conceptual level. It cannot reveal whether such benefit is significant enough in generic practical settings. To evaluate the practicality of the design, researchers have implemented it on the TinyOS [60]/MICAz platform in nesC [61] and compared CF with Standard Flooding (FLD) and Reliable Broadcast Propagation (RBP) [45]. Experiments were conducted in both indoor and outdoor multi-hop environments. For example, in an outdoor experiment, 48 MICAz nodes were deployed along a 326 m long bridge. As shown in Figure 18.4a, the reliability of RBP8 (maximum retransmit 8 times to uncovered nodes), RBP4 (maximum retransmit 4 times), CF, and standard flooding was 99.96%, 97.6%, 99.93%, and 61.96%, respectively. While achieving similar reliability as RBP8, CF reduced the number of packets transmitted by 31.2% as shown in Figure 18.4b. In addition, Figure 18.4c shows that the average dissemination delay of RBP8, RBP4, CF, and standard flooding was 4.46, 3.93, 2.85, and 2.34 s, respectively. Again, the average delay of CF was 36% less than that of RBP8.

The main reason for the performance difference is that RBP and standard flooding do not use spatial link correlation to predict the packet reception of neighboring nodes. Duplicated transmissions happen when a sender does not realize that neighboring receivers have already received the packet, so the sender retransmits the packet. Using spatial link correlation information, a sensor node can more accurately predict whether its neighbors have successfully received the packet,

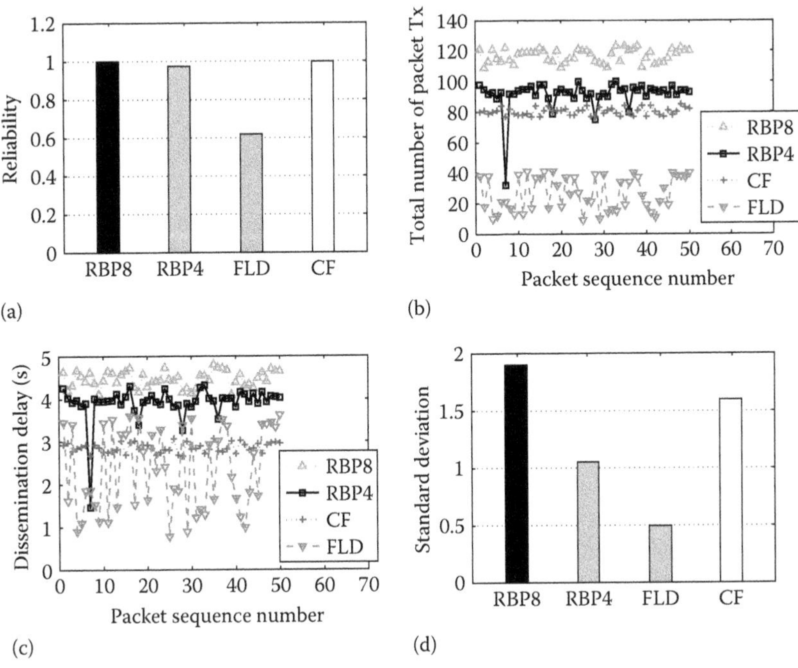

Figure 18.4 The performance of outdoor linear network experiment. (a) Reliability, (b) message overhead, (c) dissemination delay, and (d) load balance.

leading to fewer duplicated transmissions, lower congestion, shorter delays, and lower energy consumption.

18.3 Energy-Efficient Opportunistic Routing Protocol Design

Wireless networks is an active research area and many wireless routing protocols and scheduling policies have been proposed ([62–73]). Most of these routing protocols *preselect* a minimum cost single path or multiple alternate paths and use the preselected static path(s) to forward data packets. At any specific time, these routing protocols use unicast to forward data packets inside a single path even though these protocols discovered multiple alternate paths during route discovery process.

On the other hand, broadcasting protocols have already been extensively investigated ([44, 74–78]). The literature in broadcasting protocol designs can be classified into two categories: deterministic approaches and probabilistic approaches. In the deterministic approaches, a fixed node within a connected dominating set is determined as a forwarding node. These approaches are also called fixed-forwarder approaches. In these approaches, the connected dominating set is calculated by using global or local information. In a probabilistic approach, when a node receives a packet, it forwards the packet with probability p. The value of p is determined by relevant information gathered at each node. Simple probabilistic approaches predefine a single probability for every node to rebroadcast the received packet. When running these protocols in a network with different node densities, the nodes in a dense area may receive a lot of redundant transmissions. More complicated and efficient protocols, such as distance-based and location-based [75] schemes, use either area or precise position information to reduce the number of redundant transmissions.

Despite this rich literature, the existing broadcasting approaches try to disseminate packets from one node to all the other nodes inside the network by using broadcast. ExOR [79] is an influential opportunistic routing protocol for wireless mesh networks. ExOR preselects prioritized forwarding candidates and each packet carries those forwarding candidates list. Only receivers in the forwarding list forward packets in the order of forwarding priority estimated based on the proximity (measured using ETX [80]) to the destination.

Zhong et al. [81] have proposed the expected any-path transmissions system. Westphal has proposed opportunistic routing in dynamic ad hoc networks (OPRAH protocol) [82]. OPRAH prepares multiple paths for a destination and each packet carries forwarding node list as ExOR did. These opportunistic routing protocols either preselect forwarder list or prepare multiple paths and use them to forward data packets. Furthermore, these protocols mainly focus on data plane improvement with underlying link state routing protocol or AODV-style control plane protocols.

Unlike the previous approaches, the energy-efficient routing (E^2R) protocol [83] neither preselects any single or multiple alternate paths nor maintains next hop information. Instead, E^2R delivers data packets using broadcast and simultaneously utilizes all the neighboring nodes to forward data packets. Moreover, E^2R simultaneously focuses on the reduction of control overhead and the enhancement of data delivery ratio.

The key idea of E^2R is to exploit spatial diversity (i.e., broadcast nature) in wireless networks rather than specifying data delivery paths (i.e., next hops). In E^2R, the source node does not specify any particular paths. Both route metric discovery (RMD) packets and data packets are delivered through broadcast. Nodes that have better opportunities to deliver the packets are automatically selected to forward the control packets and data packets. Similar to other wireless routing protocols, E^2R operates in two phases: RMD and data delivery. In this section, we first briefly introduce these two phases and discuss the design challenges in these two phases. The detailed design is described in Section 18.3.3.

18.3.1 Route Metric Discovery Phase

In this phase, RMD packets are delivered via broadcast. There are two challenges that have to be carefully addressed. The first challenge concerns how to prevent the repeated flooding of RMD packets when the network is first constructed. At the early stage of network construction, RMD packets must be delivered to every node inside the network since the source node and intermediate nodes do not know the direction of the destination node. Controlled flooding schemes (e.g., each node always forwards the newly received RMD packet once or multiple times) can be applied. However, flooding schemes can cause many unnecessary rebroadcasts of RMD packets especially when the network density is high. To address this challenge, greedy forwarding algorithm is devised for the distribution of RMD packets. The current forwarding node's covered neighbor list is embedded in the control packets. Here, we say a node is *covered* if it has already received the packet. Whenever a node receives the packet it marks whether its neighbors have already been covered based on the covered neighbor list in the packet. Then the node sets a waiting (i.e., back-off) time based on the number of its neighbors that have not been covered. Intuitively, the node with more uncovered neighbors should have higher priority to forward the packet. Therefore, if a node has a large number of uncovered neighbors, its waiting time should be shorter.

The second challenge concerns how to reduce redundant transmissions of the route metric reply (RMREP) packet. At the stage after the destination node receives the RMD packets, the destination node tries to deliver the RMREP packet back to the source node. Similarly, controlled flooding schemes will introduce redundant transmissions. This challenge differs from the first one, because the destination node and the intermediate nodes already know the direction of the source node because they have received the RMD packet initiated from the source node. Therefore, the challenge is to utilize this knowledge to further reduce redundant transmissions. To address this challenge, researchers introduce an efficient self-suppression scheme (described in Section 18.3.3), which suppresses the forwarding of the RMREP packet based on the route metric.

18.3.2 Data Delivery Phase

After the RMD phase, the source node obtains a route metric for the new destination. The source node now needs to deliver data packets to the destination node. The challenge is how to utilize spatial diversity to improve end-to-end performance (i.e., end-to-end packet delivery ratio and delay) and reduce energy consumption (by reducing the total number of packet transmissions inside the networks).

In order to address this design challenge, researchers introduce a forwarder self-selection scheme, which is similar to the relay selection scheme used in [84]. The source node attaches the obtained route metric to data packets and broadcasts the data packets without designating forwarding nodes. Upon receiving the data packets, the nodes that have smaller route metric value than the attached route metric value are eligible to further forward the data packets. Before these nodes forward the received data packet, they wait for a small amount of time to do backoff based on their own route metric values for the destination. For example, the node with smaller route metric value will have shorter backoff time and select itself to forward the received data packets. During the backoff time interval, these nodes listen to the channel and suppress the forwarding of received data packets if they overhear data packets forwarded by a node with a smaller route metric value. When the backoff timer fires, the node updates the route metric value in the data packets by attaching its own route metric and forwards the data packets.

18.3.3 Design of E^2R

This section describes the detailed design of E^2R that contains maintenance state, RMD, and data delivery.

18.3.3.1 Maintenance State

A node enters a maintenance state after it is deployed. While in the maintenance state, every node maintains its neighboring node information and the route metric from itself to all the other nodes inside the network. Like other wireless routing protocols, every node inside the network periodically sends out HELLO messages to indicate the existence of the node. Moreover, every node uses the HELLO messages received from its neighboring nodes to update its neighboring node set ($N(i)$).

Besides neighboring node information, every node also maintains the route metric from itself to all other nodes inside the network. E^2R is compatible with all other route metrics (e.g., ETX [80] or ETT [85]) that have been proposed. For example, we can give nodes with smaller ETX values higher priority to forward received packets. Without loss of generality, in this chapter we use distance vector (e.g., hop count) as the route metric. If a node s needs to route data packets to a destination node and there is no route metric maintained at node s for that node, s will initiate the RMD process. To reduce the transmission of control messages, the establishment and maintenance of route metrics is on demand. It is triggered when the source cannot deliver data packets to the destination.

18.3.3.2 Route Metric Discovery

The RMD process includes two stages: the stage of disseminating RMD packets and the stage of propagating RMREP packets.

18.3.3.2.1 RMD Packets Dissemination

During RMD discovery, the source node originates a new RMD packet if the source node needs to route data packets to a destination node and no route metric is available to the destination node. The RMD packet contains the source node id (s), packet id (Pid), source node's covered neighbor list (CN_s), destination node id (d), route metric (R), and route metric sequence number from source to destination (S_s^d). When a node, i, receives an RMD packet, i processes the RMD packet based on the greedy forwarding algorithm (shown in Algorithm 18.1). In the first step, i updates its uncovered neighbor set ($UN(i)$) by using the source node's covered neighbor list that is embedded in the RMD packet (Line 1). Here the uncovered neighbor set of i is the set of i's neighbors that have not received the RMD packet.

Algorithm 18.1 Greedy forwarding algorithm

1: Update $UN(i)$ based on $CN(s)$
2: **if** ($i \neq s$) and ($i \neq d$) **then**
3:     // i is an intermediate node
4:     **if** *new RMD* **then**
5:         **if** $S_i^d > S_s^d$ **then**
6:             //i has a fresher route metric to destination
7:             i sends RMREP with S_i^d
8:         **else if** all neighbors in N_i received the RMD **then**
9:             drop RMD
10:         **else**
11:             // i has an out-dated route metric
12:             wait for $T_{backoff}$ period assigned based on $UN(i)$
13:             **if** a neighbor forwarded RMD and $UN(i) = \phi$ during $T_{backoff}$ **then**
14:                 drop RMD
15:             **else**
16:                 $CN(s) \leftarrow CN(i)$, $R \leftarrow R + 1$, forward RMD
17:             **end if**
18:         **end if**
19:     **end if**
20: **else if** $i = d$ and new RMD **then**
21:     // i is the destination node
22:     send RMREP
23: **end if**

If i is an intermediate node and this RMD packet is the one that i receives for the first time (Lines 2–4), there are three possible cases:

- **Case 1:** i has a fresher route metric to the destination than the source has. In other words, the sequence number of the route metric from i to the destination (S_i^d) is larger than the sequence number of the route metric from the source to the destination (S_s^d). In this case, i directly returns and RMREP packet with S_i^d(Lines 5–7).
- **Case 2:** All neighbors of i have received the RMD packet (i.e., $UN(i) = \phi$). In this case, there is no need for i to rebroadcast the RMD packet. Therefore, i drops the RMD packet (Lines 8 and 9).
- **Case 3:** i does not have a fresher route metric to the destination than the source and some neighbors of i are uncovered. In this case, i sets the backoff timer with time interval $T_{backoff}$. Here the value of $T_{backoff}$ is inversely proportional to the size of the uncovered neighbor set $UN(i)$. The larger the uncovered neighbor set $UN(i)$, the smaller the value of $T_{backoff}$. Therefore, E^2R protocol allows the node (assume node j) that has a larger number of uncovered neighbors to rebroadcast the RMD packet first. When i overhears j's rebroadcast during i's backoff time interval, i updates its uncovered neighbor set $UN(i)$. If all the neighbors of i are covered, then i drops the RMD packet (Lines 10–14). Otherwise, i will update the covered neighbor set $CN(s)$ with i's covered neighbor set $CN(i)$ and increase the value of route metric R (if the route metric is hop count, then the number of hop count is increased by 1) in the RMD packet and rebroadcast the RMD packet (Lines 15 and 16).

If i is the destination node and this RMD packet is the one that i receives for the first time, i will send back an RMREP packet (Lines 20–23).

18.3.3.2.2 RMREP Propagation

As discussed in Section 18.3.3.2.1, RMREP packets are generated either by the destination or by the intermediate node, which has a fresher route metric to the destination than the source node. The RMREP contains the destination node id (d), source node id (s), packet id (Pid), route metric (R), and route metric sequence number from source to destination (S_s^d). Here the route metric R is the number of hops from source to destination. Since the RMREP does not contain any intermediate node information, it will be unnecessarily propagated to all nodes inside the network, which will result in a large amount of energy waste. In order to address this issue, E^2R protocol introduces an efficient self-suppression scheme which contains two rules:

- **Rule 1:** If the node has rebroadcasted the source originated RMD packet, this node is eligible to forward the RMREP packet. No other node is eligible to forward the RMREP packet. This rule avoids unnecessary rebroadcasts from nodes far from both the source and the destination.
- **Rule 2:** If the route metric of the node to the destination is larger than the route metric in the RMREP packet, this node is not eligible to forward the RMREP packet, for example, if we use hop count as route metric. If the node has a larger number of hops to the destination than the source has, the source will not use this node to forward the data packet. Therefore, there is no need to let this node forward the RMREP packet.

18.3.3.3 Data Delivery

After the source receives the RMREP packet and obtains the route metric, the source needs to forward the data packets to the destination node. Unlike other wireless routing protocols (such as AODV) and forwarding methods (such as ExOR), in E^2R, the source node does not need to designate its next hop or a forwarder list within the data packets. The source node only attaches the obtained route metric to data packets and then broadcasts the data packets. In this scheme, the forwarders of the data packets are selected by the intermediate nodes themselves. It is called a forwarder self-selection scheme.

When an intermediate node i receives the data packets, i compares its own route metric value with the value of the route metric embedded in the data packets. If i's route metric value is smaller than the value of the route metric embedded in the data packets, i selects itself to be a potential forwarder of the data packets. However, i does not know whether its neighbors also received the data packets and have smaller route metric values than i. In order to handle this problem, we introduce a backoff mechanism and design the backoff time interval based on the route metric values. The smaller the value of the route metric a node has, the shorter the backoff time this node experiences. During the backoff time interval, i listens to the channel and suppresses the forwarding of the data packets if i overhears that one of its neighbors with a better route metric already forwarded the data packets. If i does not overhear its neighbors' forwarding and its backoff timer fires, i updates the route metric in the data packets with its own route metric and then rebroadcasts the data packets.

When the destination node receives the data packets, the destination node returns an ACK to the source node. The propagation of the ACK is similar to the propagation of the RMREP packet.

Researchers have performed extensive simulation with various network configurations to reveal the performance of E^2R. The results show that the E^2R protocol can provide high packet delivery ratio, low control overhead, and low packet delivery delay in unreliable environments. Moreover, by reducing the number of packet transmissions, E^2R protocol can effectively reduce the energy consumption and make it an energy-efficient routing protocol for multi-hop green wireless networks.

References

1. T. Zhu, A. Mishra, D. Irwin, N. Sharma, P. Shenoy, and D. Towsley. The case for efficient renewable energy management for smart homes. In *ACM BuildSys*, 2011.
2. T. Zhu, S. Xiao, P. Yi, D. Towsley, and W. Gong. A secure energy routing protocol for sharing renewable energy in smart microgrid. In *IEEE SmartGridComm*, 2011.
3. Z. Pei and T. Zhu. A multi-information localization algorithm in wireless sensor networks and its application in medical monitoring systems. In *WiCOM*, 2011.
4. Y. Sun, T. Zhu, Z. Zhong, and T. He. Energy profiling for mPlatform. In *SenSys*, 2009.
5. E. Uysal-Biyikoglu, B. Prabhakar, and A. Gamal. Energy-efficient packet transmission over a wireless link. *IEEE/ACM Transactions on Networking*, 10(4), 487–499, 2002.
6. G. Hackmann, O. Chipara, and C. Lu. Robust topology control for indoor wireless sensor networks. In *SenSys*, 2008.
7. A. Woo, T. Tong, and D. Culler. Taming the underlying challenges of reliable multihop routing in sensor networks. In *SenSys*, 2003.
8. C. Intanagonwiwat, R. Govindan, and D. Estrin. Directed diffusion: A scalable and robust communication paradigm for sensor networks. In *MobiCom*, 2000.
9. R. Szewczyk, A. Mainwaring, J. Anderson, and D. Culler. An analysis of a large scale habit monitoring application. In *SenSys*, 2004.

10. G. Tolle, J. Polastre, R. Szewczyk, N. Turner, K. Tu, S. Burgess, D. Gay, P. Buonadonna, W. Hong, T. Dawson, and D. Culler. A macroscope in the redwoods. In *SenSys*, 2005.
11. J. Polastre and D. Culler. Versatile low power media access for wireless sensor networks. In *SenSys*, 2004.
12. M. Buettner, G. V. Yee, E. Anderson, and R. Han. X-mac: A short preamble mac protocol for dutycycled wireless sensor networks. In *SenSys*, 2006.
13. A. El-Hoiydi and J. Decotignie. Wisemac: An ultra low power MAC protocol for the downlink of infrastructure wireless sensor networks. In *Proceedings of 9th ISCC International Symposium on Computers and Communications*, vol. 1, pp. 244–251, June 2004.
14. C. Zhou and T. Zhu. Highly spatial reusable MAC for wireless sensor networks. In *Proceedings of the 3rd IEEE International Conference on Wireless Communications, Networking and Mobile Computing (WiCOM)*, 2007.
15. C. Zhou and T. Zhu. A spatial reusable MAC protocol for stable wireless sensor networks. In *Proceedings of The 4th IEEE International Conference on Wireless Communications, Networking and Mobile Computing (WiCOM)*, 2008.
16. C. Zhou and T. Zhu. Thorough analysis of MAC protocols in wireless sensor networks. In *Proceedings of the 4th IEEE International Conference on Wireless Communications, Networking and Mobile Computing (WiCOM)*, 2008.
17. A. Arora, P. Dutta, S. Bapat, V. Kulathumani, H. Zhang, V. Naik, V. Mittal, et al. A line in the sand: A wireless sensor network for target detection, classification, and tracking. *The International Journal of Computer and Telecommunications Networking - Special issue: Military communications systems and technologies archive*, 46(5):605–634, December 2004.
18. M. Batalin, M. Rahimi, Y. Yu, D. Liu, A. Kansal, G. Sukhatme, W. Kaiser, et al. Call and response: experiments in sampling the environment. In *SenSys*, 2004.
19. N. Xu, S. Rangwala, K. Chintalapudi, D. Ganesan, A. Broad, R. Govindan, and D. Estrin. A wireless sensor network for structural monitoring. In *SenSys*, 2004.
20. X. Jiang, J. Polastre, and D. E. Culler. Perpetual environmentally powered sensor networks. In *IPSN*, 2005.
21. T. Zhu, Y. Gu, T. He, and Z.-L. Zhang. eShare: A capacitor-driven energy storage and sharing network for long-term operation. In *SenSys*, 2010.
22. T. Zhu, Y. Gu, T. He, and Z. Zhang. Achieving long-term operation with a capacitor-driven energy storage and sharing network. *ACM Transactions on Sensor Networks (TOSN)*, 8(3), Aug. 2012.
23. T. Zhu, Z. Zhong, T. He, and Z. Zhang. Energy-synchronized computing for sustainable sensor networks. *Elsevier Ad Hoc Networks Journal*, 2010. http://www.cs.binghamton.edu/~tzhu/Paper/J1_Energy-synchronized_Ad%20Hoc%20Networks_2011.pdf
24. T. Zhu, Z. Zhong, Y. Gu, T. He, and Z.-L. Zhang. Leakage-aware energy synchronization for wireless sensor networks. In *MobiSys*, 2009.
25. T. Zhu, Z. Zhong, Y. Gu, T. He, and Z. Zhang. Feedback control-based energy management for ubiquitous sensor networks. *IEICE Transactions on Communications*, E93-B(11), 2846–2854, November 2010.
26. Y. Ping, T. Zhu, B. Jiang, B. Wang, and D. Towsley. An energy transmission and distribution network using electric vehicles. In *ICC*, 2012.
27. Z. Zhong, T. Zhu, T. He, and Z. Zhang. Demo abstract: Leakage-aware energy synchronization on twin-star nodes. In *SenSys*, 2008.
28. T. Zhu, A. Mohaisen, Y. Ping, and D. Towsley. DEOS: Dynamic energy-oriented scheduling for sustainable wireless sensor networks. In *INFOCOM*, 2012.
29. Y. Gu and T. He. Data forwarding in extremely low duty-cycle sensor networks with unreliable communication links. In *SenSys*, 2007.

30. A. Malvankar, M. Yu, and T. Zhu. An availability-based link QoS routing for mobile ad hoc networks. In *Proceedings of the 29th IEEE Sarnoff Symposium*, 2006.
31. T. Zhu and M. Yu. A dynamic secure QoS routing protocol for wireless ad hoc networks. In *Proceedings of the 29th IEEE Sarnoff Symposium*, 2006.
32. J. W. Hui and D. Culler. The dynamic behavior of a data dissemination protocol for network programming at scale. In *Sensys*, 2004.
33. M. Agarwal, J. Cho, L. Gao, and J. Wu. Energy-efficient broadcast in wireless ad hoc networks with hitch-hiking. In *INFOCOM*, 2004.
34. P. Kyasanur, R. Choudhury, and I. Gupta. Smart gossip: An adaptive gossip-based broadcasting service for sensor networks. In *MASS*, 2006.
35. A. Cerpa, J. L. Wong, M. Potkonjak, and D. Estrin. Temporal properties of low power wireless links: Modeling and implications on multi-hop routing. In *MobiHoc*, 2005.
36. A. Cerpa, J. L.Wong, L. Kuang, M. Potkonjak, and D. Estrin. Statistical model of lossy links in wireless sensor networks. In *Proceedings of The Fourth International Symposium on Information Processing in Sensor Networks (IPSN)*, 2005.
37. A. Kamthe, M. A. Carreira-Perpin, and A. E. Cerpa. Multi-level Markov model for wireless link simulations. In *SenSys*, 2009.
38. G. Zhou, T. He, S. Krishnamurth, and J. Stankovic. Impact of radio irregularity on wireless sensor networks. In *MobiSys*, pp. 125–138, 2004.
39. K. Srinivasan, M. A. Kazandjieva, S. Agarwal, and P. Levis. The beta-factor: Measuring wireless link burstiness. In *SenSys*, 2008.
40. M. Albano and S. Chessa. Distributed erasure coding in data centric storage for wireless sensor networks. In *IEEE Symposium on Computers and Communications*, 2009.
41. M. Zuniga and B. Krishnamachari. Analyzing the transitional region in low power wireless links. In *IEEE SECON*, 2004.
42. S. Lin, J. Zhang, G. Zhou, L. Gu, T. He, and J. A. Stankovic. Atpc: Adaptive transmission power control for wireless sensor networks. In *SenSys*, 2006.
43. T. He, C. Huang, B. M. Blum, J. A. Stankovic, and T. Abdelzaher. Range-free localization schemes in large-scale sensor networks. In *MobiCom*, 2003.
44. T. Zhu, Z. Zhong, T. He, and Z.-L. Zhang. Exploring link correlation for efficient flooding in wireless sensor networks. In *Proceedings of the 7th USENIX conference on Networked Systems Design and Implementation (NSDI)*, 2010.
45. F. Stann, J. Heidemann, R. Shroff, and M. Z. Murtaza. RBP: Robust broadcast propagation in wireless networks. In *Sensys*, 2006.
46. V. Naik, A. Arora, P. Sinha, and H. Zhang. Sprinkler: A reliable and energy efficient data dissemination service for wireless embedded devices. In *RTSS*, 2005.
47. W. Lou and J. Wu. Double-covered broadcast (DCB): A simple reliable broadcast algorithm in MANETs. In *INFOCOM*, 2004.
48. W. Peng and X. Lu. On the reduction of broadcast redundancy in mobile ad hoc networks. In *MobiHoc*, 2000.
49. S. Ni, Y. Tseng, Y. Chen, and J. Sheu. The broadcast storm problem in a mobile ad hoc network. In *MobiCom*, 1999.
50. E. Rozner, J. Seshadri, Y. Mehta, and L. Qiu. SOAR: Simple opportunistic adaptive routing protocol for wireless mesh networks. *IEEE Transactions on Mobile Computing (TMC)*, 8(12), 1622–1635, 2009.
51. S. Biswas and R. Morris. Exor: Opportunistic multi-hop routing for wireless networks. In *SIGCOMM*, 2005.
52. M. Kurth, A. Zubow, and J. P. Redlich. Cooperative opportunistic routing using transmit diversity in wireless mesh networks. In *INFOCOM*, 2008.

53. X. Qin and R. Berry. Exploiting multiuser diversity for medium access control in wireless networks. In *INFOCOM*, 2003.
54. K. Srinivasan, M. Jain, J. Choi, T. Azim, Kim, P. Levis, and B. Krishnamachari. The kappa factor: inferring protocol performance using inter-link reception correlation. In *MobiCom*, 2010.
55. S. Guo, S. M. Kim, T. Zhu, Y. Gu, and T. He. Correlated flooding in low-duty-cycle wireless sensor networks. In *ICNP*, 2011.
56. J. Wu, W. Lou, and F. Dai. Extended multipoint relays to determine connected dominating sets in MANETs. *IEEE Transactions on Computing*, 55(3), 334–347, 2006.
57. J. Padhye, S. Agarwal, V. N. Padmanabhan, L. Qiu, A. Rao, and B. Zill. Estimation of link interference in static multi-hop wireless networks. In *IMC*, 2005.
58. R. Gummadi, D. Wetherall, B. Greenstein, and S. Seshan. Understanding and mitigating the impact of rf interference on 802.11 networks. In *SIGCOMM*, 2007.
59. B. Hull, K. Jamieson, and H. Balakrishnan. Mitigating congestion in wireless sensor networks. In *SenSys*, 2004.
60. J. Hill, R. Szewczyk, A. Woo, S. Hollar, D. E. Culler, and K. S. J. Pister. System architecture directions for networked sensors. In *Proceedings of Architectural Support for Programming Languages and Operating Systems (ASPLOS)*, pp. 93–104, 2000.
61. D. Gay, P. Levis, R. von Behren, M. Welsh, E. Brewer, and D. Culler. The nesC language: A holistic approach to networked embedded systems. In *Programming Language Design and Implementation*, 2003.
62. Y. Gu, T. Zhu, and T. He. Esc: Energy synchronized communication in sustainable sensor networks. In *ICNP*, 2009.
63. P. Jacquet, P. Muhlethaler, and A. Qayyum. Optimized link state routing protocol. In *IETF MANET*, Internet Draft, 1998.
64. D. B. Johnson and D. A. Maltz. Dynamic source routing in ad hoc wireless networks. In *Mobile Computing*. Kluwer Academic Publishers, Dordrecht, the Netherlands, 1996.
65. S. Lee and M. Gerla. Split multipath routing with maximally disjoint paths in ad hoc networks. In *Proceedings of the IEEE ICC*, 2001.
66. S.-J. Lee and M. Gerla. AODV-BR: Backup routing in ad hoc networks. In *WCNC* , 2000.
67. M. K. Marina and S. R. Das. Ad hoc on-demand multipath distance vector routing. *WCMC*, 2001.
68. V. D. Parka and M. S. Corson. A highly adaptive distributed routing algorithm for mobile wireless networks. In *INFOCOM*, 1997.
69. C. Perkins and P. Bhagwat. Highly dynamic destination-sequenced distance-vector routing (DSDV) for mobile computers. In *SIGCOMM*, 1994.
70. C. Perkins and E. Royer. Ad hoc on demand distance vector routing, mobile computing systems and applications. In *WMCSA*, 1999.
71. L. Tassiulas and A. Ephremides. Stability properties of constrained queueing systems and scheduling policies for maximum throughput in multihop radio networks. *IEEE Transactions on Automatic Control*, 37(12), 1936–1948, December 1992.
72. A. Valera, W. Seah, and S. Rao. Cooperative packet caching and shortest multipath routing in mobile ad hoc networks. In *INFOCOM*, 2003.
73. T. Zhu and M. Yu. A secure quality of service routing protocol for wireless ad hoc networks. In *IEEE GLOBECOM*, 2006.
74. Z. Haas, J. Halpern, and L. Li. Gossip-based ad hoc routing. In *INFOCOM*, 2002.
75. S. Ni, Y. Tseng, Y. Chen, and J. Sheu. The broad-cast storm problem in a mobile ad hoc network. In *MobiCom*, 1999.
76. W. Peng and X. Lu. On the reduction of broadcast redundancy in mobile ad hoc networks. In *MOBIHOC*, 2000.

77. W. Peng and X. Lu. Ahbp: An efficient broadcast protocol for mobile ad hoc networks. *Journal of Computer Science and Technology*, 16, 114–125, 2002.
78. A. Qayyum, L. Viennot, and A. Laouiti. Multipoint relaying: An efficient technique for flooding in mobile wireless networks. Technical report, 2000.
79. S. Biswas and R. Morris. Exor: Opportunistic routing in multi-hop wireless networks. In *SIGCOMM*, 2005.
80. D. D. Couto, D. Aguayo, J. Bicket, and R. Morris. High-throughput path metric for multi-hop wireless routing. In *MobiCom*, 2003.
81. Z. Zhong, J. Wang, S. Nelakuditi, and G. Lu. On selection of candidates for opportunistic anypath forwarding. *Mobile Computing and Communications Review*, 10(4), 1–2, 2006.
82. C. Westphal. Opportunistic routing in dynamic ad hoc networks. In *Proceedings of IEEE MASS*, October 2006.
83. T. Zhu and D. Towsley. E2R: Energy efficient routing for multi-hop green wireless networks. In *IEEE Conference on Computer Communications Workshops (INFOCOM WORKSHOPS)*, 2011.
84. A. Bletsas, A. Khisti, D. Reed, and A. Lippman. A simple cooperative diversity method based on network path selection. *IEEE JSAC*, 24(3), 659–672, March 2006.
85. R. Draves, J. Padhye, and B. Zill. Routing in multi-radio, multi-hop wireless mesh networks. In *Mobicom*, 2004.

Chapter 19

Agent-Driven Wireless Sensors Cooperation for Limited Resources Allocation

Sameh Abdel-Naby, Conor Muldoon, Olga Zlydareva, and Gregory O'Hare

Contents

19.1 Agents' Negotiation for Wireless Sensor Networks

In multiagent systems (MASs) [1,2], several contributions in the literature are addressing the problem of sharing resources between two or more agents within a predefined space, for example [3]. For two agents to efficiently share resources, a mutually beneficial agreement should be reached so that each is achieving part or all of the assigned goals, (e.g., monitor/report temperature). Several constraints are considered by each agent while attempting to achieve the assigned goals (e.g., time vs. number of goals).

In order for several agents/sensors to discuss the possibility of reaching a certain agreement, a common negotiation protocol must be applied, similar to the one explained in [4]. Then each agent adopts a specific negotiation strategy, such as this in [5], before engaging in a potentially beneficial interaction. The focus of these strategies vary depending on the purpose this agent or robot was designed for. For example, some negotiation strategies can be time or resources driven while others can be score-oriented.

In this section, we give an overview of agents' negotiation approaches from three different contexts. These contexts are (1) common settings, (2) service provision/acquisition, and (3) wireless networks.

19.1.1 Agents' Negotiation in Common Settings

In [6] agents, negotiation protocols are addressed with respect to three abstract domains: (1) task-oriented domains, where an agent's activity is a set of tasks to be achieved; (2) state-oriented domains, where an agent is moving from an initial state to a set of goal states; and (3) worth-oriented domains, where agents evaluate each potential state to identify its level of desirability.

In [5], a particular focus was given to agents interacting in distributed information retrieval systems arguing that the cooperation of information servers relatively increases with the advances made on agents' negotiation. Two scenarios of agent's negotiation are considered: (1) negotiation about data allocation, where autonomous agents/servers are sharing documents and they need to decide how best they could locate them and (2) negotiation about resource allocation, where the main focus is given to domains of limited resources as well as those of unlimited ones, and agents are bilaterally negotiating to share expensive or common resources.

Based on Rubinstein's model for alternating offers [7], the negotiation protocol presented in [5] is straightforward. One agent makes an offer to another that has to choose between accepting, rejecting, or opting out of the negotiation process. Each agent has its own utility function that evaluates all possible negotiation results and a strategy to decide what actions to perform at every expected situation. Although the negotiation of agents about the allocation of limited resources is similar to our scenario wherein sensors are interacting to determine the use of limited resources, still in our case sensors are expected to be more than just two, and sensors are not expected to generate offers and wait for responses.

The Contract-Net protocol presented in [4] is a high-level negotiation protocol for communicating service requests among distributed agents. R. G. Smith considers the high-level negotiation protocols as methods that lead system designers to decide "what agents should say to each other." And low-level protocols make system designers decide "how agents should talk to each other." The Contract-Net protocol assumes the simultaneous operation of both agents asking to execute tasks and agents ready to handle it. The asking agents broadcast a call for proposals, and the helping agents submit their offers and then one is granted the pending task, or the session is closed. Three points are worth highlighting in the earlier approach:

1. Linking between high-level and low-level negotiation protocols is essential when it comes to agents interacting in limited and variable resources environment. For example, when users of pocket computing devices delegate software agents to exchange and accomplish service requests on the go, the efficiency of the negotiation protocol that agents will employ is relatively increasing with the size of bandwidth a network utilizes, and the time it takes to transfer agent's requests/messages from one location to another.
2. A central decision-making situation may easily occur when a service seeker initiates the call for proposals, and it receives back all of the prospects offers, and the same agent is the only one who decides upon the termination of the negotiation process.
3. In the Contract-Net, it is always assumed that two different types of agents are interacting (e.g., buyer and seller agents), which is different in the sensors scenario because they are all of the same type.

19.1.2 Agents' Negotiation for Service Acquisition

In [8], a service-oriented negotiation model for autonomous agents was presented. Following the traditional client/server approach, scholars assumed that an autonomous agent is a "client" to another serving agent "server" that is in turn delegated to achieve a certain goal, which is acquiring or selling a service. Authors have focused their research on the reasoning model an agent will employ to identify its prospective servers, deciding about whether to perform parallel or sequential negotiations, making or accepting an offer, and abandoning a process.

Accordingly, the times to reach and execute an agreement were both considered while identifying their negotiation model. Based on a model for bilateral negotiation that was presented in [9], authors of the service-oriented negotiation model proposed a multilateral variation of it to satisfy the application domain they are interested in. However, the requirements they attempted to satisfy (i.e., privacy of information, privacy of models, value restrictions, time restrictions, and resources restrictions) are related to our research focus but of ordinary network communications.

In [10], an agent-based architecture for service discovery and negotiation was presented. The realization of three novel requirements has motivated the work of these authors; these requirements are as follows: (1) interactions of software agents are not necessarily happening in one network only and, these interactions may involve more than two types of agents (e.g., service provision agent, service acquisition agent and service evaluator agent); (2) diverse connection technologies can be utilized at different costs, which increase the complexity of a system and enable a higher level of end-user dynamicity; and (3) a service application should automatically react to changes as long as it is of end-user benefits. The scenario they used to motivate their work involves three different agents. The user agent is located on the portable device of the user, and it contacts a marketplace agent that is installed at wireless hotspot location and is responsible for maintaining a list of available Internet services providers (ISPs) that each has a representing agent called ISP agent.

An agreement is reached when the user agent succeeds to make a contract with one of the ISP agents and retrieves a configuration file that, eventually, the end user installs on its pocket device to get an Internet access through the best available ISP. The sequence of interactions among the involved agents was described, and negotiation protocols and strategies were the traditional FIPA Contract-Net [11] and FIPA-English Auction Protocol [12].

It is also worth highlighting here that four different types of auctions are widely considered in the literature of agents: (1) English, (2) Dutch, (3) first price sealed-bid, and (4) second-price sealed-bid (e.g., in [13]). These auction types share the same goal, which is granting a single item (sometimes combinatorial) to a single agent (sometimes a coalition) in a limited resources environment. An agent may participate in an auction, so one of the carried "personal" tasks can be accomplished or—*like in cooperative systems*—a learning behavior can be implemented, so agents are able to predict the future importance of this item to another agent, which is known as common value.

A multiagent negotiation method for on-demand service composition is presented in [14]. Agents here are expected to negotiate in order to reach agreements about combining different services from different providers to finally meet a consumer's expectations. The negotiation process is functioning by means of messages exchanging. When an invocation for service acquisition occurs, it is assumed that all available agents are representing specific services in a network, and they receive a message that contains a set of requirements to fulfill. If a single agent is capable of providing this service on its own, it broadcasts an OK message, if not, it transmits a Nogood message to the others asking for help. Other agents receive this help request and review their capabilities and give a response, and so on.

Using messages exchange to negotiate the acquisition of a service between entities located in a fixed wireline network is likely to be satisfactory. However, in a sensors network, where communication resources are expensive and limited, this approach will add a considerable amount of traffic, and it will increase the time a service application takes to act in response to requests.

19.1.3 Agents' Negotiation across Wireless Communications

An architecture for pervasive negotiation that uses multiagent and context-aware technologies is proposed in [15]. The main focus of this research effort is to consider specific profiles, preferences, and locations so that personalized services are transmitted to other remote devices. Authors of this chapter have considered three different agents that are involved in any negotiation scenario, these agents are (1) user agents that announce the preferences of the assigned goal, (2) supplier agents that take into consideration the preferences and the context provided to compete for service provision, and (3) negotiation agents that maintain all of the allowed negotiation strategies and mediate between the earlier two types of agents.

The negotiation mechanism proposed is influenced by semantic web approaches. The negotiation agent (i.e., the mediating agent) is coming up with the best-fitting strategy in particular situation by parsing throughout predefined negotiation ontology. This ontology is flexible and interactive so that end users and service providers are able to present their own negotiation conditions.

Intelligent information agents [16] are those capable of interacting with several distributed or heterogeneous information systems representing specific preferences in obtaining data and overcoming information overload. In [17], authors have relied on information agents, data acquisition agents, and rule-based reasoning agents to build an MAS capable of receiving data from a legacy information system—*enterprise resource planning*—and control the extracted information using artificial intelligence techniques. That effort has added the possibility of an existing information system to be customized according to the new preferences of end users without any reengineering processes.

Using ubiquitous agents, another approach was taken in [18] to allow mobile devices to access Web information systems depending on their location. Authors have used agents to represent the goals nomadic users would like to achieve, store the exact location and connection features of each user, and then migrate these agents to different information systems (or other mobile devices) to find relevant data or another information agent capable of answering user's requests. In PUMAS, the agents negotiation was implemented using standard distributed systems technique—*message passing and recommendations*—in spite of the dynamicity of mobile users and the limited resources of a mobile network.

Coalition formation is another approach to agents' negotiation. In [19], a model for coalition formation was proposed to enable each agent to select individually its allies. The model supports the formation of any coalition structure, and it does not require any extra communication or central coordination entity. Similarly, the definition of an optimal coalition in [20] is based on Pareto dominance and distance weighting algorithm. In addition, in [21], a trusted kernel-based mechanism for forming a coalition is presented, rather than on efficiency of a common task solving.

Coming from a WSN background authors of [22] are proposing an approach for developing autonomous agents that can employ economic behavior while engaged in a negotiation process. Applying this behavior let agents become able to independently take rational economic-driven decisions. Same authors described two related scenarios wherein agents are negotiating the possibility of acquiring a service. These scenarios are involving: (1) the negotiation of the transportation and communication services with the available ad-hoc network, and (2) granting the usage of the WSN to successfully agreed agents.

From electronics, another approach to automated agents' negotiation and decision making in resources-limited environment was presented in [23]. Scholars here perceive an MAS as a set of distributed radio antennas of a cellular network, and by means of agents' contracting, these agents would share the use of resources (e.g., bandwidth).

In [24], authors argue that by using reinforcement learning an agent will then be capable of employing a set of strategies that respond to the opponents' with various concessions and preferences. The protocol authors propose let agents respond to a negotiator by either giving a deadline for an agreement to be reached, or a complete rejection. The main focus of the two experiments they explained shows how an agent learns a behavior while still keeping its functionalities at the same level of performance, which they call it "not breaking down."

19.2 Agent Factory Micro Edition for Wireless Sensor Network Motes

Agent factory micro edition (AFME) [25] is an intelligent agent framework for resource-constrained devices and is based on a declarative agent programming language, which, in a similar vein to other intelligent agent platforms, is used in conjunction with imperative components. These imperative components imbue agents with mechanisms that enable them to interact with their environment; agents perceive and act upon the environment through perceptors and actuators, respectively. The word perceptor is used rather than sensor to distinguish the software component from hardware sensors.

Perceptors and actuators represent the interface between the agent and the environment and are implemented in Java. This interface acts as an enabling mechanism through which the agents are situated. AFME incorporates a variant, specifically designed for resource-constrained devices, of the agent factory agent programming language (AFAPL) [26]. AFAPL is a declarative language and is a derivative of Shoham's agent-oriented programming framework [27]. It is based on a logical formalism of belief and commitment and forms part of the agent factory framework [26], which is an open source collection of tools, platforms, and languages that support the development and deployment of MASs . In its latest incarnation, the agent factory framework has been restructured to facilitate the deployment of applications that employ a diverse range of agent architectures. As such, the framework has become an enabling middleware layer that can be extended and adapted for different application domains. The framework is broadly split into two parts:

- Support for deploying agents on laptops, desktops, and servers
- Support for deploying agents on constrained devices such as mobile phones and WSN nodes

AFME represents the latter, whereas the former is delivered through agent factory standard edition (AFSE). In the remainder of this chapter, we shall only consider AFME. An overview of the AFME control process is provided in Figure 19.1. In AFME, rules that define the conditions under which commitments are adopted are used to encode an agent's behavior. The following is an example of an AFME rule:

```
message(request,?sender, removeData(?user))>deleteRecord(?user);
```

The truth of a belief sentence (text prior to the > symbol) is evaluated based on the current beliefs of the agent. The result of this query process is either failure, in which case the belief sentence

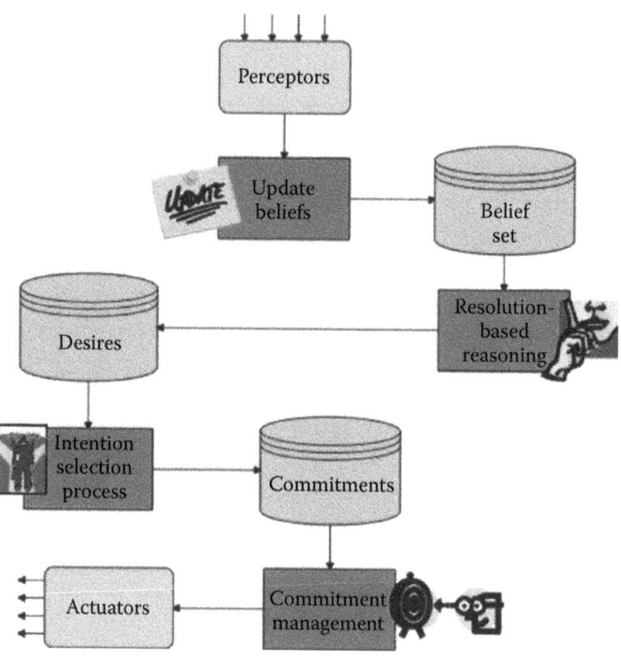

Figure 19.1 AFME control process.

is evaluated to false or to a set of bindings that cause the belief sentence to be evaluated to true. In AFAPL, the ? symbol represents a variable. In this example, if the agent has adopted a belief that it has received a message from another agent to remove user data, it adopts a commitment to delete the record related to the user. At an imperative level, a preceptor, which is written in Java, monitors the message transport service, which contains a server thread that receives incoming messages. Once a message is received, it is added to a buffer in the service. Subsequently, the perceptor adds a belief to the agent's belief set. The interpreter periodically evaluates the belief set. If the conditions for a commitment are satisfied (i.e., all of the beliefs prior to the > symbol in the rule have been adopted), either a plan is executed to achieve the commitment or a primitive action or actuator is fired. In this chapter, we shall only consider primitive actions. When an actuator is created, it is associated with a symbolic trigger. In this case, a delete record actuator, written in Java, is associated with the trigger string deleteRecord(?user). Once the commitment is activated, the ?user variable is passed to the actuator and the imperative code for deleting the file is executed. Structuring agents in this manner is useful in that it enables their behavior to be altered at a symbolic level rather than having to modify the imperative code.

In AFME, the commitment to the right of the implication (the > symbol) can take additional arguments. These arguments represent to whom the commitment is made, the time at which the commitment should be executed, the predicate for maintaining the commitment, and the utility values of the commitment. These additional arguments go beyond the scope of this discussion and shall not be described here.*

In order to facilitate communication between AFME agents in WSN applications, a wireless message transport service has been developed that can be controlled and monitored through the

* For a discussion of how these arguments are supported in AFME along with other features of AFME/AFSE, such as dynamic role adoption/retraction and the AFME intention selection process, see [29].

use of actuators and perceptors. The Sun SPOT motes [28] communicate using the IEEE 802.15.4 standard. The wireless message transport service facilitates peer-to-peer communication between agents and is based on the Sun SPOT radiogram protocol rather than TCP/IP, which is used for agents deployed on mobile phones or PDAs that have a 3G or GPRS connection. The radiogram protocol uses datagrams to facilitate communication between motes.

With the Sun SPOT radiogram protocol, the connections operating over a single hop have different semantics to those operating over multiple hops. This is due to a performance optimization. When datagrams are sent over more than one hop, there are no guarantees about delivery or ordering. In such cases, datagrams will sometimes be silently lost, be delivered more than once, and be delivered out of sequence. When datagrams are sent over a single hop, they will not be silently lost or delivered out of sequence, but sometimes they will be delivered more than once.

The radiogram protocol operates in a client–server manner. When the message transport service is created, a server thread is created to receive incoming messages. When a message is received, it is added to an internal buffer within the service. An agent will subsequently perceive messages through the use of a perceptor.

When an agent is sending a message, it attempts to open a datagram client connection. The datagram server connection must be open at the destination. With datagrams, a connection opened with a particular address can only be used to send to that address. The wireless message transport service only allows agents to send messages of a maximum size. If the content of the message is greater than the limit, it is first split into a number of sub-messages within an actuator and each sub-message is then sent using the message transport service. When all sub-messages have been received, the entire message is reconstructed within a perceptor and then added to the belief set of the agent.

One of the core features of AFME is the support for agent migration. For the Sun SPOT platform, this support is delivered through the AFME wireless migration service. Agent migration is often classified as either strong or weak. This classification is related to the amount of information transferred when an agent moves. The more information transferred, the stronger the mobility. Within AFME, support is only provided for the transfer of the agent's mental state. Any classes required by the agent must already be present at the destination. The Java platform AFME has been developed for, namely the Java Micro Edition (JME) Constrained Limited Device Configuration (CLDC). Therefore, it does not contain an API to dynamically load foreign classes. Only classes contained, and preverified, in the deployed jar can be dynamically loaded through the use of the Class.forName method. This is also one of the reasons why component deployment frameworks, such as OSGi [30], cannot be used for CLDC applications.

In the Squawk JVM, which operates on Sun SPOTs, it is possible to migrate an application to another Squawk-enabled device. Squawk implements an isolate mechanism, which can be used for a type of code migration. Isolate migration is not used in AFME. The reason for this is that isolate migration is dependent on internal details of the JVM and is therefore not really platform independent in the sense that an isolate can only be transferred to another Squawk JVM. It could not be used to transfer an application to a C or C++ CLDC JVM written for a mobile phone JVMs, for instance. Additionally, with isolates, it is necessary to migrate the entire application or platform, rather than just a single agent.

This AFME migration service uses both the Sun SPOT radiogram protocol and the radiostream protocol. The radiostream protocol operates in a similar manner to TCP/IP sockets. It provides reliable, buffered, stream-based communication between motes. This, however, comes at a cost in terms of power usage. The reason this approach is adopted for agent migration is that we wish to ensure that agent does not become corrupt or lost due to the migration process. If a message is lost or corrupt, the system can recover by resending the message. If an agent is lost or corrupt, it cannot be recovered without duplication or redundancy, which would also use up resources and would become complex to manage as agent artifacts would be scattered throughout the network.

The problem with the radiostream protocol, however, is that both the target platform and the source platform must know each other's MAC address before a connection can be established. That is, it does not adopt a client—server approach or operate in a similar manner to the radiogram protocol. In a dynamic mobile agent setting, it is unlikely that the addresses of the platforms of all source agents will be known a priori at compile time. To get around this problem, when an agent wishes to migrate to a particular platform, initial communication is facilitated through the use of datagrams. Using datagrams, the platforms exchange address and port information and subsequently construct a radiostream. Once the radiostream is established, the agent is transferred through the reliable connection and then terminated at the source. Subsequently, the stream connection is closed. At the destination, the platform creates and starts the agent.

19.3 Motivating Example

In this section, we describe a health care–oriented example that is a potentially suitable testbed for future examination to the idea of integrating agent-based negotiation models with wireless sensor networks (WSNs) through the AFME.

Considering the increasing number of older people around the world, it brings to our attention the health care monitoring systems deployed within smart houses, which reflect the world's increasing emphasis on independent living [31]. In addition, one of the current priorities for UN activities targeting aging is to support active aging, where older people play a central role through continuous participation in social, economical, and cultural aspects of world society [32]. The key goal of the active aging framework is maintaining autonomous and independent living for older population in the home environment. And, we believe that the use of Intelligent WSNs that has approaches from Artificial Intelligence implemented within will better enhance the quality of life for aging people.

The number of people over 65 with cardiovascular diseases is growing every year [33]. Recent activities of European Union are placing more industrial and academic emphasis on how to support and enhance the idea of independent living [34]. To enable a high-level patient's mobility and free him or her from being connected to the expensive hospital equipment, a number of remote monitoring systems are introduced, for example [35]. The basic principle is to track the condition of the human cardiovascular system by means of a daily transmission of medically related events and vital signs. This approach reduces the in-clinic visits by up to 43% and results in patients leading a satisfying independent and safe life [36]. It does, however, require a large amount of information to be obtained from a limited number of measuring devices and, consequently, it also requires highly reliable sensors interactions.

To increase the level of self-confidence, mobility, and autonomy of the older person, ambient assisted living (AAL) systems have been introduced, which made up of state-of-the-art sensors [37], robotics [38], and wireless communications that are all research areas that are increasingly becoming a main focus for the scientific society. The AAL's primary task is to support daily activities through natural interactions and person's behavior analysis. Recent research activities on AAL have resulted in the creation of complex frameworks such as the one presented in [39] that use system intelligence and WSN to build an awareness of the condition of the older person.

The creative way of thinking and in-depth understanding of the scientific principles with targeting social and health aspects have helped the society improve the quality of life for the elderly

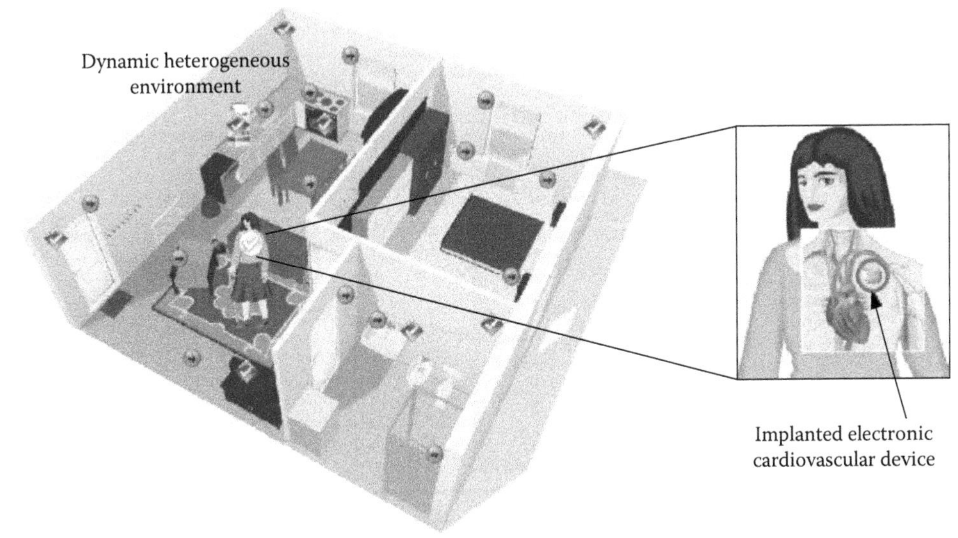

Figure 19.2 The implantable electronic cardiovascular device and smart environments.

population. As a response to today's situation on the assisted living and with the concept described in this example, the fields of wireless communications, cognitive networks, low-power systems, artificial intelligence, robotics, biomedical engineering, and health informatics are combined.

The subject of this example deals directly with the concept of achieving seamless communications of biologically driven AAL with emphasis on health care. Todays AAL systems respond based on context obtained from embedded sources as the motion, humidity, water and gas sensors, or wearable sensors. As shown in Figure 19.2, the testbed we propose will be built upon an implantable electronic cardiovascular device (IECD) integrated in to an AAL system as a complementary source of the vital signs for increasing the responsiveness level for the active health care. This testbed is expected to be effective enough to guarantee fast and accurate responses to the needs of the concerned person.

In this example, the IECD will bring significant extra information to bear on any analysis that might take place and on what responses may be forthcoming from the AAL. The vital signs received directly from the IECD can mean that steps are taken to prevent serious situations from developing. Furthermore, the focus on sustainable communication will ensure that the system will be capable of functioning irrespective of the activity of the user. The development of the seamless interactions between the core intelligent system and the biological active element is based on the wireless communications while respecting complexity and functionality.

The goal of this example is to emphasize the importance of wireless communication devices and networks, artificial intelligence and ambient systems, cognitive modeling, computational organizational theory, health informatics, implantable devices and their integration within environment, the human and system behavior analysis with respect to health care support, safety and security, independence, and mobility. Their symbiotic combination will work toward highly efficient environment that is capable of the self-adjustment and learning as a response on the patientÕs conditions (Figure 19.3). The methodology consists of a set of tools from infrastructural and behavioral analysis and modeling to implantable device testing within AAL context.

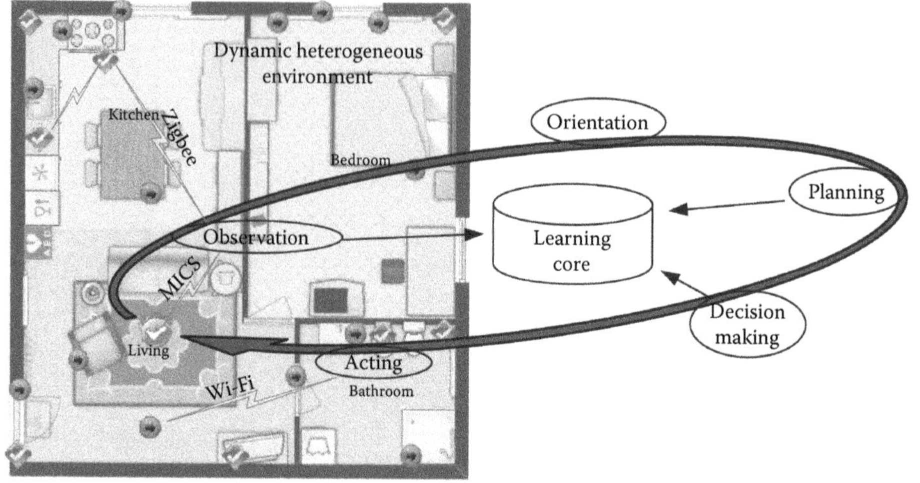

Figure 19.3 A general look on wireless dynamic heterogeneous environments.

The proposed example involves an analysis of system configuration time, latency and accuracy, system responsiveness, sensors sensitivity, vital signs and their timeliness, communication frequency range, data rate and transmission channel, elements interactions and timing, level of the context awareness as a trade-off between scalability, complexity, and security. The objective is to create a reliable wireless communication system within the AAL context with the great impact on the system responsiveness for the health care support and daily assistance.

The result of experimenting with testbed will have the potential to impact positively on the future development of wireless communications between the humans and its environment and serves as a reliable and fast link between them. A better understanding of the AAL context infrastructure and IECD mechanism is gained to provide the interconnections between them. At the same time, in-depth research and analysis of the system and human behaviors, their modeling and simulations are contributing elements of the project development to challenge the signal propagation inside the house. Medical data integration, environmental impact, and social acceptance studies will create a base of knowledge for the development of seamless communications in the AAL system.

The correct analysis of the vital signs, interrogated from IECD, and studies on the human behavior with the focus on the particular group will aim on the mobility model's development for robust communication link. Feasibility and safety will be in the focus of communication engineers to verify the efficiency of available information.

19.4 Conclusions and Future Work

In Section 19.1 we looked at the capabilities and interactions of WSNs by reflecting them through the literature of MASs that are responsible for allocating resources within complex and sometimes distributed environments. We discussed the idea of having sensors/agents negotiate to reach agreements and achieve goals. We then gave an overview of several approaches to agents' negotiation from different contexts. In Section 19.2 we provided an overview of the AFME and discussed implementation details in relation to its deployment on WSNs motes, such as the Java-enabled Sun SPOT.

In Section 19.3 we described a health care–oriented example that is potentially a suitable testbed for future examination to our idea of integrating agent-based negotiation protocols with WSNs.

In the near future, we will consider to extend our research on several related directions in order to tackle the negotiation aspect of AFME so that the efficiency of interactions between wireless sensors are enhanced. We will also be working on implementing a case-specific negotiation model for the Java-based Sun SPOT motes and examine the performance before and after the integration of the negotiation model.

References

1. G. Weiss. *Multiagent Systems: A Modern Approach to Distributed Artificial Intelligence.* The MIT Press, Cambridge, MA, March 1999.
2. M. Wooldridge. *An Introduction to MultiAgent Systems.* John Wiley & Sons, New York, June 2002.
3. S. Kraus. Negotiation and cooperation in multi-agent environments. *Artificial Intelligence Journal, Special Issue on Economic Principles of Multi-Agent Systems*, 94(1–2):79–98, July 1997.
4. R. G. Smith. The contract net protocol: High-level communication and control in a distributed problem solver. *IEEE Transactions on Computers*, C-29(12):1104–1113, 1981.
5. S. Kraus. *Strategic Negotiation in Multiagent Environments.* MIT Press, Cambridge, MA, September 2001.
6. J. S. Rosenschein and G. Zlotkin. *Rules of Encounter: Designing Conventions for Automated Negotiation Among Computers.* MIT Press, Cambridge, MA, July 1994.
7. A. Rubinstein. Perfect equilibrium in a bargaining model. *Econometrica*, 50(1):97–109, January 1982.
8. P. Faratin, C. Sierra, and N. R. Jennings. Negotiation decision functions for autonomous agents. *International Journal of Robotics and Autonomous Systems*, 24:159–182, 1998.
9. H. Raiffa. *The Art and Science of Negotiation.* Belknap Press of Harvard University Press, Cambridge, MA, March 1985.
10. E. Bircher and T. Braun. An agent-based architecture for service discovery and negotiation in wireless networks. In *Proceedings of the 2nd International Conference on Wired/Wireless Internet Communications (WWIC 2004)*, Frankfurt/Oder, Germany, pp. 295–306, February 2004.
11. FIPA TC Communication. FIPA Contract Net Interaction Protocol Specification. Technical report, The Foundation for Intelligent Physical Agents, Geneva, Switzerland, 2002.
12. FIPA TC Communication. FIPA English Auction Interaction Protocol Specification. Technical report, The Foundation for Intelligent Physical Agents, Geneva, Switzerland, 2002.
13. A. Chavez and P. Maes. Kasbah: An agent marketplace for buying and selling goods. In *The Proceedings of the First International Conference on the Practical Application of Intelligent Agents and Multi-Agent Technology*, pp. 75–90, Cambridge, MA, 1996.
14. J. Cao, J. Wang, S. Zhang, and M. Li. A multi-agent negotiation based service composition method for on-demand service. In *SCC'05: Proceedings of the 2005 IEEE International Conference on Services Computing*, pp. 329–332, Washington, DC, 2005. IEEE Computer Society.
15. O. Kwon, J. M. Shin, and S. W. Kim. Context-aware multi-agent approach to pervasive negotiation support systems. *Expert Systems with Applications*, 31(2):275–285, 2006.
16. M. Klusch. *Intelligent Information Agents: Agent-Based Information Discovery and Management on the Internet.* Springer-Verlag, Secaucus, NJ, July 1999.
17. D. Kehagias, A. L. Symeonidis, K. C. Chatzidimitriou, and P. A. Mitkas. Information agents cooperating with heterogenous data sources for customer-order management. In *SAC'04: Proceedings of the 2004 ACM Symposium on Applied Computing*, pp. 52–57, New York, 2004. ACM.

18. A. Carrillo-Ramos, J. Gensel, M. Villanova-Oliver, and H. Martin. Pumas: A framework based on ubiquitous agents for accessing web information systems through mobile devices. In *SAC'05: Proceedings of the 2005 ACM Symposium on Applied Computing*, pp. 1003–1008, New York, 2005. ACM.
19. T. Wanyama and B. H. Far. Negotiation coalitions in group-choice multi-agent systems. In *AAMAS'06: Proceedings of the Fifth International Joint Conference on Autonomous Agents and Multiagent Systems*, pp. 408–410, New York, 2006. ACM.
20. T. Scully, M. G. Madden, and G. Lyons. Coalition calculation in a dynamic agent environment. In *ICML'04: Proceedings of the Twenty-First International Conference on Machine Learning*, p. 93, New York, 2004. ACM.
21. B. Blankenburg, R. K. Dash, S. D. Ramchurn, M. Klusch, and N. R. Jennings. Trusted kernel-based coalition formation. In *AAMAS'05: Proceedings of the Fourth International Joint Conference on Autonomous Agents and Multiagent Systems*, pp. 989–996, New York, NY, 2005. ACM.
22. J. M. Anjos and L. B. Ruiz. Service negotiation over wireless mesh networks: An approach based on economic agents. In *IEEE Wireless Days: WD'08: 1st IFIP*, Dubai, United Arab Emirates, pp. 1–5, November 2008.
23. J. Bigham and L. Du. Cooperative negotiation in a multi-agent system for real-time load balancing of a mobile cellular network. In *AAMAS'03: Proceedings of the Second International Joint Conference on Autonomous Agents and Multiagent Systems*, pp. 568–575, New York, 2003. ACM.
24. S. Yoshikawa, T. Kamiryo, Y. Yasumura, and K. Uehara. Strategy acquisition of agents in multi-issue negotiation. In *Proceedings of the 2006 IEEE/WIC/ACM International Conference on Web Intelligence*, pp. 933–939, Washington, DC, 2006. IEEE Computer Society.
25. C. Muldoon, G. M. P. O'Hare, and J. F. Bradley. Towards reflective mobile agents for resource-constrained mobile devices. In *Proceedings of the 6th International Joint Conference on Autonomous Agents and Multiagent Systems*, AAMAS'07, pp. 141:1–141:3, New York, NY, 2007. ACM.
26. R. Collier, G. O'Hare, T. Lowen, and C. Rooney. Beyond prototyping in the factory of agents. In M. Marík, J. P. Müller, and M. Pechoucek, eds., *Multi-Agent Systems and Applications III*, volume 2691 of *Lecture Notes in Computer Science*, pp. 1068–1068. Springer Berlin, Heidelberg, Germany, 2003.
27. Y. Shoham. Agent-oriented programming. *Artificial Intelligence*, 60:51–92, March 1993.
28. R. Goldman. Sun Spots Wiki, January 2011. http://java.net/projects/spots/pages/Home.
29. C. Muldoon. *An agent framework for ubiquitous services*. Ph.d. dissertation, School of Computer Science and Informatics—University College Dublin (UCD), Dublin, Ireland, 2007.
30. The OSGi Alliance. OSGi service platform core specification, release 4.1, 2007. http://www.osgi.org/Specifications.
31. M. Bal, S. Weiming, Q. Hao, and H. Xue. Collaborative smart home technologies for senior independent living: A review. In *CSCWD'11, The Fifteenth International Conference on Computer Supported Cooperative Work in Design*. Lausanne, Switzerland, pp. 481–488, 2011.
32. World Health Organization (World Health Organization). Active ageing: A policy framework. Technical report WHO/NMH/NPH/02.8, Noncommunicable Diseases and Mental Health Cluster, 2002.
33. V. L. Roger, A. S. Go, D. M. Lloyd-Jones, R. J. Adams, and J. D. Berry. Heart disease and stroke statistics. *American Heart Association*, Circulation 2011:1–25, 2010.
34. P. Bamidis, M. Alborg, V. Moumtzi, and A. Koumpis. Synergy between social and health services under an ambient assisted living environment for the elderly. In *eChallenges*, Warsaw, Poland, pp. 1–8, October 2010.
35. A. Hein, S. Winkelbach, B. Martens, O. Wilken, M. Eichelberg, J. Spehr, M. Gietzelt, et al. Monitoring systems for the support of home care. *Informatics for Health and Social Care*, 35:157–176, 2010.

36. H. Burri and D. Senouf. Remote monitoring and follow-up of pacemakers and implantable cardioverter defibrillators. *Europace European Pacing Arrhythmias and Cardiac Electrophysiology Journal of the Working Groups on Cardiac Pacing Arrhythmias and Cardiac Cellular Electrophysiology of the European Society of Cardiology*, 11(6):701–709, 2009.
37. R. Paoli, F. J. Fernández-Luque, G. Doméch, F. Martínez, J. Zapata, and R. Ruiz. A system for ubiquitous fall monitoring at home via a wireless sensor network and a wearable mote. *Expert Systems with Applications*, 39(5):5566–5575, 2012.
38. M. Cirillo, L. Karlsson, and A. Saffiotti. Human-aware task planning for mobile robots. In *International Conference on Advanced Robotics ICAR 2009*, Munich, Germany, pp. 1–7, June 2009.
39. A. Bamis, D. Lymberopoulos, T. Teixeira, and A. Savvides. The behavior scope framework for enabling ambient assisted living. *Personal Ubiquitous Computing*, 14:473–487, September 2010.

Chapter 20

Event Detection in Wireless Sensor Networks

Norman Dziengel, Georg Wittenburg, Stephan Adler, Zakaria Kasmi, Marco Ziegert, and Jochen Schiller

Contents

Event detection allows for a wireless sensor network to reliably detect application-specific events based on data sampled from the environment. Events, in this context, may range from the comparatively easy to detect outbreak of a fire to the more complex stumbling and falling of a patient, or even to an intrusion into a restricted area. Distributed event detection leverages the data processing capability of the sensor nodes in order to locally extract semantically meaningful information from the sensed data. This process may involve one or multiple nodes in a given region of the deployment, and generally avoids data transmissions to the base station of the network. This in turn reduces the overall energy consumption and ultimately leads to a prolonged lifetime of the network.

In this chapter, we motivate the need for in-network data processing and event detection, and briefly review the current state of the art. We then present in detail our own approach to distributed event detection and discuss hardware platform, software architecture, and algorithms and protocols used in the detection process. Our approach has been tested extensively in several real-world deployments and we give an overview of the results from these experiments. The results illustrate how distributed event detection can achieve both high accuracy and energy efficiency, and thus mark the advantages of this approach to data processing in sensor networks.

20.1 Introduction

Event detection is a special form of in-network data processing for wireless sensor networks (WSNs) that pushes the logic for application-level decision making deeply into the network. Raw data samples from the sensors are evaluated directly on the sensor nodes in order to extract information that is semantically meaningful for the given task of the WSN. For example, a WSN that is tasked with detecting a fire will process the temperature readings from a sensor and only send an alert to the base station of the network if the readings reliably exceed a threshold value [1]. More complex examples include scenarios such as vehicle tracking [2], area surveillance [3], undersea monitoring [4], or the classification of human motion sequences [5].

In all of these deployments, sampled data are processed and evaluated close to its source, thereby reducing communication with the base station, minimizing energy consumption, and extending the lifetime of the network. As illustrated in Figure 20.1, this is achieved by programming the sensor nodes in such a way that the occurrence of deployment-specific semantic events can be established

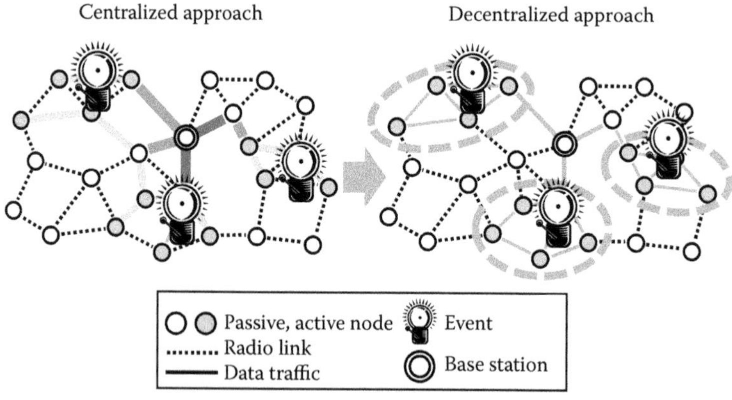

Figure 20.1 Centralized data processing vs. decentralized event detection.

directly on sensor nodes, relying only on the locally available data from the sensors. Depending on the application, a variety of sensors can be employed, including light and temperature sensors, accelerometers, microphones, or even cameras. The transformation of data from the physical sensory domain into the scenario-specific application domain stands in contrast to other techniques for in-network data processing, for example, compression and aggregation, which are agnostic to the application-level properties of the data.

In this chapter, we give an overview of current approaches to event detection in WSNs and present our own exemplary work in more detail. Our research focuses on the distributed detection of events based on one- or multi-dimensional motion patterns. A prime example of a potential application is a WSN in which sensor nodes are equipped with acceleration sensors. These sensor nodes are then mounted onto a fence and the movement of the fence measured with the goal of deciding whether a security-relevant event, for example, an intrusion into—a restricted area, has occurred. The system is trainable to detect new types of events and given the distributed nature of the algorithms employed—resilient to failures of individual nodes. We have implemented and evaluated our approach through a series of real-world deployments on several construction sites and summarize our finding in this text.

We begin our treatment of the subject matter with a review of the state of the art in Section 20.2. Afterward, we proceed with the detailed description of our platform for distributed event detection *AVS-Extrem* [6] and cover the hardware platform, the software stack, and the event detection algorithms and protocols in Section 20.3. In Section 20.4, we then present experimental results from several deployments of this system, and finally conclude in Section 20.5.

20.2 State of the Art

Since event detection is one of the key areas of application of sensor networks, there is a large body of work, and multiple deployments have already been undertaken. In this section, we give an overview of the most common system architectures and event detection algorithms.

20.2.1 Architectural Approaches

As a first step, we discuss several possible system and networking architectures and evaluation strategies for event detection WSNs and their advantages and disadvantages. The networking architecture has a high impact on the reliability and energy consumption of the sensor network and is highly dependent on the application and the complexity of the events that need to be detected.

20.2.1.1 Local Evaluation

Local evaluation describes a method according to which one sensor node gathers and processes data by itself and decides whether a certain event has occurred or not. The results are sent to a base station where an alert can be triggered or the user can be asked to confirm the findings of the sensors. One system using this method is proposed by Yang et al. [5] and is aimed at recognizing human motions. It is a body area network consisting of eight sensor nodes attached to a person's body who may perform 1 out of 12 actions. Data are delivered from an accelerometer and a gyroscope and are processed on each node. If local data processing results in a recognized event, then data of the node are transmitted to the base station and processed once again.

20.2.1.2 Centralized Evaluation

The centralized evaluation architecture is possibly the most widely used: All nodes in a WSN communicate exclusively with the central base station, which has much more computing power and energy resources than the sensors nodes. The communication can contain the sensors' raw data or the signal processing results (i.e., compression or reduction), which takes place on each of the nodes. Individual sensor nodes have no knowledge about the semantics of the collected data. The complete data are gathered and interpreted at the base station. This method has several advantages, for example, the high detection accuracy that is possible with a complete centralized overview. The main disadvantage of this approach is that the network suffers from poor performance for large numbers of nodes. Also, the system may run into energy problems as the whole network needs to constantly communicate with the base station and energy-saving techniques cannot easily be applied.

As mentioned earlier, several projects use this method to detect events in their WSNs. An exemplary implementation is presented by Wang et al. [7] who describe a habitat monitoring system that is capable of recognizing and localizing animals based on acoustics. They employ a cluster head with additional processing capabilities that may request compressed raw data from other nodes for centralized evaluation. Animals are identified by the maximum cross-correlation coefficient of an observed spectrogram with a reference spectrogram. Using reports from multiple sensor nodes, the animals are then localized in the field.

20.2.1.3 Decentralized Evaluation

The decentralized evaluation approach forms smaller clusters within the WSN. Each cluster has a special node that performs the task of a cluster head. This node has the role to classify the data and to communicate with the base station. One advantage of that architecture is that it is fault tolerant against the malfunction of single nodes or the loss of one whole cluster. If one cluster fails, the other clusters remain functional. With regard to energy awareness, this architecture also has advantages because it is possible to put clusters that are not needed or not triggered by an event into an energy-efficient idle mode.

An exemplary deployment has been conducted as part of the SensIT project. Duarte and Hu [2] evaluated several classification algorithms in a vehicle tracking deployment. After gathering the acoustic and seismic data, each sensor node classifies the events by using extracted features. The features are extracted from the frequency spectrum after performing a fast fourier transform. The evaluation comprises three classification algorithms: k-nearest neighbor, machine learning, and support vector machines. The classification result is sent to a fixed cluster head for evaluation and is combined with reports received from other nodes for tracking a vehicle.

A decentralized event detection system is also proposed by Martincic and Schwiebert [8]. Sensor nodes are grouped into cells based on their location. All nodes in a cell transmit their data samples to a cluster head that averages the results and retrieves the averages from adjacent cells. Event detection is performed on the cluster heads by arranging the collected averages in the form of a matrix and comparing it to a second predefined matrix that describes the event. An event is detected if the two matrices match.

20.2.1.4 Distributed Evaluation

The distributed evaluation method differs from the decentralized evaluation in two important points. First, all event detection and data processing take place in the network and no data except for the signaling of an event are sent to the base station. Second, there are no designated cluster head nodes in the network. Instead, all nodes are able to perform the data processing task by themselves. In contrast to the previously discussed approaches, a cluster head is not needed since a leader node is chosen dynamically every time an event occurs. The result of a detection is distributed to all nodes which also detected the event and compared with their own results. The choice of which node takes the leading role is chosen by an application-dependent algorithm and may, for instance, result in a node that is physically close to the event or has an otherwise advantageous position within the network. This leader transmits the results of the distributed evaluation to the base station.

The system presented in Section 20.3 of this chapter also falls into this category. Moreover, Li and Liu [9] propose an event detection system for coal mine surveillance to localize collapse events. The Structure-Aware Self-Adaptive principle of the 3D environment surveillance is used to detect falling nodes or changing positions of sensor nodes by using acceleration data, Received Signal Strength Indicator (RSSI) evaluations, neighbor loss, and some acoustic analysis. All nodes have to be set up with an initial known position in the mine. A beacon-based communication is periodically initiated to set up the neighborhood topology of each node. In the case of an event, the measured data are mapped with a random hash-function and transformed into a data signature file that is transmitted to the base station.

20.2.2 Algorithmic Approaches

We now shift our focus from the architectural and networking aspects to the data processing aspects of the event detection. Several methods to detect an event and to distinguish between noise have been proposed in the literature and are deployed in real-world installations of WSNs. The overall trade-off in this area is—as we will explore in this section—the weighting of algorithmic simplicity against the complexity of the events that the system is able to detect. Another interesting aspect to consider is whether a system is capable of learning events on its own, or whether expert knowledge needs to be embedded into the detection algorithm.

20.2.2.1 Threshold Usage

Current approaches to integrate event detection in WSNs often apply a threshold detection, either as a trigger for a more intensive gathering of data or as an event itself. Threshold values are suitable for a lot of applications, for example, fire detection, detection of flooding, or generally other applications in which a sensor can detect a critical boundary of the measured value. Although this method is very efficient and robust for simple detection problems, it is unable to detect more complex events such as movement patterns, faces, or speech. In a simple scenario, a possible drawback of the system is that every node that detects an exceeding threshold value will start to communicate with the base station. This can lead to an early energy shortage in the network if events occur often, or to network congestion if events occur simultaneously at the sensor nodes. These problems can be avoided by resorting to sophisticated networking architectures, as discussed previously.

A typical application for a threshold-based event system is a system to detect fire, like the one Doolin and Sitar described in [1]. Temperature and humidity sensors are fitted to sensor nodes and deployed in an area that is to be monitored. When a fire occurs, the sensors measure temperatures and humidity values that do not occur under normal environmental conditions. The nodes can then assume that a critical event has occurred and can alert the base station.

20.2.2.2 Pattern Recognition

Pattern recognition algorithms are able to process a wide variety of inputs and map them to a known set of event classes. As WSNs deal with a lot of data from different sources that are used to measure complex incidents, it is evident that pattern recognition algorithms can also be found in this domain. Pattern recognition on signals is commonly implemented by subdividing the process into several steps that can be executed by different entities of the system. During the *sampling* process, raw data are gathered, optionally preprocessed, and handed over to the *segmentation process* that detects the start and the end of the samples that belong to an event. The segmented data are then handed over to a *feature extraction* algorithm that extracts highly descriptive features from the data and creates a feature vector for the data segment. Features can be all kind of descriptive attributes of the data such as a histogram, a spectrogram, the minimum, the maximum, and so forth. The final step is the *classification* as part of which the feature vector is statistically analyzed or compared to either previously trained or fixed features like threshold values. In most cases, a prior training is necessary to deliver a sufficient set of training data that initializes the classifier. An in-depth introduction to this subject is available in Duda et al. [10] and Niemann [11].

The system presented in Section 20.3 also falls into this category. Additionally and as previously mentioned, Duarte and Hu [2] use pattern recognition algorithms in a vehicle tracking deployment. The system performs the first three steps, that is, sampling, segmentation, and feature extraction. The features in use are all based on frequency analysis of the signal. Afterward, each node classifies the event and sends the result to a fixed cluster head for evaluation and combination with reports received from other nodes.

20.2.2.3 Anomaly Detection

Anomaly detection is a term used for approaches that focus on the specific case of detecting whether a particularly unusual event has occurred. This is achieved by learning typical system behavior over time and then classifying specific events as either normal or anomalous. Approaches with this goal expand upon the principles of the two previously described approaches and incorporate techniques from the field of intrusion detection and even bio-inspired artificial immune systems.

For example, Wälchli et al. [12] designed a distributed event localization and tracking algorithm (DELTA) that is based on a small short-term memory to temporarily buffer normal state. The algorithm was deployed as part of a light sensor office surveillance with the goal of detecting and tracking persons carrying flashlights. DELTA provides the leader node with the information needed to localize and classify an event based on a simplex downhill algorithm.

A summary of all the approaches, advantages and disadvantages, and exemplary implementations is given in Table 20.1.

20.3 Exemplary Platform: AVS-Extrem

Based on the general discussion of architectures and algorithms, we now present an exemplary platform for distributed event detection, called *AVS-Extrem*: To detect an intrusion into a protected area, wireless sensor nodes equipped with accelerometers are integrated into a fence surrounding the area. The sensor nodes distinguish collaboratively between numerous event classes like opening the fence or a person climbing over the fence. In order to acquire a prototype for each event class, the sensor nodes are trained in a real-world construction site scenario. A prototype is an event class abstraction incorporating significant features that are extracted from the training data.

Table 20.1 Overview of Proposed Architectures and Detection Algorithms

Architecture	*Advantages*	*Disadvantages*	*Exemplary Implementations*
Local detection	Easy to implement and energy efficient. Low hardware requirements	Only works well for simple events, i.e., those detectable by monitoring data for exceeded thresholds or sudden alterations	Yang et al. [5]
Centralized evaluation	Full control of network from one single point. Access to all raw data	Energy inefficient and traffic intensive, which may lead to bottlenecks in data transmission	Wang et al. [7]
Decentralized evaluation	Robust to failure of single nodes due to clustering. Good energy efficiency	Special-purpose cluster head nodes may be needed. Deployment influences network topology	Duarte and Hu [2] Martincic and Schwiebert [8]
Distributed evaluation	Data are exchanged locally between nodes, events are reported once to base station. Minimal communication overhead and good energy efficiency	Hard to implement since application knowledge is distributed across the network	Li and Liu [9] Wittenburg et al. [3]
Algorithm			
Threshold usage	Easy to implement and very reliable	Only applicable to certain (simple) application scenarios	Doolin and Sitar [1]
Pattern recognition	Ability to detect very complex events	Hard to implement. Requires sufficient computational resources on nodes	Duarte and Hu [2] Wittenburg et al. [3]
Anomaly detection	Very robust because system adjusts itself to minor changes in environment	Cannot be applied to scenarios that lack normal condition occurence	Wälchli et al. [12]

The features are characteristic attributes that describe one or more patterns observed in the raw data. After the training, the extracted set of features enables to describe a class precisely and to distinguish it from other classes. This way, each sensor node can assign a newly occurred event to the appropriate event class.

20.3.1 Requirement Analysis

Our proposed platform has to meet a set of scenario-independent and scenario-dependent requirements as well as hardware/software requirements in order to achieve energy efficiency and to be applicable to a wide range of deployment scenarios. We develop our own peripheral component drivers that are integrated in our FireKernel OS. These components are application-related and intermesh the hardware and software requirements. Therefore, we, subdivide our discussion into the following two groups including details to the hardware and software.

Scenario-independent requirements: On the hardware side, we migrate away from our initial MSP430-based prototype to an ARM-based MCU with additional RAM, in order to implement more complex detection algorithms and to realize an efficient hardware and software energy-awareness control. The MCU as master and the slaves, for example, the accelerator sensor, supports several power-saving modes. To store the application-dependent settings and the prototype data, we use a nonvolatile storage.

The sensor nodes operate in a distributed and demand-based fashion, and in an environment with a dynamic topology. Hence, we employ a reactive routing protocol (cf. Section 20.3.4). Additionally, to reduce the transceiver power consumption, we implement an energy-aware and configurable Wake-On-Radio (WOR) duty-cycle. The sensor node is equipped with an energy-efficient acceleration sensor that is able to wake up the MCU when an acceleration threshold value is exceeded, otherwise the MCU remains in power-down mode (PD). Furthermore, we apply an OS-level priority-based task processing to enable an uninterrupted sampling and continuous communication. Finally, an adequately large and appropriate feature pool is gathered for the event detection by performing training intervals in order to enable an efficient feature set selection to build a prototype.

Scenario-dependent requirements: From the hardware point of view, we choose an acceleration sensor based on previous experiences with the fence monitoring application. The sensor meets the following requirements: a range of at least 8 g to avoid overload during event sensing, a 10-bit resolution with low noise level, and a sampling rate of up to 100 Hz to avoid aliasing. To cover a wide range of scenarios, we implement a self-calibrating routine for the acceleration sensor. The calibration routine takes care of the individual positional setup of the fence elements after every event.

The battery case is shuttered with a screw cap that is actuated by a spring. In rare cases, a vertical shock is able to compress the spring that disconnects the battery. This problem is solved by an additional capacitor to bypass the resulting temporary power failure of up to 60 ms. Since we target rough environments, we package the sensor node within a housing that resists severe weather conditions and heavy shocks.

20.3.2 Sensor Node Hardware

The AVS-Extrem wireless sensor nodes, as depicted in Figure 20.2a, were developed to meet the demands of motion-centric applications and localization [6]. Each node is equipped with an ARM7-based NXP LPC2387 MCU operating at a maximum speed of 72 MHz with 96 KB RAM, and 512 KB ROM, and a CC1101 radio transceiver from Texas Instrument operating at the

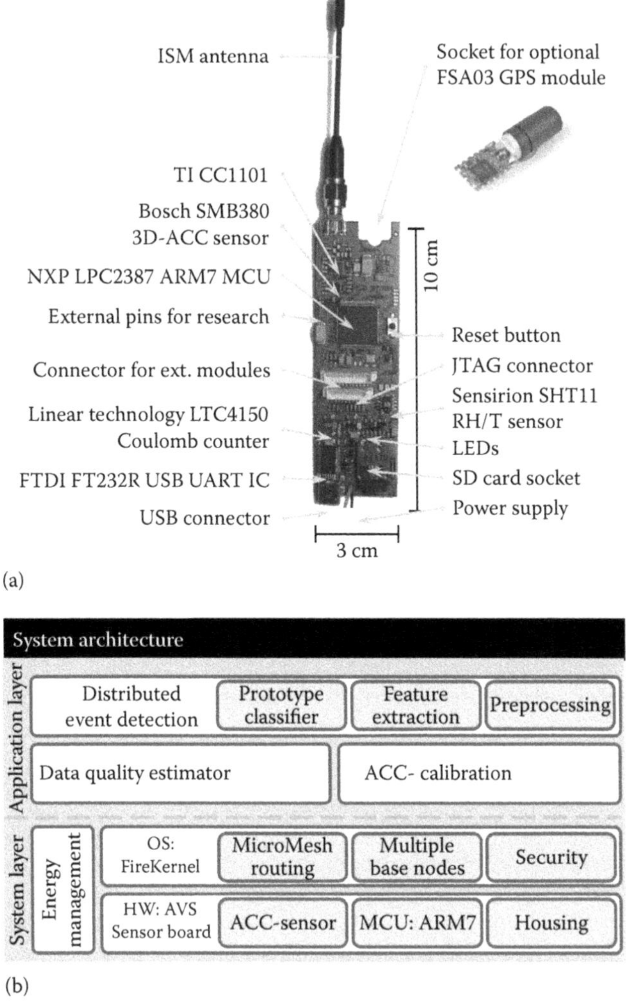

Figure 20.2 AVS-Extrem platform. (a) sensor node and (b) system architecture.

868 MHz ISM band. An LTC4150 coulomb counter is used to monitor the state of the batteries. For acquiring the movement data, a low-power and triaxial acceleration sensor is employed. We chose a Bosch SMB380 sensor with a range of up to 8 g and 10-bit resolution that adds only very little noise and has a configurable sampling rate of up to 1500 Hz. The sensor wakes up the system for further processing, once an acceleration above a user-defined threshold occurs.

In addition to the core components explained previously, the PCB also carries a temperature and humidity sensor to monitor environmental parameters. Optionally, a Falcom FSA03 GPS sensor can be employed, as well as an SD Card slot to store individually trained motion patterns and current battery status in a nonvolatile storage. Adding new external peripherals to the platform is possible by attaching them to the provided connector accessing the SPI or GPIO interface.

The sensor nodes are mounted in a very stable and fixed way utilizing a rugged housing within the fence that also contains the long-term battery supply [13]. The housing enables to conduct experiments in a highly repeatable manner with identical settings.

20.3.3 Software Stack

As shown in Figure 20.2b, the system layer of our architecture contains the AVS-Extrem sensor board, the operating system, and the energy management. All components of the system layer are briefly introduced in the subsequent sections, while the details can be found in [13].

We employ FireKernel [14] as a low-power operating system to support our concepts and to evaluate concurrent algorithms on a sensor platform that features threading and a priority-based preemptive multitasking. A flexible energy management that supports WOR and numerous MCU-based energy saving techniques has also been implemented. For secured communications, we implemented the *ParSec* security layer that employs symmetric cipher algorithms in CBC-mode in conjunction with an initialization vector and a counter to provide confidentiality, authenticity, integrity, freshness, and semantic security for the wireless communication [15].

The application layer contains an optional Dempster–Shafer-based data quality estimator that is used to establish whether the reliability of a measurement is high or low. Finally, the system comprises the distributed event detection module, which is described in Section 20.3.5.

20.3.4 Networking Protocols

Multiple base nodes (BNs) are adopted in order to provide fault-tolerant connections between arbitrary nodes in the WSN and the base station of the sensor network. Figure 20.3 depicts a configuration of a network with two base nodes. Base nodes act as proxies between the WSN and the base station and also operate as a monitor of the network. The base station is connected to all base nodes via either a wireless or a wired connection. For monitoring and data collection purposes, the base station exchanges messages with the nodes. The messages contain information about the network state such as temperature or battery charge level of a node. Furthermore, they report about occurring events in the surrounding area, such as shaking the fence or kicking against the fence. In case of failure or non-reachability of a base node, an alternative base node is selected. This way, the reliability of the network and the accessibility of the base station are improved.

The implementation of the BN principle relies on the micro mesh routing (MMR) protocol [14] to benefit from a dynamic and fault-tolerant routing protocol covering changes in the network topology. MMR is a reactive routing protocol and combines aspects of different routing protocols

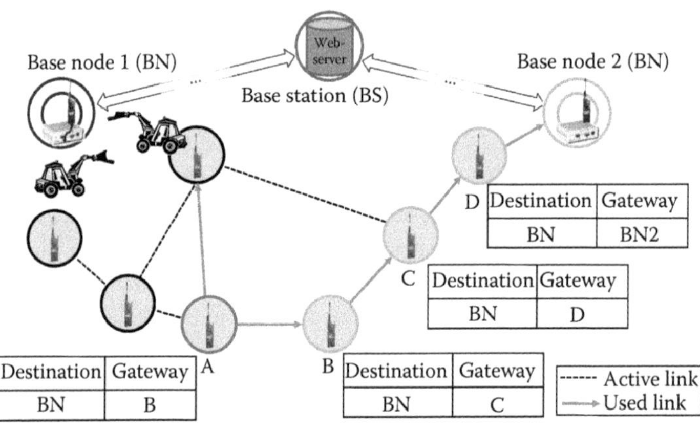

Figure 20.3 Data transmission over base nodes to base station.

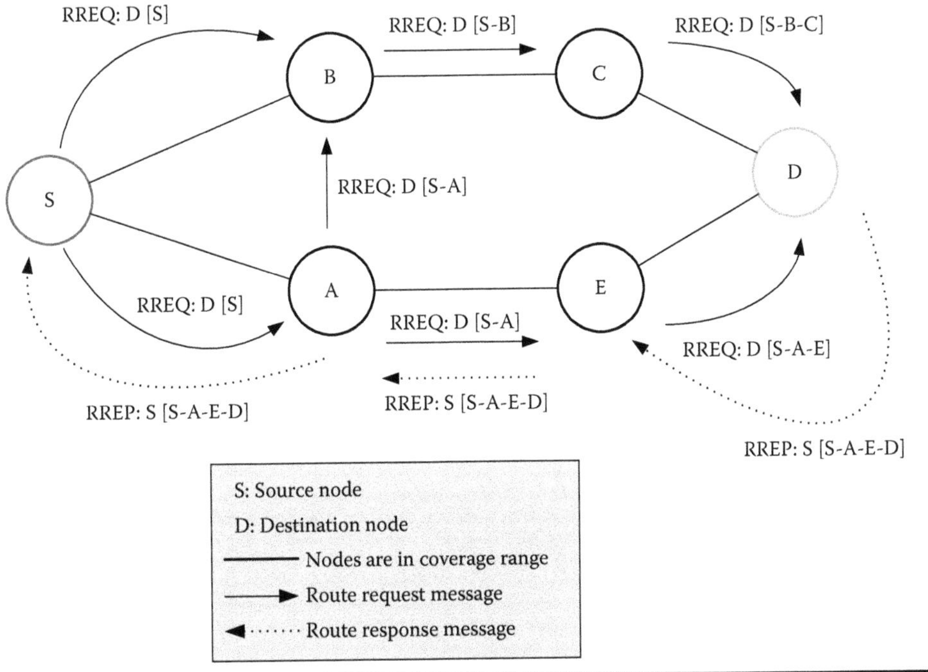

Figure 20.4 Course of action of a micro mesh route discovery process.

such as AODV and DSR. AODV-like functionality is used to take advantage of the hop-by-hop routing during the data transmission, while the principle of DSR enables to collect partial route information in the course of the discovery phase. In contrast to a strict reactive protocol, intermediate nodes analyze all forwarded packets to update their routing tables. Sequence numbers are used to prevent the formation of loops, to avoid the reception of duplicated messages, and to determine the freshness of a route.

MMR defines three message types—route request (RREQ), route reply (RREP), and route error (RERR)—and comprises two operation modes: route discovery and route maintenance. A route discovery process is triggered when a node intends to send a packet to a destination node, whose address is not available in the local routing table. In contrast to AODV, each intermediate node appends its own address to the route record, in which is concatenated a list of the hop addresses taken by the route request packet. Furthermore, the intermediate node extracts the routes from the route record of the received RREQ message and saves them in its routing table. In the case of a link failure, for example, due to the node movement, the node that detects the error and is closest to the error source initiates a route maintenance process, to notify other nodes about the link loss. The route discovery is illustrated in Figures 20.4.

20.3.5 Event Detection Algorithms and Protocols

The event detection approach implemented in our system consists of a combination of algorithms and protocols that observe and evaluate events distributively across several sensor nodes. This distributed event detection system does not rely on a base station or any other means of central coordination and processing. In contrast, current approaches to event detection in WSNs transmit raw data to an external device for evaluation or rely on simplistic pattern recognition schemes. This implies either high communication overhead or low event detection accuracy, especially for complex events (cf. Section 20.2).

In a specific deployment of our system, the nodes are equipped with accelerometers and attached to a fence at fixed internode distances. When an event occurs, a lateral oscillation will propagate through the interconnected fence elements. As a result, the nodes can sample the acceleration at different distances from the event source. According to the neighbor-relative position, which is encoded in a unique node ID, each sensor can position itself relatively to its neighbors. A node can deduce its relative position simply by comparing its own ID with the neighbor ID. The relative position of the neighbors is required for each sensor node affected by the event, in order to perform a feature fusion. If an event occurs, each affected node concatenates its own features and the received features from the neighbors based on the relative sender position. This overall process is illustrated in Figure 20.5 and described in more detail in [3].

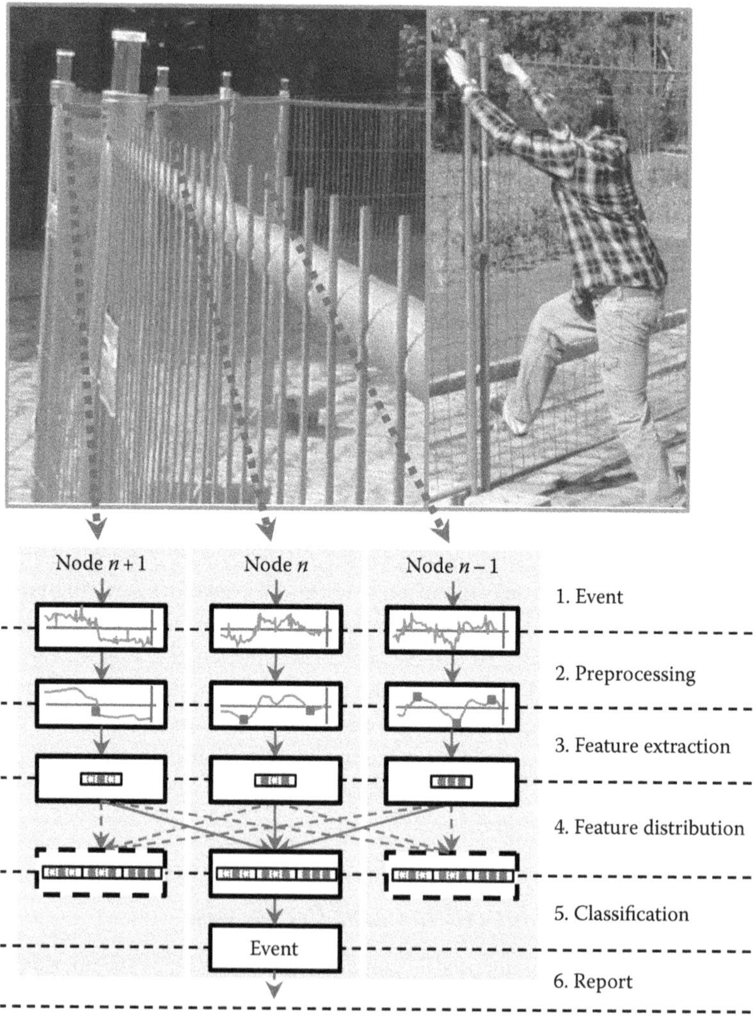

Figure 20.5 Distributed event detection process.

The system works in three distinct phases: In the *calibration phase*, the acceleration sensors are automatically calibrated before and after each event. An application-dependent calibration routine takes noise and interruptions during the calibration period into account.

The *training phase* is initiated by the user via the control station, which is used to accomplish the initial training and to calculate an optimal feature set. During the deployment, the control station may optionally also be used to serve as a base station for event reporting. The purpose of this phase is to train the WSN to recognize application-specific events by executing a set of representative training events for each event class. Every event class is individually trained by extracting a broad feature set from the sampled raw data and transmitting it to the control station. The control station performs a leave-one-out cross-validation algorithm [16] across all collected features to select an optimal subset of features. A prototype for each class is then calculated and transmitted back to each sensor node. The prototypes represent the a priori reference values needed for the classification of known patterns.

Once the training is complete, the system enters the *detection phase*. In this phase, the sensors gather raw data that are preprocessed, segmented, and normalized (cf. Figure 20.5). Only features used in the prototype vectors are extracted from the raw data, and then combined to form a feature vector. The extracted feature vector is transmitted via broadcast to all sensor nodes in the n-hop neighborhood. n is usually set to 1 for our scenario, since the radio range of a sensor node is significantly larger than the spacial expansion of the physical effects caused by an event. After receiving all required feature vectors, each node performs a feature fusion by combining the feature vectors based on a bit-mask and the relative position of the sender. In other words, each node builds its own event view using feature vectors from its neighboring nodes. Obviously, the combined feature vector is different for each node, because the respective neighboring nodes perceive the event depending on their location. Ideally, only the prototype classifier running on the node whose event view matches the trained view will classify the correct event, while the other nodes reject it. Finally, only the relevant detected events that require user interaction, for example, an intrusion or a fire, are reported to the base station of the WSN.

20.4 Experimental Results

In this section, we summarize the results from three major deployments that we conducted over the past years [3,13,17]. We first summarize the experiments related to the accuracy of the distributed event detection algorithm and then proceed to the evaluation of the energy efficiency of the system.

20.4.1 Detection Accuracy

In [3], we attached 100 ScatterWeb MSB sensor nodes to the fence elements of a construction site near our institute. We exposed the WSN to four different classes of events: shaking the fence, kicking against the fence, leaning against the fence, and climbing over the fence. For the training, we chose a region of the fence that was free of any external obstructions and trained the WSN with 15 repetitions of each event. We compared the results of this experiment with two prior experiments. In the first one, we exposed 10 sensor nodes that were attached to a fence to detect 6 different events. The system did not support autonomous training at that time. Instead, we relied on a custom-built heuristic classifier implemented in a rule-based middleware that was manually configured to classify events based on human visible patterns in the raw data. The second experiment we bring in for result comparison was part of an additional lab experiment. Here, we trained three sensor nodes to cooperatively recognize four different geometric shapes based on the acceleration data measured by the sensor. The four shapes, comprising a square, a triangle, a circle, and the capital letter U, were drawn on a flat surface by physically moving the sensor nodes and then classified using a cooperative fusion classifier. There were three persons, each moving one sensor node along these shapes for a total of 160 runs.

We collected the following metrics while exposing the system to the different classes of events (with TP = true positive, TN = true negative, FP = false positive, and FN = false negative):

- Sensitivity $= \frac{\#TP}{\#TP + \#FN}$, also called recall, corresponds to the proportion of correctly detected events.
- Specificity $= \frac{\#TN}{\#TN + \#FP}$ corresponds to the proportion of correctly ignored events.
- Positive predictive value (PPV) $= \frac{\#TP}{\#TP + \#FP}$, also referred to as *precision*, corresponds to the probability that correctly detecting an event reflects the fact that the system was exposed to a matching event.
- Negative predictive value (NPV) $= \frac{\#TN}{\#TN + \#FN}$ corresponds to the probability that correctly ignoring an event reflects the fact that the system was not exposed to a matching event.
- Accuracy $= \frac{\#TP + \#TN}{\#TP + \#TN + \#FP + \#FN}$ corresponds to the proportion of true results in the population, that is, the sum of all correctly detected and all correctly ignored events.

Figure 20.6 shows the sensitivity, specificity, PPV, NPV, and accuracy for three different deployments of our distributed event detection system. The algorithms and protocols as described in Section 20.3 are employed in the latter two deployments, i.e. in the lab prototype and in the distributed event detection deployment. We can observe that the system achieves near-perfect results under lab conditions. Also, the latest deployment performs substantially better than the initial proof-of-concept setup with the rule-based classifier. The current system achieves an overall detection accuracy of 87.1%, an improvement of 28.8% over the proof of concept.

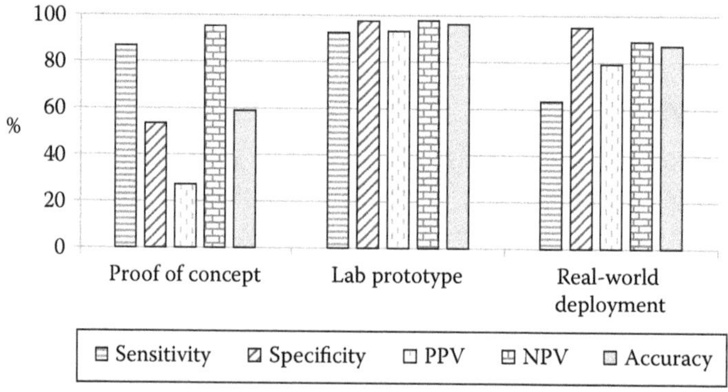

Figure 20.6 Detection quality of distributed event detection for different deployments.

20.4.2 Energy Consumption

Our experiments in [13] deal with the energy consumption of a sensor node during distributed event detection and recognition. To measure the energy consumption of the whole sensor board circuit as accurately as possible, we soldered a 10 Ω shunt resistor into the supply line which is powered by a reference voltage of 5 V. To measure the voltage of the shunt resistor a digital sampling oscilloscope (DSO) was attached. As the resistor and the voltage are known, we can calculate the value of the current and use it to calculate the electric power used by the sensor node over the time of one DSO sample. By integrating the electric power over the time of one system state, like packet transmission, sampling or IDLE mode, we can exactly measure the energy needed and use this information to approximate the energy consumption of the sensor node over a certain time. During the event detection phase, we make use of the low power modes of each of our components. In detail, the sensor nodes use the MCU PD that also shuts down all internal peripherals. The wireless transceiver makes use of the WOR mode that enables the processing of incoming data by using a duty-cycle. All sensor nodes are aware of this duty-cycle and retransmit data until it is assured that the timeslots have matched. The acceleration sensor is active and monitors the movements of the fence elements to wake up the MCU and alert the application layer in case of suspicious acceleration data that are different from the noise level.

An exemplary energy measurement is illustrated in Figure 20.7. During PD, a mean energy consumption of 9.0 mW is measured. During an event, the MCU is periodically utilized to fetch acceleration data from the acceleration sensor to the MCU (206.25 mW). This is followed by the feature extraction (350 mW) and classification (58.80 mW on average). As described in [3], a maximum of seven sensor nodes is involved in a fence event. Hence, in the phase of feature distribution one broadcast packet is sent (373.33 mW), and during classification up to six packets are received (178.5 mW) from the neighborhood. Finally, the result is calculated and sent to the base station. Afterward, the sensor node is recalibrated or, as in our example, the hysteresis function has converged and the sensor node immediately returns to the detection mode. The average time duration of an event is about 10 s. Including sampling, feature extraction, distribution, and classification, the average energy consumption for one event is, thus, about 145.4 mW.

Figure 20.8 illustrates the average energy consumption and the resulting extrapolated network lifetime. The underlying scenario for this calculation is a deployment of seven nodes in which

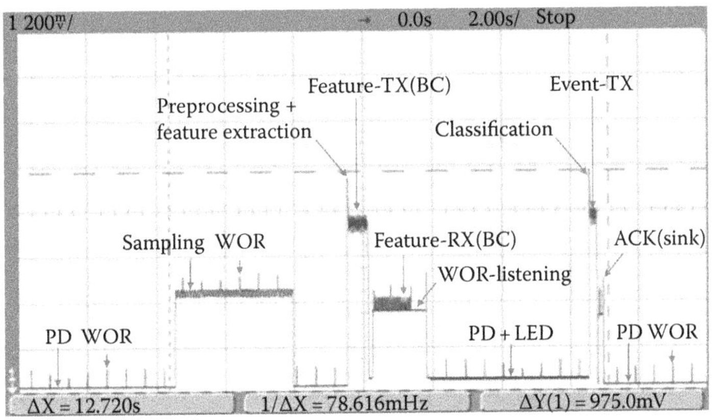

Figure 20.7 Energy consumption during event processing.

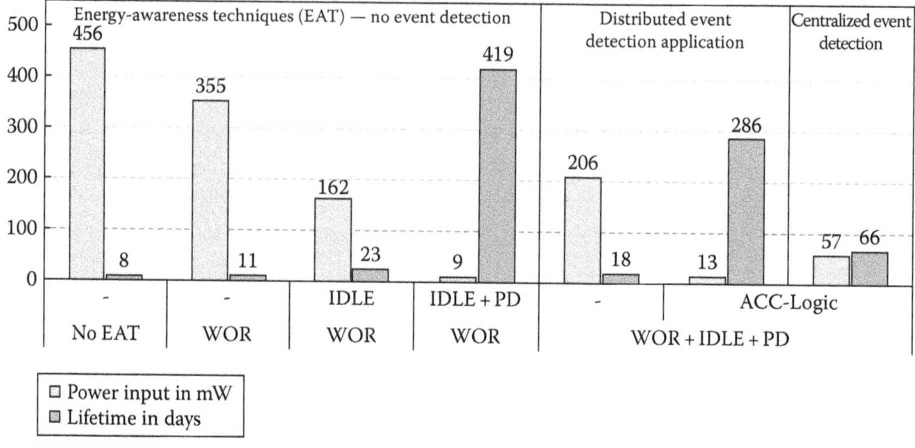

Figure 20.8 Energy consumption and network lifetime of different system configurations.

an event is generated and processed 5 times/h. For comparison, we also plot the numbers for a deployment of the same sensor network, but without any event-related activity. As can be seen in the figure, the lifetime of the network that employs centralized event detection is drastically reduced in comparison to the idle network due to the increasing energy consumption. In contrast, the distributed event detection is able to reduce the per-node energy consumption by 77%, thus increasing the lifetime of the network by 4 times to an extrapolated total of 41 weeks. By applying an acceleration-logic (ACC-Logic), we are able to wake up the MCU in the case, of an occurring event and make use of PD during the remaining time. In this way, the distributed event detection gains a lifetime improvement of about 16 times compared to a software logic calculated by the MCU. This underlines the importance of clever techniques that wake up further sensor node components only when they are required.

In conjunction, these results underline the validity of our initial statement that the distributed event detection is able to jointly achieve the otherwise conflicting goals of high-accuracy event detection and long network lifetime.

20.5 Conclusion

Distributed event detection in WSNs, as discussed in this chapter, combines highly accurate event detection with low-energy consumption. Further, given the distributed design of the classification algorithm, it is robust against failures of individual nodes as long as the network has been deployed in a sufficiently dense manner. The key advantage of this approach is, however, its generality: Since most of the architecture, that is, all components above the level of feature extraction, operates independently of the deployment-specific sensors, the system can be employed in a variety of scenarios. In addition to our main use of perimeter security, the system can also easily be trained, for example, to recognize human motions, spatial temperature distributions, or patterns in infrared readings.

With systems like the one covered in this chapter, research into event detection in sensor networks moves into a phase of incremental improvements. Depending on the difficulty of the application scenario, detection accuracies beyond the 90% mark are already feasible with current platforms. In order to push the envelope and work toward the 99% mark, careful refinements of the sensing platform are necessary, for example, the choice of sensors and their mounting on the sensor node. Furthermore, the design space of which features to use in which type of deployment needs to be explored systematically, and the training process needs to be streamlined. In conclusion, one can say that—given the overall capabilities present in current sensing platforms—*intelligent* sensor networks are just around the next corner.

Acknowledgment

This work was funded in part by the German Federal Ministry of Education and Research (Bundesministerium für Bildung und Forschung, BMBF) through the project AVS-Extrem.

References

1. D. M. Doolin and N. Sitar. Wireless sensors for wildfire monitoring. In *Proceedings of the Symposium on Smart Structures and Materials 2005*, San Diego, CA, March 2005.
2. M. F. Duarte and Y. H. Hu. Vehicle classification in distributed sensor networks. *Journal of Parallel and Distributed Computing*, 64(7), 826–838, July 2004.
3. G. Wittenburg, N. Dziengel, C. Wartenburger, and J. Schiller. A system for distributed event detection in wireless sensor networks. In *Proceedings of the Ninth ACM/IEEE International Conference on Information Processing in Sensor Networks (IPSN '10)*, Stockholm, Sweden, April 2010.
4. A. Tavakoli, J. Zhang, and S. H. Son. Group-based event detection in undersea sensor networks. In *Proceedings of the Second International Workshop on Networked Sensing Systems (INSS '05)*, San Diego, CA, June 2005.
5. A. Yang, S. Iyengar, S. S. Sastry, R. Bajcsy, P. Kuryloski, and R. Jafari. Distributed segmentation and classification of human actions using a wearable motion sensor network. In *Proceedings of the IEEE Conference on Computer Vision and Pattern Recognition Workshops (CVPRW '08)*, Anchorage, AK, June 2008.
6. N. Dziengel, M. Ziegert, Z. Kasmi, F. Hermans, S. Adler, G. Wittenburg, and J. Schiller. A platform for distributed event detection in wireless sensor networks. In *Proceedings of the First International Workshop on Networks of Cooperating Objects (CONET '10)*, Stockholm, Sweden, April 2010.
7. H. Wang, J. Elson, L. Girod, D. Estrin, and K. Yao. Target classification and localization in habitat monitoring. In *Proceedings of the IEEE International Conference on Acoustics, Speech, and Signal Processing (ICASSP '03)*, Hong Kong, China, April 2003.
8. F. Martincic and L. Schwiebert. Distributed event detection in sensor networks. In *Proceedings of the International Conference on Systems and Networks Communications (ICSNC '06)*, Tahiti, French Polynesia, October 2006.
9. M. Li and Y. Liu. Underground coal mine monitoring with wireless sensor networks. *ACM Transactions on Sensor Networks*, 5(2), March 2009.
10. R. O. Duda, P. E. Hart, and D. G. Stork. *Pattern Classification*, Wiley-Interscience, Hoboken, NJ, 2nd edn., November 2000.
11. H. Niemann. *Klassifikation von Mustern*, Springer, Berlin, Germany, 1st edn., July 1983.
12. M. Wälchli, P. Skoczylas, M. Meer, and T. Braun. Distributed event localization and tracking with wireless sensors. In *Proceedings of the Fifth International Conference on Wired/Wireless Internet Communications (WWIC '07)*, Coimbra, Portugal, May 2007.
13. N. Dziengel, M. Ziegert, S. Adler, Z. Kasmi, S. Pfeiffer, and J. Schiller. Energy-aware distributed fence surveillance for wireless sensor networks. In *Proceedings of the Seventh International Conference*

on Intelligent Sensors, Sensor Networks and Information Processing (ISSNIP '11), Adelaide, Australia, December 2011.

14. H. Will, K. Schleiser, and J. Schiller. A real-time kernel for wireless sensor networks employed in rescue scenarios. In *Proceedings of the 34th IEEE Conference on Local Computer Networks (LCN '09)*, Zürich, Switzerland, October 2009.
15. N. Dziengel, N. Schmittberger, J. Schiller, and M. Guenes. Secure communications in event-driven wireless sensor networks. In *Proceedings of the Third International Symposium on Sensor Networks and Applications (SNA '11)*, Honolulu, HI, November 2011.
16. R. Kohavi. A study of cross-validation and bootstrap for accuracy estimation and model selection. In *Proceedings of 14th International Joint Conference on Artificial Intelligence (IJCAI '95)*, Montreal, Canada, August 1995.
17. G. Wittenburg, K. Terfloth, F. L. Villafuerte, T. Naumowicz, H. Ritter, and J. Schiller. Fence monitoring—Experimental evaluation of a use case for wireless sensor networks. In *Proceedings of the 4th European Conference on Wireless Sensor Networks (EWSN '07)*, pp. 163–178, Delft, the Netherlands, January 2007.

Chapter 21

Dynamic Coverage Problems in Sensor Networks

Hristo Djidjev and Miodrag Potkonjak

Contents

21.1 Introduction

21.1.1 Historical Picture: Coverage and Tessellation

Coverage may be defined as a task where the objective is to guarantee that a set of entities of interest (e.g., points, objects, or events) is completely covered. The covering is broadly defined. For example, it may be physical or using observation points. Coverage is one of the oldest problems in mathematics and physics. For example, in 1619, Johannes Kepler, a famous German mathematician and astronomer, published his seminal book entitled *Harmonices Mundi* that included the first study on tessellation [45]. The task of tessellation is a special coverage case where the goal is to cover infinite two-dimensional space using the repetition of a single or a finite number of geometric shapes. Of course, no overlaps or gaps are allowed. Probably, the most celebrated result related to tessellation was discovered by Yevgraf Fyodorov at the end of the nineteenth century. He presented proof that all periodic tilings of the plane feature 1 of 17 unique groups of isometrics.

21.1.2 Coverage and Sensor Networks

Although coverage has a long and rich history, it only recently emerged as a premier computer science research topic. This is a confluence of technology push and application pool. The technology push was provided due to the creation of sensor network. This rapidly growing area provides means for comprehensive surveillance of both objects and area under reasonable cost and energy constraints.

The second part of the research and development impetus was provided by rapid emergence of security as one of the most important and desired system and application aspects. In a sense, coverage is the fourth wave of information security. The first was created in 1976 by the introduction of public key cryptography. It provided practical and theoretically sound techniques for ensuring privacy of data storage and data communication. The second is related to system security. In a sense, these techniques have longer and richer history than public key cryptography. Recent emphasis has been on hardware-based security and detection of malicious circuitry. The third wave aims at protection of the Internet and the WWW. Although this wave is by far the most diverse and covers issues from phishing to privacy, a significant emphasis has been on denial of service.

The fourth wave that has been just started is related to physical and social security using large-scale sensing, computation, communication, and storage resources. It is often envisioned in the form of multiple sensor network that uses (standard) wireless communication infrastructure to enable transfer of data to computational clouds. While the exact system picture has been radically changing (e.g., initially network processing of collected data was a dominant system paradigm), the frontier component (sensor networks) has been constant in all efforts.

Coverage is naturally both a sensor network canonical task as well as the basis for numerous physical and social security tasks. It has extraordinarily broad basis and numerous coverage subtasks can be defined.

The concept of coverage was introduced by Gage, who defined three classes of coverage problems: (1) blanket coverage (also known as area coverage), where the goal is to have each point of the region be within a detection distance from at least one of the sensors (static sensors, static objects coverage); (2) sweep coverage, where the goal is to move a number of sensors across the region as to maximize the probability of detecting a target (mobile sensors, static objects); and (3) barrier coverage, where the objective is to optimally protect the region from undetected penetration (static sensors, mobile objects). In addition, one can pose the fourth possible definition (mobile sensors, mobile objects). The last class of problems is not just practically very important, but also technically very challenging. Its theoretical treatment requires several probabilistic models. Its practical addressing requires sound and realistic statistical models that consider correlations.

One can also envision many other generalizations of dynamic coverage problems. For example, a number of authors considered techniques for maximizing the lifetime of the network and, therefore, the length of the pertinent coverage. Also, coverage under multiple objectives and/or multiple constraints, most often related to sensing and communication, has been a popular topic. It is important to note that technological trends may evolve so that communication ranges are much longer than sensing. Nevertheless, multiobjective coverage has tremendous practical importance. For instance, it is a natural way to address common scenarios that detection of an object or an event can be accomplished only by using sensors of different modalities and therefore properties. Another important dimension is providing guarantees of proper functioning of the coverage system in the presence of faults or security attacks.

21.1.3 Challenges in Solving Coverage Problems

We place special emphasis on the following four types of challenges.

Algorithmic challenges: Coverage problems are almost always intrinsically multidimensional. Many of them also include time dimension. Interestingly, some of the effective coverage problems can be naturally mapped into equivalent combinatorial and in particular graph formulation. For wide classes of coverage problems and, in particular, exposure problems, very often the most effective techniques involve variational calculus and its discretized realization using dynamic programming.

Finally, in some applications it is important that the algorithms have their localized versions where each sensor node contacts only a small subset of other nodes using high quality communication links in such a way that the overall global optimality is preserved completely or within a certain application bound. These types of coverage problems are most relevant in situations where one of the objectives is low-energy operation or preservation of the communication bandwidth. Also, this type of operation may be important when security is one of the important requirements. Our last remark is that probabilistic and statistical analysis of coverage algorithms is increasingly important.

Modeling challenges: There are two main aspects that require careful modeling decisions. The first is the modeling of sensitivity of sensors. Of course, for different types of sensors different types of models are more appropriate. Initially, many coverage tasks were treated under assumption that the detection is binary, for example, either an object of interest is observed or not. Consequently, much more comprehensive sensing models are introduced. For example, exposure requires that an object of interest is under surveillance in such a way that an integral of closeness over time is above a user-specified threshold. Also, directionality of some type of sensors was recognized. Of course, more and more complex models can be and should be addressed. However, as is often the case in statistics, a more complex sensing model does not imply a more realistic problem formulation and may significantly reduce (or enhance) the application domain.

The other important modeling issue is related to targeted objects and terrain. For example, in many applications, the mobility models are of prime importance. It is common to start from simple and intuitive models and keep increasing their complexity. It is interesting to mention that mobility models, unfortunately, have a long and painful history of being not just tremendously speculative, but even obviously and deeply completely counterproductive.

System challenges: It is customary that papers in top sensor networks are divided into two groups: theory and system. Not so rarely, theory papers are considered elegant and well mathematically founded but of rather low practical relevance. On the other hand, system people are primarily based on complete and demonstrated implementation that requires unacceptably high levels of abstraction and simplification. So, the first and most important system challenge is to combine useful properties of previous generation of both system and theory papers while eliminating past and some of the current problems.

Other premier system problems include low cost realization and energy efficiency. The last metric is further enhanced to include low power requirement in particular in self-sustainable coverage systems.

Security challenges: Security is one of the premier requirements in many applications and its relative role is rapidly increasing. It already ranges from privacy and trust to resiliency against hardware, software, and physical attacks. Very often, sensor networks used to ensure coverage are not attended or may even be deployed in hostile environments. Particularly interesting is the situation when two or more parties are observing each other and simultaneously aim to ensure high coverage while preserving their privacy of action. We expect that game theory techniques will be soon used in this context.

21.1.4 Focus of This Survey

In summary, coverage has a great variety of potential formulations and is a premier sensor network and emerging physical security task. In this survey we have three major objectives. The first is to survey the most popular and the most important, in terms of application coverage, tasks and proposed techniques. There are already several thousand coverage techniques. Therefore, it is not even possible to aim to be comprehensive. Instead, we focus on the most effective techniques that target most generic and pervasive coverage formulation.

The second goal is to try to establish the place of coverage in the global picture and its relationship with other sensor network, security, and system design tasks and applications. Our final target is to identify and provide a research impetus for the most important and challenging new coverage research directions.

21.2 The Coverage Problem

In this section, we discuss the importance of the coverage problem in sensor networks and briefly review the topic of static coverage. In static coverage, the goal is to place the smallest number of sensors in such a way that an area of interest is observable. In comparison, dynamic coverage addresses the situation in which either the sensors or the objects are allowed to move in the area of interest. A special case of dynamic coverage is the exposure problem in which the detection is accomplished if an integral over time of a specific sensing function is large enough to ensure detection and possibly the characterization of the pertinent object.

21.2.1 Historical Perspective

As we indicated in Section 21.1, coverage is an optimization problem, in particular with a long history in mathematics and crystallography, and more recently in robotics, computational geometry (e.g., art gallery problems), and television and wireless networks. However, the explosion of interest in coverage received a tremendous impetus with the emergence of sensor networks somewhere around the turn of the last century.

Table 21.1 provides the quantification of our claim. It shows the number of sensor coverage papers according to Google Scholar. We see that while the overall number of papers is relatively constant per year, the number of papers with words "coverage in sensor networks" has experienced consistent growth and increased by more than 30 times in the last decade even when normalized against slight growth of the overall number of papers. The overall number of papers is actually increasing every year, but nontrivial latency in paper indexing hides this growth. It also results in reporting somewhat understated growth in the number of coverage papers.

There have been several survey papers completely dedicated to coverage in sensor networks [9,16,35,41]. In addition, several ultra-popular comprehensive surveys of sensor networks devoted

Table 21.1 The Number of Sensor Coverage Papers According to the Google Scholar Database

Year	*Coverage*	*Total*
2001	190	2,670,000
2002	343	3,020,000
2003	681	3,100,000
2004	1440	3,120,000
2005	2460	2,970,000
2006	3470	3,040,000
2007	4360	2,950,000
2008	5190	2,810,000
2009	6020	2,510,000
2010	6990	2,400,000
2011	6750	3,100,000
2012	492	205,000

Note: The first column indicates year. The last two columns indicate the number of papers that address coverage in sensor networks and the total number of papers in the database, respectively. The data for 2012 include only publications indexed in January.

a substantial space to coverage [5,7,33,64,99]. Also, a large number of surveys have been published on more specific aspects of coverage [32,33] in particular using visual sensors [4,19,85] and energy-efficient coverage [6,27,38,61].

21.2.2 Applications and Architectures

Sensor networks provide a bridge between computational and communication infrastructures (e.g., the Internet) and physical, chemical, and biological worlds. The number of potential applications is unlimited. Most often, environmental, infrastructure security (e.g., pipelines and building), and military and public security are addressed. More recently, wireless health and medical applications have emerged as one of the most popular research directions.

Initially, Internet research has had a dominating impact on the wireless sensor network research. Energy has been recognized as one of the most important design metrics. In addition, there has been an emphasis on efficient usage of bandwidth. Ultra-low-power operation of wireless sensor networks was the focus of many wireless sensor network efforts. Therefore, the ultra-low-power node with very short communication ranges was accepted as the preferred architecture building block.

However, in the last several years it has been widely recognized that rapid progress in wireless mobile network provides numerous advantages. For example, mobile phone–based participatory sensing that involves human interaction has emerged as the dominant architecture paradigm.

Both applications and architectures have profound ramifications on how coverage problems are formulated and addressed. For example, the use of mobile phone infrastructure eliminated limitations and concerns about communication range that is now much higher than the sensing range of essentially all sensors. Also, the need for localized algorithms is greatly reduced and much more complex definitions of coverage that require much higher processing resources and energy can now be realistically addressed. On the other hand, latency has gained importance over throughput.

Also, each type of application requires new definitions of coverage. For example, medical applications can benefit little from traditional notions of coverage. In order to establish credible medical diagnosis, significantly more complex processing is needed that blurs distinctions between coverage and sensor fusion. It also introduces many new aspects such as sizing of sensors and its impact on coverage.

21.2.3 Real-Time Coverage

Operation in real time is essential for a majority of coverage applications that use sensor networks. Surprisingly, this topic still does not receive a proportional amount of research and effort. This is unexpected, in particular, since one of the three tracks of the most prestigious real-time conferences, Real-Time Systems Symposium, is dedicated to sensor networks. One of the first and most influential papers in this domain is by Jeong et al. [43], which addresses the problem of observing a net of actual pathways where vehicles move a specified maximal speed. Under a set of assumptions that include the maximal car density, the goal is to ensure that all intruding targets are detected before they reach any of the protection points. The objective is to maximize the lifetime of a sensor network that is used for coverage. The algorithm is based on the Floyd–Warshall algorithm to compute the all-pairs shortest paths formulation. In order to maximize the lifetime of the network, different sensors are assigned to different duty-cycle schedules. Jeong and his coauthors presented both centralized and localized algorithms for early detection of targets on a graph (i.e., highway or street network). Zahedi et al. [100] further explored the problem of trade-offs between the quality and duty-cycle (energy) of the sensors.

Trap coverage is a very interesting and natural formulation of coverage that is related to real-time detection and, in particular, latency of detection. It is also a way to address approximate coverage when the number of available sensors is pre-specified. Until now we mainly discussed coverage techniques in which complete coverage, of a targeted field is the objective. In trap coverage, holes in coverage are allowed, but only if their number and their size are below specified measures. One such measure that captures latency of detection is a time that an intruder spends in straight line travel at a specified speed before being detected. Recently, this problem has been addressed both under and not under the assumption that energy efficiency is one of the requirements [10,57].

21.2.4 Static Coverage

Although our survey is focused on dynamic coverage in sensor networks, it is important to discuss static coverage in which the goal is to cover a specific area using the smallest number of sensors. An alternative formulation is one in which the goal is to cover a maximum subarea of a given area using a specified number of sensors.

Although static coverage is probably conceptually the simplest possible formulation of any coverage problem, almost all of its instantiations are still NP-complete. For example, these instantiations can be often mapped to the dominating set problem. Interestingly, when we consider coverage of a rectangular area using disks, the complexity of the corresponding optimization is not known.

One of the first approaches to address static coverage was presented by Slijepcevic and Potkonjak [81]. They proposed two techniques: One uses simulated annealing and the other employs integer linear programming (ILP). In addition, D. Tian first as a student and later with his research group proposed a number of techniques for static coverage [89,90].

21.3 Barrier Coverage

In barrier coverage, the objective is to protect the area from unauthorized penetration. We discuss in detail several types of barrier coverage including perimeter coverage, where the objective is to cover with sensors a narrow strip along the boundary of the region; the maximum breach path problem, where the goal is to find a path that maximizes the minimum distance to any sensor; and the minimum exposure path problem, whose objective is to find a path of minimum exposure, where the exposure of the path is defined as the integral of the sensing signal along that path.

21.3.1 Perimeter Coverage

21.3.1.1 Problem Formulation

The objective of perimeter coverage is to study ways to detect an intrusion into a protected area by placing sensors near the border of the monitored region. There are two aspects of that problem: the *placement problem* asks to determine a placement of the sensors that offers optimal or near optimal protection for given resources or costs, and the *assessment problem* asks, given a placement of sensors, to evaluate how well they protect the area.

Instead of placing sensors on the boundary line, most authors consider instead placement in a *belt area*, a narrow region between two parallel lines containing the boundary, which we refer to as the *outside* and the *inside* of the belt, respectively, where sensors should be placed. If the boundary of the belt region is connected the belt is called open, and otherwise it is called closed

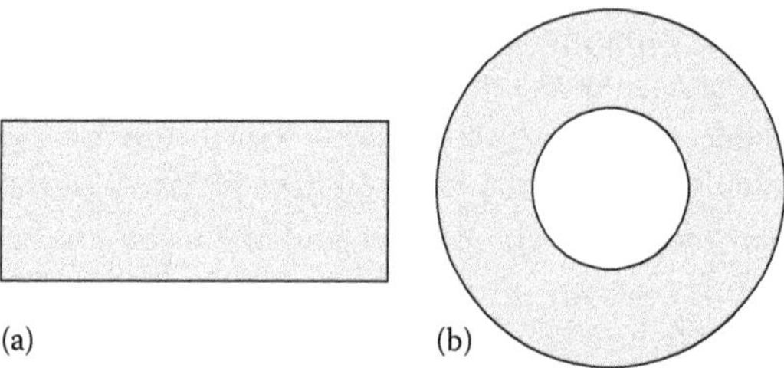

Figure 21.1 Types of belts depending on the boundary type: (a) open belt, when the boundary is connected, and (b) closed belt, when the boundary is disconnected.

(Figure 21.1). We refer to the short lines in an open-belt region connecting the outside to the inside boundary as the *left* and the *right* boundary, respectively. A belt with inside and outside boundaries l_1 and l_2, respectively, has width w, if for each point p_1 in l_1 and each point p_2 in l_2 $\text{dist}(p_1, l_2) = \text{dist}(p_2, l_1) = w$. Here, $\text{dist}(p_i, l_j)$ is defined as the minimum distance between p_i and any point in l_j.

Since any coverage of the whole area also covers the belt and the belt region is typically much smaller, it is clear that perimeter coverage is often much more cost-effective than the full-area coverage.

Kumar et al. [55,56] were one of the first to study the perimeter coverage problem in detail. They define two versions of the problem. The *weak k-barrier coverage* version considers only breaching paths with lengths equal to the belt width (called *orthogonal* paths). The rationale behind that restriction to the paths that we want to cover is that an intruder without a prior knowledge of the location of the sensors will likely choose an orthogonal path, since such a path is shortest and hence it minimizes the detection expectation. The *strong k-barrier coverage* version considers all paths crossing the complete width of the belt (called *crossing* paths) as possible breach paths. The regions are *weakly k-barrier covered* (respectively, *strongly k-barrier covered*) if every orthogonal (respectively, every crossing) path crosses the sensing region of at least k sensors. We call the maximum value of k for which the region is k-covered as the *strength* of the coverage.

21.3.1.2 Strong k-Barrier Coverage

Kumar et al. [55,56] consider two versions of the strong k-barrier coverage placement problem: a deterministic and a probabilistic one. In the deterministic version, sensors are placed on explicitly determined locations, while in the probabilistic one they are placed randomly according to a given probability distribution.

For the deterministic version, they prove that an optimal placement of the sensors in an open-belt region is on a set of k shortest paths called *separating paths* that separate the outside from the inside portion of the belt so that the sensing regions of the sensors touch or overlap inside the belt (Figure 21.2). In the case where the sensing region of each sensor is a disk of radius r, they also prove that the smallest number of sensors necessary and sufficient to cover an open-belt region is $k\lceil s/2r \rceil$, where s is the length of a shortest separating path.

For the probabilistic version of the placement problem, Liu et al. [60] show that whether a random placement of sensors in a rectangular belt yields a k-barrier coverage depends on the ratio between the length h and the width $w = w(h)$ of the belt. Specifically, if the sensors are distributed

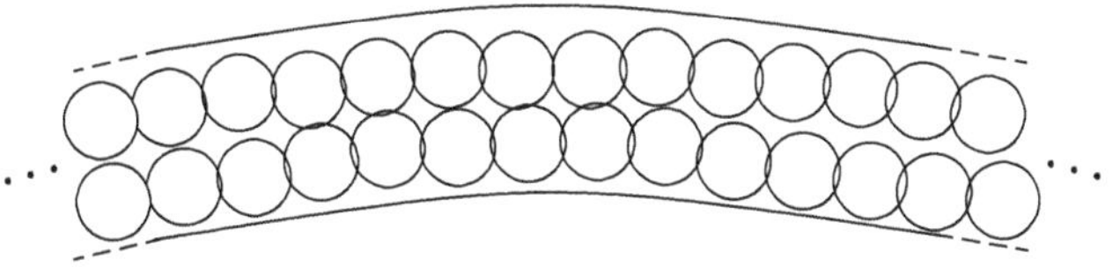

Figure 21.2 Placing the sensors on two separating paths results in a strong 2-barrier coverage of the region.

according to a Poisson point process with density λ, then if $w(h) = \Omega(\log h)$, the region is k-barrier covered with high probability if and only if the density λ of the sensors is above, a certain threshold. If, on the other hand, $w(h) = o(\log h)$, the region does not have a barrier coverage with high probability for any λ. With high probability (w.h.p.) means that the probability tends to 1 as h tends to infinity. The strength of the coverage for a fixed density γ grows proportionally with $w(h)/r$.

Another interesting question is, given a belt and the positions of a set of sensors placed in it, to determine whether the sensors provide a barrier coverage and to find the strength of such a coverage. Kumar et al. [55,56] answer that question for open-belt regions by reducing the aforementioned problem to the problem of finding a set of node-disjoint paths in a graph. They define a coverage graph G whose nodes are the sensors of the network and whose edges connect all pairs of nodes whose corresponding sensors have overlapping sensing regions. They also define two additional nodes u and v and edges between u (respectively v) and all nodes whose corresponding sensing regions intersect the left (respectively right) boundary of the belt. Using Menger's theorem [96, p. 167], they prove that k-barrier coverage by the given sensors of the belt is equivalent to the existence of k vertex-disjoint paths between u and v in G. Moreover, computing the maximum number of k vertex-disjoint paths between u and v in G can be done in time $O(k^2 n + m)$, where n and m are the number of the nodes and edges of G. However, the same proof cannot be used for the closed-belt case since Menger's theorem is not applicable to that case. The assessment problem for strong k-barrier coverage for closed-belt regions is still an open problem.

21.3.1.3 Weak k-Barrier Coverage

Weak barrier coverage allows only crossing paths that are perpendicular to the belt boundary. In [56], Kumar et al. consider sensors that are Poisson distributed with density np and ask the question which values of np produce a weak barrier coverage with high probability. We can think of the parameter n as corresponding to the total number of the sensors and p as the probability of each sensor being awake at any given time. Kumar et al. define function

$$c(s) = 2npr/(s\log(np))$$

and show that, for a belt of width $1/s$ and for any $\varepsilon \in (0,1)$, if

$$c(s) \geq 1 + \frac{(\log(\log np))^{1-\varepsilon} + (k-1)\log(\log np)}{\log(np)} \tag{21.1}$$

for sufficiently large s, then all orthogonal lines crossing the belt are k-covered with high probability as $s \to \infty$. On the other hand, if

$$c(s) \leq 1 - \frac{(\log(\log np))^{1-\varepsilon} + \log(\log np)}{\log(np)} \tag{21.2}$$

for sufficiently large s, then there exists a non-1-covered orthogonal crossing line in the belt with high probability as $s \to \infty$. Condition (21.1) is a sufficient condition for achieving k-barrier weak coverage and condition (21.2) provides a necessary condition (if the inequality is reversed) for 1-barrier weak coverage. Evidently, there is a gap between the two bounds and finding an optimal k-barrier weak coverage condition is an interesting open question.

As noted in [56], the right-hand sides of (21.1) and (21.2) tend to 1 as $s \to \infty$. Hence, asymptotically the critical value for $c(s) = 2npr/(s\log(np))$ is 1, meaning that there should be at least $\log(np)$ sensors deployed in the r-neighborhood of each orthogonal crossing line in order to produce a weak barrier coverage of the region.

In a different approach to the problem, Li et al. [58] found a lower bound on the probability for a weak k-barrier coverage, given the size of the region and the number and the distribution of the sensors. Specifically, they show that if the belt region is a rectangle with dimensions $s \times 1/s$, r is the sensing radius, the sensors are distributed according to a Poisson point process with density np, and B_k denotes k-barrier coverage, then

$$\Pr(B_k) \geq \left(1 - \sum_{j=0}^{k-1} \frac{(2nr/s)^j}{j!} e^{-2nr/s}\right)^n \cdot \left(1 - \sum_{j=0}^{k-1} \frac{(nr/s)^j}{j!} e^{-nr/s}\right)^2$$

Given the placement of the sensors, a natural question to ask is whether those sensors provide a weak k-barrier coverage. Answering that question is easier in the weak barrier coverage case than the similar question for strong barrier coverage. The reason is that, for weak coverage, the vertical positions of the sensors do not matter as only vertical paths are considered. Hence, the problem can be reduced to a one-dimensional case: just consider the projections of the sensor positions onto the line segment S defining the internal (or external) belt boundary and determine whether those projections k-cover that segment. Li et al. [58] present a simple algorithm that considers the set Q of the endpoints of all sensing intervals on S, that is, for each point x on S corresponding to a sensor projection, we add points $x - r$ and $x + r$ to Q. Then S is swept from left to right keeping track on how many sensors cover each point. The resulting algorithm has time complexity of $O(N \log N)$, where N is the number of the sensors.

21.3.1.4 Other Perimeter Coverage Results

Kumar et al. establish in [56] that it is not possible to determine locally whether a region is strongly k-barrier covered or not. This is in contrast to the full-area coverage case, where a "yes" answer is not possible, but a "no" answer is, that is, it is possible in the full coverage case to determine that a region is *not* k-covered. In order to deal with the problem of local barrier coverage, Chen et al. [20] introduce the notion of *L-local barrier coverage*. Informally, having L-local barrier coverage requires that any path contained in a box of length at most L be covered (or k-covered). Hence, L-local barrier coverage is a generalization of weak coverage for L equal to zero and to strong barrier coverage for L equal to the belt length. If L is sufficiently small, it is possible to locally determine if the region is not L-locally k-barrier covered, as proved in [20].

Chen et al. [21] use the idea of L-local barrier coverage in order to quantify the quality of k-barrier coverage. Previously, the quality measure has been binary—1 if there is k-barrier coverage and 0 if there isn't. Chen et al. define the *quality* of k-barrier coverage as the maximum value of L for which the belt is L-local k-barrier covered. If there is no such L then they define the quality as -1. They design an algorithm that computes the quality given the sensor positions and a value for k. Their algorithm also identifies weak regions that need extra sensors. The property of being able to quantify the quality of barrier coverage is analyzed from another perspective and in much more detail in the following subsections.

21.3.2 Maximum Breach Path

The maximum breach path tries to determine the least covered (the most vulnerable) path between a pair of points. In this context, a measure of how well a path p is covered is the minimum distance between any point of p to any of the sensors. The key conceptual difficulty is that there are continuously many possible paths for the intruder. Nevertheless, this is one of the first problems of coverage in sensor networks that has not only been addressed but actually solved optimally.

The key idea behind the solution is remarkably simple. The crucial step is to translate this computational geometry and continuous problem into an instance of graph theoretical problem. It is easily accomplished using the notion of a Voronoi diagram. A Voronoi diagram is a tessellation of the space using piecewise linear connected components. If we have two sensors, A and B, the line of separation between them is orthogonal to the line that connects them and passes through the middle of the distance between these two sensors. It is easy to see that during calculation of dynamic coverage, it is sufficient to consider only Voronoi diagram edges and more specifically their weight, which is equal to the distance of the closest point on the Voronoi diagram edge to either one of two sensors that define it. The justification for this observation is that if the intruder does not use for his traversal only Voronoi diagram edges, it will become closer to at least one of the sensors that are used to define the pertinent Voronoi diagram edge.

Now, in order to find if there is a breach in the system of deployed sensors of length l, all that is required is to check if there is a path in the graph that is defined on top of the Voronoi diagram, where at least one edge is not larger than a specified value. There are many ways to accomplish this task. Conceptually probably the simplest is one where we iteratively add larger and larger edges until there is a path from the starting point to the ending point. There are several important observations about this approach. One is that one can easily consider the case where different sensors have different sensitivity ranges, or even one can superimpose a grid over the area and define for each field in the grid the level of sensitivity over a single or multiple sensors. All these problems can be easily solved using dynamic programming. The much more in-depth technical presentation of these algorithms can be found in [62,65].

21.3.3 Minimum Exposure Path

As we already said several times, one of the key degrees of freedom in defining the coverage problem is related to the way in which we define the sensitivity with respect to a single or multiple sensors. The exposure is a generalization of dynamic coverage in the sense that it is asked whether it is possible to find a path through a particular field covered with sensors in such a way that the total integral of exposure over time to sensing by all relevant sensors is below the user-specified value.

There are two conceptually similar but highly different ways, in terms of implementation, to address this problem. The first one uses rasterization of the pertinent field into a particular grid or some other structure where in each field all points are sufficiently close to each other. This is easy to accomplish by decreasing the size of individual fields. For each small area, we can calculate the amount of exposure for any given period of time. Now, under the natural assumption of constant speed, we can easily use dynamic programming to find the path of minimal exposure from a starting point s to a destination point d. This task can be easily accomplished in polynomial time that depends on additional constraints that may be imposed on the definition of exposure. This solution was presented by Meguerdichian, who subsequently changed his last name to Megerian, in [63,68].

Another very interesting approach uses variational calculus to solve the exposure problem in a way that guarantees the correct solution (by Veltri et al. [91]). The key idea is to solve a small number of simplified problems such as one where very few sensors are used and to concatenate these locally optimal solutions into one that is globally optimal.

An approximation algorithm for the exposure problem with provable accuracy and polynomial running time was designed by Djidjev [29]. In this algorithm, the points are not placed on a grid covering the region (rasterization), as in the previous algorithms, but only on the edges of a Voronoi diagram for the set of the sensors. This, in effect, replaces a two-dimensional mesh by an one-dimensional mesh, significantly reducing the computational complexity of the algorithm. For any given $\varepsilon > 0$, the algorithm from [29] can find a path with exposure no more than $1 + \varepsilon$ times larger than the optimal. Hence, by reducing the value of ε, one can get paths with exposures arbitrarily close to the optimal. The running time of the algorithm is proportional to $n\varepsilon^{-2} \log n$, assuming that the Voronoi diagram does not have angles very close to zero.

21.4 Coverage by Mobile Sensors

In the mobile version of the coverage problem, the goal is to cover a region of interest with mobile sensors so that the trajectories of the sensors go through points or areas of interest at predetermined time intervals, form barriers, or relocate themselves to better static locations.

21.4.1 Sweep Coverage

Li et al. [59] consider the following problem they call the *sweep coverage* problem: There are n mobile sensors located in a region that contains m points of interest (POIs) that need to be monitored. The sensors move at the same constant speed v and a POI is considered *covered* at a given time if a mobile sensor is at that location at that time. Given a coverage scheme (schedule), a POI is considered *t-sweep covered* if it is covered at least once in every time interval of length t. The goal is to design a coverage scheme so that each of the m POIs is t-sweep covered. A more general version of the problem specifies individual sweep periods t_i for sensor t_i.

It is proved in [59] that the t-sweep coverage problem is NP-hard by reducing the traveling salesman problem to it. An even stronger result is proved in the same paper [59] that the t-sweep coverage problem cannot be approximated within a factor of less than 2 unless P = NP. It is also shown that for any $\varepsilon > 0$ there exists a polynomial time algorithm for solving the t-sweep coverage problem within a factor of $2 + \varepsilon$. The algorithm uses the $1 + \varepsilon$-approximation algorithm for the traveling salesman problem [8] to construct a short route r visiting all POIs exactly once. Then r is divided into n equal parts, one for each of the n sensors. Finally, each sensor is assigned to monitor one of the parts p_i of p by moving forward and backward along p_i. This algorithm is generalized in [59] for the case of different sweep periods for the POIs, resulting in an algorithm with an approximation ratio of 3.

21.4.2 Optimal Repositioning of Mobile Sensors

The problem of repositioning the sensors so that they provide a better barrier coverage while minimizing the distance they have to travel or the energy they need to consume is studied in [11,15,88]. Bhattacharya et al. [15] assume that n sensors are initially located in the interior of a planar region and study the problem of how to move the sensors to the boundary of the region so that the distance along the boundary between two consecutive sensor positions is the same. Hence, after repositioning the sensor positions will form a regular n-gon that is called *destination polygon.* We call the new position of each sensor the *destination* of that sensor. There are two versions of the problem:

- The *min-max* problem, aiming to minimize the maximum distance traveled by any sensor
- The *min-sum* problem, where the objective is to minimize the sum of the distances traveled by all sensor

For both problems they consider two type of regions: a unit disk and a simple polygon. We discuss first the algorithms for the min-max problem and then for the min-sum problem.

21.4.2.1 Min-Max Problem

For the min-max problem on a disk region, Bhattacharya et al. call a positive real number λ *feasible*, if all the sensors can move to the new positions on the boundary of the disk that form a regular n-gon P and the maximum distance between an old and a new position of any sensor does not exceed λ. Such polygon P is called λ-feasible. Hence, the min-max problem is equivalent to the problem of finding the minimum feasible number λ_{min} and a λ_{min}-feasible polygon. If we can construct an algorithm to check the feasibility of any number in time $T(n)$ and we know an interval containing λ_{min}, then we can do a binary search on that interval, at each step reducing twice the size of the interval containing λ_{min}. Clearly, the interval $[0, 2]$ contains λ_{min} since the distance between any two points in the disk cannot exceed its diameter. Hence, the running time of the resulting algorithm will be $T(n)\log(1/\varepsilon)$, where $\epsilon > 0$ is the required accuracy. Using a more complex binary search algorithm that uses a finite set of candidate new-position points, Bhattacharya et al. show that the exact value of λ_{min} can be found in time $O(T(n)\log n)$.

For testing feasibility of a number $\lambda > 0$ for n sensors on positions $A_1, \ldots, A_n$ inside a circle C, Bhattacharya et al. construct for each i a circle of radius λ and center A_i and consider the two intersection points of that circle with C. The resulting set Q contains $2n$ points. It is shown that, if λ is feasible, then there is a λ-feasible n-gon one of whose vertices is in Q. Hence, assuming λ is feasible, one can find a λ-feasible n-gon by checking each of the regular n-gons that contain a node in Q, whose number is at most $|Q| = 2n$. Then the problem is reduced to checking whether the vertices $B_1, \ldots, B_n$ of each of those $2n$ polygons can be mapped to distinct points among $A_1, \ldots, A_n$ so that for each i the distance between B_i and A_i is at most λ. The latter mapping problem can be solved using an algorithm due to [40] for finding a prefect matching in a bipartite graph with time complexity of $O(n^{2.5})$. The total complexity of the resulting feasibility-checking algorithm is $O(n^{3.5})$, and the resulting complexity of the min-max algorithm is $O(n^{3.5} \log n)$.

Tan and Wu [88] improve the complexity of the min-max algorithm for a disk from [15] by using a better characterization of λ_{min}-feasible polygons. Specifically, they show that if $B_1, \ldots, B_n$ are the vertices of a λ_{min}-feasible n-gon such that $|A_iB_i| \leq \lambda$ for all i, then either

1. For some i such that $|A_iB_i| = \lambda$, the line joining A_i and B_i contains the center of C
2. For some $i \neq j$, $|A_iB_i| = |A_jB_j| = \lambda$

Using this fact, one can construct a set of all n distances of type (1) and all, say m, distances of type (2). Doing a binary search on that set will yield in $O(\log(n + m))$ feasibility tests the value of λ_{min} and the corresponding n-gon. Unfortunately, in the worst case m can be of order n^3, which implies that the worst-case complexity of the resulting min-max algorithm will be $O(n^3)$. By employing a more elaborate search procedure, Tan and Wu [88] show that the complexity of their algorithm can be reduced to $O(n^{2.5} \log n)$.

For the min-max problem in a simple-polygonal region P, Bhattacharya et al. [15] show that their algorithm for disk regions can be adapted, resulting in an algorithm of time complexity $O(ln^{3/.5} \log n)$, where l is the number of the vertices of P. The additional factor of l comes from the fact that the intersection of a circle centered at a sensor and the boundary of P can consist of upto l points, unlike the disk-region problem when it consists of at most two points.

21.4.2.2 Min-Sum Problem

Unlike the min-max problem, for the min-sum version no exact polynomial algorithm is known yet, and neither is known whether the problem is NP-hard or not. The reason is that, for the min-sum problem, no characterization of the λ_{opt}-polygon is known that would allow for reducing the search space from continuous to discrete, as it is in the min-max version. Instead, it is shown in [15] that the destination of at least one sensor in any optimal n-gon belongs to a specified short segment along the circle C. Based on that fact, the corresponding segment for each of the sensors A_i is discretized by adding $O(1/\varepsilon)$ equally spaced points, each of which is then considered as a candidate of a destination for A_i. Then, for each sensor and candidate, a minimum cost-weighted matching problem is solved for a weighted graph whose nodes are the sensors A_i and the vertices of the currently considered n-gon candidate, whose edges join each sensor and each polygon vertex, and whose edge weight is equal to the Euclidean distances. The matching problem can be solved in $O(n^3)$ time using the algorithm from [54]. The complexity of the resulting min-sum approximation algorithm is $O(n^4/\varepsilon)$ and the approximation ratio is $1 + \varepsilon$. A similar approximation algorithm can be constructed for the min-sum problem for a simple-polygon region with time complexity $O(ln^5/\varepsilon)$ and approximation ratio $1 + \varepsilon$, where l is the number of the vertices of the polygon.

Tan and Wu [88] consider a special version of the min-sum problem, where the sensors are initially positioned on C. For that version, they show that an exact polynomial-time algorithm for the min-sum problem does exist, and its complexity is $O(n^4)$. Their algorithm is based on a characterization of the optimal solution that limits the search space for a destination polygons to a discrete set. Specifically, they show that in any optimal solution, there exists at least one sensor A_i whose destination is A_i, that is, that does not change its position.

21.5 Other Coverage Issues

21.5.1 Wireless Links and Connectivity

There exists a large literature on simultaneous maintenance of coverage and connectivity. As we already stated, originally the sensor research community was targeting wireless sensor nodes with ultra-low-power radios and multi-hop communication. This type of wireless links has been widely studied experimentally and using statistical generalization in terms of their transmission properties as well as quality of link vs. energy consumption properties. Unfortunately, many of these studies are to a serious extent unrealistic because it was not recognized that the radio consumption model is such that listening is often as expensive as receiving or transmission.

It has been recognized that there exists high positive and negative correlation in link qualities, both spatially and in the temporal domain. Some of the key references in these domains are [17,18,73,102]. With the change of architecture of wireless sensor networks from ultra-low-power multi-hop communication to communication using wireless phone infrastructure, many fundamental assumptions about the role of communication in coverage tasks are drastically altered. For example, in this new architecture, it is very rarely the case that communication is the bottleneck and much higher emphasis is on the use of sensors in the best possible way.

21.5.2 Multi-Objective Coverage

Multi-objective coverage is one where at least two objectives or two constraints have to be addressed during node deployment or operation. The initial literature focused on maintaining sensing coverage and connectivity in large sensor networks [92,98,102]. In this situation, the key assumption is related to the ratio of communication range and sensing domain. In particular, a very interesting situation is when these two entities are of relatively similar cardinality. These problems may not be an issue in mobile phone–based sensor networks, but multi-objective is bound to emerge as one of the most important definitions of coverage.

For example, in many security applications, it is essential that we observe the enemy while the enemy is not able to observe us. Also, it is easy to imagine that in many types of coverage one has to ensure that fundamentally different types of sensors are able to collect information (e.g., audio and visual sensors). These sensors may have not just different sensitivity ranges, but also they may or may not be directed with various angles of coverage. The key goal here is to make adequate and simple to use sensing models as well as to find which type of sensor fusion is most relevant in a particular application.

21.5.3 Localized Algorithms Coverage

Localized algorithms are those that are executed on a small number of sensor nodes that are close to each other in terms of quality of their communication links and/or in terms of sensed events. Localized algorithms are important for several reasons. They are intrinsically low energy and fault tolerant. Localized protocols usually induce much lower latency and preserve bandwidth. Finally, in very large networks they are the only practical alternative.

A comprehensive but certainly somewhat outdated survey on localized algorithms has been published in 2004 [32]. Several authors have been able to develop localized coverage algorithms that are optimal or competitive with corresponding centralized algorithms [43]. Interestingly, even algorithmic paradigms have been developed for creation of localized algorithms [70,86]. The key idea is to use as a starting point any regular centralized algorithm. The results of the pertinent centralized algorithm provide statistical knowledge about which information should be used in which way in the corresponding algorithm. The final step is to use statistical validation techniques for the evaluation of the localized algorithm. It is important to emphasize that different instances of the coverage problem should be used for the learning and testing phases. Of course, for best performance the whole procedure is reiterated in a loop until the specified level of discrepancy between the centralized and the localized algorithms is found.

21.5.4 Lifetime and Energy-Efficient Coverage

It has been realized early that energy is one of the most severe constraints in wireless sensor networks. For example, Srivastava et al. [84] recognized that in the Smart Kindergarten project, batteries have to be changed at least once per day and that in order to instrument a sufficient number of subjects (kids) for the duration of the project, one would spend millions of dollars only on batteries. Therefore, a number of approaches have been developed to maintain one or more formulations of coverage while minimizing energy consumption.

The main idea is to schedule different subsets of sensors to be active in any given point of time in such a way that each group of sensors in each subset is sufficient to guarantee the coverage objective while the number of subsets is maximized. It is related to the well-known k-coverage problem in graph theoretic literature, which is NP-complete. Interestingly, in many applications with a relatively small number of nodes (up to several hundred), one can obtain the optimal solution using ILP [47,69]. It is interesting to note that there are also a very large number of survey papers that are completely dedicated to energy-efficient strategies in wireless ad hoc and sensor networks [6,27,38,61]. In particular, a large number of heuristics have been developed to maintain network coverage using low duty-cycle sensors [26,30,44,71].

21.5.5 Fault Tolerance and Errors

There are two major sources of sensing data errors that have been widely considered. The first is that sensor measurement may provide incorrect values. The second source of error is less dangerous for the accuracy and the correctness of the evaluation of coverage and is related to missing data.

There are three main types of errors that have high impact on coverage algorithms and applications. The first is related to readings of detection sensors. The second is associated with location errors [82]; these are particularly important for mobile sensors. These two types of errors may be both in terms of missing data or incorrect measurement. The final type is related to communication using lossy links and is of the missing data nature. Note that once real-time issues are considered, a new type of error related to late-arriving data emerges. It is important to note that in more complex scenarios, new types of errors may play important roles. For example, if nodes use a sleep mode for energy conservation, errors in time synchronization may be of essential importance [37,43].

There is a tremendous amount of literature in sensor measurement data. By far, the most popular approach is to assume independent errors that follow a Gaussian distribution. A number of interesting and theoretically important results are established under these assumptions. Unfortunately, the actual properties of real errors in data essentially always have highly nonparametric distributions and rather high spatial temporal correlations. It has been demonstrated that assuming a Gaussian error distribution may result in location errors that are several orders of magnitude higher than if nonparametric models that consider correlations are used for location discovery [31]. Conceptually, the most difficult problem with error modeling is that in many applications corresponding signals are nonstationary.

There have been several efforts to accurately and realistically model errors of individual sensors [34,51] and errors and communication links of a system of sensor and wireless nodes [48,49].

There is a complex interplay between error properties and optimization techniques used for calculating or optimizing coverage. In some situations, there are readily available provably optimal solutions. For example, if the coverage problem can be optimally solved using error-free data and if an error model is Gaussian, convex programming addresses the same problem in the presence of error optimally. Unfortunately, this situation rarely has practical benefits [52]. The impact of realistic error models is discussed in detail using several sensor networks applications [82].

In many scenarios, sensor networks for coverage are deployed in hostile environments where repair is either difficult or essentially impossible. In some scenarios, the environment is harsh and may have highly negative impact on the reliability of the sensors. Essentially all scenarios in which sensor networks are used to establish coverage are not attended by humans. Therefore, it has been recognized that there is a need for fault-tolerant coverage.

The most natural and the most popular way to ensure fault tolerance is through the use of redundancy [24]. In particular, k-cover algorithms simultaneously provide both energy efficiency and fault tolerance [1,47,53]. Interestingly, a much more efficient approach can be derived when tolerance is treated within the framework of sensor fusion [50,75].

21.5.6 Dealing with Uncertainty

Coverage under uncertainty in terms of locations of nodes has been widely studied [13,25,36,72,87]. Many of these efforts use mathematically sophisticated concepts (e.g., homology) or verification techniques. We expect that soon other uncertainty degrees of freedom will be addressed. For example, probabilistic or, even better, statistical guarantees of the coverage quality in the presence of uncertainty about the actual actions of other side (attacker, intruder), will be essential in many applications. One such potential framework to address these issues is the use of game theory.

21.5.7 Visual Coverage

One of the key predecessors of coverage is tasks in computational geometry such as art gallery observation by a limited number of agents. It is assumed that an agent can detect object at an arbitrary distance unless the object is hidden by a wall. The problem asks to deploy the smallest number of art gallery employees in such a way that there does not exist any area of the gallery that is not observed by at least one employee. In many security applications, as well as in entertainment applications, visual information is of the ultimate importance. Therefore, in the last 5 years, visual coverage emerged as one of the most popular topics. There are several surveys that treat this important problem in great detail [19,85].

In addition, there is a survey by Georgia Institute of Technology researchers that covers multimedia wireless sensor networks that is concerned with both data acquisition and data transmission [4]. The main conceptual difference between the standard definition of coverage and visual coverage is that cameras are subject to directional field of view and that they have rather large but nevertheless limited sensing range. A very important assumption is about the ability to rotate camera as required by tracking or coverage needs. As a consequence of these intricate sensor models, very intriguing and challenging optimization problems arise. It is surprising that a significant number of these can be solved in provably optimal ways using polynomial time complexity algorithms [2,3,14,39,42,78,83,97,101].

21.5.8 Security

Security is one of the most important parameters in many mobile and unattended system. In addition to papers published at many wireless, sensor, and security conferences, even dedicated conferences for wireless security attract a large number of submissions. Essentially, all security issues related to system security directly apply to coverage in sensor networks. It is not surprising that security of coverage results is of high importance. After all, coverage problems are very often directly related themselves to security applications. There are a large number of surveys on security in sensor networks [22,28,76].

In addition, there are at least two security challenges that are specific for sensor networks and coverage. The first is issue of physical attacks. Usually, security attacks require sophisticated mathematical, software, or system techniques. Therefore, it can be undertaken only by experts in these fields and significant efforts. However, reading of sensors can be easily altered using corresponding source of excitation. For example, one can easily increase the temperature of a sensor or alter the speed of acoustic signal propagation using dust. These type of attacks can easily result in greatly incorrect distance, location, or other measurements [23]. The development of techniques that mitigate or even better eliminate such impacts are of high importance.

The second issue is that in addition to the correct measurements one need to ensure that each of the measurements is collected by a sensor deployed by trusted party at exact location where the sensor is initially deployed at exactly the time when it claimed that data are collected. Recently, several such solutions that utilize the notion of public physical unclonable function (PPUF) [12] have been developed [67,77]. The key idea is to combine challenges and/or GPS as inputs to one or more PPUFs. The characteristics of PPUF are such that any attempt to separate or replace them destroys their characteristics and therefore security properties.

21.5.9 Emerging Directions

Initial efforts on coverage in sensor networks have formulated and solved several canonical problems. There are exponentially many new formulations that consider more and more issues or accept more complex and detailed sensing models as well as object movement. While many of them are interesting and technically challenging, there is still an ongoing search for killer applications of large and profound practical importance. Also, several basic problems such as static coverage with respect to static objects are still not completely answered.

There are too many new applications for any survey or even book to cover. Due to space limitations, we just very briefly go through two new applications: mobile wireless health [46] and energy harvesting [66,93–95]. In addition, we briefly discuss the related and intriguing emerging topic of local sensing using global sensors [79,80].

We illustrate issues in coverage problems using a very small crosscut of wireless health research, specifically, medical shoes. Medical shoes are instrumented with a large number of sensors that record pressure below each small area of a soul and several other types of sensors (e.g., accelerators) [74,75]. These remarkably simple systems are capable of facilitating remarkable broad sets of diagnoses and of supporting a wide spectrum of medical treatments. However, these systems are rather expensive and have high energy budgets. It has been recently demonstrated that both can be reduced by more than an order of magnitude by using the notion of semantic coverage. Semantic coverage does not detect all events, but only ones that are relevant for medical purposes [93–95]. Therefore, in a sense it provides a natural bridge between coverage and general sensor fusion that is driven by applications.

We use the term "global sensors" for large sensors that simultaneously sense multiple locations. Probably the best illustration is one where a single sensor is used to sense pressure from any of *k* keyboards. At first this approach to coverage of events (one where any single key of a keyboard senses pressure) may sound counterintuitive. However, it results in great energy sensing. For example, if we just want to detect if any key is activated when we have standard one key–one sensor scheme, we need as many sensor readings as there are keys. However, if each sensor covers *k* keys, this requirement is reduced by a factor of *k* times. Judicious placement of such global sensors can ensure complete coverage of keys while reducing energy requirements by more than an order of magnitude [79,80]. Although the first algorithms have been proposed and they are very effective, we still know rather little of advantages and limitations of the use of global sensors for local sensing.

21.6 Conclusion

We have surveyed the history, state of the art, and trends of coverage in sensor networks. Since comprehensive and complete coverage is out of the question due to the tremendous amount of research, we placed emphasis on the most important conceptual and practical issues. Even then, only a small slice of research results are covered. Nevertheless, we hope that this chapter will help practitioners and facilitate starting research in obtaining a better global picture of coverage in sensor networks.

References

1. Z. Abrams, A. Goel, and S. Plotkin. Set k-cover algorithms for energy efficient monitoring in wireless sensor networks. In *Proceedings of the 3rd International Symposium on Information Processing in Sensor Networks, IPSN*'04, pp. 424–432, New York, 2004. ACM.
2. J. Adriaens, S. Megerian, and M. Potkonjak. Optimal worst-case coverage of directional field-of-view sensor networks. In *2006 3rd Annual IEEE Communications Society on Sensor and Ad Hoc Communications and Networks, 2006. SECON '06*, vol. 1, pp. 336–345, Reston, VA, Sept. 2006.
3. J. Ai and A.A. Abouzeid. Abouzeid, coverage by directional sensors in randomly deployed wireless sensors networks. *Journal of Combinatorial Optimization*, 11:21–41, 2006.
4. I.F. Akyildiz, T. Melodia, and K.R. Chowdhury. A survey on wireless multimedia sensor networks. *Computer Networks*, 51:921–960, Mar. 2007.
5. I.F. Akyildiz, W. Su, Y. Sankarasubramaniam, and E. Cayirci. A survey on sensor networks. *Communications Magazine, IEEE*, 40(8):102–114, Aug. 2002.
6. G. Anastasi, M. Conti, M.D. Francesco, and A. Passarella. Energy conservation in wireless sensor networks: A survey. *Ad Hoc Networks*, 7:537–568, May 2009.
7. T. Arampatzis, J. Lygeros, and S. Manesis. A survey of applications of wireless sensors and wireless sensor networks. In *Proceedings of the 2005 IEEE International Symposium on Intelligent Control, 2005. Mediterranean Conference on Control and Automation*, pp. 719–724, Washington, DC, June 2005.
8. S. Arora. Polynomial time approximation schemes for euclidean tsp and other geometric problems. In *FOCS*, pp. 2–11, Los Alamitos, CA, IEEE Computer Society, 1996.
9. N.A.A. Aziz, K.A. Aziz, and W.Z.W. Ismail. Coverage strategies for wireless sensor networks. *World Academy of Science, Engineering and Technology*, 50:145–150, 2009.

10. P. Balister, Z. Zheng, S. Kumar, and P. Sinha. Trap coverage: Allowing coverage holes of bounded diameter in wireless sensor networks. In *INFOCOM 2009*, Rio de Janeiro, Brazil, pp. 136–144, *IEEE*. April 2009.
11. D. Ban, W. Yang, J. Jiang, J. Wen, and W. Dou. Energy-efficient algorithms for *k*-barrier coverage in mobile sensor networks. *International Journal of Computers, Communications and Control*, V(5):616–624, December 2010.
12. N. Beckmann and M. Potkonjak. Hardware-based public-key cryptography with public physically unclonable functions. In *Information Hiding*, Springer, Heidelberg, Germany, pp. 206–220, 2009.
13. Y. Bejerano. Simple and efficient k-coverage verification without location information. In *INFOCOM 2008. The 27th Conference on Computer Communications* pp. 291–295, *IEEE.*, Phoenix, AZ, April 2008.
14. P. Bender and Y. Pei. Development of energy efficient image/video sensor networks. *Wireless Personal Communications*, 51:283–301, 2009. 10.1007/s11277-008-9643-6.
15. B. Bhattacharya, B. Burmester, Y. Hu, E. Kranakis, Q. Shi, and A. Wiese. Optimal movement of mobile sensors for barrier coverage of a planar region. In *Proceedings of the 2nd International Conference on Combinatorial Optimization and Applications*, COCOA 2008, pp. 103–115, Berlin, Heidelberg, Germany, Springer-Verlag, 2008.
16. M. Cardei and J. Wu. Coverage in wireless sensor networks. In M. Ilyas and I. Mahgoub, eds., *Handbook of Sensor Networks*. CRC Press, Boca Raton, FL, 2004.
17. A. Cerpa, J.L. Wong, L. Kuang, M. Potkonjak, and D. Estrin. Statistical model of lossy links in wireless sensor networks. In *Information Processing in Sensor Networks, 2005. IPSN 2005. Fourth International Symposium on*, Los Angeles, CA, pp. 81–88, April 2005.
18. A. Cerpa, J.L. Wong, M. Potkonjak, and D. Estrin. Temporal properties of low power wireless links: Modeling and implications on multi-hop routing. In *Proceedings of the 6th ACM International Symposium on Mobile ad hoc Networking and Computing, MobiHoc '05*, pp. 414–425, New York, 2005. ACM.
19. Y. Charfi, N. Wakamiya, and M. Murata. Challenging issues in visual sensor networks. *Wireless Communications*, 16:44–49, April 2009.
20. A. Chen, S. Kumar, and T.H. Lai. Local barrier coverage in wireless sensor networks. *IEEE Transactions on Mobile Computing*, 9(4):491–504, April 2010.
21. A. Chen, T.H. Lai, and D. Xuan. Measuring and guaranteeing quality of barrier coverage for general belts with wireless sensors. *ACM Transactions on Sensor Networks*, 6:2:1–2:31, January 2010.
22. X. Chen, K. Makki, N. Yen, and K. Pissinou. Sensor network security: A survey. *IEEE Communications Surveys and Tutorials*, 11:52–62, 2009.
23. Y. Chen, K. Kleisouris, X. Li, W. Trappe, and R.P. Martin. The robustness of localization algorithms to signal strength attacks: A comparative study. In *International Conference on Distributed Computing in Sensor Systems (DCOSS)*, San Francisco, CA, June 2006.
24. B. Cărbunar, A. Grama, J. Vitek, and O. Cărbunar. Redundancy and coverage detection in sensor networks. *ACM Transactions on Sensor Networks*, 2:94–128, February 2006.
25. V. de Silva and R. Ghrist. Coverage in sensor networks via persistent homology. *Algebraic and Geometric Topology*, 7:339–358, 2007.
26. J. Deng, Y.S. Han, W.B. Heinzelman, and P.K. Varshney. Scheduling sleeping nodes in high density cluster-based sensor networks. *Mobile Networks and Applications*, 10:825–835, December 2005.
27. I. Dietrich and F. Dressler. On the lifetime of wireless sensor networks. *ACM Transactions on Sensor Networks*, 5:5:1–5:39, February 2009.
28. D. Djenouri, L. Khelladi, and N. Badache. A survey of security issues in mobile ad hoc and sensor networks. *IEEE Communications Surveys and Tutorials*, 7:2–28, 2005.
29. H.N. Djidjev. Approximation algorithms for computing minimum exposure paths in a sensor field. *ACM Transactions of Sensor Networks*, 7:23:1–23:25, October 2010.

30. C.F. Hsin and M. Liu. Network coverage using low duty-cycled sensors: Random coordinated sleep algorithms. In *Third International Symposium on Information Processing in Sensor Networks, 2004. IPSN 2004*, Berkeley, CA, pp. 433–442, April 2004.
31. J. Feng, L. Girod, and M. Potkonjak. Location discovery using data-driven statistical error modeling. In *IEEE Infocom*, Barcelona, Spain, pp. 1–10, 2006.
32. J. Feng, F. Koushanfar, and M. Potkonjak. *Localized Algorithms for Sensor Networks*, pp. 1–17. CRC Press, Boca Raton, FL, 2004.
33. J. Feng, F. Koushanfar, and M. Potkonjak. *Sensor Network Architecture*, pp. 1–26. CRC Press, Boca Raton, FL, 2004.
34. J. Feng and M. Poktonjak. Transitive statistical sensor error characterization and calibration. In *IEEE Sensors*, Irvine, CA, pp. 572–575, 2005.
35. A. Ghosh and S.K. Das. Coverage and connectivity issues in wireless sensor networks. In *Mobile, Wireless, and Sensor Networks: Technology, Applications, and Future Directions*. JohnWiley & Sons, New York, 2006.
36. R. Ghrist and A. Muhammad. Coverage and hole-detection in sensor networks via homology. In IPSN 2005. *Fourth International Symposium on Information Processing in Sensor Networks, 2005*, Los Angeles, CA, pp. 254–260, April 2005.
37. Y. Gu, T. Zhu, and T. He. Esc: Energy synchronized communication in sustainable sensor networks. In *ICNP*, Princeton, NJ, 2009.
38. S. Halawani and A.W. Khan. Sensors lifetime enhancement techniques in wireless sensor networks-a survey. *Journal of Computing*, 2:34–47, 2010.
39. X. Han, X. Cao, E.L. Lloyd, and C.C. Shen. Deploying directional sensor networks with guaranteed connectivity and coverage. In *5th Annual IEEE Communications Society Conference on Sensor, Mesh and Ad Hoc Communications and Networks, 2008 SECON '08*, San Francisco, CA, pp. 153–160, June 2008.
40. J.E. Hopcroft and R.M. Karp. An $n^{5/2}$ algorithm for maximum matchings in bipartite graphs. *SIAM Journal of Computing*, 2(4):225–231, 1973.
41. C.-F. Huang and Y.-C. Tseng. A survey of solutions to the coverage problems in wireless sensor networks. *Journal of Internet Technology*, 6(1):1–8, 2005.
42. C. Istin and D. Pescaru. Deployments metrics for video-based wireless sensor networks. *Transactions on Automatic Control and Computer Science*, 52(66)(4):163–168, 2007.
43. J. Jeong, Y. Gu, T. He, and D. Du. Visa: Virtual scanning algorithm for dynamic protection of road networks. In *INFOCOM 2009, IEEE*, Rio de Janeiro, Brazil, pp. 927–935, April 2009.
44. A. Kansal, A. Ramamoorthy, M.B. Srivastava, and G.J. Pottie. On sensor network lifetime and data distortion. In *Proceedings. International Symposium on Information Theory, 2005 ISIT 2005*, Adelaide, Australia, pp. 6–10, September 2005.
45. J. Kepler. *The Harmony of the World*. Translated by Dr. Juliet Field, The American Philosophical Society, 1997.
46. J. Ko, C. Lu, M.B. Srivastava, J.A. Stankovic, A. Terzis, and M. Welsh. Wireless sensor networks for healthcare. *Proceedings of the IEEE*, 98(11):1947–1960, November 2010.
47. F. Koushanfar, A. Davare, M. Potkonjak, and A. Sangiovanni-Vincentelli. Low power coordination in wireless ad-hoc networks. In *Proceedings of the 2003 International Symposium on Low Power Electronics and Design, 2003. ISLPED '03*, pp. 475–480, August 2003.
48. F. Koushanfar, N. Kiyavash, and M. Poktonjak. Interacting particle-based model for missing data in sensor networks: Foundations and applications. In *IEEE Sensors*, Daegu, Korea, pp. 888–891, 2006.
49. F. Koushanfar and M. Poktonjak. Markov chain-based models for missing and faulty data in mica2 sensor motes. In *IEEE Sensors*, Irvine, CA, pp. 576–579, 2005.
50. F. Koushanfar, M. Potkonjak, and A. Sangiovanni-Vincentelli. On-line fault detection of sensor measurements. In *Proceedings of IEEE Sensors, 2003*, Toronto, Canada, vol. 2, pp. 974–979, October 2003.

51. F. Koushanfar, A. Sangiovanni-Vincentelli, and M. Potkonjak. Error models for light sensors by non-parametric statistical analysis of raw sensor measurements. In *IEEE Sensors*, Vienna, Austria, pp. 1472–1475, 2004.
52. F. Koushanfar, S. Slijepcevic, M. Potkonjak, and A. Sangiovanni-Vincentelli. Error-tolerant multi-modal sensor fusion. In *IEEE CAS Workshop on Wireless Communications and Networking*, Pasadena, CA, 2002.
53. F. Koushanfar, A. Davare, D.T. Nguyen, A. Sangiovanni-Vincentelli, and M. Potkonjak. Techniques for maintaining connectivity in wireless ad-hoc networks under energy constraints. In *ACM Transaction on Embedded Computer System*, vol. 6, July 2007.
54. H.W. Kuhn. The Hungarian method for the assignment problem. *Naval Research Logistic Quarterly*, 2:83–97, 1955.
55. S. Kumar, T.H. Lai, and A. Arora. Barrier coverage with wireless sensors. In *Proceedings of the 11th annual international conference on Mobile computing and networking MobiCom '05*, pp. 284–298, New York, 2005. ACM Press.
56. S. Kumar, T.H. Lai, and A. Arora. Barrier coverage with wireless sensors. *Wireless Networks*, 13:817–834, December 2007.
57. J. Li, J. Chen, S. He, T. He, Y. Gu, and Y. Sun. On energy-efficient trap coverage in wireless sensor networks. In *2011 IEEE 32nd Real-Time Systems Symposium (RTSS)*, pp. 139–148, 29 2011–December 2 2011.
58. L. Li, B. Zhang, X. Shen, J. Zheng, and Z. Yao. A study on the weak barrier coverage problem in wireless sensor networks. *Computer Networks*, 55(3):711–721, 2011.
59. M. Li, W. Cheng, K. Liu, Y. He, X. Li, and X. Liao. Sweep coverage with mobile sensors. *IEEE Transactions on Mobile Computing*, 10(11):1534–1545, November 2011.
60. B. Liu, O. Dousse, J. Wang, and A. Saipulla. Strong barrier coverage of wireless sensor networks. In *Proceedings of the 9th ACM international symposium on Mobile Ad Hoc Networking and Computing, MobiHoc '08*, pp. 411–420, New York, 2008. ACM.
61. S. Mahfoudh and P. Minet. Survey of energy efficient strategies in wireless ad hoc and sensor networks. In *Seventh International Conference on Networking, 2008. ICN 2008*, Cancun, Mexico, pp. 1–7, April 2008.
62. S. Megerian, F. Koushanfar, M. Potkonjak, and M.B. Srivastava. Worst and best-case coverage in sensor networks. *IEEE Transactions on Mobile Computing*, 4(1):84–92, January–February 2005.
63. S. Megerian, F. Koushanfar, G. Qu, G. Veltri, and M. Potkonjak. Exposure in wireless sensor networks: theory and practical solutions. *Wireless Network*, 8:443–454, September 2002.
64. S. Megerian and M. Potkonjak. *Wireless Sensor Networks*. John Wiley & Sons, Inc., New York 2003.
65. S. Meguerdichian, F. Koushanfar, M. Potkonjak, and M.B. Srivastava. Coverage problems in wireless ad-hoc sensor networks. In *Proceedings of the Twentieth Annual Joint Conference of the IEEE Computer and Communications Societies INFOCOM 2001*, Anchorage, AK, vol. 3, pp. 1380–1387, 2001. *IEEE*.
66. S. Meguerdichian, H. Noshadi, F. Dabiri, and M. Potkonjak. Semantic multimodal compression for wearable sensing systems. In *Sensors, 2010*, Kona, HI, pp. 1449–1453, November 2010. *IEEE*.
67. S. Meguerdichian and M. Poktonjak. Security primitives and protocols for ultra low power sensor systems. In *IEEE Sensors*, Limerick, Ireland, pp. 1225–1228, 2011.
68. S. Meguerdichian, F. Koushanfar, G. Qu, and M. Potkonjak. Exposure in wireless ad-hoc sensor networks. In *Proceedings of the 7th Annual International Conference on Mobile Computing and Networking, MobiCom '01*, pp. 139–150, New York, 2001. ACM.
69. S. Meguerdichian and M. Potkonjak. Low power 0/1 coverage and scheduling techniques in sensor networks. Technical Report 030001, Department of Computer Science, University of California, Los Angeles, CA, 2003.

70. S. Meguerdichian, S. Slijepcevic, V. Karayan, and M. Potkonjak. Localized algorithms in wireless ad-hoc networks: Location discovery and sensor exposure. In *Proceedings of the 2nd ACM International Symposium on Mobile Ad Hoc Networking and Computing, MobiHoc '01*, pp. 106–116, New York, 2001. ACM.
71. V.P. Mhatre, C. Rosenberg, D. Kofman, R. Mazumdar, and N. Shroff. A minimum cost heterogeneous sensor network with a lifetime constraint. *IEEE Transactions on Mobile Computing*, 4(1):4–15, January–February 2005.
72. A. Muhammad and A. Jadbabaie. Decentralized computation of homology groups in networks by gossip. In *American Control Conference, 2007. ACC '07*, New York, NY, pp. 3438–3443, July 2007.
73. S. Myers, S. Megerian, S. Banerjee, and M. Potkonjak. Experimental investigation of IEEE 802.15.4 transmission power control and interference minimization. In *4th Annual IEEE Communication Society Conference on Sensor, Mesh and Ad Hoc Communications and Network, 2007. SECON '07*, San Diego, CA, pp. 294–303, June 2007.
74. H. Noshadi, F. Dabiri, S. Meguerdichian, M. Potkonjak, and M. Sarrafzadeh. Behavior oriented data resource management in medical sensing systems. In *ACM Transactions on Sensor Networks*, 2013.
75. H. Noshadi, F. Dabiri, S. Meguerdichian, M. Potkonjak, and M. Sarrafzadeh. Energy optimization in wireless medical systems using physiological behavior. In *Wireless Health 2010, WH '10*, pp. 128–136, New York, 2010. ACM.
76. A. Perrig, J. Stankovic, and D. Wagner. Security in wireless sensor networks. *Communications ACM*, 47:53–57, 2004.
77. M. Poktonjak, S. Meguerdichian, and J.L. Wong. Trusted sensors and remote sensing. In *IEEE Sensors*, Kona, HI, pp. 1104–1107, 2010.
78. M. Rahimi, S. Ahmadian, D. Zats, R. Laufer, and D. Estrin. Magic of numbers in networks of wireless image sensors. In *Workshop on Distributed Smart Cameras*, Boulder, CO, pp. 71–81, 2006.
79. M. Rofouei, M. Sarrafzadeh, and M. Potkonjak. Detecting local events using global sensing. In *IEEE Sensors*, Limerick, Ireland, pp. 1165–1168, 2011.
80. M. Rofouei, M. Sarrafzadeh, and M. Potkonjak. Efficient collaborative sensing-based soft keyboard. In *International Symposium on Low Power Electronics and Design*, Fukuoka, Japan, pp. 339–344, 2011.
81. S. Slijepcevic and M. Potkonjak. Power efficient organization of wireless sensor networks. In *IEEE International Conference on Communications, 2001. ICC 2001*, Helsinki, Finland, vol. 2, pp. 472–476, 2001.
82. S. Slijepcevic, S. Megerian, and M. Potkonjak. Location errors in wireless embedded sensor networks: Sources, models, and effects on applications. *SIGMOBILE Mobile Computing and Communication Review*, 6:67–78, June 2002.
83. S. Soro and W.B. Heinzelman. On the coverage problem in video-based wireless sensor networks. In *2nd International Conference on Broadband Networks, 2005. BroadNets 2005*, Boston, MA, pp. 932–939 vol. 2, October 2005.
84. M. Srivastava, R. Muntz, and M. Potkonjak. Smart kindergarten: Sensor-based wireless networks for smart developmental problem-solving environments. In *Proceedings of the 7th Annual International Conference on Mobile Computing and Networking, MobiCom '01*, pp. 132–138, New York, 2001. ACM.
85. S. Soro and W. Heinzelman. A survey of visual sensor networks. *Advances in Multimedia*, vol. 2009, Article ID 640386, 21 pages, 2009.
86. E.-S. Sung and M. Potkonjak. Optimized operation for infrastructure-supported wireless sensor networks. In *7th Annual IEEE Communications Society Conference on Sensor Mesh and Ad Hoc Communications and Networks (SECON), 2010*, Boston, MA, pp. 1–9, June 2010.
87. A. Tahbaz-Salehi and A. Jadbabaie. Distributed coverage verification in sensor networks without location information. In *47th IEEE Conference on Decision and Control, 2008. CDC 2008*, Cancun, Mexico, pp. 4170–4176, December 2008.

88. X. Tan and G. Wu. New algorithms for barrier coverage with mobile sensors. In *Proceedings of the 4th International Conference on Frontiers in Algorithmics, FAW 10*, pp. 327–338, Berlin, Heidelberg, Germany, 2010. Springer-Verlag.
89. Di Tian and Nicolas D. Georganas. A coverage-preserving node scheduling scheme for large wireless sensor networks. In *Proceedings of the 1st ACM International Workshop on Wireless Sensor Networks and Applications, WSNA '02*, pp. 32–41, New York, 2002. ACM.
90. D. Tian, Z. Lei, and N.D. Georganas. Configuring node status in a two-phase tightly integrated mode for wireless sensor networks. *International Journal of Ad Hoc Ubiquitous Computing*, 2:175–185, February 2007.
91. G. Veltri, Q. Huang, G. Qu, and M. Potkonjak. Minimal and maximal exposure path algorithms for wireless embedded sensor networks. In *Proceedings of the 1st International Conference on Embedded Networked Sensor Systems, SenSys '03*, pp. 40–50, New York, 2003. ACM.
92. Y.-C. Wang, C.-C. Hu, and Y.-C. Tseng. Efficient deployment algorithms for ensuring coverage and connectivity of wireless sensor networks. In *Proceedings of the First International Conference on Wireless Internet, 2005*, Budapest, Hungary, pp. 114–121, July 2005.
93. J. B. Wendt, S. Meguerdichian, H. Noshadi, and M. Potkonjak. Energy and cost reduction in localized multisensory systems through application-driven compression. In *Data Compression Conference*, Snowbird, UT, 2012.
94. J. B. Wendt, S. Meguerdichian, and M. Potkonjak. *Small is Beautiful and Smart*, pp. 1–17. Science Publishers, 2012.
95. J. B. Wendt and M. Potkonjak. Medical diagnostic-based sensor selection. In *IEEE Sensors*, Limerick, Ireland, pp. 1507–1510, 2011.
96. D. B. West. *Introduction to Graph Theory*. Prentice Hall, Englewood Cliffs, NJ, 2nd edn., 2001.
97. C.-H. Wu and Y.-C. Chung. A polygon model for wireless sensor network deployment with directional sensing areas. *Sensors*, 9(12):9998–10022, 2009.
98. G. Xing, X. Wang, Y. Zhang, C. Lu, R. Pless, and C. Gill. Integrated coverage and connectivity configuration for energy conservation in sensor networks. *ACM Transaction on Sensor Networks*, 1:36–72, August 2005.
99. J. Yick, B. Mukherjee, and D. Ghosal. Wireless sensor network survey. *Computer Networks*, 52:2292–2330, August 2008.
100. S. Zahedi, M.B. Srivastava, C. Bisdikian, and L.M. Kaplan. Quality tradeoffs in object tracking with duty-cycled sensor networks. In *IEEE 31st Real-Time Systems Symposium (RTSS), 2010*, San Diego, CA, pp. 160–169, 30 2010-December 3, 2010.
101. J. Zhao, S.-C. Cheung, and T. Nguyen. Optimal camera network configurations for visual tagging. *IEEE Journal on Selected Topics in Signal Processing*, 2(4):464–479, Aug. 2008.
102. G. Zhou, T. He, S. Krishnamurthy, and J.A. Stankovic. Impact of radio irregularity on wireless sensor networks. In *Proceedings of the 2nd International Conference on Mobile Systems, Applications, and Services, MobiSys '04*, pp. 125–138, New York, 2004. ACM.

Chapter 22

Self-Organizing Distributed State Estimators

Joris Sijs and Zoltan Papp

Contents

Distributed solutions for signal processing techniques are important for establishing large-scale monitoring and control applications. They enable the deployment of scalable sensor networks for particular application areas. Typically, such networks consists of a large number of vulnerable components connected via unreliable communication links and are sometimes deployed in harsh environment. Therefore, dependability of sensor network is a challenging problem. An efficient and cost-effective answer to this challenge is provided by employing runtime reconfiguration techniques that assure the integrity of the desired signal processing functionalities. Runtime reconfigurability has a thorough impact on system design, implementation, testing/validation, and deployment. The presented research focuses on the widespread signal processing method known as state estimation with Kalman filtering in particular. To that extent, a number of distributed state estimation solutions that are suitable for networked systems in general are overviewed, after which robustness of the system is improved according to various runtime reconfiguration techniques.

22.1 Introduction

Many people in our society manage their daily activities based on knowledge and information about, for example, weather conditions, traffic jams, pollution levels, oil reservoirs, and energy consumptions. Sensor measurements are the main source of information when monitoring these surrounding processes. Moreover, a trend is to increase the amount of sensors, as they have become smaller, cheaper, and easier to use, so that large-area processes can be monitored with a higher accuracies. To that end, sensors are embedded in a communication network creating a so-called sensor network, which typically consists of *sensor nodes* linked via a particular network topology (Figure 22.1). Each sensor node combines multiple sensors, a central processing unit (CPU), and a (wireless) communication radio on a circuit board. Sensor networks have three attractive properties for system design: they require low maintenance, create "on-the-fly" (ad hoc) communication networks, and can maintain large amounts of sensors.

Nowadays, sensor nodes are commercial off-the-shelf products and give system designers new opportunities for acquiring measurements. Although they make sensor measurements available in large quantities, solutions for processing these measurements automatically are hampered by limitations in the available resources, such as energy, communication, and computation.

Energy plays an important role in remotely located processes. Such processes are typically observed by severely energy-limited sensor nodes (e.g., powered by battery or energy scavenging) that are not easily accessible and thus should have a long lifetime. Some applications even deploy sensor nodes in the asphalt of a road to monitor traffic or in the forest to collect information on habitats. See, for example, the applications described in [1,2] and recent surveys on sensor networks in [3–6]. To limit energy consumption, one often aims to minimize the usage of communication and computational resources in sensor nodes. However, there are other reasons why these latter two resources should be used wisely.

Limited communication mainly results from upper bounds on the network capacity, as it was established in the Shannon–Hartley theorem for communication channels presented in [7].

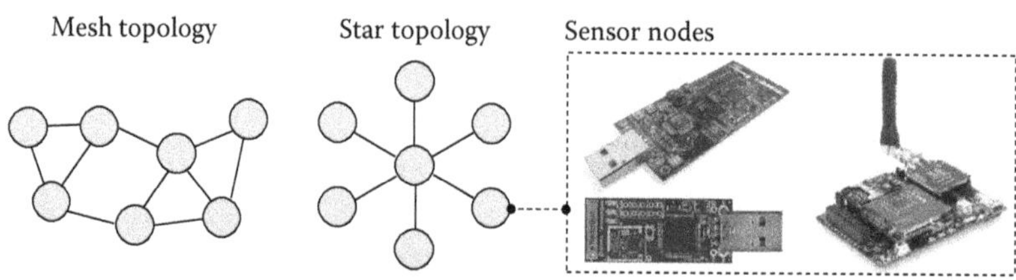

Figure 22.1 Sensor nodes in a mesh and star network topology with some examples of nodes: Tmote-Sky (top-left), G-node (bottom-left), and Waspmote (right).

It shows that the environment in which nodes communicate influences the amount of data that can be exchanged without errors. In addition, communication is affected by package loss as well, which occurs due to message collision (i.e., simultaneous use of the same communication channel by multiple transmitters). Hence, a suitable strategy for exchanging data is of importance to cope with the dynamic availability of communication resources.

Computational demand is related to the algorithms performed in sensor networks for processing the measurements. The established centralized solutions, where measurements are processed by a single node, fail for large-scale networks even when communication is not an issue: With an increasing amount of sensor nodes, the computational load of a centralized solution will grow polynomially, up to a point that it is no longer feasible or highly inefficient. To that extent, non-centralized solutions are explored that aim to make use of local CPUs that are already present in each node.

A straightforward consequence of the resource limitation, the scale, and the often-hostile embedding environment is that fault-tolerance and/or graceful degradation are critical requirements for large-scale distributed systems. This means that the sensor network should be able to cope with situations that emerge from common operational events, such as node failures, sensor degradation, and power loss. Building in redundancy to cover the anticipated failure modes may result in complex, prohibitively expensive implementations. Instead, dynamical system architectures are to be realized via runtime reconfiguration, as it realizes a networked system that can follow the changes in the internal and external operational conditions and assure optimal use of available resources.

Limitations of the earlier-mentioned resources are important design parameters. Depending on the sensor network application at hand, suitable trade-offs must be made to enable a feasible and practical deployment. One of these trade-offs is the local processing–communication trade-off. This encourages the local processing of the sensor measurements rather than communicating them, since exchanging 1 bit typically consumes much more energy than processing 1 bit. Hence, *centralized* methods for processing measurements are unpractical, due to their significant impact on the communication requirements. To solve this issue, *distributed* signal processing methods are increasingly studied. Such methods seek for a more efficient use of the spatially distributed computation and sensing resources according to the network topology. The signal processing method addressed in this chapter is state estimation.

Well-studied state estimation methods are the Kalman filter (KF) for linear processes, with extensions known as the extended KF and unscented KF for nonlinear processes, see, for example, [8–10]. Apart from their centralized solutions, some distributed implementation are found in [11–19]. Typically, these distributed solutions perform a state estimation algorithm locally in each node and thereby compute a local estimate of the global state vector. Note that these distributed solutions can thus be regarded as a network of state estimators. However, they were not designed to cope with the unforeseen operational events that will be present in the system, nor address deliberate reconfigurations of a sensor network during operation.*

Therefore, the contribution of this chapter is to integrate solutions on distributed Kalman filtering with a framework of self-organization. To that extent, each node not only employs a state estimator locally but additionally performs a management procedure that supports the network of state estimators to establish self-organization. The outline of this chapter is as follows. First, we address the used notation, followed by a problem description in Section 22.3. Section 22.4

* For example, a reduction of the sampling time of nodes that run out of battery power, so to save energy and increase their lifetime.

then presents several existing solutions on distributed Kalman filtering, with its required resources in Section 22.5, for which a supportive management procedure is designed in Section 22.6. The proposed network of self-organizing state estimators is further analyzed in Section 22.7 in an illustrative example, while concluding remarks are summarized in Section 22.8.

22.2 Notation and Preliminaries

$\mathbb{R}$, $\mathbb{R}_+$, $\mathbb{Z}$, and $\mathbb{Z}_+$ define the set of real numbers, nonnegative real numbers, integer numbers, and nonnegative integer numbers, respectively. For any $C \subset \mathbb{R}$, let $\mathbb{Z}_C := \mathbb{Z} \cap C$. The notation 0 is used to denote zero, the null-vector, or the null-matrix of appropriate dimensions. The transpose, inverse (when it exists), and determinant of a matrix $A \in \mathbb{R}^{n\times n}$ are denoted by $A^\top$, A^{-1}, and $|A|$, respectively. Further, $\{A\}_{qr} \in \mathbb{R}$ denotes the element in the qth row and rth column of A. Given that $A, B \in \mathbb{R}^{n\times n}$ are positive definite, denoted by $A \succ 0$ and $B \succ 0$, then $A \succ B$ denotes $A - B \succ 0$. $A \succeq 0$ denotes that A is positive semi-definite. For any $A \succ 0$, $A^{\frac{1}{2}}$ denotes its Cholesky decomposition and $A^{-\frac{1}{2}}$ denotes $(A^{\frac{1}{2}})^{-1}$. The Gaussian function (Gaussian in short) of vectors $x, \mu \in \mathbb{R}^n$ and matrix $\Sigma \in \mathbb{R}^{n\times n}$ is denoted by $G(x, \mu, \Sigma)$, for which $\Sigma \succ 0$ holds. Any Gaussian function $G(x, \mu, \Sigma)$ can be illustrated by its corresponding ellipsoidal sub-level-set $\mathcal{E}_{\mu,\Sigma} := \{x \in \mathbb{R}^n | (\mu - x)^\top \Sigma^{-1}(\mu - x) \le 1\}$. See, Figure 22.2 for a graphical explanation of a sub-level-set.

22.3 Problem Formulation

Let us consider a linear process that is observed by a sensor network with the following description.

Networked System The network consists of N sensor nodes, in which a node $i \in \mathcal{N}$ is identified by a unique number within $\mathcal{N} := \mathbb{Z}_{[1,N]}$. The set $\mathcal{N}_i \subseteq \mathcal{N}$ is defined as the collection of all nodes $j \in \mathcal{N}$ that have a direct network connection with node i, that is, node i exchanges data with node j.

Process Each node $i \in \mathcal{N}$ observes a perturbed, dynamical process according to its local sampling time $\tau_i \in \mathbb{R}_{>0}$. Therefore, the discrete-time process model of node i, at the k_ith sampling instant, yields

$$\begin{aligned} x[k_i] &= A_{\tau_i} x[k_i - 1] + w[k_i - 1], \\ y_i[k_i] &= C_i x[k_i] + v_i[k_i]. \end{aligned} \tag{22.1}$$

The state vector and local measurement are denoted as $x \in \mathbb{R}^n$ and $y_i \in \mathbb{R}^{m_i}$, respectively, while process-noise $w \in \mathbb{R}^n$ and measurement-noise $v_i \in \mathbb{R}^{m_i}$ follow the Gaussian distributions

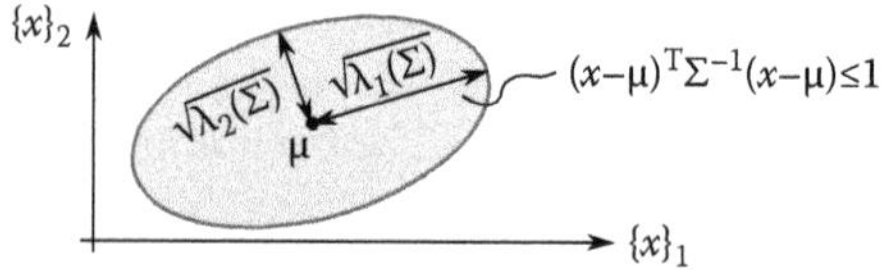

Figure 22.2 An illustrative interpretation of the sub-level-set $\mathcal{E}_{\mu,\Sigma}$.

$p(w[k_i]) := G(w[k_i], 0, Q_{\tau_i})$ and $p(v_i[k_i]) := G(v_i[k_i], 0, V_i)$, for some $Q_{\tau_i} \in \mathbb{R}^{n\times n}$ and $V_i \in \mathbb{R}^{m_i\times m_i}$. A method to compute the model parameters A_{τ_i} and Q_{τ_i} from the corresponding continuous-time process model $\dot{x} = Fx + w$ is the following:

$$A_{\tau_i} := e^{F\tau_i} \quad \text{and} \quad Q_{\tau_i} := B_{\tau_i} cov\big(w(t-\tau_i)\big) B_{\tau_i}^{\top},$$
$$\text{with} \quad B_{\tau_i} := \int_0^{\tau_i} e^{F\eta} d\eta.$$

The goal of the sensor network is to compute a local estimate of the global state x in each node i. Note that the process model is linear and both noises are Gaussian distributed. As such, it is appropriate to assume that the local estimate is Gaussian distributed as well, that is, $p_i(x[k_i]) := G(x[k_i], \hat{x}_i[k_i], P_i[k_i])$ for some *mean* $\hat{x}_i[k_i] \in \mathbb{R}^n$ and *error-covariance* $P_i[k_i] \in \mathbb{R}^{n\times n}$. This further implies that one can adopt a distributed KF solution in the sensor network for state estimation, for example, [11,13–19]. Such solutions typically compute a local estimate of x in each node i based on y_i and on the data exchanged by its neighboring nodes $j \in \mathcal{N}_i$. Existing methods on distributed Kalman filtering present an a priori solution on what data should be exchanged, at what time, and with which nodes. Hence, for a given sensor network, a matched (static) estimation procedure is derived per node under predefined conditions. Such static estimation procedures are infeasible when deploying large-scale networked systems. Broken communication links, newly added nodes to an existing network, node failures, and depleted batteries are just a few examples of operational events likely to occur in large-scale sensor networks. Solutions should thus be in place that enables the (data processing) sensor network to cope with these configuration changes by reconfiguring its own operation in runtime. These topics are often addressed by methods that establish a self-organizing network, in which a feasible solution for unforeseen system changes is sought for during the operation of a network rather than during its design time.

Therefore, this chapter investigates a self-organization sensor network with the purpose of estimating the state vector of large-area processes (Figure 22.3). More specifically, the problem addressed is to integrate state-of-the-art results in distributed Kalman filtering with applicable solutions for establishing a self-organizing networked system. The (modified) Kalman filtering algorithms performed in the different nodes interact with each other via a management layer "wrapped around" the KF. The management layer is responsible for parameterization and topology control, thus assuring coherent operational conditions for its corresponding estimator. Note that this warrants a two-way interaction between the modified KF and the management layer. Let us present the state of the art in distributed Kalman filtering, next, before addressing the solutions that establish a self-organizing networked system.

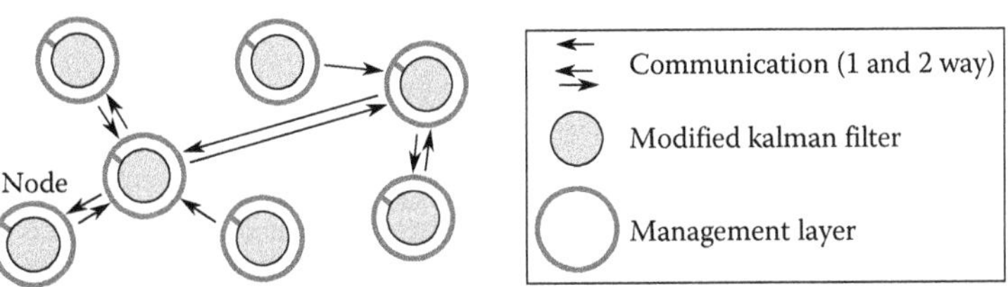

Figure 22.3 A network of Kalman filters with supporting management layer to realize the self-organizing property of the network.

22.4 Distributed Kalman Filtering

The linear process model of (22.1) is characterized by Gaussian noise distributions. A well-known state estimator for linear processes with Gaussian noise distributions is the KF, formally introduced in [9]. Since many distributed implementations of the KF make use of its original algorithm, let us define the Kalman filtering function $f_{\mathrm{KF}} : \mathbb{R}^n \times \mathbb{R}^{n\times n} \times \mathbb{R}^{n\times n} \times \mathbb{R}^{n\times n} \times \mathbb{R}^m \times \mathbb{R}^{m\times n} \times \mathbb{R}^{m\times m} \to \mathbb{R}^n \times \mathbb{R}^{n\times n}$. Different nodes will employ this function. Therefore, let us present a generalized characterization of f_{KF} independent of the node index i. To that end, let $y[\mathrm{k}] \in \mathbb{R}^m$ denote a measurement sampled at the synchronous sampling instants $\mathrm{k} \in \mathbb{Z}_+$ with a sampling time of $\tau \in \mathbb{R}_{>0}$ according to the following description:

$$y[\mathrm{k}] = Cx[\mathrm{k}] + v[\mathrm{k}], \quad p(v[\mathrm{k}]) = G(v[\mathrm{k}], 0, V). \tag{22.2}$$

Then, a characterization of the Kalman filtering function, which computes updated values of the state estimates $\hat{x}[\mathrm{k}]$ and $P[\mathrm{k}]$ based on $y[\mathrm{k}]$ in (22.2), yields

$$(\hat{x}[\mathrm{k}], P[\mathrm{k}]) = f_{\mathrm{KF}}(\hat{x}[\mathrm{k}-1], P[\mathrm{k}-1], A_\tau, Q_\tau, y[\mathrm{k}], C, V), \tag{22.3}$$

$$\begin{aligned} \text{with } M &= A_\tau P[\mathrm{k}-1] A_\tau^\top + Q_\tau; \\ K &= MC^\top (CMC^\top + V)^{-1}; \\ \hat{x}[\mathrm{k}] &= A_\tau \hat{x}[\mathrm{k}-1] + K(y[\mathrm{k}] - CA_\tau \hat{x}[\mathrm{k}-1]); \\ P[\mathrm{k}] &= (I_n - KC)M. \end{aligned} \tag{22.4}$$

The KF is a successful and well-studied state estimator. See, for example, some assessments presented in [20–22]. Its success is based on three aspects:

- Measurements are included iteratively.
- The estimation error $x - \hat{x}$ is asymptotically unbiased and attains the minimal quadratic value of the error-covariance P.
- The Kalman filtering algorithm is computationally tractable.

Therefore, when distributed solutions for state estimation became apparent, the Kalman filtering strategy was often the starting point for any novel distributed state estimator. Moreover, many of the ideas explored in distributed Kalman filtering are easily extendable toward distributed state estimation in general. A summary of these ideas is given in the next sections, as it facilitates in the decision on how to compute a node's local estimate $p_i(x)$.

The overview on distributed Kalman filtering distinguishes two different approaches. In the first approach, nodes exchange their local measurement, while in the second approach nodes share their local estimate (possibly additional to exchanging local measurements). This second approach was proposed in recent solutions on distributed Kalman filtering, as it further improves the estimation results in the network. For clarity of exposition, solutions are initially presented with synchronized sampling instants $\mathrm{k} \in \mathbb{Z}_+$, that is, each node i has the same sampling instant $\tau \in \mathbb{R}_+$. After that, modifications are given to accommodate asynchronous sampling instants $\mathrm{k}_i \in \mathbb{Z}_+$ and local sampling times $\tau_i \in \mathbb{R}_+$.

22.4.1 Exchange Local Measurements

22.4.1.1 Synchronized Sampling Instants

First solutions on distributed KFs proposed to share local measurements. See, for example, the methods presented in [11,23–25]. Local measurements are often assumed to be independent (uncorrelated). Therefore, they are easily merged with any existing estimate in a particular node. To reduce complexity even further, most methods do not exchange the actual measurement but rewrite y_i, C_i, and V_i into an *information form*, that is,

$$z_i[\mathrm{k}] := C_i^\top V_i^{-1} y_i[\mathrm{k}] \quad \text{and} \quad Z_i[\mathrm{k}] := C_i^\top V_i^{-1} C_i, \quad \forall i \in \mathcal{N}. \tag{22.5}$$

Established terms for $z_i[\mathrm{k}] \in \mathbb{R}^n$ and $Z_i[\mathrm{k}] \in \mathbb{R}^{n\times n}$ are the *information vector* and *information matrix*, respectively. They are used in an alternative KF algorithm with equivalent estimation results but different computational complexity, known as the *information filter*. To that extent, let us introduce the information filtering function $f_{\mathrm{IF}} : \mathbb{R}^n \times \mathbb{R}^{n\times n} \times \mathbb{R}^{n\times n} \times \mathbb{R}^{n\times n} \times \mathbb{R}^m \times \mathbb{R}^{m\times m} \to \mathbb{R}^n \times \mathbb{R}^{n\times n}$, for $z[\mathrm{k}] := C^\top V^{-1} y[\mathrm{k}]$ and $Z[\mathrm{k}] := C^\top V^{-1} C$ as the information form of the generalized measurement $y[\mathrm{k}]$ expressed in (22.2), that is,

$$(\hat{x}[\mathrm{k}], P[\mathrm{k}]) = f_{\mathrm{IF}}(\hat{x}[\mathrm{k}-1], P[\mathrm{k}-1], A_\tau, Q_\tau, z[\mathrm{k}], Z[\mathrm{k}]), \tag{22.6}$$

$$\begin{aligned} &\text{with } M = A_\tau P[\mathrm{k}-1] A_\tau^\top + Q_\tau; \\ &P[\mathrm{k}] = (M^{-1} + Z[\mathrm{k}])^{-1}; \\ &\hat{x}[\mathrm{k}] = P[\mathrm{k}](M^{-1} A_\tau \hat{x}[\mathrm{k}-1] + z[\mathrm{k}]). \end{aligned} \tag{22.7}$$

Notice that a node i can choose between f_{KF} and f_{IF} for computing a local estimate of x. This choice depends on the format in which nodes share their local measurement information, that is, the normal form (y_i, C_i, V_i) or the information form (z_i, Z_i), as well as the computational requirements of f_{KF} and f_{IF}. In addition, note that when the original KF is employed by a node i, that is, $(\hat{x}_i[\mathrm{k}], P_i[\mathrm{k}]) = f_{\mathrm{KF}}(\hat{x}_i[\mathrm{k}-1], P_i[\mathrm{k}-1], A_\tau, Q_\tau, \bar{y}_i[\mathrm{k}], \bar{C}_i, \bar{V}_i)$, then $\bar{y}_i[\mathrm{k}]$ is constructed by stacking $y_i[\mathrm{k}]$ with the received $y_j[\mathrm{k}]$ column wise,* for all $j \in \mathcal{N}_i$. However, the distributed KF proposed in [11] showed that the administration required to construct $\bar{y}_i$, $\bar{C}_i$, and $\bar{V}_i$ can be simplified into an addition when local measurements are exchanged in their information form instead. This implies that each node i performs the following function, which is also schematically depicted in Figure 22.4, that is,

$$\begin{aligned} &(\hat{x}_i[\mathrm{k}], P_i[\mathrm{k}]) = f_{\mathrm{IF}}(\hat{x}_i[\mathrm{k}-1], P_i[\mathrm{k}-1], A_\tau, Q_\tau, \bar{z}_i[\mathrm{k}], \bar{Z}_i[\mathrm{k}]), \\ &\text{with } \bar{z}_i[\mathrm{k}] = z_i[\mathrm{k}] + \sum_{j\in\mathcal{N}_i} z_j[\mathrm{k}] \quad \text{and} \quad \bar{Z}_i[\mathrm{k}] = Z_i[\mathrm{k}] + \sum_{j\in\mathcal{N}_i} Z_j[\mathrm{k}]. \end{aligned} \tag{22.8}$$

This simple, yet effective, distributed KF triggered many novel extensions. For example, to reduce communication requirements by quantization of the measurement values, as presented in [17], or to estimate only a part of global state vector x in a node i, for example, [26,27]. However, a drawback when exchanging measurements is that node i receives localized data from

---

* Parameters $\bar{C}_i$ and $\bar{V}_i$ can be constructed similar to $\bar{y}_i$.

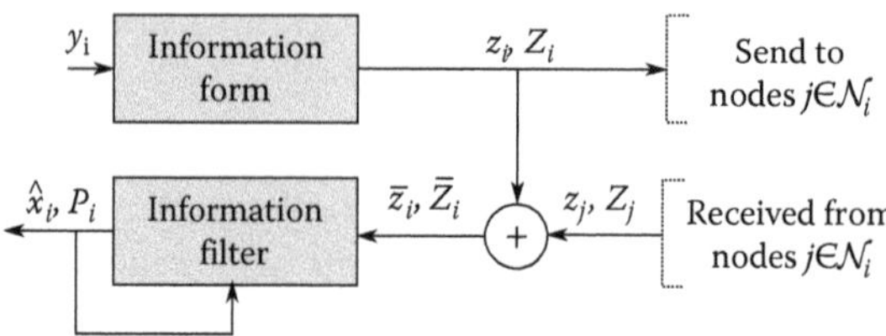

Figure 22.4 Schematic setup of a node's local algorithm for estimating the state x according to a distributed KF where local measurements are exchanged in their information form.

the neighboring nodes $j \in \mathcal{N}_i$. Hence, only a part of the measurements produced by the sensor network is used for computing $\hat{x}_i$ and P_i. A solution to exploit more measurement information, as proposed in [14,15], is to attain a consensus on local measurements. This means that, before f_{IF} is performed, each node i first employs a distributed consensus algorithm on $z_i[\mathrm{k}]$ and $Z_i[\mathrm{k}]$, for all $i \in \mathcal{N}$. Some popular consensus algorithms are found in [28–31]. However, they require that neighboring nodes exchange data multiple times in between two sampling instants. Due to this demanding requirement, distributed KFs with a consensus on local measurements are not very popular. Other extensions of the distributed KF presented in (22.8) take into account that the sampling instants of individual nodes can differ throughout the network. As this is also the case for the considered network, let us discuss the extension for asynchronous measurements next.

22.4.1.2 Asynchronous Sampling Instants

The assumed sensor network of Section 22.3 has different sampling instants per node. This means that the k_ith sample of node i, which corresponds to its local sampling instant $t_{\mathrm{k}_i} \in \mathbb{R}_+$, will probably not be equal to the time $t \in \mathbb{R}_+$ at which a neighboring node $j \in \mathcal{N}_i$ sends $\left(z_j(t), Z_j(t)\right)$. To address this issue, let us assume that node i received $\left(z_j(t), Z_j(t)\right)$ at time instant $t \in \mathbb{R}_{(t_{\mathrm{k}_i-1}, t_{\mathrm{k}_i}]}$. Then, this received measurement information is first "predicted" toward the local sampling instant t_{k_i}, so that it can be used when node i runs its local estimation function f_{IF}. The results of [32] characterize such a prediction, for all $j \in \mathcal{N}_i$ and $t \in \mathbb{R}_{(t_{\mathrm{k}_i-1}, t_{\mathrm{k}_i}]}$, as follows:

$$\begin{aligned} z_j[\mathrm{k}_i|t] :=& \left(A_{t_{\mathrm{k}_i}-t}^{-\top}\right) z_j(t) + \left(\Phi^\top \Sigma^{-1} A_{t_{\mathrm{k}_i}-t}^{-\top}\right)\hat{x}_i(t) \\ & - \Phi\left(\Phi_{P_i} + Q_{t_{\mathrm{k}_i}-t}^{-1} + \Phi_{Z_j}\right)^{-1} A_{t_{\mathrm{k}_i}-t}^{-\top}\left(\hat{x}_i(t) - z_j(t)\right), \\ Z_j[\mathrm{k}_i|t] :=& \Phi_{Z_j} + \Phi_{P_i}\left(\Phi_{P_i}^\top + Q_{t_{\mathrm{k}_i}-t}^{-1}\right)\Phi_{P_i} - \Phi\left(\Phi + Q_{t_{\mathrm{k}_i}-t}^{-1}\right)^{-1}\Phi^\top, \end{aligned} \tag{22.9}$$

in which $\Phi_{P_i} := A_{t_{\mathrm{k}_i}-t}^{-\top} P_i^{-1}(t) A_{t_{\mathrm{k}_i}-t}^{-1}$, $\Phi_{Z_j} := A_{t_{\mathrm{k}_i}-t}^{-\top} Z_j^{-1}(t) A_{t_{\mathrm{k}_i}-t}^{-1}$ and $\Phi := \Phi_{P_i} + \Phi_{Z_j}$. Further, note that a node $j \in \mathcal{N}_i$ may have send multiple data packages in between t_{k_i-1} and t_{k_i} with local measurement information, for example, when node j has a smaller sampling time than node i.

The (predicted) measurement of (22.9) in information form can directly be used by an information filter. This means that the values of $\hat{x}_i[\mathrm{k}_i]$ and $P_i[\mathrm{k}_i]$ are updated at the local sampling instant t_{k_i} of node i according to an algorithm that is similar to the one presented in (22.8), that is,

$$
\begin{aligned}
(\hat{x}_i[\mathrm{k}_i], P_i[\mathrm{k}_i]) &= f_{IF}(\hat{x}_i[\mathrm{k}_i-1], P_i[\mathrm{k}_i-1], A_{\tau_i}, Q_{\tau_i}, \bar{z}[\mathrm{k}_i], \bar{Z}[\mathrm{k}_i]),\\
\text{with } \bar{z}[\mathrm{k}_i] &= z_i[\mathrm{k}_i] + \sum_{j\in\mathcal{N}_i} z_j[\mathrm{k}_i|t], \quad \forall t \in \mathbb{R}_{(t_{\mathrm{k}_i-1}, t_{\mathrm{k}_i}]},\\
\bar{Z}[\mathrm{k}_i] &= Z_i[\mathrm{k}_i] + \sum_{j\in\mathcal{N}_i} Z_j[\mathrm{k}_i|t], \quad \forall t \in \mathbb{R}_{(t_{\mathrm{k}_i-1}, t_{\mathrm{k}_i}]}.
\end{aligned}
\tag{22.10}
$$

Note that the earlier-mentioned information filter assumes that local measurement are exchanged in the information form. A solution when nodes exchange local measurements in their normal form, that is, (y_i, C_i, V_i), is to employ the Kalman filtering function f_{KF} for each time instant $t \in \mathbb{R}_{(t_{\mathrm{k}_i-1}, t_{\mathrm{k}_i}]}$ at which a new measurement is received. Such a procedure could reduce the computational demands of a node, since the prediction formulas of (22.9) are complex. Nonetheless, incorporation of local measurements $y_j(t)$ that are not sampled at the predefined sampling instants t_{k_i} requires much attention from the management layer of the individual node i. A more natural solution to this problem is obtained in distributed KFs that exchange local estimates instead of local measurements, which are presented next.

22.4.2 Exchange Local Estimates

22.4.2.1 Synchronous Sampling Instants

The main advantage of exchanging local estimates is that measurement information spreads through the entire network, even under the condition that nodes exchange data only once per sampling instant. However, since local estimation results are exchanged, note that nodes require a method that can merge multiple estimates of the same state x into a single estimate. Various solutions of such methods are found in literature. However, before addressing these methods, let us start by presenting the generalized estimation algorithm performed by each node i that corresponds to this type of distributed KF solutions.

Typically, solutions of distributed KF that exchange local estimates first merge the local measurement $y_i[\mathrm{k}]$ with the previous local estimate $p_i(x[\mathrm{k}-1])$ via a KF and thereby, compute the updated estimate $p_i(x[\mathrm{k}])$. This updated local estimate is then shared with neighboring nodes, due to which node i will receive the local estimate of nodes $j \in \mathcal{N}_i$. It will be shown that not every solution requires to share both the locally estimated mean as well as its corresponding error-covariance. Therefore, let us introduce set of received means at node i as $\mathcal{X}_i \subset \mathbb{R}^n$ and a corresponding set of received error-covariances as $\mathcal{P}_i \subset \mathbb{R}^{n\times n}$, that is,

$$
\mathcal{X}_i[\mathrm{k}] := \left\{\hat{x}_j[\mathrm{k}] \in \mathbb{R}^n | j \in \mathcal{N}_i\right\}, \tag{22.11}
$$

$$
\mathcal{P}_i[\mathrm{k}] := \left\{P_j[\mathrm{k}] \in \mathbb{R}^{n\times n} | j \in \mathcal{N}_i\right\}. \tag{22.12}
$$

The earlier-mentioned information of the local estimation results at neighboring nodes, together with the node's own local estimate, that is, $\hat{x}_i[\mathrm{k}]$ and $P_i[\mathrm{k}]$, will be used as input to a merging function. More precisely, let us introduce this merging function $\Omega : \mathbb{R}^n \times \mathbb{R}^{n\times n} \times \mathbb{R}^n \times \mathbb{R}^{n\times n} \to \mathbb{R}^n \times \mathbb{R}^{n\times n}$, which results in the merged Gaussian estimate $p_{i^+}(x[\mathrm{k}]) := G(x[\mathrm{k}], \hat{x}_{i^+}[\mathrm{k}], P_{i^+}[\mathrm{k}])$, as follows:

$$
(\hat{x}_{i^+}[\mathrm{k}], P_{i^+}[\mathrm{k}]) = \Omega(\hat{x}_i[\mathrm{k}], P_i[\mathrm{k}], \mathcal{X}_i[\mathrm{k}], \mathcal{P}_i[\mathrm{k}]). \tag{22.13}
$$

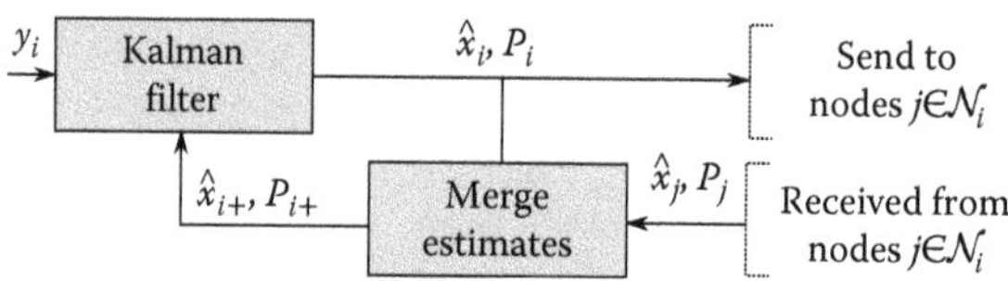

Figure 22.5 Schematic setup of a node's local algorithm for estimating the state x according to a distributed KF where local estimates are exchanged.

Then, the generalized local algorithm performed by a node $i \in \mathcal{N}$ for estimating the state, which is also depicted in the schematic setup of Figure 22.5, yields

$$\begin{aligned}(\hat{x}_i[\mathrm{k}], P_i[\mathrm{k}]) &= f_{\mathrm{KF}}(\hat{x}_{i^+}[\mathrm{k}-1], P_{i^+}[\mathrm{k}-1], A_{\tau}, Q_{\tau}, y_i[\mathrm{k}], C_i, V_i);\\ &\text{share } (\hat{x}_i[\mathrm{k}], P_i[\mathrm{k}]) \text{ with all } j \in \mathcal{N}_i;\\ &\text{collect } (\hat{x}_j[\mathrm{k}], P_j[\mathrm{k}]) \text{ for all } j \in \mathcal{N}_i;\\ (\hat{x}_{i^+}[\mathrm{k}], P_{i^+}[\mathrm{k}]) &= \Omega(\hat{x}_i[\mathrm{k}], P_i[\mathrm{k}], \mathcal{X}_i[\mathrm{k}], \mathcal{P}_i[\mathrm{k}]).\end{aligned} \tag{22.14}$$

Note that a suitable strategy for the merging function $\Omega(\cdot,\cdot,\cdot,\cdot)$ is yet to be determined. Literature indicates that one can choose between three types of strategies—consensus, fusion, and a combination of the two. A detailed account on these three strategies is presented next, by starting with consensus.

Consensus strategies aim to reduce conflicting results of the locally estimated means $\hat{x}_i$, for all $i \in \mathcal{N}$. Such an objective makes sense, as $\hat{x}_i$ in the different nodes i of the network is a local representative of the same global state x. Many distributed algorithms for attaining a consensus (or the average) were proposed, which all aim to diminish the difference $\hat{x}_i[\mathrm{k}] - \hat{x}_j[\mathrm{k}]$, for any two $i, j \in \mathcal{N}$. See, for example, the distributed consensus methods proposed in [28–31]. The general idea is to perform a weighted averaging cycle in each node i on the local and neighboring means. To that extent, let $W_{ij} \in \mathbb{R}^{n\times n}$, for all $j \in \mathcal{N}_i$, denote some weighting matrices. Then, a *consensus* merging function $\Omega(\cdot,\cdot,\cdot,\cdot)$ is typically characterized as follows:

$$\begin{aligned}(\hat{x}_{i^+}[\mathrm{k}], P_{i^+}[\mathrm{k}]) &= \Omega(\hat{x}_i[\mathrm{k}], P_i[\mathrm{k}], \mathcal{X}_i[\mathrm{k}], \mathcal{P}_i[\mathrm{k}]),\\ \text{with } \hat{x}_{i^+}[\mathrm{k}] &= \Big(I_n - \sum_{\hat{x}_j[\mathrm{k}]\in\mathcal{X}_i[\mathrm{k}]} W_{ij}\Big)\hat{x}_i[\mathrm{k}] + \sum_{\hat{x}_j[\mathrm{k}]\in\mathcal{X}_i[\mathrm{k}]} W_{ij}\hat{x}_j[\mathrm{k}],\\ P_{i^+}[\mathrm{k}] &= P_i[\mathrm{k}].\end{aligned} \tag{22.15}$$

Note that the previously mentioned consensus merging function is limited to the means and that the error-covariance of a node is not updated, due to which $\mathcal{P}_i[\mathrm{k}]$ can be the empty set. Further, most research on consensus methods concentrates on finding suitable values for the weights W_{ij}, for all $j \in \mathcal{N}_i$. Some typical examples of *scalar* weights were proposed in [28,31], where $d_i := \sharp\mathcal{N}_i$ (number of elements within the set $\mathcal{N}_i$) and $\epsilon < \min\{d_1, \ldots, d_N\}$, that is,

Nearest neighboring weights $\quad W_{ij} := (1 - d_i)^{-1}, \qquad \forall j \in \mathcal{N}_i;$

Maximum degree weights $\quad W_{ij} := (1 - \epsilon)^{-1}, \qquad \forall j \in \mathcal{N}_i;$

Metropolis weights $\quad W_{ij} := \big(1 + \max\{d_i, d_j\}\big)^{-1}, \quad \forall j \in \mathcal{N}_i.$

An analysis on the effects of these weights, when they are employed by the consensus function in (22.15), was presented in [28,31]. Therein, it was shown that employing *nearest neighboring weights* in (22.15) results in a bias on $\lim_{k\to\infty} \hat{x}_{i^+}[k]$. This is prevented by employing *maximum degree weights* or *metropolis weights*. However, *maximum degree weights* require global information to establish ϵ in every node, which reduces its applicability in sensor networks.

Employing a consensus strategies for merging the local estimates of neighboring nodes is very popular in distributed KFs. As a result, many extensions of the preceding solution are found in literature. A common extension is to perform the averaging cycle not only on the means $\hat{x}_i[k]$ and $\hat{x}_j[k]$, as characterized in (22.15), but also on the error-covariances $P_i[k]$ and $P_j[k]$ of neighboring nodes. See, for example, the distributed KF proposed in [33] and a related solution presented in [34]. It is worth to point out that an in-depth study on distributed KFs with a consensus on local estimates is presented in [35]. Therein, it is shown that minimization of the estimation error by jointly optimizing the Kalman gain K of f_{KF} and the weights W_{ij} of Ω is a non-convex problem. Hence, choosing the value of the Kalman gain K affects the weights W_{ij}, for all $j \in \mathcal{N}_i$, which raised new challenges. A solution for joint optimization on K and W_{ij} was introduced in [36] as the *distributed consensus information filter*. However, a drawback of any consensus method is that the local error-covariance $P_i[k]$ is not taken into account when deriving the weights W_{ij}, for all $j \in \mathcal{N}_i$. The error-covariance is an important variable that represents a model for the estimation error $cov(x[k] - x_i[k])$. Therefore, merging two local estimates $p_i(x[k])$ and $p_j(x[k])$ in line with their individual error-covariance implies that one can choose the value of W_{ij} such that the result after merging, that is, $p_{i^+}(x[k])$, is mainly based on the local estimate with the least estimation error. This idea is in fact the fundamental difference between a consensus approach and a fusion strategy. In fusion, both error-covariances $P_i[k]$ and $P_j[k]$ are explicitly taken into account when merging $p_i(x[k])$ and $p_j(x[k])$, as it is indicated in the next alternative merging function based on fusion.

Fusion-consensus strategies is a label for characterizing some initial fusion solutions that are based on the fusion strategy *covariance intersection*, which was introduced in [37]. Fusion strategies typically define an algorithm to merge two prior estimates $p_i(x[k])$ and $p_j(x[k])$ into a single, "fused" estimate. Some fundamental fusion methods presented in [25,38] require that correlation of the two prior estimates is available. In (self-organizing) sensor networks, one cannot impose such a requirement, as it amounts to keeping track of shared data between all nodes in the network. Therefore, this overview considers fusion methods that can cope with unknown correlations. A popular fusion method for unknown correlations is *covariance intersection*. The reason that this method is referred to as a fusion-consensus strategy is because the fusion formula of *covariance intersection* is similar to the averaging cycle of (22.15) in consensus approaches. The method characterizes the fused estimate as a convex combination of the two prior ones. As an example, let us assume that node i has only one neighboring node j. Then employment of *covariance intersection* to characterize $\Omega(\cdot,\cdot,\cdot,\cdot)$ of (22.14) as a fusion function, for some $W_{ij} \in \mathbb{R}_{[0,1]}$, yields

$$P_{i^+}[k] = ((1 - W_{ij})P_i^{-1}[k] + W_{ij}P_j^{-1}[k])^{-1},$$

$$\hat{x}_{i^+}[k] = P_{i^+}[k]((1 - W_{ij})P_i^{-1}[k]\hat{x}_i[k] + W_{ij}P_j^{-1}[k]\hat{x}_j^{-1}[k]).$$

Note that the preceding formulas indicate that the error-covariance $P_i[k]$ and $P_j[k]$ are explicitly taken into account when merging $\hat{x}_i[k]$ and $\hat{x}_j[k]$. Moreover, even the weight W_{ij} is typically based on these error-covariances, for example, $W_{ij} = \text{tr}(P_j[k])(\text{tr}(P_j[k]) + \text{tr}(P_{i(l-1)}))^{-1}$ with some other examples found in [39–41]. As a result, the updated estimate $p_{i^+}(x[k])$ computed by the merging function Ω will be closer to the prior estimate $p_i(x[k])$ or $p_j(x[k])$ that is "the most accurate one," that is, with a smaller error-covariance. An illustrative example of this property will be given later on. For now, let us continue with the merging function in case node i has more than one neighboring node. Fusion of multiple estimates can be conducted recursively according to the order of arrival at a node. Therefore, the merging function $\Omega(\cdot,\cdot,\cdot,\cdot)$ based on fusion method *covariance intersection* has the following characterization:

$$
\begin{aligned}
&(\hat{x}_{i^+}[k], P_{i^+}[k]) = \Omega(\hat{x}_i[k], P_i[k], \mathcal{X}_i[k], \mathcal{P}_i[k]),\\
&\text{with:} \qquad \text{for each received estimate } (\hat{x}_j[k], P_j[k]), \text{ do}\\
&\qquad \Sigma_i = \left((1 - W_{ij})P_i^{-1}[k] + W_{ij}P_j^{-1}[k]\right)^{-1};\\
&\qquad \hat{x}_i[k] = \Sigma_i\left((1 - W_{ij})P_i^{-1}[k]\hat{x}_i[k] + W_{ij}P_j^{-1}[k]\hat{x}_j^{-1}[k]\right);\\
&\qquad P_i[k] = \Sigma_i;\\
&\quad \text{end for}\\
&\quad \hat{x}_{i^+}[k] = \hat{x}_i[k], \quad P_{i^+}[k] = P_i[k].
\end{aligned}
\tag{22.16}
$$

Although *covariance intersection* takes the exchanged error-covariances into account when merging multiple estimates, it still introduces conservatism. Intuitively, one would expect that $p_{i^+}(x[k])$ is more accurate than $p_i(x[k])$ and $p_j(x[k])$, for all $j \in \mathcal{N}_i$, as prior estimates of neighboring nodes are merged. A formalization of this intuition is that $P_{i^+}[k] \preceq P_i[k]$ and $P_{i^+}[k] \preceq P_j[k]$ should hold for all $j \in \mathcal{N}_i$. One can prove that *covariance intersection* does not satisfy this property, due to which an alternative fusion method is presented next.

Fusion strategies aim to improve the accuracy after fusion, for which the basic fusion problem is the same as previously mentioned, that is, merge two prior estimates $p_i(x[k])$ and $p_j(x[k])$ into a single, "fused" estimate $p_{i^+}(x[k])$, when correlations are unknown. Some existing fusion methods are found in [42–44]. In this survey the *ellipsoidal intersection* fusion method of [42,43] is presented, since it results in algebraic expressions of the fusion formulas. In brief, *ellipsoidal intersection* derives an explicit characterization of the (unknown) correlation a priori to deriving algebraic fusion formulas that are based on the independent parts of $p_i(x[k])$ and $p_j(x[k])$. This characterization of the correlation, for any two prior estimates $p_i(x[k])$ and $p_j(x[k])$, is represented by the *mutual covariance* $\Gamma_{ij} \in \mathbb{R}^{n\times n}$ and the *mutual mean* $\gamma_{ij} \in \mathbb{R}^n$. Before algebraic expressions of these variables are given, let us first present the resulting merging function $\Omega(\cdot,\cdot,\cdot,\cdot)$ when *ellipsoidal intersection* is employed in this function for fusion:

$$
\begin{aligned}
&(\hat{x}_{i^+}[k], P_{i^+}[k]) = \Omega(\hat{x}_i[k], P_i[k], \mathcal{X}_i[k], \mathcal{P}_i[k]),\\
&\text{with:} \qquad \text{for each received estimate } (\hat{x}_j[k], P_j[k]), \text{ do}\\
&\qquad \Sigma_i = (P_i^{-1}[k] + P_j^{-1}[k] - \Gamma_{ij}^{-1})^{-1};\\
&\qquad \hat{x}_i[k] = \Sigma_i\left(P_i^{-1}[k]\hat{x}_i[k] + P_j^{-1}[k]\hat{x}_j^{-1}[k] - \Gamma_{ij}^{-1}\gamma_{ij}\right);\\
&\qquad P_i[k] = \Sigma_i;\\
&\quad \text{end for}\\
&\quad \hat{x}_{i^+}[k] = \hat{x}_i[k], \quad P_{i^+}[k] = P_i[k].
\end{aligned}
\tag{22.17}
$$

The mutual mean γ_{ij} and mutual covariance Γ_{ij} are found by a singular value decomposition, which is denoted as $(S, D, S^{-1}) = svd(\Sigma)$ for a positive definite $\Sigma \in \mathbb{R}^{n\times n}$, a diagonal $D \in \mathbb{R}^{n\times n}$ and a rotation matrix $S \in \mathbb{R}^{n\times n}$. As such, let us introduce the matrices $D_i, D_j, S_i, S_j \in \mathbb{R}^{n\times n}$ via the singular value decompositions $(S_i, D_i, S_i^{-1}) = svd(P_i[\mathrm{k}])$ and $(S_j, D_j, S_j^{-1}) = svd(D_i^{-\frac{1}{2}} S_i^{-1} P_j[\mathrm{k}] S_i D_i^{-\frac{1}{2}})$. Then, an algebraic expression of γ_{ij} and Γ_{ij}, for some $\varsigma \in \mathbb{R}_+$ while $\{A\}_{qr} \in \mathbb{R}$ denotes the element of a matrix A on the qth row and rth column, yields

$$
\begin{aligned}
D_{\Gamma_{ij}} &= \mathrm{diag}_{q\in\mathbb{Z}_{[1,n]}}\left(\max[1, \{D_j\}_{qq}]\right),\\
\Gamma_{ij} &= S_i D_i^{\frac{1}{2}} S_j D_{\Gamma_{ij}} S_j^{-1} D_i^{\frac{1}{2}} S_i^{-1},\\
\gamma_{ij} &= \left(P_i^{-1} + P_j^{-1} - 2\Gamma^{-1} + 2\varsigma I_n\right)^{-1}\\
&\quad\times \left((P_j^{-1} - \Gamma^{-1} + \varsigma I_n)\hat{x}_i + (P_i^{-1} - \Gamma^{-1} + \varsigma I_n)\hat{x}_j\right).
\end{aligned}
$$

A suitable value of ς follows: $\varsigma = 0$ if $|1 - \{D_j\}_{qq}| > 10\epsilon$, for all $q \in \mathbb{Z}_{[1,n]}$ and some $\epsilon \in \mathbb{R}_{>0}$, while $\varsigma = \epsilon$ otherwise. The design parameter ϵ supports a numerically stable result of *ellipsoidal intersection.*

This completes the three alternatives that can be employed by the merging function $\Omega(\cdot,\cdot,\cdot,\cdot)$. Before continuing with an extension of this merging function toward asynchronous sampling instants, let us first present an illustrative comparison of the two fundamentally different approaches. An illustration of this comparison is depicted in Figure 22.6, which is established when $p_i(x[\mathrm{k}])$ and $p_j(x[\mathrm{k}])$ are either the result of a fusion or a consensus approach. The consensus result is computed with the averaging cycle of (22.15) and $W_{ij} = 0.1$. Recall that only the means $\hat{x}_i[\mathrm{k}]$ and $\hat{x}_j[\mathrm{k}]$ are synchronized and not their error-covariances. The fusion result is computed with *ellipsoidal intersection* of (22.17). Let us further point out that Figure 22.6 is not included to decide which method is better. It is merely an example to illustrate the goal of consensus (reduce conflicting results) with respect to the goal of fusion (reduce uncertainty).

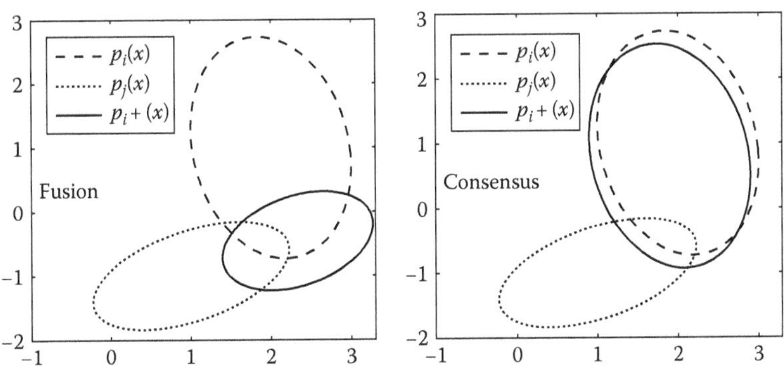

Figure 22.6 A comparison of consensus versus fusion. Note that PDFs are represented as ellipsoidal sub-level-set, that is, $G(\theta, \mu, \Sigma) \rightarrow \mathcal{E}_{\mu,\Sigma}$. A graphical characterization of such a sub-level-set is found in Figure 22.2, though let us point out that a larger covariance Σ implies a larger area size of $\mathcal{E}_{\mu,\Sigma}$.

22.4.2.2 Asynchronous Sampling Instants

The assumed networked system of Section 22.3 has different sampling instants per node. This implies that the k_ith sample of node i, which corresponds to the sampling instant $t_{k_i} \in \mathbb{R}_+$, will probably not be equal to the time $t \in \mathbb{R}_+$ at which a neighboring node $j \in \mathcal{N}_i$ sends $(\hat{x}_j(t), P_j(t))$. Compared to exchanging measurements, asynchronous sampling instants can be addressed more easily for distributed KF solutions that exchange local estimates. More precisely, the received variables $(\hat{x}_j(t), P_j(t))$ should be predicted from time t toward the sampling instant t_{k_i}, that is,

$$\begin{aligned} \hat{x}_j[k_i|t] &:= A_{t_{k_i}-t}\hat{x}_j(t), & \forall j \in \mathcal{N}_i, t \in \mathbb{R}_{(t_{k_i-1}, t_{k_i}]}, \\ P_j[k_i|t] &:= A_{t_{k_i}-t}P_j(t)A_{t_{k_i}-t}^\top + Q_{t_{k_i}-t}, & \forall j \in \mathcal{N}_i, t \in \mathbb{R}_{(t_{k_i-1}, t_{k_i}]}. \end{aligned} \tag{22.18}$$

Then, solutions of distributed Kalman filtering that are in line with the setup depicted in (22.14) can cope with asynchronous sampling instants by redefining $\mathcal{X}_i[k_i]$ and $\mathcal{P}_i[k_i]$ as the collection of the preceding predicted means $\hat{x}_j[k_i|t]$ and error-covariances $P_j[k_i|t]$, for all $j \in \mathcal{N}_i$.

This completes the overview on distributed Kalman filtering, in which nodes can adopt a strategy that exchanges local measurements or local estimates. Next, existing self-organization methods are presented, though an analysis of the required resources for estimation is studied first.

22.5 Required Resources

The distributed KFs presented in the previous section are typically proposed for static sensor networks. However, the focus of this chapter is to extent those methods for sensor networks that have to deal with changes in the networked system. To cope with these changes, nodes must be able to adapt the conditions of their local estimation algorithm, or even choose a local algorithm that is based on a different type of distributed KF. In order to carry out these reconfiguration processes, certain design decisions should be made in runtime depending on the available resources (e.g., how to reassign the KF tasks in case of node failures, what type of KF algorithms are feasible to run under given communication constraints, etc.). Therefore, this section presents a summary of the required resources for the different distributed KF strategies. Important resources in sensor networks are communication and computation. Let us start by addressing the communication demand of a node i. Section 22.4 indicates that there are three different types of data packages that a node can exchange, that is, the local measurement $y_i \in \mathbb{R}^{m_i}$ in normal form or information form, and the local estimate of $x \in \mathbb{R}^n$. The resulting communication demands of node i that correspond to these different data packages are listed in Table 22.1.

Next, let us indicate the computational demand of a node i by presenting the algorithm's complexity of the different functionalities that can be chosen to compute $p_i(x)$. This complexity involves the number of floating points operations depending on the size of local measurements $y_i \in \mathbb{R}^{m_i}$ and state vector $x \in \mathbb{R}^n$. To that extent, the following properties on the computational complexities of basic matrix computations are used:

- The summation/subtraction of $A \in \mathbb{R}^{q\times r}$ with $B \in \mathbb{R}^{q\times r}$ requires $O(qr)$ operations.
- The product of $A \in \mathbb{R}^{q\times r}$ times $B \in \mathbb{R}^{r\times p}$ requires $O(qrp)$ operations.
- The inverse of $A \in \mathbb{R}^{q\times q}$ invertible matrix requires $O(q^3)$ operations.
- The singular value decomposition of $A \in \mathbb{R}^{q\times q}$ requires $O(12q^3)$ operations.

Table 22.1 Communication Demand in the Amount of Elements (Floating Points) That Is Exchanged by Each Node Depending on the Data Shared

Exchanged Data	*Communication Demand*
(y_i, C_i, V_i)	$m_i^2 + m_i + nm_i$
(z_i, Z_i)	$n^2 + n$
$(\hat{x}_i, P_i)$	$n^2 + n$

Table 22.2 Computational Demand in the Amount of Floating Point Operations Depending on the Employed Functionality, Where $M_i := m_i + \sum_{j \in \mathcal{N}_i} m_j$

Functionality	*Computational Demand*
f_{KF}	$\approx O(4n^3 + 3M_i n^2 + 2nM_i^2 + M_i^3)$
f_{IF}	$\approx O(3n^3 + M_i n^2 + nM_i^2)$
Ω consensus of (22.15)	$\approx O(3n + M_i + 1)$
Ω fusion-consensus of (22.16)	$\approx O(3n^3 + 9n^2)$
Ω fusion of (22.17)	$\approx O(31n^3 + 7n^2)$

Then, the resulting computational complexity of the Kalman filtering functions f_{KF} and f_{IF} and of the three merging functions Ω, that is, characterized by a consensus, fusion-consensus, and fusion strategy, are listed in Table 22.2.

The next section makes use of Tables 22.1 and 22.2 to decide what type of data should be exchanged between neighboring nodes and which functionalities should be followed in the local estimation algorithm of a node.

22.6 Self-Organizing Solutions

The design challenge of any embedded system is to realize given functionalities, in this case the ones of the local estimation algorithm, on a given hardware platform while satisfying a set of nonfunctional requirements, such as response times, dependability, power efficiency, etc. Model-based design has been proven to be a successful methodology for supporting the system design process. Model-based methodologies use multiple models to capture the relevant properties of the design (when the required functionalities are mapped onto a given hardware configuration), for example, a model of the required functionalities, temporal behavior, power consumption, and hardware configuration. These models can then be used for various purposes, such as automatic code generation, architecture design, protocol optimization, system evolution, and so on. Important for the design process are the interactions between the different models, which can be expressed as constraints, dependencies, etc. In this section, a model-based design methodology is followed to assure *dependability* for state estimation in a sensor network via runtime reconfiguration.

To illustrate the model-guided design process for distributed signal processing let us consider an example. Two fundamental models for system design are emphasized here: the task model

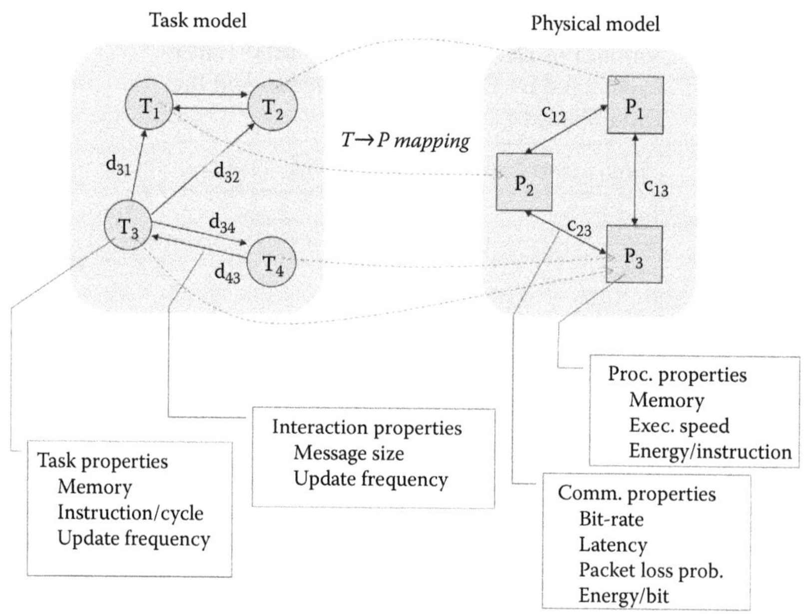

Figure 22.7 Modeling of signal processing and implementation.

(capturing the required functionalities) and the physical model (capturing the hardware configuration of the implementation). For the sake of simplicity, a particular hardware configuration and communication topology is assumed; the question to answer is how the required functionalities can be realized on the given configuration, as shown in Figure 22.7.

The task model in this Figure is represented as directed graph wherein the signal processing components (tasks) are represented by the vertices of the graph, while their data exchange (interactions) are represented by the edges. Both the tasks as well as the interactions are characterized by a set of properties, which typically reflect nonfunctional requirements or constraints. These properties are used to determine system-level characteristics, and thus the feasibility of certain design decisions can be tested (see details later). The tasks run on a connected set of processors, represented by the physical model of the system. The components of the physical model are the computing nodes, that is, consisting of processor, memory, communication and perhaps other resources, and the communication links. During the system design, the following steps are carried out (typically it is an iterative process with refinement cycles [45], but the iterations are not considered here):

- Select the algorithms for the processing realized by the tasks.
- Compose the task model.
- Select the hardware components for the physical model.
- Select a communication topology.
- Establish the mapping between the task model and the physical model.

The design process involves a particular mapping that defines the assignment of a task T_r to a processor P_q; that is, it determines which task runs on which node. * Obviously, the memory and

* We assume that nodes are equipped with a multitasking runtime environment, consequently multiple tasks can be assigned to a single node.

execution time requirements define constraints when assigning the tasks to nodes. Further, data exchange between tasks makes the assignment problem more challenging in distributed configurations, as a task assignment also defines the use of communication links, and the communication links have limited capabilities (indicated by the attached property set in Figure 22.7). After every refinement cycle, according to the steps listed earlier, the feasibility of a resulting design should be checked. For example, an assignment of T_3 to P_3 and T_4 to P_1 may yield an unfeasible design if the interaction d_{34} imposes too demanding requirements on the communication link c_{13}, that is, high data exchange rate or large data size. On the other hand, assigning both T_3 and T_4 to P_3 may violate the processing capability constraint on P_3. Changing the hardware configuration and/or using less demanding algorithms (and eventually accepting the resulting lower performance) for implementing T_3 or T_4 could be a way out.

Note that the design process results in a sequence of decisions, which lead to a feasible system design. Traditionally, the design process is "offline" (*design time*), that is, it is completed before the implementation and deployment of the system itself. The task model, the hardware configuration, and their characteristics are assumed to be known during this design time, and the design uncertainties are assumed to be low. Under these conditions, a model-based optimization can be carried out, delivering an optimal architecture ready for implementation. Unfortunately, these assumptions are overly optimistic in a wide spectrum of application cases.

(Wireless) sensor networks deployed for monitoring large-scale dynamical processes are especially vulnerable. Sensor deterioration, node failure, unreliable communication, depleted batteries, etc., are not exceptions but common events in normal operation. These events result in changes in the system configuration, as it is captured by the physical model, due to which implementations relying on static designs may fail to deliver according to the specifications. A possible work-around is to build redundancy into the system and thereby, to implement fault-tolerance. In this case, the top-level functionalities remain intact until a certain level of "damage" is reached. This approach usually leads to complex and expensive implementations—unacceptable for the majority of applications. The components are "underutilized" in nominal operation, while power consumption is increased due to the built-in redundancy. The other approach is to accept the fact that maintaining a static configuration is not feasible and make the system such that it "follows" those changes and "adjusts" its internals to assure an implementation of the assigned functionalities as far as it is feasible. The resulting behavior typically manifests "graceful degradation" property, that is, until damage reaches a certain level the set of functionalities and their quality can be kept; beyond that level the system loses noncritical functionalities and/or the quality of running functionalities is reduced due to a shortage of resources. Realizing this latter approach has significant impact both on system design and on the runtime operation of the system. Conceptually, the system design process is not completely finished in design time, instead a set of design alternatives are provided for execution. During operation—depending on the health state of the configuration and the conditions of the embedding environment—a selection is made automatically to assure an optimal use of available resources, that is, providing the highest level of the functionalities under the given circumstances. In the next section, typical solutions for implementing this latter approach are overviewed.

22.6.1 Approaches to Runtime Adaptivity

Evolution of large-scale networked embedded systems in general and (wireless) sensor networks in particular poses a number of technical challenges on the design, implementation, testing, deployment, and operation processes [46]. Considering the reconfiguration as a "vehicle" to implement such evolution, the reconfiguration of the functionalities on the available hardware can be carried out at four different stages of the system's life cycle:

1. Design time—configuration redesign, new code base, etc.
2. Load time—new functionalities are implemented via code update.
3. Initialization time—during system (or component) startup, the optimal design alternative is selected and parameterized depending on a "snapshot" of the context.
4. Runtime—reconfiguration is performed while the system is in use.

Here, only the *runtime* reconfiguration variant of the evolution is considered with special emphasis on the needs of distributed Kalman filtering.

In case of runtime reconfiguration, the reconfiguration process is triggered by observation of changes in the embedding environment of the system or in the system itself, for example, realizing node failure or a low battery status. The "trajectory" for reconfiguration is not predefined but is a result of an optimization process attempting to maximize the "usefulness" of the system as defined by a performance criterion. The concept of the reconfiguration process is illustrated in Figure 22.8.

The process relies on the model-based approach as introduced previously. The relevant models of the system, such as the task model, physical model, temporal model, etc., are formalized and stored in an efficiently accessible way in a database represented by the *models* block. The *constraints* block represents the dependencies in the models and between models. During operation of the signal processing systems, the *MONITOR* collects information about several aspects of the operation. Goals of the operation may change depending on, for example, different user needs. Changes in the observed phenomenon may cause that the models assumed in design time have become invalid. Similarly, internal changes in the system configuration should be recognized, such as broken communication and sensor failure. The *MONITOR* functionality checks if the observed changes result in violating certain constraints of the systems or a significant drop in performance. If the *MONITOR* concludes that under current circumstances the system cannot perform as requested, then the reconfiguration process is initiated. The central component is the

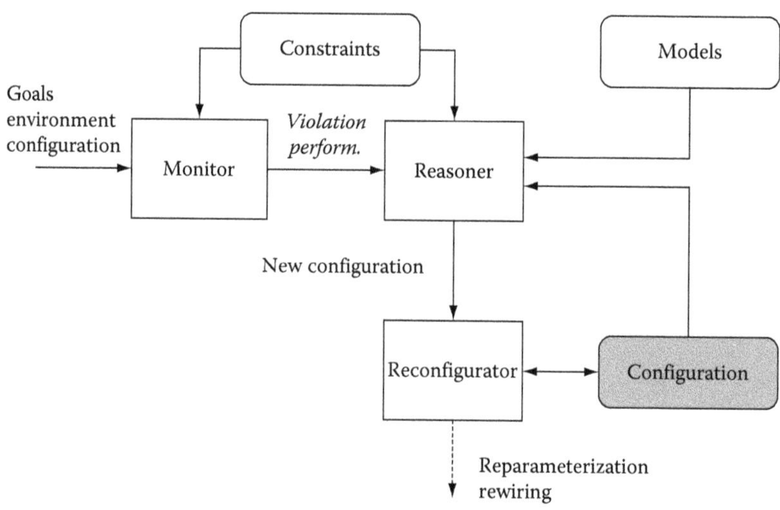

Figure 22.8 Reconfiguration process.

REASONER, which, based on the models, constraints, and the actual findings, determines a new configuration that satisfies all constraints and provides an acceptable performance. It should be emphasized the *REASONER* may carry out not only pure logical reasoning but also other types of search and optimization functions depending on the representation used to describe the models, goals, and so on. The new configuration is passed to the *RECONFIGURATOR* functionality to plan and execute the sequence of operations for "transforming" the old into the new configuration in runtime.*

Note that the reconfiguration process of Figure 22.8 runs on the same embedded monitoring system that is used for signal processing. An efficient implementation of this runtime reconfiguration should address three challenges:

- Representation: What are the right formalisms to describe the models and their interaction? To what extent should the models be made part of the running code? What is an efficient model representation in runtime?
- Monitoring: How can we collect coherent information about the health state of the system, even in case of failures? How can we deduct the potentially disruptive situations, that is, which should trigger reconfiguration actions, from the raw observation set?
- Reasoning: What are the efficient algorithms, which are matching with the model representation, to resolve the conflicts rising from changes in the environment and/or in the system configuration? What are the chances for a distributed solution of the reasoning process?

There are no ultimate answers to these questions. The application domains have crucial impact on the optimal representation and reasoning, as well as on the resources that are required to run the reconfiguration process itself. Consequently, a thorough analysis of the application in hand, its typical failure modes, the dependability requirements, and other relevant aspects of the system in its environment jointly identify the proper selection of techniques for setting up a suitable runtime reconfiguration process.

The research area of runtime reconfigurable systems design is quickly evolving. Established domains as self-adaptive software systems [47] and dynamically reconfigurable hardware systems [48,49] provide fundamental contributions. In the following, a few characteristic approaches are briefly addressed. A reconfiguration methodology based on model integrated computing (MIC) was introduced in [50]. Therein, the designer describes all relevant aspects of the system as formal models. A meta-modeling layer supports the definition of these relevant aspects that are to be modeled and generates the necessary model editors, that is, carries out model analysis, verification, etc. The program synthesis level consists of a set of model interpreters, which according to the supplied models and constraints generate program code. The reconfiguration is triggered by changes in the models or constraints, which initiates a new model interpretation cycle. Though MIC provides a flexible way to describe and implement reconfigurable systems, the model interpretation is a computationally demanding step and may seriously limit the applicability in real-time cases. Alternatively, a model-oriented architecture with related tools for runtime reconfigurable systems was presented in [51]. This approach uses variability, context, reasoning, and architecture models

* The operations for "transforming" the configuration act on the program modules implementing the task graph and on a "switchboard" realizing the flexible connections among the tasks. Consequently, the program modules should implement a "standard" application programming interface (API), which allows for a function independent, unified configuration interface to software components. This way parameter changes in the signal processing functions and in the connections between these functions can be carried out irrespective of the actual functions involved in the processing.

to capture the design space. In runtime, the interactions among the event processor, goal-based reasoner, aspect model weaver, and the configuration checker/manager components will carry out the reconfiguration. The approach is well suited for coping with a high number of artifacts but the real-time aspect is not well developed. A formalization of the reconfiguration as a constraint satisfaction problem was proposed in [52–54]. The design space is (at least partially) represented and its design constraints are explicitly stated. These methodologies implement a "constraint-guided" design space exploration to find feasible solutions under the observed circumstances. In parallel, a suitable performance criterion is calculated to guide the reconfiguration process to optimal solution. The method described in [54] is also capable of hardware/software task migration and morphing. Different reconfiguration solutions were developed for service-oriented architectures (SOAs). For example, the reconfiguration method introduced in [55] extends the "traditional" discover–match–coordinate SOA scheme with a hierarchical service overlay mechanism. This service overlay implements a composition functionality that can dynamically "weave" the required services from the available service primitives. In [56], a solution is proposed that follows an object-centric paradigm to compose the compound services. By modeling the service constraints, an underlying constraints satisfaction mechanism implements the dynamic service configuration. A different approach was presented in [57], which describes a model-based solution to validate at runtime that the sensor network functionalities are performed correctly, despite of changes in the operational conditions. It models the application logic, the network topology, and the test specification, which are then used to generate diagnostic code automatically. Though the solution does not address the *REASONING* functionality of Figure 22.8, it delivers low false-negative detection rates, that is, it covers the *MONITOR* functionality effectively.

22.6.2 Implementation of Runtime Reconfiguration

The runtime reconfiguration brings in an extra aspect of complexity, which is "woven" into the functional architecture of the system and thus makes the testing and validation extremely challenging. To keep the development efforts on a reasonable level, both design and implementation support are needed. Many of the runtime reconfiguration approaches cited propose an architectural methodology, design tool set, and runtime support, for example, [46,50,52,54–58]. A common feature of these efforts is to support the system developer with application-independent reconfiguration functionalities, which can be parameterized according to the concrete needs of the application at hand. They also attempt to "separate concerns" when feasible, that is, try to make the design of the functional architecture and the reconfiguration process as independent as possible, while still maintaining clear interactions between them. Typically, the corresponding reconfiguration functionalities manifest themselves in an additional software layer between the "nominal" real-time executive layer, such as TinyOS [59] or Contiki [60], and the application layer. See Figure 22.9 for more details. The (application independent) monitoring and reconfiguration functionalities in this figure receive the application specific information from "outside" in the form of models. Conceptually they are "interpreters." As such, they realize a virtual machine dedicated to a certain type of computational model, for example, rule based inference, finite sate machine, constraint satisfaction, and so on. They read-in the application specific "program," which is represented by the *reconfiguration rules* component in Figure 22.9, and interpret its code in the context of the data received from the *MONITOR* function. For example, if the reconfiguration process is based on a rule-based representation of the application specific knowledge,* then the *REASONER*

* This type of formalism will be used in the case study later in the chapter.

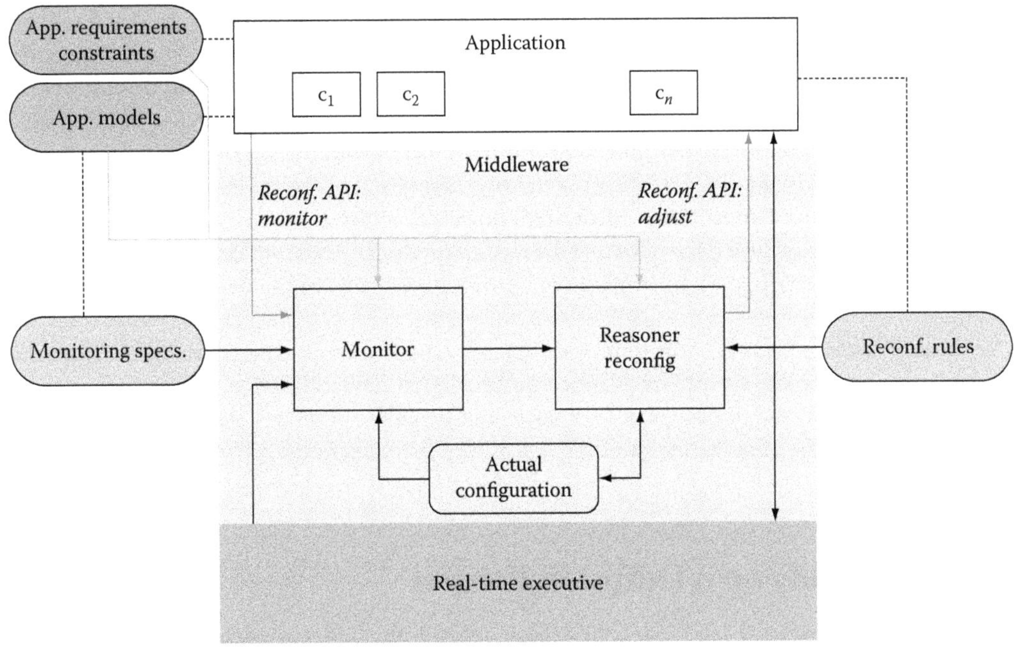

Figure 22.9 Middleware for runtime reconfiguration.

implements a forward changing (data driven) inference engine [61], in which using the *actual configuration* and the data received from *MONITOR* as fact base. The inference process results in derived *events* and *actions*, which define the reconfiguration commands issued for the application layer. The application program is characterized by (multi-aspect) models, requirements, and constraints that are created by the designer according to, for example, [57]. For efficiency reasons, the models created by a designer are rarely used directly by the reconfiguration process. Instead, after thorough compile-time checking, these models are translated to a "machine friendly" format to enable resource-aware access and transformations. The models can also be used for automatic code generation and synthesis to create the application code if the appropriate tools are available [50]. The monitoring functionality of Figure 22.9 (equivalent to the one in Figure 22.8) defines the set of observations that a reconfiguration process should take into consideration. Typically, this monitoring should cover the operational characteristics of an application, for example, sensor noise level and estimator variance, combined with the health state of its execution platform, for example battery energy level and quality of communication channels. The reconfiguration rules then define the "knowledge base" of the reasoner/reconfigurator, which is also depicted in Figure 22.9, for example, they determine how the recognized changes in operational conditions are handled. Note that reconfiguration rules do not necessarily refer to rule-based knowledge base but that the format and content of the "rules" are determined by a reasoning procedure, for example, constraint satisfaction, graph matching, first-order logic, etc. Further note that the application layer of Figure 22.9 uses a number of reconfigurable components ($c_1 \dots c_n$) to implement the required application-level functionality. These components should implement a unified API, so that the middleware layer is able to retrieve information for its monitoring purposes and for executing its reconfiguration commands.

It should be emphasized that in certain applications the reconfiguration decisions could rely on system-wide information. In these cases, the monitoring and reconfiguration activities inherently involve communication, resource scheduling, etc. This adds an extra layer of complexity to the systems, for example, implementing distributed snap-shot algorithms, leader election and distributed reasoning/planning, that may demand resources beyond the capabilities of the nodes. A work-around is to give up the fully distributed implementation of the reconfiguration and assign the most demanding functionalities to (dedicated) powerful nodes, as proposed in [50,52,54]. The monitoring information is then forwarded to the reconfiguration node(s) where a new configuration is determined. The reconfiguration commands are transferred back to the nodes for synchronized execution.

In the next section, the role of runtime reconfiguration will be demonstrated. It follows from the inherent network topology properties assumed in distributed state estimation that reconfiguration decisions are based on the information from local and neighboring nodes. As such, a distributed implementation of runtime reconfiguration is feasible, even on nodes of moderate computing capabilities.

22.7 Case Study on a Diffusion Process

The results of the presented self-organizing sensor network for state estimation are demonstrated and evaluated in a spatiotemporal 2D diffusion process. The goal of the sensor network is to follow the contaminant's distribution profile in time (i.e., the concentration distribution in space and time of a particular chemical compound) in the presence of wind. To that extent, let us consider an area of 1200×1200 meters containing a contaminant source. As time passes, the contaminant spreads across the area due to diffusion and wind. To simulate the spread, let us divide the area into a grid with a grid-size of 100 meters. The center of each grid-box is defined as a grid-point. Then, the spread of the contaminant is represented by the concentration level $\rho^{(q)} \in \mathbb{R}_+$ at the q-th grid-point $q \in [1, 144]$. This concentration level $\rho^{(q)}$ depends on the corresponding levels at neighboring grid-points, which are denoted as q_n for north, q_s for south, q_e for east and q_w for west. See Figure 22.10 for a graphical representation of these grid-points relative to the q-th grid-point. Further, the *continuous-time* process model of $\rho^{(q)}$, for some $a, a_n, a_s, a_e, a_w \in \mathbb{R}$, yields

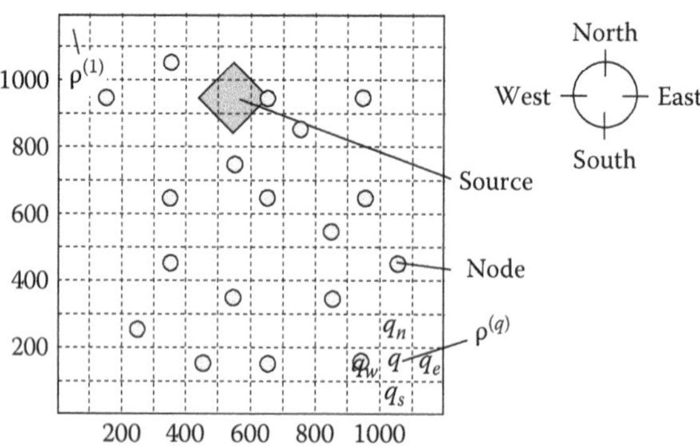

Figure 22.10 The monitored area is divided into a grid. Each grid-point q has four neighbors q_n, q_s, q_e, and q_w, that is, one to the north, south, east, and west of grid-point q, respectively. The chemical matter produced by the source spreads through the area due to diffusion and wind.

$$\dot{\rho}^{(q)} = a\rho^{(q)} + a_n\rho^{(q_n)} + a_s\rho^{(q_s)} + a_e\rho^{(q_e)} + a_w\rho^{(q_w)} + u^{(q)}, \quad \forall q \in \mathbb{Z}_{[1,144]}.$$

The variable $u^{(q)} \in \mathbb{R}_+$ in (22.7) parameterizes the production of chemical matter by a source at grid-point q and follows $u^{(18)} = 75$, $u^{(29)} = 75$, $u^{(30)} = 100$, $u^{(31)} = 100$, and $u^{(42)} = 175$ for all time $t \in \mathbb{R}_+$, while $u^{(q)} = 0$ for all other $q \in \mathbb{Z}_{[1,144]}$. The remaining parameters are chosen to establish a northern the wind direction, that is, $a = \frac{-12}{800}$, $a_n = \frac{1}{800}$, $a_s = \frac{2}{800}$, $a_e = \frac{7}{800}$, and $a_w = \frac{2}{800}$.

A sensor network is deployed in the area to reconstruct the concentration levels at each grid-point based on the local measurements taken by each node.

Communication The network consists of 18 sensor nodes that are randomly distributed across the area, see also Figure 22.10. It is assumed that the sensor nodes communicate only with their direct neighbors, that is nodes with a 1-hop distance, and that their position is available.

Process Neither the wind direction nor values of the contaminant source are available to the nodes. Therefore, the process model that is used by the local estimation algorithms of the different nodes is a simplified diffusion process in *continuous-time*, that is,

$$\dot{\rho}^{(q)} = \alpha\rho^{(q)} + \alpha_n\rho^{(q_n)} + \alpha_s\rho^{(q_s)} + \alpha_e\rho^{(q_e)} + \alpha_w\rho^{(q_w)} + w^{(q)},$$

with $\alpha = \frac{-12}{800}$, $\alpha_n = \frac{3}{800}$, $\alpha_s = \frac{3}{800}$, $\alpha_e = \frac{3}{800}$, and $\alpha_w = \frac{3}{800}$. The unknown source and model uncertainties are represented by process noise $w^{(q)} \in \mathbb{R}$, for all $q \in \mathbb{Z}_{[1,144]}$. A suitable characterization of this noise, that is, to cover unknown source values $u^{(q)}$ between -150 and 150, is given by the *continuous-time* PDF $p(w^{(q)}(t)) = G(w^{(q)}(t), 0, 2 \cdot 10^3)$. Further, the state is defined as the collection of all concentration levels, that is, $x := \left(\rho^{(1)} \;\; \rho^{(2)} \cdots \rho^{(144)}\right)^{\top}$. The model parameters A_{τ_i} and Q_{τ_i} of the *discrete-time* process model in (22.1) are characterized with the initial sampling time of $\tau_i = 10$ s, for all nodes $i \in \mathcal{N}$. To determine the other process model parameters, that is, C_i and V_i, it is assumed that each sensor node i measures the concentration level at its corresponding grid-point, that is, $y_i[k_i] = \rho^{(q)}[k_i] + v_i$, for some $q \in \mathbb{Z}_{[1,144]}$ and $p(v_i[k_i]) = G(v_i[k_i], 0, 0.5)$, for all $i \in \mathbb{Z}_{[1,18]}$. The real concentration levels at the three time instants $t = 140$, $t = 240$, and $t = 340$ are illustrated in Figure 22.11.

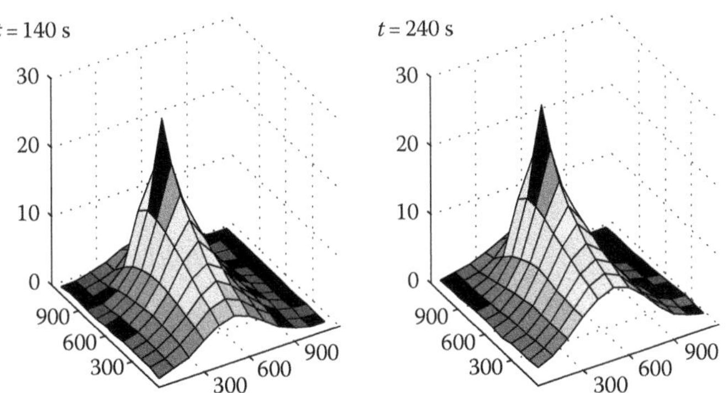

Figure 22.11 The simulated concentration levels at the different grid-points for two instances of the time $t \in \mathbb{R}_+$.

The objective of the sensor network is to determine the contaminant distribution by estimating the state x in multiple nodes of the network. This is carried out in two types of sensor networks: a hierarchical network and a fully distributed one. In each configuration, unforseen events occur indicating node breakdown and batteries depleting below critical energy level. The nodes must adapt their local state estimating functionalities to recover from lost neighbors and/or to reduce their energy consumption so that batteries do not get depleted. Let us start this analysis with the hierarchical network.

22.7.1 Hierarchical Sensor Network

In a *hierarchical* sensor network, nodes are given specific tasks prior to its deployment. Basically, the network consists of multiple subnetworks, as it is illustrated in Figure 22.12a. In each subnetwork, nodes exchange their local measurements with the center node of that particular subnetwork (denoted with dashed lines). The center node computes a local estimate based on these received measurements via f_{KF}, after which this estimate is shared with the center nodes of other subnetworks (denoted with the solid lines). The received estimates are then fused with the local estimate according to the merging function f_{ME} and the fusion method *ellipsoidal intersection* of (22.17).

Two events will occur in this network, followed by the corresponding action as it is implemented in the reconfiguration process of each node. The reconfiguration is local: Operational events are monitored locally and the reconfiguration actions influence only the node that issued the request for action. A rule-based representation formalism is used to define the "knowledge base" of the reconfiguration functionality. As such, the *REASONER* component of the middleware implements a forward chaining rule interpreter, that is, if *event* then *action* [61–63]. For clarity of the illustrative example, we do not attempt a rigorously formal description of the knowledge base but only the "style" of the rule-based representation is shown.

- At $t = 150$ s, nodes 1, 3, and 8 will cross their critical energy level;
  If *the critical energy level is crossed,* then *lower the node's local sampling time from* 10 *s to* 20 *s.*
- At $t = 250$ s, node 5 will break down. To detect whether a state estimating node, breaks down, nodes within each subnetwork exchange acknowledgments or heartbeat messages are used to indicate normal operational mode.

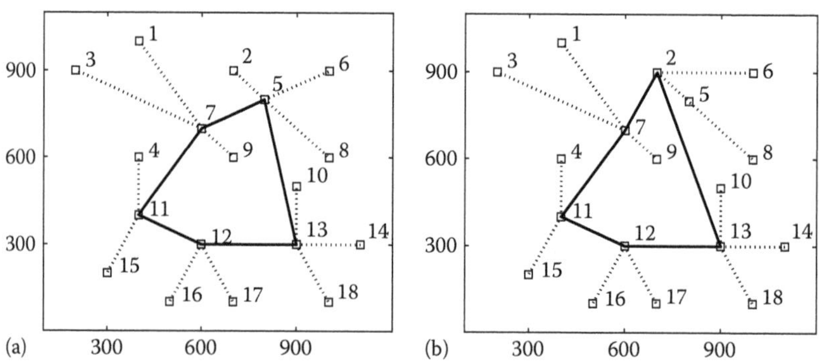

Figure 22.12 Network topology in a hierarchical network. (a) Initial topology and (b) topology after 250 s.

If *the acknowledgment of the state-estimating node is not received,* then *check the energy levels of all other nodes in the corresponding subnetwork. The node with the largest energy level takes over the responsibility for estimating the state, according to an algorithm that is similar to the node that broke down. Also, reestablish the connection with the other subnetworks.*

As an example, the rule set in the following text shows the handling of the #2 event

```
Rule_2a:
    IF   NEIGHBOR(?x) & TIMEDOUT(?x) & ?x.function = centerfun
    THEN set(go_for_newcenter,TRUE)

Rule_2b:
    IF   go_for_newcenter & NEIGHBOR(?x) &
         !TIMEDOUT(?x) & max(?x.power) = self.power
    THEN exec(assign,centerfun), exec(broadcast,centerfun_msg)
```

Figure 22.12a depicts the network topology prior to the event that node 5 breaks down, while Figure 22.12b illustrates this topology after the event (assuming that the battery of node 2 has the highest energy level). This figure indicates that node 2 has become responsible for estimating the state and thereby, replaces node 5 that broke down at $t = 250$ s. Further, Figure 22.13 depicts a particular estimation error, for which the estimation error of single node i is defined as $\Delta_i := (x - \hat{x}_i)^\top (x - \hat{x}_i)$. More specifically, the figure presents the difference in the estimation error of a network *not* effected by operational event with the estimation error in a network that *is* effected by the previously presented operational event. The reason that the figure depicts the results of node 7 is because this node is affected by both events.

Before Figure 22.13 is analyzed, let us denote the hierarchical network in the *presence* of the aforementioned operational events as the reconf-case and the hierarchical network in the *absence* of operational event as the ideal-case. Then the figure indicates that the results of the reconf-case and the ideal-case are equivalent until the two operational events occur, which is expected as both network cases are similar until 150 s. After that time, the estimation error of node 7 in the reconf-case increases with respect to the ideal-case. This is due to the fact that nodes 1, 3, and 8 double their local sampling times from $t = 150$ on and thus, node 7 will receive twice as less

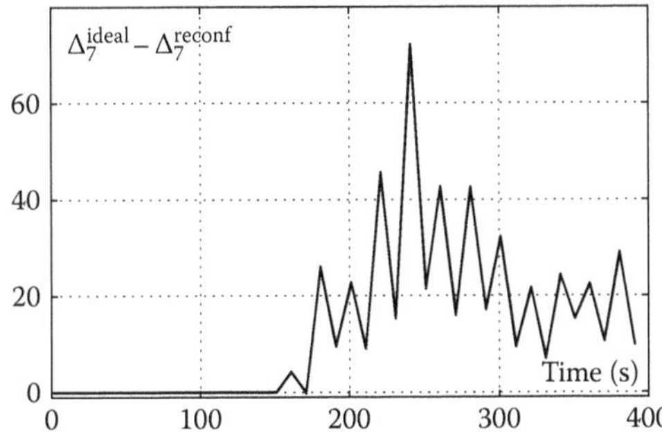

Figure 22.13 The difference in the estimation error of node 7 for a network that is *not* effected by operational events (Δ_7^{ideal}) with the estimation error in a network that *is* effected by the previously presented operational event (Δ_7^{reconf}).

measurement information from nodes 1 and 3. This leads to an increase in estimation error of node 7 compared to the ideal-case. Further, note that this error decreases when local measurement information from nodes 1 and 3 is received, that is, at the time instants 170, 190, 210, ..., 370, 390. At these instances node 7 receives two more local measurements, that is, y_1 and y_3, which is not the case at the other sampling instants as nodes 1 and 3 doubled their local sampling time. After the second operational event, that is, node 5 breaks down at $t = 250$, the difference in the estimation error of node 7 for the reconf-case with respect to the ideal-case decreases (on average). This behavior can be explained from the fact that node 2 has become a direct neighbor of node 7, while this node 2 was indirect neighbor via node 5 prior to $t = 250$. Since node 2 is closer to the contaminant source, node 7 obtains an improved estimation result when node 2 is its direct neighbor rather than an indirect one.

22.7.2 Distributed Sensor Network

The distributed sensor network reflects an ad hoc networked system. This means that the nodes establish a mesh-network-topology, as it is depicted in Figure 22.14a. Since there is no hierarchy in this network, each node estimates the local state by performing the distributed KF of (22.14): the local measurement is processed by f_{KF} to compute a local estimate of the state, which are then shared with neighboring nodes as input to the merging function Ω employing the state fusion method *ellipsoidal intersection* of (22.17).

Two events will occur in this network, followed by the corresponding action as it is implemented in the management layer of each node.

- At $t = 150$ s, nodes 1, 3, and 8 will cross their critical energy level.
  If *the critical energy level is crossed,* then *lower the node's local sampling time from* 10 *s to* 20 *s.*
- At $t = 250$ s, nodes 5 and 11 will break down. Nodes detect that another node has broken down, since no new local estimates are received from that node.
  If *a node breaks down and the network has lost its connectivity,* then *establish a network connection with other nodes until this connectivity is reestablished. In case this means to increase the communication range to larger distances, decrease the sampling time accordingly.*

Figure 22.14a depicts the network topology prior to the event that nodes 5 and 11 break down, while Figure 22.14b illustrates the topology and after the event. This figure indicates that the

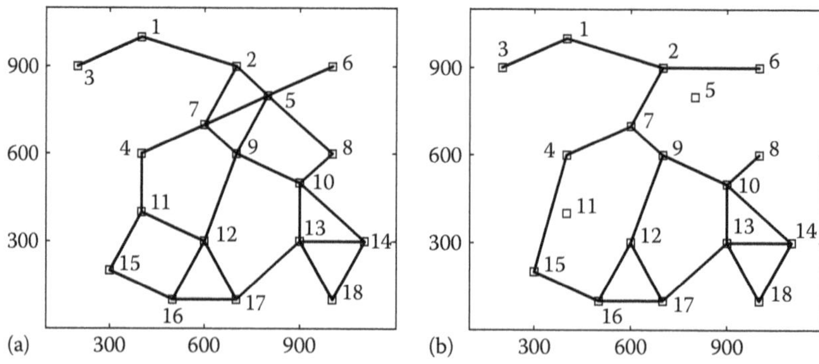

Figure 22.14 Network topology in a distributed network. (a) Initial topology and (b) topology after 250 s.

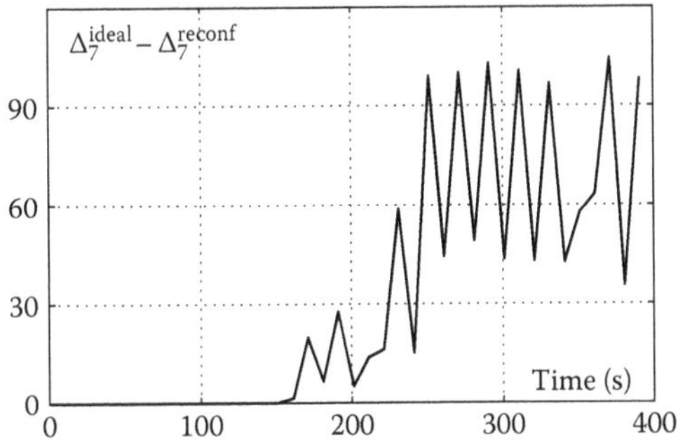

Figure 22.15 The difference in the estimation error of node 7 for a network that is *not* effected by operational events (Δ_7^{ideal}) with the estimation error in a network that *is* effected by the previously presented operational event (Δ_7^{reconf}).

sensor network establishes connectivity, also after the event of a node breaking down. However, the nodes 6 and 15 will have to exchange data with nodes that are far away. Therefore, these node will lower their local sampling time to 20 s. Further, Figure 22.15 depicts the same estimation error as Figure 22.13, only then for a distributed network. This means that the figure presents the difference in the estimation error of a network *not* effected by operational event with the estimation error in a network that *is* effected by the previously presented operational event. The reason that the figure depicts the results of node 7 is because this node is affected by both events.

Before Figure 22.15 is analyzed, let us denote the distributed network in the *presence* of the aforementioned operational events as the reconf-case and the distributed network in the *absence* of operational event as the ideal-case. Then, the figure indicates a similar behavior compared to the hierarchical network that was previously discussed; for example, the results of the reconf-case and the ideal-case are equivalent until the first operational event occurs, after which the error of the reconf-case increases with respect to the ideal-case. Also, the estimation results of node 7 have an "up-down" type of behavior, which is due to the action undertaken by nodes 1 and 3 to double their local sampling times. As such, node 7 receives an updated estimate from nodes 1 and 3 after every other of its local sampling instants. The difference between the estimation error of node 7 in the reconf-case increases even further with respect to the ideal-case after the second operational event, i.e., nodes 5 and 11 break down at $t = 250$.

Both the illustrative case studies of a hierarchical and a distributed sensor network indicate that the state is estimated by multiple nodes in the network, even in the presence of unforeseen operational events. As such, adopting a self-organizing method in large-scale and ad hoc sensor networks improves the robustness of state estimation within the network.

22.8 Conclusions

Ad hoc sensor networks typically consist of a large number of vulnerable components connected via unreliable communication links and are sometimes deployed in harsh environment. Therefore, dependability of networked system is a challenging problem. This chapter presented an efficient and cost-effective answer to this challenge by employing runtime reconfiguration techniques additional to a particular signal processing method (Kalman filtering). More precisely, a distributed Kalman filtering strategy was presented in a self-organizing sensor networks. This means that each node computes a local estimate of the global state based on its local measurement and on the data exchanged by neighboring nodes. The self-organizing property was implemented via a runtime reconfiguration process, so to have a sensor network that is robust to external and internal system changes, for example, nodes that are removed or added to an existing network during operation.

Firstly, a brief overview of existing solutions for distributed Kalman filtering was presented. The corresponding algorithms were described with equivalent input and output variables. As a result, nodes could choose which of the algorithms is currently best suitable for estimating the state vector, while taking into account the available communication and computational resources. This further enabled nodes to select what information is to be shared with other nodes, that is, local measurements or local estimates, and how the received information is merged with the local estimate. Secondly, the system architecture was addressed, such that challenging design issues could be separated from the actual implementation of a (self-organizing) distributed KF. To that extent, an overview of typical reconfiguration approaches was given with an emphasize on the interactions between the signal processing and hardware/communication aspects of system design. After that, the self-organizing property of the proposed distributed KF was assessed in a diffusion process for two types of sensor networks, that is, a hierarchical network and a fully distributed one. In both cases, the network was able to cope with unforeseen events and situations. Or differently, employing runtime reconfiguration in the nodes of the sensor network implements a kind of self-awareness with the ability to create corrective actions and thus assuring that data processing functionalities are never used beyond their scope of validity.

References

1. Z. Papp, J. Sijs, and M. Lagioia. Sensor network for real-time vehicle tracking on road networks. In *Proceedings of the 5-th International Conference on Intelligent Sensors, Sensor Networks and Information Processing*, pp. 85–90, Melbourne, Australia, 2009.
2. R. Szewczyk, E. Osterweil, J. Polastre, M. Hamilton, A. Mainwaring, and D. Estrin. Habitat monitoring with sensor networks. *ACM Communications*, 47:34–40, 2004.
3. I.F. Akyildiz, W. Su, Y. Sankarasubramaniam, and E. Cayirci. Wireless sensor networks: A survey. *Elsevier, Computer Networks*, 38:393–422, 2002.
4. C. Y. Chong and S. P. Kumar. Sensor networks: Evolution, opportunities and challenges. In *Proceedings of the IEEE*, volume 91, pp. 1247–1256, 2003.
5. J. M. Kahn, R. H. Katz, and K. S. J. Pister. Next century challenges: Mobile networking for smart dust. In *Proceedings of the 5-th ACM/IEEE International Conference on Mobile Computing and Networking*, pp. 271– 278, Seattle, WA, 1999.
6. F. L. Lewis. Wireless sensor networks. In *Smart Environments: Technologies, Protocols, Applications (Chapter 2)*, Lecture Notes in Computer Science. Wiley, New York, 2005.
7. C. E. Shannon and W. Weaver. *The Mathematical Theory of Communication*. The University of Illinois Press, Urbana, IL, 1949.
8. S. J. Julier and J. K. Uhlmann. A new extension of the kalman filter to nonlinear systems. In *Prceedings of AeroSense: The 11-th International Symposium on Aerospace/Defense Sensing, Simulation and Controls*, pp. 182–193, Orlando, FL, 1997.
9. R.E. Kalman. A new approach to linear filtering and prediction problems. *Transactions of the ASME Journal of Basic Engineering*, 82(D):35–42, 1960.
10. B. Ristic, S. Arulampalam, and N. Gordon. *Beyond the Kalman filter: Particle Filter for Tracking Applications*. 2002.

11. H.F. Durant-Whyte, B.Y.S. Rao, and H. Hu. Towards a fully decentralized architecture for multi-sensor data fusion. In *1990 IEEE International Conference on Robotics and Automation*, pp. 1331–1336, Cincinnati, OH, 1990.
12. S.C. Felter. An overview of decentralized Kalman filters. In *IEEE 1990 Southern Tier Technical Conference*, pp. 79–87, Birmingham, U.K., 1990.
13. F. Garin and L. Schenato. *Networked Control Systems*, volume 406 of *Lecture Notes in Control and Information Sciences*, chapter A survey on distributed estimation and control applications using linear consensus algorithms, pp. 75–107. Springer, Berlin, Germany, 2011.
14. S. Kirti and A. Scaglione. Scalable distributed Kalman filtering through consensus. In *Proceedings of the IEEE International Conference on Acoustics, Speech and Signal Processing*, pp. 2725–2728, Las Vegas, NV, 2008.
15. R. Olfati-Saber. Distributed Kalman filtering for sensor networks. In *Proceedings of the 46-th IEEE Conference on Decision and Control*, pp. 5492–5498, New Orleans, LA, 2007.
16. A. Ribeiro, G. B. Giannakis, and S. I. Roumeliotis. SOI-KF: Distributed Kalman filtering with low-cost communications using the sign of innovations. *IEEE Transactions on Signal Processing*, 54(12):4782–4795, 2006.
17. A. Ribeiro, I. D. Schizas, S. I. Roumeliotis, and G. B. Giannakis. Kalman filtering in wireless sensor networks: Reducing communication cost in state-estimation problems. *IEEE Control Systems Magazine*, 4:66–86, 2010.
18. J. Sijs, M. Lazar, P.P.J. Van de Bosch, and Z. Papp. An overview of non-centralized Kalman filters. In *Proceedings of the IEEE International Conference on Control Applications*, pp. 739–744, San Antonio, TX, 2008.
19. A. Speranzon, C. Fischione, K.H. Johansson, and A. Sangiovanni-Vincentelli. A distributed minimum variance estimator for sensor networks. *IEEE Journal on Selected Areas in Communications*, 26(4):609–621, 2008.
20. B.D.O. Anderson and J. B. Moore. *Optimal filtering*. Prentice-Hall, Englewood Cliffs, NJ, 1979.
21. M. S. Grewal and A. P. Andrews. *Kalman Filtering: Theory and Practise*. Routledge, U.K., 1993.
22. G. Welch and G. Bishop. An introduction to the kalman filter, 1995.
23. H.R. Hashmipour, S. Roy, and A.J. Laub. Decentralized structures for parallel Kalman filtering. *IEEE Transaction on Automatic Control*, 33(1):88–93, 1988.
24. M.F. Hassan, G. Salut, M.G. Sigh, and A. Titli. A decentralized algorithm for the global Kalman filter. *IEEE Transaction on Automatic Control*, 23(2):262–267, 1978.
25. J.L. Speyer. Computation and transmission requirements for a decentralized Linear-Quadratic-Gaussian control problem. *IEEE Transactions on Automatic Control*, 24(2):266–269, 1979.
26. U.A. Khan and J.M.F. Moura. Distributed Kalman filters in sensor networks: Bipartite fusion graphs. In *IEEE 14-th Workshop on Statistical Signal Processing*, pp. 700–704, Madison, WI, 2007.
27. A.G.O. Mutambara and H. F. Duranth-Whyte. Fully decentralized estimation and control for a modular wheeled mobile robot. *International Journal of Robotic Research*, 19(6):582–596, 2000.
28. A. Jadbabaie, J. Lin, and A. Morse. Coordination of groups of mobile autonomous agents using nearest neighbor rules. *IEEE Transaction on Automatic Control*, 48(6):988–1001, 2003.
29. A. Tahbaz Salehi and A. Jadbabaie. Consensus over ergodic stationary graph processes. *IEEE Transactions on Automatic Control*, 55:225–230, 2010.
30. L. Xiao and S. Boyd. Fast linear iterations for distributed averaging. *Systems and Control Letters*, 53(1):65–78, 2004.
31. L. Xiao, S. Boyd, and S. Lall. A scheme for robust distributed sensor fusion based on average consensus. In *Proceedings of the 4-th International Symposium on Information Processing in Sensor Networks*, pp. 63–70, Los Angelos, CA, 2005.

32. V. Hasu and H. Koivo. Decentralized kalman filter in wireless sensor networks - case studies. In K. Elleithy, T. Sobh, A. Mahmood, M. Iskander, and M. Karim, eds., *Advances in Computer, Information, and Systems Sciences, and Engineering: Proceedings of IETA 2005, TeNe 2005 and EIAE 2005*, pp. 61–68, the Netherlands, 2006. Springer.
33. W. Ren, R. Beard, and D. Kingston. Multi-agent Kalman consensus with relative uncertainty. In *Proceedings of the American Control Conference*, pp. 1865–1870, Portland, OR, 2005.
34. D. W. Casbeer and R. Beard. Distributed information filtering using consensus filters. In *Proceedings of the American Control Conference*, pp. 1882–1887, St. Louis, MO, 2009.
35. R. Carli, A. Chiuso, L. Schenato, and S. Zampieri. Distributed Kalman filtering based on consensus strategies. *IEEE Journal in Selected Areas in Communications*, 26(4):622–633, 2008.
36. R. Olfati-Saber. Kalman-Consensus filter: Optimality, stability, and performance. In *Proceedings of the 48-th IEEE Conference on Decision and Control*, pp. 7036 – 7042, Shanghai, China, 2009.
37. S. J. Julier and J. K. Uhlmann. A non-divergent estimation algorithm in the presence of uknown correlations. In *Proceedings of the American Control Conference*, pp. 2369–2373, Piscataway, NJ, 1997.
38. Y. Bar-Shalom and L. Campo. The effect of the common process noise on the two-sensor fused-track covariance. *IEEE Transactions on Aerospace and Electronic Systems*, AES-22(6):803–805, 1986.
39. D. Franken and A. Hupper. Improved fast covariance intersection for distributed data fusion. In *Proceedings of the 8-th International Conference on Information Fusion*, pp. WbA23:1–7, Philadelphia, PA, 2005.
40. U. D. Hanebeck, K. Briechle, and J. Horn. A tight bound for the joint covariance of two random vectors with unknown but constrained cross-correlation. In *Proceedings of the IEEE Conference on Multisensor Fusion and Integration for Intelligent Systems*, pp. 85–90, Baden-Baden, Germany, 2001.
41. W. Niehsen. Information fusion based on fast covariance intersection filtering. In *Proceedings of the 5-th International Conference on Information Fusion*, pp. 901–905, Annapolis, MD, 2002.
42. J. Sijs and M. Lazar. Distributed Kalman filtering with global covariance. In *Proceedings of the American Control Conference*, pp. 4840–4845, San Francisco, CA, 2011.
43. J. Sijs and M. Lazar. State fusion with unknown correlation: Ellipsoidal intersection. *Automatica (in press)*, 2012.
44. Y. Zhuo and J. Li. Data fusion of unknown correlations using internal ellipsoidal approximations. In *Proceedings of the 17-th IFAC World Congress*, pp. 2856–2860; 2008. Seoul, South Korea.
45. A.T. Bahill and B. Gissing. Re-evaluating systems engineering concepts using systems thinking. *IEEE Transactions on Systems, Man, and Cybernetics, Part C: Applications and Reviews*, 28(4):516–527, Nov 1998.
46. G. Karsai, F. Massacci, L.J. Osterweil, and I. Schieferdecker. Evolving embedded systems. *Computer*, 43(5):34–40, May 2010.
47. B. H. C. Cheng, R. D. Lemos, H. Giese, P. Inverardi, J. Magee, J. Andersson, B. Becker et al. Software engineering for self-adaptive systems: A research roadmap. In B. H. C. Cheng, R. de Lemos, H. Giese, P. Inverardi, and J. Magee, eds., *Software Engineering for Self-Adaptive Systems*, volume 5525 of *Lecture Notes in Computer Science*, pp. 1–26, Springer, Berlin, Germany, 2009.
48. E. L. de Souza Carvalho, N. L. V. Calazans, and F. G. Moraes. Dynamic task mapping for mpsocs. *IEEE Design and Test*, 27:26–35, September 2010.
49. R. Hartenstein. A decade of reconfigurable computing: A visionary retrospective. In *Design, Automation and Test in Europe, 2001. Conference and Exhibition 2001. Proceedings*, pp. 642–649, 2001.
50. G. Karsai and J. Sztipanovits. A model-based approach to self-adaptive software. *Intelligent Systems and Their Applications, IEEE*, 14(3):46–53, May/Jun 1999.
51. B. Morin, O. Barais, J.-M. Jzquel, F. Fleurey, and A. Solberg. Models at runtime to support dynamic adaptation. *IEEE Computer*, pp. 46–53, October 2009.

52. S. Kogekar, S. Neema, B. Eames, X. Koutsoukos, A. Ledeczi, and M. Maroti. Constraint-guided dynamic reconfiguration in sensor networks. In *Proceedings of the 3rd International Symposium on Information Processing in Sensor Networks*, IPSN '04, pp. 379–387, New York, 2004. ACM.
53. P. J. Modi, H. Jung, M. Tambe, W.-M. Shen, and S. Kulkarni. A dynamic distributed constraint satisfaction approach to resource allocation. In *Proceedings of the 7th International Conference on Principles and Practice of Constraint Programming*, CP '01, pp. 685–700, London, U.K., 2001. Springer-Verlag.
54. T. Streichert, D. Koch, C. Haubelt, and J. Teich. Modeling and design of fault-tolerant and self-adaptive reconfigurable networked embedded systems. *Eurasip Journal on Embedded Systems*, p. 15, 2006.
55. S. Kalasapur, M. Kumar, and B. Shirazi. Seamless service composition (sesco) in pervasive environments. In *Proceedings of the First ACM International Workshop on Multimedia Service Composition*, MSC '05, pp. 11–20, New York, 2005. ACM.
56. X. D. Koutsoukos, M. Kushwaha, I. Amundson, S. Neema, and J. Sztipanovits. *OASiS: A Service-Oriented Architecture for Ambient-Aware Sensor Networks*, volume 4888 LNCS, pp. 125–149. 2007.
57. Y. Wu, K. Kapitanova, J. Li, J. A. Stankovic, S. H. Son, and K. Whitehouse. Run time assurance of application-level requirements in wireless sensor networks. In *Proceedings of the 9th ACM/IEEE International Conference on Information Processing in Sensor Networks*, IPSN '10, pp. 197–208, New York, 2010. ACM.
58. J.C. Georgas, A. van der Hoek, and R.N. Taylor. Using architectural models to manage and visualize runtime adaptation. *Computer*, 42(10):52–60, Oct. 2009.
59. P. Levis, S. Madden, J. Polastre, R. Szewczyk, A. Woo, D. Gay, J. Hill, M. Welsh, E. Brewer, and D. Culler. Tinyos: An operating system for sensor networks. In *Ambient Intelligence*, 2004. Springer Verlag, New York.
60. A. Dunkels, B. Grnvall, and T. Voigt. Contiki—a lightweight and flexible operating system for tiny networked sensors. In *Proceedings of the First IEEE Workshop on Embedded Networked Sensors (Emnets-I)*, Tampa, FL, November 2004.
61. S. Russell and P. Norvig. *Artificial Intelligence: A Modern Approach*. Prentice-Hall, Englewood Cliffs, NJ, 2nd edn., 2003.
62. D. Chu, L. Popa, A. Tavakoli, J. M. Hellerstein, P. Levis, S. Shenker, and I. Stoica. The design and implementation of a declarative sensor network system. In *SenSys'07*, pp. 175–188, 2007.
63. A. D. Jong, M. Woehrle, and K. Langendoen. Momi: Model-based diagnosis middleware for sensor networks. In *Proceedings of the 4th International Workshop on Middleware Tools, Services and Run-Time Support for Sensor Networks*, MidSens '09, pp. 19–24, New York, 2009. ACM.

Chapter 23

Low-Power Solutions for Wireless Passive Sensor Network Node Processor Architecture

Vyasa Sai, Ajay Ogirala, and Marlin H. Mickle

Contents

23.1 Introduction

Conventional wireless sensor networks (WSNs) are generally made up of a set of autonomous multifunctional sensor nodes distributed over a specific environment. These sensor nodes are used to collect environment data and transfer these data to the user through the network that can include Internet segments. Besides collecting data, a node may also need to perform computations on the measured data. In general, deployment of conventional sensor networks for environmental monitoring is mainly limited due to the active life span of the onboard non-rechargeable power source. As the sensors are battery powered, it becomes difficult to periodically monitor the manual replacement of the batteries. There have been a lot of research efforts in the direction of prolonging the limited lifetime of WSNs through efficient circuit, architecture, and communication techniques [1,2]. In summary, the use of a WSN system is strictly limited by the battery life of the sensor nodes [36]. There is a need for a new sensor network paradigm that is not based on an enhanced lifetime of a conventional WSN but is developed for a network that is free of any battery constraints.

A wireless passive sensor network (WPSN) is a non-disposable and a cost-efficient system that operates based on the incoming received power [3–7]. This system is considered to be an efficient and a novel solution for energy problems in WSN [3]. The concept to remotely feed a sensor node on the power from an external radio frequency (RF) source has led to the emergence of the WPSNs. This concept was first introduced to power a passive RF identification (RFID) tag. It is well known that passive RFID design blocks form the basis for passive sensor node (PSN) architectures [8]. PSN operating frequencies fall under the same industrial, scientific and medical (ISM) frequency bands as most RFID applications. The latest trend in environmental monitoring application is to have sensor nodes operating at power levels low enough to enable the use of energy harvesting techniques [9,10]. This facilitates the deployed system, in theory, for continuous sensing for a considerable extended period of time reducing recurring costs.

Building blocks of typical wireless PSN architecture consist of a sensing unit, communication unit, a processing unit, and a power source as shown in Figure 23.1 [4,5]. The sensing unit in most cases consists of a sensor (s) and an analog-to-digital converter (ADC) as components. A sensor is a device generally used to measure some physical quantity such as temperature, light, etc. The ADC is used to convert the received analog data signal into a digital signal so as to be processed by the microcontroller. The processing unit consists of a low-power microcontroller and a storage block. The microcontroller processes data and controls and coordinates other component functionalities. The communication unit consists of an RF transceiver module that transmits and receives data to/from other devices connected to the wireless network. The power unit mainly delivers the RF–DC converted power to the rest of node units and also stores additional power based on availability.

The major differences in the architectures of a conventional WSN node and a WPSN node are the hardware of the power unit and the transceiver [4]. The power unit of the conventional WSN generally consists of a battery along with a support block called the power generator. The power unit for a WPSN node is basically an RF-to-DC converter–capacitor network. The converted DC power is used to wake up and operate the node or is kept in a charge capacitor for future usage.

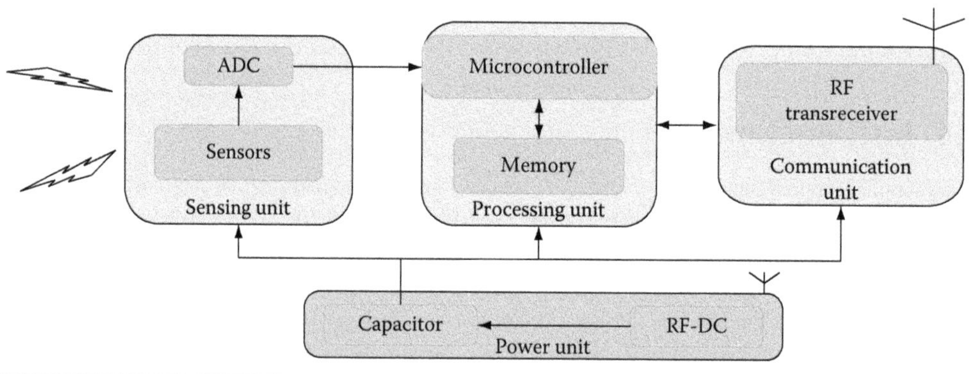

Figure 23.1 General WPSN node architecture.

A short-range RF transceiver, typically a major power-consuming unit on the node, is used in a conventional WSN as compared to a much simpler transceiver for modulated backscattering in the WPSN node [4,11].

To minimize the power consumed at the sensor node, the simplest of all solutions is to eliminate any or at least most signal processing at the node by transferring sampled data from all nodes to a central server. But there exist several applications in real-time control and analysis of continuous data sampling processes, where real-time signal processing and conditioning are implemented on discrete time sampled data. In many cases, the sensor data at each node must be preprocessed or conditioned before it can be handled by the processor. In such a scenario, where each sensor node must include a processor, there is need for application based dedicated processor hardware implementations that improve power efficiency and allow fine-grained design optimization. This chapter introduces a low-power conceptual design for a distributed architecture of a single passive sensor processor.

Typically, the smaller the area of the processor used in WPSN node architecture, the lower is the price. Using small-area processors, which require minimum power to operate, to provide greater read distances is significant in this scenario. WPSN being an emerging research area, there is little documentation on all the power-efficient scenarios applicable to passive sensor devices. In Refs. [3,4,6,7] efficient antenna designs, low-power transceivers were introduced for WPSNs. Not only is it important to have energy-efficient front-end and power unit designs, but there is also a need to have low-power novel processor designs that allow greater ranges for WPSN nodes. This chapter forms the basis for the low-power passive distributed sensor node architecture providing an increased operating range of the passive device.

Consider an intelligent sensor network topology with a sink node in the center communicating with several nodes around it. The server and a single sensor combination can be viewed as a single instruction single data (SISD) processor or the intelligent sensor network as a whole can be viewed as a single instruction multiple data (SIMD) processor. The power consumption of such an SIMD system depends on the hardware complexity of the passive node processor, which in turn depends on the instruction set (IS) supported by the architecture.

An intelligent combination of circuit techniques, applications, and architecture support is required to build a low-power sensor node system. This chapter introduces and elaborates the key concepts of the low-power design of the WPSN node processor. Using the 8051-ISA as an example for the distributed design concept, application-based customization of the sensor processor architecture is also elucidated in the later sections.

23.2 Low-Power Circuit Techniques in WSNs

Energy has become a critical aspect in the design of modern wireless devices and especially in WSNs. There is a need for a new architecture that takes into account such factors especially for passive or battery-operated device applications. Energy is defined in general as the sum of switching energy plus the leakage current energy.

The energy consumption equation is given as follows [1,12]:

$$E_{total} = V_{dd}(\alpha C_{sw} V_{dd} + I_{leakage} \Delta t_{op}). \tag{23.1}$$

Switching activity for 1 s is represented by α and the amount of time required to complete an operation is denoted by Δt_{op} as in the Equation 23.1. V_{dd}, $I_{leakage}$, and C_{sw} shown in Equation 23.1 represent the supply voltage, leakage current, and switching capacitance, respectively.

Conventional WSNs employ a variety of low-power design techniques, and a short overview of these approaches is presented in the following paragraphs. The following circuit techniques are most commonly classified under the following categories, which are used to minimize the power consumption in sensor networks [1].

Asynchronous designs are increasingly becoming an integral part of numerous WSNs [13–16] due to their low-power advantages. These designs are characterized by the absence of any globally periodic signals that act as a clock. In other words, these designs do not use any explicit clock circuit and hence wait for specific signals that indicate completion of an operation before they go on to execute the next operation. Low-power consumption, no clock distribution, fewer global timing issues, no clock skew problems, higher operating speed, etc., are advantages of asynchronous designs over synchronous designs.

Power supply gating is also a low-power circuit technique widely used to reduce the subthreshold leakage current of the system [17]. This process allows unused blocks in the system to be powered down in order to reduce the leakage current. This technique was used in the Harvard sensor network system [18].

A subthreshold operation technique allows supply voltages (V_{dd}) lower than threshold voltages (V_{th}) to be used for lowering the active power consumption. This technique was first used in the complete processor design for WSNs from University of Michigan [19–21].

The aforementioned techniques can also be extended to the WPSN based on the requirements of the application and the power available to a sensor node. The focus of this chapter will be on low-power solutions to wireless passive distributed sensor node architectures. The novel low-power techniques described in the following sections are applicable not only to WPSN but also to RFID systems and RFID sensor networks (RSNs).

23.3 Novel Low-Power Data-Driven Coding Paradigm

Wireless digital transmission systems are known to use different data encoding techniques especially the variable pulse width encoding and Manchester encoding techniques. Many RF applications, such as RFID passive tags, RSNs, sensors, serial receivers, etc., use this type of encoding. Most well-known receiver decoder designs use an explicit clock to decode (Manchester or Pulse Interval) encoded data. On receiving encoded data from the transmitter, the clock is extracted from it. A classical decoding process for pulse width modulated signals is by oversampling with a clock [22,23]. The received signal is sampled at a much higher bit rate clock than the received signal in order to decode it as shown in Figure 23.2. *Symbol-1* or *symbol-0* of the received encoded data stream can be identified by counting the number of clock pulses within each symbol as shown in Figure 23.2 (2 for *symbol-0* and 4 for *symbol-1*). The well-known architecture of this classical decoding scheme is shown in Figure 23.3. This architecture basically consists of high-frequency oscillator, fast-clocked counter, and a comparator. The major disadvantage of using high clock rate driven decoders is the significant increase in the power consumption at the receiver side [23–26].

23.3.1 Pulse Width Coding Scheme

A novel explicit clock-less coding scheme for communication receivers is shown in Figure 23.4 for reducing the power consumption on the receiver side [24,27,37]. Pulse width coding (PWC) data shown in Figure 23.4 represent the encoded input demodulated serial data as "01100110." In Figure 23.4, the PWC data signal is sampled at every rising edge of the delayed version of

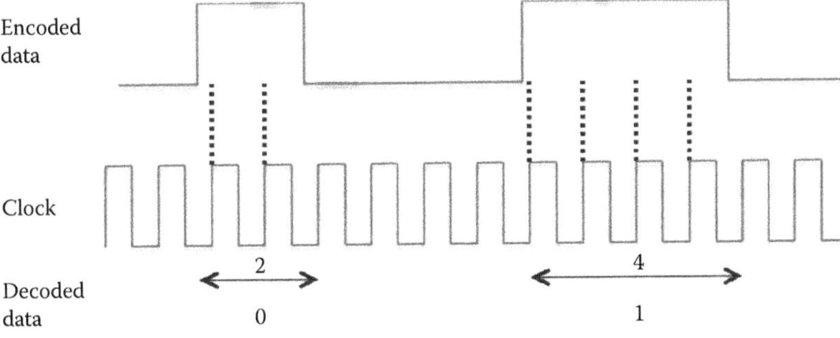

Figure 23.2 Conventional decoding scheme.

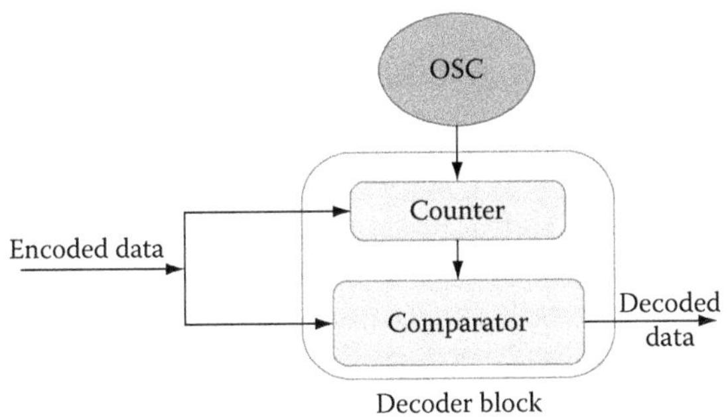

Figure 23.3 Conventional decoder architecture.

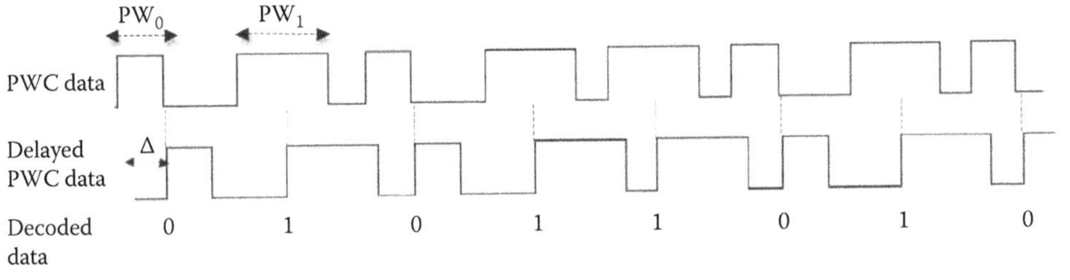

Figure 23.4 PWC scheme.

the PWC data signal in order to differentiate "1" and "0" for completing the decoding process. We can clearly see from Figure 23.2 that the decoded output bit is "1" whenever both the signals are high; otherwise it is a "0." The most important power parameter in this decoding scheme is the delay (Δ). The minimum possible delay required is about PW_0 and the maximum delay required is less than PW_1. In other words, for a successful PWC decoding, we need delay (Δ) to satisfy the condition: $PW_0 < \Delta < PW_1$. The decoding mechanism proposed in Ref. [24] for this scheme is an extremely simple, low-power, and clock-less circuit realized using Complementary metal-oxide-semiconductor (CMOS) digital chip design techniques.

The PWC scheme can be implemented and easily integrated to other well-known synchronous and asynchronous design variants that have high-power-consuming decoder modules in serial data communication receivers. A well-known direct application is in the symbol decoding process of the passive RFID tag and RSN node systems while the encoding remains unchanged at the transmitter.

The delay generally in hardware translates to a buffer element. A buffer element is generally built using even number of inverters. Using an optimized library, there is a possibility to further optimize the design schematic generated and lower the power values for inverters that are used to interpret large delay values. Another alternative to designing low-power inverters is to individually model them based on the choice of parameters such as width, length, and target technology of the metal-oxide-semiconductor (MOS) layout designs [29]. The inverter can be designed from the transistor level using the CAD (computer-aided design) layout tools. This would also give the designer the flexibility to alter the width of individual P-type metal-oxide-semiconductor (PMOS) and N-type metal-oxide-semiconductor (NMOS) transistors to generate the necessary delays within the circuit [30,31] conforming to the low-power requirements.

23.3.2 Data-Driven Decoder Architecture

The PWC scheme introduced in the previous Section can be realized using the explicit clock-less architecture shown in Figure 23.5. This data-driven architecture does not use any explicit clock to drive its components such as the shift register and the comparator. The delayed encoded input acts as a clock to trigger the components of the decoder thus eliminating the need for any explicit clock. This architecture was successfully simulated using the standard CAD tools as a low-power and low-area data-driven decoder [24]. In Ref. [24], it has been reported that the post-layout power consumption of the data-driven chip was about one-fourth the power consumed by the conventional decoder design for the same data rate of 40 kHz. The cell area of the data-driven decoder is 69% smaller than that of the conventional design. Elimination of the high-frequency oscillator, fast-clocked counter, and an explicit clock has contributed to this significant reduction in both power consumption and the area occupied by the data-driven chip.

An example passive RFID post-layout CMOS design was successfully simulated to operate at very low power using a custom low-power asynchronous computer [28]. This design uses the data-driven decoder with an integrated counter to it as the major low-power component of the entire architecture. The power consumption of the post-layout simulation results of both the synchronous/asynchronous RFID designs is illustrated in Figure 23.6a for different data transmission rates. The switching power consumption comparisons for different data rates are also illustrated in Figure 23.6b. There is a consistent linear increase in switching power for the synchronous design when compared to the asynchronous design as the data rate increases. In other words, the switching power of the asynchronous design is lower when compared to the synchronous design at each data rate. These results are very encouraging especially at the typical data rate corroborating the

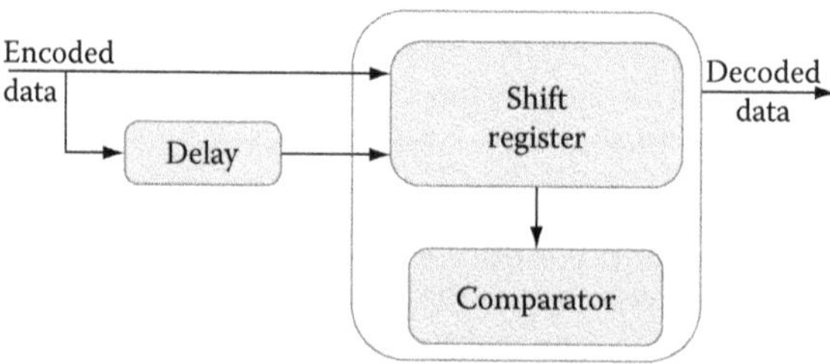

Figure 23.5 Novel data-driven decoder architecture.

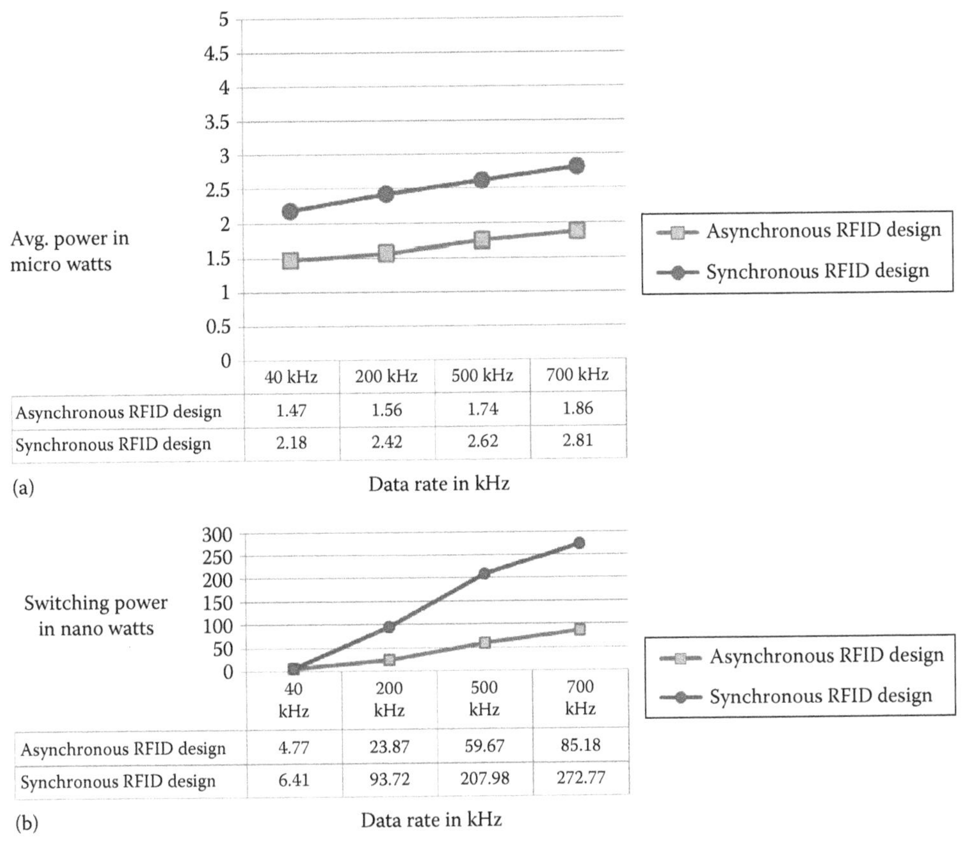

	40 kHz	200 kHz	500 kHz	700 kHz
Asynchronous RFID design	1.47	1.56	1.74	1.86
Synchronous RFID design	2.18	2.42	2.62	2.81

	40 kHz	200 kHz	500 kHz	700 kHz
Asynchronous RFID design	4.77	23.87	59.67	85.18
Synchronous RFID design	6.41	93.72	207.98	272.77

Figure 23.6 (a) Asynchronous versus synchronous average power comparisons. (b) Asynchronous versus synchronous switching power comparisons.

concept that a passive RFID tag can be designed using an asynchronous design to significantly reduce power requirements and thereby increasing its read range. The same concept can also be applied to sensor node architecture especially for the case where the data-driven decoder can be integrated with a low-power processing unit. The low-power processing unit design concept will be discussed in the next section.

23.4 Distributed Architecture Design for a WPSN Node Processor

Microcontroller design choice for a sensor node leads to a trade-off between speed and energy efficiency. In most cases, power constraints dominate, which in turn leads to significant computational constraints. In general, a microcontroller consists of a controller, volatile memory for data storage, ROM/EPROM/EEPROM, parallel I/O interfaces, clock generator, serial communication interfaces, etc. In Ref. [6], a general-purpose low-power 16 bit programmable microcontroller (MSP430F2132) is used for managing the entire node. The microcontroller design can be tailor-made for applications to further reduce the power requirements at the node. A significant contribution toward achieving a low-power sensor node processor is introduced in this chapter that highlights customizing a processor based on its subset instruction set architecture (ISA) for the specific target application [27].

SIMD is a well-known class of parallel computers in Flynn's taxonomy. SIMDs have the ability to perform the same operation on multiple data simultaneously for processors with multiple processing units. The need for synchronization between processors is not required. The proposed architecture presents a PSN(s) that can be replicated to produce an SIMD architecture. The passive units are powered and controlled by RF energy that enables convenient reconfiguration due to the ability to address nodes individually or in groups that can be simply and conveniently changed using RF communications. Thus, bits within the passive node processors can be set to perform or ignore commands thus allowing dynamic reconfigurability of the units composing the SIMD.

Consider an intelligent sensor network topology with a base station (sink node) in the center communicating with several wireless PSNs around it as shown in Figure 23.7 [27]. In a WPSN, a PSN is passive and is powered by the impinging RF wave, which is also used for communication, from a sink node. The major change with respect to architecture of a PSN is only the processing unit as shown in Figure 23.1. The sink node acts as the RF source assumed to have unlimited power that feeds the PSN with RF power. The sink node transmits RF power to the randomly deployed PSN nodes for processing, sensing, and data collection activities. This sink node wirelessly transmits commands to the PSN that executes these commands and responds back to the sink node. A PSN can be implemented as a CMOS chip that provides logic to respond to commands from a sink node. Thus, the sink node and the PSN combination can be viewed as a complete processor or as multiple processing units [38]. This will form the basis of our distributed concept that will be introduced in the following paragraphs.

The sink node (Control and Memory [C&M]) is an RF equipped control and storage base station, and the PSN processor is an execution unit with minimal storage capacity (e.g., registers) as shown in Figure 23.8. The sink node is allowed the flexibility to be a classical von Neumann– or Harvard-type architecture that consists of an interrogation control unit along with a program and memory units. Commands will be stored on the powered sink node that transmits the commands wirelessly to the PSN. The intent is to keep the PSN processor as simple as possible so as to maintain

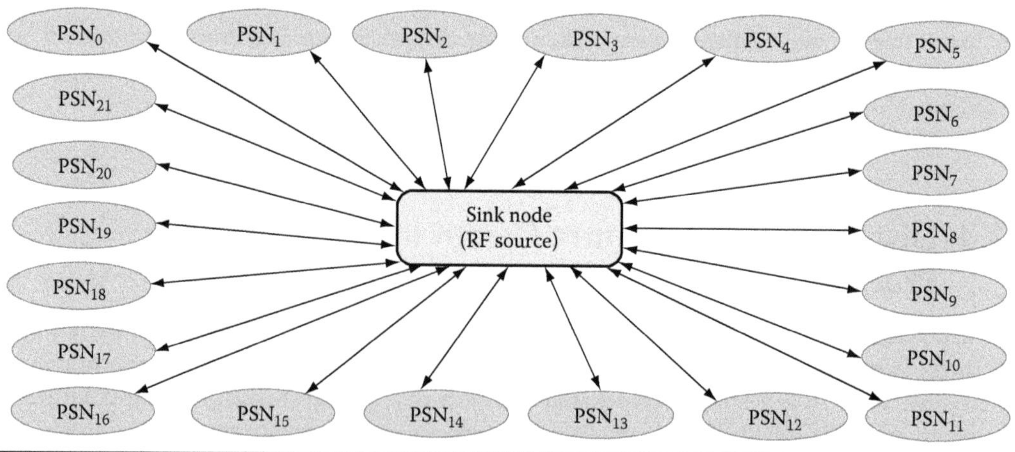

Figure 23.7 WPSN topology with PSN fed by an RF source–sink node.

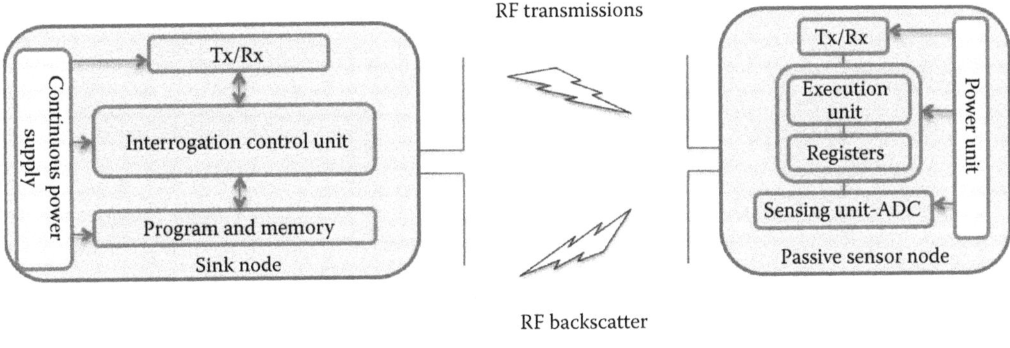

Figure 23.8 High-level distributed WPSN node processor architecture.

low-power requirements and/or extend read range from the sink node. Any unnecessary complexity on the PSN processor will be moved onto the powered sink node. This would significantly reduce the hardware on the PSN side, thus reducing the overall power consumption of the node.

Many other circuit design techniques can be applied to the sensor node architectures to reduce power consumption. One of the main power reduction techniques that can be employed in the proposed PSN processor design is to eliminate the use of an explicit clock overhead [28]. The entire circuitry of the PSN processor is to be asynchronous with the remote command execution controlled by the sink node making the system programmable and reconfigurable. The design uses a clock-less data-driven symbol decoder introduced earlier as a low-power component instead of the conventional input clocked data decoding process used at the sensor nodes [24]. Another technique commonly used to reduce power consumption is scaling down the supply voltage of the system or part of the system [1]. Any combination of these energy-efficient techniques can be used in addition to the distributed architecture concept based on application-specific requirements.

23.4.1 Exploring the 8051 Microcontroller and Its ISA for WPSN Applications

The choice of the Intel 8051 (i8051) is justified by the fact that it is still one of the most popular embedded processors. Furthermore, due to its small size and low cost, it has numerous applications where power efficiency is necessary. The most commonly used 8051 microcontroller in sensor nodes will be considered as an example for exploring its ISA and its application to the proposed conceptual distributed design.

The 8051 is an 8 bit microcontroller that includes an IS of 255 operation codes. The 8051 architecture consists of five major blocks, namely, control unit, ALU, decoder, ROM, and RAM. Based on the distributed design concept introduced as shown in Figure 23.8, the WPSN node processor consists of two major blocks with respect to 8051: 8051 compatible execution unit and the minimum number of temporary storage registers required. The execution unit is mainly an 8051 ALU. The number of instruction supported by the execution unit depends on the target application.

The sink node will transmit the program instructions to the WPSN node that executes these instructions and returns the results back to the sink node. The WPSN node, for example, will have the capabilities to perform functions like OR, XOR, AND, ADD, etc., that are compatible with 8051 depending on the application. This sink node and the WPSN node together form a complete processor. As the program to be executed by the WPSN node is stored in the sink node, the need

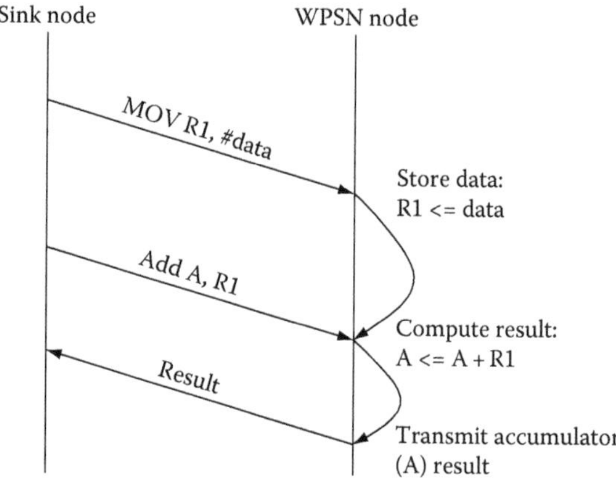

Figure 23.9 Sequence diagram for an ADD operation.

for program memory at the passive node is eliminated. There still may be a need for local scratch pad memory at the WPSN node although the number of bytes is drastically reduced in order to satisfy the power requirements. The WSPN node executes the instructions wirelessly as issued by the sink node.

Figure 23.9 represents a high-level sequence diagram for an ADD operation. Let us consider an ADD operation: ADD A, R1 (A = A + R1), where R1 denotes one of the eight (R0–R7) 8 bit 8051 working registers for a selected register bank and A denotes the 8-bit accumulator register. The sink node sends out the R1 values to load and store it in the temporary storage on the WPSN node processor unit. On receiving the ADD instruction, the passive node processor's execution unit performs the addition operation on the already existing value in the accumulator and the new R1 value. The computed result on the accumulator register is sent back to the sink node. The sink node will contain main memory that acts as the major storage area for the majority of data items.

Sensor applications require special-purpose hardware suitable to cater to a different set of requirements. Characteristics of the target applications and the utility of the sensors make it important to choose applicable hardware for sensor networks on a case-by-case basis. The power requirements of a WPSN limit the requirements needed for different applications used. Some of the well-known basic core algorithms form a class of simple applications such as the sum-array (sum of all values in a list), Top10 (finds top 10 values in a list), majority consensus (finds the majority values in a list), min-max finder (finds minimum and maximum values in a list), Binary search (typical search algorithm for a sorted list), Matrix Multiplication (matrix multiplication for small size matrices), etc. [32].

Generally, sensor networks employ only data filtering at the node so that every sensor sample need not be transmitted on the radio so as not to consume all the wireless bandwidth available to the network. By transmitting only necessary sensor data readings over the radio allows saving the available stored energy on the node. Let us consider a simple application scenario, for example, using sum-array application using 8051-ISA. The amount of temporary storage and the ALU capabilities of the WPSN node processor will be chosen to maintain low-power requirements. Assume that the only temporary memory space available for the execution unit is the register R7–R0 of a selected single register bank of 8051. The major function is an ADD operation and hence the choice of the arithmetic instructions that would be part of the execution unit on the WSPN node are ADD A, Rn; ADDC A, Rn. The minimal data transfer instruction necessary would be the MOV A, Rn;

MOV Rn, A; and MOV Rn, #DATA (8 bit). The WSPN node processor will support only those features required to interface and communicate with the sink node. Therefore, the branch, comparison, load, and store instructions will be implemented on the sink node side rather than on the passive side. This ISA will be compatible with the i8051-ISA. Additional instructions can be added to enhance the capability of the execution unit and also depending on the application.

To arrive at an energy-efficient computation solution, there is always a trade-off between the communications from and computation on a sensor node. Hence the choice of a design for a sensor node architecture depends not only on the low-power technique but also on the application space.

23.5 Data-Driven Architecture Design Flow Methodology

A data-driven architecture is a design paradigm that uses no explicit clock to drive its components [24]. Either data or local signals are used to drive components of the processor. This type of an asynchronous design uses no global periodic signals to synchronize its operations. Lack of strong support of commercial CAD tools is a major hurdle for synthesis of explicit clock-less designs. Asynchronous (very-high-speed integrated circuits) hardware description language (VHDL) designs generally are known to use non-synthesizable delay constructs such as wait, delay, etc., for their implementation in the absence of a global periodic signal for synchronization. Standard VHDL compilers are not known to synthesize VHDL code that implements an asynchronous design. Based on the conventional hardware descriptor languages, most asynchronous design methodologies [33–35] that have been proposed are not accessible to standard high-level design tools. The high-level data-driven design flow will be described in this section that requires minimum changes to a traditional synchronous flow [24,27,39].

Step 1: The data-driven design is first written in VHDL along with the necessary non-synthesizable delay constructs. This VHDL design is then simulated using Mentor Graphic's ModelSim. A customized test bench is used to verify the correct functionality of the design in ModelSim.

Step 2: A synthesized netlist is generated for the data-driven design using the Synopsys Design Compiler upon the successful verification of the VHDL design in step 1.

The "dc_shell" command interface provides a script execution environment based on Tool Command Language (TCL). The basic directives of a TCL script include setup environment variables, constraints, basic compilation directives, etc. The major modification will be to eliminate any clock in the script. All the statements that involve the non-synthesizable VHDL delay constructs are identified and removed. Synthesizable delay commands need to be separately inserted into the TCL script during the synthesis process. The available delay commands are set_max_delay and set_min delay. These commands have several options, generally needs a -from set of start points, -to set of end points along with a fixed target delay value. The start and end points refer to specific cells and their corresponding input and output pins in the schematic of the design. Identification of the necessary start/end points in the design is the key for the accurate working of the design as insertion of these delay changes the timing graph of the design.

The next step is to compile, simulate, and verify the generated design netlist Verilog file along with the target technology library using ModelSim.

Step 3: After the successful post-synthesis verification, the design layout along with the post place-and-route netlist is generated using a VLSI Cadence layout tool known as Cadence Encounter. The post place-and-route design netlist is simulated and verified for the expected operation using ModelSim along with a delay file format. After the successful post-layout simulation of the netlist, the layout verification of the data-driven design is said to be complete.

The final step is to estimate the power consumed by the data-driven design using Cadence Encounter. A switching activity file is generated from the initial test bench using ModelSim. During the final stage of the place-and-route process, the activity file is used in the power rail analysis option available with Encounter to produce the power report.

The advantage of using data-driven designs is not only for achieving low power but also for allowing flexibility to implement explicit clock-less designs using standard CAD tools. The data-driven symbol decoder introduced in Ref. [24] was implemented using this methodology. This methodology can also be applied to the distributed WPSN 8051-node architecture when implemented as a data-driven design. The design flow used for data-driven designs can also be extended to a variety of design paradigms such as synchronous, asynchronous, globally asynchronous locally synchronous, and globally synchronous locally asynchronous [24].

23.6 Conclusion

Conventional WSNs are a disposable system as they are dependent on the limited lifetime of their batteries. WPSN system of passive nodes does not face this problem as they are remotely powered by an RF source, but do have limited ranges. This chapter discusses novel low-power solutions to increase the range of WPSN nodes. This chapter illustrates the importance of the data-driven architecture using a novel clock-less symbol decoder architecture and its low-power applications that include synchronous and asynchronous design variants that have high-power-consuming modules in communication receivers. A detailed power analysis of the post-layout simulation power results of the data-driven symbol decoder and the conventional clocked symbol decoder for passive RF receiver systems has been analyzed. This chapter introduces the elements and concepts of the design of a WPSN node processor as a distributed architecture that operates remotely and wirelessly from the sink node. A high-level asynchronous design flow that can be used to implement the data-driven design using synchronous CAD tools is also discussed. This design flow will provide the reader with sufficient guidelines to design and implement application-specific data-driven processor architectures.

This research has the potential to realize WPSN node applications for environmental, structural, and medical fields especially while providing the basis for a programmable, reconfigurable, and a low-power passive processing unit for distributed computing.

Currently, our research work is focused on developing a low-power distributed 8051-SIMD architecture as a clock-less design based on the concepts described in this chapter.

References

1. M. Hempstead, M. J. Lyons, D. Brooks, and G.-Y. Wei, Survey of hardware systems for wireless sensor networks, *ASP Journal of Low Power Electronics*, 4(1), April 2008.
2. I. F. Akyildiz et al., A survey on sensor networks, *IEEE Communications Magazine*, 40(8), 102–114, August 2002.
3. M. T. Isik and O. B. Akan, PADRE: Modulated backscattering- based passive data retrieval in wireless sensor networks, *Proceedings of IEEE WCNC*, Budapest, Hungary, pp. 1–6, April 2009.
4. O. B. Akan, M. T. Isik, and B. Baykal, Wireless passive sensor networks, *IEEE Communications Magazine*, 47(8), 92–99, August 2009.
5. A. Bereketli and O. B. Akan, Communication coverage in wireless passive sensor networks, *Communications Letters, IEEE*, 13(2), 133–135, February 2009.

6. R. D. Fernandes, N. B. Carvalho, and J. N. Matos, Design of a battery-free wireless sensor node, *EUROCON—International Conference on Computer as a Tool (EUROCON)*, 2011 IEEE, Lisbon, Portugal, pp. 1–4, April 27–29, 2011.
7. R. D. Fernandes, A. S. Boaventura, N. B Carvalho, and J. N. Matos, Increasing the range of wireless passive sensor nodes using multisines, *2011 IEEE International Conference on RFID-Technologies and Applications (RFID-TA)*, Sitges, Spain, pp. 549–553, September 15–16, 2011.
8. M. Philipose et al., Battery-free wireless identification and sensing, *Pervasive Computing*, 4(1), 37–45, January–March 2005.
9. R. Amirtharajah and A. P. Chandrakasan, Self-powered signal processing using vibration-based power generation, *IEEE Journal of Solid-State Circuits*, 33, 687–695,1998.
10. Y. Ammar, A. Buhrig, M. Marzencki, B. Charlot, S. Basrour, K. Matou, and M. Renaudin, Wireless sensor network node with asynchronous architecture and vibration harvesting micro power generator, *Proceedings of the 2005 Joint Conference on Smart Objects and Ambient Intelligence: Innovative Context-Aware Services: Usages and Technologies*, Grenoble, France, pp. 287–292, ACM Press, October 2005.
11. H. Stockman, Communication by means of reflected power, *Proceedings of the I.R.E.*, 36, 1196–1204, 1948.
12. L. P. Alarcon, T. T. Liu, M. D. Pierson, and J. M. Rabaey, Exploring very low energy logic: A case study, *ASP Journal of Low Power Electronics*, 3, 223–233, 2007.
13. V. Ekanayake, C. Kelly, and R. Manohar, An ultra low power processor for sensor networks, *Proceedings of the 11th International Conference on Architectural Support for Programming Languages and Operating Systems*, Boston, MA, pp. 27–36, October 2004.
14. V. Ekanayake, C. Kelly, and R. Manohar, BitSNAP: Dynamic significance compression for a low-energy sensor network asynchronous processor, *Proceedings of the 11th International Symposium on Asynchronous Circuits and Systems*, New York, pp. 144–154, March 2005.
15. C. Kelly, V. Ekanayake, and R. Manohar, SNAP: A sensor-network asynchronous processor, *Proceedings of the 9th International Symposium on Asynchronous Circuits and Systems*, Vancouver, British Columbia, Canada, pp. 24–33, May 2003.
16. L. Necchi, L. Lavagno, D. Pandini, and L. Vanzago, An ultra-low energy asynchronous processor for wireless sensor networks, *Proceedings of the 12th IEEE International Symposium on Asynchronous Circuits and Systems*, Grenoble, France, p. 78, March 13–15, 2006.
17. M. Powell, S.-H. Yang, B. Falsafi, K. Roy, and T.N. Vijaykumar, Gated-Vdd: A circuit technique to reduce leakage in deep-submicron cache memories, *International Symposium on Low Power Electronics and Design (ISLPED)*, Rapallo, Italy, pp. 90–95, June 2000.
18. M. Hempstead, N. Tripathi, P. Mauro, G.-Y. Wei, and D. Brooks, An ultra low power system architecture for sensor network applications, *The 32nd Annual International Symposium on Computer Architecture (ISCA)*, Madison, WI, pp. 208–219, June 2005.
19. S. Hanson, B. Zhai, M. Seok, B. Cline, K. Zhou, M. Singhal, M. Minuth, J. Olson, L. Nazhandali, T. Austin, D. Sylvester, and D. Blaauw, Performance and variability optimization strategies in a sub-200 mV, 3.5 pJ/inst, 11 nW subthreshold processor, *IEEE Symposium on VLSI Circuits (VLSI-Symp)*, Kyoto, pp. 152–153, June 2007.
20. L. Nazhandali, B. Zhai, J. Olson, A. Reeves, M. Minuth, R. Helfand, S. Pant, T. Austin, and D. Blaauw, Energy optimization of subthreshold-voltage sensor network processors, *The 32nd Annual International Symposium on Computer Architecture (ISCA)*, Madison, Wisconsin, pp. 197–207, June 2005.
21. B. Zhai, L. Nazhandali, J. Olson, A. Reeves, M. Minuth, R. Helfand, S. Pant, D. Blaauw, and T. Austin, A 2.60pJ/Inst subthreshold processor for optimal energy efficiency, *IEEE Symposium on VLSI Circuits (VLSI-Symp)*, Honolulu, HI, pp. 154–155, June 2006.

22. N. Cho, S.-J. Song, S. Kim, S. Kim, and H.-J. Yoo, A 5.1-uW 0.3-mm2 UHF RFID tag chip integrated with sensors for wireless environmental monitoring, *IEEE European Solid State Circuits Conference (ESSCIRC)*, Grenoble, France, pp. 279–282, September 2005.
23. N. Cho et al., A 8-uw, 0.3-mm RF-powered transponder with temperature sensor for wireless environment monitoring, *IEEE International Symposium on Circuits and Systems*, 5, 4763–4766, May 2005.
24. V. Sai, A. Ogirala, and M. H. Mickle, Low power data driven symbol decoder for a UHF passive RFID tag, *Journal of Low Power Electronics (JOLPE)*, 8(1), February 2012.
25. M. Minhong, RFID radio circuit design in CMOS, *Ansoft*, Technical Report, September 2006.
26. S.-J. Kim, M.-C. Cho, J. Park, K. Song, Y. Kim, and S. H. Cho, An ultra low power UHF RFID tag front-end for EPCglobal Gen2 with novel clock-free decoder, *IEEE International Symposium Circuits and Systems*, Seattle, WA, pp. 660–663, May 2008.
27. V. Sai, A low power distributed architecture: A passive remote execution processor, PhD dissertation (in process), unpublished.
28. V. Sai, A. Ogirala, and M. H. Mickle, Low power RFID design using custom asynchronous passive computer, *Journal of Low Power Electronics (JOLPE)*, 6(4), 469–627, December 2010.
29. J. M. Rabaey, A. Chandrakasan, and B. Nikolic, *Digital Integrated Circuits*, 2nd edn., Upper Saddle River, NJ: Prentice Hall, 2002.
30. A. Shebaita and Y. Ismail, Lower power, lower delay design scheme for CMOS tapered buffers, *Design and Test Workshop (IDT)*, 1–5, 2009.
31. D. Sharma and R. Mehra, Low power, delay optimized buffer design using 70 nm CMOS technology, *International Journal of Computer Applications*, 22(3), 13–18, 2011.
32. S. Mysore, B. Agrawal, and F. T. Chong, Timothy sherwood, exploring the processor and ISA design for wireless sensor network applications, *Proceedings of the 21st International Conference on VLSI Design*, Hyderabad, India, pp. 59–64, January 4–8, 2008.
33. C. H. (Kees) van Berkel, M. B. Josephs, and S. M. Nowick, Special issue on asynchronous circuits and systems, *Proceedings of the IEEE*, 87(2), February 1999.
34. S. Hauck, Asynchronous design methodologies: An overview, *Proceedings of the IEEE*, 83(1), 69–93, January 1995.
35. T. Nanya, Challenges to dependable asynchronous processor design, in *Logic Synthesis and Optimization*, T. Sasao, ed., Dordrecht, the Netherlands: Kluwer Academic Publishers, 1993.
36. M. W. Chiang, Z. Zilic, J.-S. Chenard, and K. Radecka, Architectures of increased availability wireless sensor network nodes, *Proceedings of the International Test Conference on International Test Conference*, pp. 1232–1241, Charlotte, NC, October 26–28, 2004.
37. V. Sai, A. Ogirala, and M. H. Mickle, A low-power pulse width coding scheme for communication receiver systems, Communications, ACTA Press, 1(1):1–4, 2012.
38. V. Sai, A. Ogirala, and M. H. Mickle, A Low-Power Wireless Distributed Dynamic Network Design Concept: FFT Processor Architecture, Communications, ACTA Press, 1(1), 2012.
39. V. Sai, A. Ogirala, and M. H. Mickle, Implementation of an Asynchronous Low-Power Small-Area Passive RFID Design Using Synchronous Tools For Automation Applications, *Journal of Low Power Electronics (JOLPE)*, 8(4), August 2012.

Chapter 24

Fusion of Pre/Post-RFID Correction Techniques to Reduce Anomalies

Peter Darcy, Prapassara Pupunwiwat, and Bela Stantic

Contents

24.1 Introduction

Radio frequency identification (RFID) technology refers to the use of multiple tags being attached to various items that are scanned and recorded in a database. The process itself eliminates mundane tasks such as conducting manual inventory checking and counting or using a barcode scanner to count the items to be purchased at a supermarket checkout. Modern integrations of this technology include baggage tracking at airports, pet owner identification, and tagging objects in stores to enforce security by alerting management when an item has left the facility without the tag being deactivated. Despite the wide-scale adoption and advantages of RFID, several issues exist that introduce a level of unreliability resulting in the technology being used only in a fraction of its potential applications.

Due to the RFID technology relying on radio signals automatically collecting tag data, certain anomalies are also recorded lowering the quality of the captured observations. These anomalies are classified as data that are captured but are not meant to be, the false-positives, and the data that are not present in the data set when it is meant to be, false-negatives. False-positives usually arise when either an observation is captured twice as a duplicate reading when it is only meant to be recorded once, or in the case of data being captured by a reader when they are not within its normal range. False-negatives usually occur due to external forces such as interference from water or metal, or tags interfering and colliding with each other on the air interface.

Previous methodologies have been proposed in past literature to correct the anomalies at the different stages of the RFID capturing process: the physical, middleware, and deferred stages. Common physical solutions involve altering the location or conditions at which the data are physically captured such as modifying the orientation of the objects or attaching additional tags to increase the likelihood of scanning the object. Middleware solutions involve the use of algorithms such as anti-collision techniques or filters to correct data as they are read into the reader. Finally, deferred solutions employ algorithms to process captured observations after they have been stored within the data warehouse. Each of these methodologies, however, either introduces an amount of artificially created anomalies or lacks the intelligence to correctly decipher the correct approach to eliminate the anomalies.

To reduce the amount of unintentional artificial anomalies and increase the intelligence of the correction approach, we have introduced various novel techniques both before and after the storage of captured data. In this chapter, we examine various techniques used in different ways. This includes both probabilistic and deterministic anti-collision algorithms, and various machine-learning algorithms designed specifically to identify and correct both false-positive and false-negative anomalies. With regards to the tree-based pre-processing approach, we reduce the tag starvation problem resulting in missed readings having an increased chance of being recorded and compare it to already used state-of-the-art methodologies. Within the machine-learning deferred approaches, we examine the use of a Bayesian Network, Neural Network, and Non-Monotonic Reasoning classifiers in a novel application designed to both detect and correct an anomaly where possible, and discover the highest achieving techniques.

24.2 Background

RFID technology refers to the use of multiple tags being attached to various items that are scanned and recorded in a database. Modern integrations of this technology include baggage tracking at airports, pet owner identification, and tagging objects in stores to enforce security by alerting management when an item has left the facility without the tag being deactivated.

Due to the RFID technology relying on radio signals automatically collecting tag data, certain anomalies are also recorded lowering the quality of the captured observations. These anomalies are classified as data that are captured but are not meant to be, the false-positives, and the data that are not present in the data set when it is meant to be, false-negatives. After significantly reducing these issues, it will possible to integrate RFID technology into numerous commercial sectors resulting in an increase in both efficiency and effectiveness of business processes.

To combat the highly ambiguous anomalies in RFID data sets, intelligent anti-collision protocols, such as deterministic and probabilistic anti-collision algorithms, along with classifiers, such as Bayesian networks, neural networks, and non-monotonic reasoning, may be employed. Deterministic method begin the identification process by issuing a prefix until it gets matching tags. It will then continue to ask for additional prefixes until all tags within the region are found. In contrast, probabilistic methods allow tags to respond at randomly generated times. If a collision occurs, colliding tags will have to identify themselves again after waiting for a random period of time. Bayesian networks operate by finding the highest probable conclusion when given a list of observations. Similarly, the neural network has a list of observations input into it as a feature set and the trained network will determine which conclusion is to be drawn. Finally, non-monotonic reasoning (NMR) operates by having a set of rules defined by the user including precedence of these rules and uses them to determine its conclusion.

24.2.1 Radio Frequency Identification

RFID has had a long history commencing with its utilization during the Second World War to its modern usage [1]. The basic architecture of RFID itself consists of a tag, reader, and middleware to perform advanced analysis on the data that make it practical for use in many applications with beneficial outcomes. There are several problems that arise when using the passive tags due to the nature of the system, in particular, the amount of unreliable readings in the raw data.

24.2.1.1 System Architecture

The system architecture of an RFID system contains four important components [2]: an RFID tag, an RFID reader, the RFID middleware, and the database storage. For a diagram representing the flow of information in this system architecture, see Figure 24.1.

The RFID tag is the simplest, lowest level component of the RFID system architecture. These tags come in three types—passive, semi-passive, and active. The tag itself is made up of three different parts: the Chip that holds the information the tag is to dispense, the antenna that is used to transmit the signal out, and the packaging that houses the chip and antenna and may be applied to the surface of other items [2]. Passive tags are the most error prone but, due to the lack of a battery, also the most cost effective and long lasting [3].

Electromagnetic pulses emitted from the readers allow the passive tag enough energy to transmit its identification back [4]. In comparison, the semi-passive tag has a battery source attached to it. However, the battery is only utilized to extend the readability range as it will use the reader's pulse to transmit its information resulting in a shorter life span but increased observation integrity. The final tag is the active tag that utilizes a battery to not only extend its range but also to transmit its identification number. From its heavy reliance of the battery, this tag has the highest cost and shortest life span of all the tags currently available. The active tag also has advanced features not capable by the other types such as communicating with other tags within proximity [2]. Even today, there are novel and emerging technologies to reduce the production cost even further such as the *chipless RFID system tags* [5,6] *and readers* [7], which will be integrated into future applications [8,9] such as space exploration [10] and airport baggage tracking [11].

The RFID readers are the machines used to record the tag identifiers and attach a timestamp of the observation. They do this by emitting a wave of electromagnetic energy that then interrogates the tags until they have responded. These devices have a much greater purpose when needing to

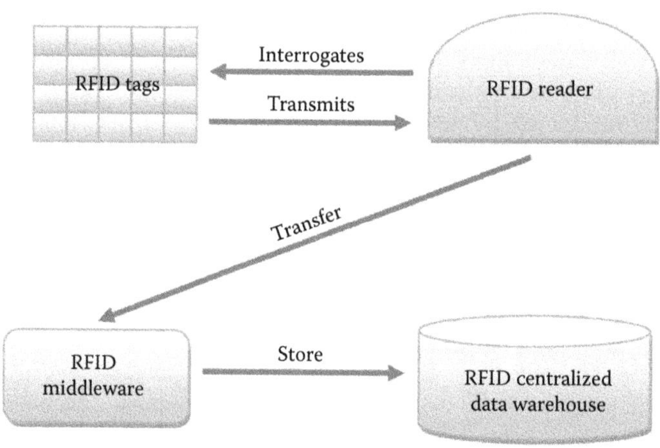

Figure 24.1 Flow of information between the different components of the RFID system architecture.

interrogate passive and semi-passive tags as they also provide the power necessary to transmit the information back. Readers like the tags come in a variety of types such as the handheld reader and the mounted reader. The mobile handheld tags are used for mainly determining which objects are present within a group, for example, when needing to stocktake items within a supermarket. In comparison, the mounted readers are static in geographical locations and used primarily to track items moving through their zones such as mounted readers to observe all items on a conveyer belt.

The middleware, also commonly known as the savant or edge systems, is the layer at which the raw RFID readings are cleaned and filtered to make the data more application friendly. It receives information passed into it from the readers and then applies techniques such as anti-collision and smoothing algorithms to correct simple missing and duplicate anomalies [12,13]. The filtrated observational records including the tag and reader identifiers along with the timestamp the reading was taken are then passed onto the Database storage.

At the end of the data capture cycle, observational records are passed into a common area where all readings are achieved. This component is known as the database storage and is used to hold all information that is streamed from the readers. In most cases, due to the massive amount of interrogation required to read every tag constantly, this can result in a massive flood of data. For example, it has been stated that Wal-Mart has generated 7TB of data daily from its integration of RFID systems [14]. Having all information stored in a central database also allows for higher level processes such as data cleaning, data mining, and analytical evaluations.

24.2.1.2 Format of Observations

The format of the data recorded in the database after a tag has been read consists of three primary pieces of information: the electronic product code (cEPC), the reader identifier that made the observation, and the timestamp that contains the time the reading occurred. Table 24.1 contains information typically stored in the database storage.

The EPC is a unique identification number introduced by the Auto-ID center and given to each RFID tag. It is made up of a 96 bit, 25 character-long code containing numbers and letters. The number itself is made up of a header for 8 bits, EPC manager for 28 bits, object class (COC), for 24 bits, and serial number (SN) for 36 bits [15]. Ward and Kranenburg state that a possible alternative to using the EPC is to employ IPv6, which is the advanced version of Internet addresses. These will take over the current system, which is IPv4 [15]. It is estimated that since IPv6, will have 430 quintillion Internet addresses as opposed to the current 4 billion address limit, there will be enough addresses for all items being tracked with RFID.

Table 24.1 Table Populated with Sample RFID Data Containing Information about EPC, Reader, and Timestamp

EPC	*Reader*	*Timestamp*
030000E500023C000431BA3	001	2008-07-29 14:05:08.002
030000E500023C000431BA3	003	2008-07-29 14:32:12.042
030000E500023C000431BA3	002	2008-07-29 14:45:54.028
030000E500023C000431BA3	004	2008-07-29 15:02:06.029
030000E500023C000431BA3	007	2008-07-29 15:18:49.016

The reader identifier attribute is the unique identifier of the reader so that the analyzer will be informed of which reader took the EPC reading. If the reader is static in its location as well, such a position of the reading may be derived from a simple query in the database later using this value. Knowledge of the geographical location of each unique reader identifier may also provide additional information needed for future business processes. The Timestamp contains a temporal reading used to identify the date and time that the tag passed within the vicinity of the reader. For example, 2008-07-29 14:05:08.002 would be stored as a timestamp.

24.2.1.3 RFID Anomalies

There are certain characteristics associated with the nature of RFID technology [16,17]. These challenges include low-level data, error-prone data, high data volumes, and its spatial and temporal aspects. With regards to the error-prone data, RFID observations suffer from three various anomalies that are recorded along with actual RFID readings. The first is a wrong reading in which data are captured where they should not be. The second is duplicate readings in which a tag is observed twice rather than once. The third is the missed readings that occur when a tag is not read when and where the object it is attached to should have been physically within proximity.

24.2.2 Collision Handling in RFID Data Streams

RFID collision handling is one of the most heavily researched topics because it is a very important step to determine a quality of captured data. The better quality of data at the earlier stage of data processing means less complex algorithms are needed for RFID event process and database management. This section explains the type of each collision and surveys on existing *deterministic* and *probabilistic anti-collision* methods.

24.2.2.1 RFID Collision Types

Simultaneous transmissions in RFID systems lead to collisions as the readers and tags typically operate on the same channel. Three types of collisions are possible: *reader-to-reader* collision, *reader-to-tag* collision, and *tag-to-tag* collision [18].

- *Reader-to-reader collisions:* Interference occurs when one reader transmits a signal that interferes with the operation of another reader and prevents the second reader from communicating with tags in its interrogation zone. Reader-to-reader collision can be easily avoided by determining the appropriate reader's deployment that prevents direct signal interference between two or more readers.
- *Reader-to-tag collisions:* Interference occurs when one tag is simultaneously located in the interrogation zone of two or more readers, where more than one reader attempts to communicate with that tag at the same time.
- *Tag-to-tag collisions:* Tag collision in RFID systems happens when multiple tags are energized by the RFID reader simultaneously and reflect their respective signals back to the reader at the same time. This problem is often seen whenever a large volume of tags must be read together in the same reader zone. The reader is unable to differentiate these signals.

24.2.2.2 Deterministic Anti-Collision Protocols

Deterministic methods can be classified into a *memory* tree-based algorithm and a *memoryless* tree-based algorithm. In the memory algorithm, which can be grouped into "tree splitting," "binary search," and "bit arbitration," the reader's inquiries and the responses of the tags are stored and managed in the tag memory. This results in an equipment cost increase especially for RFID tags. In contrast, in the memoryless algorithm, the responses of the tags are not determined by the reader's previous inquiries. The tags' responses are determined only by the present reader's inquiries so that the cost for the tags can be minimized. "Query Tree" (QT) is classified as memoryless algorithm.

Depending on the number of tags that respond to the interrogator, there are three cycles of communication between tag and reader in deterministic approaches.

- *Collision cycle: Collision cycle* occurs when the number of tags that respond to the reader is more than one. The reader cannot identify the ID of tags.
- *Idle cycle: Idle cycle* occurs when there is no response from any tag to the reader. This type of cycle is unnecessary and should be minimized.
- *Successful cycle: Successful cycle* happens when exactly one tag responds to the reader and the reader can identify the ID of that tag.

For the tree-based *anti-collision*, we focus on QT-based protocols because it is the most acceptable and effective *anti-collision* technique for passive UHF tags [19]. The QT [20] is a data structure for representing prefixes that are sent by the RFID reader. The QT algorithm consists of loops, and in each loop, the reader issues a query with specific prefixes, and the matching tags respond with their information. If only one tag replies, the reader successfully recognizes the tag. If more than one tag tries to respond to reader's query, tag collision occurs and the reader cannot get any information about the tags. The reader, however, can recognize the existence of tags to have ID that matches the query. To further identify collided tags, the QT algorithm tries to query with 1-bit longer prefixes in the next round of identification. By extending the prefixes, the reader can recognize all the tags.

Figure 24.2 displays an example of a QT procedure. An identification process starts at Level one of tree, where QT uses tag IDs to split a tag set. It can be seen that *Tag 1010* is successfully identified in the first round because from all three tags, only *Tag 1010* has "1" for the first bit of string. In the second round of identification, *idle cycle* was created, as there was no tag starting with "00" for the first two bits. In the third round of identification, the other two tags, *Tag 0100* and *Tag 0111*, are successfully identified.

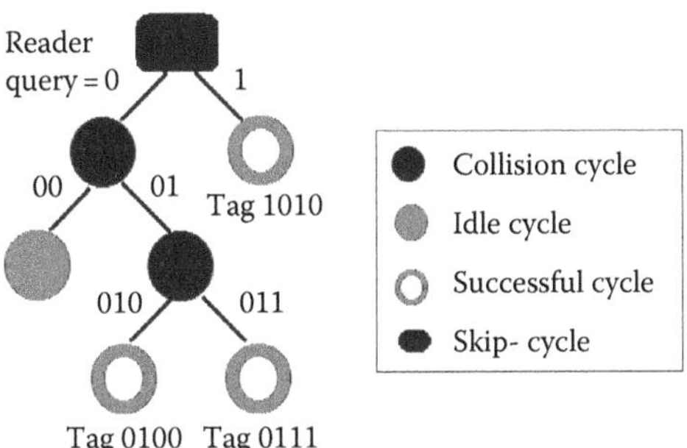

Figure 24.2 Query tree memoryless-based anti-collision protocol.

24.2.2.3 Probabilistic Anti-Collision Protocols

In a probabilistic approach, tags respond to readers at randomly generated times. If a collision occurs, colliding tags will have to identify themselves again after waiting a random period of time [19]. When we mentioned the probabilistic *anti-collision* approach in RFID, we usually refer to the ALOHA-based approach, which is the most widely used type of *anti-collision*. "Slotted ALOHA" [21], which initiates discrete time-slots for tags to be identified by reader at the specific time, was first employed as an *anti-collision* method in an early days of RFID technology. The principle of slotted ALOHA techniques is based on the "pure ALOHA" introduced in early 1970s [22], where each tag is identified randomly. To improve the performance and throughput rate, different *anti-collision* schemes were suggested in the past literature. "Framed-slotted ALOHA" technique is the most improved ALOHA-based technique currently applied in many applications. The most accepted framed-slotted ALOHA technique is the "dynamic framed-slotted ALOHA (DFSA)."

In DFSA, each tag in an interrogation zone selects one of the given N slots to transmit its identifier; and all tags will be recognized after a few frames. Each frame is formed of specific number of slots that is used for communication between the readers and the tags. To determine the *frame-size*, it gathers and uses information such as number of *successful slots*, *empty slots*, and *collision slots* from the previous round to predict the appropriate *frame-size* for the next identification round [23–26]. DFSA can identify the tag efficiently because the reader adjusts the *frame-size* according to the estimated number of tags. However, the *frame-size* change alone cannot sufficiently reduce the tag collision when there are a number of tags because it cannot increase the *frame-size* indefinitely. DFSA has various versions depending on different tag estimation methods used. There have been several researches to improve the accuracy of *frame-size* by implementing *frame-size estimation* techniques [27–30]. According to the DFSA protocol, the reader picks a tag within an interrogation zone by the command "Select," then issues "Query," which contains a "Q" parameter to specify the *frame-size* (frame-size $F = 2^Q - 1$). Each selected tag will pick a random number between 0 and $2^Q - 1$ and places it into its slot counter. The tag, which picks zero as its slot number, will respond and backscatter its EPC to the reader. Then, the reader issues the "queryrep" or "queryadjust" command to initiate another slot [31,32].

Similar to the tree-based *anti-collision*, there are three kinds of slot in ALOHA-based *anti-collision*, as shown in Figure 24.3: (1) *empty slot* where there is no tag reply, (2) *successful slot* where there is only one tag reply, and (3) *collision slot* where there is more than one tag reply. The term "initial Q" refers to the first "Q" or *frame-size*, which applies to a specific identification cycle.

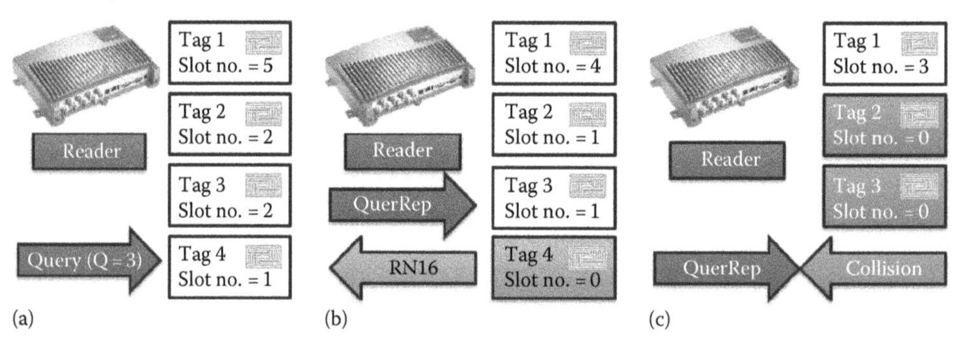

Figure 24.3 (a) Empty slot, (b) successful slot, and (c) collision slot in EPC class 1 generation 2 protocol.

In Figure 24.3a, the reader first initiates a "query" and broadcasts the signal to nearby tags. Since there is no tag that picks zero as its slot counter, the slot is counted as an *empty slot*. Figure 24.3b shows that after the first "query" was sent, each tag deducted its slot counter by one. The reader then sends "QueryRep" to tags in close proximity; and any tag that has zero as its slot counter replies. When there is only one tag that responds, a *successful slot* occurs and the tag replies to the reader with its RN16. Figure 24.3c demonstrates that when two tags respond to the reader at the same time, a *collision slot* occurs and in this case, no information is transmitted.

24.2.3 Classifiers

While anti-collision techniques can be applied to filter the incoming data, it is not able to restore highly ambiguous missing reading and no wrong anomalies. Thus, a highly intelligent approach should be utilized to correct these anomalies after it has been stored inside the database. One such highly intelligent approach is the integration of classifiers to correctly determine if an anomaly is present and then the actions to correct the information to maintain a high level of integrity.

24.2.3.1 Bayesian Networks

Bayesian networks refer to a network designed to find the highest probable solution to any given problem. This is usually performed by determining the product of evidence found in a situation and comparing it with other possible causes until the greatest probable outcome is discovered. When expressing the mechanics of any Bayesian network, there are three common mediums: a joint distribution equation, an influence diagram, and a Bayesian network table. To demonstrate the idea of a Bayesian network, we have developed an example scenario in which a network is developed to determine the cause of a tree falling down (human or nature) when given such attributes as council markings and weather. The specific rules for this example are that there is a very high chance for the council to cut down the tree if this scenario is coupled with fine weather. However, if the weather is stormy, there is less chance of the tree being cut down by humans:

$$P(X_1, \ldots, X_n) = \prod_{i=1}^{n} P(X_i | (X_1, \ldots, X_{i-1})) \tag{24.1}$$

The mathematical equation is a formula designed to express the process utilized in determining the percentage of likelihood of a cause being true. The information depicted in the equation will then be translated into a table. This table consists of the evidence vs. the causes in which a percentage is given to each case for the true and false outcomes of each scenario. From this table, all the percentages are multiplied together and a percentage score is given to each of the causes. A Bayesian network will then conclude that the most probable cause is the cause with the highest achieving percentage.

24.2.3.2 Artificial Neural Network

As seen in Figure 24.4, an artificial neural network (ANN) is a classifier designed to emulate the learning behavior of the brain. It does this by creating a fixed amount of neurons that are trained to deliver a certain output when fed various input. The entire process has actually been based on the biological neuron. Dendrites will receive information that is passed to the cell body whose objective is to pass the information into the axon when certain requirements are met and, thus,

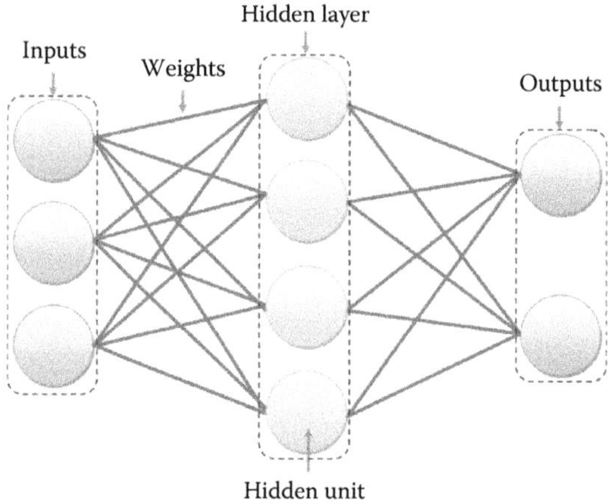

Figure 24.4 A high-level interpretation of how a neural network is designed with its three main layers: the input, hidden, and output layers.

to dendrites of other neurons via the synapse connection. The crucial difference between a digital neuron and its biological counterpart is that there is a computational limit to the amount of hidden units that may be present within a network. Unfortunately, technology has not advanced enough to effectively and efficiently emulate the amount of neurons the human brain possesses, which is estimated to be between 10 billion and 1 trillion [33].

The ANN consists of three main layers: the *input layer*, *hidden layer(s)*, and the *output layer*[34]. The processes include receiving inputs that are modified at a central sum area. The neuron will then apply an activation function such as the hard limiter, in which it is either assigned +1 or −1 if the value is positive or negative, respectively, or sigmoidal functions, which is displayed in Equation 24.2, to derive a value for the output. With regard to training the network, there are several techniques available such as the *back-propagation* (BP) [35,36] and *genetic algorithms* [37,38], which are both considered to be leaders with regard to the configuration of ANN:

$$f(x) = 1/(1 + exp(-x)) \tag{24.2}$$

When attempting to configure the neural network weights, one method that exists is to utilize training algorithms. Two dominant training algorithms that have been proven to excel in network training are the back-propagation algorithm and the evolutionary neural network. BP relies on the concept of training the network by propagating error back through the network via modifying the weights after the output has been calculated [35]. The algorithm uses either a predetermined limited amount of iterations or the root-mean-square (RMS) error threshold of the calculated output as stopping criteria [36].

The evolutionary neural network training algorithm in contrast to BP utilizes the theory of genetic evolution to train the network weights. Similar to the genetic algorithm process of training a Bayesian network, all the weights are added into a chromosome as genes to be manipulated according to the fittest output obtained [37]. The weights are initialized as small random numbers which are checked for either obtaining a high enough score or if the amount of generation limit has been reached. In the case where neither of the stopping criteria has been met, the algorithm will examine each chromosome in relation to achieving the correct output. A certain amount of the unfit chromosomes is then destroyed within the population and is replaced with child chromosomes of two of the fitter chromosomes. The child chromosomes can be created by varied means such as one-point, two-point, or uniform crossover [38]. Mutation will then be applied to a certain small percent of the population to ensure that the network avoids problems such as network paralysis or local minima [39].

24.2.3.3 Non-Monotonic Reasoning

NMR refers to a deterministic logic used to decipher the solution when given a number of relevant pieces of evidence. NMR is set apart from classic monotonic reasoning in that, in contrast of arriving at one conclusion for any given problem, NMR will consider a number of outcomes and will eliminate or add them as extra information that is readily available. In particular, we have investigated the clausal defeasible logic (CDL) as the proof algorithm to arrive at a conclusion as it has been designed specifically to be implemented in a computer [40].

A language called "decisive programming language" (DPL), proposed by [41], has been employed to illustrate scenarios which use CDL. Within DPL, several symbols are used to represent different relationships of the entities preceding the relation, the antecedent, and the entities positioned subsequently after the relation, the conclusion. The first symbol is the strict rule relation that is represented as "$\rightarrow$." It dictates that this rule is certain with no possible ambiguity involved. The second symbol is the defeasible rule symbol "$\Rightarrow$" denoting a relationship in which it is defeasible to say the former entity will result in the latter entity. The third symbol is the warning rule symbol "$\leadsto$" that describes when the former entity cannot disprove the latter (usually the negative of the latter). Other symbols that are used include the priority relation "$>$," which dictates that the former rule is greater than the latter rule, and the negative symbol "$\sim$," which turns the following variable into its negative counterpart. Although it is true that one conclusion must be drawn for any given situation, CDL has several levels of confidence, represented in formulae that may be used to obtain a different correct answer. These different formulae include the following:

- **μ:** This formula uses only certain information to obtain its conclusion.
- **π:** This formula allows conclusions in which ambiguity is propagated.
- **β:** This formula does not allow any ambiguity to be used in obtaining its conclusion.
- **α:** A formula in which any conjunction of the π and β formulae are used to reach its conclusion.
- **δ:** The disjunction of π and β are used to draw conclusions.

As discussed by [42], other strengths that set CDL apart from other reasoning algorithms are its ability to uses team defeat, failure-by-looping, and discovering the loops in a given reasoning system within a set number of steps. This logical engine has already been tested and implemented in two different scenarios to allow a robot dog play soccer and inside a robot designed as a means of alarming individuals of an emergency in an elderly care situation [43].

24.3 Anti-Collision Techniques

In this section, we introduce a deterministic joined Q-ary tree [44] with the intended goal to minimize memory usage queried by the RFID reader. Most implementations of tree-based algorithms are deployed with older type of EPC class 1, which has limited memory and capability. We also introduce the probabilistic cluster-based technique (PCT) [3] *anti-collision* method to improve the performance of tag recognition process in probabilistic anti-collision algorithm. The remaining of this section comprises the explanation of our proposed joined Q-ary tree and PCT and the comparative analysis of both techniques.

24.3.1 Deterministic Anti-Collision Approaches

This section comprises the explanation on EPC encoding schemes, the typical scenarios discussion, and the foundation of joined Q-ary tree.

24.3.1.1 EPC Encoding Schemes Analysis

The most common type of encoding is the general identifier 96 (GID-96) bits scheme, which is independent of any existing identity specification or convention and can be used in most events. The GID is defined for a 96-bit EPC and is independent of any existing identity specification or convention. In addition to the *header* that guarantees uniqueness of the encoding type, the GID is composed of three fields: the *general manager number* (GMN), OC, and SN, as shown in Table 24.2.

In order to manage and monitor the traffic of RFID data effectively, the *EPC pattern* is usually used to keep the unique identifier on each of the items arranged within a specific range [45]. The *EPC pattern* does not represent a single tag encoding, but rather refers to a set of tag encodings. For instance, the GID-96 includes three fields in addition to the *header* with a total of 96-bits binary value. 25.1545.[3456–3478].[778–795] is a sample of the *EPC pattern* in decimal, which later will be encoded to binary and embedded onto tags. Thus, within this sample pattern, the *header* is fixed to 25 and the GMN is 1545, while the OC can be any number between 3456 and 3478, and the SN can be anything between 778 and 795.

24.3.1.2 Warehouse Distribution Scenarios

For deterministic anti-collision approaches, we examine specific scenarios based on the assumption that items tend to move and stay together through different locations especially in a large warehouse. We focus on crystal warehouse scenario using GID-96 bits encoding scheme, which can be classified into four different scenarios: (1) unique item level, (2) unique container level, (3) Unique company level, and (4) unique warehouse level.

Table 24.2 GID-96 Includes Three Fields in addition to the Header, with a Total of 96-Bits Binary Value

GID-96	*Bit*	*Maximum Decimal/Binary*
Header (H)	8	0011 0101
General manager number (GMN)	28	268,435,455
Object class (OC)	24	16,777,215
Serial number (SN)	36	68,719,476,735

Note: Only "H" is shown in binary, while the rest are shown in decimal.

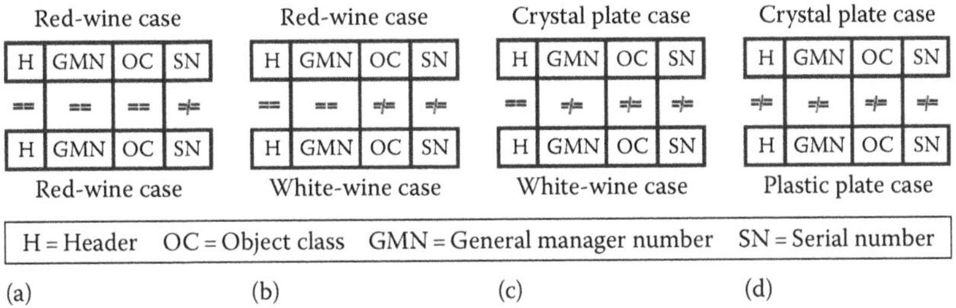

Figure 24.5 Crystal warehouse scenario: (a) unique item level, (b) unique container level, (c) unique company-level, and (d) unique warehouse level.

Unique item-level scenario: This scenario occurs when two collided tags (GID-96 encoding) are captured and they have the same *encoding scheme* (*header*), same *GMN*, same *OC*, but different *SN*. By using the crystal warehouse scenario example from Figure 24.5a, it can be seen that two collided tags are captured with the same *encoding scheme*, GMN, and OC. We believe that both tags are each attached to two different cases of red wine.

Unique container-level scenario: This scenario takes place when two collided tags are captured and they have the same *header*, same *GMN*, different *OC*, and different *SN*. Figure 24.5b shows that crystal red-wine glasses and crystal white-wine glasses are packed in different case and pallet because they are different type of wine glasses. Within this scenario, each case of wine glasses will have a unique *SN* attached to it, with different *OC* for each pallet of white wine or red wine.

Unique company-level scenario: This scenario is illustrated in Figure 24.5c. Two collided tags are captured and they have the same *header*, and unique *GMN*, *OC*, and *SN*. We believe that one tag is attached to the crystal plate case, while the other tag is attached to the white-wine case. We can assume that there are two different companies producing separate crystal ware, and that the wine glasses and plates are from different companies but share the same warehouse because they are both crystal.

Unique warehouse-level scenario: This scenario occurs when two collided tags are captured and they have different *header*, *GMN*, *OC*, and *SN*. We can assume that all items are from different companies that use different encoding schemes. For example, Figure 24.5d shows that two wine glasses with different sculpture, one made from crystal and the other from plastic, are allocated in the same warehouse. The unique warehouse-level scenario will not be discussed any further in this chapter because we are only looking at a large warehouse distribution where most items move together as a group. Therefore, most items from the same type of manufacturing will stick together until they are deployed to smaller retailer.

24.3.1.3 Joined Q-ary Tree

The joined approach [44] is a combination of Q-ary trees, specifically 2-ary and 4-ary trees, which have been identified to be the best Q-ary trees [46,47]. The joined Q-ary tree employs the right combination of Q-ary trees for each specific scenario. Assuming that most items from the warehouse have massive movements, the first few bits of the EPC will be identical and the remaining bits will be very similar. In order to optimize the performance of the joined Q-ary tree, the right separating point (SP) between the two Q-ary trees needs to be configured. This procedure will further reduce the accumulative bits from the reader's queries and improve the robustness of the overall identification process.

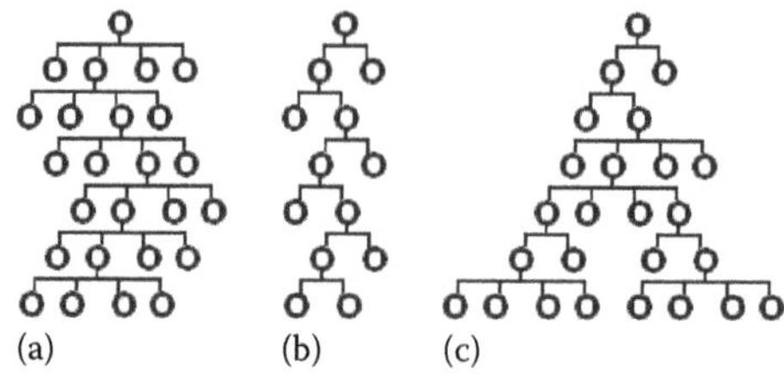

Figure 24.6 Sample of: (a) a Naive 4-ary tree; (b) a naive 2-ary tree, and (c) a joined Q-ary tree.

Figure 24.6 shows the example of (a) Naive 2-ary, (b) Naive 4-ary, and (c) joined Q-ary tree. Joined Q-ary tree bonded both 2-ary and 4-ary trees together and applied to specific bits of EPC depending on how identical or unique they are.

24.3.1.4 EPC Bits Prediction and Classification

In warehouse distribution environment according to unique item-level and unique container-level scenarios, it is known that the first 36-bits of EPC (header and GMN) are definitely identical. However, 24-bits of *OC* can be both identical and unique for all tags, depending on how many pallets existed within one interrogation zone. For example, if there are 5 pallets of 12 cases each in the interrogation zone, there will be 5 different *OC* and 60 unique *SN* for all 60 items (cases).

Since *OC* involved 24-bits of EPC (allow 16,777,215 unique tags) but only 5 unique *OC* are needed, we must calculate a certain number of *unique bits* needed in order to apply the right Q-ary tree. This also applies to *SN* that contains 36-bits of string. Assuming that the *EPC pattern* is used, not all 36-bits of these strings will be unique.

Table 24.3 shows a formal structure for bits classification of GID-96 bits EPC. It can be seen that the *identical bits* of EPC always equal to 36-bits for the first 36-bits of EPC. This includes 8-bits of *header* and 28-bits of *GMN*, which are always the same for all tags. For OC, 24-bits are available where *unique bits within object class* (UOC) can be predicted using Equation 24.3. In addition, *unique bits within serial number* (USN) with 36-bits can also be predicted using the same equation.

Our method is executed based on the assumption that the approximate number of tags (pallets, cases) is known prior to the identification process. This information is needed for *unique bits* calculation: UOC and USN from Table 24.3. However, in most circumstances, number of tags is

Table 24.3 Formal Structure of Bits Classification of EPC GID-96 Bits

	Length	*Identical*	*Unique*
Header	8	8	0
General manager number	28	28	0
Object class	24	24 - UOC	UOC[a]
Serial number	36	36 - USN	USN[b]

[a] UOC the number of unique bits within object class.
[b] USN is the number of unique bits within the serial number.

usually unknown until the first query is issued by the reader. Therefore, UOC and USN of joined Q-ary tree can be initially set to zero and after the first round of identification, these two parameters can be computed.

Joined Q-ary tree adaptively adjusts their tree branches at specific *SP*. These SP are configured according to *identical bits* and *unique bits* within an EPC data. In order to calculate the estimated number of *unique bits* within an EPC, we need the average number of tags within an interrogation zone, and then to apply the following equation:

$$B = \log_2(N) \tag{24.3}$$

where

N is the number of tags
B is the *unique bits* of EPC

24.3.2 Probabilistic Anti-Collision Approaches

This section comprises the mathematic fundamental for *probabilistic anti-collision* schemes, and the foundations of the proposed PCT.

24.3.2.1 Mathematic Fundamental for ALOHA-Based Tag Estimation

In the framed-slotted ALOHA-based *probabilistic* scheme, to estimate the number of present tags, binomial distribution is a good fundamental method. For a given initial Q in a frame with F slots and n tags, the expected value of the number of slots with occupancy number x is as follows:

$$a_x = n \times C_n^x \left(\frac{1}{F}\right)^x \left(1 - \frac{1}{F}\right)^{n-x}$$

Therefore, the expected number of empty slot e, successful slot s, and collision slot c is given by the following equations:

$$\begin{cases} e = a_0 = F\left(1 - \frac{1}{F}\right)^n \\ s = a_1 = n\left(1 - \frac{1}{F}\right)^{n-1} \\ c = a_k = F - a_0 - a_1 \end{cases}$$

Thus, the system efficiency (E) is defined as the ratio between the number of successful slot and the frame-size, as per the following equation:

$$E = \frac{s}{F} = \frac{n\left(1 - \frac{1}{F}\right)^{n-1}}{F} = n\frac{1}{F}\left(1 - \frac{1}{F}\right)^{n-1}$$

It has been proven that the highest efficiency can be obtained if the frame-size F is equal to the number of tags n, provided that all slots have the same fixed length:

$$F(optimal) = n$$

Therefore, we make the assumption that by keeping the number of tags close to the available *frame-size*, the optimal performance efficiency can be obtained. According to literatures, it is possible to achieve the theoretically optimal efficiency of 36.8% in ALOHA-based systems.

24.3.2.2 Probabilistic Cluster-Based Technique

The PCT [3] employs a dynamic probabilistic algorithm concept and uses a group-splitting rule to split *backlog* into group if the number of unread tags is higher than the maximum *frame-size*.

The PCT approach first estimates the number of *backlog*, or the remaining tags, within the interrogation zone. If the number of *backlog* is larger than the specific *frame-size*, it splits the number of *backlog* into a number of groups and allows only one group of tags to respond. PCT approach derived new rules using particular equations, according to the optimal system efficiency obtained for specific number of tags. We first conducted an experiment to acquire optimal *frame-size* for specific number of tags as shown in Figure 24.7. It can be seen that the optimal system efficiency achieved by the probabilistic ALOHA method is approximately 38% and the optimal number of tags is close to the maximum *frame-size*. Efficiency is calculated as shown in Equation 24.4:

$$Efficiency = \left(\frac{S}{S + C + E}\right) \tag{24.4}$$

where

S is the number of successful slots
C is the number of collision slots
E is the number of empty slots

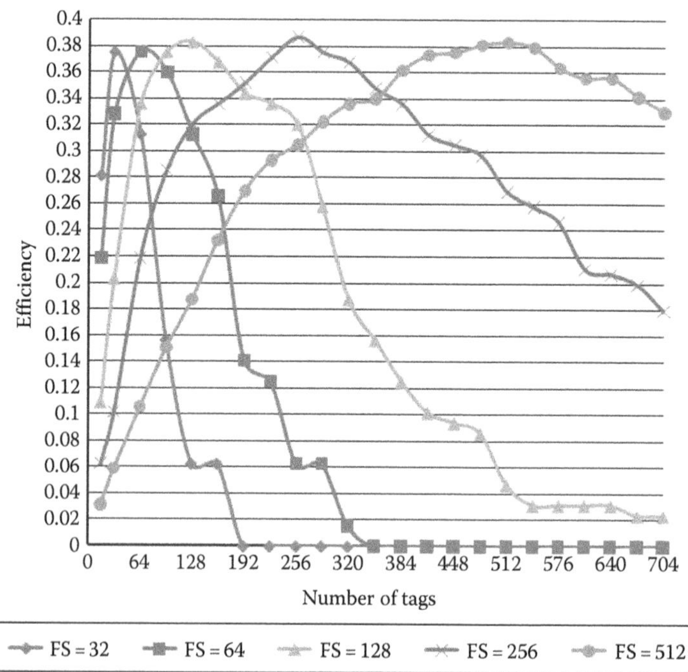

Figure 24.7 Performance efficiency of different frame-size on different number of tags.

24.3.2.3 PCT Rules

The PCT [3] method employs a dynamic probabilistic algorithm concept and uses the group-splitting rule to split *backlog* into group if the number of unread tags is higher than the maximum *frame-size*.

The PCT approach first estimates the number of *backlog*, or the remaining tags, within the interrogation zone. If the number of *backlog* is larger than the specific *frame-size*, it splits the number of *backlog* into a number of groups and allows only one group of tags to respond. The reader then issues a "Query," that contains a "Q" parameter to specify the *frame-size* (frame-size F(min) = 0; F(max) = $2^Q - 1$). Each selected tag in the group will pick a random number between 0 and $2^Q - 1$ and place it into its slot counter. Only the tag that picks zero as its slot counter responds to the request. When the number of estimated *backlog* is below the threshold, the reader adjusts the *frame-size* without grouping the unread tags. After each read cycle, the reader estimates the number of *backlog* using the PTES algorithm and adjusts its *frame-size*.

Table 24.4 shows the PCT rule. For instance, if the number of *backlog* equals 900 tags, the PCT algorithm will split the unread tags into three groups of Q8 ($2^8 - 1 = 256$).

Table 24.4 PCT Rule—Number of Unread Tags, Optimal Frame-Size (FSA and FSB), and Number of Groups (A and B)

PCT Rule				
Backlogs	*FS A*	*Group A*	*FS B*	*Group B*
....	....	....	....	....
1233 to 1408	256	4	–	–
1057 to 1232	256	3	128	1
881 to 1056	256	3	–	–
705 to 880	256	2	128	1
529 to 704	256	2	–	–
353 to 528	256	1	128	1
177 to 352	256	1	–	–
89 to 176	128	1	–	–
45 to 88	64	1	–	–
23 to 44	32	1	–	–
12 to 22	16	1	–	–
6 to 11	8	1	–	–
....	....	....	....	....

24.3.3 Comparative Analysis of Deterministic and Probabilistic Techniques

In this chapter, we have empirically compared the performance of the joined Q-ary tree against the PCT *anti-collision* approach because our *deterministic* and *probabilistic* methods have outperformed existing techniques in their own grounds [3,44,48]. The joined Q-ary tree uses less resource, has no complexity in implementation, and needs low reader power and memory consumption, because it does not need to keep memory during identification. On the other hand, the PCT works well in arbitrary situation, minimizes resource used, and increases system efficiency, without the need for complex implementation. We believe that this comparative analysis is necessary to identify the best overall method for specific circumstances.

24.3.3.1 Data Set

For joined Q-ary tree *anti-collision* approach, there are 10 pallets of inventories in test case A, with each pallet containing 100 cases/tags, giving a total of 1000 tags. Similarly, test case B also contains 1000 tags, but each pallet only holds 50 cases/tags.

- *Test case A*: joined Q-ary tree with 100 tags per pallet (Joined(100))—10 pallets, 100 cases each, total 1000 tags
- *Test case B*: joined Q-ary tree with 50 tags per pallet (Joined(50))—20 pallets, 50 cases each, total 1000 tags

For *probabilistic anti-collision* approach, we considered different number of tags, from 100 to 1000 tags. For each identification round, optimal tunable *initial Q* is applied.

24.3.3.2 Results

From the empirical study, we have investigated the performance of our proposed joined Q-ary tree and PCT. Figure 24.8a illustrates that the difference in performance between each method increased with the increased number of tags, this has particularly become visible when examining 1000 tags. The overall number of slot results have shown that the joined Q-ary tree with 100 tags per pallet (joined(100)) has obtained the minimal number of slots throughout the whole experiment, which also obtains the shortest identification time required. In contrast, the joined Q-ary tree with 50 tags per pallet (joined(50)) performed poorly compared with the joined(100) and PCT. These results has proven that the selection of the *EPC pattern* has a large impact on the performance of the joined Q-ary tree. When the chosen *EPC pattern* involved has a very small group of tags (such as 50 tags per pallet), the performance of joined Q-ary tree cannot be optimized.

Figure 24.8b shows the performance efficiency of all methods. It can be seen that the Joined(100) achieved close to 47% efficiency once the number of tags reach 1000. Additionally, we can see that the performance efficiency of both the joined(100) and joined(50) methods keeps increasing in accordance to the number of tags. In contrast, the PCT cannot achieve a performance efficiency higher than 38%. By examining Figure 24.8, it can be assumed that the efficiency of the joined Q-ary tree will increase slowly once the number of tags within the interrogation zone becomes very high. For the joined(50), if the number of tags keeps increasing, it is possible that the performance efficiency will achieve the same level as PCT.

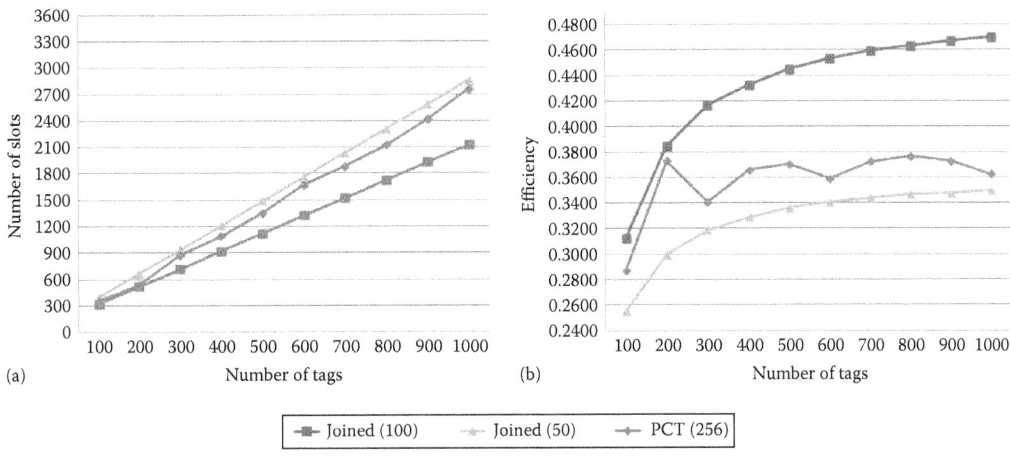

Figure 24.8 Comparative analysis of joined Q-ary tree vs. PCT: (a) number of slots comparison and (b) performance efficiency.

From the comparative analysis, we have identified certain properties of importance for *anti-collision* methods in general. For *deterministic* methods, we have discovered that there are impacts from similar *EPC patterns*, the number of tags within one group of the *EPC pattern*, and the overall number of tags within the interrogation zone. For *probabilistic* methods, we have determined that the performance of the *anti-collision* technique depends on the initial *frame-size* (or the Q value) specification, the accuracy of *backlog* prediction techniques, and the overall number of tags within the interrogation zone.

24.4 Deferred Cleaning Approaches

In the following section, we discuss the cleaning of the stored RFID observations after they have been placed inside the data warehouse. Unlike the anti-collision techniques, it is not only possible to restore missing records in the database, but also to eliminate wrong and duplicate data. Thus, we have divided this section into both false-negative and false-positive cleaning sections to correct missing and wrong/duplicate anomalies, respectively. In each of the methodologies, we apply a Bayesian network, neural network, and non-monotonic reasoning classifiers, and then compare each approach to determine the highest performing methodology.

24.4.1 False-Negative Cleaning

To accurately correct missing readings, we have constructed an advanced data analysis methodology coupled with high-level intelligence to correctly decipher the most likely candidates of observations to be returned into the data set. Specifically, the concept we have introduced will intelligently analyze the missing data anomaly and use the Bayesian network [49,50], neural network [51], and non-monotonic reasoning [52–54] classifiers to find the correct observations to load back into the database. This will include an outline the motivation and scenario considered in this work followed by a description of the system architecture of our approach. These discussions will be followed by the database structure that houses the RFID observations and all assumptions made toward our methodology. We then present the results obtained from our experimental evaluation before summarizing our findings.

24.4.1.1 System Architecture

We have divided our system's architecture into three core components. The first is designed to analyze the data where the missed reading occurred which we have named the *analysis phase.* The data discovered in this analysis phase are then passed onto the *intelligence phase* where the correct permutation is selected. After the resulting data set has been chosen, the *loading phase* will complete the program's cycle by inputting the information back into the data warehouse.

Analysis phase: The analysis phase consists of the tool locating missed readings and identifying essential data about the anomaly. The first process is to divide the tags into "tag streams" (Definition 24.1) as seen in Figure 24.9. These tag streams include chronicle information relating only to one individual tag. From these tag streams, certain information is ascertained relating to the nature of the false negative anomaly. This includes finding the reader locations of the observations two readings before and directly before the anomaly (a and b, respectively) and the two readings directly after the reader (c and d, respectively). Additionally, the shortest path between readings b and c using the map data is found. The total missing readings calculated via the number of missing timestamps (n), and the amount of observations within the shortest path (s), is then calculated.

Definition 24.1 Tag stream: We define tag streams as individually analyzed streams for one tag from the mass amount of readings.

The analytical information obtained from our approach includes detecting if readers a and b are equal ($a == b$); determining if readers b and c are relatively close to each other according to the map data ($b \leftrightarrow c$); discovering if the readers b and c are equal ($b == c$); finding if readers c and d are equal ($c == d$), and discovering if n is equal to, less than, or greater than s minus two ($n == (s - 2)$), ($n < (s - 2)$), ($n > (s - 2)$). The reason as to why we subtract two from value s is that the shortest path will include the values of b and c which are not necessarily a part of the missing gaps of knowledge. All of these analytical Boolean variables are then passed on to the correction phase which utilize it to seek out the most ideal imputed reader values. We utilize four main arithmetic operations to obtain this binary analytical information. These include the equivalent symbol $==$, the less $<$ and greater than $>$ symbols, and the $\leftrightarrow$ symbol we have elected to represent geographical proximity. The rationale as to why s is always having two taken away from it lies in the fact that the shortest path always includes the boundary readers b and c, which are not included within the n calculation.

Figure 24.9 A visual representation of how we analyze one tag at a given moment within a tag stream.

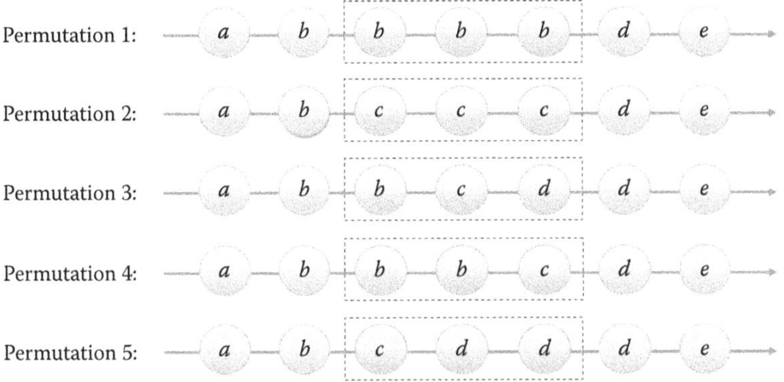

Figure 24.10 An illustration of what reader values are placed into each false-negative anomaly for each of the five different permutations.

Intelligence phase: The intelligence phase occurs when the various permutations of the missing data are generated as candidates to be restored in the data set. The five different permutations that are been generated and depicted in Figure 24.10 are described as follows:

- Permutation 1: All missing values are replaced with the reader location of observation *b*.
- Permutation 2: All missing values are replaced with the reader location of observation *c*.
- Permutation 3: The shortest path is slotted into the middle of the missing data gap. Any additional missing gaps on either end of the shortest path are substituted with values *b* for the left side and *c* for the right.
- Permutation 4: The shortest path is slotted into the latter half of the missing data gap. Any additional missing gaps on the former end are substituted with value *b*.
- Permutation 5: As the anti-thesis of Permutation 4, the shortest path is slotted into the former half of the missing data gap. Additional missing gaps found at the latter end of the missing data gap are substituted with value *c*.

After the data analysis and permutation formations are completed, all relevant information found in the analytical phase are treated as a feature set definition and passed into either the *Bayesian network*, *neural network*, or the *non-monotonic reasoning engine*. The various classifiers will then return the permutation that has been found to best suit the missing data. With regard to the Bayesian network, all weights inside the network are first created with small random numbers. After this, we utilize a genetic algorithm to train these weights based on a training algorithm of previously correct permutations based on each variation of the feature set definitions. The resulting network configuration from this training will then provide the optimal permutation of readings to be inserted back into the data warehouse.

The ANN accepts seven binary inputs to reflect the analysis data ($a == b$, $c == d$, etc.) and has five binary outputs to reflect that the permutations that have been found. The ANN also includes nine hidden units found in one hidden layer. We specifically chose this configuration as we have found that there should be a larger number of hidden nodes than inputs and one layer would be sufficient for our network at this moment. If we chose to extend the complexity of our system, we may wish to also either increase the number of hidden units, layers, or both.

We have also applied a momentum term and learning rate whose values are 0.4 and 0.6, respectively, to avoid local minima and network paralysis. Each input and output value will not be

1 and 0 as this may not yield a very high classification rate. Instead, we will use the values of 0.9 and 0.1, respectively. We will set the stopping criteria as both the RMS error threshold when it reaches below 0.1 and 1000 iterations. We have also utilized the sigmoidal activation function to derive our outputs.

The NMR classifier is based on the rules we have created as displayed in Tables 24.5 through 24.9. Each of the rules present within the tables are combinations found from the analytical data joined by "and" statements ($\bigwedge$) which have been gathered within the analysis phase. Within the logic engine build, the precedence of the rules corresponds to the larger number of the rule (e.g., rule 17 will beat rule 4 in Table 24.6). In the event that more than one permutation has been found to be ideal in a given situation, we use the following hierarchical weighting: permutation 3 > permutation 1 > permutation 2 > permutation 4 > permutation 5. In the unlikely case where no conclusions have been drawn from the NMR engine, permutation 3 will be elected as the default candidate due to it having perfect symmetry within the imputed data. This ordering has been configured to be the most accurate conclusion assuming that the amount of consecutive missed readings are low due to the randomness of the anomalies.

Table 24.5 Table Depicting the Non-Monotonic Reasoning Rules Used to Create the Permutation 1 Logic Engine

Rule No.	*Rule*	*Conclusion*
1	$b == c$	~Perm1
2	$b == c \bigwedge n < (s-2)$	~Perm1
3	$b == c \bigwedge n == (s-2)$	~Perm1
4	$a == b$	Perm1
5	$c == d \bigwedge n == (s-2)$	~Perm1
6	$c == d \bigwedge n > (s-2)$	~Perm1
7	$c == d \bigwedge b \leftrightarrow c \bigwedge n == (s-2)$	~Perm1
8	$c == d \bigwedge b \leftrightarrow c \bigwedge n > (s-2)$	~Perm1
9	$a == b \bigwedge b \leftrightarrow c$	Perm1
10	$a == b \bigwedge b \leftrightarrow c \bigwedge n == (s-2)$	Perm1
11	$a == b \bigwedge b \leftrightarrow c \bigwedge b == c \bigwedge \sim c == d \bigwedge n == (s-2)$	Perm1
12	$c == d$	~Perm1
13	$c == d \bigwedge b \leftrightarrow c$	~Perm1
14	$a == b \bigwedge b \leftrightarrow c \bigwedge n > (s-2)$	Perm1
15	$a == b \bigwedge b == c \bigwedge b \leftrightarrow c \bigwedge \sim c == d \bigwedge n > (s-2)$	Perm1
16	$\sim b \leftrightarrow c$	~Perm1
17	$n < (s-2)$	~Perm1

Table 24.6 Table Depicting the Non-Monotonic Reasoning Rules Used to Create the Permutation 2 Logic Engine

Rule No.	*Rule*	*Conclusion*
1	$b == c$	~Perm2
2	$b == c \wedge n < (s-2)$	~Perm2
3	$b == c \wedge n == (s-2)$	~Perm2
4	$c == d$	Perm2
5	$a == b \wedge n == (s-2)$	~Perm2
6	$a == b \wedge n > (s-2)$	~Perm2
7	$a == b \wedge b \leftrightarrow c \wedge n == (s-2)$	~Perm2
8	$a == b \wedge b \leftrightarrow c \wedge n > (s-2)$	~Perm2
9	$b \leftrightarrow c \wedge c == d$	Perm2
10	$b \leftrightarrow c \wedge c == d \wedge n == (s-2)$	Perm2
11	$\sim a == b \wedge b \leftrightarrow c \wedge b == c \wedge c == d \wedge n == (s-2)$	Perm2
12	$a == b$	~Perm2
13	$a == b \wedge b \leftrightarrow c$	~Perm2
14	$b \leftrightarrow c \wedge c == d \wedge n > (s-2)$	Perm2
15	$\sim a == b \wedge b == c \wedge b \leftrightarrow c \wedge c == d \wedge n > (s-2)$	Perm2
16	$\sim b \leftrightarrow c$	~Perm2
17	$n < (s-2)$	~Perm2

Table 24.7 Table Depicting the Non-Monotonic Reasoning Rules Used to Create the Permutation 3 Logic Engine

Rule No.	*Rule*	*Conclusion*
1	$a == b \wedge c == d$	Perm3
2	$\sim a == b \wedge \sim c == d$	~Perm3
3	$a == b \wedge c == d \wedge n > (s-2)$	Perm3
4	$\sim a == b \wedge \sim c == d \wedge \sim n > (s-2)$	~Perm3
5	$n == (s-2)$	Perm3
6	$b == c$	~Perm3
7	$a == b \wedge c == d \wedge n == (s-2)$	Perm3
8	$n < (s-2)$	~Perm3

Table 24.8 Table Depicting the Non-Monotonic Reasoning Rules Used to Create the Permutation 4 Logic Engine

Rule No.	*Rule*	*Conclusion*
1	$c == d$	~Perm4
2	$b == c$	~Perm4
3	$a == b$	Perm4
4	$\sim a == b$	~Perm4
5	$n > (s-2) \wedge a == b$	Perm4
6	$\sim n > (s-2)$	~Perm4
7	$\sim a == b \wedge \sim n > (s-2)$	~Perm4

Table 24.9 Table Depicting the Non-Monotonic Reasoning Rules Used to Create the Permutation 5 Logic Engine

Rule No.	*Rule*	*Conclusion*
1	$a == b$	~Perm5
2	$b == c$	~Perm5
3	$c == d$	Perm5
4	$\sim c == d$	~Perm5
5	$n > (s-2) \wedge c == d$	Perm5
6	$\sim n > (s-2)$	~Perm5
7	$\sim c == d \wedge \sim n > (s-2)$	~Perm5

Loading phase: The loading phase consists of the selected permutation being uploaded back into the data storage at the completion of the *intelligence phase*. The user will have the opportunity to either elect to load the missing data into the current data repository or to copy the entire data set and only modify the copied data warehouse. This option would effectively allow the user to revisit the original data set in the event that the restored data are not completely accurate.

24.4.1.2 Experimental Evaluation

Within the following section, we have included a thorough description of the setup of the experimentation used in our methodology. First, we discuss the environment used to house the programs. This is followed by a detailed discussion of our experimentation including the four experiments we performed and their respective data sets used. These experiments include the training of the three various classifiers and then taking the highest performing configurations to compare against each other.

Environment: Our methodology has been coded in the C++ language and compiled with Microsoft Visual Studio C++. The code written to derive the lookup table needed for the non-monotonic reasoning data has been written in Haskell and compiled using Cygwin Bash shell. All programs were written and executed on Dell machine with the Windows XP service pack 3 operating system installed.

Experiments: We have conducted four experiments to adequately measure the performance of our methodology. The first set of experimentations we conducted involved finding a Bayesian network that performs the highest clean on RFID anomalies. After this, we investigated the highest performing neural network configuration in our second experiment. The third experiment conducted was to determine which of the CDL formulae performs most successfully when attempting to correct large amounts of scenarios. The training cases used in the each of these experiments consisted of various sets of data consisting of ambiguous false negative anomaly cases.

The fourth experiment we conducted was designed to test the performance of our selected highest performing Bayesian network, neural network, and non-monotonic reasoning logic configurations to determine which classifier yielded the highest and most accurate clean of false-negative RFID anomalies. The reason as to why these techniques were selected as opposed to other related work is that only other state-of-the-art classifying techniques may be compared with respect to seeking the select solution from a highly ambiguous situation.

The fourth experiment testing sets included four data repositories consisting of 500, 1,000, 5,000, and 10,000 ambiguous false-negative anomaly cases. We defined our scoring system as if the respective methodologies were able to return the correct permutation of data that had been previously defined. All data within the training and testing set have been simulated to emulate real RFID observational data.

Database structure: To store the information recorded from the RFID reader, we utilize portions of the "Data model for RFID applications" DMRA database structure found in Siemens Middleware software [55]. Additionally, we have introduced a new table called MapData designed to store the map data crucially needed within our application. Within the MapData table, two reader IDs are stored in each row to dictate if the two readers are geographically within proximity. The structures of the two tables we are using in experimenting include the following:

```
OBSERVATION(Reader_id, Tag_value, Timestamp)
MAPDATA(Reader1_id, Reader2_id)
```

Assumptions: We have made three assumptions that are required for the entire process to be completed. The first assumption is that the data recorded will be gathered periodically. The second assumption we presume within our scenario is that the amount of time elected for the periodic readings is less than the amount of physical time needed to move from one reader to another. This is important as we base our methodology around the central thought that the different readings will not skip over readers that are geographically connected according to the MapData. The final assumption we make is that all readers and items required to be tracked will be enclosed in a static environment that has readers which cover the tracking area.

24.4.1.3 Results and Analysis

To thoroughly test our application, we devised four different examinations which we have labeled the *Bayesian network*, *neural network*, *non-monotonic reasoning*, and *false-negative comparison experiments*. In the Bayesian network false-negative experimentation, we conducted various investigations into both static and dynamic configurations to attempt to achieve the highest cleaning performance. Similar to the Bayesian network experiment, the neural network false-negative experimentation included finding the highest training algorithm to configure the network and obtain the highest performing clean. The non-monotonic reasoning false-negative experiment compared the cleaning rate of each of the CDL formulae. The highest performing non-monotonic reasoning setup was then compared to Bayesian and neural network approaches to find the highest performing cleaning algorithm.

24.4.1.3.1 Bayesian Network False-Negative Experiment

We decided to augment the Bayesian Network with a genetic algorithm to train the network based on a test set we developed. The training experiment consisted of the utilization of the genetic algorithm with 250 generation iterations to determine the fittest chromosome. To this end, we trained and compared the genes of chromosomes where 100–1000 chromosomes were in the population and incremented by 100 in each sequential experiment. The data set used to determine the fitness of the chromosomes contained every permutation possible with the analysis and its correct permutation answer. From the results, we have found that the chromosome which was the fittest resulted when 500 chromosomes were introduced into the population. Unlike the other experimentation we have performed in this research, the Bayesian network has been measured by the number of inserts that were exactly correct as opposed to measuring how correct the entire data set is. This results in the Bayesian network achieving a relatively low cleaning rate as permutations that are not an exact match with the training results will be counted as incorrect when the resulting imputed data may actually be correct in the data set.

24.4.1.3.2 Neural Network False-Negative Experiment

The neural network configuration experiment has the goal of seeking out the training algorithm that yields the highest clean rate. The two training methods used for comparison are the BP and genetic algorithms. The training set utilized in this experiment is comprised of every possible combination of the inputs and their respective outputs that amount to a total of 128 entries. We have conducted tests upon three different training algorithm setups: The first is the BP algorithm and the other two are genetic algorithms that use 20 and 100 chromosomes to find the optimized solution. Additionally, we conducted each experiment in three trials to further generalize our results. The algorithm that had the hardest time finding the correct configuration was the BP algorithm when it iterated for 50 and 100 times, earning it 1.56% classification rate. The trainer that performed the best was the genetic algorithm, both using 20 and 100 chromosomes in every test and the iteration number excluding the 20 chromosome configuration which lasted for 5 generations.

The analysis we performed on these results consisted of graphing the average of our findings into a bar graph to illustrate the difference in algorithms. On average, the neural network genetic algorithm performed the best, obtaining an 87.5% cleaning rate. Unfortunately, as discovered previously in the results, the BP algorithm performed the weakest within the three algorithms. We believe the poor results of the BP algorithm was due directly to over-training the network. We noticed that there was also a particularly low result when attempting to train this algorithm for 50 attempts in trial 2. However, it wasn't until the 100 iteration training that we could clearly see the effects of the training routine on the average cleaning rate [51].

24.4.1.3.3 NMR False-Negative Experiment

We created the third experiment with the goal of determining which of the five CDL formulae would be able to clean the highest rate of highly ambiguous missing RFID observations. We did this by comparing the cleaning results of the μ, α, π, β, and δ formulae on various training cases. There were three training sets in all with 100, 500, and 1000 ambiguous false-negative anomalies. Additionally, at the completion of these experiments, the average was determined for all three test cases and was used to ascertain which of the five formulae would be used within the significance experiment.

We have found that the highest average achieving formula has been found to be α (Alpha). This is probably due to the fact that it discovers cases in which both the β and π formulae agree upon, thereby increasing the intelligence of the decision. Also of note is that the disjunction of β and π formulae shown within δ achieves a relatively high average cleaning rate. The lowest performing average cleaning rate has been found to be β, which is probably due to its nonacceptance of ambiguity when drawing its conclusion. We believe it is crucial for the cleaner to have a low level of ambiguity when drawing its conclusions as the problem of missed readings needs a level probability to infer what readings need to be replaced. As stated previously, we have chosen the α formula as the highest performing cleaner to be used within the classifier comparison experiment [54].

24.4.1.3.4 False-Negative Comparison Experiment

The goal of our fourth experimental evaluation was designed to put three classifiers through a series of test cases with large amounts of ambiguous missing observations. The three different classifications techniques included our NMR engine with CDL using the α formula compared against both Bayesian and neural networks with the highest performing configurations obtained from previous experimentations. We designed the experiment to have an abnormally high amount of ambiguous false-negative anomalies consisting of 500 and 1000 test cases to thoroughly evaluate each approach. Following the conclusion of these experiments, we derived the average of each technique to find the highest performing classifier [54].

The results of this experimentation shown in Figure 24.11 have shown that the neural network obtained a higher cleaning average than that of both the Bayesian network and NMR classifiers with 86.68% accuracy. The lowest performing cleaner was found to be the Bayesian network.

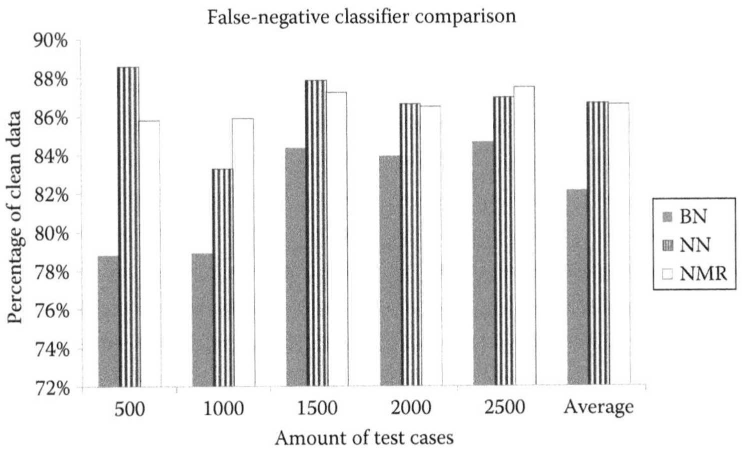

Figure 24.11 The revised results of the Bayesian network, neural network, and non-monotonic reasoning classifiers when attempting to clean an evenly spaced amount of false-negative anomalies.

From these findings, we have determined that the Neural Network has provided the highest accuracy when attempting to clean ambiguous false-negative anomalies. It is interesting to note that the probabilistic approach has actually outperformed the deterministic methodology in the case of imputing missing data. We believe that this is due to the fact that there would be a level of ambiguity and probability needed to be introduced when correcting missing data due to the lack of information available to the classifier. In this case, a methodology that attempts to investigate the validity of observations not normally considered by deterministic approaches would yield a higher cleaning rate.

24.4.2 False-Positive Cleaning

In this section, we have modified our RFID anomaly management system to identify and eliminate false-positive observations. We will first review the motivation and architecture of the novel concept. This includes the feature set definition, Bayesian network, neural network, non-monotonic reasoning, and loading phases. Next, we provide details of the ideal scenario to design our system most effectively and any assumptions we have made to ensure that the algorithms operates correctly will be listed. Finally, we provide an experimental evaluation of each of the classifier performance and discuss which provides the highest accuracy when correcting wrong and duplicate anomalies.

24.4.2.1 System Architecture

The design of our system has been broken into three sections: the *feature set definition phase*, the *classifier phase*, and the *modification phase*. The feature set definition phase is the first process that is conducted within our application in which raw data are searched and sorted to find suspicious readings and the circumstances surrounding each of these observations. The classification is where the system deviates between three different classifiers—the *Bayesian network*, *neural network*, or NMR. Each classifier has one goal, which is to determine if the flagged reading should be deleted or kept within the data set. This decision is based solely upon the input gathered from the *feature set definition phase*. After each classifier has determined the validity of the observation, it will then pass the decision onto the *modification phase* that will either delete or keep the value being passed to it.

24.4.2.1.1 Feature Set Definition Phase

The first stage of the program is the *feature set definition phase* whose main goal is to analyze the data to discover suspicious readings and investigate key characteristics surrounding the flagged observation. Initially, this phase breaks the tag readings into tag streams designed to analyze the route of one tag. A tag will be flagged suspicious if the difference in timestamps exceed the user-defined duration it should take to reach the location, or if the geographical locations of the readers are not within proximity. To determine the geographic validity of the readers, the program utilizes a table named MapData that is constructed by the user and reflects the geographic layout of all adjacent readers within the static environment. As illustrated in Figure 24.12, there are five observational values that are ascertained: a, b, x, c, and d. The values of observations a and b are the readings taken two or one positions, respectively, before the suspicious reading x. The c and d readings are the observations that have been recorded once and twice after the suspicious reading. From all these observations, the timestamp and the location are all recorded and used in further analysis.

Figure 24.12 Graphical representation of a tag stream with observations that are to be examined highlighted as *a*, *b*, *x*, *c*, and *d*.

After the values of a, b, x, c, and d with their respective timestamps and locations have been found, the feature set definition phase further investigates key characteristics of the data. The characteristics comprise 10 different binary mathematical operations; however, additional characteristics may be added by the user. Each of the characteristics contains spatial and temporal information regarding the observations before and after the suspicious readings. With regards to the proximity of timestamps, we have utilized the value of half a second. This time value may be altered to better suit the application for which it is designed. The characteristics we discovered are as follows:

- $b.loc \leftrightarrow x.loc$
- $c.loc \leftrightarrow x.loc$
- $b.time == x.time$
- $c.time == x.time$
- $b.loc == x.loc$
- $c.loc == x.loc$
- $a.loc \leftrightarrow x.loc$
- $d.loc \leftrightarrow x.loc$
- $b.time \leftrightarrow x.time$
- $c.time \leftrightarrow x.time$

The data used in these characteristics are five different values that have two sub-values each. The five values include the observations of a, b, x, c, and d, where each have the time (time) and location (loc) for each value stored. The characteristics our methodology uses in analysis include when values are within certain proximity which is represented as $\leftrightarrow$, or are equivalent which is represented as $==$. It is important to note that the function which states that the two values are within proximity of each other have different meanings between the location and time. With regards to the location information (loc), the proximity is determined by the "MapData" information, whereas the time proximity refers to the temporal interval between two observations being within the user-defined time value of each other (i.e. two observations being with 5 seconds of each other). After all these characteristics have been gathered, they are passed onto the various classifying methodologies as inputs to determine whether or not the flagged item should remain within the data set.

24.4.2.1.2 Classifier Phase: Bayesian Network

The first option that the *classifier phase* can utilize is the Bayesian network. In this example, we have considered the Bayesian network to have 10 inputs that correspond to the analytical characteristics found at the end of the *feature set definition phase*. Using these 10 inputs based on the weights it has obtained through training the Bayesian network, it will determine whether the flagged observational reading should be kept within the database. This will result in one output known as the *keep_value* which will be set to either true or false. This *keep_value* output will be passed to the *modification phase* at the end of this process at which point the entire application will repeat each time a suspicious reading is encountered. We also set all binary input numbers from 0 and 1 to 0.1 and 0.9, respectively, to allow for higher mathematical functions to benefit from avoiding a multiplication of zero or one.

We have chosen a genetic algorithm to train the Bayesian network weights based on the various test cases that may arise. The genetic algorithm will have the population of the chromosomes to determine the ideal number of chromosomes utilized for training purposes. The mutation rate of the genetic algorithm being utilized will be 1% for the top 10% chromosomes with regard to fitness and 5% for all other chromosomes. After the best weight configuration has been determined, the network will be utilized to compare it against the neural network and non-monotonic reasoning approaches.

24.4.2.1.3 Classifier Phase: Neural Network

ANN is the second option we have chosen for the classifier phase that utilizes weighted neurons to determine the validity of the flagged value. Like the Bayesian network, this ANN will use the 10 inputs gathered from the *feature set definition phase* to pass through the network and obtain 1 output. The network comprises a single hidden layer with 11 hidden nodes resulting in 121 weights between all the nodes. We specifically wanted to choose more hidden units than inputs and only one layer as we have found that multilayered networks do not necessarily enhance the performance of the classifier.

We have also set the momentum and learning rates to 0.4 and 0.6, respectively, and have utilized a sigmoidal activation function. Additionally, as with the Bayesian network, we shall use the numbers 0.1 and 0.9 rather than the binary numbers of 0 and 1, respectively. Two prominent training algorithms have been utilized to properly configure the neural network. The first is the BP algorithm while the second is the genetic algorithm, that has also been utilized within the Bayesian network. Both algorithms will use a limited amount of iterations as stopping criteria for the training.

24.4.2.1.4 Classifier Phase: Non-Monotonic Reasoning

The final classifier we have utilized within our implementation is NMR logic engines. The actual algorithm utilizes a series of rules that we have created based upon the input analysis variables obtained from the *feature set definition phase*. From this, the logic engines determine the correct course of action to either keep the value or not based on the different levels of ambiguity we enforce. The rules utilized within the logic engine may be examined in Table 24.10. The four symbols that are used to interact with the values within the rules are the logic AND operator $\wedge$, the negative operator $\sim$, the equal operator $==$, and our use of the double arrow $\leftrightarrow$ to illustrate proximity between the two analysis variables.

As a default case where neither keep_val nor ~keep_val are encountered, the logic engine will keep the flagged reading to avoid artificially introduced false-negative observations. Additionally, the order in which they have been written in this document is also the order of priority with regards to finding the conclusion.

Table 24.10 Table Containing all the Rules and Respective Conclusions Utilized in the Non-Monotonic Reasoning Engines

Rule No.	*Rule*	*Conclusion*
1	c.time $\leftrightarrow$ x.time $\bigwedge \sim$ c.loc == x.loc	keep_val
2	b.time $\leftrightarrow$ x.time $\bigwedge \sim$ b.loc == x.loc	keep_val
3	$\sim$ b.loc $\leftrightarrow$ x.loc $\bigwedge \sim$ c.loc $\leftrightarrow$ x.loc $\bigwedge \sim$ a.loc $\leftrightarrow$ x.loc $\bigwedge \sim$ d.loc $\leftrightarrow$ x.loc	~keep_val
4	$\sim$ b.loc $\leftrightarrow$ x.loc $\bigwedge \sim$ c.loc $\leftrightarrow$ x.loc	~keep_val
5	a.loc $\leftrightarrow$ x.loc $\bigwedge$ d.loc $\leftrightarrow$ x.loc	keep_val
6	b.loc $\leftrightarrow$ x.loc $\bigwedge$ c.loc $\leftrightarrow$ x.loc	keep_val
7	b.loc $\leftrightarrow$ x.loc $\bigwedge$ c.loc $\leftrightarrow$, x.loc $\bigwedge \sim$ b.time == x.time $\bigwedge \sim$ c.time == x.time	keep_val
8	c.time == x.time	~keep_val
9	b.time == x.time	~keep_val
10	b.time == x.time $\bigwedge$ c.time == x.time	~keep_val
11	b.loc $\leftrightarrow$ x.loc $\bigwedge$ c.loc $\leftrightarrow$ x.loc $\bigwedge \sim$ b.time == x.time $\bigwedge \sim$ c.time == x.time	keep_val
12	b.loc == x.loc $\bigwedge$ c.loc == x.loc $\bigwedge$ b.time == x.time $\bigwedge$ c.time == x.time	~keep_val

24.4.2.1.5 Modification Phase

After each intelligent classifier has determined whether or not to keep or delete the flagged reading, it will pass it to the *modification phase*. After the decision has been received, the application will then delete the identified value in the original data warehouse.

24.4.2.2 Experimental Results and Analysis

In order to investigate the applicability of our concepts, we conducted four experiments. The first three were dedicated to finding the optimal configuration of each classifier, whereas the last focused on the comparison of the three classifiers. In this section, we describe the database structure, assumptions, and environment in which we conducted these experiments and an analysis of the experiments. Furthermore, we describe the experimental evaluation and present the results obtained. The first three are designed to determine the highest achieving configuration of the Bayesian network, neural network, and non-monotonic reasoning classifiers. In the fourth experiment, the highest achieving classifiers have been compared against each other to find which one achieves the highest cleaning rate. All experimentation was performed with an identical database structure and computer as the false-negative experimentation.

24.4.2.2.1 Environment

As outlined earlier, there are four main experiments that were conducted using the methodology. The first experiment was designed to test the highest performing genetic algorithm when training the Bayesian network. For this experiment, the amount of chromosomes in the population was manipulated to find the highest performing number. The second experiment was designed to discover which training algorithm of either the BP or genetic algorithm obtained the highest cleaning rate. For this experiment, the amount of chromosomes were modified and compared with the BP algorithm to determine the highest achieving algorithm. The third experiment was designed to determine which formulae achieved the highest cleaning rate within the NMR approach.

We specifically chose only to examine classifier techniques as the related work is not comparable due to either it not being able to clean ambiguous data or not using an automated process. The last experiment which was conducted took the highest achieving configurations of each of the classifiers and compared each methodology against the other. Four data sets were utilized for this experimentation, the first three were training sets in which 500, 1000, and 5000 scenarios were used to train the algorithms and find the optimal configuration. Each training set contained different scenarios to avoid the risk of over-fitting the classifiers.

The second data set was three testing sets in which 1,000, 5,000, and 10,000 randomly chosen scenarios were selected and passed to the application to have the anomalies eliminated. Each of these testing sets contained feature set definitions generated within our sample scenario. After each of the training and testing experiments have been conducted, the average of cleaning rate of the experiments has been derived for each technique and used to identify the highest achieving method.

24.4.2.2.2 Bayesian Network Experiment

For our first experiment, we conducted an investigation into the optimal amount of chromosomes that are needed to clean the false-positive anomalies. To accomplish this, we created three Bayesian networks that have been configured using a genetic algorithm with 10, 50, and 100 chromosomes. Each network was trained for 10 generations to breed and optimize the configuration. With regard to the set of data being used for training, we used three different "training cases" comprising 500, 1000, and 5000 false-positive anomaly scenarios. After these experiments were completed, the average of the three training cases was then extracted and, subsequently, used to determine the amount of chromosomes that were needed to achieve the highest cleaning rate.

From our results, we have found that the configuration that used 10 chromosomes to train the network obtained the highest average cleaning rate. As a result, the Bayesian network using a genetic algorithm with 10 chromosomes will be utilized in the final experiment in which all three classifiers are compared. The lowest achieving configuration using 100 chromosomes tested upon 500 training cases was the Bayesian network. The highest achieving configuration has been found to be the configurations with 10 and 50 chromosomes against 500 and 1000 training cases, respectively.

24.4.2.2.3 Neural Network Experiment

The second experiment we conducted was in relation to determining the highest performing network configuration for a neural network to clean anomalous RFID data. To do this, we trained the weights of the networks using the BP and the genetic algorithms with 10 (GA-10), 50 (GA-50), and 100 (GA-100) chromosomes present. The performance of the resulting networks is determined based upon the correctness of the classification from three Training cases using 500, 1000, and 5000 false-positive anomaly scenarios. Each configuration had been trained by 10 iterations or generations before the training experiment commenced. The main goal of this experiment was to determine the highest average achieving network trainer; thus, the average of the three training cases has also been found.

From the experimental results, we have derived a general observation that the performance of the network is vastly improved with 50 and 100 chromosomes using the genetic algorithm. The highest performing average of the neural network has been found to be the genetic algorithms when trained with both 50 and 100 chromosomes. As such, we decided to use the genetic algorithm with 100 chromosomes as the attempt to clean 500 training cases performed the highest. The lowest performing cleaning algorithm was the BP algorithm when attempting to clean 1000 training cases.

24.4.2.2.4 NMR Experiment

The main goal of the third experiment was to derive the highest performing NMR formula from the five different options used in CDL. With this in mind, the μ (Mu), α (Alpha), π (Pi), β (Beta), and δ (Delta) formulae were each trained using three training cases containing 500, 1000, and 5000 false-positive scenarios each. Like the previous two experiments, the average of each performing algorithm was ascertained and used to determine which of the five formulae would be utilized to proceed onto the final experiment.

From the results, we have found that the highest performing formulae are μ, α, and π. In contrast, the β and δ formulae both performed the least cleaning. With regard to the final experimentation, we have chosen the π formula as it performed the highest and is the most likely to continue to perform highly. The reasons as to why we rejected the μ and α formulae lie in the fact that the μ formula is strict in that it only accepts factual information and the α formula is connected directly to the β. Hence, we determined that the π formula would be superior to the other formulae.

24.4.2.2.5 Comparison Experiment

The goal of the fourth experiment was to determine which of the three highest performing classifier techniques would clean the highest percentage of a large amount of false-positive RFID anomalies. The three classifiers used in this experiment included the Bayesian network trained by a genetic algorithm with 10 chromosomes (BN), the neural network trained by genetic algorithm with 100 chromosomes (NN), and the π of the NMR. The classifiers were all chosen based upon the high performance found within the first three experiments previously discussed. Both of the Bayesian and neural networks had been trained for 10 generations before these tests were conducted. As opposed to the previous experiments, we determined that three "testing cases" containing 1000, 5000, and 10,000 randomly chosen false-positive scenarios would be utilized to determine the highest performing classifier. To ascertain the highest performance, the average of each of the three test cases has been found from the results.

The results of this experiment are depicted in Figure 24.13 where the amount of test cases and classifier has been graphed against the percentage of correctness. From these results, it can be seen that the NMR Engine achieves the highest average cleaning rate among other classifiers. The highest performing classifier has been found to be the NMR when attempting to clean

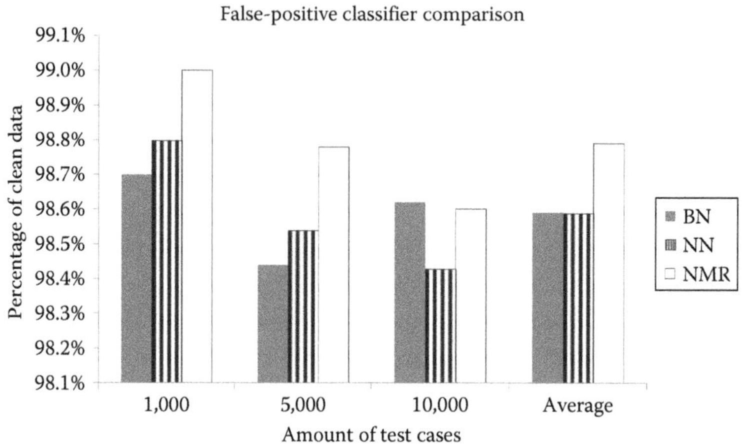

Figure 24.13 Experimental results of the fourth experiment designed to find the highest performing classifier when faced with a large amount of false-positive anomalies.

1,000 test cases, whereas the lowest achieving classifier has been found to be the neural network when attempting to clean 10,000 test cases.

The NMR engine outperformed the other classifiers in dealing with false-positive data due to the fact that it is a deterministic approach. The Bayesian and neural networks, by contrast, rely on a probabilistic nature to train their respective networks. The major drawbacks of this system are that it is specifically tailored for the static RFID cleaning problem; however, we believe that the same concept may be applicable to any static spatial-temporal data enhancement case study. With regard to applying our methodology to other applications where the environment is dynamic, the feature-set definition and NMR will need greater complexity to accommodate the change in anomalies. Although the test cases utilized in experimentation were small in comparison to the immense amount of RFID readings that get recorded in real-world systems, we believe our methodology would behave similarly upon larger data sets.

24.5 Conclusion

In this chapter, we have discussed the issues associated with anomalies present in captured RFID data and presented solutions to improve its integrity. First, we proposed deterministic and probabilistic anti–collision approaches, which increased the efficiency of the system performance. We also performed a comparative analysis of our two proposed deterministic and probabilistic anti–collision methods, and identified the benefits and disadvantages of each approach. We then proposed deferred cleaning approaches to be applied after the filtering of the data is complete to correct any ambiguous anomalies still present in the stored observations. For postcapture cleaning, we have integrated the Bayesian network, neural network, and non-monotonic reasoning classifiers to introduce a high level of intelligence to combat the ambiguous anomalies.

First, we proposed a deterministic anti–collision algorithm using combinations of Q-ary trees, with the intended goal to minimize memory usage queried by the RFID reader. By reducing the size of queries, the RFID reader can preserve memories, and the identification time can be improved. We then introduced the probabilistic group-based anti–collision method to improve the overall performance of the tag recognition process and provide a sufficient performance over existing methodologies. We also performed a comparative analysis of our proposed deterministic joined Q-ary tree and PCT, and identified the benefits and disadvantages of each approach for specific circumstances. Empirical analysis shows that the joined Q-ary tree method can achieve higher efficiency if the right EPC pattern is configured. However, for arbitrary situations where EPC pattern cannot be found, it is more preferable to use a probabilistic approach rather than the deterministic method.

After the filtering has been completed with the anti–collision algorithms, we introduced the use of intelligent classifiers to discover ambiguous anomalies and decide upon the action to be taken to correct them. With regards to false-negative anomaly detection and correction, we have found that the highest performing classifier is the neural network. In contrast, we found that the NMR classifier achieved the highest cleaning rate when correcting the false-positive anomalies. From the results, we have seen that it is much easier to return a database to the highest integrity when attempting to clean false-positive anomalies as opposed to false-negatives. Additionally, we have found that the NMR classifier was able to achieve the highest false-positive anomaly clean as its deterministic nature makes it ideal to clean wrong and duplicate data, while probabilistic techniques would introduce an additional level of ambiguity. With regards to the false-negative anomalies, the neural network classifier gained the highest cleaning rate due to it being able to introduce a limited amount of ambiguity needed to find the ideal missing values.

References

1. H. Stockman. Communication by means of reflected power. In *Insistute of Radio Engineers (IRE)*, New York, pp. 1196–1204, 1948.
2. S. S. Chawathe, V. Krishnamurthy, S. Ramachandran, and S. E. Sarma. Managing RFID Data. In *VLDB*, Toronto, Canada, pp. 1189–1195, 2004.
3. P. Pupunwiwat and B. Stantic. Resolving RFID data stream collisions using set-based approach. In *The Sixth International Conference on Intelligent Sensors, Sensor Networks and Information Processing (ISSNIP)*, pp. 61–66, Brisbane, Queensland, Australia, 2010. IEEE.
4. Á. Barbero, E. Rosnes, G. Yang, and Ø. Ytrehus. Constrained codes for passive RFID communication. In *Information Theory and Applications Workshop (ITA)*, San Diego, CA, pp. 1–9, 2011.
5. S. Preradovic and N. Karmakar. Design of short range chipless RFID reader prototype. In *Intelligent Sensors, Sensor Networks and Information Processing (ISSNIP 2009)*, Melbourne, Victoria, Australia, pp. 307–312, 2009.
6. S. Preradovic, N. Karmakar, and M. Zenere. UWB chipless tag RFID reader design. In *IEEE International Conference on RFID-Technology and Applications (RFID-TA)*, Guangzhou, China, pp. 257–262, 2010.
7. N. C. Karmakar, S. M. Roy, and M. S. Ikram. Development of smart antenna for RFID reader. In *IEEE International Conference on RFID*, Las Vegas, Nevada, USA, pp. 65–73, 2008.
8. D. Giusto, A. Iera, G. Morabito, L. Atzori, S. Tedjini, E. Perret, V. Deepu, and M. Bernier. Chipless tags, the next RFID frontier. In *The Internet of Things*, pp. 239–249. Springer, New York, 2010.
9. S. Preradovic and N. Karmakar. 4th generation multiresonator-based chipless RFID tag utilizing spiral EBGs. In *2010 European Microwave Conference (EuMC)*, Paris, France, pp. 1746–1749, 2010.
10. J. Pavlina and D. Malocha. Chipless RFID SAW sensor system-level simulator. In *2010 IEEE International Conference on RFID*, Orlando, Florida, USA, pp. 252–259, 2010.

11. R. B. Ferguson. Logan airport to demonstrate baggage, passenger RFID tracking. eWeek, July 2006. Available from: http://www.eweek.com/c/a/Mobile-and-Wireless/Logan-Airport-to-Demonstrate-Baggage-Passenger-RFID-Tracking/ (accessed on September 28, 2011).
12. S. R. Jeffery, M. N. Garofalakis, and M. J. Franklin. Adaptive cleaning for RFID data streams. In *VLDB*, Seoul, Korea, pp. 163–174, 2006.
13. D.-H. Shih, P.L. Sun, D. C. Yen, and S.M. Huang. Taxonomy and survey of RFID anti-collision protocols. *Computer Communications*, 29(11):2150–2166, 2006.
14. E. Schuman. Will users get buried under RFID data? [online]. eWeek, Nov 2004. Available from: http://www.eweek.com/c/a/Enterprise-Applications/Will-Users-Get-Buried-Under-RFID-Data/ (accessed on September 28, 2011).
15. M. Ward, R. van Kranenburg, and G. Backhouse. RFID: Frequency, standards, adoption and innovation. Technical report, JISC Technology and Standards Watch, 2006.
16. R. Cocci, T. Tran, Y. Diao, and P. J. Shenoy. Efficient data interpretation and compression over RFID streams. In *ICDE*, Cancun, Mexico, pp. 1445–1447. IEEE, 2008.
17. R. Derakhshan, M. E. Orlowska, and X. Li. RFID data management: Challenges and opportunities. In *IEEE Conference on RFID*, Texas, USA, pp. 175–182, 2007.
18. S. Jain and S. R. Das. Collision avoidance in a dense RFID network. In *WiNTECH '06: Proceedings of the 1st International Workshop on Wireless Network Testbeds, Experimental Evaluation and Characterization*, pp. 49–56, New York, 2006. ACM.
19. D. K. Klair, K.-W. Chin, and R. Raad. A survey and tutorial of RFID anti-collision protocols. *Communications Surveys Tutorials, IEEE*, 12(3):400–421, 2010.
20. C. Law, K. Lee, and K.-Y. Siu. Efficient memoryless protocol for tag identification. In *Proceedings of the 4th International Workshop on Discrete Algorithms and Methods for Mobile Computing and Communications, DIALM '00*, pp. 75–84, New York, 2000. ACM.
21. C.-H. Quan, W.-K. Hong, and H.-C. Kim. Performance analysis of tag anti-collision algorithms for RFID systems. In *Emerging Directions in Embedded and Ubiquitous Computing*, volume 4097, pp. 382–391, Seoul, South Korea, 2006. Springer, Berlin, Germany.
22. N. Abramson. The ALOHA system—another alternative for computer communications. In *Proceedings of Fall Joint Computer Conference, AFIPS Conference*, pp. 281–285, Houston, TX, 1970.
23. M. R. Devarapalli, V. Sarangan, and S. Radhakrishnan. AFSA: An efficient framework for fast RFID tag reading in dense environments. In *QSHINE '07: The Fourth International Conference on Heterogeneous Networking for Quality, Reliability, Security and Robustness Workshops*, pp. 1–7, New York, 2007. ACM.
24. J. Ding and F. Liu. Novel tag anti-collision algorithm with adaptive grouping. *Wireless Sensor Network (WSN)*, 1(5):475–481, 2009.
25. X. Fan, I. Song, and K. Chang. Gen2-based hybrid tag anti-collision Q algorithm using chebyshev's inequality for passive RFID systems. In *PIMRC 2008. IEEE 19th International Symposium on Personal, Indoor and Mobile Radio Communications, 2008*, pp. 1–5, Cannes, France, 2008.
26. S.-R. Lee and C.-W. Lee. An enhanced dynamic framed slotted ALOHA anti-collision algorithm. In *Emerging Directions in Embedded and Ubiquitous Computing*, volume 4097, pp. 403–412, Seoul, Korea, 2006. Springer Berlin, Germany.
27. C. W. Lee, H. Cho, and S. W. Kim. An adaptive RFID anti-collision algorithm based on dynamic framed ALOHA. *IEICE Transactions*, 91-B(2):641–645, 2008.
28. S.-R. Lee, S.-D. Joo, and C.-W. Lee. An enhanced dynamic framed slotted ALOHA algorithm for RFID tag identification. In *MOBIQUITOUS '05: Proceedings of The Second Annual International Conference on Mobile and Ubiquitous Systems: Networking and Services*, pp. 166–174, Washington, DC, 2005. IEEE Computer Society.

29. P. Pupunwiwat and B. Stantic. A RFID explicit tag estimation scheme for dynamic framed-slot ALOHA anti-collision. In *The Sixth Wireless Communications, Networking and Mobile Computing (WiCOM)*, pp. 1–4, Chengdu, China, 2010. IEEE.
30. P. Pupunwiwat and B. Stantic. Dynamic framed-slot ALOHA anti-collision using precise tag estimation scheme. In H. T. Shen and A. Bouguettaya, eds., *The Twenty-First Australasian Database Conference (ADC)*, volume 104 of *CRPIT*, pp. 19–28, Brisbane, Queensland, Australia, 2010. ACS.
31. Z. Wang, D. Liu, X. Zhou, X. Tan, J. Wang, and H. Min. anti-collision scheme analysis of RFID system. *Auto-ID Labs White Paper.*, 2007. http://www.autoidlabs.org/single-view/dir/article/6/281/page.html
32. L. Zhu and P. T.-S. Yum. The optimization of framed aloha based RFID algorithms. In *MSWiM '09: Proceedings of the 12th ACM International Conference on Modeling, Analysis and Simulation of Wireless and Mobile Systems*, Canary Islands, Spain, pp. 221–228, 2009. ACM.
33. R. W. Williams and K. Herrup. The control of neuron number. *Annual Review of Neuroscience*, 11(1):423–453, 1988.
34. W. S. Mcculloch and W. Pitts. A logical calculus of the ideas immanent in nervous activity. *Bulletin of Mathematical Biophysic*, 5:115–133, 1943.
35. D. Rumelhart, G. Hinton, and R. Williams. Learning representations by back-propagating errors. *Nature*, 323:533–536, 1986.
36. M. Blumenstein, X. Y. Liu, and B. Verma. An investigation of the modified direction feature for cursive character recognition. *Pattern Recognition*, 40(2):376–388, 2007.
37. J. Holland. *Adaptation in Natural and Artificial Systems.* University of Michigan Press, Cambridge, MA, 1975.
38. A. Rooij, R. Johnson, and L. Jain. *Neural Network Training Using Genetic Algorithms.* World Scientific Publishing Co., Inc., River Edge, NJ, 1996.
39. D. Cha, M. Blumenstein, H. Zhang, and D.-S. Jeng. A neural-genetic technique for coastal engineering: Determining wave-induced seabed liquefaction depth. In *Engineering Evolutionary Intelligent Systems*, Berlin, Germany, pp. 337–351. 2008.
40. G. Antoniou, D. Billington, G. Governatori, and M. J. Maher. Embedding defeasible logic into logic programming. *Theory and Practice of Logic Programming*, 6(6):703–735, 2006.
41. D. Billington, V. E. Castro, R. Hexel, and A. Rock. Non-monotonic reasoning for localisation in robocup. In Claude Sammut, ed., *Proceedings of the 2005 Australasian Conference on Robotics and Automation*, Sydney, Australia, December 2005.
42. D. Billington. An introduction to clausal defeasible logic [online]. David Billington's Home Page, Aug 2007. Available from: http://www.cit.gu.edu.au/~db/research.pdf (accessed on September 28, 2011).
43. D. Billington. Propositional clausal defeasible logic. In *European Conference on Logics in Artificial Intelligence (JELIA)*, Dresden, Germany, pp. 34–47, 2008.
44. P. Pupunwiwat and B. Stantic. Joined Q-ary tree anti-collision for massive tag movement distribution. In B. Mans and M. Reynolds, eds., *The Thirty-Third Australasian Computer Science Conference (ACSC)*, volume 102 of *CRPIT*, pp. 99–108, Brisbane, Queensland, Australia, 2010. ACS.
45. P. Darcy, P. Pupunwiwat, and B. Stantic. *The Challenges and Issues facing Deployment of RFID Technology*, pp. 1–26. Deploying RFID—Challenges, Solutions and Open Issues. 2011.
46. P. Pupunwiwat and B. Stantic. Performance analysis of enhanced Q-ary tree anti-collision protocols. In *The First Malaysian Joint Conference on Artificial Intelligence (MJCAI)*, vol. 1, pp. 229–238, Kuala Lumpur, Malaysia, 2009.
47. P. Pupunwiwat and B. Stantic. Unified Q-ary tree for RFID tag anti-collision resolution. In A. Bouguettaya and X. Lin, eds., *The Twentieth Australasian Database Conference (ADC)*, volume 92 of *CRPIT*, pp. 47–56, Wellington, New Zealand, 2009. ACS.

48. P. Pupunwiwat, P. Darcy, and B. stantic. Conceptual selective RFID anti-collision technique management. In *International Conference on Mobile Web Information Systems (MobiWIS 2011)*, Niagara Falls, Canada, pp. 827–834, 2011.
49. P. Darcy, B. Stantic, and A. Sattar. Augmenting a deferred Bayesian network with a genetic algorithm to correct missed RFID readings. In *Malaysian Joint Conference on Artificial Intelligence (MJCAI 2009)*, Kuala Lumpur, Malaysia, pp. 106–115, 2009.
50. P. Darcy, B. Stantic, and A. Sattar. Improving the quality of RFID data by utilising a Bayesian network cleaning method. In *Proceedings of the IASTED International Conference Artificial Intelligence and Applications (AIA 2009)*, Innsbruck, Austria, pp. 94–99, 2009.
51. P. Darcy, B. Stantic, and A. Sattar. Applying a neural network to recover missed RFID readings. In *Australasian Computer Science Conference (ACSC 2010)*, Brisbane, Queensland, Australia, pp. 133–142, 2010.
52. P. Darcy, B. Stantic, and R. Derakhshan. Correcting stored RFID data with non-monotonic reasoning. *Principles and Applications in Information Systems and Technology (PAIST)*, New Zealand, 1(1):65–77, 2007.
53. P. Darcy, B. Stantic, and A. Sattar. A fusion of data analysis and non-monotonic reasoning to restore missed RFID readings. In *Intelligent Sensors, Sensor Networks and Information Processing (ISSNIP 2009)*, Melbourne, Victoria, Australia, pp. 313–318, 2009.
54. P. Darcy, B. Stantic, and A. Sattar. Correcting missing data anomalies with clausal defeasible logic. In *Advances in Databases and Information Systems (ADBIS 2010)*, Novi Sad, Serbia, pp. 149–163, 2010.
55. S. Liu, F. Wang, and P. Liu. A temporal RFID data model for querying physical objects. Technical Report TR-88, TimeCenter, Denmark, 2007.

Chapter 25

Radio Frequency Identification Systems and Sensor Integration for Telemedicine

Ajay Ogirala, Shruti Mantravadi, and Marlin H. Mickle

Contents

25.1 RFID for Telemedicine

25.1.1 Telemedicine: Is It Necessary?

Cost associated with individual health care and the health-care industry is one of the most debated and discussed topics in recent times. According to the U.S. Centers for Medicare & Medicaid Services, the average cost per patient per stay has increased from $1851 in the year 1980 to $8793 in the year 2005 [1]. The average annual expenditure per consumer for health care in 1 year increased from $1500 (approximate value) in the year 1990 to $2800 (approximate value) in the year 2007 [2] (Figures 25.1 and 25.2).

The direct effect of the rising health-care costs can be observed in Figures 25.3 through 25.5, which project the trends in hospital emergency room visits, hospital inpatient days, and hospital outpatient days per 1000 population, respectively, during 1999–2009 [3]. The trend graphs clearly show decrease in inpatient numbers and an increase in outpatient numbers. A rise in emergency

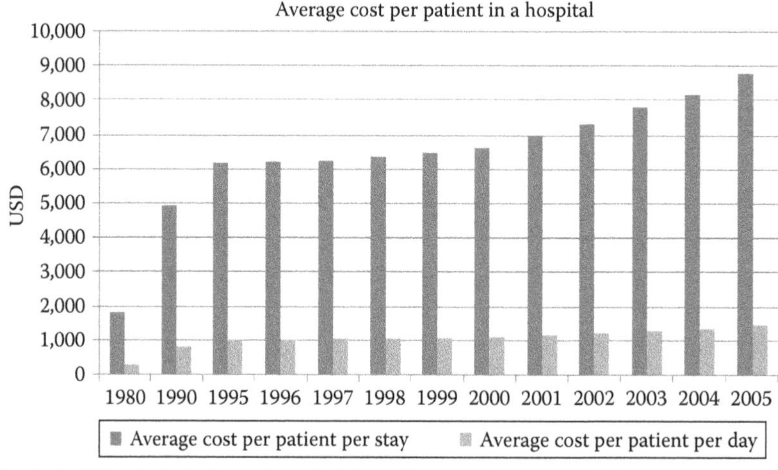

Figure 25.1 Average cost per patient per stay trend (1980–2005).

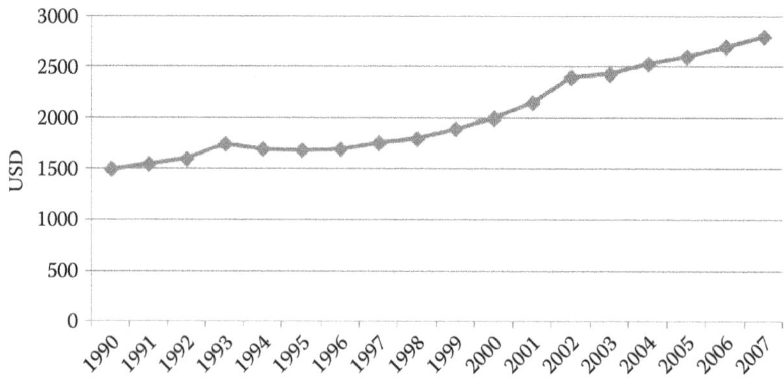

Figure 25.2 Average annual expenditure per consumer for health care in a year (1990–2007).

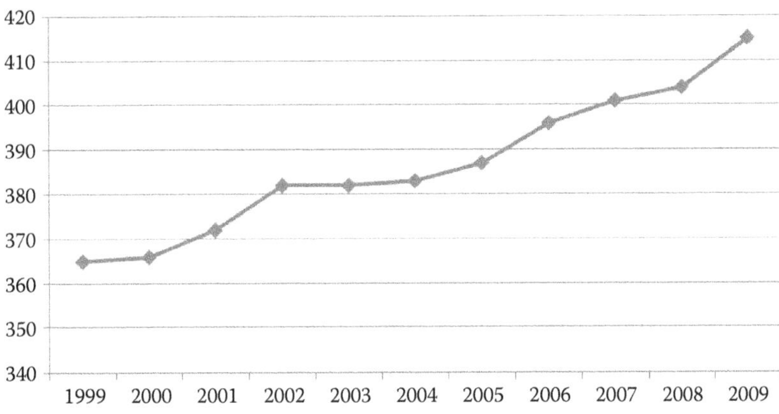

Figure 25.3 Emergency room visits per 1000 population (1999–2009).

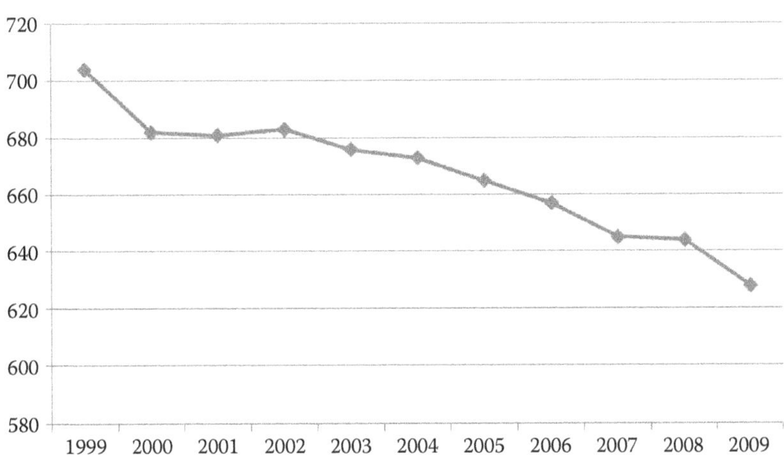

Figure 25.4 Inpatient days per 1000 population (1999–2009).

room visits also suggests that patients are unwilling to visit hospitals unless absolutely necessary and/or life threatening.

According to the U.S. Department of Health and Human Services, the total health expenditures as a percent of gross domestic product of the United States increased from 5.1% in the year 1960 to 15.3% in the year 2005 [4,5]. This percentage is the highest among several developed and developing countries. The total health expenditures as a percent of per capita health expenditures in the United States increased from $147 in the year 1960 to $6410 in the year 2007.

With the exponentially increasing cost of patient care and liability in medical institutions, there is ongoing research to consult and treat patients via remote methods. The current trends in health-care costs and developments in wireless sensors lead to a reasonable advancement toward telemedicine.

Radio frequency identification (RFID) technology, primarily developed to replace and overcome the limitations of the bar code technology, is rapidly finding applications in several industries and

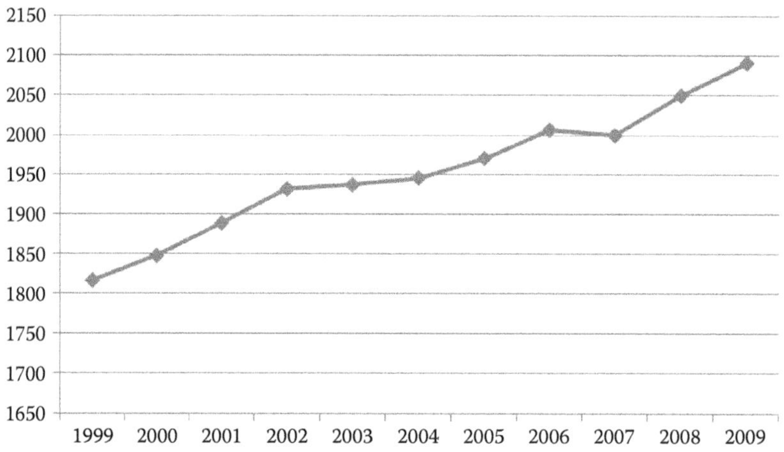

Figure 25.5 Outpatient days per 1000 population (1999–2009).

particularly in the health-care industry. Many hospitals across the globe are adapting to this latest wireless technology by integrating the available biomedical sensors with commercially available RFID networks to create an intelligent wireless health-care network that in recent times is more generally known as mHealth (Mobile Health) or telemedicine. It is widely believed and aspired that telemedicine is the solution for growing health-care needs, cost, and resources in developed as well as developing countries.

25.1.2 Fundamentals of RFID

RFID technology is the latest in wireless communication technology that was originally intended for item tracking applications. The hardware modules of the RFID technology can be broadly classified into the interrogator (also called the reader) and the tag. The interrogator is analogous to the bar code reader and the tag is analogous to the bar code sticker. The primary advantage of RFID over traditional bar code technology is in terms of speed, memory, and beyond line-of-sight communication. To explain further, RFID can identify more tags in unit time compared to the bar code technology. There is more memory on an RFID tag compared to the bar code sticker enabling detailed information of the item to be stored on the tag attached to the item and eliminating the necessity of referring a database to store and/or retrieve additional information. The most important advantage of RFID over bar code is the ability to communicate beyond line of sight between the interrogator and the tag.

To simplify the application of the technology, RFID technologies allow the transmission of a unique serial number wirelessly, using radio waves. The two fundamental parts of the system that are necessary to do this are the RFID tag and the RFID interrogator. The interrogator requests all tags in its range to identify themselves. The tags respond by transmitting their unique serial number. Attaching an RFID tag to a physical object allows the object to be "seen" and monitored by existing computer networks and office administration systems.

RFID can be broadly classified into active RFID and passive RFID. Active RFID tags are powered by an onboard battery. Passive RFID tags, depending on the distance from the interrogator, either harvest energy from the interrogator-transmitted electromagnetic energy (near field) or communicate with the interrogator by modulating and reflecting the incident electromagnetic energy (far field). It is important to note that most RFID communication protocols are reader initiated and therefore the tags have to wait for interrogator initiation to transfer or request for data. Passive RFID technology is typically used for item management applications and active RFID has more specific applications like cargo tracking, military applications, and sensors [6–8].

The RFID interrogator can be broadly divided into the following blocks from a system design engineer's perspective:

- Antenna
- Power circuit
- Modulator and demodulator
- Data processor
- Memory
- Host interface

The RFID tag (irrespective of the power source) can be broadly divided into the following blocks from a system design engineer's perspective:

- Antenna
- Power and switching circuit
- Modulator and demodulator
- Data processor
- Memory
- Attached sensors

Let us try to understand the fundamental design challenges for the RFID system without diving too deep into the details. The major difference in designing an interrogator and a tag is that the interrogator is not constrained by the amount of power available as it is typically powered by an AC outlet. Another major difference is that the dimensions of an interrogator are typically more flexible. This flexibility greatly affects the degree of freedom available to the engineer in designing the antenna and electronic circuitry of the tag. Memory for the passive tag is typically ROM type for obvious power constraints, and this restricts the flexibility and speed of operation of such RFID systems. When understanding the design challenges of RFID systems, it is absolutely essential to mention the Friis transmission equation (25.1) that is overlooked by most literature available:

$$\frac{P_r}{P_t} = G_t\left(\theta_t, \varphi_t\right) G_r\left(\theta_r, \varphi_r\right) \left(\frac{\lambda}{4\pi R}\right)^2 \left(1 - |\Gamma_t|^2\right)\left(1 - |\Gamma_r|^2\right) \left|a_t \cdot a_r^*\right| e^{-\alpha R} \tag{25.1}$$

where

P_r is the power received
P_t is the power transmitted
G_t is the gain of transmitting antenna
G_r is the gain of receiving antenna
θ and φ is the inclination angle and azimuth angle measured in spherical coordinate system
λ is the wavelength of electromagnetic energy transmitted
R is the distance between transmitting and receiving antennas
Γ_t is the reflection coefficient of transmitting antenna

Γ_r is the reflection coefficient of receiving antenna
a_t is the polarization vector of transmitting antenna
a_r is the polarization vector of receiving antenna
e is the antenna efficiency
α is the absorption coefficient of transmission medium

The Friis transmission equation relates the power transmitted by the transmitting antenna to the power received by the receiving antenna as a function of the antenna gain, frequency of operation, antenna reflection coefficient, antenna polarization, and medium. It is important to understand that the entire RFID system can operate as designed only if the received power (at interrogator or tag) is adequate to decipher. As clearly equated by the Friis equation, this received power is highly sensitive to the polarization of the interrogator and tag antennas. While the available literature portrays the varied applications of RFID, the reader has to note that all that is possible in theory and textbooks cannot be practiced due to current limitations in the manufacturing process of RFID tags that directly affect the cost and dimension of available tags.

The phenomenon that influence the received electromagnetic energy at the receiver include but are not limited to multipath, diffraction, fading, Doppler shift, noise (internal and external), interference, and ducting [9,10].

25.1.3 Current Applications of RFID

Typical applications of RFID include the following:

- Automotive security, automotive location, automotive passive entry systems
- Highway toll booths, traffic congestion detention and avoidance
- Livestock tracking, wild animal tracking, pet tracking
- Asset tracking in multiple and varied industries
- Contactless payment and shopping
- Supply chain management—one of the most widely used and primary RFID application

Current applications of RFID in health care include [11–14] the following:

- Item tracking in hospitals and operating rooms
- Patient tracking in hospitals
- Data tracking
- Drug tracking
- Crash cart tracking
- Nurse tracking
- Remote wireless continuous arrhythmia detection
- Implant monitoring
- Smart patient rooms

25.1.4 RFID Standards and Spectrum Utilization

The number and use of standards within RFID and its associated industries are quite complex and undocumented. They involve a number of bodies and are in a continuous process of development. Standards have been produced to cover four key areas of RFID application and use:

- Air interface standards (basic communication between reader and tag)
- Data content and encoding (numbering schemes)
- Conformance
- Interoperability (as an extension of conformance)

It is important to note that there are no published RFID standards that particularly define the physical layer communication protocol or frequencies for biomedical or telemedicine applications. The available market solutions adopt a combination of RFID and data communication protocols or resort to developing custom proprietary protocols for application-specific solutions. The available data communication protocols apart from the RFID standards are Zigbee, Bluetooth, IEEE 802.11a/b/g/n, GSM, CDMA, and GPRS.

While application-specific custom communication protocols for sensor networks are understandable during this phase of telemedicine evolution, the development of standards by a global entity will become an absolute necessity for expedited growth and sustenance of telemedicine in near future.

There are several U.S. and international standard bodies involved in the development and definition of RFID technologies including the following:

- International Organization of Standards
- EPCglobal Inc.
- European Telecommunication Standards Institute (ETSI)
- Federal Communications Commission (FCC)

RFID communication technologies are governed by the ISO/IEC 18000 family of standards for their physical layer of communication. The various parts of ISO/IEC 18000 describe air interface communication at different frequencies in order to be able to utilize the different physical behaviors. The various parts of ISO/IEC 18000 are developed by ISO/IEC JTC1 SC31, "Automatic Data Capture Techniques." Conformance test methods for the various parts of ISO/IEC 18000 are defined in the corresponding parts of ISO/IEC 18047. Performance test methods are defined in the corresponding parts of ISO/IEC 18046. A list of parts in the ISO/IEC 18000 family is as follows:

- ISO/IEC 18000-1—Generic parameters for the air interface globally accepted
- ISO/IEC 18000-2—Communication at frequencies below 135 kHz
- ISO/IEC 18000-3—Communication at 13.56 MHz frequency
- ISO/IEC 18000-4—Communication at 2.45 GHz frequency
- ISO/IEC 18000-6—Communication at frequencies between 860 and 960 MHz
- ISO/IEC 18000-7—Communication at 433.92 MHz frequency

Within a given frequency band, the nonideal communication range will vary depending on the environment and the factors in Friis equation. Table 25.1 provides a reference to the different RFID communication standards and their typical characteristics at a glimpse.

The power that can be transferred across variable depth through the human skin and tissue is not equal over a wide range of frequencies. Several human torso models have been developed [15–19], and the loss across several frequencies is available in literature. As a rule of thumb, from all the different human torso simulators, it is observed that there is a greater loss as the carrier frequency increases [20]. This is mainly because ultrahigh frequency (UHF) and higher

Table 25.1 RFID—A Classification

	Low Frequency (LF)	*High Frequency (HF)*	*Ultrahigh Frequency (UHF)*
Typical frequency band	30–300 kHz	3–30 MHz	300 MHz–3 GHz
Typical RFID communication frequency	125 and 134.2 kHz	13.56 MHz	433.92 MHz, 860–960 MHz, 2.45 GHz
Approximate read range	<1 m	<2 m	433.92 MHz—up to 100 m 2.45 GHz—1–10 m 860–960 MHz—0.5–5 m
Typical data rate	<1 kbps	<25 kbps	433.92 MHz—up to 30 kbps 2.45 GHz—up to 100 kbps 860–960 MHz—up to 30 kbps
Typical power source	Passive RFID	Passive RFID	433.92 MHz—active RFID 2.45 GHz—active and passive RFID 860–960 MHz—passive RFID
Important characteristic near water and metal	These signals penetrate water but not metal	These signals penetrate water but not metal	These signals can neither penetrate water nor metals
Typical application	Animal ID, automobile wireless access	Smart labels, contactless travel cards, and security cards	Document tracking, inventory tracking, and military applications. Most widely used RFID frequencies

frequencies are extremely sensitive to water and humidity. The human skin and tissue absorb the electromagnetic (EM) radiation at these frequencies contributing to higher losses (radiated as heat) in the body medium. This encourages the use of low frequency (LF) and high frequency (HF) in health-care industry when penetration into the body is necessary. But the disadvantage is that the data rate and the communication range at these frequencies are very limited. The communication frequency of sensor network for telemedicine is therefore very crucial.

25.1.5 RFID Communication Protocols Extended

Today, most office buildings, educational institutions, and commercial stores are hot spots, where several wireless systems can be connected to the Internet [21] via a wireless interface. These systems transfer data according the IEEE 802.11 communication protocol at 2.45 GHz. The IEEE 802.11 protocol (only as an example), while providing high data rates, includes recurring packet header and footer information and an initial high-volume data exchange between the wireless system and the wireless router to authenticate and initiate further data transfer [22]. These protocol requirements along with the addition of multiple layers for ease of communication and added overhead make the IEEE 802.11 and other high long-range communication protocols such as GSM, CDMA, and GPRS unattractive for communication where the time available for data transfer is critical, i.e., in the order of microseconds. For example, communicating with automobiles on a highway traveling at high speeds is a time-sensitive application. Another time-sensitive application is data acquisition from multiple sensors in a network that needs frequent calibration. Applications where the transmitter and receiver cannot be active for extended intervals of time due to power limitations [23–25] require a simple yet secure protocol for quick data transfer that is also standardized.

An attractive alternate is, for example, the EPCglobal™ Class-1 Generation-2 protocol [26] popularly known as the Gen-2 protocol. This protocol, originally developed to communicate with only passive RFID tags, is now evolving into a communication link with greater application potential [27], one of them to maintain an intermittent-connection wireless network. An intermittent-connection wireless network is usually a star topology, where the connection between the central server and the system(s) is not continuous. Further, by replacing or modifying the wireless front end of the system with a suitable alternate, it is possible to extend this technology to replace or assist wired communication links such as the Intel invented USB. To simplify, the internationally accepted Gen-2 protocol can be used as a communication protocol, not only between passive RFID tags and interrogators, but also between other short-ranged wireless sensor networks. The same concept is also applicable to other RFID communication protocols such as ISO 18000-2 [28] and ISO 18000-3 [29] and is currently being researched [27].

25.2 Sensors for Telemedicine

25.2.1 Biomedical Sensors versus Environmental Sensors

The fundamental difference between environmental sensors and biomedical sensors is because of the change in behavior of electric and electromagnetic signals in the atmosphere and inside and near the human body. The electromagnetic energy reacts with the moisture content in the human body, and this is the main concern when designing, implanting, and communicating with biomedical sensors.

Biomedical sensor design and dispatch differs from traditional environmental sensors in terms of material safety, communication, replacement, maintenance, lifetime, power, size, interference, and noise. The list is not all inclusive but a collection of the basic engineering challenges.

The materials used in biomedical sensors that are either implantable or wearable have to be human friendly and nontoxic even after long-term exposure and contact. The different fluids and enzymes in the human body react differently with a variety of metals, ceramics, polymers, and plastics. Some materials are absorbed by the body after prolonged exposure. Common allergies to safe materials are also of concern.

As mentioned before, the biggest challenge in designing biomedical sensors is the communication with the sensor. Wired communication is generally not preferred not just for aesthetic and convenience reasons but also for safety and health reasons. Wireless communication is greatly challenged by the moisture content in the human body. The absorption of electromagnetic energy by the human body increases as the frequency in the electromagnetic spectrum increases. Selection of the frequency for communication is therefore critical. Communication with implanted sensors is typically using volume conduction technique. The biocompatibility of this technique is still in research. Sensors that are passively powered have to be designed so that they can reliably operate at low power levels as the electromagnetic energy available for the sensor to harvest energy from is limited compared to the environment sensor. The transmitted energy from the sensor has to be sufficiently high so that it can account for the absorption by the human body and also be within human tolerance limits.

The size of the sensor is also of concern, whether it is implantable or wearable. Implantable sensors are limited in size for obvious reasons. Wearable sensors have to be small enough not to cause inconvenience to the patient even after long-term use.

Noise among nonsynchronized biomedical sensors is of concern from an application layer perspective. The electromagnetic energy and its harmonics interfering with the electrophysiology of the human heart and brain are still under research and not absolutely established but are definitely of concern.

Examples of environmental sensors (not biomedical sensors) are sensors with applications in environmental monitoring, agriculture, biological processes detection and monitoring, food processing, and pharmacological industries.

Examples of biomedical sensors include active implantable medical devices, cardiac rhythmic management devices, and vital sign(s) monitors.

25.2.2 Classification of Biomedical Sensors

A biomedical sensor can be defined as a transducer for measuring a physiological variable. Examples of physiological variables include body temperature, blood flow, blood velocity, electromyographic (EMG) signals, electroencephalogram (EEG) signals, and electrocardiogram (ECG) signals. Biomedical sensors are the basic building blocks of diagnostic medicine and therefore communicating with patients and biomedical sensors at great distances via existing or existential data networks is inevitable for advancement of telemedicine. Among latest trends in health care, self-testing is on the rise. This trend is driven by the desire of the patients and physicians alike to have the ability to perform instantaneous diagnosis and displacement of external and lengthy diagnosis model into the point-of-care model.

Biomedical sensors can be classified based on their application *in vitro* or *in vivo* measurements. Sensors used primarily in laboratories and diagnostic clinics can be classified under the *in vitro* category. *In vitro* sensors can be further classified into physiological sensors and pathological sensors. Physiological sensors measure electrolytes, enzymes, and biochemical metabolites in blood and pathological sensors, as the name suggests, measure or detect pathogens in the blood. Biomedical sensors for measuring pressure, flow, and concentration of gases are used *in vivo* (Figure 25.6).

Biomedical sensors can also be classified based on the quantity of measurement and can be characterized into physical, electrical (bio-potential) and bio-analytical (chemical), gaseous, and optical sensors. It has to be noted that in all electronic sensors, the quantity being measured is converted into an electrical signal (either voltage or current) by a transducer. For example, an oximeter converts bold SpO_2 data into current signal using light at a particular infrared frequency. According to the aforementioned classification, an oximeter can be classified as a gaseous sensor but not as an electrical sensor. Each sensor type will be explained briefly.

Physical sensors are typically used to measure the change in position of an object or medium. They are used in measuring or quantifying changes in dimensions, pressure, force, or temperature.

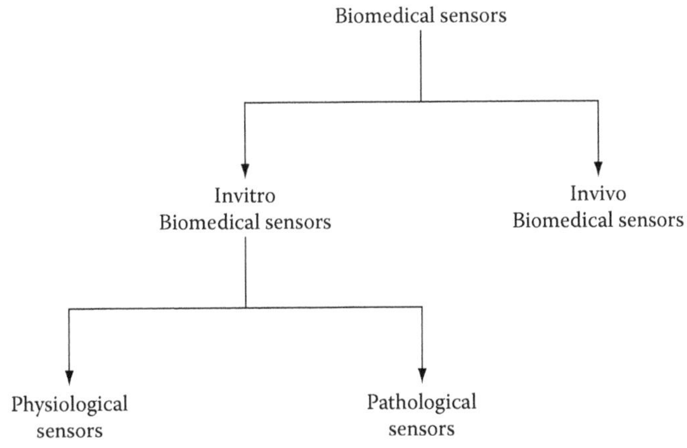

Figure 25.6 Classification of biosensors.

Examples of apparatus that can be classified as physical sensors include blood pressure monitor, electromagnetic blood flow monitor, and thermometers.

The purpose of electrical or bio-potential sensors is to measure the ionic potential generated inside the human body. Examples of apparatus that can be classified as bio-potential sensors are ECG, EMG, and EEG.

Bio-analytical sensors are primarily used to measure concentration and traces of enzymes and bacteria. These sensors can be further classified as enzyme-based sensors and microbial-based biosensors. Most enzymes react only with specific chemicals preset in simple form or as a complex compound. The action of specific enzymes can be used to construct a wide variety of biosensors. A typical example is a glucose sensor that uses the enzyme glucose oxidase. Microbial sensors are used for controlling biochemical processes in environmental, agricultural, food, and pharmaceutical applications. These sensors typically involve the assimilation of organic compounds by the microorganism, followed by a change in respiration activity or the production of specific electrochemically active metabolites such as hydrogen, carbon dioxide, or ammonia produced by the microorganism.

Knowledge of patient's arterial blood gases such as oxygen and carbon dioxide is important in medicine to sustain the patient using mechanical ventilation or chemical drugs. There are several chemical, optical, and temperature transducers to quantify the percentage of oxygen and carbon dioxide available in blood.

Optical sensor makes use of dispersion and diffraction of certain frequencies or visible, ultraviolet, and infrared spectrum. The change in absorbance, reflection, scattering, polarization, or refractivity of light through a biological medium is converted into quantifiable data by a transducer. These sensors are used to measure the health or healing rate of implanted tissue among several other applications.

25.2.3 Sensors for Oral Telemedicine

As any other field of life science, dentistry has been continuously evolving. Among many improvisations in the field that have enhanced the speed, accuracy, and treatment modalities, sensors have also carved a niche. With the miniature dimensions of the sensors, they barely require any camouflage and are both complacent and effective in the modern-day treatment.

The role of oral sensors in telemedicine has been established for over a decade. With the advancements in the production cycle of microelectromechanical system (MEMS) devices, the role of sensors in oral medicine for telemedicine is once again a topic of interest for the industry and academics.

In orthodontic treatments, relapse generally occurs due to mild inaccuracies in the angulations of teeth, which in turn is caused by obscure application of pressures that cause angulations of teeth. Sensors aid in continuous monitoring of the pressures causing angulations and dictating precise time for altering the pressure(s) in order to cause concrete tooth movements and hence successful treatment.

Sensors are used in protecting infants from sleep apnea by quantizing their breathing and respiration. A continuous monitoring helps the management in pediatric centers to ensure the safety of each infant individually with minimal effort and prevent sudden infant deaths.

A brief literature survey elicits that sensors in the oral cavity for long-term use primarily include orthodontic sensors and oral airflow sensors [30,31]. Sensors in development include Ph sensors for cavity detection and for monitoring the osseointegration of oral implants and the levels of various microorganisms in post-flap surgeries and graft placement surgeries.

25.3 Telemedicine Models

Telemedicine network can be broadly classified into a real-time telemedicine network or a continuous monitoring telemedicine network. As the names suggest, a real-time network will be primarily used when the patient and the medical personnel communicate with each other in real time. The continuous monitoring model will be used when a patient requires constant or frequent monitoring of his/her vitals or any other critical information as seen fit by the medical personnel for his/her particular condition. A third model called the hybrid model can also be used for a telemedicine network. As the name suggests, this model will be a combination of the real-time model and the continuous monitoring model. In the hybrid model, it is possible to monitor the patient's vitals continuously and, when required, communicate with the patient and make changes to the sensor implanted or attached to the patient. The different modules of the two models and further explanation follow in the next two sections.

25.3.1 Real-Time Model

As explained before, this model is adopted when the medical personnel and the patient communicate with each other in real time. The communication here is not just the audio or visual data of the patient explaining his symptoms but also the vitals or any other biological information captured and transmitted by the sensor to the medical personnel via the telemedicine network. The medical personnel can in real-time adjust the sensor(s) and/or other medical electronic equipment attached or implanted into the patient and get immediate feedback from the patient.

Consider the following example for a real-time model in telemedicine to adjust an implanted cardiac defibrillator (ICD). An ICD is necessary for patients with a heart condition where the heart needs help to pace (or beat) at a consistent pace. The ICD provides the beat (electrical signal) when the heart fails to do so. It also slows the beating heart by providing an electric shock to the heart when it is pacing (beating) too fast, which is life threatening to the patient. Patients with implanted ICDs have to make regular visits to the cardiac clinic to adjust the settings on the ICD. The medical personnel adjust the ICD so that it can maintain a healthy heart rate based upon

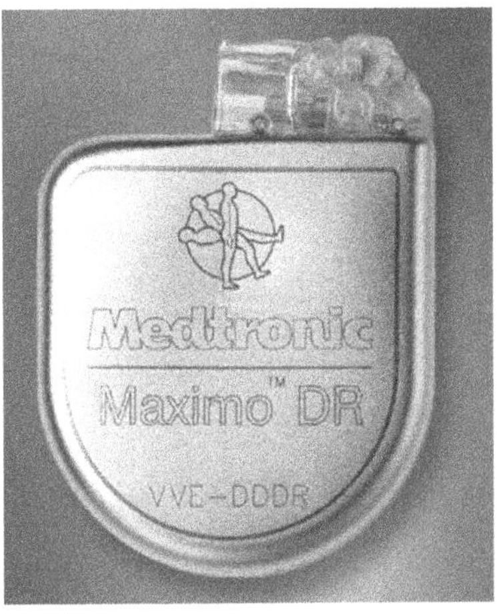

Figure 25.7 Medtronic ICD.

not only the medical tests and ECG results but also the experiences and activities of the patient (Figure 25.7).

A telemedicine communication network will greatly benefit the patient and the medical personnel in this case. The patient can benefit from saving the time, energy, and resources from making frequent visits to a cardiac clinic. The medical personnel can benefit from not having to go through the hustle of accommodating a patient at every visit (cost efficient).

A typical real-time model will be explained where the data link between the sensor and the existing data networks is an RFID system. It is possible to complete the network by using other communication protocols and systems such as a Wi-Fi link between the sensor and the existing data networks. The sensor can either be implanted inside the patient or be worn by the patient. The transceiver design on the sensor will greatly depend upon the sensor being inside the body or outside. It is well documented that the human body, because of the moisture content, will absorb the electromagnetic energy making it a challenge to transmit modulated electromagnetic waves through it. When the sensor is implanted within the patient's body, the transceiver is typically a separate entity to the sensor that is worn by the patient. Data communication between the sensors is wired or by using volume conduction techniques [32,33] (Figure 25.8).

A real-time model will consist of the patient equipped with a sensor. The sensor is either integrated or interfaced to an RFID tag. The RFID tag communicates with an RFID interrogator. It is also possible for one RFID reader to communicate with several RFID tags each associated with a different patient. Such networks are used in a hospital setting. The data collected by the RFID interrogator is transmitted to a secure terminal that can be accessed only by authorized medical personnel. The data transmitted by the real-time model is private, highly sensitive, and privileged information. This model incurs the same security threats as other RFID and WAN networks [34,35]. There are several documented solutions to make these networks stronger, and these solutions can be adopted as required to the real-time telemedicine communication model.

The model can include an optional communication channel between the patient and the medical personnel that may or may not be considered part of the real-time model. This communication channel can be as simple as a telephone or postal mail to communicate the session summary.

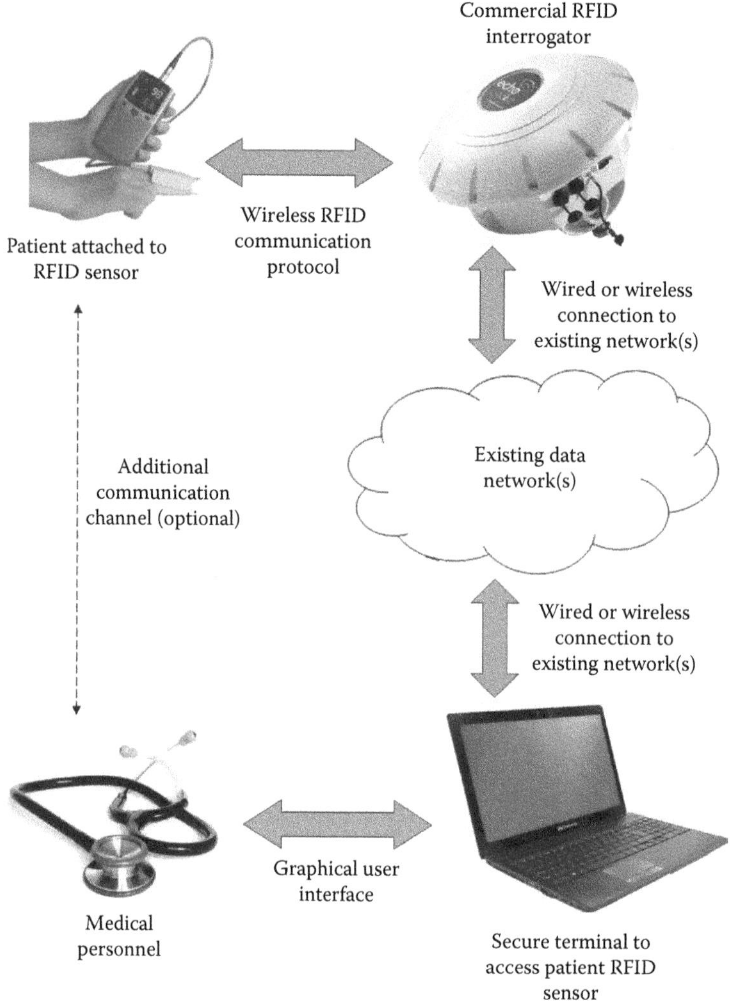

Figure 25.8 Real-time model for telemedicine network.

25.3.2 Continuous Monitoring Model

This model is adopted when it is necessary to monitor the patient's conditions continuously. The sensor can continuously (or at regular intervals) capture required data and store it in memory. The sensor will transmit the data to the RFID interrogator when it is in range. In this model, the memory required depends on the frequency at which the sensor acquires data and the frequency of availability of the interrogator for the sensor to transmit the information.

Consider the following example for a continuous model in telemedicine to monitor the vitals of a patient suffering with or prone to pneumonia. The vitals can include blood oxygen (SpO_2%), body temperature, and the beats-per-minute (bpm) of the patient. These vitals can be measured by using a pulse oximeter. The pulse oximeter can be wearable (called finger pulse oximeter) or implanted inside the patient. The pulse oximeter acquires the necessary vitals and transmits the data for the perusal of the medical personnel. In this case, the medical personnel do not have to

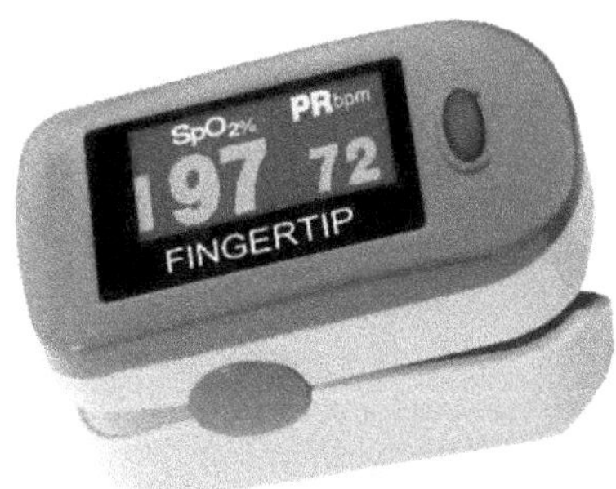

Figure 25.9 Finger pulse oximeter.

alter the settings of the pulse oximeter at regular intervals. A continuous monitoring model can be applied for arrhythmia detection. For patients suspected with arrhythmia, an external sensor is attached for typically 24–48 h. The maximum time for which the sensor can be attached to the patient is limited by the maximum memory available on the sensor. By transmitting the data acquired by the arrhythmia sensor via the telemedicine network, the patient can be monitored for longer time. A continuous monitoring mode typically transfers data in one direction (Figure 25.9).

The continuous monitoring telemedicine communication network is beneficial to the patient as he can be provided quality health care from the comforts of his home. The medical personnel can benefit from having an extra bed at the hospital.

Again, a typical continuous monitoring model will be explained using an RFID communication link. Other options are available including Wi-Fi.

The model consists of one or more patients, each equipped with a sensor integrated with an RFID tag. The tag transfers the sensor data to the RFID interrogator. The interrogator transfers the data to a secured hospital server via existing data networks. The data can be analyzed by the medical personnel as necessary. Since the data communication is typically one way, an additional communication channel becomes necessary for communication between the patient and the medical personnel. This can be a simple telephone line or an audiovisual communication channel (Figure 25.10).

25.3.3 ISO 18000-6c and ISO 18000-7 for Telemedicine Sensors

The internationally accepted RFID standards like the EPC™ Radio-Frequency Identity Protocols Class-1 Generation-2 UHF RFID Protocol and the ISO/IEC 18000-7 standard are developed with provisions to communicate and control the sensor attached to the RFID tag. Without replicating the entire standard, this part of the chapter will discuss the provision in the two aforementioned standards to communicate with optional attached sensors to RFID tags.

A typical Gen-2 tag goes through the singulation process to enter the open or secured state of operation. In this state, there is provision in the standard to communicate with the attached sensors using custom commands. A total of 256 custom commands are available in the Gen-2 standard. Each custom command is 16 bits. The operation code of custom commands starts from 1110000000000000 and end at 1110000011111111. Custom commands can be used only as point-to-point command by the interrogator to communicate with only one tag. The memory of a typical Gen-2 tag is mapped into four banks. The four different banks are Reserved, EPC, TID, and User with bank codes, 00, 01, 10, and 11, respectively. The size of the user memory bank and

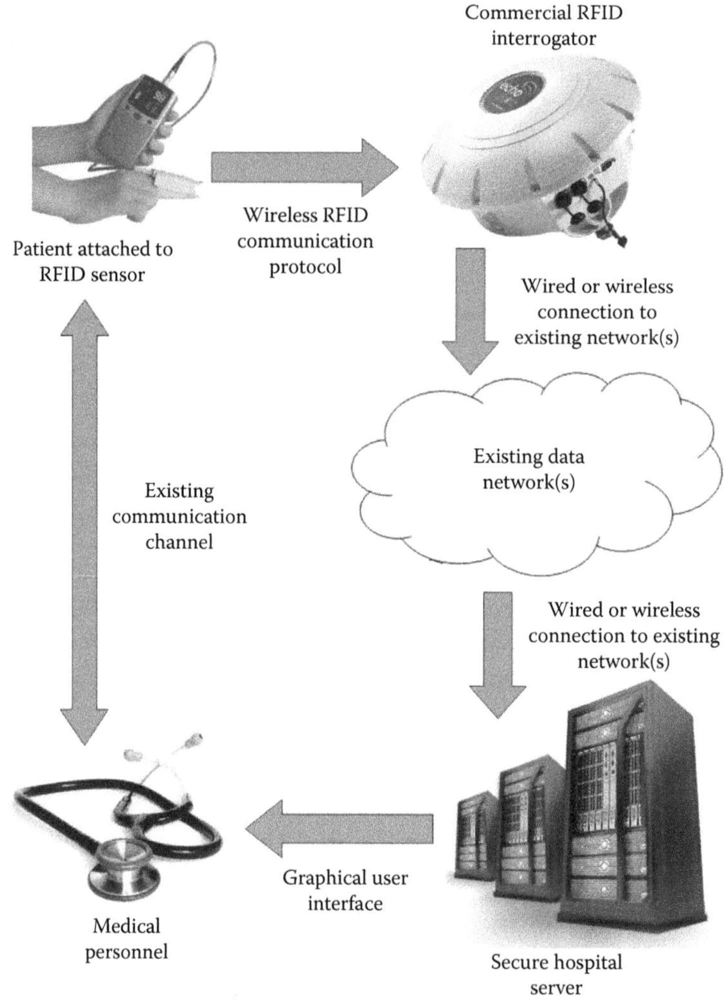

Figure 25.10 Continuous model for telemedicine network.

its organization is not limited by the protocol. The user memory bank can also be shared by the sensor(s).

A typical ISO 18000-7 active tag goes through the wake-up and collection process to enter the awake state. In this state, there is provision in the standard to communicate with the attached sensors using the custom commands. The number of custom commands available is not clearly described in the standard that is still young. Each command (custom or mandatory) is 8 bits. According to the current standard, any operation code not defined by the ISO 18000-7 standard can be used as a custom command. Sensor data have to be stored in the UDB Application Extension Block format. Since the data are stored in the tag's memory bank (organization is not stringent), it can be retrieved by using the available memory read commands or custom commands. The sensor data format is specified by the IEEE 1451 [36] and ISO/IEC 24753 [37] standards. The physical interface between the RFID tag and the sensor is described in IEEE 1451.7 and ISO/IEC 24753 standards.

25.4 Current Research in Telemedicine

The current research in telemedicine at the RFID Center of Excellence, University of Pittsburgh, includes developing a finger pulse oximeter with remote monitoring capability, implantable Doppler viscometer with remote monitoring capability, and a pacemaker programmer with remote control capability. This section will introduce the functionality of each of the aforementioned electronics and their scope in telemedicine.

25.4.1 Finger Pulse Oximeter

The design is an extension of a commercially available finger pulse oximeter. For the prototype development, the MD300C2 oximeter from Choice Med is selected. The commercial oximeter is interfaced to the CC2510 mini development board, an RF System-on-Chip (SoC) solution from Texas instruments.

Finger pulse oximeters use an infrared and a red LED to transmit light through the fingertip of a patient. The light that exits the fingertip of the patient falls on a photo-sensor. The relative attenuation of the two different frequency ranges of light (infrared and red) at the photo-sensor corresponds to the oxygen saturation and heart rate of the patient.

The finger pulse oximeter has an onboard microcontroller unit (MCU) that is used to both drive the infrared and red LEDs and measure the analog voltages produced by the photo-sensor. After calculating the oxygen saturation level and heart rate, these values are sent to an LCD screen via a serial peripheral interface (SPI) bus from the MCU.

The SPI data from the MCU to the LCD are transmitted by the CC2510 to another CC2510 that replicates the data as seen by the finger pulse oximeter. For proof of concept, the initial design transmits the SPI data from the MCU to the LCD via the Texas Instruments (TI) proprietary communication protocol at 2.45 GHz. The typical data rate required to transmit the necessary information is between 100 and 250 kbps. Using compression algorithms and sampling the data in regular time intervals, the required data rate can be significantly decreased.

Figure 25.11 shows the prototype finger pulse oximeter with remote monitoring capability. The data as measured by the oximeter on the right are replicated by the oximeter on the left.

The design was extended by incorporating a Wi-Fi link into the system as shown in Figure 25.12. The MS300C2 commercial pulse oximeter from Choice Med was interfaces to the MatchPort NR Embedded Ethernet Device Server from Lantronix via the CC2510 mini development board, an RF SoC solution from Texas instruments. In this system, the data are transmitted from the oximeter on the right to the oximeter on the left via the Internet.

Figure 25.11 Wearable pulse oximeter for telemedicine with TI's link.

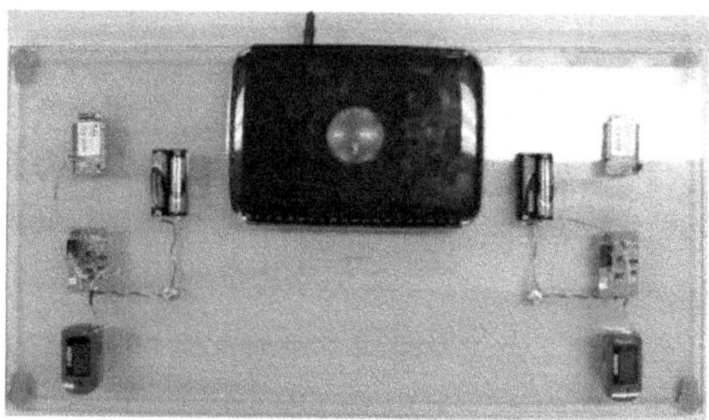

Figure 25.12 Wearable pulse oximeter for telemedicine with Wi-Fi link.

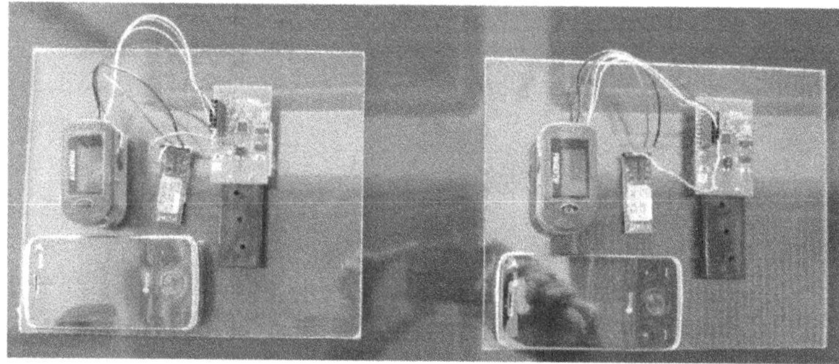

Figure 25.13 Wearable pulse oximeter for telemedicine with Bluetooth link.

Further development of the system includes incorporating a bluetooth link into the system as shown in Figure 25.13. The MS300C2 commercial pulse oximeter from Choice Med was interfaced to the Bluetooth Mate Gold (WRL-09358) bluetooth adaptor from SparkFun Electronics via the CC2510 mini development board, an RF SoC solution from Texas instruments. In this system, the data as seen by the oximeter in the right can be transmitted via the cellular network to the oximeter on the left.

The telemedicine systems shown in Figures 25.11 through 25.13 are examples of continuous model. The information about the patient's vitals is regularly monitored by the medical personnel. The data flow is from the patient to the medical personnel.

25.4.2 Implantable Doppler Flowmeter

The Doppler flowmeter is an implantable continuous Doppler device capable of wirelessly transmitting blood flow information in real time to a remote receiver. As part of the University of Pittsburgh's Telemedicine initiative, this implantable Doppler flowmeter allows for remote monitoring of vitals (Figure 25.14).

25.4.3 Remote-Controlled Pacemaker Programmer

In this system, a commercial pacemaker and implantable cardiac defibrillator from Medtronic, the Medtronic CareLink 2090, is interfaced to the MatchPort NR Embedded Ethernet Device Server from Lantronix via the CC2510 mini development board, an RF SoC solution from Texas

Figure 25.14 Implantable Doppler flowmeter.

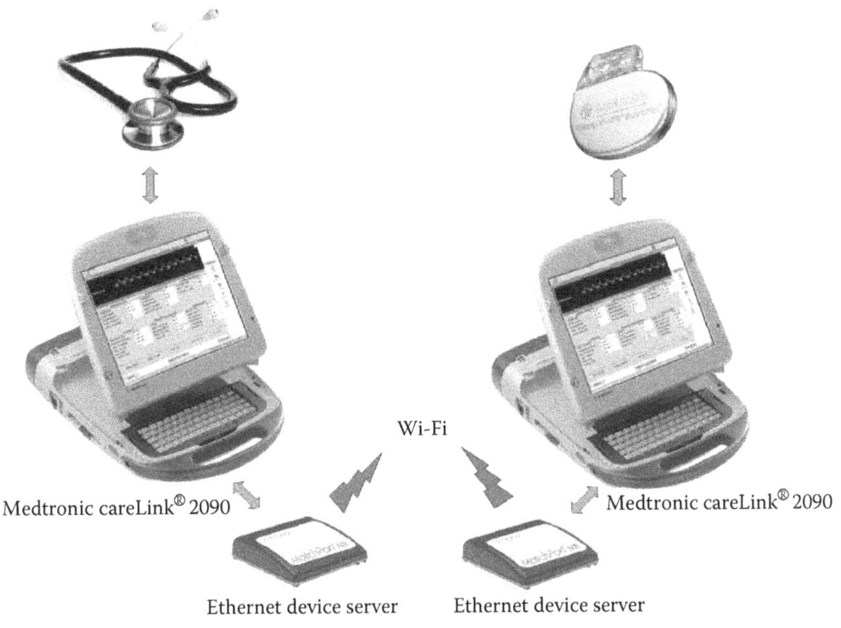

Figure 25.15 Prototype version of Medtronic CarelinK® 2090 with remote control capability.

instruments. This system is shown in Figure 25.15. The programmer on the left is operated by the medical personnel. The operations are transmitted to the programmer on the right via the Internet. The programmer on the right replicates the operation performed by medical personnel on the programmer on the left. Any interactions between the programmer on the right and the pacemaker implanted in the patient are replicated by the programmer on the left.

The future developments of this system will include a video link between the patient and the medical personnel.

This system is an example of a real-time model in telemedicine. The data flow both ways, i.e., from the medical personnel to the patient and from the patient to the medical personnel.

To summarize, tele-medicine has the innate ability to significantly improve the quality of healthcare [38–43]. Continuous monitoring networks of today, is only the beginning of a network that is yet unimaginable in terms of potential, infrastructure and research.

References

1. Project Americal.org, Data Compilation ID 319—Health Care: Hospitals: Cost per Patient, Original Data Source from U. S. Centers for Medicare & Medicaid Services, 2008. Available Online at http://www.project.org/info.php?recordID=319
2. Project Americal.org, Data Compilation ID 79—Health Care: Spending, Original Data Source from U. S. Centers for Medicare & Medicaid Services, 2008. Available Online at http://www.project.org/info.php?recordID=319
3. Statehealthfacts.org, Kaiser Family Foundation, Data Compilation, Original Data Source from AHA Annual Survey, April 2011. Available Online at http://www.statehealthfacts.org/index.jsp
4. Health, United States, 2007 with Chartbook on Trends in the Health of Americans, U.S. Department of Health and Human Services, Centers for Disease Control and Prevention, National Center for Health Statistics, 2007.
5. Health, United States, 2008 with Special Feature on Health of Young Adults, U.S. Department of Health and Human Services, Centers for Disease Control and Prevention, National Center for Health Statistics, 2008.
6. A. Ogirala, Automated testing of wireless link standards, Master's thesis, Department of Electrical and Computer Engineering, University of Pittsburgh, Pittsburgh, PA, 2007.
7. A. Murari, Interoperability as a standard's ingredient for active RFID systems, Master's thesis, Faculty of Computing, Engineering and Mathematical Sciences, University of the West of England, Bristol, U.K., 2008.
8. A. Ogirala, Development of automated test analysis, methodology and procedure for interoperability measure in ISO 18000-7 active RFID, PhD dissertation, Department of Electrical and Computer Engineering, University of Pittsburgh, Pittsburgh, PA, 2009.
9. S. Haykin and M. Moher, *Modern Wireless Communication*, Pearson Education, Inc., Upper Saddle River, NJ, 2005.
10. D. Tse and P. Viswanath, *Fundamentals of Wireless Communication*, Cambridge University Press, Cambridge, U.K., 2005.
11. S.K. Elizabeth and N.D. Alberto, A surprise wireless remote transmission, *Heart Rhythm*, 5(7), 1092, July 2008.
12. A. Mauricio, S.D. Budimir, N.K. Marko, F. Sina, and N. Andrea, P6–82: A novel portable wireless multi-channel ECG and intracardiac recording system, *Heart Rhythm*, 3(5 Suppl), S328–S329, May 2006.
13. M. Loredana and P.R. Renato, Remote monitoring of implantable devices: The European experience, *Heart Rhythm*, 6(7), 1077–1080, July 2009.
14. R. Maurizio, Z. Markus, E. Wunderlich, C. Morais, B. Dominique, A. Hartmann, S. Winter, and S. Massimo, Early detection of atrial arrhythmia through home monitoring technology. The Home-PAT trial, *Heart Rhythm*, 2(5 Suppl), S319–S320, May 2005.
15. ANSI/AAMI PC69, *Active Implantable Medical Devices: Electromagnetic Compatibility—EMC Test Protocols for Implantable Cardiac Pacemakers and Implantable Cardioverter Defibrillators*, 2nd edn., January 2006.

16. D.L. Hayes, P.J. Wang, D.W. Reynolds, M. Estes III, J.L. Griffith, R.A. Steffens, G.L. Carlo, G.K. Findlay, and C.M. Johnson, Interference with cardiac pacemakers by cellular telephones, *New England Journal of Medicine*, 336, 21, 1473–1479, May 1997.
17. S. Futatsumori, Y. Kawamura, T. Hikage, T. Nojima, B. Koike, H. Fujimoto, and T. Toyoshima, In vitro assessment of electromagnetic interference due to low-band RFID reader/writers on active implantable medical devices, *Journal of Arrhythmia*, 25(3), 142–152, 2009.
18. V. Barbaro, P. Bartolini, G. Calcagnini, F. Censi, B. Beard, P. Ruggera, and D. Writters, On the mechanisms of interference between mobile phones and pacemakers: Parasitic demodulation of GSM signal by the sensing amplifier, *Physics in Medicine and Biology*, 48(11), 1661–1671, June 2003.
19. J. Rahbek, Comparison of the RF immunity of operational amplifiers, in *Proceedings of Zurich 12th International EMC Symposium*, 8B3, Zurich, Switzerland, pp. 43–46, 1997.
20. C. Gabriel, Compilation of the dielectric properties of body tissues at RF and microwave frequencies, Brooks Air Force Base, Books AFB, TX, Technical Report Al/oe-tr-1996–0037, 1996.
21. E.H. Qi, M. Meylemans, and M. Hattig, Augmenting wireless LAN technology for Wi-Fi PAN, *Conference on Signals, Systems and Computers, 2009 Conference Record of the Forty-Third Asilomar*, Pacific Grove, California, USA, pp. 321–324, November 1–4, 2009.
22. IEEE Standard for Information Technology—Telecommunication and information exchange between systems—Local and metropolitan area networks—Specific requirements. Part 11: Wireless LAN Medium Access Control (MAC) and Physical Layer (PHY) Specifications, June 2007.
23. M.H. Mickle et. al., Methods and apparatus for reducing power consumption of an active transponder, United States Patent Number US 7375637 B2, May 20, 2008.
24. M.H. Mickle et. al., Methods and apparatus for switching a transponder to an active state, and asset management systems employing same, United States Patent Number US 7876225 B2, January 25, 2011.
25. M.H. Mickle et. al., Methods and apparatus for reducing power consumption of an active transponder, United States Patent Number US 7525436 B2, April 28, 2009.
26. EPC Radio-Frequency Identity Protocols Class-1 Generation-2 UHF RFID Protocols for Communications at 860 MHz–960 MHz, Ver 1.0.9, EPCGlobal 2004.
27. Phase IV Engineering Inc., Document Number 61-100015-00 Rev2.1, Innovative Wireless Sensing, Wireless Sensor Overview 2010. Available online at: http://www.phaseivengr.com/p4main/LinkClick.aspx?fileticket=mKTdbJeUenY%3d&tabid=159&mid=908
28. ISO 18000-2, Information technology—Radio frequency identification for item management—Part 2: Parameters for active air interface communications below 135 kHz, 2009.
29. ISO 18000-3, Information technology—Radio frequency identification for item management—Part 3: Parameters for active air interface communications at 13.56 MHz, 2008.
30. K.R. Magliocca and J.I. Helman, Obstructive sleep apnea diagnosis medical management and dental implications, *Journal of the American Dental Association*, 136(8), 1121–1129, August 1, 2005.
31. N. Yoshida, Y. Koga, A. Saimoto, T. Ishimatsu, Y. Yamada, and K. Kobayashi, Development of a magnetic sensing device for tooth displacement under orthodontic forces, *IEEE Transactions on Biomedical Engineering*, 48(3), 354–360, March 2001.
32. B.L. Wessel, P. Roche, M. Sun, and R.J. Sclabassi, Optimization of an implantable volume conduction antenna, *Engineering in Medicine and Biology Society, 2004. IEMBS '04. 26th Annual International Conference of the IEEE*, San Francisco, USA, Vol. 2, pp. 4111–4114, September 1–5, 2004.
33. M. Sun, Q. Liu, W. Liang, B.L. Wessel, P.A. Roche, M. Mickle, and R.J. Sclabassi, Application of the reciprocity theorem to volume conduction based data communication systems between implantable devices and computers, *Engineering in Medicine and Biology Society, 2003. Proceedings of the 25th Annual International Conference of the IEEE*, Cancun, Mexico, Vol. 4, 3352–3355, September 17–21, 2003.

34. S. Rao, N. Thanthry, and R. Pendse, RFID security threats to consumers: Hype vs. reality, *41st Annual IEEE International Carnahan Conference on Security Technology*, 2007, Ottawa, ON, Canada, pp. 59–63, October 8–11, 2007.
35. M.S. Siddiqui, and C.S. Hong, Security issues in wireless mesh networks, *International Conference on Multimedia and Ubiquitous Engineering, 2007 (MUE '07)*, Seoul, Korea, pp. 717–722, April 26–28, 2007.
36. IEEE Draft Standard for a Smart Transducer Interface for Sensors and Actuators—Transducer to Microprocessor Communication Protocols and Transducer Electronic Data Sheet (TEDS) Formats, IEEE P1451.2/D20, February 2011, pp. 1–28, March 18, 2011.
37. *Information technology—Radio Frequency Identification (RFID) for Item Management—Application Protocol: Encoding and Processing Rules for Sensor and Batteries*, ISO/IEC 24753:2011, 1st edn., April 2012.
38. J.D. Enderle and J.D. Bronzino, *Introduction to Biomedical Engineering*, 3rd edn., Elsevier Inc., Amsterdam, the Netherlands, 2012.
39. J. Moore and G. Zouridakis (eds), *Biomedical Technology and Devices Handbook*, CRC Press, Boca Raton, FL, 2004.
40. R.C. Fries (ed), *Handbook of Medical Device Design*, Marcel Dekker, Inc., New York, 2001.
41. M. Kutz (ed), *Biomedical Engineering and Design Handbook, Volume 1: Fundamentals*, 2nd edn., McGraw-Hill, New York, 2009.
42. M. Kutz (ed), *Biomedical Engineering and Design Handbook, Volume 2: Applications*, 2nd edn., McGraw-Hill, New York, 2009.
43. S.B. Miles, S.E. Sharma, and J.R. Williams (eds), *RFID Technology and Applications*, Cambridge University Press, Cambridge, U.K., 2008.

Chapter 26

A New Generation of Intrusion Detection Networks

Jerry Krill and Michael O'Driscoll

Contents

26.1 Introduction

Various means exist today to monitor, ensure the safety of, and control access to public and private areas, both large and small. Such means include video monitoring, infrared (IR) moving object detectors, and "electric eye" tripwire approaches with IR signals across key pathways. The concepts presented are alternative means to monitor a security zone in ways that are difficult to counter and yet are highly automated and relatively inexpensive. The concepts, in combination, also appear scalable and, thus, can have commercial applications for small or very large security businesses in addition to some special applications such as oil pipelines.

In the last 4 years, we have published several papers that investigate and evolve design concepts of highly automatic, autonomous intrusion detection and tracking sensor networks. Our first approach was to develop a very inexpensive option of tiny, expendable sensors (called "pebbles") that emit extremely low-power microwave tones to cue their neighbors, exhibit swarming behavior, and are scalable, perhaps to hundreds of square miles.

This initial work investigated the viability of using simple tones for signaling and combining tones from multiple nodes to track detected objects. We will briefly discuss the architecture and implementation of a low-cost prototype using MICAz motes with acoustic sensors as surrogates. The prototype was used to demonstrate and validate the swarming behavior of a field of "pebbles" in a physical environment and to explore the effect of various sensing parameters and network configurations.

We have also developed the design concept and requirements for a higher cost, more sophisticated, second concept featuring multifunction array lidars (MFALs) with novel electronic scanning laser array apertures allowing near-simultaneous, interleaved functions including detection, tracking, and cueing. The network of array lidars provides collaborative automatic detection, tracking, acquisition cueing, and intruder-type identification, as well as free space optics (FSO) communication. A lidar consists of four optical phased arrays, each with about 1 million radiating elements in a 1 cm^2 aperture that enable electronic beam steering analogous to microwave array antennas. The four arrays are on four sides at the top of a square cross-section mast and provide full 360° azimuth coverage. A small number of such MFAL nodes can perform target detection over a significant area, but this approach is not considered as easily scalable as the first concept of swarming pebbles.

Finally, we merge the two approaches to optimize the high performance of the lidar approach and the cost effectiveness of the swarming pebble network. The concept is to use a very low-cost field of pebbles for wide area coverage that cues the more precise inner zone MFAL network. The pebbles will provide coverage in the wide open, exterior spaces, inherently scalable to cover up to hundreds of square km, while the MFALs will be focused on the inner layer and choke points of a few hundred square meters or a few square km in area. We develop requirements and provide the calculations on the feasibility of this combined approach, including automated cueing and hand-off-between pebbles and MFAL zones. We also briefly discuss the idea of mobile patrolling MFAL mini-vehicles that follow or converge on the intruders.

26.2 Tone-Based Swarming Detection Network

26.2.1 Concept

Emergence and swarm intelligence have been widely discussed topics over the last several years. An attractive concept is to engineer a collective intelligence demonstrated vividly in natural swarming, such as the construction of termite mounds where swarms are the prime movers and no organizing principles are evident. For such swarms to be effective, however, certain characteristics appear necessary: diversity of perspective, independence, decentralization, and aggregation. In short, there must be exchange of information with multiple viewpoints, and the solution must be an aggregation of these views.

The swarming sensor network that we have conceived involves several features:

- Swarming—The ability of the network to focus in an ad hoc manner based on a collective response to inputs to each individual component.
- Distributed intelligence—A collection of elements appearing to respond intelligently, although no single element possesses the information or orchestrates a response.
- Inferential signaling without protocol—The nontraditional conveyance of information that may consist of inferences from simple signals or behaviors, i.e., "body language" from near neighbors. In our case, we will use a microwave tone as the signal.
- Chaotic behavior—The system seemingly behaving in a manner that borders on unpredictable or unstable.
- Emergence—Effects and behaviors appearing at higher levels that are not explicitly evident in lower-level component designs.
- Multiple systems—including a sensor system, signaling system, and remote sensor field monitor.

This system concept was conceived to approach the simplest, most economic, and lowest risk means of establishing a sensor network.

Figure 26.1 illustrates the essential elements of the network. A large number of nodes, or "pebbles," is randomly distributed, for example, air dropped or individually placed, with the condition that there is sufficient density so each pebble is within overlapping sensor and communication range of its near neighbors. The pebble field surrounds an installation, pipeline, or building to be protected, for example. Although disguised as pebbles, even if they are discovered, they are far too numerous to gather.

Each sensor node contains (a) a small, specific sensor, for example, acoustic, radio frequency (RF), chemical, optical, or biological; (b) a power supply; (c) a microwave communications transceiver (two-way communications); (d) a controller chip; (e) a suitable container to disguise and/or otherwise protect the sensor node components; and (f) an optional solar array. One or more remote directive transceivers monitor the spectral power density levels and transmission locations of this "pebble field" and, as a design option, can transmit over a "control" channel to modify some number of nodes' programming, for example, detection thresholds.

The pebble nodes' sensors are passive, i.e., nonemitting for minimum power consumption. All the pebbles could contain the same type of sensor, but, more generally, they could contain a mix

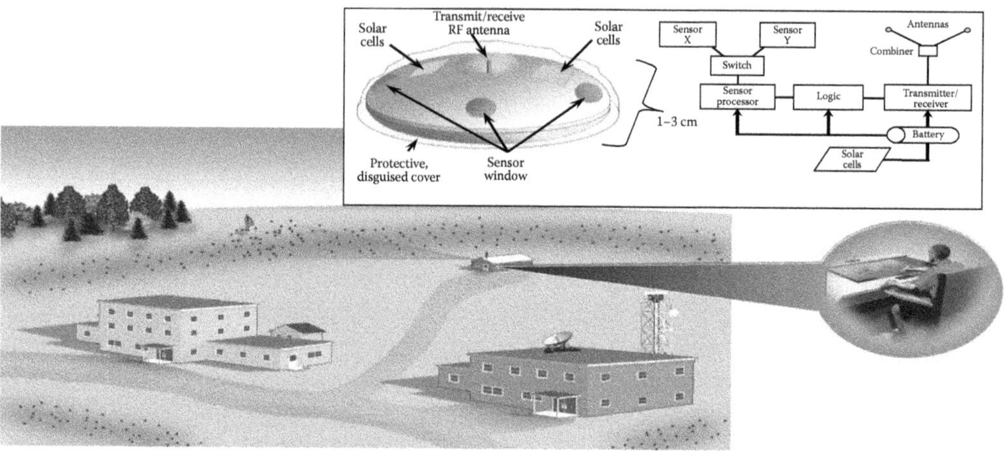

Figure 26.1 Elements of the network.

of different sensor types, perhaps even two or more sensor types per pebble node. The sensors can be preset to search for specific characteristics, such as to detect certain vibration frequencies of the human voice or visual motion indicative of humans, or they may be set to detect any sound above average background or any moving object, etc. The detection threshold can also be set. The set threshold would have a reasonably insensitive "cold detection" value to ensure a low false alarm rate. However, the sensor could be cued to become more sensitive if it receives a cueing signal from its node neighbors, indicating that one or more neighbors have detected an intruder. The requisite settings will be discussed later.

Assume one pebble senses an event according to its preset cold detection threshold. It will emit a weak communication signal in all directions but with intentionally limited range to save energy and to only reach its neighbors, which would likely also have a reasonable probability of detection (P_D) if the event is real and not a false alarm.

As each sensor makes a subsequent associated detection, it continues to emit a simple signal, such as a tone, that can be received by its neighbors. If a correlation of sensor events to an intruder incident begins to transpire, the activity of the sensors will naturally increase the total power density in that area at the cue tone frequency via their communications transmissions. They will also be inherently collaborating and, by their near-neighbor interactions, producing a swarming behavior. As long as a sensor node continues to detect, it will send a signal to support continuation of the swarming activity. If detections begin to wane, the swarming signals will diminish.

From a distance, a directional receiver tuned to the frequency band(s) or tone(s) of the pebble nodes scans the sensor field to monitor activity. The directional antenna may, for example, determine pebble signal activity above normal at a particular direction. The strength of the received signal would be proportional to the strength of pebble detection activity, implying a firm indication of detecting an intruder. The angle of reception, for example, from a direction-finding antenna, indicates the approximate location of the intruder. Multiple remote directional receivers could be operated for triangulation to further localize the swarm activity. A similar localization approach would be to have pebbles at different locations radiating at different tonal frequencies. The remote receiver would be able to link the sensor net activity to a command and control (C2) node for action, if indicated. For example, a swarm activity indicating a high confidence of detection could result in a decision to intercept the detected intruder. The remote receiver could be manned and

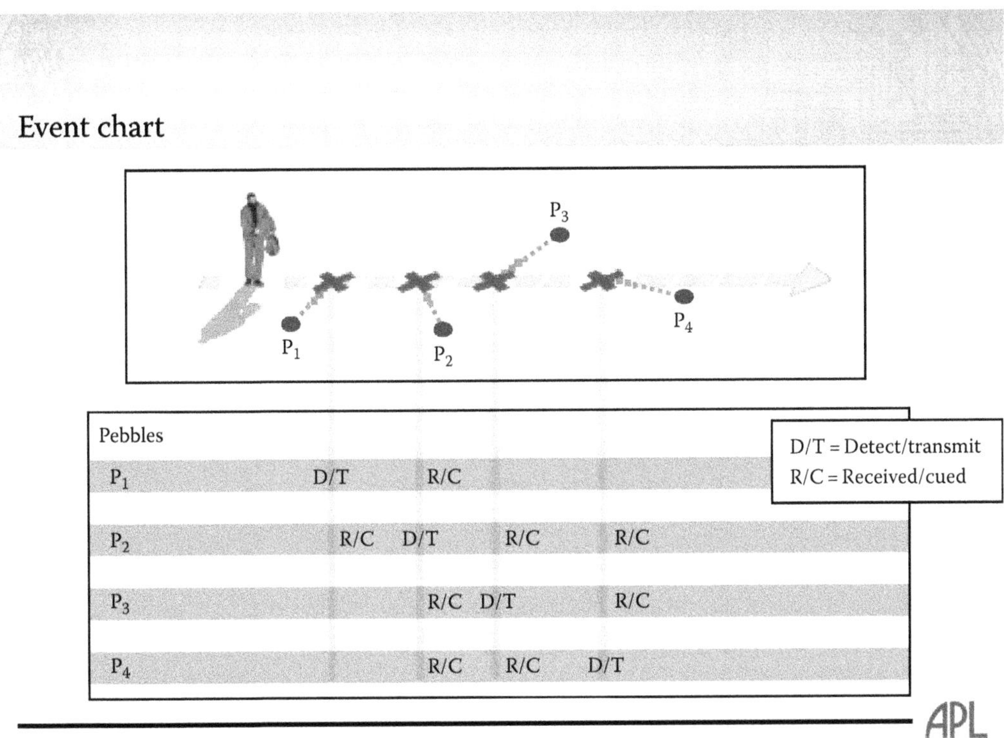

Figure 26.2 Event chart.

the C2 decision made at that location, or it could be unmanned and operated remotely via a communications link to a security office. Additional options with this type of operation include cueing of video cameras and tripping an audible or silent alarm if activity of the nodes exceeds a threshold, enabling a response by security forces.

Figure 26.2 is an event chart to further illustrate the intruder detection by a field of signaling sensor pebbles. Shown are several pebbles, P1 through P4, each with a sensor, such as a microphone or IR detector, as well as a microwave signal cue tone communications transceiver. As an intruder passes within the sensor range of pebble P1, illustrated by the "X" on the left, P1 detects above a preset threshold (D) and transmits a cue tone (T). The tone from P1 is received (R) by pebble P2, which, in turn, cues its detection threshold to a lower, more sensitive setting (C). Sometime later, as the intruder passes into the detection range of pebble P2, the pebble detects (D) with even less signal than for P1 (second "X" from the left). It then sends out a tone that is received by pebbles P1, P3, and P4. Pebble P3, which is now set at the cued detection threshold, detects (D) and transmits a tone (T) that is received by Pebbles P2 and P4, and so on. A remote receiver may be set to detect the tones of, say, five transmitting pebbles. Further, scanning with high antenna gain can allow tracking of the intruder's progress as new pebbles detect and transmit tones, in turn.

26.2.2 Analysis

Our analysis of the swarming network concept is based on an existing design. We selected the "Mica mote" for consideration because the design exists and could be modified to accommodate this concept [1,2]. Although, perhaps, larger, more costly, and higher power than may ultimately

Table 26.1 Characteristics for Pebble Nodes and Remote Receiver

	Pebble Node	*Remote Receiver*
Noise figure	2 dB	2 dB
Bandwidth (tone)	50 kHz	50 kHz
Carrier frequency range	2.4 GHz	2.4 GHz
Received signal to noise required (with noncoherent integration)	12 dB	12 dB
Receive antenna loss	3 dB	3 dB
Antenna gain	−8 dBi	
Transmit power	−5 dBm (−30 dBm excursion)	
Receive aperture		0.025 m^2 (0.25 m^2 excursion)

be desired for pebbles with this concept, it represents a design that may be adaptable to inferential swarming behavior. We begin by discussing communications connectivity calculations for the mote (i.e., pebble) and then describe blockage detection, as a variant of this concept, in Section 26.2.3.

The mote design description of [1,2] indicates up to a 30 m communication range at a moderately high data rate (hundreds of kilobits per second) using the Bluetooth protocol at 2.4 GHz. Based on these characteristics and on power consumption information and, further, recognizing that we are merely communicating narrowband tones, we postulate the design characteristics for the mote-based pebbles and a remote receiver in Table 26.1. We consider nearly omnidirectional sensor and communication antennas rather than more complex sector sensors and antennas. We also postulate expected propagation loss values between pebbles and the remote receiver, assuming potential foliage effects, of up to 15 dB when well within the radio horizon and up to 35 dB at the radio horizon [3]. The radio horizon range depends on receiver heights, for example, 7.1 km between a surface pebble and a 3 m high remote receiver antenna for standard propagation conditions.

Table 26.1 includes two levels of pebble node transmit power, −5 dBm of the present mote design and an excursion to a much lower power of −30 dBm representing a potential advanced, very low-power design. Also, two remote receiver antenna apertures are considered: a significant gain, directional antenna of a 0.5-m-by-0.5-m area and a smaller, lower gain antenna with a 16 cm side dimension.

Figure 26.3 provides example results from our parametric calculations of the minimum number of "pebbles" detectable by a remote receiver for the combinations of pebble transmit powers and receiver apertures versus range. We assume the pebbles are placed randomly but well within each other's reception range to induce swarming response. Our calculations confirm maximum communications range between pebbles of about 30 m for −5 dB power. If we assume that each transmitting pebble has a transmit power of P_{peb} and a transmit gain of G_{peb}, and there are N transmitting pebbles, then for noncoherent combining the collection of transmitting pebbles has an average effective radiated power of $NP_{peb}G_{peb}$. The signal-to-noise ratio (SNR) at a distant receiver is then

$$\frac{\mathrm{S}}{\mathrm{N}} = \frac{NP_{peb}G_{peb}A_{rcv}\sqrt{tB}}{4\pi R^2(kT_{sys}B)L_P L_s}$$

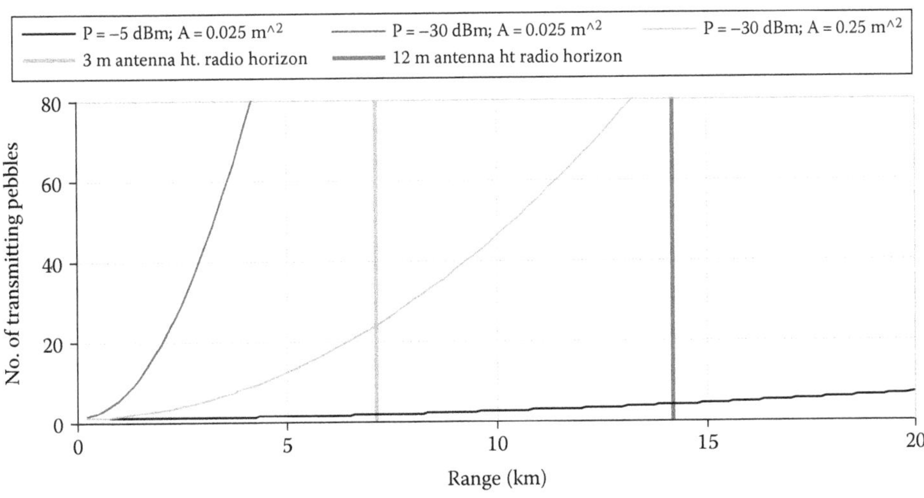

Figure 26.3 Number of transmitting pebbles versus range for various pebble transmitting powers and receiver apertures.

where

A_{rcv} is the receive antenna aperture

R is the range to the receive antenna

$kT_{sys}B$ is the noise power where k is Boltzmann's constant, T_{sys} is the system noise temperature and B is the receiver bandwidth

$\sqrt{tB}$ is the S/N improvement factor due to noncoherent integration over a time t

$L_P L_s$ is the propagation loss and other system losses, respectively

In Figure 26.3, we assume the parameters of Table 26.1 and a receiver noncoherent integration time of 1 s. For ranges well within the radio horizon, we assume a propagation loss of 15 dB to address fading and foliage effects. The propagation loss will increase significantly at the radio horizon and beyond. From Figure 26.3, for a pebble transmit power of −30 dBm and a distance of 5 km, a remote receiver with a 3 m antenna height would detect 12 pebbles or more with a 0.25 m^2 aperture. This implies that out of perhaps thousands of pebbles, at least 12 would need to radiate, indicating intruder activity, before the remote receiver would detect any response. Note that our assumption of noncoherent power combining requires more than a few pebbles to be radiating. Experimental verification of the propagation loss and incoherent tone integration will be discussed later. Whereas the minimum number of pebbles might ensure a minimum of false alarms, it may also be insufficient to ensure adequate intruder detection sensitivity, for example, if the pebbles are sufficiently separated and sparse so a human intruder would only trigger a smaller number of pebbles at any time. Thus, swarming network configuration analysis was performed to determine the requisite pebble density and remote receiver dynamic detection range for the intruder detection sensor sensitivity, as will be described in a later section. From our parametric calculations, we conclude that, for a pebble transmit power of −5 dBm, we can detect a few pebbles with a remote receiver with a reasonable antenna aperture under significant propagation loss from 1 to 20 km in range. For a much lower pebble transmit power (−30 dBm) (e.g., to reduce cost and detectability by an adversary intruder), a remote receiver with a significant antenna aperture could detect the beginning with a few dozen pebbles out to several kilometers.

Observe that this system is potentially scalable from a few such pebbles to literally millions over many square kilometers with no design or settings changes to the network other than perhaps to connect a larger network of multiple remote monitor stations. Also observe that the concept features three interactive systems. First, are the sensors themselves: one or more hosted in the pebble chassis. These individual sensors could be quite sophisticated in terms of miniaturization and filtering mechanisms, for example, to ensure requisite detection and false alarm probabilities, P_D and P_{FA} etc., respectively, and to ensure distinguishing of the intruder of interest from false identifications such as wild animals or tumbling debris. The sensors also must provide adequate detection range to ensure overlapping coverage with neighboring pebbles' sensors. The second system in the concept is the signaling network. It must have adequate range and detection/false alarm performance and maintain continual connectivity among near neighbors. The third system is the remote monitor for sensing swarming behavior, with requisite swarming and false swarming statistics that can also transmit commands to reset or change the state of the pebble field.

As we have mentioned, sensor detection of an intruder could be accomplished by a number of means, including passive audio detection or passive IR detection [4]. We begin by stipulating that a detection system is designed to provide 12 dB or greater SNR out to a range of 10 m from the intruder location. Signal detection theory can then be used to determine probabilities for detection and false alarms. For 12 dB or greater sensor SNR, the "cold" probability of detection is greater than 0.93 for a P_{FA} of 10^{-4}. If the sensor has been cued, we assume it lowers the detection threshold to increase the P_D. For example, for a 12 dB SNR, the detection threshold can be lowered to achieve a P_D of 0.995 for a P_{FA} of 10^{-2}. If the detection is increased to 20 m, the same sensor system provides a 6 dB SNR, assuming that free-space spreading propagation conditions are in place. In this case, the cold P_D is 0.1 for a P_{FA} of approximately 10^{-4}, and the cued P_D is 0.5 for a P_{FA} of 10^{-2}. These passive sensor detection results are summarized in Table 26.2.

Figure 26.4 illustrates the range relationships among adjacent pebbles using standard P_D versus P_{FA} curves, for example [5].

In the network system that signals to cue neighboring pebbles upon declaring a cold detection by a pebble, the cue is accomplished by emitting a RF tone that is detected by a neighboring pebble. To help prevent a "swarming instability" in which all pebbles are eventually cued and yield an excessive false alarm rate, cue tones last only a few seconds (we have initially selected 5 s) before turning off and will not be reinitiated unless another detection is made. For a RF tone-based cueing system designed to provide 12 dB or greater SNR out to a range of 10 m, signal detection theory can then be used to determine probabilities for detection and false alarms (according to classical P_D versus P_{FA} curves). For 12 dB or greater SNR, the P_D is greater than 0.93 for a P_{FA} of 10^{-4}. If we assume that the tone signal variation with range corresponds to free-space spreading, the design would then provide a 6 dB SNR at a range of 20 m, which corresponds to a P_D of $\sim$0.1 for a P_{FA} of 10^{-4}. For free-space propagation conditions, nearest neighbors will have a

Table 26.2 Zone Range Requirements for Pebbles in a Passive Detection Sensor System

	Passive Detection at 10 m (12 dB SNR)	*Passive Detection at 20 m (6 dB SNR)*
Cold detection	$P_D > 0.9$, $P_{FA} = 10^{-4}$	$P_D < 0.1$; $P_{FA} = 10^{-4}$
Cued detection	$P_D > 0.99$; $P_{FA} = 10^{-2}$	$P_D < 0.5$; $P_{FA} = 10^{-2}$

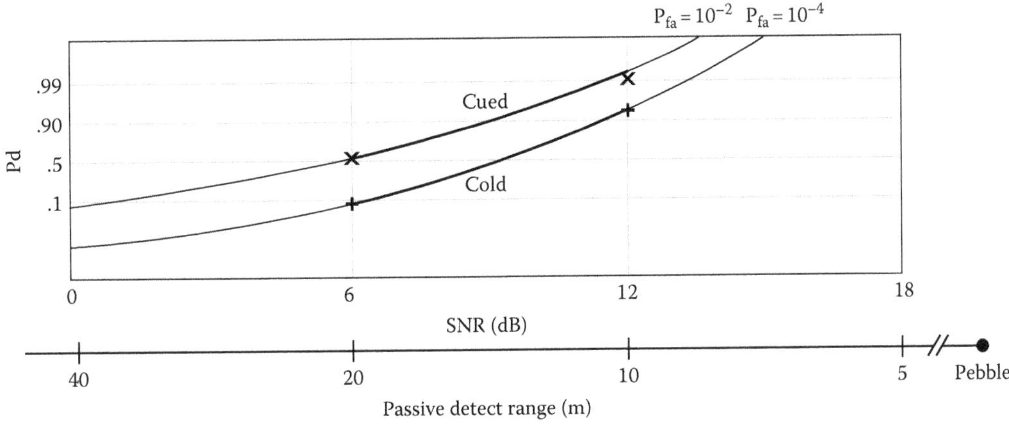

Figure 26.4 Pebble sensor range performance.

Table 26.3 Zone Range Requirements for Near-Neighbor Signal Cueing System

	10 m (12 dB SNR)	20 m (6 dB SNR)
Communication tone detection	$P_D > 0.9$, $P_{FA} = 10^{-4}$	$P_D < 0.1$; $P_{FA} = 10^{-4}$

high P_D, whereas the next closest neighbors will have a significant reduction in detection probability. More severe propagation conditions would further reduce the detection probability beyond the nearest neighbors. This analysis specifies the influence of neighbors. Table 26.3 summarizes these requirements. Figure 26.5 illustrates the near-neighbor signaling range and associated received signal in a free-space environment.

Experiments were performed, which provide the basis for the propagation conditions assumed earlier, demonstrated free-space spreading for elevated modules and more severe attenuation with range for modules located on the ground. The results are illustrated in Figure 26.6.

A straightforward analysis showed, and associated experimental results confirmed, that a remote receiver receiving the incoherent sum of five or more pebbles represents in its directive beam an acceptable likelihood of intruder presence, as will be discussed.

The zone requirements for the remote monitor could be designed independently from specifying its range from the pebble field. They are

- P_D for at least five pebbles >0.9 with associated P_{FA} less then 10^{-4} within the spatial dimensions of the directive antenna beam
- Localization of transmitting pebbles (assuming detecting "swarms" of at least five near-neighbors) to within 100 m circular error probable of centroid

For given ranges between a monitor and the pebble field boundaries, these requirements accommodate derivation of required monitor receiver sensitivity, antenna gain, and antenna beam splitting accuracy (e.g., via con-scan or monopulse techniques [6]).

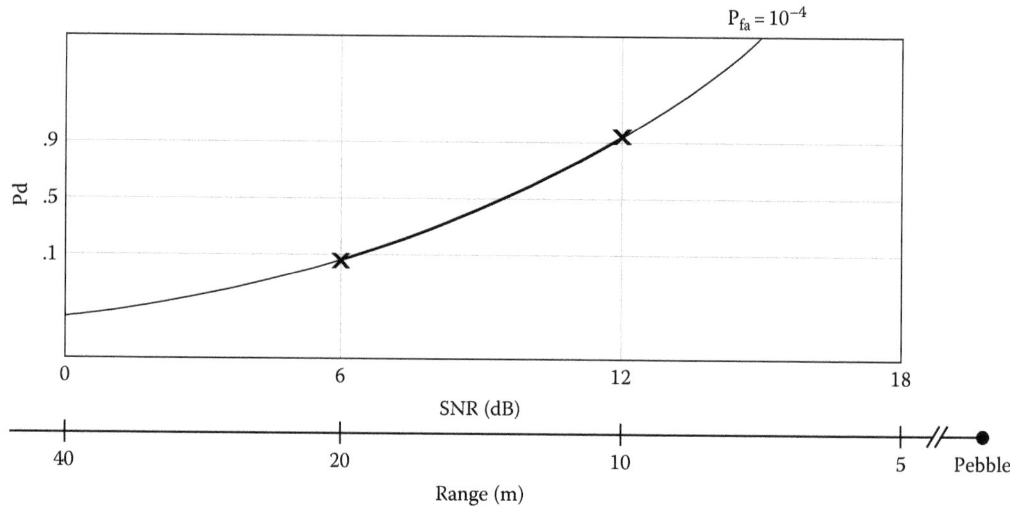

Figure 26.5 Pebble signaling range performance.

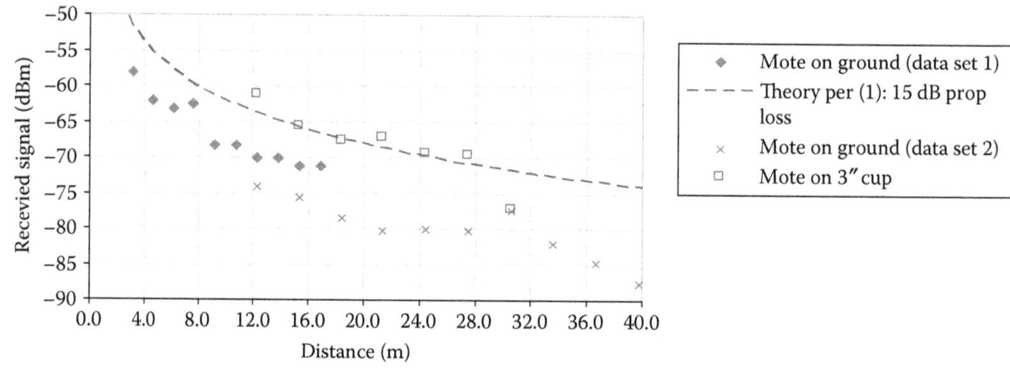

Figure 26.6 Mote-to-receiver connectivity.

26.2.3 Intruder Blockage Detection Analysis

A novel sensor was explored analytically, which is a rudimentary, yet difficult to counter, intruder detection mechanism: "intruder blockage detection." For this concept, each pebble is set not only to receive a communication tone, but also a blockage-sensing tone at a different frequency. We intermix into the pebbles several "illuminator" pebbles that continually transmit a low-power, blockage-sensing tone. All other pebbles are set to receive the communication tone as well as the blockage-sensing tone. The concept is that pebbles will receive rather constant blockage-sensing tone signal power levels unless an intruder passes through the path between the transmitting pebble and receiving pebble.

During the passage of an intruder, which blocks the blockage-sensing transmission to a receiving pebble, the receiving pebble will detect a significant drop in received signal power for a short period. If that occurs, a potential detection is declared, and the pebble emits the 2.4 GHz communication tone. Upon receiving the communication tone, the neighboring pebbles increase the sensitivity of their blockage signal triggering threshold so they would detect the loss of signal more readily, i.e., they are cued to "listen" more carefully.

Note that countering blockage detection in the microwave band between several illuminator pebbles and many randomly placed receiving pebbles would likely be difficult. Further mitigation could be in the form of pebble receiver detection of attempts to "jam" the transmission frequency or provision for randomized tone hopping or modulation that would be difficult for an intruder to mimic.

We performed some preliminary diffraction calculations to determine whether there is adequate blockage signal loss from a human intruder for detection at representative pebble distances. We modeled a human intruder as an infinitely long, vertical cylinder that is 0.3 m in diameter. The cylinder's complex permittivity is that of saltwater to approximate the permittivity of the human body. For such a simple shape, vertical signal polarization causes a deeper shadow than horizontal polarization by 2–3 dB. However, because this model ignores irregularities in human shape and composition as well as irregularities in the surface and due to nearby obstacles, which would tend to weaken the polarization effect on the blockage, we consider as a worst-case horizontal polarization. Figure 26.7 plots blockage loss versus distance from the obstacle for 3, 10, and 20 GHz blockage-sensing tones.

The signal drop at 2.5 m distance is about 3, 6, and 8 dB for 3, 10, and 20 GHz, respectively. Thus, it appears that using 20 GHz as the blockage-sensing tone provides more effective blockage detection, i.e., a sufficient change in signal against a typical environment for reliable detection by a receiving pebble without excessive false alarms.

Figure 26.8 illustrates the idealized blockage "shadow" in two dimensions for 20 GHz. For this calculation, we used a parabolic equation computation method described in [3]. The blockage signal loss appears to be significant at 5–6 dB, even 10 m behind the blocking cylinder, and some loss at 3–4 dB even occurs at an assumed maximum inter-pebble communications distance of 30 m. We therefore conclude that a blockage detection capability may be effective against a human intruder near 20 GHz. One can set a signal power threshold in the remote receiver requiring some minimum number of pebbles to transmit a communications tone to conclude that there may be an intruder, thus further reducing the prospects for a false alarm. The stability of the pebble node network must

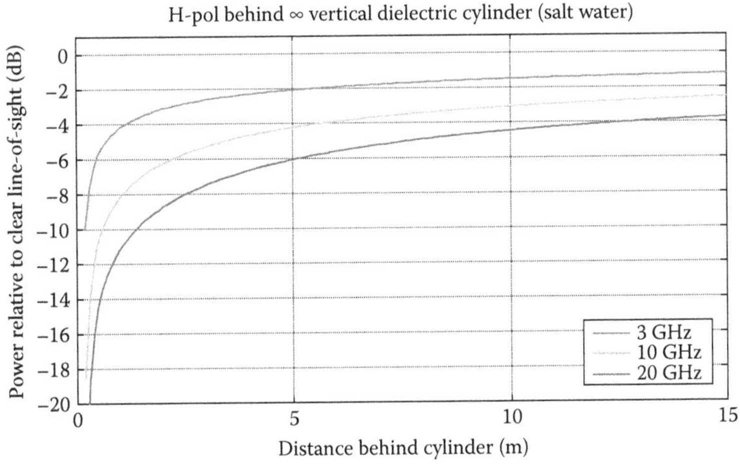

Figure 26.7 Power reduction versus distance behind a 0.3 m diameter blocking cylinder for horizontal polarization.

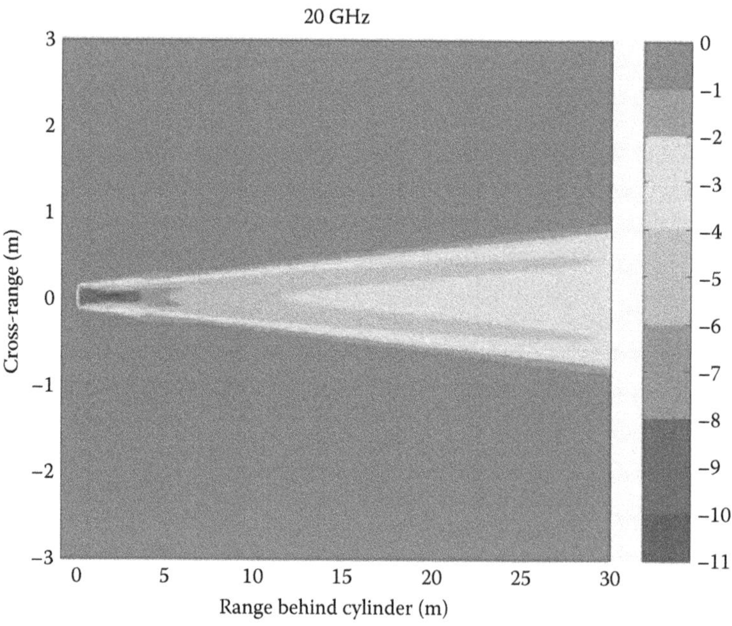

Figure 26.8 Two-dimensional plot of power reduction near a 0.3 m diameter blocking cylinder for horizontal polarization.

be maintained so cueing for greater pebble detection sensitivity does not cause the network to go unstable, in which sensitized pebbles continue to detect false alarms after the triggering blockage event has ceased. Greater network stability may be achieved with a timeout feature in which the transmissions of the detecting pebbles cease after, for example, 5 s and the detection threshold is reset to the "cold detection" value. The timeout approach would also conserve node power.

A preliminary detection and false alarm analysis was performed for the intruder blockage detection approach. A noncentral chi-square distribution was used to model the received signal plus noise power. For received signals 30 dB above thermal noise power, a single pebble cold detection threshold set to detect a drop in signal level of 4–6 dB (below the 30 dB level) will yield a very high P_D and very low probability of false alarm. Additional pebble detections correlated with the first detection would not appreciably improve the detection performance, but would indicate intruder movement through the pebble field. Figure 26.9 plots probability versus SNR. The ascending line indicates the probability that the 20 dB signal plus noise exceeds the SNR. The descending line indicates the probability that the signal reduced by 6 dB is less than the SNR. For received signals 20 dB above the noise, a 6 dB cold detection threshold would be set to provide a P_D of about 90% and a false alarm rate of 10^{-4}. If the pebble then cues neighboring pebbles to reduce their threshold to detect a 4 dB drop in signal level, the reduced threshold would be more likely to trigger cued detections, and these detections will serve to improve detection performance (and indicate intruder movement). If the cold threshold were retained by the neighboring pebbles, rather than the cued threshold, further cold detections would not occur as readily and, as a result, would not provide intruder movement indication.

The cueing mechanism for reducing the detection threshold for 20 dB signal to noise may not provide better detection performance than other strategies, such as reporting any events beyond a very low threshold (like 2 dB) and taking M out of N as a basis for declaring a detection. However,

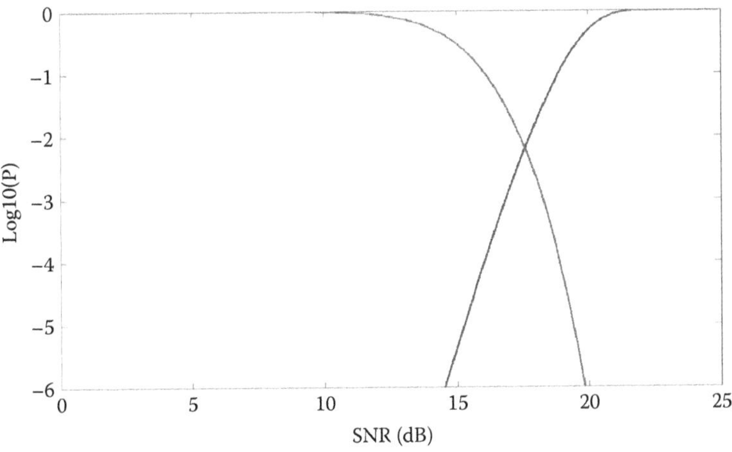

Figure 26.9 Probability versus SNR for 20 dB signal case.

given that the pebbles need to minimize transmission time and power for energy conservation, the cueing approach may prove optimal.

Note that multiple illumination frequencies may be needed to reduce interference associated with a pebble receiving the combined signal of multiple blockage-sensing illuminators. For example, if a pebble is receiving signals at comparable strengths and at the same frequency from two or more illuminators, intruder blockage from one of the illuminators could be masked, or jammed, by the signals of the unblocked illuminators. If neighboring illuminators operate on different frequencies and each of the detection pebbles is tuned to only one of the illumination frequencies, or, alternately, could be tuned to discriminate different illuminations, this interference problem could be alleviated.

For the desired detection performance, we mentioned previously that an illuminator pebble must provide 20 dB signal to noise at 20 GHz to a receiving pebble at approximately 10 m range. We considered the feasibility of a continuously transmitting illuminator from a power consumption viewpoint. We estimate that a −15 dBm transmit power is sufficient (assuming omnidirectional 20 GHz antennas, a 100 kHz receive bandwidth, 3 dB receive noise figure, 3 dB losses on transmit and receive, and a 15 dB propagation loss). For an overall efficiency of less than 5%, we estimate that the total power consumption could be on the order of 1 mW.

The mote design description in [1] indicates a 3 W-h battery, which would indicate up to 3000 h of continuous operation of an illuminator. A 1 cm^2 solar panel that can generate 10 mW of power in full sunlight [1] would extend operation. A pulsed system could also be considered to minimize power consumption. Such a system would increase complexity, requiring clock synchronization between the illuminating and receiving pebbles. Finally, because the pebbles are considered expendable, periodic replacement of blockage-sensing illuminators with depleted batteries would likely be economical.

26.2.3.1 Network Sensitivity Analysis Results

The basic assumption of the swarming pebbles concept is that inter-pebble and pebble field-to-remote receiver propagation losses are approximately constant and predictable. Then pebble transmit power can be set to only allow near-neighbor pebbles to receive cue tones. It may turn out that under certain conditions, such as multipath, obstacle blockages, and propagation ducting, actual propagation loss is highly variable over seconds to minutes by tens of decibels. The following paragraphs describe a bounding analysis that was performed of what would occur in a swarming pebble field for extreme propagation conditions. Then, additional network design features are proposed for consideration in the prototypes to accommodate propagation variations while retaining network stability and performance.

Consider a pebble field with 20 m separation over about a 4 km^2 area. This could be represented by 10,000 pebbles covering approximately a 2 × 2 km^2 pebble field or a long rectangle, say 0.25 × 16 km, along a pipeline or on the periphery of a utility complex such as a power plant. Assume also that the pebbles reset to their cold, uncued detection thresholds every 10 s so that we do not need to consider cumulative probabilities (and power is conserved). Propagation loss variations can greatly alter network performance. For example, if a total swing of ±18 dB of propagation would occur, a 20 m nominal communication range between pebbles would increase to 106 m or reduce to 2.5 m. Case 1 is an idealization of the former case, and case 2 considers an extreme of the latter case. Consider the following limiting cases:

- Case 1: Perfect propagation—In this case, for example, strong ducting, if one pebble makes a cold detection and emits a cueing tone, all other pebbles receive the cue and set their more sensitive cued detection thresholds. From aforementioned text, a cold detection threshold is assumed with $P_D \cong 0.9$ and $P_{FA} \cong 10^{-4}$. A cold detection cues all other 9999 pebbles to the cued threshold with $P_D \cong 0.99$ and $P_{FA} \cong 10^{-2}$.
- Case 2: No Propagation—In this case, for example, extreme blockages, if a pebble makes a cold detection and emits a cueing tone, no other pebbles receive the cue. Therefore, all 10,000 pebbles remain at the cold detection threshold of $P_D \sim 0.9$ and $P_{FA} \cong 10^{-4}$.
- Case 3: Intermediate Threshold—In this case, assume all pebbles retain a single cold threshold of $P_D \cong 0.95$ and $P_{FA} \cong 10^{-3}$, with no cueing. This represents a potential alternative state if either case 1 or 2 is determined.

For each case in Table 26.4, the longer term average number of false alarm detections per 10,000 pebbles is shown in the second column. Assuming a searching directional antenna for the remote receiver that covers 0.1 of the pebble field area (0.4 km^2) at any time, column 3 indicates the average number of false alarms per antenna beam position. If the remote receiver is set to detect, at

Table 26.4 Performance Analysis Results

Case	*No. of False Alarms per 10,000 Pebbles*	*No. of False Alarms in Remote Receiver Antenna of Beam Covers 1/10 Pebble Area*	*Probabilities of Five Pebbles That Can Be Detected per Beam (Cold plus Cued)* P_D	P_{FA}
1	100	10	~0.9	10^{-12}
2	1	~0	~0.6	10^{-20}
3	10	~1	~0.8	10^{-15}
Nom. op. perf.	1	~0	~0.9	10^{-12}

a minimum, 5–10 pebbles, then 5–10 false alarms among pebbles in a beam would cause a remote receiver reception to falsely indicate an intruder. Columns 4 and 5 illustrate the P_D for five pebbles within a beam and P_{FA} with a cold detection plus four cued detections (for case 1) and five cold detections in cases 2 and 3. Table 26.4 also shows nominal operation performance described in this chapter in addition to the three cases.

Case 1 indicates that a cold intruder detection made by a pebble that cues all other pebbles would yield a P_{FA} of 10^{-12} and P_D of 0.9. However, if the pebble making the cold detection sends a cue signal that is received by all pebbles in the field, the resulting per-pebble P_{FA} of 10^{-2} implies that during the course of the 10 s interval about 10 false alarms per beam would light up all beam positions and prevent localization of the intruder.

Conversely, case 2 with essentially no propagation would result in only cold detections. An intruder would undergo a series of five cold detections with a cumulative P_D of 0.6 and P_{FA} of 10^{-20}. This is not a very high P_D.

If, under conditions of uncertain propagation, all pebbles are ordered, via the remote monitor, to set cold thresholds of a P_{FA} of 10^{-3} and a P_D of 0.95, and no cueing was allowed, an intruder would be detected with five cold detections with cumulative P_D of 0.8 and P_{FA} of 10^{-15}.

These cases suggest several possible options in pebble network design. Common to all three cases is the need to, in some way, measure propagation conditions as they likely vary over seconds, minutes, or hours. Measurements could take the form of either direct measurement of propagation loss or signal strength or monitoring the false alarm density as indication of propagation effects. To first order, local effects such as specular multipath from buildings or blockage from shrubs or ridges are thought not to have an overall negative sensitivity impact on performance. Blind spots (poor propagation) or enhanced sensitivity zones (enhanced signals) may influence when a detection is made or whether an intruder track is maintained consistently, but overall network performance is likely essentially maintained.

In the interest of only minimal design feature additions to maintain low cost, the following design options are discussed.

26.2.3.2 Test Tones and Gain and Sensitivity Control

Periodically, all pebbles could be commanded by the remote monitors to send brief tones at a test frequency (other than the frequency of the cue tone). Depending on received signal strengths, the pebbles would change the receiver sensitivity or the gain of the transmitter or receiver amplifier. The remote receiver monitors themselves could also adjust receiver sensitivity during these pebble calibration transmissions to ensure a detection threshold for the prescribed number of pebble tones per beam (e.g., 5–10) based on the test tones. The cost of adding the circuit is expected to be negligible. The most likely drawback is power consumption. However, the test tone could occur in much less than 1 s.

26.2.3.3 False Alarm Monitoring

In a significant size field of, for example, 10,000 pebbles, false alarm statistics per monitor beam position may be gathered by the remote receiver. If a remote monitor never detects random false alarms per beam position (perhaps via a test receiver channel with higher sensitivity than the normal monitor channel), it is likely that either nominal performance or case 2 performance is in effect. However, if the receiver is consistently detecting two or more apparently random false alarms in more than one beam position, this would be indicative of low propagation loss (case 1). One option under this condition would be for the remote receiver to command the pebbles in that area to a nominal case 3 condition. Although detection performance is somewhat lower than the nominal case, it might be adequate.

A no-cost alternative could be to have a person walk through a pebble field on occasion or trigger a few dispersed test pebbles to emit tones to test the response of the field.

The advantage of the false alarm monitoring approach is that no additional design feature would be required for the pebbles except the ability to receive a command to change to a case 3 threshold setting.

26.2.4 Swarming Simulation

A simulation was developed to gain further first-order insights into the swarming behavior. The simulation that was created to perform some initial investigations into the system performance is not unlike Mitch Resnick's "Star Logo" simulation [7] that mimics the behavior of slime mold, except that the sensors in our simulation are not allowed to move. Basically, a large number of pebble sensors is deployed, each with a very simple rule set governing its behavior, such that the collective behavior could indicate "intrusions" that could be remotely monitored.

The simulation models the case of 20,164 pebble sensors distributed in an evenly spaced grid over 1 km^2 (for a grid spacing of 7 m). A mouse interface allows insertion of a moving intruder through the grid. At each time step, the state of each sensor is updated to reflect intruder detections, communication among nearby sensors, false alarms, etc. The result is effectively a cellular automaton with the mouse-controlled intruder as an additional external stimulus.

Each sensor's cued "alert state" is indicated by gray levels. On the simulation display, each white pixel indicates a "cold" sensor that is not detecting or receiving, and each black pixel indicates a sensor that has detected an intruder and is emitting a tone to be received by nearby sensors that in turn respond by temporarily decreasing their detection threshold to the more sensitive cued setting, indicated by the gray pixels.

At each time step, the process of updating the state of each cell in the automaton—that is, each sensor in the grid—consists of a single Bernoulli trial. The probability of "success" (i.e., detection or false alarm) for a sensor is determined by two factors: the current detection threshold state of a sensor ("cold" or "cued") and a unit step function (detection or false alarm) of range to the intruding target, if one exists. This detection range is fixed at 10 m. Given the resulting probability of success, a uniform pseudorandom number in the unit interval determines whether the sensor has a detection or false alarm. In the event of a detection or false alarm, the detection threshold state of all sensors within communication range (also fixed at 10 m) is lowered to the "cued" state.

Some performance parameters of the sensor are adjustable by slider controls. The default "cold" probabilities of detection and false alarm are 0.9 and 10^{-4}, respectively. In the "cued" state, with a lowered detection threshold, the default probabilities of detection and false alarm are 0.99 and 10^{-2}, respectively. The detection and communication ranges of each sensor are fixed at 10 m, the effect being that the neighborhood of influence of each sensor is fixed to a subset of eight surrounding sensors.

Figure 26.10 shows by black and gray dots, pebble detections (black) and those receiving a cue (gray). The dashboard on the right allows the user to adjust the key zone design parameters of cold and cued pebble communication tone detection and false alarm probabilities over one and several orders of magnitude, respectively. Also shown is a sliding adjustment of cue tone persistence time from 0 to 5 s.

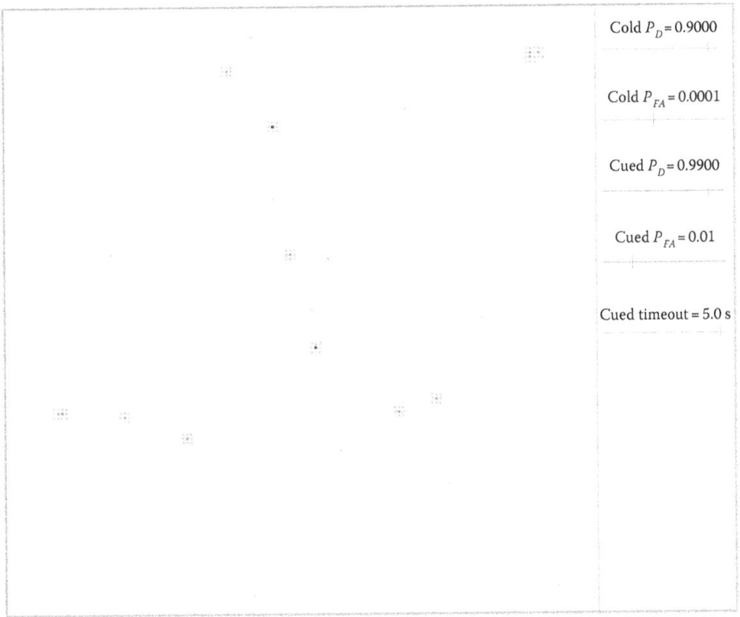

Figure 26.10 Simulation false alarm snapshot at zone requirements to first order.

The simulation was useful in testing the stability of the network. Figure 26.10 illustrates a typical picture of the display with random false alarm detections for a pebble packing density of 7 m separation. The lack of correlation (less than five pebbles detecting in a local area) indicates that a remote receiver with a directive antenna beam that scans over the pebble field would not receive the requisite signal level to expect an intruder.

Figure 26.11 illustrates an intruder moving through the sensor field and the sensors having detected the intruder, leaving a persistent trail of transmitting pebbles for a directive remote receiver to detect. If the remote receiver display retains the pebble transmission history and its beam directivity is sufficiently focused, then the track of an intruder could be followed.

Figures 26.12 and 26.13 illustrate instabilities of the sensor field. Figure 26.12 shows the false alarms if the cold detection false alarm probability is increased by a factor of 10 from 10^{-4} to 10^{-3}. Enough individual false alarms occur that a directive remote receiver could be receiving a spatially correlated signal for a false intruder alert, as shown. Figure 26.13 is the case for which the cold detection false alarm probability is retained at the nominal 10^{-4}, but the false alarm probability for a cued detection is increased by a factor of 10 from 10^{-2} to 10^{-1}. Swarm "clouds" appear within seconds due to a high percentage of cued false detections. Moving these false alarm probabilities, shown in Figures 26.12 and 26.13, back to their nominal values causes the sensor field to calm down to the picture of Figure 26.9 in a matter of seconds.

Other insights gained from the sensor net model include the following:

- The density, or average distance of near-neighbor pebbles, should not be too dense or cued "swarming" instability such as Figure 26.13 can commence from an intruder detection. Too little density will provide insufficient internode interaction to respond to an intruder with sufficient cued detection.

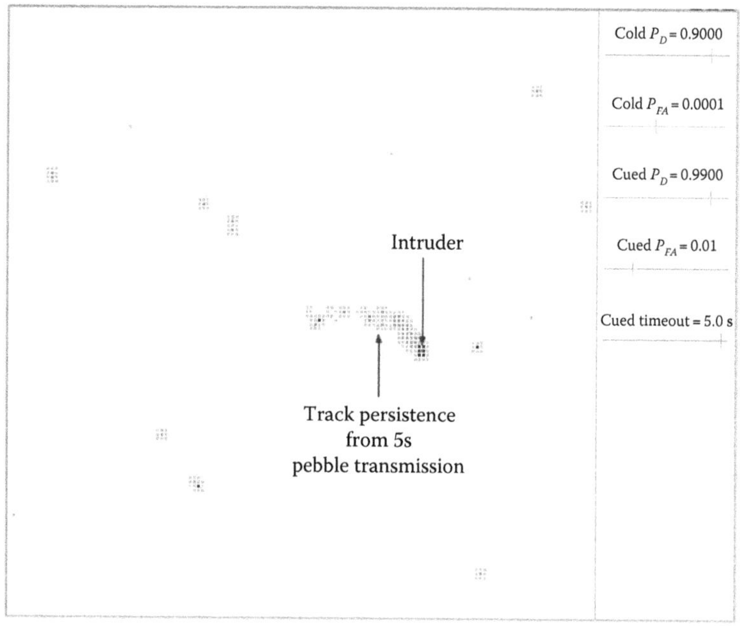

Figure 26.11 Simulation of false alarms and intruder track for nominal zone requirements.

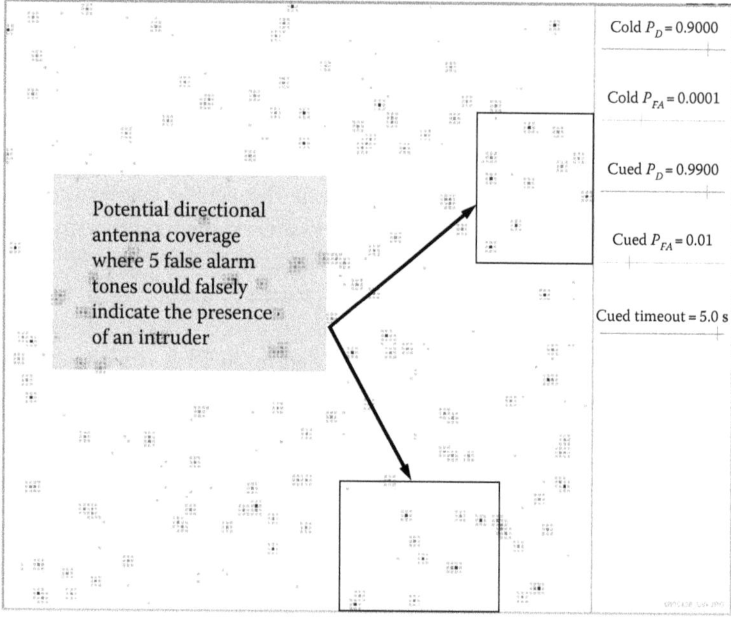

Figure 26.12 Simulation false alarm snapshot at 10 times higher rate than zone requirement for cold detection.

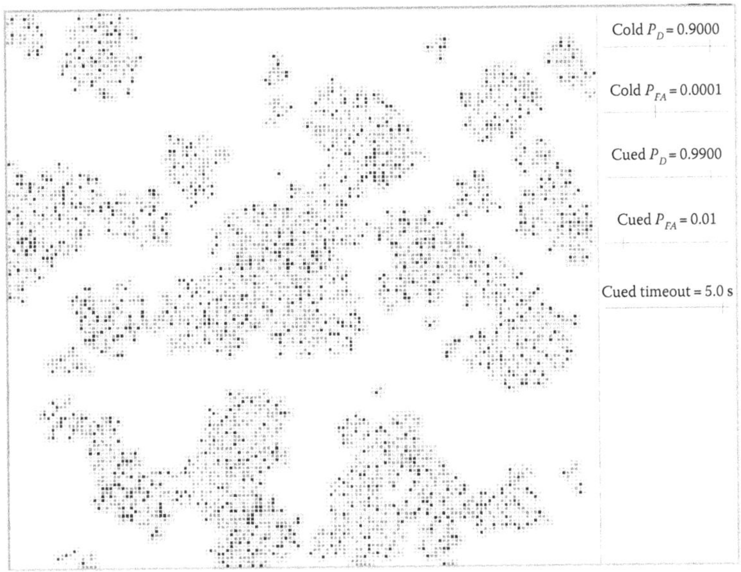

Figure 26.13 Simulation false alarm snapshot at 10 times higher rate than zone requirements for cued detection.

- Reducing the tone timeout from 5 to 1 or 2 s cleans up the false alarm picture (Figure 26.10) but further reduces the persistence of the track picture at the remote receiver.

Whereas this simple system features stationary pebbles exhibiting simple behavior, additional features can be conceived that would exhibit further emergent behavior.

- Pebble motion toward the detected intruder to further maintain sensor contact, i.e., kinematic swarming could be incorporated. This would require some damping mechanism, such as slow speed, so that one intruder does not empty the field in some locations allowing another intruder to enter undetected due to insufficient pebble density.
- The feature of identification via algorithms or a mix of sensor types for correlating could narrow the types of intruders of interest, for example, humans or motor vehicles versus wild animals. A zone condition for probabilities of correct and false identifications would need to be added.
- Transmission (spraying) of tagants at the intruder by detecting sensor pebbles, such as small RF tags, IR emitters, or phosphorescent dust, would allow for subsequent human or other sensor tracking beyond the pebble field.

26.2.5 Architecture

As described earlier, the objective intrusion detection system consists of "pebbles," small, simple sensor nodes dispersed throughout the environment being monitored, and a base station, or remote receiver, located at a command post within transmission range of the pebble field. While the approach ultimately relies on the eventual development of small, extremely discreet, inexpensive devices, our prototype implementation uses widely available mote hardware to investigate swarming sensor network behavior in the real world. We utilize functionality of the motes that is envisioned to be feasible in these small devices, which could be produced in quantity.

26.2.5.1 Preliminary Hardware Design

Critical to the affordability and scalability of the intruder sensor network concept is pebble technology that supports small size, low power consumption, and low cost. To support the swarming network concept, each pebble would contain a sensor, a microwave transceiver for two-way tone-based communications, an antenna, a power supply, a controller, and a suitable container to disguise and protect the sensor node components. A variety of sensor techniques could be used for intruder detection. The microwave blockage signal detection concept identified earlier would require the addition of a microwave receiver and antenna, both at 20 GHz or higher.

Over the past few years, a number of different motes have been designed and built. Network experiments described in this chapter use a MICAz Mote [8]. The MICAz Mote has a form factor of 5.7 × 3.18 × 0.64 cm and uses two size AA batteries. Although larger and more expensive than required for this application, each Mica mote contains a radio and microprocessor/controller that are significantly more complex than needed for our application. Recently, a team of researchers at the University of California Berkeley developed the 2.5 × 2.0 mm Spec Mote [9]. The Spec Mote is a fully working single-chip mote, which has a RISC core and 3 kB of memory, uses a 902.4 MHz radio, and has been shown to communicate about 12 m with a data rate of 19.2 kbps. The Spec Mote shows the potential for integrated circuit (IC) technology to facilitate reduced size and volume production costs.

Although similar in technology, existing motes do not perform the functions needed by the proposed concept, particularly the need for both a two-way tone-based signaling channel and potentially also a blockage detection receiver. Conversely, existing motes are more functionally complex than needed for the swarming network concept. To further establish size, performance, and cost feasibility of the intruder detection pebble, a design concept is being developed, as described in the following paragraphs.

There are two major functional aspects of the proposed pebble: (1) a sensor mechanism and (2) a tone-based RF signaling mechanism. Key to realizing the low size and cost is to realize the RF signaling mechanism on an RF IC. The RF signaling mechanism has two primary modes of operation. In the receive mode, the pebble listens to detect, cue, or alert tones from other nearby pebbles. When an alert tone is received, the pebble lowers its sensor detection threshold. When the pebble's sensor makes a detection, the pebble is switched to a transmit mode in which it transmits an alert tone for a fixed amount of time. The RF signaling mechanism requires the following functions: a detector for power sensing of the RF alert tone, a tone generator, a timer, registers and comparators and very simple control logic, a small antenna for RF transmission and reception, amplifiers including a low-noise amplifier on receive and possibly a power amplifier on transmit, and a power source.

The receive detector, transmit tone generator, control circuits, and any amplifiers would be included on a single RF IC. The frequency is assumed to be 2.4 GHz. RSSI circuitry is used for the receive detector and a voltage-controlled oscillator (VCO) is used for the tone generator circuitry. An off-chip crystal provides suitable frequency stability and drift for the tone signals. Stability is an important parameter because it drives the noise bandwidth and integration time that can be used at the remote receiver. Using a VCO alone to drive the antenna provides -10 dBm of power to the antenna. Higher-power levels can be achieved by adding a power amplifier, but at the cost of higher

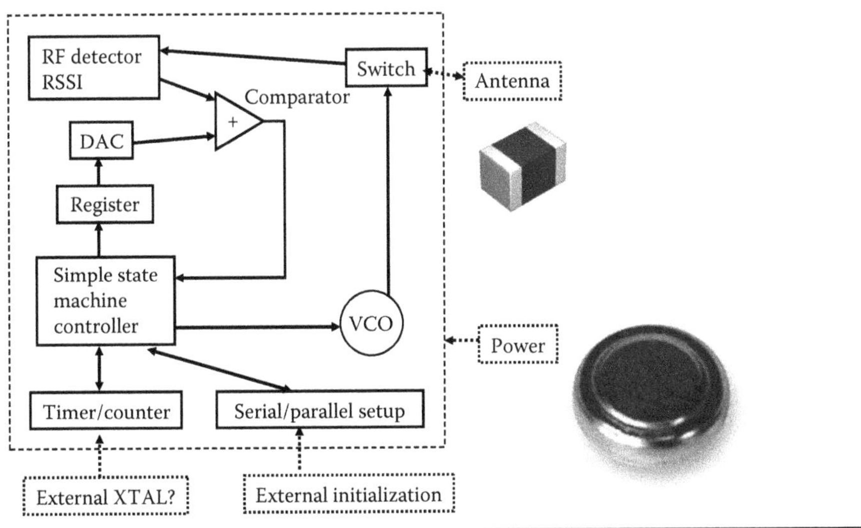

Figure 26.14 RFIC block diagram.

power consumption during transmit. Using function circuit blocks and the Jazz Semiconductor Multi-Project SiGe process, a preliminary RFIC design is shown in Figure 26.14. Chip cost is estimated to be less than $2 in volume production.

Several options are available for the antenna, including wire antennas and surface-mount antennas. Significant recent development has addressed the miniaturization of board-level omnidirectional antennas at 2.4 GHz for wireless applications. One example is the NanoAnt technology from Laird Technologies [10]. This technology operates in the 2.4 GHz Bluetooth band and is designed for high-volume surface-mount attachment through use of pick-and-place processing. This antenna measures 2.5 × 2.0 × 2.0 mm and reportedly costs $1.10 in high volume. This and other antennas will be evaluated as part of our concept development effort.

A preliminary board layout containing a chip antenna, the RFIC, and a temperature-controlled crystal oscillator was developed and provided a size of 14 × 13 mm. The board size may have to be increased pending study of the ground planes necessary for chip-based antennas. The pebble concept would include a 3 V cell battery mounted on the bottom side of the RF signaling board. Three-volt coin batteries are available in a variety of capacities and cost under $1 in quantity.

A simple rugged container is envisioned to be very inexpensive. The simplest sensor, a microphone manufactured by Horn Industrial Company, is available for $0.16 in batches of 1000 or more from Digi-Key Corporation [11]. The upper bound volume cost per unit for a pebble with a simple microphone is therefore less than $5. To add a 20 GHz circuit, as the illumination blockage detection would require, a receive-only RFIC circuit with a somewhat smaller size chip could cost about $1.50. A separate antenna would be required. The bounded volume per unit cost of the illumination sensor pebble would therefore be expected to be less than $7.50.

26.2.5.2 Prototype Software for Audio Microphone Sensors in MICAz Motes

The software running on the pebbles in the intrusion detection network prototype using MICAz motes is described later. The pebbles can be in one of three different timed modes: (1) time synchronization, (2) swarming, and (3) data transmit.

The time synchronization mode ensures that timers kept by all of the motes in the network are synchronized in order to get consistent real-time data. One mote in the network is specially programmed to act as the master mote and the rest act as slave motes. In time synchronization mode, the master mote waits a specified period of time, transmits a packet, and immediately enters swarming mode. The slave motes wait to receive this packet, upon which they immediately enter swarming mode as well. The time synchronization mode is the initialization mode for the network, prompting all motes to enter the swarming mode at the same time. While not essential to the swarming behavior and the longer-term pebbles implementation, time synchronization in the prototype allows us to accurately correlate event and power data over time among all nodes.

The swarming mode produces the "instinctive" actions of each node that contribute to the ultimate swarming behavior of the network. This mode, which can be modeled as a state machine, implements the actual intrusion detection using the acoustic sensors on the motes and simple tone-based communication. There are four different states in this mode as shown in Figure 26.15. The mote's LEDs provide a visual indication of its state.

State 1 is the normal detection mode, in which the pebble is not emitting any cue tones and its microphone detection threshold is set to the normal, baseline level. Once the pebble detects a sound event, indicated in Figure 26.15 by the "Sensor event" transition out of State 1, the pebble enters State 3, in which it emits a 2.48 GHz unmodulated cue tone. The pebble returns to State 1 after t_c seconds (the Cue Tone Timer period), indicated by the "Cue tone timer fired" transition between State 3 and State 1 in the figure.

If a pebble in State 1 detects a radio event—i.e., its RSSI readings are above the specified threshold, indicating that a neighboring pebble is emitting the cue tone—it moves into State 2 and lowers its microphone detection threshold. The pebble initiates a low threshold timer upon detecting a radio event. If the pebble still has not detected a sound when the low threshold timer fires (t_{lt} seconds after the radio event), the pebble returns to State 1. If, on the other hand, the pebble detects a sound exceeding the lower threshold, the sensor event causes a transition from State 2 to State 4, in which the pebble begins transmitting the cue tone as well.

In State 4, the pebble will continue sound detection at the lower threshold level while emitting the cue tone. Similar to the behavior in State 3, the pebble initiates the cue tone timer when it

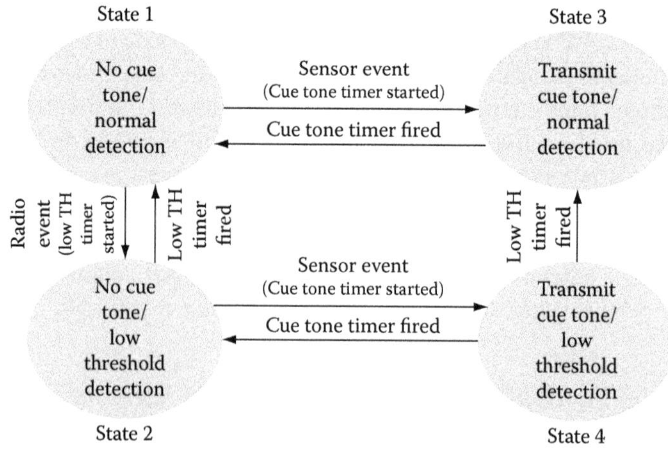

Figure 26.15 Pebble software state diagram.

begins emitting the tone. Should the cue tone timer fire before the low threshold timer, the pebble will return to State 2. If, instead, the low threshold timer fires, the pebble will move into State 3 and reset the detection threshold to the normal value.

The swarming mode can be set to run for any amount of time, depending on the length of the experiment. After the swarming mode completes, the mote enters the data transmit mode, during which each mote sends its event and time data as packets to the base station. The motes send their data one after another using a timer offset based on their ID (identity) number to prohibit transmission overlap. The MICAz mote at the base station receives the data packets and forwards them to the base station computer over the serial port. The motes' LEDs indicate which mote is sending packets during the data transmit mode.

26.2.6 Initial Experiments

To assess the impact of propagation loss in a tone-based network, a series of tests were conducted to collect RF data in real-world settings. The experiments used the MICAz platform [8] from Crossbow Technology, Inc., to emulate the pebbles. These programmable motes (short for remotes) have a simple, low-power microcontroller and communicate via a Chipcon CC2420 radio chip. The CC2420 transceiver has features that enable a realistic simulation of the swarming pebbles network. The transceiver can be programmed to transmit an unmodulated sine wave at approximately 2.4 GHz and has the receive signal strength indicator (RSSI) circuitry to report approximate received signal levels. The specifications of the MICAz radio are shown in Table 26.5.

For our experiments, we configured the MICAz motes to transmit a 2.48 GHz tone and measured the effective radiated power (ERP) and the transmit frequency of several motes to characterize performance. We measured the ERP of 13 motes in an anechoic chamber. The ERP of 12 of the motes varied from −2.6 to 1.7 dBm; one mote was found to be significantly lower in ERP. We measured the frequency across 6 motes and found the tone frequencies to be tightly grouped within a 30 kHz span. The excellent repeatability across motes of the tone frequencies is conducive to using a small noise bandwidth to maximize detection range at a remote receiver. The measured performance of these motes compares well to the pebble performance we assumed in our connectivity analyses in Section 26.2.2 [12], where we assumed a receiver noise bandwidth of 50 kHz.

The first experiment conducted was to measure the mote-to-mote connectivity, which relates to the required mote spacing for the swarming network concept. In this experiment, a MICAz Mote was configured to transmit a 2.48 GHz tone and the connectivity to another MICAz Mote was measured using the mote's RSSI. The RSSI measurement is not a precise measurement; the mote's radio specification quotes an RSSI accuracy of ±6 dB. The experiments were conducted on a flat grass field that contained nearby structures and fencing. A picture of the field is shown in Figure 26.16.

Table 26.5 MICAz Radio Specifications

Parameter	*Specification*
Frequency band	2400–2483.5 MHz
Transmit data rate	250 kbps
Transmit power	−24 – 0 dBm
Receive sensitivity	−90 dBm (minimum), −94 dBm (typical)

Figure 26.16 Grass field where experiments were conducted.

One set of measurements was taken with both transmit and receive motes placed on the grass, and another set of measurements was taken with transmit and receive motes placed on 0.61 m-high tripods. The preliminary results show that elevated motes without the presence of blockage structures can detect alert tones over relatively large separations consistent with free-space propagation and little fading. The results for motes placed on the grass are much more limiting. These results show a significant attenuation, roughly on the order of 3 dB/m. Results to date suggest that embedding pebbles in a dielectric layer (e.g., Styrofoam) to elevate them off the ground could provide tailorable propagation range, depending on the layer thickness.

An important aspect of the swarming network concept is the ability to achieve sufficient connectivity to the remote receiver that is monitoring the sensor field. Factors that affect the connectivity include the following:

- Physical limitations on pebble transmit power due to the need for low power consumption and electrically small antennas
- Propagation effects such as multipath, fading, and blockage
- The expectation that the tone signals emitted by the individual pebbles are not adding coherently at the remote receiver due to variations in their transmit signal phase and electrical path lengths to the receiver
- The variation of the aggregate emitted signal due to the dynamic nature of the network

The connectivity experiment used a MICAz mote configured to transmit a tone at 2.4 GHz and measured the signal received by a horn antenna and spectrum analyzer at various separation distances. Again, the measurements were taken on a grass field as shown in Figure 26.6 for three sets of data.

The first two sets were taken on different dates and used a MICAz mote on the grass. For the first set of data, the horn was approximately 0.6 m high; for the second set of data, the receive horn was 1.35 m high. For the third set of data, the transmitting mote was placed on a 7.6 cm high plastic cup. The second set of data for the mote on the ground indicates reduced performance. Because these two sets of data were taken on separate dates with some differences in equipment, there could be various reasons for the difference (e.g., variations in mote transmit power and variations in the measurement setup). The original connectivity calculations allocated a 15 dB propagation loss to

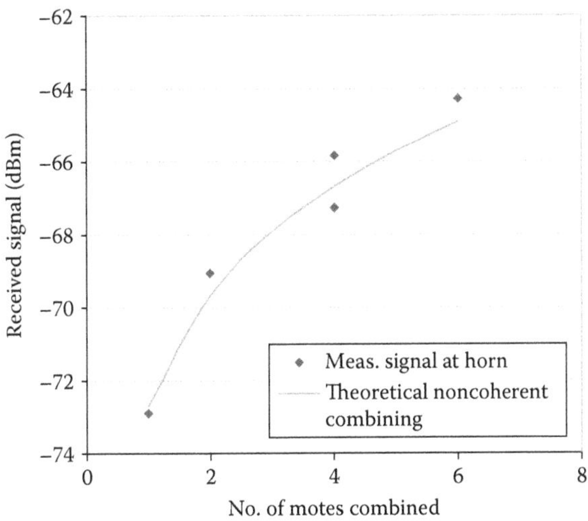

Figure 26.17 Noncoherent combining experiment.

account for the effects of fading, foliage, etc. The theoretical curve is consistent with the mote elevated 7.6 cm and is optimistic relative to a mote resting on the ground.

An experiment was conducted to test the assumption that tone signals transmitted from a collection of pebbles will add noncoherently at a remote receiver even for just a few radiating pebbles. For this experiment, conducted in an anechoic chamber, the receive signal from one, two, four, and then six transmitting motes was measured (Figure 26.17). The tone signals from the motes themselves are noncoherent. Two measurements were taken with four transmitting motes; the mote locations were varied for these two measurements. The motes were located in a line with separations of several centimeters and at a distance of 10 m from the receive horn. Figure 26.17 indicates that the motes exhibited noncoherent combining as expected.

26.2.7 Prototype Evaluation

Once we completed the implementation and simulation, we defined experiments to answer the following questions:

1. Does the pebble field exhibit swarming behavior in response to stimulus?
2. Is the remote receiver able to detect signals from the swarming field?
3. How does the configuration of the pebble field affect the network?
4. How reliable is the intrusion detection (e.g., what is the probability of detecting an event and how often are false events detected)?

We conducted experiments in a mostly empty rectangular room with a concrete floor on which we directly placed the motes. Characterization tests included raising the motes from the floor by placing them on cups; however, we found that in our space, the increased transmission range caused higher false-positive detections. We tested pebble fields containing 8 and 25 motes arranged in several configurations. The remote receiver station was situated approximately 20 ft from the pebble field such that all pebbles fell within the line of sight of the receiver's directional antenna.

26.2.7.1 Configurations

We conducted initial experiments using eight pebbles in two different geometrical configurations that are shown in Figure 26.18a and b. In configuration (a), the pebbles are in a grid layout with the antenna placed 25 ft away from the pebble field's edge. In configuration (b), the pebbles are in a line with the antenna placed 20 ft away from the pebble at one end of the line. The pebbles are labeled with the numbers 0–7, which we use to identify them later in Section 26.2.7.2 for each configuration. We simulate an intruder with a constant sound tone (we refer to this as the "stimulus tone") played while moving in the indicated direction. Using a constant stimulus tone and well-defined path allows us to perform controlled experiments. Future work will vary the simulated intruder's sound level and mobility pattern.

All mote configurations used the same parameter settings. The cue tone must be low enough that only immediately nearly pebbles can detect it to reduce false alarms. The level must be high enough, however, that the receiver antenna can detect cue tones. We performed preliminary characterization tests in the indoor testing environment to measure power levels and distances required for detection. These tests indicted the appropriate tone power level, pebble spacing, sensitivity thresholds, and the distance to the receiver antenna for our specific setup. Table 26.6 list the mote parameter settings that we used in all of our experimental configurations.

Note that the threshold levels are relative to the mote's running average of raw RSSI and microphone readings and are not physical values that trigger events.

26.2.7.2 Results

The first configuration we examine is the eight-pebble grid shown in Figure 26.18a. Figures 26.19 and 26.20 show the results from one individual run of the experiment using the grid configuration. The goal of our experiments using the eight-pebble field was to validate the swarming network concept in a real implementation and physical environment. To do this, we first examine the events occurring at each pebble during the experiment.

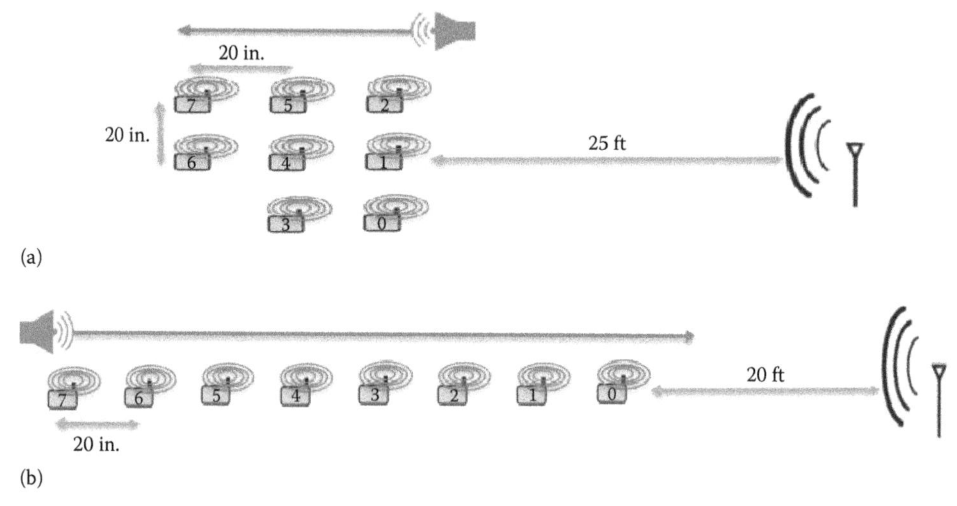

Figure 26.18 Experiment configurations. (a) Grid configuration. (b) Line configuration.

Table 26.6 Mote Parameters Used in Experiments

Cue tone timer value	8 s
Low threshold timer value	20 s
Cue tone power level	25 dBm
Mote cue tone detection threshold	5
Microphone threshold (normal)	20
Microphone threshold (lowered)	10

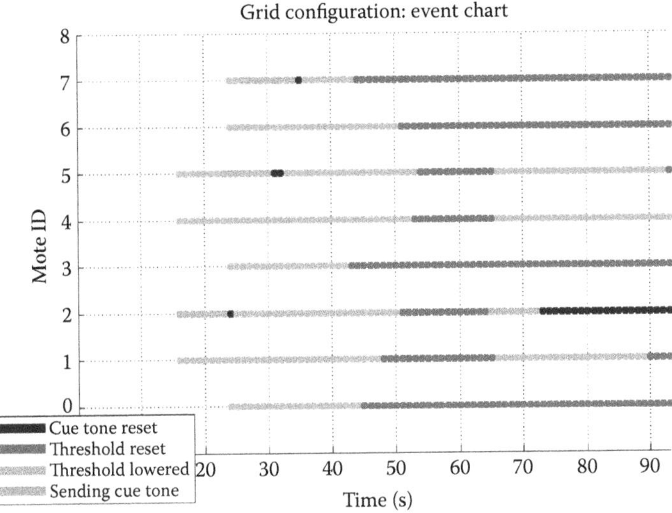

Figure 26.19 Event chart for grid configuration.

Figure 26.19 shows an event chart that displays the event and time data collected by each pebble during a representative experiment run. The mote IDs on the event chart refer to the numbering of the pebbles in Figure 26.18a. In this chart, we can see a successful swarming network that is responding to stimulus. For example, when we initiated the stimulus tone, Mote 2 detected a sensor event and transmitted the cue tone, which caused its neighbors, motes 1, 4, and 5 to lower their thresholds. As the stimulus tone source moved in the direction indicated in Figure 26.18a, Mote 5 and eventually Mote 7 also detected sensor events and began emitting cue tones, causing their neighbor motes' thresholds to be lowered.

Figure 26.20 shows the power level received over time at the antenna during the same experiment run, as collected from the spectrum analyzer by the automated LabVIEW program. The peaks in the power level, which can be correlated to the event–time data from the motes, show that the remote receiver can successfully detect the tones emitted by the motes. Note that the second power level rise, caused by Mote 2 emitting a tone, is an example of a false detection. This shows the importance of detection thresholds both at the pebbles and the base station.

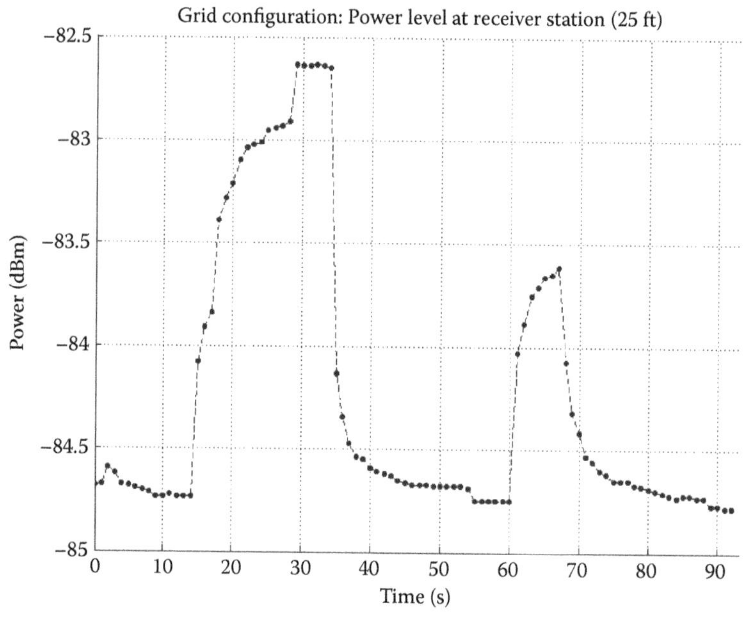

Figure 26.20 Power levels at receiver station for grid configuration.

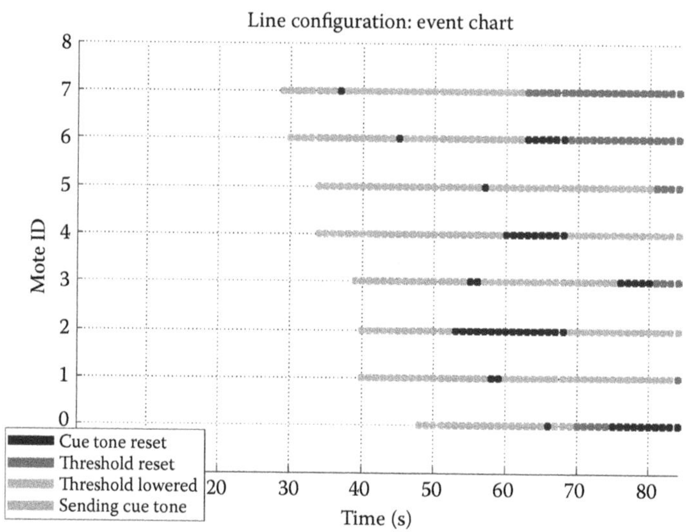

Figure 26.21 Event chart for line configuration.

Our next series of experiments used the eight-pebble line configuration shown in Figure 26.18b. Figures 26.21 and 26.22 show a representative set of results from an experimental run using the line configuration.

The stimulus tone was started near Mote 7 and moved along the line of motes toward Mote 0. The swarming network behavior can once again be observed—Mote 7 emits the first cue tone,

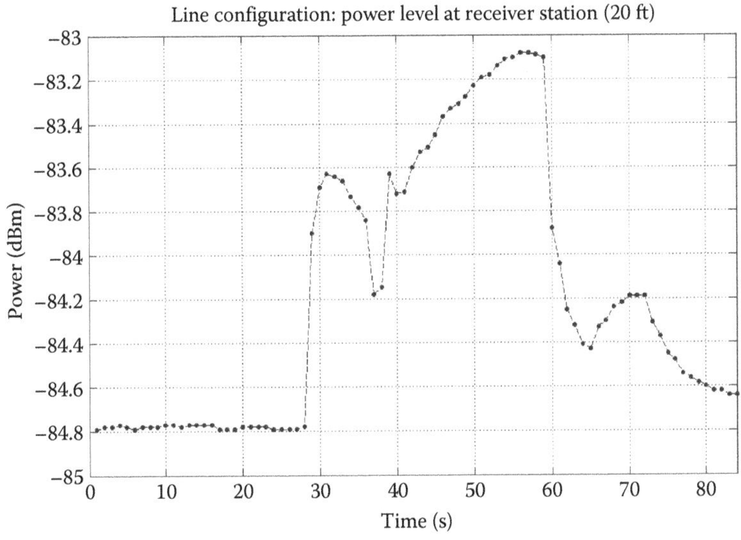

Figure 26.22 Power levels at receiver station for line configuration.

lowering the threshold of its neighbor Mote 6, which then detects the stimulus tone and begins to emit a cue tone, with the others following as the stimulus tone source moves through the field.

The eight-pebble experiments successfully demonstrated swarming behavior; however, the reliability of the intrusion detection is difficult to fully evaluate.

Inconsistencies in the sensitivity of the acoustic sensors of the motes make it difficult to accurately control the sound detection, especially with as few as eight motes. Some microphones are more sensitive than others, so a mote that is farther away might detect a sound that a nearer mote may not have detected. However, with a larger number of motes and a larger pebble field, this problem may not be severe enough to be noticeable.

These issues, while MICAz mote-specific and unrelated to the swarming concept itself, give light to the various factors that must be addressed when implementing a swarming sensor network in a real environment. Even as technology advances, the sensor nodes will always be limited by size, power, and cost factors that will affect their reliability and consistency. This provides support to the idea that a large number of motes are necessary for the swarming concept to successfully detect intruders. This also reinforces the notion that tuning of parameters in the sensor devices is not a trivial aspect of using this type of behavior for intrusion detection.

26.3 Multifunction Array Lidar Network for Intruder Detection and Tracking

26.3.1 Baseline Concept

With the limitations noted in the swarming pebble discussion [12,13], it was recognized that even with automated wide area intruder detection that swarming pebbles provide, potentially intense operator monitoring and inspection would most likely be required to verify the identity of the intruder, for example, hostile intent versus a wild animal. This led us to consider an alternate approach that utilizes a new device—a thin, phased-array laser aperture used for a MFAL that is highly analogous to microwave phased array antennas used in multifunction array radar (MFAR). The concept offers fully automatic detection-intruder type determination and precision tracking without requiring operator monitoring or intervention. The design, performance, and fabrication characteristics of this new optical phased array concept were summarized in a recent paper [14].

Compared to visible light imaging, passive IR imaging provides the possibility of operation both day and night by exploiting the thermal radiance of objects of interest. Using IR cameras on the ground, tracking the three-dimensional (3D) position of moving objects is accomplished by correlating images of two or more cameras and computing the distance to each object using stereoscopic range estimation algorithms. When multiple objects must be tracked, correlation of images becomes difficult because each camera may have a different view of the objects unless cameras are located very close to each other. Also, stereoscopic range estimation accuracy varies with the range to the object. The lidar system described here provides the 3D position and identification of the basic shape (for identification) of each object with one sensor, without correlation or estimation algorithms, and with high range accuracy independent of range. It also allows multiple lidars to collaborate to mitigate obstacle blockages and countermeasures such as lidar blinding.

Note that this is a concept study to explore the potential of an MFAL network. The authors emphasize that the flat, thin optical phased array, although they believe to be feasible, has yet to be built.

The concept of the MFAL was enabled by the recent conception of an extremely thin, true optical phased array. This optical phased array has allowed, for the first time, a direct extension of the capabilities embodied in modern MFARs, such as the U.S. Navy's Aegis system [15], and in sensor netting systems, such as the U.S. Navy's Cooperative Engagement Capability (CEC) [16].

Figure 26.23 illustrates a concept for a network of MFAL nodes. Each node provides part of a surveillance fence for azimuthal surveillance coverage, and some elevation coverage, using four arrays with 1 cm^2 aperture areas on the vertical faces of a cube. Three MFAL nodes are shown monitoring the area around a building complex. A fourth node is shown on the rooftop of the smaller building. This node serves as the remote monitor as well as a backup surveillance node.

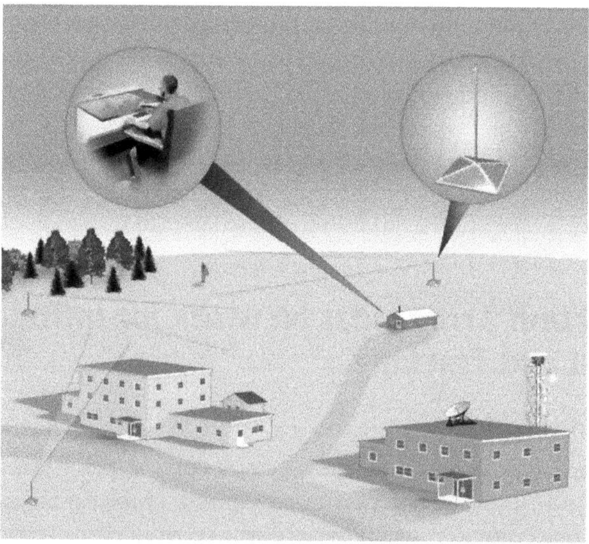

Figure 26.23 Intruder detection swarming network concept.

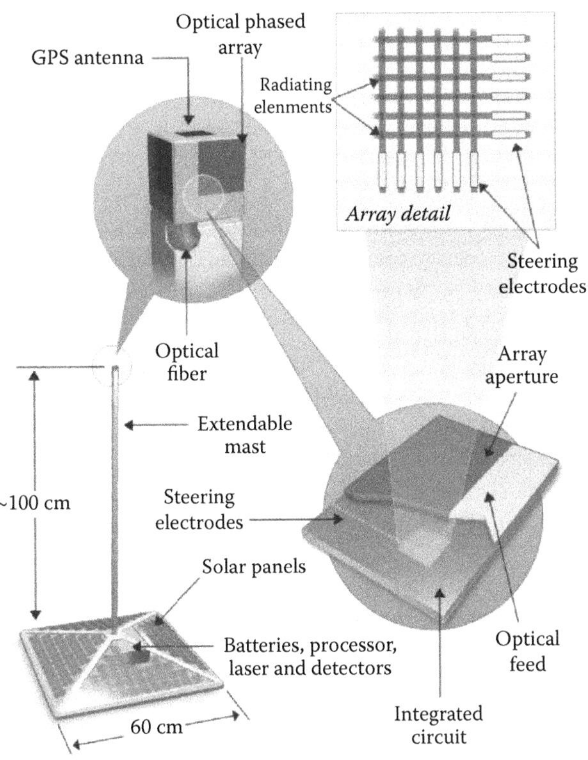

Figure 26.24 Illustration of MFAL node details.

As a remote monitor, it receives data from the other nodes via FSO links established by their optical arrays. Each array can be used for either lidar surveillance or FSO communication at any given time. Shown in an inset of the figure is a command monitor. The monitor could be automatic with an alarm to alert an operator if an intruder is detected and/or identified.

Figure 26.24 illustrates the configuration of an MFAL node consisting of a pedestal base, extendable mast, and the optical array "cube." The extendability of the mast ensures obstacle clearance and adequate range to the optical horizon. As shown in the figure, the pedestal base incorporates solar arrays, power supply, processing and control electronics, a Global Positioning System (GPS) receiver, and the laser transmitters and receivers. The cube consists of four optical phased-array apertures and a GPS antenna. The insets illustrate the structure of an optical array including functional sections and magnified features at the micron scale.

Each face of the cube consists of an optical layer stacked on an IC layer. The optical layer is composed of an array of crossed optical waveguides (dark gray and labeled "Array Aperture" in the figure), as well as a cascade of multimode interference (MMI) splitters (light gray and labeled "Optical Feed") [14]. Each array is fed along its lower edge by edge-coupled optical fibers leading to the base that carry optical power from the transmitter and to the receiver.

Scattering occurs at the intersections of the array. The waveguides are fabricated in an electro-optic (EO) material, and electrodes on each waveguide control the phase of the light at each intersection and thus the beam steering. For an $N \times N$ array, there are $2N$ steering electrodes. These electrodes are located on the back side of the optical layer where they can be controlled by the IC layer, which translates steering directions from the beam controller in the pedestal base into voltages for the steering electrodes. A ground plane is established by placing a single, large, L-shaped electrode on top of the optical layer over all the steering electrodes.

The array's rows and columns are spaced pseudorandomly to minimize cross-coupling while providing adequate array element density and sidelobe control [14]. The initial design was based on using polymethyl methacrylate as the EO material. This material yields an average waveguide period of 9 μm. For $N = 1000$, this spacing yields a total array size of 9 × 9 mm. The steering electrodes and MMI splitters occupy about 5 mm in each direction on two sides of the array, leading to a face size of about 15 × 15 mm.

Notional electronic processing and control components in the base include ICs on multi-chip modules for the lidar functions and the communications functions, memory modules, a GPS receiver for time synchronization, and location fixes used for the lidar alignment. Figure 26.25 shows the modes of the MFAL.

Figure 26.26 illustrates the basic operation of the MFAL network in the case of an approaching intruder. Node N1 searches in a pseudorandom search pattern with a 0.01° width beam. The description following the figure uses MFAR and sensor network terminology found, for example, in [6].

When an intruder has been detected, nominally at 0.4 km in an unobstructed view, node N1 detects motion A against a 3D background "clutter map" and transitions to track dwells B that includes a contiguous pattern of beams to scan the cross-sectional extent of the object and identify its nature (e.g., human, animal, vehicle), and, in so doing, determine whether it appears to be in

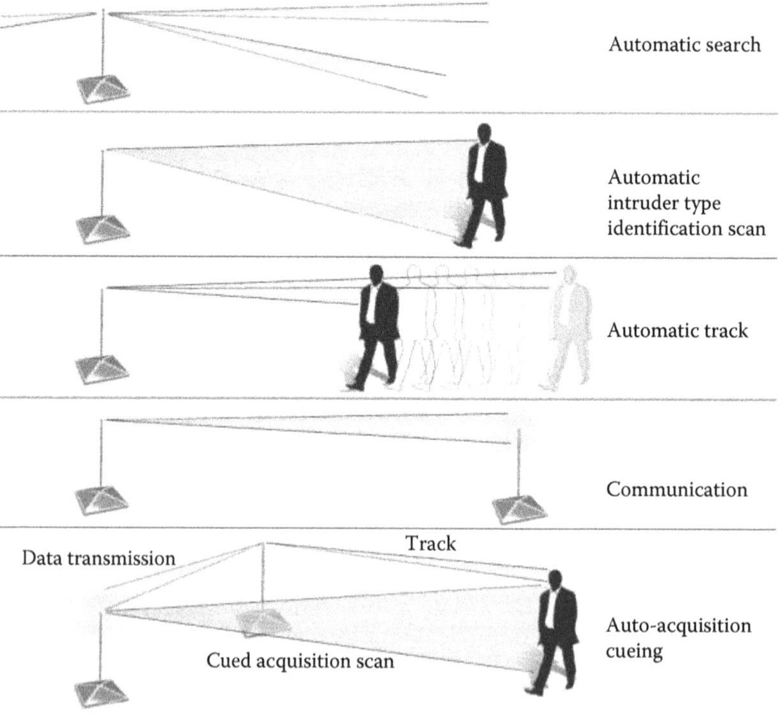

Figure 26.25 Modes of the network nodes.

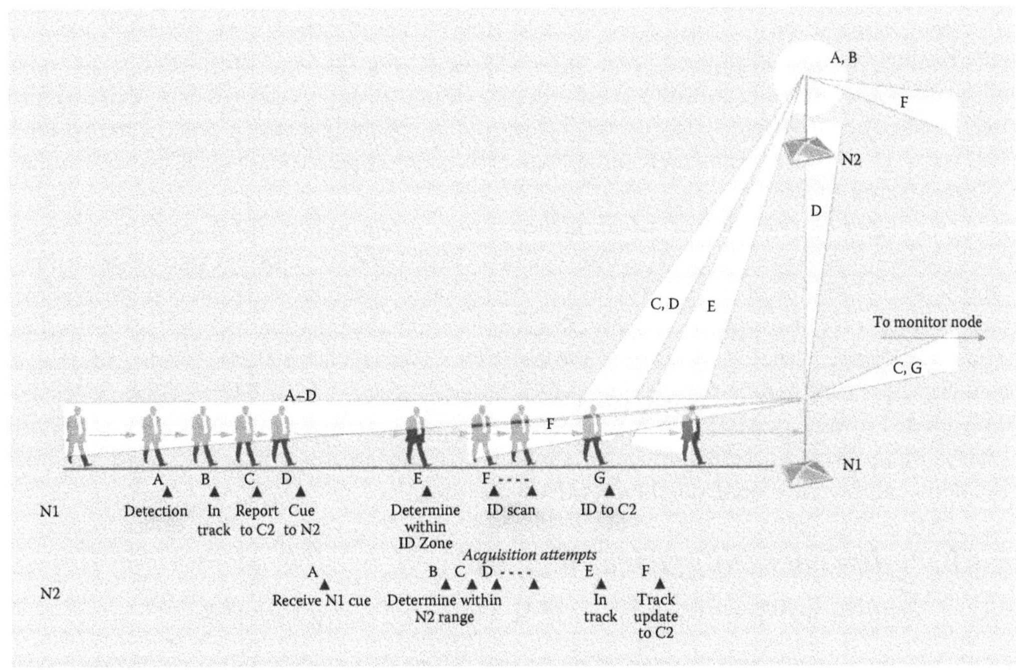

Figure 26.26 Network operation with approaching intruder.

the shape of a real object or a false lidar return. The track state centered on the intruding object is transmitted via FSO C to the central monitor node for which node N1 forms a beam in the monitor's direction, and the monitor receives during a periodic communication reception timeslot. The monitor receives the node identification, node location, and intruder coordinates from the most recent lidar measurement, along with a track state vector for correlation with the returns of other nodes. The tracking node also sends the same information to its nearest neighbors (e.g., N2) D at beam locations provided by the monitor node. At a predetermined, periodic reception timeslot, the neighboring node N2 receives the data from the tracking node N1. The monitor and neighboring nodes (e.g., node N2) receive periodic track updates from the tracking node, N1, once per second. This cueing information, together with a priori knowledge of the node positions and their respective sensor orientations, is then used for the track handover process. The neighboring node N2 monitors the track, and when the intruder is determined to be sufficiently close for possible detection, an autonomous attempt is made to acquire the target. Should the node N2 successfully acquire the target with sufficient track quality, it assumes the track to free up the original tracking node (informed via the communication link). Alternatively, both nodes can retain the track and report separately to the monitor node for maximum track certainty.

Multiple beams are required to cover the rather large ambiguity as a result of GPS-based lidar alignments relative to the highly accurate range and angle accuracies of the lidars themselves. If there is no return at the direction of the received remote track, the neighboring node continues to send out periodic acquisition beam patterns at the latest track location until it either receives a return or times out. Upon receiving its own lidar return, the neighboring node then sends a series of beams to develop its own track scan E and reports the track updates to the monitor and neighboring nodes F.

Tracking multiple intruders by each node is possible because the track mode need only provide dwells for limited target directions, thus minimally impacting the lidar timeline. For example, limiting the system to one detection acquisition or two acquisition cues will only decrease the search rate by approximately 10%. Although this condition slightly increases a particular volume search coverage period, it is tolerable on average because every search frame will not contain an interleaved acquisition.

One advantage of a network of MFAL nodes versus a single node is that if an intruder passes near obstacles where one node's lidar is obstructed, another node at a different vantage point may be able to see around the obstacle so that the combination of tracks received and combined at the remote monitor may be complete even if no single node can maintain a continuous track. Selection of node locations can therefore be made on the basis of uncovering obstructions. Another advantage is if an intruder attempts to jam a lidar node with an optical emitter, other lidar nodes would likely not be blinded at the same time due to their diverse locations; however, if they were, the strobes of the jamming could be reported to the remote monitor for triangulation. They would also indicate such anomalous behavior occurring in the surveillance area.

26.3.2 Preliminary Design and Analysis

26.3.2.1 Preliminary Network and MFAL Design Characteristics

This section provides design features and parameter values, as well as example calculations that support the design concept. Table 26.7 presents MFAL network detection and track parameter values and nominal response times. These represent system requirements for an operational mission to detect, track, and identify the types of intruders with nominally one sensor node per km^2. We believe that the processing and data speeds support hundreds of track updates and detections per second per node.

Table 26.8 presents the specific design parameters of the MFAL. These parameter values and the associated range calculation are based on commercially available laser and receiver components.

At these powers and pulse widths, care must be taken to avoid damaging the optical components because of the small mode-field area (MFA) of the waveguides. The fibers connecting the lasers to the array faces have a MFA of about 80 μm^2, yielding a fluence of about 2 J/cm^2, well below the fiber's damage threshold of about 50 J/cm^2. The array itself is a polymeric EO material, with the input power divided equally among all of the waveguides. However, as initially conceived in [14], the cascading splitter network that divides this power was also polymeric. Coupling directly into a single polymer waveguide on each surface would result in a fluence of about 40 J/cm^2, far in excess of the polymer's approximate 1 J/cm^2 damage threshold. To solve this problem, the optical layer

Table 26.7 Characteristics of MFAL Functions

Detection range	400 m at 10% reflectivity for human walker
FSO communication range	1 km
Volume search coverage	360° azimuth −0.1° to +0.4° elevation
Volume search coverage period	<1 s for 2% of volume, 50 s all beam positions
Transition to track	<5 s
Track update rate	1 Hz

Table 26.8 Design Parameter Values

Energy per pulse	1.6 μJ
Pulse width	4 ns
Pulse repetition interval	25 μs
Pulse integration	Four pulses noncoherent
Target dwell time	100 μs
Range resolution	0.6 m
Wavelength	1550 nm
F-number	3
Beamwidth	0.01°
Pass band of elastic channel	0.1 nm
Quantum efficiency	0.2
Receiver electronic bandwidth	20 MHz
Preamplifier current noise density	2.12 e−12 A $Hz^{1/2}$
Amplifier noise factor	1
Non-multiplied dark current	0.2e nA
Multiplied dark current	0.2e nA
Detector noise factor	20
Detector current gain	100
Target area (person)	11148 cm^2(0.7 × 1.9 m)
Target reflectance	0.1
Array loss	6 dB

will be composed of multiple materials, with silica waveguides used for the first two stages of the cascading MMI splitter network (to go from 1 to 100 waveguides per side of the array) and polymer used for the last stage of the cascade (to go from 10 to 1000 waveguides) and the array itself.

The patterned silica waveguides will be butt-coupled directly to the polymer waveguides lithographically. All waveguides will be fabricated on the same substrate.

The communications mode requires orders of magnitude less power than the lidar modes, assuming both transmit and receive directivity. Therefore, the lidar modes dominate the power requirement of 30 W. The solar cells covering the 0.64 m^2 base, as shown in Figure 26.24, provide 90 W of power during peak lighting conditions, thrice that required to operate the system, allowing the batteries to be charged, during operation, at twice the discharge rate. Lithium polymer batteries with capacity sufficient for 3 days of sunless operation, 2500 W-h, easily fit within the base. Alternately, the MFAL could operate on line power if greater reliability were required or if a higher-power laser, enabling longer range, was employed.

As will be discussed later, each array face has its own lidar transmitter and receiver, totaling four per node; so, for lidar modes, each array operates in parallel and independently. We also assume a single 250 mW fiber communication transmitter/receiver that can be switched into the arrays, one array at a time, according to a specific time window reserved for FSO communications.

Because lidar and communications signals are in the infrared region of the spectrum, at 1550 nm wavelength, ranges are affected by significant dust and fog. Lidar range will likely be impaired before communications range. For example, at a fog visibility of 100 m visibility, the preceding MFAL range would be reduced from nearly 400 to 56 m at 1550 nm wavelength. One all-weather compensation approach would be to integrate and intermingle the all-weather, but less-precise, swarming microwave network, described previously with the FSO MFAL network. If weather-based FSO impairment (i.e., loss of FSO reception) is observed at the monitor node operating both the MFAL and the swarming microwave networks, the monitor could command the MFAL network to shut down and the swarming network to start up as a backup capability, thereby conserving MFAL power.

26.3.2.2 MFAL Surveillance, Acquisition, Tracking, and Identification

In describing an example array search and track strategy, we apply the parameter values of Tables 26.7 and 26.8; specifically, an aperture beamwidth of 0.01°, a 4 ns pulse length yielding approximately 0.6 m range resolution, and a pulse repetition frequency and 4-pulse noncoherent integration yielding a lidar dwell time (to send pulses and receive echoes per beam position) of 100 μs. Further, because each array is connected to an independent laser source and receiver, all four arrays can transmit and receive in parallel. Finally, for such short detection ranges (0.4 km calculated for 10% object reflectivity), one wishes to effectively search the volume and update tracks every second.

For each array that scans 90° in azimuth and from −0.1° to +0.4° in elevation per second, 450,000 contiguous beams would be required. Further, at 100 μs per beam dwell, it would take 45 s per volume scan at all beam positions. In contrast, only 10,000 dwells are achieved per second at 100 μs dwell intervals. To accomplish the volume update with many fewer beams (10,000 versus 450,000), we recognize that the cross-range beam coverage at 500 m range is only 8.7 cm. Therefore, for detection of objects of interest that are, for example, 0.5 m wide and 2 m tall, we can choose to only transmit every fifth contiguous beam position in azimuth and every tenth beam position in elevation. In this manner, every transmitted azimuth beam position center will be separated by about 44 cm, and each transmitted elevation beam position will be separated by 87 cm (Figure 26.27).

Thus, a human, larger animal, or vehicle would be covered by at least one of the sparsely distributed beams. As a further hedge against an intruder slipping through the sparsely sampled volume, the system changes which of the continuous beam positions are covered with pulses every second, so that all beam positions have been covered over each 45 s with 10,000 available beams. By using only 2% of the 450,000 beam positions, we will therefore require 9000 per second per array to adequately search the volume per second, with all beam positions covered every 50 s (Figure 26.28).

For detection, we develop a 3D clutter map, storing whether an echo was received (a "1") or not (a "0") at each pulse resolution cell for each of the 450,000 beam positions (only partly updated per second, but all ranges and positions updated every 50 s). We use a clutter map rather than Doppler detection to enable use of off-the-shelf, low-cost lidar systems. Even for expected detections of only 400 m, we assume detections could occur, for example, with higher reflectivity

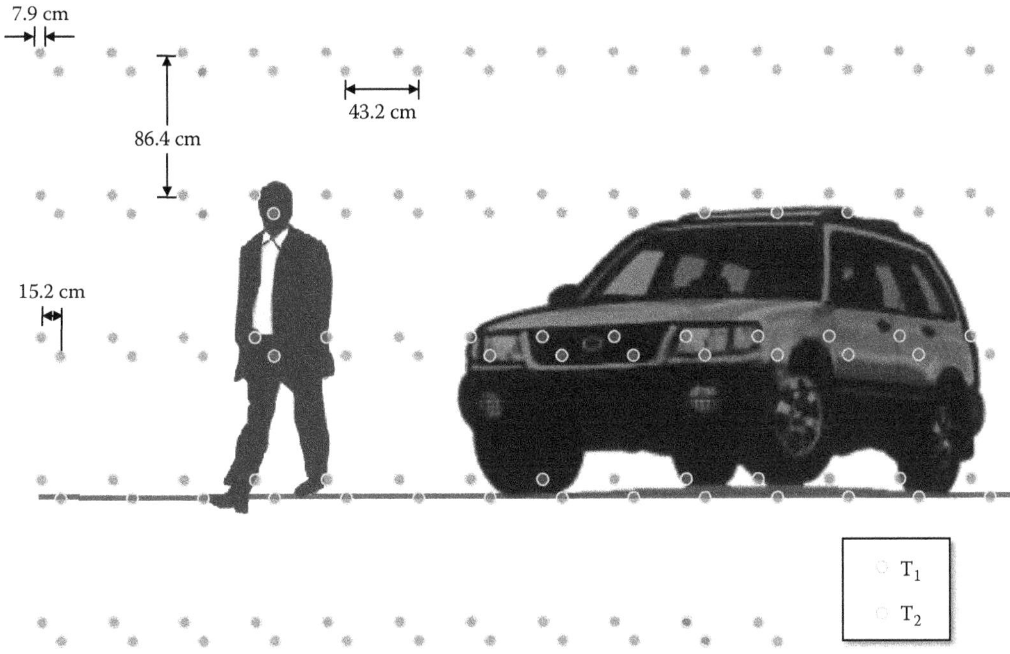

Figure 26.27 Beam pattern for two 1 s frames at 0.5 km range.

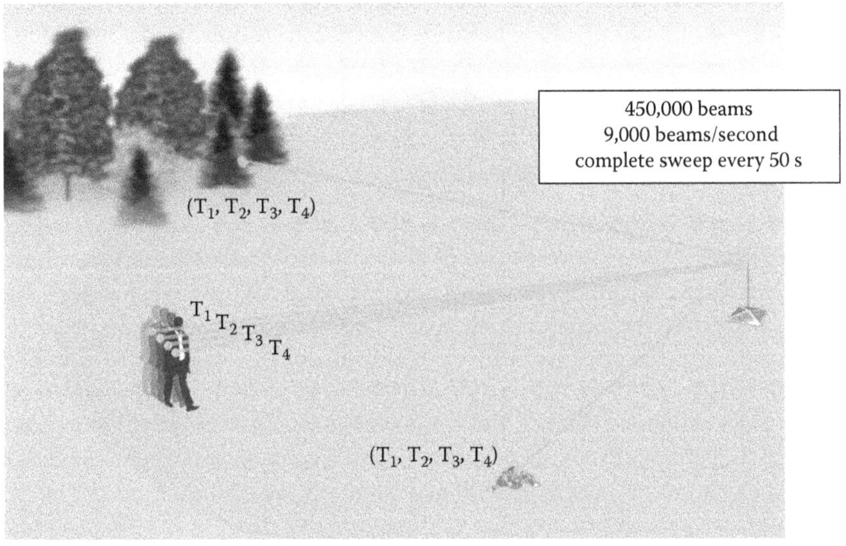

Figure 26.28 Beam patterns and timing.

objects, out to 1 km. Then, 1500 range resolution cells result per beam position and 450,000 beam positions per array and four arrays result in a clutter volume of 2.7 Gb. The memory of 2% of the memory cells is updated every second, with all cells updated each 50 s. The detection algorithm would determine physical motion by detecting changes in a number of contiguous cells over time (Figure 26.28). Once it is determined that a grouping of cells of comparable range and angle have changed, the MFAL is directed to scan a 20 × 20 beam pattern of every other beam position over a 40 × 40 beam position area at that location as a priority interrupt from the search pattern. This would cover a 3.5 × 3.5 m cross-range at 500 m. If a significant portion of beam positions in this pattern receives echoes at about the same ranges, detection is declared. During each second from then on, a track update beam is scheduled at the center of the detected beam pattern. Updates are entered into a track filter for each track. If a track update return is not received over several seconds, another 20 × 20 beam acquisition is attempted. Therefore, of the 9000 array dwells identified for search each second, a multiple of 400 beams will be interrupted for each transition to track.

When a detection has been determined and a track initiated, the node transmits the track state data as well as the estimated cross-sectional area of the intruder to the nearby nodes and to the remote monitor. At the remote monitor node, the cross-sectional area is an indication of intruder size, and the track velocity can indicate whether the intruder is potentially a vehicle traveling beyond human speed. The information is used by other MFAL nodes to cue an acquisition of the intruding object. This is of value in further verifying the detection and to maintain track by other nodes if the intruder passes out of sight or behind obstacles from the originally detecting node. Because of GPS position uncertainty much larger than lidar beam and range accuracy, a receiving node will provide special monitoring of a beam pattern covering the indicated location out to 5–10 m on each side in azimuth, depending on ambiguity calculations for the target and lidar geometry. In this special region, the clutter map detector is set to high detection and corresponding false alarm probabilities in that area based on the acquisition message of a neighboring node. If the cued node makes a detection and transition to a tracking process, it will send a message to neighboring nodes, indicating a detection associated with the track state received by the cueing node that sent the track state. In this way, a basic swarming behavior is established similar to the swarming intruder network in [13].

26.3.2.3 MFAL FSO Network Operations

Because we have 10,000 dwells available and use 9000 for detection, acquisition, and tracking, as previously described, we shall reserve 1000 dwells, or 0.1 s per second available for each array, to communicate track and identification data via the FSO channel. Note that whereas we assume four lidar sources and receivers, one for each array, we assume only one communication source receiver, shared among the four arrays, with each individual array transmitting data only in the directions of other nodes. At 100 Mbps, a total of 10 Mb can be transmitted and/or received at a time. Assuming a simple error detection and correction code of 12 bits per information bit, this translates to 0.8 Mb of data per second per node. If 32 bits are used per word, then 26,000 words per second could be sent or received from each node in the allocated 0.1 s window each second.

More detail is now provided concerning the unique FSO network. Typical successful FSO networks assume the need for high gain in transmission beams and reception beams via precision gimbaled mirror alignment, plus highly agile automatic turbulence correction and automatic gain control (AGC). FSO communication in the present concept requires no mechanical gimbals because the optical phased arrays provide electronic beam pointing. Also, at 1 km of communication range between neighboring nodes, it is sufficient that the transmitter aperture gain be used but significant receiver aperture gain is not needed. Finally, because of the short communication distances and low data rates, atmospheric turbulence compensation and advanced AGC features are not required.

A variety of network operations schemes can be devised for directive transmission and reception using combinations of time-division multiplexing (TDM) and wavelength-division multiplexing (WDM). A concept is offered here as evidence that a practical approach is feasible. For network initiation, each node is initially set for omnidirectional reception in which the arrays are spoiled to an approximate 0 dB gain (±3 dB). The monitor sends out interrogation beams, at low data rates, indicating GPS time and location and a node-responsive time window. The monitor beam sweeps in a 360° azimuth "interrogation" pattern. As individual nodes receive the interrogation, generally at different times, they respond with high gain transmit beams pointing toward the monitor node location within the indicated time window, during which time the monitor arrays are set to omnidirectional gain. A number of potential interrogation response WDM channels are available from which each node is randomly assigned for transmission of a response. The monitor can receive multiple responses at different WDM channels simultaneously for channel decoding during the response time windows. Alternatively, all nodes could use the same wavelength and, using GPS time synchronization, a TDM structure could be implemented in which each node takes turns communicating. It is expected that FSO communications will include appropriate FSO error detection and correction coding and a commercial data encryption product. The monitor will transmit interrogation beams followed by a listening time window for several cycles to ensure all nodes have responded. The monitor will then individually transmit to each node, all of which are set to omni receive, with a table of the locations of all reporting nodes in the network. From this point onward, both transmission and reception will be directional.

During each second, with GPS time synchronization and position alignment, each node will perform surveillance, tracking, cued acquisition, and intruder type identification functions for the first 0.9 s and provide the remaining 0.1 s for transmission and reception of neighboring node acquisition cue track updates. During the 0.1 s window, all nodes will set their arrays to receive from their nearest neighbors in anticipation of a potential acquisition cue message, and those nodes with track cue data will transmit to immediate neighbors via their directive apertures. This time is also reserved for reporting detections, track updates, and identification images to the monitor node. The monitor node may also transmit a command to individual nodes or all nodes during the timeslot each second that is reserved for acquisition cues.

26.3.3 Potential Next Steps

The initial concept description has been provided for a new type of intruder detection system analogous to military microwave phased-array radar and communications systems. The description includes functions, timing, component descriptions, and initial calculations. Its feasibility hinges on the successful implementation and cost effectiveness of the new optical phased-array aperture originally described by Papadakis et al. [14] and further articulated here. Clearly, the prototype development of the phased array is the principal next step. Software functions, signal processing, sensor node alignment, communication protocols, and timing structure have been successfully developed for microwave systems and are not considered high-risk areas. Further, the application of interest is much less complex than that of the military microwave systems and would be expected to be straightforward to design [17].

As observed, the MFAL concept could be applied to other problems such as vehicle collision avoidance and control, short-range inter-vehicle communication, and even surveillance and communication inside buildings where line-of-sight internode communications are possible.

26.4 Intersection between the Swarming "Pebbles" and the MFAL Surveillance Zone

The two types of intrusion detection networks are very different; in fact, intentionally designed to have opposing attributes:

1. Simple nodes versus highly advanced nodes
2. Many randomly placed nodes, scalable to a wide area versus carefully placed and focused on a specific area

Such diversity can provide a complementary means to detect intruders; however, having the networks interact is not straightforward. The swarming pebbles provide imprecise tracking of intruders and cannot distinguish the type of intruder using such simple sensors as microphones and IR detectors. In contrast, the MFAL zone is very precise and can determine type of intruder. Therefore, without design changes, the two networks cannot automatically cue and alert each other in the same or adjacent region.

There are, however, advantages to designing them to interact. We will consider interaction in three scenarios in which a very large swarming pebbles field extends at the perimeter of a monitored area and the precise network of MFALs monitors an inner, high-priority perimeter, as shown in Figure 26.29.

26.4.1 Intruder Leaving the Swarming Pebbles Field and Entering the MFAL Zone

As described earlier, the track of an intruder from the monitor center is relatively imprecise. A remote monitor antenna beam with a 0.25 m^2 aperture would be expected to cover a rather large area, for example, 0.25 km^2 from a range of less than 1 km at 2.5 GHz. For 10 GHz pebble frequency, the same size aperture could cover a 0.25 $mile^2$ spot from about 10 km. Scanning the antenna could "centroid" the composite pebble signal as it scans to further reduce the ambiguity

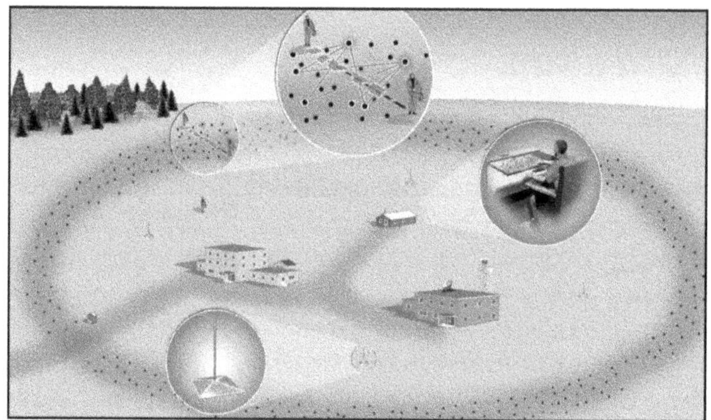

Figure 26.29 Configuration of an outer zone of swarming pebbles and an inner zone of MFAL nodes. Scenario A: Intruder leaving the swarming pebbles field and entering the MFAL zone. Scenario B: Intruder leaving the MFAL zone and entering the pebble field. Scenario C: Intruder in the midst of a surveillance zone containing both MFALs and pebbles.

area by standard techniques. However, regardless of approach, it is likely that the intruder track would not be more accurate than 100–200 m. Such low accuracy is helpful for large surveillance regions. However, it is too imprecise to cue an MFAL as an intruder tracked by pebbles leaves the pebble field and enters the MFAL zone. There are some design measures that could provide an advantage to the MFAL zone over merely cold detection by MFALs with no input from the pebble field.

As described earlier, an MFAL will scan every fifth beam position over 360° azimuth, and every tenth beam in elevation, during every second. Each MFAL maintains a 3D "clutter map" and automatically monitors for consistent movement by pixel groupings. If such behavior is detected over a number of dwells, a moving object is declared to have been detected, and scanning and track update dwells are then scheduled from then on. For fast-moving vehicles, it is likely that an MFAL would detect such movement in only two rotations, i.e., within 2 s, because the pixel motion would be obvious. For slow-moving humans, an MFAL might require four rotations, or 4 s, before enough pixel movement is determined to declare a "detection."

Even if the remotely monitoring command and control unit of the pebble field can send a track accurate to within 200 m for an approaching intruder, the closest MFAL would not be able to use the track to cue with sufficient accuracy to provide a significant time advantage. However, the MFAL could reset its 1 s rotation cycle to span the 200 m cross-range, thereby potentially saving a fraction of a second. And if the presence of the cue could be used to allow a more sensitive pixel movement detection threshold, then perhaps the MFAL could make a detection one cycle (second) sooner, say after three cycles rather than four. Therefore, direct cueing from the pebbles to the MFALs does not significantly enhance performance.

If, however, the pebbles also produced a 1550 nm tone in addition to the microwave tone monitored by its neighbors and the central monitor antenna, then they could provide a beacon to cue the MFAL with high accuracy. For this design case, suppose that those pebbles bordering the MFAL zone carried optical tone emitters in the same wavelength band of the MFAL communications transceivers, but at a wavelength separate from those of the communications channels. The MFAL could be designed to receive the tone and perform a cued acquisition.

To accomplish this, a design feature would also need to be made to the beam scheduling. As presently designed, the MFAL would not detect the tone until the MFAL trained a beam in the precise direction sometime within the 50 s full coverage. Otherwise, for every fifth beam per second, there would only be a 20% chance of alignment of beam and tone per 1 s scan cycle. A new mode would need to be added to the MFAL to allow it to passively sweep the horizon for beacon detections with a spoiled beam width with wider azimuthal spread. A lower gain, broader azimuthal laser beam would still allow adequate received signal strength and yet allow for faster scanning of the 360° of azimuth. If the azimuthal beamwidth were spoiled from the nominal 0.01°–0.1°, it is possible that within 1 ms the beam could be scanned continually through the full 360° horizon, to receive, detect, and locate any optical pebble tone. The beacon could then be used to self-cue the MFAL to acquire the emerging intruder with the beam spread mentioned in the previous section. Adding such a feature only requires additional functionality to be programmed into the MFAL control element. As for the pebbles, at least those near the boundary with the MFAL zone would need to be specially manufactured to emit not only microwave cue tones, but also optical cue tones of sufficient strength for nearby MFALs.

In summary, the required design changes for pebble field cueing of MFALs are

- Optical tone capability in pebbles near the boundary allowing dual microwave and optical transmission upon detection of an intruder

- Optical cue tone beacon reception via the MFAL communication transceiver when the scanning beam is pointed at the pebble
- Spoiling of the MFAL receive beam in the horizontal direction to facilitate rapid scanning of the horizon for beacon signals
- Addition of a beacon scanning mode in the MFAL to scan 360° for pebble beacon tones once per second

With these design changes, an intruder moving from the pebble field into the MFAL field could be acquired and identified by the nearest MFAL in less than 2 ms, which is the time of a passive scan (1 ms) plus the time to send out the 20 by 20 track beam acquisition pattern (0.4 ms) during a single one-second time sweep. This latency of less than 2 ms to less than 1 s, depending on beacon scanning timing, is in contrast to uncued detection and transition to track of an MFAL of 2–4 s. Such a cueing design measure is likely most useful for high-speed intrusion such as a vehicle.

26.4.2 Intruder Leaving the MFAL Zone and Entering the Pebble Field

In this case, the intruder track is very accurate and could, in principle, be used to cue a limited set of pebbles in the direction that the intruder is headed. Again, as for the aforementioned case, special design measures would be needed for cueing to minimize tracking discontinuity during the time required to reestablish a pebbles track. For example, without track continuity, it is possible that an intruder would penetrate tens of meters into the pebbles field before a new pebbles track was established. If the intruder changes direction abruptly during that time, it is possible that the new track would not be considered the same object as the previous precise MFAL track because of the tracking gap in combination with a change of direction and accuracy differences.

Beginning with the simplest design option of the MFAL sending its precise track coordinates to the swarming pebbles field command center, the monitor antenna could send a command signal to the nearest pebbles to the MFAL track to cause them to trigger to the greater cued detection sensitivity. This would increase the probability of early detection as the intruder enters the field to minimize the tracking gap. However, with such a large command/monitor antenna footprint, nominally 0.25 km^2, so many pebbles could be commanded to cued detection sensitivity that a false alarm instability could be triggered, as mentioned earlier.

For the selected cold and cued detection settings of the pebbles identified previously, it is desired that only pebbles within about plus or minus 10 m of the last MFAL coordinates for the intruder be set to cued detection sensitivity as the intruder enters the pebble field. To match such a narrow 20 m wide sector covered by the MFAL beam from 500 m range would require a microwave antenna beam at the MFAL location to cue a narrow swath of pebbles with only a 2.3° azimuthal beam width. Even at the relatively high microwave frequency of 10 GHz, this would require an unacceptably large antenna.

Therefore, as for the previous case, an optical solution is warranted. By implementing a simple optical receiver and wide field-of-view aperture in the pebbles at the edge of the pebble field bordering the MFAL zone, they would be able to directly receive a precise optical cue tone. The MFAL could emit a cueing frequency via its communication transmitter at a uniquely different wavelength from its radar and communications band, as for the previous situation. There would be sufficient gain for the MFAL to spoil its 0.01° beam to, perhaps, 0.1°, as mentioned earlier, and briefly (within a millisecond) transmit the cue tone for a sequence of beams over the 2.3° area where the MFAL track enters the pebble field. This could serve to only cue those pebbles most likely to detect the intruder, and perhaps maintain track continuity.

26.4.3 Intruder in the Midst of a Surveillance Zone Containing Both MFALs and Pebbles

As identified earlier, it may be of advantage to extend the pebble field into the MFAL zone for the purpose of ensuring all-weather tracking capability. In this manner, when fog or rain attenuation reduces the MFAL range, the pebble field tracks can maintain surveillance coverage in the inclement weather. The pebble field track accuracy is orders of magnitude less accurate than the MFAL track accuracy, however. Therefore, even when both colocated networks are in operation simultaneously prior or after the inclement weather, it is only likely that the tracks of one will automatically correlate at a command/control center with the tracks of the other network if intruding objects are separated more widely than the accuracy of the less accurate tracking network, the pebble field. Because of this, it is probably the most expedient to only use the MFAL tracks in good weather and only resort to pebbles field tracks in severely degraded weather.

26.5 Autonomous Mobile Tracking Sensor Nodes

Especially for the MFAL network, the individual terminals are sufficiently compact to make them attractive candidates to mount on small autonomous vehicles. A notional view of such a vehicle is shown in Figure 26.30, using an unattended ground vehicle chassis. In this way, the mobile MFAL nodes could also be designed to automatically run an intercept course to engage an intruder for purposes such as (a) forcing a halt to the intruder, (b) providing a warning, or (c) furnishing a video/audio feed to the command center to visually identify and communicate. One of the issues that would arise is that there would need to be net-wide coordination. A completely autonomous, uncoordinated set of MFAL nodes, all capable of moving toward an intruder, or multiple intruders, could leave coverage gaps in the network while they intercepted.

Coordination could be performed via the optical communications channels. Coordination options could include (a) the first node to detect an intruder moves to make the interception or (b) the unit that has the shortest time required to make the intercept along the intruder trajectory moves to complete the interception. This could be a centrally commanded operation, or, alternatively, each node could perform its own intercept calculations and communicate to others to confirm its

Figure 26.30 Example of mobile MFAL vehicle.

intent to intercepts. It may be best to include several such intercept logic options as one may be more useful than another depending on the behavior of the intruder. For example, a high-speed vehicle that is moving faster than an intercepting node can travel would motivate the first-to-intercept logic over first-to-detect.

The network would also be required to have reserve coverage capability to compensate for a node moving out of position to make the intercept. Alternative approaches include (a) packing the sensor nodes more tightly than otherwise required, so there is more sensor overlap or (b) placing reserve mobile nodes in the zone to fill in for gapped positions. In the latter alternative, there would need to be a sufficient number of reserve nodes so that the time lag for the closest such reserve unit to move into a gapped location is acceptable without compromising coverage continuity.

Acknowledgments

The authors would like to thank the following people who were coauthors with them on one or more of the referenced papers: Donald A. Day, Kenneth W. O'Haver, I.-J. Wang, D. G. Lucarelli, Eric R. Farmer, Amy Vaduthalakuzky, A. B. Dalton, R. A. Nichols, M. C. Gross, S. J. Papadakis, G. F. Ricciardi, J. S. Peri, and I. N. Bankman.

References

1. D. E. Culler and H. Malder, Smart sensors to network the world, *Scientific American*, 290(6), 85–91, 2004.
2. J. L. Hill and D. E Culler, Mica: A wireless platform for deeply embedded networks, *IEEE Micro*, 22, 12–24, November–December 2002.
3. M. H. Newkirk, J. Z. Gehman, and G. D. Dockery, Advances in calculating electromagnetic field propagation near the earth's surface, *Johns Hopkins APL Technical Digest*, 22(4), 2001.
4. J. A. Krill, K. W. O'Haver, M. J. O'Driscoll, I.-J. Wang, and D. G. Lucarelli, Prototype design and experimental results for an intruder detection swarming network, *Forth International Conference on ISSNIP*, December 15–18, 2008, Sydney, New South Wales, Australia.
5. *Electronic Warfare and Radar Systems Engineering Handbook*, Naval Air Systems Command, TP8347 Rev 2, Washington, DC, April 1999.
6. M. Skolnik, *Radar Handbook*, 3rd edn., McGraw Hill, New York, January 2008.
7. S. Johnson, *Emergence*, Scribner, New York, 2004.
8. Crossbow Technology, Micaz Wireless Measurement System, 2007. [Online] Available: http://www.openautomation.net/uploadsproductos/micaz_datasheet.pdf
9. D. Pescovitz, A Big Radio in a (Very) Small Package, College of Engineering, University of California, Berkeley, 2003. [Online] Available: http://coe.berkeley.edu/labnotes/0403/spec.html
10. Laird Technologies Antenna Division Unveils Nanoant^{TM} Antenna For Handheld, Wireless Devices, 2006. [Online] Available: http://www.lairdtech.com/NewsItem.aspx?id=965
11. Microphone OMNI Direct 6.0X5.0MM, Digi-Key Corporation, Part Number 359-1012-ND. http://www.digikey.com
12. J. A. Krill, D. A. Day, M. J. O'Driscoll, and K. W. O'Haver, Swarming network for intruder detection, presented at the *Third International Conference on Intelligent Sensors, Sensor Networks and Information Processing (ISSNIP)*, 3–6 December 2007, Melbourne, Australia.
13. A. Vaduthalakuzhy, A. B. Dalton, J. A. Krill, K. W. O'Haver, M. J. O'Driscoll, and R. A. Nichols, Implementation and evaluation of swarming sensor network for intrusion detection, *Fifth International Conference on ISSNIP*, December 7–10, 2009, Melbourne, Victoria, Australia.

14. S. J. Papadakis, G. F. Ricciardi, M. C. Gross, and J. A. Krill, A flat laser array aperture, *Proceedings of SPIE*, 7666, 76661S-76661S-10, 2010.
15. C. Whitcomb, The story of aegis, *Naval Engineers Journal*, Special edn., 121(3), 2009.
16. M. J. O'Driscoll and J. A. Krill, Cooperative engagement capability, *Naval Engineers Journal*, 109, 43–57, March 1997.
17. J. A. Krill and M. J. O'Driscoll, Multifunction array lidar for intruder detection and tracking, presented at the *Sixth International Conference on Intelligent Networks and Information Processing (ISSNIP)*, 5–7 December 2012, Brisbane, Australia.

Index

A

B

D

E

F

J

K

L

M

N

Q

R

S

U

W

X

Z

Milton Keynes UK
Ingram Content Group UK Ltd.
UKHW051928141024
449569UK00027B/1394